Lecture Notes in Production Engineering

Lecture Notes in Production Engineering (LNPE) is a new book series that reports the latest research and developments in Production Engineering, comprising:

- Biomanufacturing
- Control and Management of Processes
- Cutting and Forming
- Design
- Life Cycle Engineering
- Machines and Systems
- Optimization
- Precision Engineering and Metrology
- Surfaces

LNPE publishes authored conference proceedings, contributed volumes and authored monographs that present cutting-edge research information as well as new perspectives on classical fields, while maintaining Springer's high standards of excellence. Also considered for publication are lecture notes and other related material of exceptionally high quality and interest. The subject matter should be original and timely, reporting the latest research and developments in all areas of production engineering. The target audience of LNPE consists of advanced level students, researchers, as well as industry professionals working at the forefront of their fields. Much like Springer's other Lecture Notes series, LNPE will be distributed through Springer's print and electronic publishing channels. To submit a proposal or request further information please contact Anthony Doyle, Executive Editor, Springer (anthony.doyle@springer.com).

Mathias Liewald · Alexander Verl ·
Thomas Bauernhansl · Hans-Christian Möhring
Editors

Production at the Leading Edge of Technology

Proceedings of the 12th Congress of the German Academic Association for Production Technology (WGP), University of Stuttgart, October 2022

Springer

Editors
Mathias Liewald
Institut für Umformtechnik
Universität Stuttgart
Stuttgart, Germany

Alexander Verl
ISW
Universität Stuttgart
Stuttgart, Germany

Thomas Bauernhansl
IFF
Universität Stuttgart
Stuttgart, Germany

Hans-Christian Möhring
Institut für Werkzeugmaschinen
Universität Stuttgart
Stuttgart, Germany

ISSN 2194-0525 ISSN 2194-0533 (electronic)
Lecture Notes in Production Engineering
ISBN 978-3-031-18317-1 ISBN 978-3-031-18318-8 (eBook)
https://doi.org/10.1007/978-3-031-18318-8

This Springer imprint is published by the registered company Springer Nature Switzerland AG
The registered company address is: Gewerbestrasse 11, 6330 Cham, Switzerland

Preface

On behalf of the German Academic Association for Production Technology (WGP), Frankfurt/Germany, four belonging institutes agreed to organize its 12th Annual Plenary congress entitled "Production at the Leading Edge of Technology" in Stuttgart during October 11–14, 2022. The Institutes for Metal Forming Technology (IFU), for Control Engineering of Machine Tools and Manufacturing Units (ISW), for Industrial Manufacturing and Management (IFF) and for Machine Tools (IfW) cordially welcome researchers and guests to present recent scientific findings in the areas of productions technologies related to our Academic Association. With this, the Association invites researchers coming from its institutes and from industry to the University of Stuttgart. We are looking forward to share exciting and fruitful discussions with experts from industry and research!

Research in production permanently shifts the boundaries of what is feasible. Contributing presentations do show production processes that advance into new areas of production in terms of fundamental findings, methodology, more efficient use of resources or interdisciplinary. But where does the search for new borders lead to? Which borders do we still have to cross, which ones do we prefer not to cross?

The focus of congress is put on production processes in actually limiting fields related to extreme velocity, size, accuracy, methodology, use of resources and interdisciplinarity. Challenges from the fields of metal forming, cutting processes and machine tools, automated assembly and robotics, process planning and management sciences as well as interdisciplinary research topics will be addressed in the congress program. The program therefore summarizes peer-reviewed contributions from production science and industrial research. Scheduled sessions provide an overview of current trends in production research and give an insight into ongoing research recently conducted in the German Academic Association for Production Technology.

We wish all participants an interesting and inspiring Annual Plenary congress and are looking forward to welcoming you to Stuttgart!

Prof. Mathias Liewald MBA

Prof. Thomas Bauernhansl

Universität Stuttgart
Institut für Industrielle Fertigung
und Fabrikbetrieb IFF

Prof. Alexander Verl

Prof. Hans-Christian Möhring

September 2022
Stuttgart, Germany

Organization

Universität Stuttgart
Institut für Umformtechnik
Prof. Dr.-Ing. Dr. h. c. Mathias Liewald MBA
Maximilian Bachmann, M.Sc.

Universität Stuttgart
Institut für Steuerungstechnik der Werkzeugmaschinen und Fertigungseinrichtungen
Prof. Dr.-Ing. Alexander Verl
Xenia Günther

Universität Stuttgart
Institut für Industrielle Fertigung und Fabrikbetrieb
Prof. Dr.-Ing. Thomas Bauernhansl
Dipl.-Ing. Jörg Siegert
Frank Herbrig

Universität Stuttgart
Institut für Werkzeugmaschinen
Prof. Dr.-Ing. Hans-Christian Möhring
Sibylle Krug, M.A.

Contents

Recent Developments in Manufacturing Processes

Development of a Temperature-Graded Tailored Forming Process for Hybrid Axial Bearing Washers

J. Peddinghaus[1](✉), Y. Faqiri[2], T. Hassel[2], J. Uhe[1], and B.-A. Behrens[1]

[1] Institut für Umformtechnik und Umformmaschinen (Forming Technology and Machines), Leibniz University Hannover, Garbsen, Germany
peddinghaus@ifum.uni-hannover.de

[2] Institut für Werkstoffkunde (Materials Science), Leibniz University Hannover, Garbsen, Germany

Abstract. The Tailored Forming process developed in the presented research enables the production of axial bearing washers with AISI 1022M (C22.8/1.0460) base material and AISI 52100 (100Cr6/1.3505) cladding on the rolling contact surface. By limiting the use of the bearing steel to the highly loaded surface, significant amounts of alloyed steel can be saved. The cladding is applied through plasma transferred arc (PTA) welding and subsequently formed to improve its properties. The challenge in developing a hot upsetting process lies in the high difference in flow stress of the two materials, since the harder bearing steel is merely pressed into the softer base without sufficient deformation. In order to equalise the flow stress of both materials, an adapted temperature gradient is induced over the washer height before upsetting. Due to this, a higher cladding temperature is set while the base material remains significantly cooler. This is realised by means of local inductive heating of the cladding and different transfer times to the upsetting process. The process variants are applied in an automated forging cell and subsequently evaluated in metallographic analysis of cross sections after welding and after forming. The results show the most favourable material properties after forming when local inductive heating of the cladding is simultaneously combined with cooling of the base material and the transfer time between the heating stage and forming is minimized.

Keywords: Tailored forming · Upsetting · Plasma transferred arc welding · Inductive heating

1 Introduction

The load collective of machine components is rarely uniform and homogeneous throughout the entire component. The choice of material can therefore be locally adapted to the specific load spectrum in different sections of the part. This approach not only enables increase in performance but potentially also a significant improvement in terms of economic factors. When combining different steel materials, the use of high performance

M. Liewald et al. (Eds.): WGP 2022, LNPE, pp. 3–12, 2023.
https://doi.org/10.1007/978-3-031-18318-8_1

alloyed steels can be limited to highly loaded sections, with high requirements. Using unalloyed steels in less loaded zones, economical cost savings as well as ecological savings in problematic alloying elements can be achieved depending on the use case at hand.

A promising application area of hybrid components is roller bearings. While the surface is subjected to high tribological loads, the requirements for the core beneath are lower. Significant savings can be achieved with unalloyed steel for the unloaded core, especially when upscaling this concept to large components for heavyweight applications. In order to achieve such a hybrid bearing component, a high-performance material is applied to a simpler steel base by means of deposition welding. The tribological performance of the steel cladding is highly dependent on its microstructure. After the deposition process, the properties of the applied material are however often not ideal for bearing applications. The tailored forming approach, developed in the Collaborative Research Centre (CRC) 1153 is therefore a promising strategy for reinforced hybrid bearing components. The components are joined first, and then formed in a hot forming process, thermomechanically enhancing the joint zone and both of the applied materials in order to achieve properties locally adapted to the local load conditions in the different sections of the part after subsequent heat treatment.

2 State-of-the-Art

The present work focuses on the tailored forming process chain for bearing applications based on an axial bearing washer as a demonstrator component established in prior research. The alloyed steel AISI 52100 is welded onto the unalloyed steel base (AISI 1022M) in a PTA process and is then formed near-net-shape in an upsetting process. The forming process is followed by machining and heat treatment [1].

PTA welding is a thermal coating process which is used primarily to produce wear- and corrosion-resistant coatings. A high-energy plasma arc is used between a non consumable tungsten electrode and the base material, which melts the surface of the base material. The powder is melted in the arc and a substance-to-substance bond between the cladding and base material is formed. The advantages of this process are a low dilution rate between the cladding and base material, a small heat affected zone and deposition rates up to 10 kg/h. Further benefits are a high degree of automation and reproducibility [2].

The disadvantage of welded layers lies in the microstructure and welding defects such as pores resulting from the solidification of the cladding after welding. Coors et al. found out, that pores significantly influence the performance of hybrid bearing washers [3]. The microstructure after welding is characterised by large grains with significant agglomerations of carbides. Large grains have a significant negative effect on the bearing performance [4] but the size can be reduced through subsequent heat treatment. In terms of carbides, a small size and fine distribution improves performance [5]. The improvement of carbide distribution is however limited in heat treatment [6]. In order to achieve a favourable microstructure, a forming process following the welding process can be beneficial for the performance of the resulting hybrid bearing. Doblesky et al. showed that when forming a hybrid friction welded component, the deformation is concentrated in the weaker material [7]. Further research on hybrid components showed that hot forming can have a beneficial effect on the joining zone resulting from the prior friction welding process [8]. The tailored forming approach was transferred to the combination of deposition welding for enhanced surface properties in combination with a subsequent forming process by Lehnert et al. who found it to be a promising strategy for steel-steel combinations [9]. PTA Welding was identified as a suitable deposition process for subsequent forming of a nickel alloy on a steel base by Landgrebe et al. [10].

In order to achieve deformation in a cladding layer, the flow stress of the materials must be approximated. This can be achieved by inducing an adapted temperature profile in the component during forming. A higher temperature in the cladding reduces the flow stress while a lower temperature increases flow stress of the softer material. A heating method common in forging which can be adapted for local applications is induction. Guo et al. found out, that forming results are uniform and predictable even with inhomogeneous inductive heating in pipe bending [11]. The applicability of local induction heating in bulk forming of crank shafts was successfully tested and modelled by Song and Moon [12]. Induction heating is characterised by the skin effect, which limits the induction of heat to the surface zones of a heated component. The heating of the remaining part is based on conduction form the shell into the core of the material [13].

Locally adapted temperature profiles through inductive heating have also been studied for hybrid components, such as impact extruded steel-aluminium shafts [14]. The temperature gradient mainly effected by heating and transfer time, geometry and material properties showed a strong effect on the forming results [14]. For tailored forming of a laser-cladded bevel gear consisting of a cylindrical core of AISI 1022M with a shell of AISI HNV3, an axial temperature gradient of 200 °C was found to be necessary to guarantee form filling [15]. This material combination was also applied in prior work on the axial bearing washer geometry, which this paper deals with. The results showed, that inert atmosphere is necessary to prevent decarburisation in furnace heating [1]. When applying the bearing steel AISI 52100, further results showed, that welding defects were not eliminated through forming with homogeneous heating and upsetting. These defects were the cause for premature failure of the resulting bearing washers, limiting their performance [3]. Further development of a tailored forming process for sufficient deformation of the cladding layer is therefore necessary.

The presented work aims at further developing the tailored forming application for hybrid bearing applications. The main open question is how a high-performance steel cladding such as AISI 52100 on a softer steel base can be sufficiently deformed to improve microstructural properties.

3 Materials and Methods

In this work, a novel inductive heating strategy for tailored forming of hybrid axial bearing washers was developed through iterative improvement.

Prior to forming, the AISI 52100 cladding of appr. 5 mm was applied to a base plate made of AISI 1022M with a diameter of 140 mm and a height of 10 mm in a robot-guided deposition PTA process. The welding process, developed iteratively in order to minimise pore formation. The weld seam is applied in a circular pattern where the torch oscillates stationary at one point while the specimen is rotated 360°. The oscillation increases the dynamics of the weld pool, which causes degassing of the melt, preventing pores. The welding process starts from the centre of the disc towards the edge of the disc. During this movement, the input of plasma and transport gas as well as the powder and current input are constantly increased until the desired welding parameters are reached and the torch reaches the outer starting point. After the specimen rotation is complete, the torch moves out tangentially from the circular track. The welding process takes 5 min and 25 s, and the disc heats up to 800 °C. In order to keep the dilution of the materials constant, the welding current is reduced dynamically during the welding process. Before subsequent forming, the unclean welding surface of cladding was grinded down using an angle grinder.

The forming process is carried out in an automated forging cell with a 40 kJ screw press. The final height of 9 mm after forming is set through limiting mechanical stoppers. The global strain in the washer resulting of the forming process is defined by its initial and its final height. The distribution of the strain over the height of washer is however dependent on the local flow stress of the materials. At a forming temperature of 900 °C and a strain rate of 50 1/s, material data generated in the scope of the CRC 1153 showed a flow stress of AISI 52100 is appr. 325 MPa while AISI 1022M exhibits a lower flow stress of appr. 275 MPa. In order to evaluate the transferability of the developed processes to other materials, a second cladding material consisting of a mixture of the powder alloys Rockit 706 ® (Höganäs Sweden AB) [16] and Rockit 401 ® (Höganäs Sweden AB) [17] (1:5 ratio). Rockit 401 ® is a ferritic steel with a high chromium content and is characterised by good wear and corrosion resistance. Rockit 706 ® is a martensitic steel with a high hardness, wear resistance and high impact strength. Material data shows similar flow stress of appr. 275 MPa at a higher temperature of 1050 °C and lower strain rate of 1 1/s.

With the applied process chain, the reinforcing layer thickness after forming reaches appr. 3 mm through the component. Since bearing loads are mostly limited to below 1 mm in their depth of penetration for the given dimensions, the joining zone and the softer base material are not subdue to any significant dynamic stress. Therefore, in contrast to other tailored forming components, no specific requirements must be formulated for the joint properties or the base material support.

Approach A: local heating and cooling:

In order to achieve a temperature gradient over the height of the washer, an inductor was designed to locally heat the cladding surface. During surface heating of the cladding above, the washer was placed on a water-cooled plate to remove heat conducted within the part from the base material and further enhance the temperature gradient. To guarantee contact to the cooling plate, the bottom AISI 1022M surface was machined prior to heating. The inductor geometry was individually custom-designed to the ring geometry of the washer before forming. The inductive heating process was powered by a high frequency generator with a frequency range in the range of 40 to 100 kHz and a heating capacity of 40 kW. The local heating effect is based on the skin effect, limiting the introduced heat to the cladding surface.

The entire process is carried out in an automatic forging cell including the press, an industrial robot and a heating station as displayed on the left in Fig. 1. A robot-gripper adapted to the ring geometry picks up the ring and places it on the cooling plate beneath the inductor in the heating station. The setup of the heating station is displayed in the right in Fig. 1. The ring is heated in a sequence of adaptable segments with individually defined power input. After heating, the ring is removed by the robot and placed in the forging press before being upset between two flat dies. The minimal transfer time between heating and upsetting was 7 s and crucial for the temperature gradient during forming. The heat gradient equalises through heat conduction during this time.

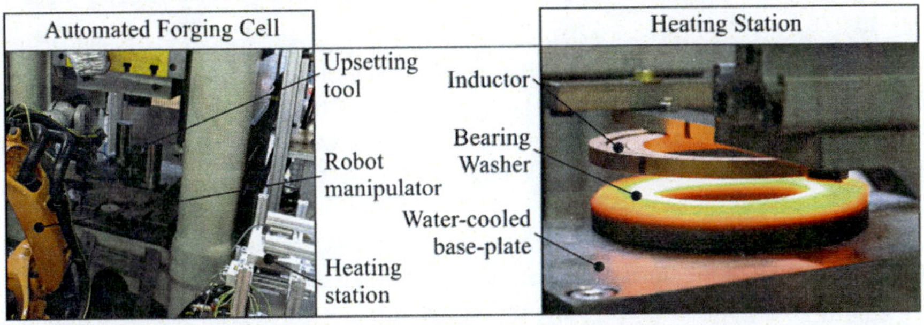

Fig. 1. Test setup for forming with heating strategy A in the automated forging press

Heat conduction is highly dependent on the temperature gradient, which means, that the cladding temperature instantly decreases when the heat input ends if the base is too cold. The cladding cools even more during transfer and is formed at low temperature. Initial tests showed that moderate heating of the base material leads to a higher cladding temperature and temperature gradient during forming. The heating profile determined consisted of initial heating at 75% power for 10 s, a slower heating segment at 50% for 7 s to allow some heat to be conducted into the base and a final heating stage for 10 s with 100% power to maximize the temperature gradient. The temperature profile was recorded throughout the heating process with a thermal imaging camera.

Approach B: local heating and cooling in forging process.

To further increase the temperature gradient in forming, the transfer time must be shortened beyond the limits of the robot handling system. The heating station was therefore integrated into the tooling system in the forging press and fitted with a linear actuator as displayed in Fig. 2. After heating, the specimen is directly pushed onto the tool for forming. The transfer time was reduced to 3 s using this adapted upsetting process, significantly increasing the temperature gradient in the forming process. The heating process was recorded through thermal imaging as well.

Tests with the initial inductive heating strategy also showed, that the temperature was also highly dependent on the height of the cladding surfaces varying between 11.5 and 13.5 mm. Minor deviations in height between the specimens had a significant influence on the temperature profile when using the identical heating power sequence. Before heating and forming, the specimens for this strategy were therefore machined on the cladding side as well to a uniform height of 13 mm.

Fig. 2. Test setup for forming with heating strategy B in the automated forging press

A comparison of the different processes in terms of heat treatment is of relevance, since this state is characteristic for the application as bearing washers, with focus on material hardness. For heat treatment, the specimens were quenched in oil after being heated to 850 °C for 45 min and subsequently tempered at 150 °C for 1 h, according to parameters tailored to the cladding material AISI 52100.

In order to evaluate the performance of the developed process strategies, the microstructure of the resulting washers was characterised. The microstructure is analysed after welding and after forming in untreated and heat-treated state to evaluate the influence of the forming process. In order to analyse the microstructure, each washer was separated into halves, of which one was heat treated. Specimens were extracted through wet abrasive cut-off grinding, prepared and etched with 2% nitric acid solution to visualise and analyse the microstructure through light microscopy.

4 Results and Discussion

Thermal imaging is not precise, but can provide an extensive overview of the temperature distribution in the process. The results of the heating processes are displayed in Fig. 3 and show, a significant temperature gradient of over 500 °C directly after heating (left). The temperature peak of up to nearly 1200 °C is reached on the cladding, while the base material is significantly colder with temperatures below 700 °C. The heating profile of the faster strategy B in the moment before impact of the upsetting die shows, that the peak temperature is reduced to below 1100 °C and the base material has heated up to appr. 750 °C, decreasing the temperature gradient to appr. 340 °C and resulting in a lower flow stress of appr. 120 MPa on the cladding surface and appr. 160 MPa flow stress at the base both for 0.1 true strain and 10 MPa strain rate. These results confirm that low transfer time is required for a high temperature gradient and therefore a sufficient cladding deformation. Strategy B is therefore the most promising approach for upsetting of the cladding. The developed strategies showed no effect however for the applied powder mixture with Rockit 706 ® and Rockit 401 ® as the deformation was limited to the base material. This can be lead back to high flow stress of the material at high temperatures and at high strain rates, as it is the case in the process at hand with a screw press. This material is therefore not analysed further.

In order to compare the different strategies with regard to the resulting bearing washer, the microstructure development is analysed. A comparison of the specimen after

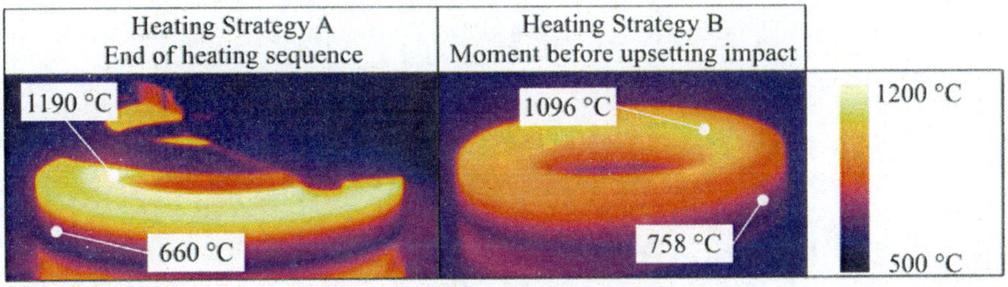

Fig. 3. Thermal imaging results

welding and after forming with the different heating strategies is shown in Fig. 4. For all specimens, the cladding microstructure is finer after forming, but is not quantifiable with the applied methods due to the martensite content. The etched image for strategy B shows a different appearance than the other specimens. The base material appears darker while the cladding is not affected by the etching solution. The discrepancy in etching reaction occurred for all specimens after forming with strategy B. The heat-treated halves of the same specimens showed microstructure consistent with the other strategies. It is therefore not clear what the cause for this behaviour is. Further analyses will be carried out in order to characterise the occurring mechanisms.

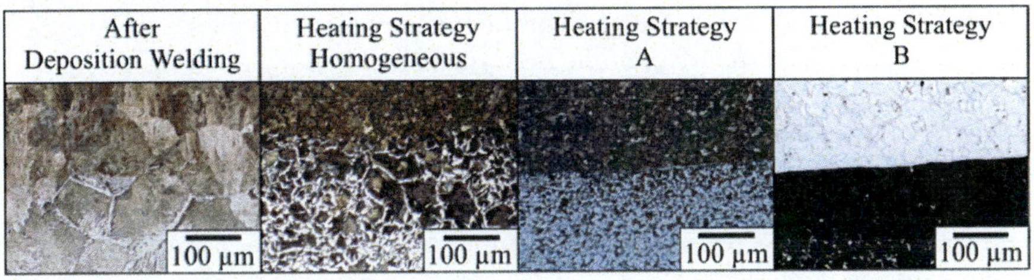

Fig. 4. Joint zone analysis for different specimen after welding or forming with different heating strategies etched with 2% nitric acid solution

The segregation between the two materials is more distinct for the local heating strategies than for homogeneous heating. The less pronounced boundary between the materials can be explained with the longer heating time of 20 min which allows diffusion and therefore the equalisation of concentration gradients of the materials. The cladding produced with strategy A is characterised by finer microstructure due to the enhanced deformation in the cladding layer through local heating. The finer structure is a promising result for the performance of the bearing washers.

In order to evaluate the influence of the forming process on the cladding microstructure after heat treatment, the strategies A and B are compared with specimens heat-treated directly after deposition welding. The results are displayed in Fig. 5 in different magnifications to evaluate the microstructure and the carbide distribution. The microstructure is refined compared to the non-heat-treated specimen displayed above. The microstructure after welding and heat treatment appears to be slightly coarser than in formed and heat treated specimens.

Fig. 5. Cladding layer microstructure for different specimen welded or formed with different heating strategies after subsequent heat treatment etched with nitric acid solution

The welded and heat-treated microstructure is also characterised by a dendritical microstructure which can be explained by the directional solidification of the weld pool in the cladding layer. The carbides are dot-shaped in appearance and are concentrated between the dendrites. They are driven in front of the solidification front causing agglomerations in the resulting cladding layer. This microstructure with inhomogeneous carbide distribution has an adverse effect of the performance of bearing components.

After forming with heating strategy A, the overview in the top centre shows that the dendritical microstructure remains visible but is significantly distorted through the applied process. The image below, of the microstructure with higher magnification shows a finer distribution of the dark dot-shaped carbides throughout the grains with sporadic agglomerations. The specimen produced with heating strategy B shows no agglomeration of carbides in the overview in the top right of Fig. 5. The detailed image shows evenly distributed dark sections, which cannot clearly be identified as carbide agglomerations, since the dotted carbides do not increase in concentration in their proximity. The carbides appear to be more evenly distributed through the microstructure as a result of forming with heating strategy. The applied methodology allows for a qualitative comparison of the resulting microstructures of the different process strategies. A quantitative assessment of properties such as grain size or carbide distribution however requires further research and in-depth analysis methods and is planned as the next step. The qualitative analysis is a sufficient method to detect tendencies in process behaviour which allows the identification of the most promising strategies.

Due to their temperature-stable nature, carbides are limited in their mobility and can only be dispersed through recrystallisation and breaking up of grain boundaries of the surrounding metal matrix. The changes in carbide distribution therefore correlate with the deformation of the cladding material. As described in Chap. 2, a fine distribution of carbides is also beneficial for bearing performance. It can therefore be summarised, that the implementation of a high temperature gradient through the local inductive heating strategy and the reduction in transfer time allows an increase in deformation of the cladding material and as a result can potentially improve the bearing performance.

5 Conclusion and Outlook

The development of a process chain for tailored forming of a hybrid bearing washer is the main objective of the presented research. In order to achieve sufficient properties in the resulting deposition welded component, an adapted local heating strategy is required for the upsetting process. An induction heating process was developed to limit heat input into the base material. Two strategies were developed, primarily differing in transfer time. The results showed, that a higher temperature gradient was achieved through a reduction of transfer time. Metallographic analysis showed, that this increase results in finer grain and carbide distribution. The etching behaviour for strategy B strongly deviates from the remaining analysed specimen and requires further research. The microstructure appeared to be the most promising for strategy B. The potential for the developed strategy is however limited, since the additionally applied powder mixture could not be formed despite temperature gradient. Based on the presented results, further analyses to quantify and validate the findings, such as grain size and carbide distribution analysis will be carried out and life testing will be performed to evaluate and qualify the tailored forming approach for bearing applications.

Acknowledgments. The research results presented in this paper were obtained within the Collaborative Research Centre 1153 "Process chain to produce hybrid high performance components by tailored forming"—252662854 in the subprojects B2, A4 and T1. The authors thank the subproject A2 for supplying the heat treatment of the hybrid components, subproject C1 for providing the material data and the German Research Foundation (DFG) for the financial support of this project.

References

1. Behrens, B.-A., et al.: Manufacturing and evaluation of multi-material axial-bearing washers by tailored forming. Metals **9**(2), 232 (2019)
2. Schuler, V., Twrdek, J.: Praxiswissen Schweißtechnik: Werkstoffe, Prozesse, Fertigung, 6th edn. Springer Fachmedien Wiesbaden, Wiesbaden (2019)
3. Coors, T., et al.: Investigations on tailored forming of AISI 52100 as rolling bearing raceway. Metals **10**(10), 1363 (2020)
4. Weinzapfel, N., Sadeghi, F., Bakolas, V.: An approach for modeling material grain structure in investigations of Hertzian subsurface stresses and rolling contact fatigue. ASME J. Tribol. **132**(4), 041404 (2010)
5. Parker, R.J., Zaretsky, E.V.: Rolling-element fatigue lives of through-hardened bearing materials. ASME J. Lubrication Tech. **94**(2), 165–171 (1972)
6. Stickels, C.A.: Carbide refining heat treatments for 52100 bearing steel. Metall. Mater. Trans. B **5**, 865–874 (1974)
7. Domblesky, J., Kraft, F., Druecke, B., Sims, B.: Welded preforms for forging. J. Mater. Process. Technol. **171**(1), 141–149 (2006)
8. Behrens, B.-A., et al.: Experimental investigations on the state of the friction-welded joint zone in steel hybrid components after process-relevant thermo-mechanical. AIP Conf. Proc. **1769**, 130013 (2016)

9. Lehnert, T., Sterzing, A., Mauermann, R., Schubert, N., Krüger, L., Keßler, A., Wolf, G., Wagler, H.: Development of innovative combustion chamber components for large marine engines. In: Proceedings of the 22nd International ESAFORM Conference on Material Forming, 040016 (2019)
10. Landgrebe, D., Krüger, L., Schubert, N., Jentsch, E., Lehnert, T.: Resource-efficient development of thermally highly resistant engine components of hybrid metal composites—experiments and numerical analysis. Proc. Eng. **207**, 884–889 (2017)
11. Guo, X., et al.: Numerical simulations and experiments on fabricating bend pipes by push bending with local induction-heating process. Int. J. Adv. Manuf. Technol. **84**(9–12), 2689–2695 (2015)
12. Song, M.C., Moon, Y.H.: Coupled electromagnetic and thermal analysis of induction heating for the forging of marine crankshafts. Appl. Therm. Eng. **98**, 98–109 (2016)
13. Rudnev, V., Loveless, D., Cook, R.L.: Handbook of induction heating. In: Manufacturing Engineering and Materials Processing, 2nd edn, p. 61 (2017)
14. Behrens, B.-A., Wester, H., Schäfer, S., Büdenbender, C.: Modelling of an induction heating process and resulting material distribution of aerial distribution of a hybrid semi-finished product after impact extrusion. In: 24th International Conference on Material Forming, Liège, Belgique (2021)
15. Behrens, B.-A., et al.: Microstructural evolution and mechanical properties of hybrid bevel gears manufactured by tailored forming. Metals **10**(10), 1365 (2020)
16. Höganäs AB: Rockit® 606/706—Combat Impact and Abrasive Wear (2021). https://www.hoganas.com/globalassets/download-media/sharepoint/brochures-and-datasheets-all-documents/rockit_rockit-606-706_2653hog.pdf
17. Höganäs AB: Rockit® 401—Sustainable Solution to Replace Hard Chrome Plating (2021). https://www.hoganas.com/globalassets/download-media/sharepoint/brochures-and-datasheets-all-documents/rockit_rockit-401-sustainable-solution-to-replace_2275hog.pdf

Study on the Compressibility of TiAl48-2-2 Powder Mixed with Elemental Powders

A. Heymann(✉), J. Peddinghaus, K. Brunotte, and B.-A. Behrens

Institute of Forming Technology and Machines, Leibniz University Hannover, Hannover, Germany
heymann@ifum.uni-hannover.de

Abstract. Many metallic powder materials can be processed fast and cost-effectively using the conventional powder metallurgy method. For this purpose, the metal powder is pressed into a compact and then sintered in a furnace to produce a finished component. Gamma titanium aluminides are an exception to this. Due to their brittleness, they cannot be compacted in classical die pressing. A promising approach is the addition of elemental significantly more ductile alloy powder. The aim of this work is to investigate the influence of the admixture of elemental powder on the compressibility and the properties after the sintering process. Within the scope of the work, commercially available pre-alloyed TiAl48-2-2 (GE48) powder, which is applied e.g. for turbine blades in aircraft, is used. The powder alloy is mixed with elemental titanium, aluminium, chromium and niobium powder according to its composition and then pressed to a compact. Selected samples are sintered and metallographically characterised. By varying the pressing load and the proportion of elemental powder, as well as the proportion of elemental powder mixtures, the influence on the compaction behaviour and the mechanical properties is investigated. It is possible to produce compacts with sufficient mechanical properties by adding specific proportions of different elemental powders depending on the element and the compaction parameters. The results show a significant dependence of the relative density and tensile splitting strength on the proportion and type of elemental powder added.

Keywords: Powder compacting · TiAl powder · Mechanical properties

1 Introduction

The group of gamma-based titanium aluminides (γ-TiAl) is a new structural material for lightweight construction. They are suitable for use in the automotive industry, e.g. for valves and turbocharger wheels, and in the aviation industry, e.g. as turbine blades. In addition, they can be used as a substitute material for titanium and nickel-based alloys due to their high heat resistance, good corrosion resistance, high stiffness and high specific strength as well as their low density of 3.9 to 4.2 g/cm^3 [1, 2].

The conventional powder metallurgical (PM) process route is one of the most economical PM routes. It enables the production of near-net-shape products by consolidating

M. Liewald et al. (Eds.): WGP 2022, LNPE, pp. 13–23, 2023.
https://doi.org/10.1007/978-3-031-18318-8_2

and shaping metal powder into compacts, followed by sintering to adjust the microstructure and thus the mechanical properties [3]. On an industrial scale, the powder is usually processed in dies by double-sided pressing. It is filled into the die and compacted evenly from both sides by two opposing punches. Existing bonds between the individual powder particles are broken and cavities are filled by particle rearrangement. When the pressing load is increased further, the powder particles first become elastic and then deform plastically to the point of breaking. This further fills cavities and ultimately increases the density. Afterwards, the compact is ejected by one of the punches and sintered in a furnace [4]. The properties of the pressed part are decisive for the final properties of the sintered component. On the one hand, they are influenced by the material itself, the particle geometry and particle size, and on the other hand by pressing load, ejection pressure and the lubricant used [5, 6]. Lubricant can be introduced into the pressing process by mixing lubricant into the powder or by direct lubrication of the powder-contacting tools. The latter has the advantages of higher compact densities and consequently sintering densities, more homogeneous properties and shorter process times, as no powder-lubricant mixture has to be prepared and delubrication during sintering is unnecessary [7]. More homogeneous properties can be obtained by using pre-alloyed powder [8]. However, the processing of pre-alloyed TiAl powder within the conventional PM route has so far led to insufficient results [3]. The brittleness of the material has been identified to impair plastic deformation and thus material compaction [9]. This is caused on the one hand by the low number of sliding dislocations and on the other hand by impurities such as oxygen. Interstitially dissolved oxygen leads to solid solution hardening, which can widely reduce the ductility of the material [10, 11]. As a result, the processing of TiAl powder has so far been carried out primarily by hot isostatic pressing (HIP) and by the field-assisted sintering technique (FAST) [12]. A potential solution for processing the pre-alloyed TiAl powder within the conventional PM route is the admixture of ductile metal powder in order to increase compressibility [13]. Compressibility is the relation between the green density and the pressure applied [14]. Gethin et al. [15] showed that the addition of ductile powder to a brittle base powder significantly improves compression during pressing. In 1967, Heckel [16] reported an increase in the relative density of iron, copper, niobium and tungsten with rising pressing pressure. Tiwari et al. [17] were able to achieve an analogous result when processing an elemental powder mixture of aluminium and iron. Solimanjad et al. [18] investigated the influence of particle size and pressing load after die compaction, after sintering and after a subsequent heat treatment on the relative density and tensile strength. They were able to conclude that there is an optimum pressing load at which density and tensile strength are at a maximum. Dixit et al. [19] showed the dependence of hardness on pressing pressure for copper powder. Although the relative density increases with rising pressing load, there is an optimal load with respect to the hardness of the compact. Kumar et al. [20] were able to show the influence of the sintering time on the resulting density and hardness for two different aluminium matrix composites. So far, the addition of alloying elements in the form of elemental powder to a pre-alloyed alloy in powder form has not been investigated. Consequently, the influence of different powders in varying proportions on the compaction result is still completely unknown. Hence, the influence on the component after sintering is also unknown. This knowledge forms the basis for qualifying TiAl for the conventional PM route.

In this paper, the consolidation of commercial pre-alloyed TiAl48-2-2 (GE48) powder is investigated. For this purpose, the powder is mixed with different proportions of the alloying elements in the form of elemental powder and compacted with varying compaction pressures. The aim is to investigate the influence of different elemental powders and their proportions on the consolidation. Subsequently, selected powder mixtures are sintered to determine the final component properties. The evaluation of the results is based on the relative density and the tensile splitting strength of the compacts as well as on optical microscopic images and hardness measurements of the sintered samples.

2 Materials and Methods

The investigations were based on commercial pre-alloyed TiAl48-2Cr-2Nb (GE48) powder with a particle size in the range of 45 to 150 µm. The alloying elements of the alloy were selected as elemental powders. Titanium grade 2 and Aluminium 99.5 with a spherical particle size of 50 to 150 µm, as well as non-spherical chromium (chromium content $\geq$ 99.3 wt%) with 63–150 µm and niobium chromium (niobium content: 77–83 wt%, chromium content $<$ 17.5 wt%) up to 300 µm particle diameter were used. The niobium-chromium powder used was assumed to contain 80 wt% niobium in all mixtures in order to ensure comparability. For the preparation of the powder mixtures, GE48 powder was used as the starting material and first mixed with different proportions of aluminium or titanium powder, as well as aluminium and titanium powder according to the ratio present in GE48, in order to investigate the fundamental influence of the main alloying elements on compressibility and compact strength. Subsequently, all alloying elements were mixed according to the ratio present in GE48 and then added in different proportions to the GE48 powder. This resulted in a mixture of pre-alloyed GE48 and different proportions of elemental "GE48" powder. To ensure repeatability, the mixture was mixed in a 3D turbular shaker mixer of Willy A. Bachofen GmbH at 50 rpm for 1 h. The mixing ratios that enable successful compact production were determined within the scope of preliminary tests. The powders were mixed as described above and pressed at 300 and 600 MPa, with the aim of obtaining a pressed green body in one piece. The combinations summarised in Table 1 were investigated.

Table 1. Experimental design with the sample compositions investigated

Basis GE48 (wt%)		0	20	30	40	50	60
Admixtures (wt%)	Ti	100	80				
	Al	100	80	70	60	50	40
	Ti/Al	64.6/35.4	45.1/24.9	38.6/21.4			
	Ti/Al/Cr/Nb	59.6/33.0/2.6/4.8	47.7/26.4/2.1/3.8	41.7/23.1/1.8/3.4s	35.7/19.8/1.6 /2.9		

The subsequent conventional die pressing was carried out on a single-sided hydraulic press from MSE Teknoloji LTD with pressures of 300 and 600 MPa. In order to achieve homogeneous compaction, a die floating on springs was used. The test setup is shown in Fig. 1. Before each pressing process, the contact surface between tool and powder was spray-coated with graphite to ensure a damage-free ejection of the compact [21]. In order to achieve the best possible comparability, all cylindrical compacts were pressed with a height and diameter of 20 mm as target dimensions.

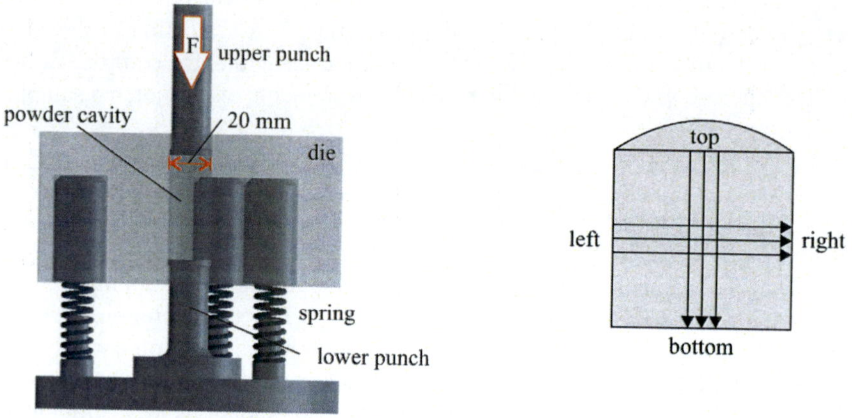

Fig. 1. Experimental setup (left) and position of the hardness measurements carried out (right)

In order to calculate the relative density, the density of the compact was first determined by measuring its actual height and diameter as well as its weight. Then the density of the compact was divided by the density of the solid material of 3.97 g/cm^3 (GE48) or the respective overall solid densities of the mixed materials [21]. The tensile splitting strength was used to evaluate the bond between the powder particles. It was determined experimentally in accordance with DIN EN 12390-6:2009 [22]. For this purpose, the cylindrical sample was placed on a metal plate and loaded radially in the centre of a cylindrical test sample until failure occurred while recording the force. The tensile splitting strength f_{ct} was obtained from the maximum force F, the sample diameter D, and the height of the sample h according to the following formula [22]:

$$f_{ct} = (2F)/(\pi Dh) \tag{1}$$

The PM route was completed by subsequently sintering the compacts. Additional samples with admixed titanium, aluminium, chromium and niobium powder according to Table 1 were pressed and sintered in an oven of Gero Ofenbau GmbH at 550 °C under low vacuum for 3 h. The subsequent evaluation was carried out by means of microstructural analysis. For this, the samples were prepared metallographically and etched according to Kroll (3 ml HF + 6 ml HNO_3 + 100 ml H_2O). In addition, Vickers hardness tests, according to DIN EN ISO 6507-1 (HV1, test load 9.807 N), were carried out by taking 20 hardness values from the left, top, right or bottom of the samples' cross-sections (Fig. 1, right).

3 Results and Discussion

Figure 2 shows the relative densities with the respective standard deviations of the compacts produced from GE48 after mixing different proportions of elemental titanium or aluminium (left) and titanium and aluminium (Ti-Al) as well as titanium, aluminium, chromium and niobium (Ti-Al-Cr-Nb, right). Most standard deviations are too small to be recognisable in both diagrams. At 600 MPa pressing load, 40% aluminium, 80% titanium, 70% Ti-Al and 60% Ti-Al-Cr-Nb are required for the successful production of a compact. At 300 MPa, 10% more of each of the elemental powders are required. An exception is the powder mixture of GE48 and titanium, for which a compact in one piece could not be produced with 300 MPa pressing load. All curves have in common that the relative density increases with decreasing GE48 content. The highest relative densities are achieved with the addition of aluminium. With increasing aluminium content, the relative density increases from about 82% (300 MPa pressing load) or 84% (600 MPa pressing load) to about 98% regardless of the pressing load. Thus, there is a pressing load after which the relative density does not change significantly [23]. The influence of the admixture of titanium on the relative density is significantly greater than with aluminium. Within a 20% difference in concentration, the relative density increases from 72% to about 86% at 100% titanium. A high fluctuation in the results is observed there. One possible reason could have been different loads during ejection. These can be attributed to fluctuating friction values on the die wall, which may have been caused by an uneven application of lubricant. Then, the higher ejection loads may have led to an expansion of the pressed pellets.

Nevertheless, the maximum relative density of titanium of 86 ± 5% is below the maximum density of aluminium of 98 ± 1%. Due to the same test conditions for both added materials, the influence of pressing load, particle size, particle geometry and lubricant can be excluded. A possible explanation could lie in the different material properties of titanium and aluminium. Aluminium has a lower yield strength and better ductility than titanium and can therefore be pressed to higher densities. The yield strength of GE48 is significantly higher and the ductility significantly lower than that of titanium or aluminium, which means that the relative density of the mixtures increases as the content of GE48 decreases [14].

This observation is confirmed by similar behaviour when pressing different proportions of Ti-Al. At 600 MPa pressing load and depending on the content of powder added, the relative densities are between those with aluminium added and those with titanium added. At 300 MPa pressing load, the relative densities are significantly lower [24]. The additional admixture of niobium and chromium had no significant influence on the respective relative densities. A possible reason could be that the amounts added were simply too small to have a visible influence. However, an intact compact could be produced with a 10% lower proportion of elemental mixture (Ti-Al-Cr-Nb) at both pressures tested.

In contrast to pressing with admixed aluminium, there is no mixing ratio of the other powder mixtures at which a similar relative density is achieved regardless of the pressure. Consequently, the comparatively low density values could also be increased by further raising the pressing load [23].

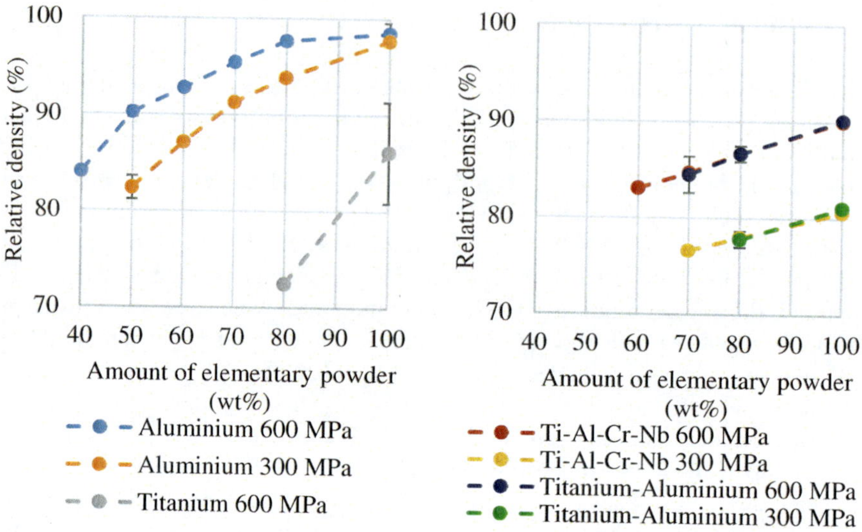

Fig. 2. Relative density as a function of different proportions of elemental titanium or aluminium powder (left) and titanium, aluminium, chromium and niobium (Ti-Al-Cr-Nb) or titanium and aluminium powder (right)

Figure 3 illustrates the results with the respective standard deviations of the tensile splitting tests as a function of different amounts of titanium or aluminium (left) and Ti-Al as well as Ti-Al-Cr-Nb. As in Fig. 2, most of the standard deviations are too small to be visible in both diagrams. Analogous to the relative density, the tensile splitting strength of the compacts tends to increase with rising amounts of elemental powder. Exceptions to this are the strengths at 100% aluminium and 100% titanium at 600 MPa pressing load. At these two values, the fluctuations are comparatively high. Uneven lubricant application may have led to increased pressing loads during ejection due to locally increased friction factors, resulting in optically non-visible pre-damage. Furthermore, the tensile splitting strengths at 40% (600 MPa pressing load) and 50% (300 MPa pressing load) aluminium are comparatively low with about 0.5 MPa, so that these admixture proportions represent the lower process limit above which an intact compact is producible. The tensile splitting strengths of Ti-Al of 3.7 ± 0.7 MPa to 5.6 ± 0.4 MPa (600 MPa pressing load) and from 0.8 ± 0.1 MPa to 2.1 ± 0.4 MPa (300 MPa pressing load) tend to be below the equivalent values of the samples with aluminium admixture, analogous to the corresponding relative densities. At 300 MPa pressing load, the strength values of the samples mixed with Ti-Al-Cr-Nb are similar to the values for the Ti-Al mixtures. In contrast, the tensile splitting strength values of the Ti-Al-Cr-Nb blended samples pressed at 600 MPa are below those of the Ti-Al blended samples. An exception is the strength for 100% Ti-Al-Cr-Nb, which is 7.9 ± 0.3 MPa and thus clearly above the maximum determined value for aluminium 6.2 ± 0.1 MPa. Consequently, the strongest particle bond was present with this composition and at this pressing load. One possible explanation could be the different particle geometries. At 100% of the elemental mixture, the highest proportion of angular particles is present. These enable better particle bonding compared to the spherical particles at e.g. 100% aluminium ultimately manifesting itself in a higher tensile splitting strength [25].

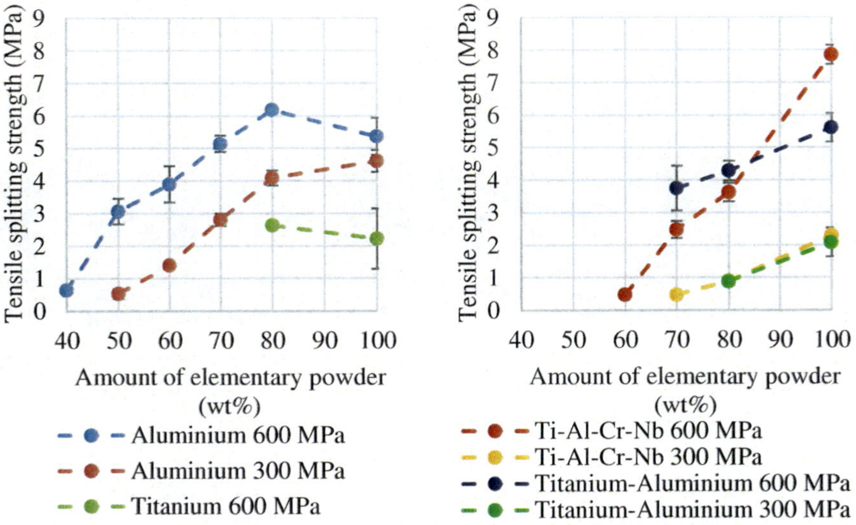

Fig. 3. Tensile splitting strength as a function of different proportions of elemental titanium or aluminium powder (left) and titanium, aluminium, chromium and niobium (Ti-Al-Cr-Nb) or titanium and aluminium powder (right)

Figure 4 shows two exemplary microstructures of sintered 60% Ti-Al-Cr-Nb with 40% GE48 on the left and 100% Ti-Al-Cr-Nb mixture on the right. The spherical powder particles with a visible structure in it are titanium, the grey ones without recognisable structure are GE48 powder, and the light grey angular particles are chromium, while the darker angular particles are niobium. At 40% GE48 (60% Ti-Al-Cr-Nb, left), significantly more pores (dark areas) are visible than at 0% GE48 (100% Ti-Al-Cr-Nb, right), which is consistent with the determined relative densities; these are significantly lower at 60% Ti-Al-Cr-Nb admixture than at 100% admixture. Apart from the decrease in the GE48 content and the simultaneous increase in the other elements, no major difference is visible. With the exception of the aluminium powder, the powder particles are in their original state. Consequently, they show no signs of diffusion processes after sintering. Usually, the temperatures of pressureless solid phase sintering are just below the melting point of a material [26]. Due to the great differences in the melting points of the individual components, e.g. 660 °C for aluminium and 2477 °C for niobium, complete solid phase sintering cannot be achieved.

The high fluctuation of the hardness values confirms incomplete sintering. Figure 5 shows the hardness curves for 100% Ti-Al-Cr-Nb as an example. The hardness curves of the other samples do not differ significantly in the hardness values.

The fluctuations tend to decrease with decreasing GE48 content, as the proportion of pores also diminishes. The hardness values averaged over the sample's width (horizontal) and the sample's height (vertical) according to Fig. 1 (right) are shown in Table 2. The

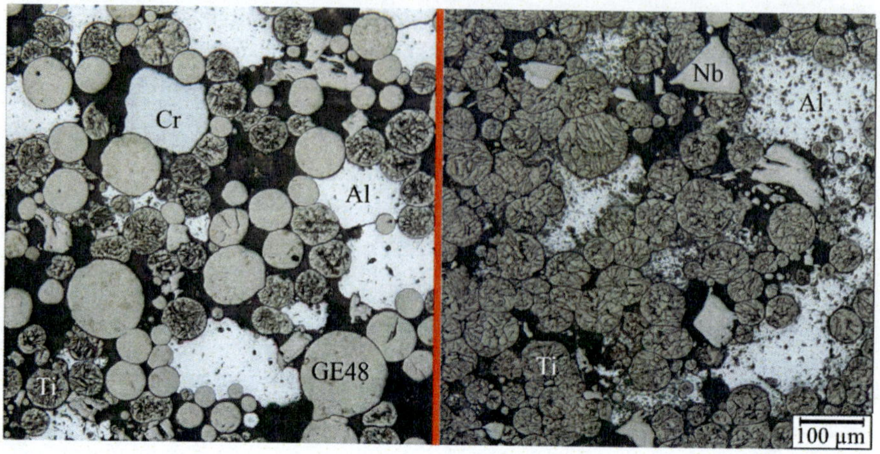

Fig. 4. Microstructure of 40% GE48 with 60% Ti-Al-Cr-Nb (left) and 100% Ti-Al-Cr-Nb (right) sintered for 3 h at 550 °C in low vacuum

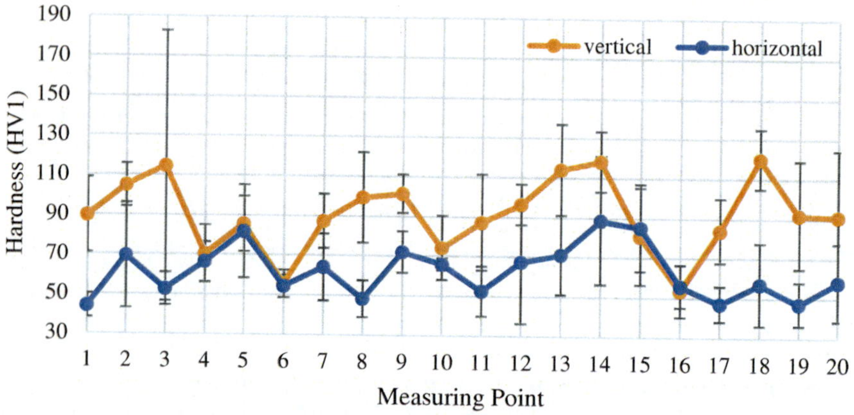

Fig. 5. Hardness profiles in vertical and horizontal direction from top or left to bottom or right according to Fig. 1 (right) of the 100% Ti-Al-Cr-Nb sample, sintered at 550 °C

hardness also tends to decrease with decreasing GE48 content. This is probably due to the higher initial hardness of GE48 compared to the elemental powders. With decreasing GE48 content, there is also more e.g. softer aluminium powder in the sample, which ultimately reduces the hardness. However, the maximum hardness determined is about 106.5 HV1, which is significantly lower than that of the solid material sintered in the FAST process (about 300 to 326 HV1) and again confirms incomplete sintering [27].

4 Summary and Outlook

Within the scope of the investigations, the influence of the admixture of different proportions of elemental powders of the main alloying elements and all alloying elements

Table 2. Hardness in HV1 averaged over the sintered sample width (vertical) and sample height (horizontal) for the sintered samples

	40% GE48	30% GE48	20% GE48	0% GE48
Horizontal (HV1)	99.9 ± 29.9	76.3 ± 23.1	75.0 ± 23.3	62.4 ± 15.4
Vertical (HV1)	106.5 ± 26.3	97.7 ± 22.2	105.6 ± 24.2	91.0 ± 20.0

of GE48 on the processing in conventional die compaction could be examined. It was demonstrated that as the proportion of GE48 decreases, the relative density and tensile splitting strength generally increase. Considering the tensile splitting strength as an indicator of particle bonding, it could be shown that the bond of the individual particles improves with an increasing proportion of elemental powders. This is attributable to the different powders used: compacting is generally easier with a ductile material with low yield strength (aluminium) than with a comparatively less ductile material with high yield strength (titanium) or a brittle material (GE48). The mixtures of both elements showed analogous behaviour. Depending on the pressing load, significantly lower proportions of aluminium, 40% at 600 MPa and 50% at 300 MPa, are required for the successful production of a pressed part than with titanium, 80% at 600 MPa. With the addition of titanium and a pressing load of 300 MPa, it was not possible to produce an intact compact. A possible remedy could be to increase the pressing load. The samples sintered at 550 °C with different admixtures of elemental titanium, aluminium, chromium and niobium could not be completely sintered according to the light microscope images. Only the aluminium participated in the sintering process, while the other components remained in their initial powder state. The strong fluctuation of the hardness values confirms this. The hardness ranges from about 62.5 to 106.5 HV1. In a further study, the influence of different sintering temperatures on the resulting microstructure and mechanical properties can be considered. In order to allow all elements to participate in the sintering process, solid phase sintering can be carried out under pressure or it can be sintered in the liquid phase of the aluminium, which is then sintered at significantly higher temperatures.

Acknowledgements. Funded by the Deutsche Forschungsgemeinschaft (DFG, German Research Foundation)—Project-ID 394563137—SFB 1368.

References

1. Liu, B., Liu, Y., Zhang, W., Huang, J.S.: Hot deformation behavior of TiAl alloys prepared by blended elemental powders. Intermetallics **19**(2), 154–159 (2011)
2. Gerling, R., Clemens, H., Schimansky, F.P.: Powder metallurgical processing of intermetallic gamma titanium aluminides. Adv. Eng. Mater. **6**(1–2), 23–38 (2004)
3. Henriques, V.A.R., Galvani, E.T., Cairo, C.A.A., Graça, M.L.A., Dutra, A.C.S.M.: Microstructural investigation of routes for gamma titanium aluminides production by powder metallurgy. Powder Metall. Titanium **II**(704), 204–213 (2016)

4. Bohr, D., Petersen, T., Brunotte, K., Behrens, B.-A.: Influence of friction-reducing powder-compaction tool coatings on green-compact properties. In: Behrens, B.-A., Brosius, A., Drossel, W.-G., Hintze, W., Ihlenfeldt, S., Nyhuis, P. (eds.) WGP 2021. LNPE, pp. 349–356. Springer, Cham (2022). https://doi.org/10.1007/978-3-030-78424-9_39
5. Hong, S.-T., Hovanski, Y., Lavender, C.A., Weil, K.S.: Investigation of die stress profiles during powder compaction using instrumented die. J. Mater. Eng. Perform. **17**(3), 382–386 (2008)
6. Solimanjad, N., Larsson, R.: Die wall friction and influence of some process parameters on friction in iron powder compaction. Mater. Sci. Technol. **19**(12), 1777–1782 (2013)
7. Simchi, A.: Effects of lubrication procedure on the consolidation, sintering and microstructural features of powder compacts. Mater. Des. **24**(8), 585–594 (2003)
8. Dai, H., et al.: Iron based partially pre-alloyed powders as matrix materials for diamond tools. Powder Metall. **58**(2), 83–86 (2015)
9. Yamaguchi, M., Inui, H., Ito, K.: High-temperature structural intermetallics. Acta Mater. **48**(1), 307–322 (2000)
10. Morris, M.A.: Dislocation mobility, ductility and anomalous strengthening of two-phase TiAl alloys: effects of oxygen and composition. Intermetallics **4**(5), 417–426 (1996)
11. Kawabata, T., Abumiya, T., Izumi, O.: Effect of oxygen addition on mechanical properties of TiAl at 293–1273 K. Acta Metall. Mater. **40**(10), 2557–2567 (1992)
12. Wang, Y.H., Lin, J.P., He, Y.H., Wang, Y.L., Chen, G.L.: Microstructures and mechanical properties of Ti–45Al–8.5Nb–(W,B,Y) alloy by SPS–HIP route. Mater. Sci. Eng.: A **489**(1), 55–61 (2008)
13. Ransing, R.S., Gethin, D.T., Khoei, A.R., Mosbah, P., Lewis, R.W.: Powder compaction modelling via the discrete and finite element method. Mater. Des. **21**(4), 263–269 (2000)
14. St-Laurent, S., Chagnon, F., Thomas, Y.: Study of Compaction and Ejection Properties of Powder Mixes Processed by Warm Compaction
15. Gethin, D.T., Lewis, R.W., Ransing, R.S.: A discrete deformable element approach for the compaction of powder systems. Modell. Simul. Mater. Sci. Eng. **11**(1), 101 (2002)
16. Heckel, R.W.: Density-pressure relationships in powder compaction. Trans. Metall. Soc. AIME (221), 671–675 (1961)
17. Tiwari, S., Rajput, P., Srivastava, S.: Densification Behaviour in the Fabrication of Al-Fe Metal Matrix Composite Using Powder Metallurgy Route. ISRN Metallurgy 2012 (2012)
18. Anand, S.S., Mohan, B.: Effect of particle size, compaction pressure on density and mechanical properties of elemental 6061Al alloy through powder metallurgical process. Int. J. Mater. Eng. Innov. **3**(3/4), 259 (2012)
19. Dixit, M., Srivastava, R.K.: Effect of compaction pressure on microstructure, density and hardness of Copper prepared by Powder Metallurgy route. IOP Conf. Ser.: Mater. Sci. Eng. **377**(1), 12209 (2018)
20. Kumar, N., Bharti, A., Saxena, K.K.: A re-investigation: effect of powder metallurgy parameters on the physical and mechanical properties of aluminium matrix composites. Mater. Today: Proc. **44**, 2188–2193 (2021)
21. Behrens, B.A., Brunotte, K., Bohr, D.: Experimental investigation of endogenous lubrication during cold upsetting of sintered powder metallurgical components. Tribol. Manuf. Process. Joining Plastic Deformation **II**(767), 163–170 (2018)
22. DIN EN 12390-6:2010-09. Testing Hardened Concrete—Part 6: Tensile Splitting Strength of Test Specimens. Beuth Verlag GmbH (2010)
23. Schatt, W.: Pulvermetallurgie. Technologien und Werkstoffe. Springer, Heidelberg (2007)
24. Solay Anand, S., Mohan, B.: Effect of particle size, compaction pressure on density and mechanical properties of elemental 6061Al alloy through powder metallurgical process. Int. J. Mater. Eng. Innov. **3**(3–4), 259–268 (2012)

25. Bourcier, D., et al.: Influence of particle size and shape properties on cake resistance and compressibility during pressure filtration. Chem. Eng. Sci. **144**, 176–187 (2016)
26. Gökçe, A., Findik, F., Kurt, A.O.: Effects of sintering temperature and time on the properties of Al-Cu PM alloy. Pract. Metallography **54**(8), 533–551 (2017)
27. Behrens, B.-A., Brunotte, K., Peddinghaus, J., Heymann, A.: Influence of dwell time and pressure on SPS process with titanium aluminides. Metals **12**(1), 83 (2022)

Concept for In-process Measurement of Residual Stress in AM Processes by Analysis of Structure-Borne Sound

J. Groenewold(✉), F. Stamer, and G. Lanza

wbk Institute of Production Science, Karlsruhe Institute of Technology (KIT), Kaiserstr. 12, 76131 Karlsruhe, Germany
jork.groenewold@kit.edu

Abstract. Process-induced residual stress is a major challenge in today's additive manufacturing (AM) processes, such as powder bed fusion by laser beam melting of metal. After the AM process, the exact stress state is usually unknown, and parts often require heat treatment to relieve residual stress. In-process measurement of residual stress is currently not possible. This paper presents a concept to derive the measurement of the residual stress by analyzing the structure-borne sound induced during the AM process. The first step of the concept is to integrate a device into a build plate to set a defined mechanical load during the manufacturing process. Then, samples can be fabricated on this build plate in several steps. By applying mechanical load with the device, the stress state in the samples can be changed between the fabrication steps. During this stepwise fabrication process, the structure-borne sound signal is recorded. Subsequently, the correlation between the stress states and the acoustic process emissions is analyzed using FFT, STFT and cross-spectral analyses. The overall goal is to establish a model to determine residual stress in AM components by evaluating the acoustic process emissions.

Keywords: Additive manufacturing · Residual stress · Structure-Borne sound · Quality control

1 Introduction

Due to the high freedom of design and the processing of a wide range of materials, additive manufacturing is becoming increasingly widespread, so that the technology has changed from "rapid prototyping" to series production of components [1]. In powder bed fusion (PBF), a coater applies thin layers of powder to a build plate or previous layers in a build chamber under an inert gas atmosphere or in a vacuum, and then an energy source (e.g. laser) selectively melts the powder. During this process, powder particles bond with each other and with previously manufactured layers. Unused powder is removed after the process and recycled [1].

One challenge in PBF processes is the formation of residual stresses during the manufacturing process. The local melting of metal powder leads to temperature gradients and eventually to the formation of thermal residual stresses in the component, which can

M. Liewald et al. (Eds.): WGP 2022, LNPE, pp. 24–31, 2023.
https://doi.org/10.1007/978-3-031-18318-8_3

lead to cracking in the build-up process or to component distortion [2, 3]. In addition, residual stresses have effects on the fatigue strength, static strength, chemical resistance, and anisotropy of the manufactured components [4, 5].

2 State of the Art

Kruth et al. [3] have identified several influencing factors for the reduction of residual stresses in PBF processes. For example, component residual stresses can be reduced by specific settings of process parameters, such as laser power, scanning speed, build temperature and exposure strategy, as well as by the appropriate use of support structures. Despite these factors, it has not yet been possible to completely suppress the formation of residual stresses and the associated defect patterns [3].

A variety of technologies exist for characterizing the residual stress state of components, such as X-ray diffraction or mechanical methods, such as the contour method [6]. These methods can only be used off-process. Another option for measuring residual stresses is the use of ultrasound [7]. This technology is based on the acousto-elastic effect, which describes the relationship between the propagation velocities of ultrasonic waves in bodies as well as the prevailing stress state [8]. The propagation velocity is measured, for example, via time-of-flight measurements of ultrasonic pulses. For this purpose, an ultrasonic pulse is coupled into the sample with a piezoelectric transmitter and the transmitted signal is measured with a receiver. The propagation velocity can be inferred from the time difference between the input and output signal and the sample geometry. Roy et al. and Holoch et al. have carried out extensive investigations for the characterization of the propagation velocities of sound waves [9, 10].

In laser-based manufacturing processes, structure-borne sound is emitted in the component as a result of exposure to the laser [11]. Eschner et al. have carried out investigations into structure-borne sound in laser-based additive manufacturing (PBF-L/M). For this purpose, a structure-borne sound sensor was integrated into a PBF-L/M system. Through this, sound emissions occurring during the manufacturing process could be recorded and correlated with the formation of defects [12, 13]. Furthermore, the recorded signals can be assigned to an area within the manufactured component by matching them with the laser trajectory.

3 Objective and Approach

The central hypothesis of this work is that the structure-borne sound emitted during the PBF-L/M process correlates with the resulting residual stress. Assuming this hypothesis is true, residual stress could be measured time and cost efficiently improving the industrialization of the PBF-L/M technology. Based on this hypothesis and motivation, the objective of this work is to develop a concept pursuing the goal of verifying the described correlation. The general approach of this concept is to measure and compare the structure-borne sound signals during additive manufacturing on test specimens with different stress states. The basic experimental setup is shown in Fig. 1.

To implement this idea, the proposed concept consists of three steps. The first step of this concept handles the creation of a defined mechanical load causing different stress

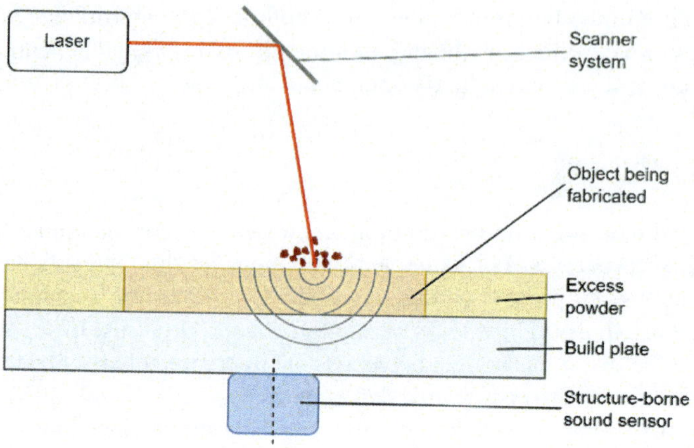

Fig. 1. Experimental setup for measuring structure-borne sound

states. In order to create this mechanical load, a device for setting a mechanical load on test specimens has to be designed. In addition, suitable test specimens need to be designed. The second step describes the execution of experiments to record structure-borne sound data. In the third step, the data obtained is analyzed to check whether there is a correlation between the stress state and the recorded structure-borne sound signal.

3.1 Device for Setting a Mechanical Load and Design of Test Specimens

The purpose of the device is to introduce a defined mechanical load into the test specimens and thus create a defined stress condition. For the design of the fixture and the test specimens, several requirements must be considered. First, the mechanical load must be changeable without having to open the process chamber. Fortunately, there is an access under the build plate that can be used if the device is advantageously designed. Second, the device must be integrated within the specified dimensions of the build plate and without protruding parts so as not to obstruct the coater of the additive manufacturing machine. Finally, the sensor measuring the structure-borne sound must be integrated under the build plate.

Considering these requirements, the device shown in Fig. 2 was designed. As geometry for the test specimens a cut-open ring was chosen. By spreading this ring with a wedge, which is tightened with a screw using a torque wrench, the mechanical load can be easily increased. In a simulation, a tensile stress of approximately 190 MPa is generated at a displacement of 1.3 mm, which corresponds to a load of 100 N. Due to this design of the device and specimen, no force flow is introduced into the build plate, unlike, for example, a design with flat bending specimens, so that the entire device can be built very compactly and with low complexity.

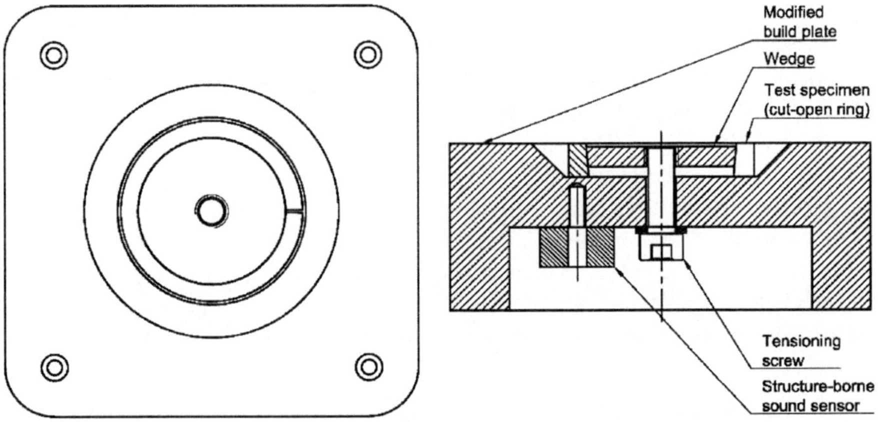

Fig. 2. Top view (left) and cross-sectional view (right) of the concept for the device for setting a mechanical load

3.2 Experiments

The aim of the experiments is to collect data of structure-borne sound signals under different stress conditions in order to analyze them later and determine possible correlations. It is to be expected that the actual additive manufacturing process will influence the recorded signal, e.g. through the powder, possibly formed pores or the change in geometry. For this reason, experiments are carried out first in which no powder is used and thus there is no influence of the additive manufacturing process. In further experiments the additive manufacturing process is carried out conventionally, i.e. with powder. In order to additionally investigate an influence of the heat input of the laser on the measurement results, the tests are performed with two load curves: from no load to maximum load and vice versa. This way, possible influences of the experiment order can be eliminated. In preparation for the experiments, the milled specimens are first stress-relief heat treated to relieve any residual stresses that may be present. The planned experiments are listed in Table 1 and are described below.

Table 1. Experimental plan

Experiment Nr	Specimen Nr	With (✓) / without (–) powder	Load curve
1	1	–	
2	2	–	
3	3	✓	
4	4	✓	

For the experiments without powder, a print job with one layer with the geometry of the specimen is created and the machine is prepared by inserting the presented device including the previously manufactured specimen. Subsequently, the print job is executed without applying powder and the structure-borne sound signal is recorded. The next step is to increase the mechanical load by one tenth of the maximum load, for example. This changes the stress state in the specimen. After that, the print job is executed a second time, and the structure-borne sound signal is also recorded. These steps are repeated ten times until the maximum load is reached. The maximum load can be determined by simulating the stress distribution in the test specimen, as shown in Fig. 3. In consideration of the mechanical properties, a maximum load of 100 N is selected so that the specimens deform only elastically and plastic deformation is prevented.

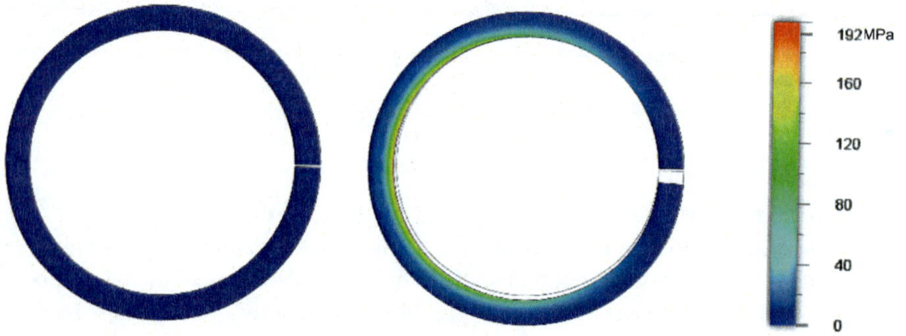

Fig. 3. Simulation of the 1st principal stress in the ring as an effect of no load (left) and a mechanical load of 100N (right)

For the experiments with powder, the influence of the additive manufacturing process is considered. For this purpose, a print job with a test specimen of several layers is created. Then a few layers are produced on the milled test specimen while it is not under load. After that, the manufacturing process is stopped, the load is increased and the manufacturing process continues. During this step-by-step manufacturing process, the structure-borne sound signal is recorded. All experiments are then followed by a comparative study of the signals recorded in the individual process steps, which will be discussed in the next section.

After the experiments, measures are taken to validate the results. In order to determine the actual stress state, one measure is to investigate the stresses that occur using reference measurement techniques. The X-ray diffraction method can be used for this purpose. In order to characterize the repeatability of the stresses generated by the proposed device, a measurement is performed with all samples applying the same mechanical load and the results are then compared. The sonic velocity of additively manufactured components is influenced not only by the residual stress state present but also by other effects, such as porosity. In order to analyze the influence of these effects, computer tomographic as well as metallographic investigations of the built-up test specimens are to be carried out as a second measure [14].

3.3 Analysis

During the experiments, data sets for the studies with and without an additive manufacturing process are recorded for different load conditions. The objective is now to check whether there is a correlation between the stress state and the recorded structure-borne sound. For this purpose, the signals from low stress states are compared with those from high stress states.

A simple way to perform this comparison is to compare the acquired signal under the different load settings of the instrument for each test, e.g. a setting with no load and a setting with maximum load for a first test. A second method of extracting data from different stress states is based on the idea that the stress within the test specimen is not constant, but has a spatial distribution. For example, the simulation shown in Fig. 3 indicates a high tensile stress at the inner edge and a low stress at the outer edge. By pre-processing the structure-borne sound signal with software developed in preliminary work at the institute (soon to be published), the signals can be related to the location where they originated. Based on this assignment and in consideration of the calculated stress in the simulation, the signals generated under different stress conditions can be compared.

Regardless of the method used to extract data from different stress states, the structure-borne sound signals must be analyzed and compared. For this purpose, an existing data processing setup is used, which was developed in previous work [13]. The data processing setup consists of several steps. First, the data is decoded from a proprietary file format and then converted to the frequency domain using Short-time Fourier transform (STFT) [12, 13]. Noise is still present in the resulting spectrogram, for example due to stepper motors installed in the system. For this reason, noise suppression is performed by applying a difference mask that subtracts the noise from the spectrogram [12, 13].

After the data has passed through this processing setup, it can be analyzed. If the hypothesis of this work is true, it is expected that a change can be seen in the processed data, e.g. a shift of frequencies or differences in amplitude. Therefore, visualizations of spectrograms from different stress states are compared manually, e.g., by subtracting them from each other, to check for obvious anomalies or features. In addition, more sophisticated methods such as the DBSCAN algorithm can be used. The DBSCAN algorithm is an unsupervised learning method that is capable of grouping data based on their density and which works well despite the presence of noise [15]. Another method for comparing spectrograms of different stress states is to perform a cross-spectral analysis. This method is analogous to linear regression, but functions for data in the frequency domain. It therefore allows to reveal correlations in spectra [16].

If the experiments and analysis show a correlation between stress states and the acquired structure-borne sound signal, it is desirable to extend this concept into a more sophisticated model predicting residual stress.

3.4 First Results

In a first test, we set up a build plate as shown in Fig. 1 and focused the laser on exactly centered on the build plate. Without applying any powder, the laser was then switched on and the structure-borne sound signal was recorded. This signal was then processed using STFT. The results are shown in Fig. 4.

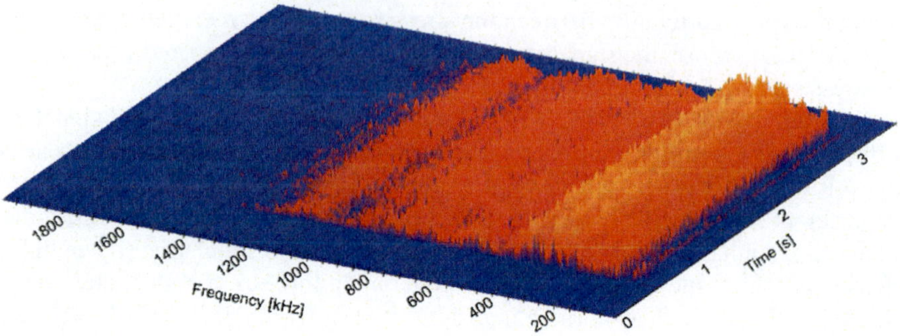

Fig. 4. STFT spectrogram (with noise suppression) of the structure-borne sound signal generated by melting of material

It can clearly be seen in Fig. 4 that it is possible to detect a signal of material melting using the structure-borne sound sensor. Further investigations are now required to determine whether this signal correlates with the residual stresses at the location of the interaction.

4 Conclusion and Outlook

This work addresses the development of new, intelligent sensor technology for additive manufacturing in order to make the manufacturing process robust for use in highly versatile production systems. The residual stresses that arise in additive manufacturing have been a hurdle so far as the process parameters have to be adjusted empirically until the component meets the required properties. This is in contradiction to the idea of highly flexible production that additive manufacturing promises.

If the results are positive, there is an opportunity for follow-up projects that address research into the measurement of residual stress states and the spatially resolved measurement of residual stresses using the structure-borne sound signal. In perspective, this can be used to selectively adjust process parameters and control residual stress states. In-process measurement of residual stresses during the additive manufacturing process would reduce process uncertainties, cost of treatment as well as time and, thus, enable the cost efficient production of higher-quality products.

References

1. Gibson, I., Rosen, D., Stucker, B.: Additive Manufacturing Technologies. 3D Printing, Rapid Prototyping and Direct Digital Manufacturing. Springer, New York, Heidelberg, Dodrecht, London (2015)
2. Mercelis, P., Kruth, J.-P.: Residual stresses in selective laser sintering and selective laser melting. Rapid Prototyping J. (2006). https://doi.org/10.1108/13552540610707013

3. Kruth, J.-P., Deckers, J., Yasa, E., Wauthlé, R.: Assessing and comparing influencing factors of residual stresses in selective laser melting using a novel analysis method. In: Proceedings of the Institution of Mechanical Engineers, Part B: Journal of Engineering Manufacture (2012). https://doi.org/10.1177/0954405412437085
4. Habschied, M., de Graaff, B., Klumpp, A., Schulze, V.: Fertigung und Eigenspannungen*. HTM J. Heat Treatment Mater. (2015). https://doi.org/10.3139/105.110261
5. Acevedo, R., Sedlak, P., Kolman, R., Fredel, M.: Residual stress analysis of additive manufacturing of metallic parts using ultrasonic waves: state of the art review. J. Market. Res. (2020). https://doi.org/10.1016/j.jmrt.2020.05.092
6. Prime, M.B.: Cross-sectional mapping of residual stresses by measuring the surface contour after a cut. J. Eng. Mater. Technol. (2001). https://doi.org/10.1115/1.1345526
7. Bergman, R.H., Shahbender, R.A.: Effect of statically applied stresses on the velocity of propagation of ultrasonic waves. J. Appl. Phys. (1958). https://doi.org/10.1063/1.1723035
8. Li, Z., He, J., Teng, J., Wang, Y.: Internal stress monitoring of in-service structural steel members with ultrasonic method. Materials (Basel, Switzerland) (2016). https://doi.org/10.3390/ma9040223
9. Roy, S., Gebert, J.-M., Stasiuk, G., Piat, R., Weidenmann, K.A., Wanner, A.: Complete determination of elastic moduli of interpenetrating metal/ceramic composites using ultrasonic techniques and micromechanical modelling. Mater. Sci. Eng. A (2011). https://doi.org/10.1016/j.msea.2011.07.029
10. Holoch, J., Czink, S., Spadinger, M., Dietrich, S., Schulze, V., Albers, A.: SLM-Topo—Prozessspezifische Topologieoptimierungsmethode für im Selektiven Laserschmelzen gefertigte Leichtbaustrukturen. Industrie 4.0 Management **36**, 45 (2020)
11. Saifi, M., Vahaviolos, S.: Laser spot welding and real-time evaluation. IEEE J. Quantum Electron. (1976). https://doi.org/10.1109/JQE.1976.1069104
12. Eschner, N., Weiser, L., Häfner, B., Lanza, G.: Development of an Acoustic Process Monitoring System for Selective Laser Melting (SLM) (2018)
13. Eschner, N., Weiser, L., Häfner, B., Lanza, G.: Classification of specimen density in Laser Powder Bed Fusion (L-PBF) using in-process structure-borne acoustic process emissions. Addit. Manuf. (2020). https://doi.org/10.1016/j.addma.2020.101324
14. Englert, L., Czink, S., Dietrich, S., Schulze, V.: How defects depend on geometry and scanning strategy in additively manufactured AlSi10Mg. J. Mater. Process. Technol. (2022). https://doi.org/10.1016/j.jmatprotec.2021.117331
15. Ester, M., Kriegel, H.P., Sander, J., Xiaowei, X.: A Density-Based Algorithm for Discovering Clusters in Large Spatial Databases with Noise, CONF-960830. AAAI Press, Menlo Park, CA (United States). https://www.osti.gov/biblio/421283-density-based-algorithm-discovering-clusters-large-spatial-databases-noise
16. Martinson, D.G.: Cross-spectral analysis. In: Martinson, D.G. (ed.) Quantitative Methods of Data Analysis for the Physical Sciences and Engineering, pp. 406–424. Cambridge University Press, Cambridge, UK, New York, NY (2018)

Characterisation and Modelling of Intermetallic Phase Growth of Aluminium and Titanium in a Tailored Forming Process Chain

N. Heimes[1](✉), H. Wester[1], O. Golovko[2], C. Klose[2], H. J. Maier[2], and J. Uhe[1]

[1] Institut für Umformtechnik und Umformmaschinen (Institute of Forming Technology and Machines), Leibniz Universität Hannover, 30823 Garbsen, Germany
heimes@ifum.uni-hannover.de

[2] Institut für Werkstoffkunde (Materials Science), Leibniz Universität Hannover, 30823 Garbsen, Germany

Abstract. The combination of aluminium ($AlSi_1MgMn$) and titanium (Ti6Al6-4V) allows producing components with high lightweight potential and at the same time high strength and chemical resistance. Upon joining of dissimilar materials, intermetallic phases (IMP) can form. These are comparatively hard and brittle and represent a weak point in the hybrid component. Along the process chain for manufacturing a hybrid bearing bushing made of $AlSi_1MgMn$ and Ti6Al-4V by co-extrusion, die forging and heat treatment, the joining zone is exposed to high thermal loads. As a result, the individual process steps can lead to the growth of IMP reducing the compound's quality. In order to investigate the formation and the growth of IMP at process-relevant temperatures and contact times in detail, experimental analogy tests were carried out. Subsequently, the specimens were examined by scanning electron microscopy. Due to the constant temperature and the respective contact time, the diffusion coefficient was calculated from the determined phase thickness using the Einstein-Smoluchowski equation. This allowed describing the diffusion coefficients as a function of temperature and implementing them into a finite element model via a subroutine. To validate the subroutine, further tests were carried out and the calculated phase thickness was validated with experimentally determined phase thickness, which exhibited good correlation.

Keywords: Tailored forming · Aluminium-titanium compound · Intermetallic phases · Finite element method

1 Introduction

Demands on technical components in terms of functionality, weight and resource-saving production are continuously rising. One way to meet the increased requirements is the use of hybrid components. However, manufacturing of innovative hybrid components also requires adapted process chains. One promising possibility is the use of pre-joined semi-finished goods, called Tailored Forming [1]. One of the challenges in the use

M. Liewald et al. (Eds.): WGP 2022, LNPE, pp. 32–41, 2023.
https://doi.org/10.1007/978-3-031-18318-8_4

of pre-joined hybrid semi-finished products made of dissimilar materials is the brittle microstructure that can develop in the joining zone caused by the thermomechanical treatment within the process chain.

The presented research focuses on the combination of aluminium ($AlSi_1MgMn$) and titanium (Ti6A6-4V) in a coaxial arrangement, which offers the possibility of producing components with high lightweight potential combined with high strength and chemical resistance. In hybrid components, the joining zone of the respective materials represents a potential weak point. Therefore, the goal is to achieve a compound strength that matches or exceeds the strength of the weaker material partner ($AlSi_1MgMn$). The process chain in this study is composed of a Lateral Angular Co-Extrusion (LACE) process, a die forging process and a final heat treatment and machining to produce a hybrid bearing bushing. The LACE process is used to extrude $AlSi_1MgMn$ and Ti6Al-4V into a hybrid semi-finished product. The pre-joined hybrid semi-finished products are later formed into a bearing bushing using the die forging process. Heat treatment provides a final adjustment of the T6 state of the aluminium. In previous work, the respective process steps were numerically modelled for the material combination $AlSi_1MgMn$/20MnCr5 [2]. For the new material combination $AlSi_1MgMn$ and Ti6Al-4V, the focus was on the numerical representation of the growth of the intermetallic phase (IMP) over all individual process steps. In the case of the material combination of $AlSi_1MgMn$/20MnCr5, it was determined that the IMP reached a critical thickness in a short time, and thus the bond strength was reduced [2]. The phase growth for the material combination $AlSi_1MgMn$/Ti6Al-4V is to be investigated numerically along the process chain in order to be able to adapt the experimental process control. For this purpose, it is necessary to determine the parameters influencing the phase growth and to consider the growth of the IMP via a subroutine in a finite element method (FE) model.

A few results on co-extrusion of aluminium and titanium have already been published by different research groups, mostly using prepared extrusion blocks. Engelhardt et al. reinforced an aluminium extrusion block in the centre with a flat section of titanium grade 1 to produce hybrid sheets [3]. After reinforcing an aluminium extrusion block with twelve titanium grade 2 pins, Arunkumar et al. fabricated reinforced cylindrical profiles [4]. In another study, Grittner et al. inserted a Ti6Al-4V core into an $AlSi_1MgMn$ extrusion block and extruded it into a coaxially reinforced profile [5]. Besides mechanically joined extrusion blocks, Grittner et al. additionally examined cast extrusion blocks. Here, the aluminium alloys used were cast around a Ti6Al-4V core [5]. In all cited studies, due to the joint heating phase (400–480 °C at 2–3 h) of the reinforced extrusion blocks, IMP could be detected in the extruded profile, and the phases were significantly larger in the profiles made from cast extrusion blocks compared to the mechanically joined extrusion blocks. An asymmetric rectangular profile made of Al99.5/$AlSi_1MgMn$ and titanium grade 2 was produced by Grittner et al. by means of a LACE process using conventional extrusion blocks. However, no IMP could be detected in the profile after the LACE process. By means of a subsequent heat treatment (540 °C up to 4 h), IMP could be determined [6]. The specimens were subsequently annealed at 540 °C up to 16 h and the growth of the IMP was analysed as well and the bond strength was determined in tensile tests. After heat treatment, the tensile specimen of the combination Al99.5/titanium grade 2 failed in the aluminium base material whereas the combination of $AlSi_1MgMn$/titanium grade 2 failed in the joining zone [7]. Xue et al. numerically investigated the manufacturability of rectangular Ti6Al4V/AA1050 profiles by non-equal channel lateral co-extrusion. Conventional extrusion blocks were used and experimentally implemented [8]. The profiles produced showed no IMP at a extrusion block temperature of 480 °C, but IMP could be detected when the temperature was increased to 520 and 580 °C [9].

Summarizing, the state of the art shows that it is possible to join titanium and aluminium by co-extrusion and heat treatment. A detailed description of the onset of phase growth or the phase growth at a certain temperature is not known yet. As a result, the phase growth as a function of temperature cannot be accurately described yet.

Behrens et al. determined the IMP thickness for $AlSi_1MgMn$ and 20MnCr5 by means of analogy experiments [10] as a function of temperature as well as contact time and were thus able to describe the diffusion growth of the IMP using the Einstein-Smoluchowski equation [11]. The developed model was subsequently implemented into the FE-program Forge NxT 2.1 via a user-subroutine and validated by Uhe on the basis of the analogy experiments as well as LACE experiments [12]. In the present study, the phase growth as a function of temperature and time are determined for $AlSi_1MgMn$ and Ti6Al-4V. In addition, the respective diffusion coefficients are calculated. The determined diffusion coefficients are transferred into a subroutine and implemented in the FE program Forge NxT 3.0. Finally the subroutine was validated via further experimental tests.

2 Material and Method

Analogy tests were carried out with a quenching and forming dilatometer (TA Instruments). The test sequence of Uhe [12] and Behrens et al. [10, 11] was extended by a holding phase without loading representing the heat treatment process. In Fig. 1a the extended test sequence is shown. The experimental setup represented in Fig. 1b) consists of two Ti6Al-4V cylinders and one $AlSi_1MgMn$ cylinder, which are inserted into a steel sleeve. Prior to the experiments, the cylindrical specimens were ground and polished on the end surfaces to remove oxide layers in the joining zone. After reaching a vacuum of 3.5×10^{-4} mbar, the specimen was heated to test temperature via inductive heating and held for 4 min to achieve a homogeneous temperature distribution. Subsequently, the $AlSi_1MgMn$ specimen was formed, followed by a holding phase with superimposed loading, resulting in a hybrid specimen. After unloading, the extended holding phase without mechanical loading took place, representing the slow cooling during extrusion and heat treatment. Finally, the hybrid specimen was cooled to room temperature using nitrogen inert gas. In order to describe the formation and the individual growth of the IMP along the process chain, the boundary conditions were defined considering the individual processes. The lower test temperature is defined by the LACE process as 450 °C [12]. Peak temperatures in the joining zone in die forging of hybrid workpieces can reach up to 590 °C. The maximum holding time is dictated by the heat treatment and can reach up to 4 h. Due to co-extrusion and subsequent cooling in air, a minimum time of 15 min was selected.

The significantly shorter contact times under load, during die forging as well as in the welding chamber in the LACE process are represented by the forming phase in the

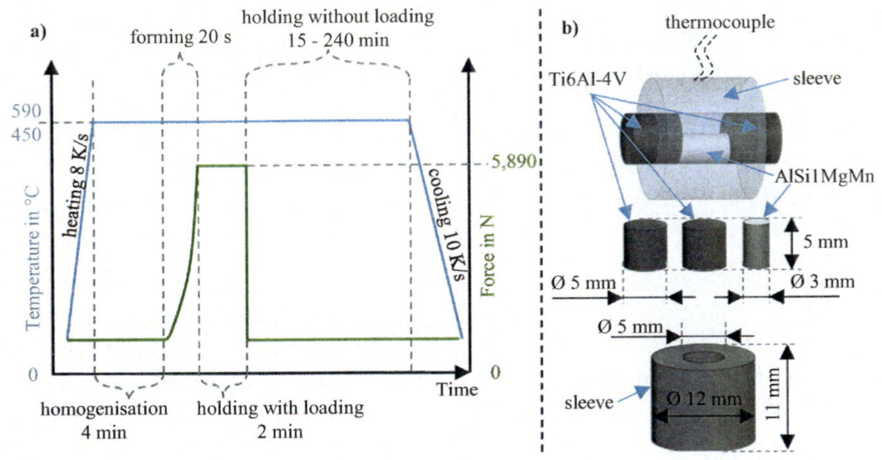

Fig. 1. a) Schematic force-time and temperature-time curves, b) schematic experimental setup of the analogy experiments

analogy test. Derived from this, the test matrix, as shown in Table 1, was investigated and each parameter combination was repeated three times. Grey coloured combinations were used for model generation, orange coloured combinations for validation.

Table 1. Test matrix of the performed analogy experiments

Time in min \ Temperature in °C	450	500	525	550	575	590
15						
30						
45						
60						
120						
240						

After the analogy tests, the specimens were embedded in conductive material and metallographically prepared along the longitudinal section. In the longitudinal section, two joining zones could be examined for each specimen. The analysis of the formation of an IMP in the joining zones was carried out using scanning electron microscopy (SEM) with a Supra 40VP from Zeiss. Specifically, five backscattered electron (BSE) images were taken on each side of the joining zone at a distance of 1 mm per specimen. This allows for a detailed recording of the phase growth over the entire joining zone. The BSE images were subsequently analysed using a MATLAB script developed by Herbst et al. [13]. In general, BSE images contain different shades of grey according to the average atomic number of the material. Lighter elements appear darker and heavy elements brighter. IMPs are composed of at least two different elements, so that the average atomic number often lies between the atomic numbers of the other elements. In the mentioned MATLAB script by Herbst et al., the pixels of the intermetallic phase are detected by classifying the grey tones and line by line evaluation [13].

In the solid state, IMP grow and form via diffusion processes, which can be described via Fick's 1st and 2nd law [14]. However, the concentration gradient of the diffusing materials must be known. Another possibility to describe diffusion is the Einstein-Smoluchowski equation. Here, the diffusion path d is described by the diffusion coefficient D and the diffusion time t, see Eq. 1.

$$d = \sqrt{2 * D * t} \tag{1}$$

In the following, the diffusion path d is equated with the determined IMP thicknesses from the SEM analyses. Thus, the diffusion coefficient can be calculated for each temperature and time investigated. For the calculation, the average phase width is used, so that an average diffusion coefficient is generated. The resulting database is subsequently numerically modeled in order to be able to use the experimentally determined data within FE simulation. Using the fitted data, the diffusion coefficient can be calculated and used in the simulation for each temperature in order to map the phase growth. Details on the practical implementation of the subroutine is given in the work of Uhe [12] and Behrens et al. [11].

3 Results and Discussion

The recorded BSE images were classified based on the appearance of the IMP: continuous IMP (grey), discontinuous IMP (orange) and not evaluable (blue) (Fig. 2). Continuous IMPs are characterised by an IMP that extends over both joining zones of the specimen. In the group of discontinuous IMPs, insular IMP occur that do not show a coherent phase, but which can be clearly identified as IMP. The last group contains all parameter combinations that cannot be clearly detected as IMP or where no IMP were present. In some images, it was not possible to clearly determine whether an IMP, precipitates or alloying elements of the $AlSi_1MgMn$ are present in the joining zone area, due to the resolution of the BSE images. A higher resolution could be achieved by focused ion beam lamellae or transmission electron microscope images, but these were beyond the scope of the study given the large amount of specimens analysed. In Fig. 2 on the right examples of BSE images representing the three categories mentioned above are shown. By means of this classification, process windows were created over the investigated parameter space, from which it can be determined if a formation of the IMP is probable.

The individual process windows are plotted in Fig. 2, whereby the two validation points (crosses) were not taken into account. Both validation points show IMP according to their position in the process window. At 525 °C the IMP is discontinuous and at 575 °C a continuous IMP occurs. Outside the temperature limits of the test matrix, it can be assumed that the adjacent classification ranges still apply. For an extension of the ranges along the time axis, further investigations need to be carried out, especially for shorter times. In general, it can be concluded from the diagram that with increasing temperature

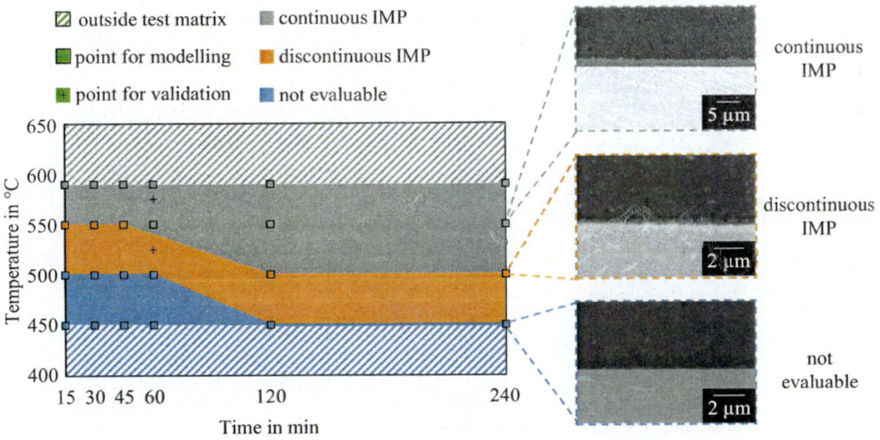

Fig. 2. Process windows for the formation of IMP classes and examples for each classification

and time, the IMP tend to form a constant phase seam. In this case, the influence of temperature is stronger than the influence of time.

The influence of temperature on the formation and the growth of IMP is shown in Fig. 3 at a constant time of 60 min. At a temperature of 450 °C there are clearly no IMP Fig. 3a. If the temperature is increased to 550 °C, the presence of IMP could not be clearly detected on the basis of the SEM images Fig. 3b. However, the evaluation of all BSE images for this parameter combination did not show a clear formation of an IMP. Therefore, (a) and (b) were both assigned to the non-evaluable group, see Fig. 2.

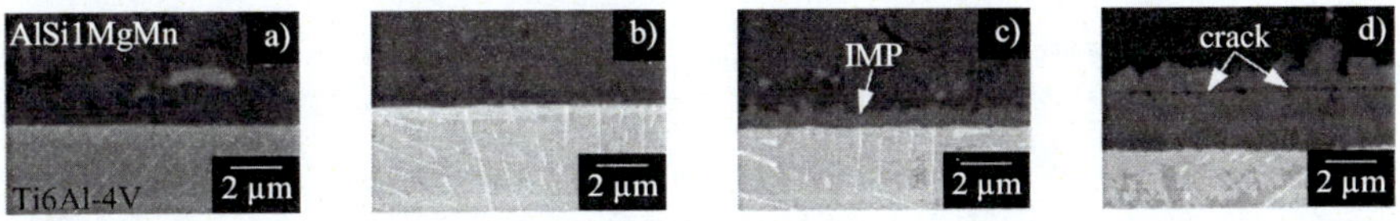

Fig. 3. Formation of the IMP for different temperatures at a holding time without loading of 60 min a) 450 °C, b) 500 °C, c) 550 °C, d) 590 °C

A further increase of the temperature to 550 °C (Fig. 3c) and to 590 °C (Fig. 3d) results in formation of a continuous IMP. The phase width at 590 °C has more than doubled compared to 550 °C. In addition, it is noticeable that at 590 °C a crack has appeared within the phase, which is recognisable as a thin black line. The crack might have been formed by the specimen preparation during grinding and polishing or by the very different coefficients of thermal expansion of $AlSi_1MgMn$ and Ti6Al6-4V. Cracking at 590 °C is seen as a first indication that a critical width of IMP has been reached. Since in the analogy test the cooling of the specimen is carried out without additional force, a tensile stress is induced in the joining zone because the coefficient of thermal expansion of $AlSi_1MgMn$ is 2.8 times higher than that of Ti6Al6-4V. Cracks were detected in all specimens that were joined at 590 °C. The growth of the IMPs over time at a temperature of 550 °C is shown in Fig. 4. After 30 min, insular IMPs can be seen in both the $AlSi_1MgMn$ and the Ti6Al6-4V (Fig. 4a). After 60 min, these islands merge to form a continuous phase seam (see Fig. 4b). If the time increased to 120 min Fig. 4c and 240 min Fig. 4d, the IMP grows further.

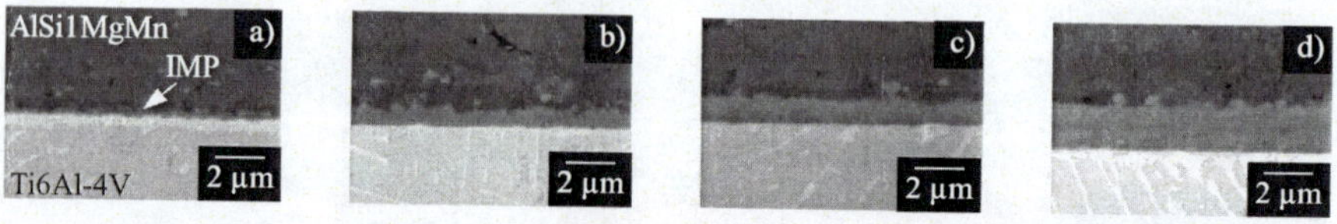

Fig. 4. Formation of the IMP for different holding times without loading at a temperature of 550 °C **a** 30 min, **b** 60 min, **c** 120 min, **d** 240 min

In Fig. 3d the individual grains of the IMP are visible. These are more globular, whereas the grain structure in Fig. 4c, d appears more stem-like. In addition, it is noticeable that the IMP initially form islands in the $AlSi_1MgMn$ (see Fig. 3b). Subsequently, insular phases also form in Ti6Al6-4V (Fig. 4a). These islands grow together on both sides to form a continuous IMP. Further growth of the IMP seems to take place primarily in Ti6Al6-4V (see Figs. 3c, d and 4c, d). A similar formation and growth behaviour of intermetallic phase seams is described by Ryabov [14] and Springer et al. [15] for aluminium-steel compounds.

The results of the analysis of the BSE images are summarised in Fig. 5a. According to the classification introduced in Fig. 2, no analysis could be carried out for 450 and 500 °C with contact times of less than 120 min. IMP thicknesses could be determined for all other parameter combinations.

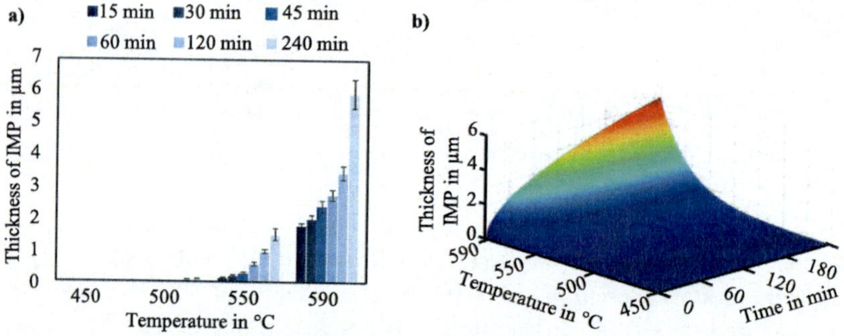

Fig. 5. a) Experimentally determined thickness of the IMP providing mean values with standard deviation, b) calculated thickness of the IMP according to Eq. 2

This confirms the observation from Figs. 3 and 4 that increasing the temperature has a significantly stronger influence on the phase growth than the contact time. The thickest IMP were obtained at 590 °C and 240 min with an average of 5.9 µm. Subsequently, the diffusion coefficient for each parameter combination was determined from the mean values of the respective parameter combinations by solving Eq. 1 with respect to the diffusion coefficient. It should be noted that the time considered for calculation of the diffusion coefficient is the sum of the holding time with loading and the holding time without loading, since the diffusion process starts directly after the phase formation. The calculated diffusion coefficients are almost constant over time at a given temperature. Uhe also concluded in her investigations that the diffusion coefficient is constant over time [12]. Therefore, an average diffusion coefficient was calculated for each temperature, which is shown in Table 2.

Table 2. Mean diffusion coefficients depending on the temperature

Temperature T in °C	450	500	550	590
Mean diffusion coefficient D in m^2/s	0	3.9E-19	3.5E-17	1.0E-15

These diffusion coefficients were fitted using an exponential approach to calculate a specific diffusion coefficient for each temperature within the parameter space. Substituting the temperature-dependent diffusion coefficient in Eq. 1 yields Eq. 2. Thus, the IMP thickness can be calculated as a function of time t and temperature T using Eq. 2, see Fig. 5b.

$$d = \sqrt{2 * \alpha * e^{\beta * T} * t}; \quad \begin{matrix} \alpha = 4.491e - 38 \\ \beta = 0.087 \end{matrix} \tag{2}$$

Equation 2 was subsequently implemented as a subroutine into the commercial FE software Forge NxT 3.0. To validate the equation, the analogy test was built up and the two parameter combinations 525 and 575 °C were simulated for a holding time of 60 min (see Fig. 6a). A very good agreement was achieved between the simulated and the experimental values for 575 °C. The simulated phase width of 1.4 μm is within the scatter of the experimental values with a deviation from the mean value of 2.9%. At 525 °C, however, the simulation overestimates the mean phase width by 37.8%. Furthermore, the calculated value of 0.16 μm is also outside the scatter of the measured values, see Fig. 6b. The larger deviation at 525 °C could be attributed to the smaller number of data points of the IMP, since the IMP are discontinuous in this case.

With the developed approach, the resulting phase width can be predicted numerically over the entire process chain and the process parameters can be determined so that the phase width does not exceed a critical width. If the phase width becomes too large, the IMP weakens the composite material produced in the final component. The critical width of the IMP for AlSi1MgMn and Ti6Al-4V can be determined according to the method published by Uhe by tensile tests on specimens with defined phase width and correlation with the obtained strengths. For the IMP between AlSi1MgMn and 20MnCr5, an optimum bond strength at a phase width of 1.1 μm was determined. Higher and lower phase widths lead to reduced strength [12]. The same behaviour is expected for the combination AlSi1MgMn and Ti6Al-4V. However, the optimum phase width for achieving high bond strengths still has to be determined by further tests. This can shorten the design time of tailored forming process chains and specifically influence the properties of the composite zone.

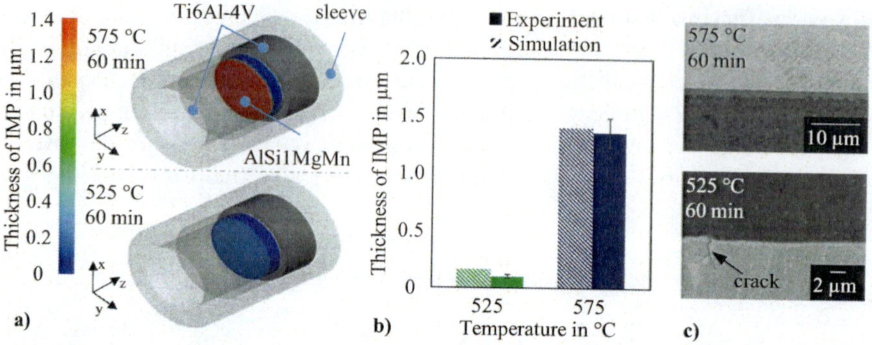

Fig. 6. Validation of the subroutine by comparing the simulated and experimental results: a) simulated IMP thickness, b) comparison of experimental and simulated IMP thickness, **c** BSE images of the validation specimens

4 Summary and Outlook

In this study, the formation and growth of IMP between $AlSi_1MgMn$ and Ti6Al-4V was investigated using analogy experiments and SEM analyses with the aim of numerically describing the phase growth over the process chain for the production of a hybrid bearing bushing. The BSE images of the joining zone were classified in terms of their evaluability and process windows were defined in which IMP occur. IMP thickness was determined to increase with increasing time and temperature, with temperature being the dominant factor. Diffusion coefficients were calculated using the Einstein-Smoluchowski equation and then approximated as a function of temperature using an exponential approach. This approximation was implemented via a subroutine in the commercial FE software Forge NxT 3.0 and validated with analogy tests. Validation shows a deviation of 2.9% for 575 °C and a deviation of 37.8% for 525 °C, which is due to the smaller database available for this condition.

In the future, the subroutine will be applied and validated along the process chain for the production of a hybrid bearing bushing so that a numerical design of the process boundary conditions can be carried out in order to set the IMP thickness specifically. In addition, the bond strength will be investigated as a function of the IMP thickness of the specimens in the tensile test and also implemented in the FE model.

Acknowledgements. The results presented in this paper were obtained within the Collaborative Research Centre 1153/2 "Process chain to produce hybrid high performance components by Tailored Forming" in the subproject A01, funded by the Deutsche Forschungsgemeinschaft (DFG, German Research Foundation)—252662854. The authors thank the German Research Foundation (DFG) for financial support of this project.

References

1. Behrens, B.-A., Uhe, J.: Introduction to tailored forming. Prod. Eng. Res. Devel. (2021)

2. Behrens, B.-A., Maier, H.J., Poll, G., Wriggers, P., Aldakheel, F., Klose, C., Nürnberger, F., Pape, F., Böhm, C., Chugreeva, A., Coors, T., Duran, D., Thürer, S.E., Herbst, S., Hwang, J.-I., Matthias, T., Heimes, N., Uhe, J.: Numerical investigations regarding a novel process chain for the production of a hybrid bearing bushing. Prod. Eng. Res. Devel. (2020)
3. Engelhardt, M., Grittner, N., Senden genannt Haverkamp, H. von, Reimche, W., Bormann, D., Bach, F.-W.: Extrusion of hybrid sheet metals. J. Mater. Process. Technol. (2012)
4. Arunkumar, S., Alphin, M.S., Kennedy, Z.E., Sriraman, N.: Development of a co-extruded Al-Ti bimetal composite. Mater. Tehnol. (2022)
5. Grittner, N., Striewe, B., Hehl, A. von, Bormann, D., Hunkel, M., Zoch, H.W., Bach, F.W.: Co-Extrusion of Aluminium-Titanium-Compounds. KEM (2011)
6. Grittner, N., Striewe, B., Hehl, A. von, Engelhardt, M., Klose, C., Nürnberger, F.: Characterization of the interface of co-extruded asymmetric aluminum-titanium composite profiles. Mat.-wiss. u. Werkstofftech (2014)
7. Striewe, B., Grittner, N., Hehl, A. von, Nürnberger, F.: Heat Treatment of Titanium-Aluminum-Compounds Made by Co-Extrusion of Asymmetric Compound Profiles. MSF (2015)
8. Xue, X., Sun, K., Tian, M., Liao, J.: Analysis of forming-induced distortion of dissimilar Ti6Al4V/AA1050 laminate made by non-equal channel lateral co-extrusion. Int. J. Adv. Manuf. Technol. **110**(5–6), 1627–1640 (2020). https://doi.org/10.1007/s00170-020-05933-3
9. Liao, J., Tian, M., Xue, X.: Interface properties of dissimilar Ti-6al-4v/Aa1050 composite laminate made by non-equal channel lateral co-extrusion and heat treatment. SSRN J. (2022)
10. Behrens, B.-A., Klose, C., Thürer, S.E., Heimes, N., Uhe, J.: Numerical modeling of the development of intermetallic layers between aluminium and steel during co-extrusion. In: Proceedings of the 22nd International ESAFORM Conference on Material Forming ESAFORM 2019. Vitoria-Gasteiz, Spain, 8–10 May 2019, p. 40029. AIP Publishing (2019)
11. Behrens, B.-A., Maier, H.J., Klose, C., Wester, H., Thürer, S.E., Heimes, N., Uhe, J.: Characterization and modeling of intermetallic phase formation during the joining of aluminum and steel in analogy to co-extrusion. Metals (2020)
12. Uhe, J.: Numerische und experimentelle Untersuchungen zum Verbundstrangpressen unter Berücksichtigung der intermetallischen Phasenbildung. Berichte aus dem IFUM, 2021, Band 06. TEWISS Verlag, Garbsen (2021)
13. Herbst, S., Dovletoglou, C.N., Nürnberger, F.: Method for semi-automated measurement and statistical evaluation of iron aluminum intermetallic compound layer thickness and morphology. Metallogr. Microstruct. Anal. (2017)
14. Ryabov, V. R.: Welding of aluminium alloys to steels. Welding Surfacing Rev. **9**(3) (1998)
15. Springer, H., Szczepaniak, A., Raabe, D.: On the role of zinc on the formation and growth of intermetallic phases during interdiffusion between steel and aluminium alloys. Acta Materialia 96 (2015)

Model Based Prediction of Force and Roughness Extrema Inherent in Machining of Fibre Reinforced Plastics Using Data Merging

Wolfgang Hintze, Alexander Brouschkin(✉), Lars Köttner, and Melchior Blühm

Institute of Production Management and Technology (IPMT), Hamburg University of Technology (TUHH), Denickestraße 17, 21073 Hamburg, Germany
alexander.brouschkin@tuhh.de

Abstract. Planning of machining operations for fibre reinforced plastics components today entails expensive trials in order to meet quality, productivity and cost requirements. The use of existing data for modeling and simulation has so far been severely limited due to the lack of universal process models that capture fundamental mechanisms in a process-independent approach and thus allow data to be merged across different cutting processes. Recently, a universal model describing the engagement conditions in oblique cutting of unidirectional FRPs has been developed. The model closes the gap described and builds the basis for cross-technology data merging from different cutting operations, which has been common practice for homogenous materials for a long time. In case of mostly thin FRP components and often poor clamping conditions the generated forces in cutting operations are crucial as they may lead to dynamic process instabilities and to unfavorable part deflections impeding part precision. Furthermore, the quality of the machined surface depends on the engagement conditions, which usually change significantly during machining. Force and quality data from different sources and across various cutting processes and FRP materials were merged using the universal engagement model to reveal generally applicable relationships. These will enable faster, more reliable and more cost-efficient planning of cutting operations for FRP components in the future.

Keywords: Fibre reinforced plastics · Cutting · Modelling · Data-based modeling: *Production at the leading edge of technology*

1 Introduction

Due to the high importance of energy and resource efficiency, the production of lightweight structures made of fibre-reinforced plastics (FRP) is increasing. This is reflected, for instance, in the demand for carbon fibre reinforced plastics (CFRP), which is measured in terms of tonnage produced, amounted to 128.5 kt in 2018, recording an average growth of 12.2% p.a. since 2010 [1]. Despite the COVID-19 pandemic, the leading CFRP manufacturers are expanding their production capacities [2]. In the production of lightweight structures, machining of FRP is an important manufacturing step in the aerospace and automotive industry.

M. Liewald et al. (Eds.): WGP 2022, LNPE, pp. 42–51, 2023.
https://doi.org/10.1007/978-3-031-18318-8_5

Compared to metals, with homogeneous and isotropic material properties, machining of FRP is more complicated. Thus, a change in the machining conditions such as continuous tool wear or a change in the clamping situation of the mostly thin-walled workpieces requires an adjustment of the machining parameters [3, 4].

In the literature, most studies focus on single FRP machining processes such as drilling [5], milling [6–10], turning [11], and planing [5, 6]. Thus, on the one hand, large amounts of measurement data of different machining processes are available. On the other hand, due to the focus on specific processes, there is a lack of process-independent models to describe the machining of FRP. Process-independent models are known from metal machining. For example *Kienzle* recognized a dependency of the cutting force on the chip cross section and introduced a specific cutting energy k_c [12]. Budak et al. later developed the most common model for calculating dynamic milling forces for metals based on predicting the cutting coefficients for oblique cutting from orthogonal cutting data and discretization of the cutting edges [13]. This method has been used to establish mechanistic models for predicting the cutting forces in FRP-machining [10, 14].

The cutting forces and quality of machined surfaces depend on the fibre orientation, so that the description of the engagement conditions to the fibre is important. Generally, this is defined by the fibre cutting angle θ_L included between fibre direction and cutting direction [8]. This is sufficient to describe an orthogonal cut, but insufficient to describe spatial engagement conditions in oblique cutting e.g. at setting angles $\kappa_r \neq 90°$. For a process-independent description of arbitrary engagement conditions *Hintze* introduced two spatial engagement angles θ_0, φ_0 [15].

Previous studies have shown that the cutting force components (F_c, F_f, F_p) behave π-periodically with θ_L. In the range $30° > \theta_L > 60°$, the forces pass through a minimum. While F_f and F_p reach the maximum between $120° > \theta_L > 180°$, the cutting force reaches it at about $\theta_L = 90°$ [5, 6, 10, 11, 16]. Likewise, optical examinations of the machined surface revealed material defects, such as fibre pull-out and breakage during drilling, which can be attributed to the engagement conditions of the tool with respect to the fibre direction [17, 18]. For example, the measured roughness of the machined surfaces (R_a, R_z, R_t) reaches a significant maximum in the range $30° > \theta_L > 60°$, while for all other angles it is constant at $R_z <$ fibre diameter [5, 7–9].

In this article, the above mentioned approach for the description of arbitrary spatial engagement conditions on the basis of the angles θ_0, φ_0 is applied to existing measurement data from the literature in order to reveal process-independent correlations with respect to process forces and roughnesses.

2 Universal Model for Engagement Conditions

2.1 Description of Engagement Conditions with Reference Angles

In most machining processes the oblique cut is applied, which is defined with the setting angle κ_r and the cutting edge inclination angle λs. Due to the anisotropy of the FRP material, the reference of the fibre to the machining directions is of great importance, as it is significant for the cutting forces and the surface quality. A model presented in [15] describes the spatial conditions of engagement in relation to the fibre direction independent of the machining process.

The model defines the rectangular coordinate system of the FRP S^L with the vectors $\overrightarrow{e_{||}}$, $\overrightarrow{e_{\perp 1}}$ and $\overrightarrow{e_{\perp 2}}$, which can be transformed into the rectangular coordinate system S^{SO} with the vectors $\overrightarrow{e_r}$, $\overrightarrow{e_s}$, $\overrightarrow{e_o}$.

$$S^L = \left\{\overrightarrow{e_{||}}, \overrightarrow{e_{\perp 1}}, \overrightarrow{e_{\perp 2}}\right\} \quad (1)$$

$$S^{SO} = \left\{\overrightarrow{e_r}, \overrightarrow{e_s}, \overrightarrow{e_o}\right\} \quad (2)$$

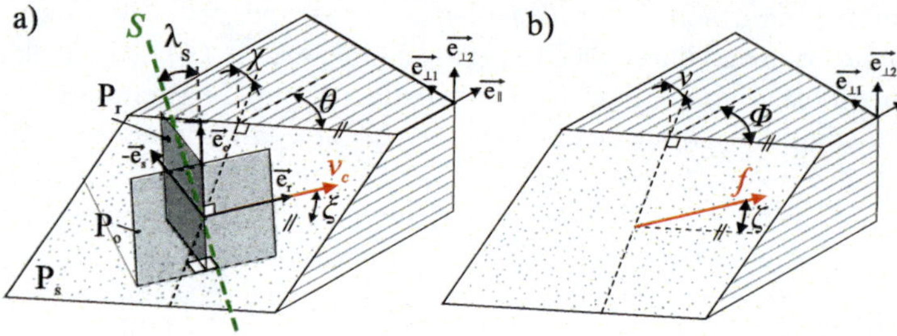

Fig. 1. Description of cutting (a) and feed (b) direction in oblique cut of orthotropic materials

As shown in Fig. 1a, the vector $\overrightarrow{e_{||}}$ is the fibre direction, $\overrightarrow{e_{\perp 2}}$, is a perpendicular vector to $\overrightarrow{e_{||}}$ in laminate thickness direction, vector $\overrightarrow{e_{\perp 2}}$, is a vector perpendicular to the previous two. $\overrightarrow{e_r}$ is the normal vector for the tool reference plane and the cutting direction, $\overrightarrow{e_s}$, is the normal vector for the cutting plane and $\overrightarrow{e_o}$, is the normal vector for the orthogonal plane. The coordinate system S^L can be transformed into S^{SO}. For the transformation the angles θ, χ and ξ are introduced as shown in Fig. 1a. The transformation is described by the following matrix.

$$S^{SO} = S^L T \quad (3)$$

$$T = \begin{bmatrix} \cos\theta & \sin\theta & 0 \\ -\sin\theta & \cos\theta & 0 \\ 0 & 0 & 1 \end{bmatrix} \cdot \begin{bmatrix} 1 & 0 & 0 \\ 0 & \cos\chi & \sin\chi \\ 0 & -\sin\chi & \cos\chi \end{bmatrix} \cdot \begin{bmatrix} \cos\xi & 0 & -\sin\xi \\ 0 & 1 & 0 \\ \sin\xi & 0 & \cos\chi \end{bmatrix} \quad (4)$$

For the completeness of the description of the engagement conditions, λ_S describes the inclination of the cutting edge in the cutting plane as usual.

Similar to the engagement conditions, the feed direction can be referenced with the corresponding angles Φ, ν and ς as shown in Fig. 1b. For a complete description of the tool movement, the engagement angle φ indicates the position of the cutting edge during rotational movement of the tool or workpiece. The Table 1 shows the value ranges of the reference angles for different machining operations.

2.2 Description of Engagement Conditions with Spatial Angles

Considering the orthotropy of the material, it is reasonable to assume that the same cutting conditions lead to the same cutting results with respect to the fibre direction.

Table 1. Value range of reference angles for common machining processes

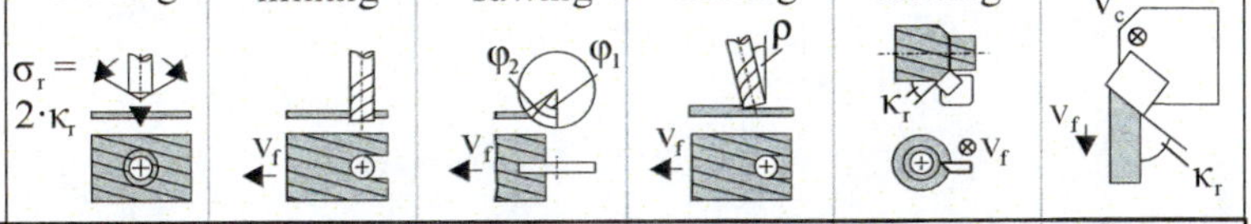

	Counter-sinking	Peripheral milling	Circular sawing	Face milling	Logitudinal turning	Planing
θ°	0 ...180	0 ...180	Const.	Const.	0 ...180	Const.
ξ°	0	0	90	ρ	0	0
φ°	90	0 ...180	$\varphi_1 - \varphi_2$	0 ...180		
Φ°		0 ...180	0 ...180	0 ...180	0 ...180	N/A
ν°	κ_r	0	90	90	$90 - \kappa_r$	N/A
ζ°	0	0	0	90	0	N/A
	Major cutting edge					
χ°	κ_r	$90 - \kappa_r$	$90 - \varphi$	$90 - \kappa_r$	$90 - \kappa_r$	$90 - \kappa_r$
	Minor cutting edge					
χ°	κ_r'		κ_r'	κ_r'	κ_r'	

With reference to the fibre direction, the engagement conditions can be described independently of the machining process. For this purpose, the spatial engagement angle φ_0 and the spatial fibre cutting angel θ_0 are introduced. As shown in Fig. 2, φ_0 is measured between $-\overrightarrow{e_s}$ and $\overrightarrow{e_{||}}$ while θ_0 is measured between $\overrightarrow{e_r}$ and $\overrightarrow{e_{||}}$. Thus it holds:

$$0^\circ \leq \varphi_0 \leq 90^\circ \tag{5}$$

$$90^\circ - \varphi_0 \leq \theta_0 \leq 90^\circ + \varphi_0 \tag{6}$$

The definition range of these angles corresponds to the white triangle in the θ_0, φ_0-diagram in Fig. 3. The spatial angles change depending on the engagement angle φ. Figure 3 shows how the spatial angles φ_0 and θ_0 behave for different machining processes. That way cross-technology data can be merged from different cutting operations. However the influence of λ_S is omitted in this diagram.

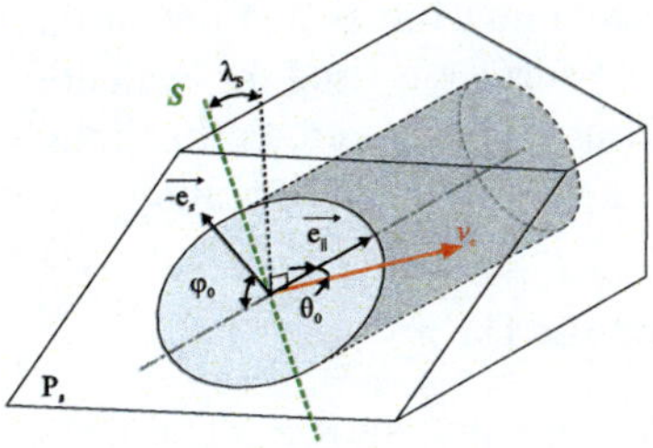

Fig. 2. Description of cutting direction with spatial engagement angles θ0, φ0

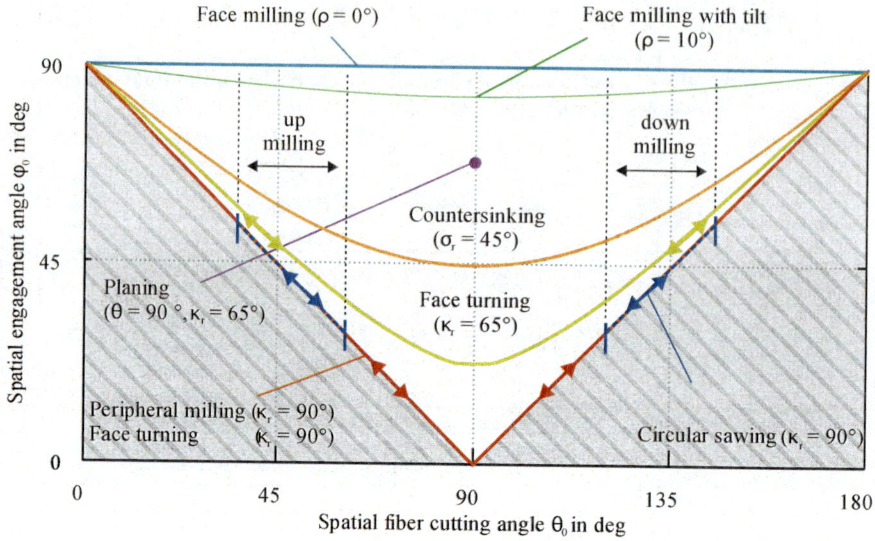

Fig. 3. θ_0, φ_0-diagram for common cutting operations

3 Merging of Measurement Data

The benefit of the presented universal model is that it enables the revelation of generally valid technological interrelationships for the machining of unidirectional fibre-reinforced plastics across different machining processes and conditions. With regard to the achievable component accuracy and surface quality, the cutting force and surface roughness are of particular importance. Accordingly, force and roughness results from different processes and machining operations are combined to find generally valid relationships.

3.1 Used Measurement Data

Table 2 lists the sources, processes and key process parameters used. The material used in these publications was unidirectional CFRP with an epoxy matrix. The cutting edge radius of the tools during the investigations was 5–90 μm. To better determine the relationships between the machining forces and the surface qualities, the force that is perpendicular to the machined surface is used for evaluation. It depends on the cutting process which force component is considered. For the roughness, R_z is evaluated, since this is most frequently used in the evaluated literature.

3.2 Procedure for Merging the Data

To analyse and compare the force and surface data, the spatial angles φ_0 and θ_0 are determined from the described test runs. The results can then be presented in the θ_0, φ_0-diagram. To ensure comparability of the forces and roughness, which are strongly dependent on various influences such as tool geometry and wear, the values of a test series are normalised to the mean value of $\varphi_0 = 90°$, $\theta_0 = 180°$ for each test series separately. The

Table 2. Parameters of the collected data

Source	Process	v_c (m/min)	h (mm)	ρ (°)	γ_f (°)	α_f (°)	κ_r (°)	λ_s (°)	F_{cN0} (N)	R_{z0} (µm)
[5]	Planing	5	0.03	–	0	12	30 … 90	0	97	8.1
[6]		5	0.01	–	30	20	90	0	100	–
			0.019						114	–
			0.03						151	–
[8]	Face milling	500	0.01	0	0	–	$\kappa_r' =$ 0°	5	99	2.8
		500	0.01	1	0	–	$\kappa_r' =$ 0°	5	24	3.5
[5]	Countersinking	7	0.025	0	30	20	45	−5 … −10	76	–
							55		46	–
							65		42	–
[6]	Peripheral milling	5	0.02	0	0	25	90	0	175	–
			0.03						197	–
[16]		4	0.02	0	0	12	90	0	348	–
			0.03						376	–
[7]		798	0.06	0	0	12	90	0	–	3.3
[9]		94.25	0.08	0	5	10	90	48	–	4.2
[11]	Face turning	90	0.03	-	10–30	7–21	90	0	Min … Max [1]	–

values for planning [6], where this value did not exist, was normalized to $\varphi_0 = 82.5°$, $\theta_0 = 7.5°$. The according mean values of F_{CN0} and R_{z0} are mentioned in Table 2.

The normalised quantities are plotted in a contour plot with the axes φ_0 and θ_0. The values between the test series were linearly interpolated so that the entire definition range of both angles is represented.

4 Results and Discussion

4.1 Process Forces

Figure 5 shows the contour plot of the normalized force, which was interpolated with an R^2 of 0.89. Along the edges of the contour plot, the force curve is similar to those in the literature. In the θ_0, φ_0-diagramm, a minimum of the normalised force is in the range $40° > \theta_0 > 60°$ and $65° > \varphi_0 > 75°$. A maximum of the force occurs in the range $140° > \theta_0 > 160°$ and $55° > \varphi_0 > 65°$. It is worth mentioning that the forces are negative at the minimum, which indicates that the tool is being pulled into the cut (Fig. 4).

Karpat presented a model to describe the dependence of machining forces on the fibre cutting angle θ_L in peripheral milling as a simple sine function (Eq. 7). Its validity

[1] Voss examined eight different tool geometries with two different wear states. Each tool variation was normalised to its own value.

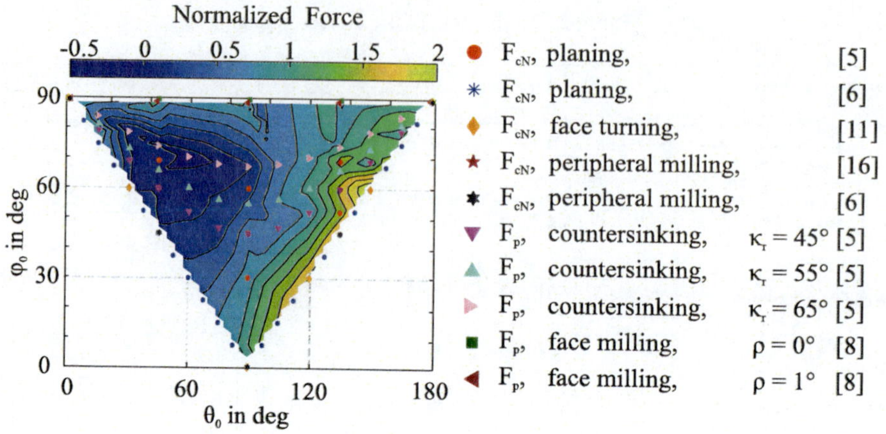

Fig. 4. Normalized forces in θ_0, φ_0-diagram

for the new approach with the angles θ_0 and φ_0 was verified by replacing θ_L with θ_0 and φ_0. Figure 5 shows the *Karpat* equation in a coloured strip, as it is only valid for this area of peripheral milling and describes it qualitatively well. However, it is unsuitable for describing spatial engagement conditions for any cutting processes.

$$F_r = F_{cN} = a_p \cdot h \cdot K_{Tc}(\theta_L) = a + b \cdot \sin(2 \cdot \theta_L + \mathrm{c}) \tag{7}$$

4.2 Surface Quality: Roughness

Figure 6 shows the contour plot of the normalised R_z, which was interpolated with an R^2 of 0.64. Along the edges of the contourplot. In the $\theta_0.\varphi_0$-plane, a maximum of the normalised R_z is in the range $45° > \theta_0 > 75°$ and $45° > \varphi_0 > 75°$. In the remaining range of the spatial angles, normalized R_z is constant with a value slightly lower than the fibre diameter.

Data with low λ_s were selected for this work, so it is noticeable that the data from Li et al. [9] with a λ_s of 48° gives similar roughness to the data with $\lambda_s = 0°$.

4.3 Discussion

When comparing the force and roughness data, the occurrence of high roughness at low forces is striking. This indicates a pull-in of the tool into the cut, which leaves a stepped finished surface due to a periodic fibre breakout, which is documented in the literature [5, 17]. In contrast, low roughness occurs with positive forces as observed by face milling [8]. Fibre-shearing, the dominant failure mechanism in face milling, is characterized by low positive forces and low roughness. It is promoted by repetitive cutting edge engagement of the machined surface in face milling without tilt ($\rho = 0°$) [8] according to $\varphi_0 = 90°$.

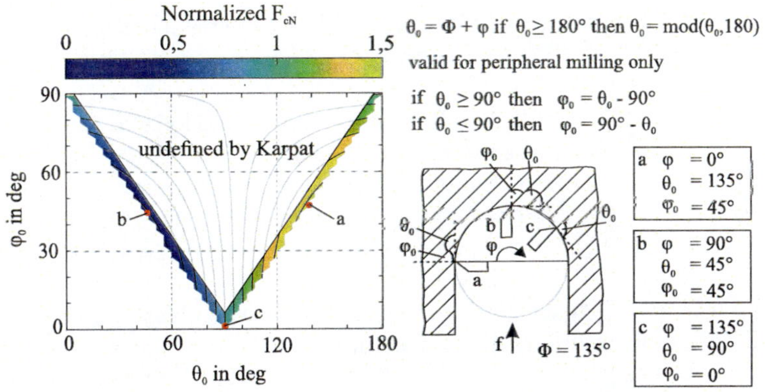

Fig. 5. Normalized forces and fit to *Karpat*-model

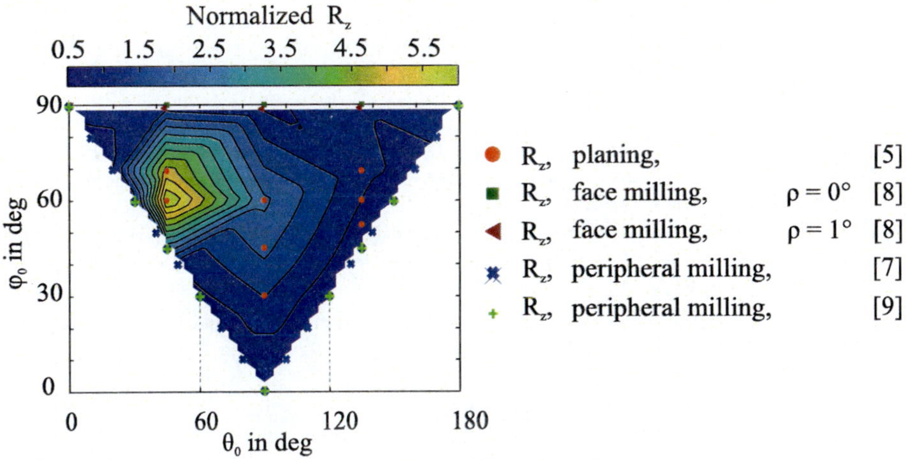

Fig. 6. Normalized R_z in θ_0, φ_0-diagram

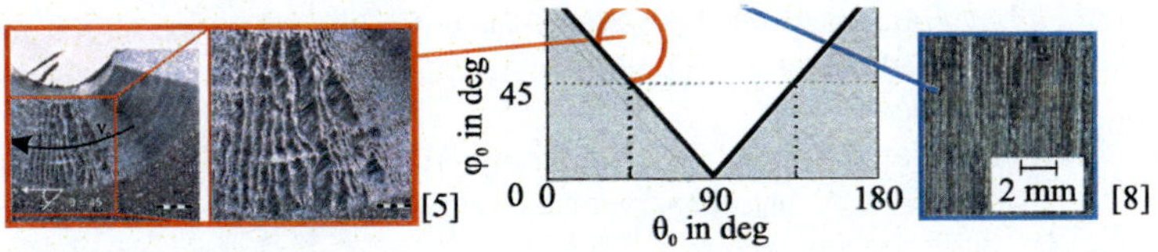

Fig. 7. Failure mechanism in θ_0, φ_0-diagramm

5 Summary and Outlook

With a new method presented in [15] for the description of engagement conditions based on spatial engagement angles, characteristic parameters, which are evaluated from different machining processes, can be compared in a unique model leading to similar qualitative results.

- Data from several publications can be compared with the novel model covering any cutting operations relevant for unidirectional CFRP.
- Regardless of the individual cutting process minimum forces occur in the range of angels $40° > \theta_0 > 60°$ and $65° > \varphi_0 > 75°$, maximum forces at $140° > \theta_0 > 160°$ and $55° > \varphi_0 > 65°$ and maximum roughness R_z in the range $45° > \theta_0 > 75°$ and $45° > \varphi_0 > 75°$
- Low and negative F_{cN} lead to an increase in roughness, which can be explained by tool pull-in and the resulting fibre break-outs.
- Thus, the novel model is able to predict cutting conditions detrimental with regard to process forces and surface quality for any cutting operation on unidirectional FRP.

The consideration of the engagement conditions with $\lambda s \neq 0°$ and their correlation to cutting forces and surface qualities is the subject of ongoing research.

References

1. Sauer M.: Composites-Marktbericht 2019 - Der globale CF- und CC-Markt 2019 - Marktentwicklungen, Trends, Ausblicke und Herausforderungen - veröffentlichte Kurzfassung. Carbon Composites e.V. (2019)
2. Sauer M.: Composites-Marktbericht 2020 - Der globale CF- Produktkapazitäten - Marktentwicklungen, Trends, Ausblicke und Herausforderungen - veröffentlichte Kurzfassung. Carbon Composites e.V. (2021)
3. Körkel, F.: Zerspanbarkeitsbewertung von Faserverbundkunststoffen bei der Fräsbearbeitung dünnwandiger Bauteile in der Großserie. Dissertation. Technische Universität Hamburg-Harburg, Hamburg (2015)
4. Klotz, S.: Dynamische Parameteranpassung bei der Bohrungsherstellung in faserverstärkten Kunststoffen unter zusätzlicher Berücksichtigung der Einspannsituation. Dissertation. Karlsruher, Institut für Technologie, Karlsruhe (2017)
5. Schütte, C., Bohren und Hobeln von kohlenstofffaserverstärkten Kunststoffen unter besonderer Berücksichtigung der Schneide-Faser-Lage. Dissertation. Technische Universität Hamburg-Harburg, Hamburg (2014)
6. Brügmann, F.: Bauteilqualität und Werkzeugverschleiß beim Fräsen von CFK-Gelege unter räumlichen Eingriffsbedingungen. Dissertation. Technische Universität Hamburg, Hamburg (2018)
7. Kindler, J.: Werkstückqualität und Standzeitoptimierung von Zerspanwerkzeugen bei der Umrissbearbeitung von kohlenstofffaserverstärkten Kunststoffen. Dissertation. Technische Universität Hamburg-Harburg, Hamburg (2010)
8. Hintze, W., Hartmann, D., Schubert, U.: Stirnfräsen von CFK zur Fügeflächenherstellung und Reparaturvorbereitung. ZwF – Zeitschrift für wirtschaftlichen Fabrikbetrieb **107**(6), 462–467 (2012)
9. Li, H., Qin, X., Huang, T., Liu, X., Sun, D., Jin, Y.: Machining quality and cutting force signal analysis in UD-CFRP milling under different fiber orientation. Int. J. Adv. Manuf. Technol. **98**, 2377–2387 (2018). https://doi.org/10.1007/s00170-018-2312-3,2018
10. Karpat, Y., Bahtiyar, O., Değer, B.: Mechanistic force modeling for milling of unidirectional carbon fiber reinforced polymer laminates. Int. J. Mach. Tools. Manuf. **56**, 79–93 (2012). https://doi.org/10.1016/j.ijmachtools.2012.01.001

11. Voß, R.: Fundamentals of Carbon Fibre Reinforced Polymer (CFRP) Machining. Dissertation.ETH Zürich, Zürich (2017)
12. Kienzle, O., Victor, H.: Spezifische Schnittkräfte bei der Metallbearbeitung. Werkstattstechnik und Maschinenbau **47**(5), 224–225 (1957)
13. Budak E., Altintas Y., Armarego E.: Prediction of milling force coefficients from orthogonal cutting data. J. Eng. Industry. **118**(2), 216 (1996). https://doi.org/10.1115/1.2831014
14. Chandrasekharan, V, Kapoor, S, DeVor, R.: A mechanistic model to predict the cutting force system for arbitrary drill point geometry. J. Manuf. Sci. Eng. **120**, 563 (1998)
15. Hintze, W.: CFK-bearbeitung—Trenntechnologien für Faserverbundwerkstoffe und den hybriden Leichtbau, Springer Vieweg Verlag (2021). https://doi.org/10.1007/978-3-662-63265-9
16. Hartmann, D.: Delamination an Bauteilkanten beim Umrissfräsen kohlenstofffaserverstärkter Kunststoffe. Dissertation. Technische Universität Hamburg-Harburg, Hamburg (2012)
17. Jawahir, I., Brinksmeier, E., M'Saoubi, R., Aspinwall, D., Outeiro, J., Meyer, D., Umbrello, D., Jayal, A.: Surface integrity in material removal processes: recent advances. CIRP Ann. **60**(2), 603–626 (2011). https://doi.org/10.1016/j.cirp.2011.05.002
18. Hintze, W., Clausen, R., Schütte, C., Kroll, K.: Evaluation of the total cutting force in drilling of CFRP - a novel experimental method for the analysis of the cutting mechanism. Prod. Eng. Res. Devel. **2018**, 1–10 (2018)

Mechanisms for the Production of Prestressed Fiber-Reinforced Mineral Cast

M. Engert[1(✉)], K. Werkle[1], R. Wegner[2], and H.-C. Möhring[1]

[1] Institute for Machine Tools (IfW), University of Stuttgart, Holzgartenstr. 17, 70174 Stuttgart, Germany
Michelle.Engert@ifw.uni-stuttgart.de

[2] Institute for Textile and Fiber Technologies (ITFT), University of Stuttgart, Pfaffenwaldring 9, 70569 Stuttgart, Germany

Abstract. Prestressed, fibre-reinforced mineral cast has the potential to replace steel in the field of structural components for machine tools. High damping properties, a low density compared to steel and adjustable creep properties coupled with a low CO_2 equivalent and a low primary energy requirement make the hybrid material a suitable material for the future. However, the pre-stressing of the reinforcing fibres requires high tensile forces corresponding to around 5% of the compressive strength of the mineral cast. The installation situation, which also schedules integrating the mechanisms for a subsequent readjustment of the prestressing into the machine tool, requires a minimum installation space for the prestressers with a maximum base area of 25×25 mm^2 and a minimum height. The problematic clamping conditions of the carbon fibers, which should only be loaded along the fibre direction due to their low transverse strength, require a novel tension mechanism. In this paper, specially developed pretensioning mechanisms for the carbon fibers used as reinforcement are investigated and the different mechanisms are compared to each other.

Keywords: Composite · Residual stress · Carbon fibres

1 Introduction

The topics of energy and resource efficiency have occupied scientific research for years. One possible approach to improve the CO_2 balance of machine tools is the use of mineral cast, also known as polymer concrete. Mineral cast has been used successfully for several years in machine tools, especially for machine beds. Mineral cast or reactive resin concrete in general, is a mixture of mineral aggregates and a binder resin, similar to concrete used in the construction industry [1]. In other words, it is not a cement matrix but a plastic matrix that holds the aggregates together in the concrete. This sounds like a small difference, but it is a significant one, as the polymer binder gives the material its special properties. Here it convinces with its excellent damping properties, which allow vibrations to decay up to five times faster than a machine bed made of a welded steel construction [1]. In this way, tool wear can be reduced and the achievable surface quality

M. Liewald et al. (Eds.): WGP 2022, LNPE, pp. 52–60, 2023.
https://doi.org/10.1007/978-3-031-18318-8_6

of the workpieces can be increased at the same time. Furthermore, low thermal expansion coefficients make mineral cast less sensitive to temperature fluctuations than comparable steel or cast iron designs. In previous research work, mineral cast was already used as a structural material for function integrated modular fixtures [2] and as damping filler within a material hybrid machine tool slide [3].

The adaptation to structural components is of particular interest due to the manifold advantages. However, the high creep tendency [4] of the material and the lower tensile strength compared to steel [5] form an obstacle. Possible optimization options are already known from the literature. Reis, for example, has already been able to show that an increase in compressive strength of 16% is possible by using short carbon fibers [6]. At the same time, he was able to show that an increase in impact strength of 200% and an increase in Young's modulus of 39% are possible when glass fibers are used [7]. Barbuta et al. were able to demonstrate that the use of natural cellulose fibers is unsuitable for improving the mechanical properties of mineral cast [8]. In summary, non of the reinforcing components mentioned so far is able to increase the tensile strength of mineral cast far enough to make it suitable for a use as a structural component in a machine tool. One possible solution is the integration of already prestressed carbon fibers into the mineral cast. This creates residual compressive stresses during the curing process. The aim is to use these residual compressive stresses and the high tensile strengths of the cast-in carbon fibers to create a new hybrid material that can withstand the tensile loads of a structural component of a machine tool. Under this approach, the pre-stressing mechanisms used for the fibers are of particular importance. Known mechanisms for prestressing and holding fibers are mainly from the construction industry and do not meet the constraints of a prestress of 2.8 kN with a minimum installation space of 25 mm × 25 mm. Many of the mechanisms are designed to prestress Carbon fiber reinforced plastic bars or laminations. In addition, none of the in the literature presented prestressing mechanisms can meet the small installation space requirements. In this paper, the results of the development of compact approaches for prestressing carbon fibers are reviewed. For this purpose, the basic principles of frictional and material bonding were taken up and supplemented by the basic idea of frictional bonding.

2 Development of New Prestressing Elements for Carbon Fibers

Based on fundamental considerations, four different attachment types for carbon fibers have been identified.

Figure 1 shows a schematic representation of the attachment methods. The upper left-hand side shows a mechanism that is based on force fitting bonding. The lower left side shows a frictional mechanism. On the right-hand side, firmly bonds are shown. In the upper right part, a bionic approach of a knot hole is shown, and on the lower right side, a combination of frictional connection of the fibers combined with a firmly connection. The friction-locked mechanism has a similar operation as the mechanism presented by Kerrouche [9]. Again, the prestressing is to be done by means of wedges. However, in order to be able to tension a roving instead of a rod, they are placed opposite each other in a plane.

The frictional mechanism consists of two pins held in a housing. The carbon fiber roving is wrapped around the metal pins in a similar way to the bollards known from

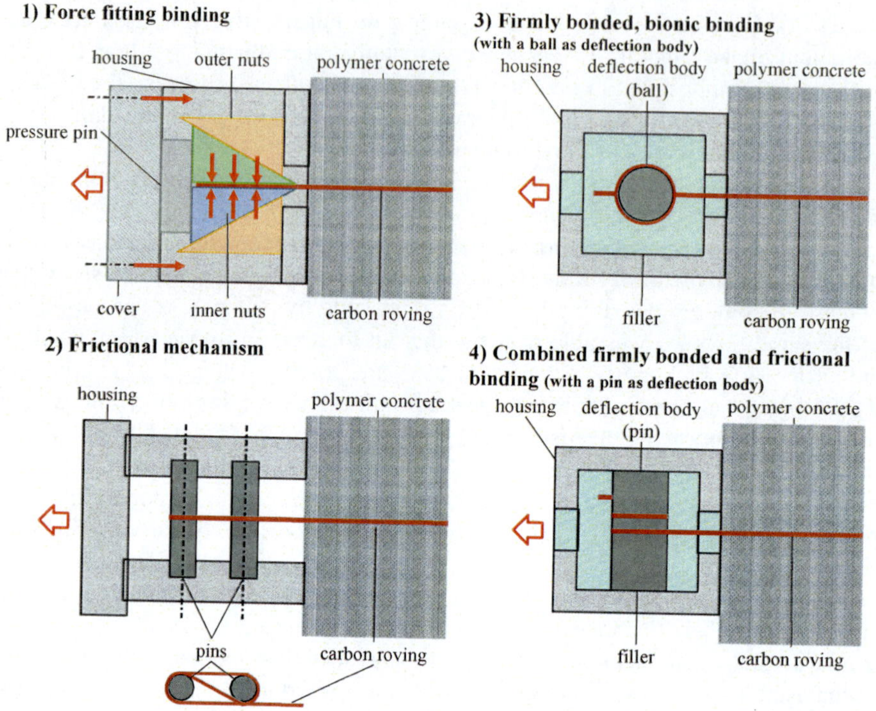

Fig. 1. Schematic representation of the planned connection types (1) force-fit connection, (2) friction-fit connection, (3) material-fit bionic connection, (4) combined friction-fit and material-fit connection

marine applications. The two firmly bonded mechanisms differ in their deflection bodies. These are a ball and a pin. Both deflection bodies are located in a housing filled with an epoxy resin filler. Since only the filler has to be replaced for each test, the resource requirement is negligible The use of the ball in the midst of the carbon fiber is intended to mimic the regrowth and thickening of branches found in nature. In contrast, the combined frictional and firmly bonded connection with a wrapped and molded pin represents an oversized connection strategy.

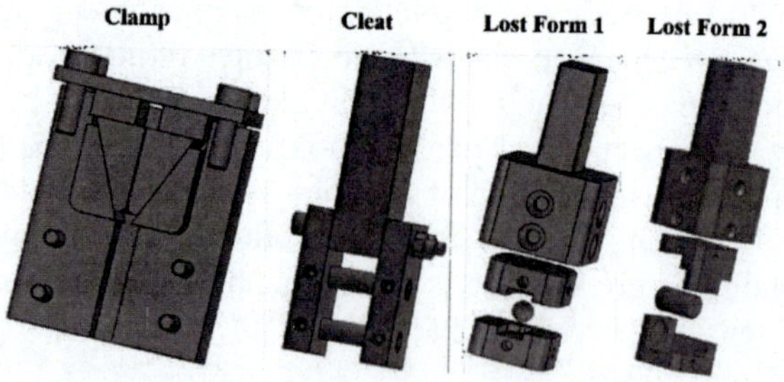

Fig. 2. Developed mechanisms for prestressing carbon fibres embedded in mineral cast

Based on these schematic representations, four prestressing mechanisms are developed (Fig. 2) and analyzed using the finite element method. Ansys Workbench 2021 is used as the simulation software. For this purpose, a carbon fiber is prestressed between two prestressing mechanisms of the same type. A parameter analysis is used to calculate the maximum tensile forces that can be achieved. According to the manufacturer, the maximum tensile strength of the GRAFIL 34-700 Carbon fibers used from Mitsubishi Chemical Carbon Fiber and Composites is 3700 MPa [10]. For the cleat and the clamp, friction with a friction coefficient of 0.3 [11] is assumed between the roving and the steel components. For the simulation of the clamp, friction with a friction coefficient of 0.15 [12] is assumed between the wedges and their contacting steel parts.

The simulative examination of the clamp in Fig. 3 shows that, from a tensile force of 7400 N, failure of the roving occurs due to tearing between the two clamping points. The maximum elongation in the fiber is 3.2%. In addition, the simulation provides evidence of high compressive forces on the carbon fiber in the area of the clamping point, which could lead to damage to the fiber and its failure.

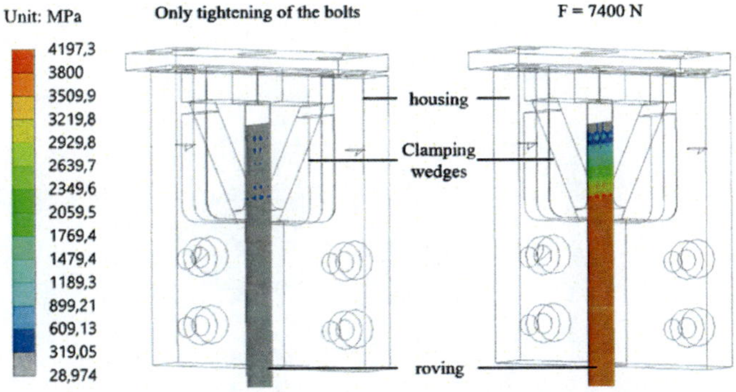

Fig. 3. Tensions in the clamping area when prestressing the carbon roving with the clamp

The results of the simulative observation of the cleat from Fig. 4 show that with this type of prestressing mechanism the single-wrapped outer side starts to slide over the pins. As there is less movement on the other side of the pins a tension builds up. The maximum stress of 3700 MPa is already exceeded at a tensile force of 320 N.

The simulation of Lost Form 1 shows that, from a tensile force of 3000 N, a stress increase occurs in the filling material in the area of the ball (Fig. 5) and thus a matrix fracture occurs. The maximum elongation of the fiber is 0.67% and is not critical. However, this effect is only noticeable at forces above the required 2800 N.

The last type of connection considered is the Lost Form 2 with wrapped pin. In the simulative analysis, the carbon fiber fails between the clamping points at a tensile force of 37000 N and above. An elongation of 1.9% can be reached. The simulation provides an indication of a pull-out of the binding material out of the form (Fig. 6). This is to be observed and, if necessary, verified in the course of the experimental investigation.

Based on the simulation results, Lost Form 2 appears to be the most suitable for the application at hand, as it can bear the highest forces. In addition, it shows no indication

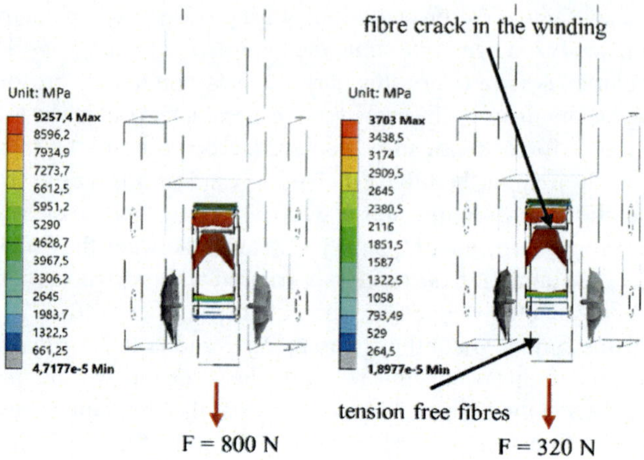

Fig. 4. Tensions with different tensile forces and internal windings

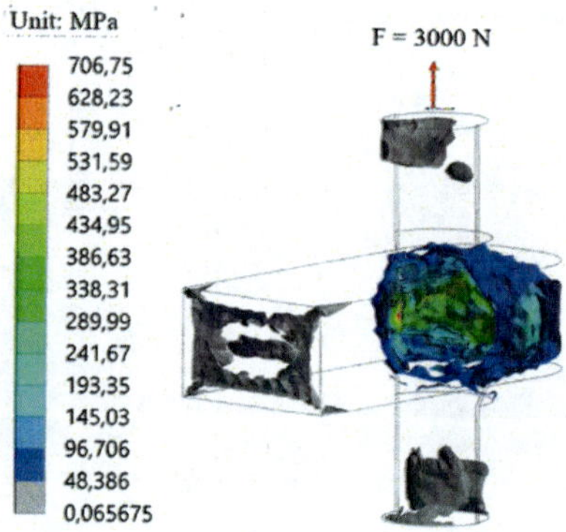

Fig. 5. Tensions in the area of the ball in the filling material of Lost Form 1

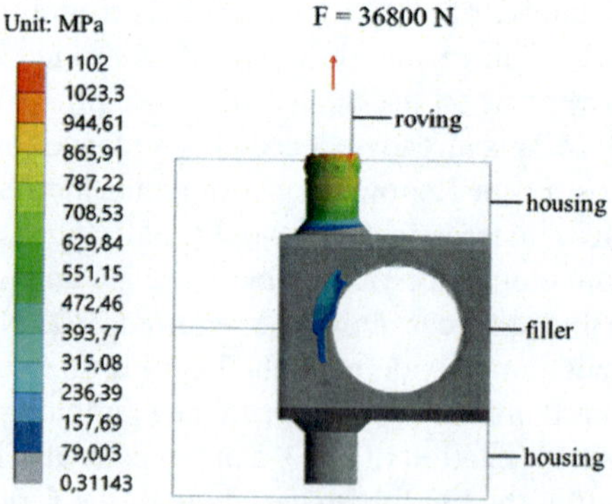

Fig. 6. Tensions at the bond in Lost Form 2

of a possible failure of the prestressing mechanism or of possible damage to the carbon fiber roving.

3 Experimental Results

The prestressing mechanisms are clamped in pneumatic sample grips of a Zwick Roell type 8497 hydraulic universal testing machine with measuring length of 100 mm. The mechanisms and the roving's clamped between them are subjected to a tensile load at a constant test speed of 2 or 5 mm/min. All the tests presented below were carried out seven times each for statistical validation. However, for better overview, only four curves are shown.

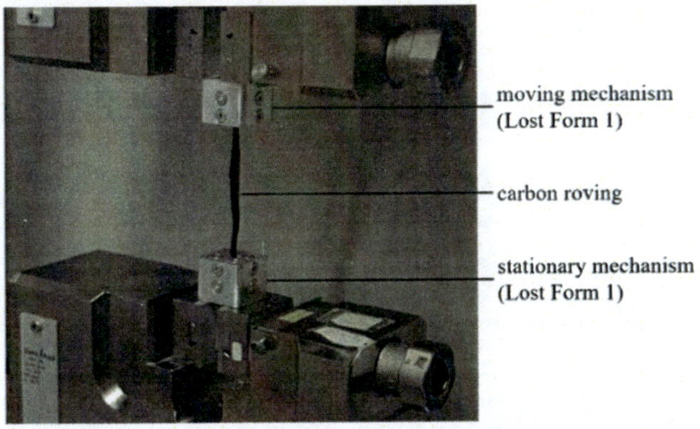

Fig. 7. Experimental Setup at the example of the Lost Form 1

The use of the machine finished clamping wedges results in a maximum reached prestressing force of 497.4 N (sample 3). The minimum reached prestressing force achieved is 274.6 N (sample 1). On average, a prestressing force of 412.1 N can be achieved (Fig. 8). The most frequent cause of failure, as already shown in the simulation, is a failure of the roving at the clamping point like shown in Fig. 7 on the right side. Alternatively using crosshatched finished wedges, damage to the fiber at the clamping point already occurs at prestressing forces of 230.7 N on average.

For the tests with the cleat, the rovings are wrapped around the pins ten times. On average, a prestressing force of 1251.9 N can be achieved. However, there is a very large scatter. The maximum prestressing force is 1679.6 N, while the minimum prestressing force is 659.9 N. The main reason for failure is tearing of the fibers between the pins. This means that the cleat does not achieve the desired pretensioning forces of 2800 N. In addition, the results differ from those from the simulation, in which failure already occurs at a pretensioning force of 320 N. This can probably be attributed to the tenfold winding of the roving around the pins in the experimental investigation in contrast to the single winding used in the simulation.

Another problem when using the cleat as a prestressing mechanism is the up to 60 mm long prestressing travel, which is due to the slipping of the fibers. In addition,

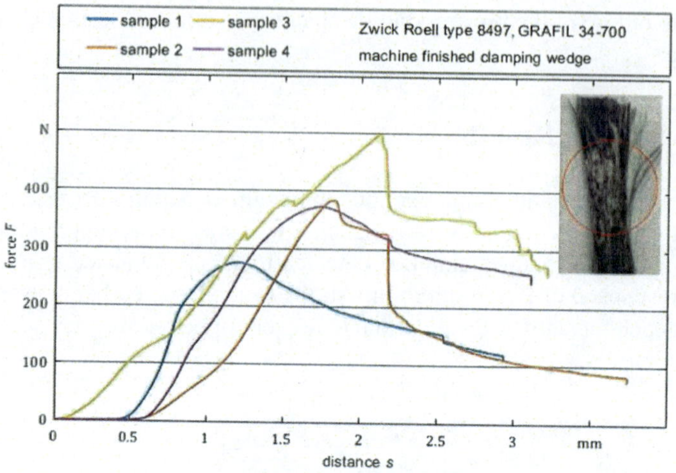

Fig. 8. Results of the tensile tests of the clamp

the required prestressing distance varies in the individual measuring runs (see Fig. 9). This precludes simultaneous prestressing by means of several mechanisms.

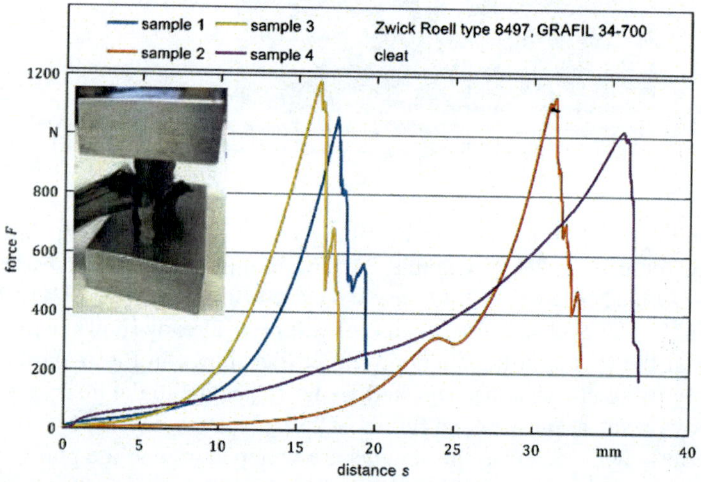

Fig. 9. Damaged fibre after applying a pre-tensioning force to the fibre clamped in the cleat, force-displacement diagram over four selected measurement runs when clamped in the cleat.

When the rovings are prestressed using Lost Form 1, forces of 1251.9 N can be achieved on average. The most frequent reason for failure is a fracture of the resin in the area of the ball. The required forces of 2800 N cannot be achieved with this prestressing mechanism (see Fig. 10). In addition, there is a large scatter of the measured values. This is probably due to a not ideal bond. This could also explain the deviating maximum prestressing forces compared to the simulation.

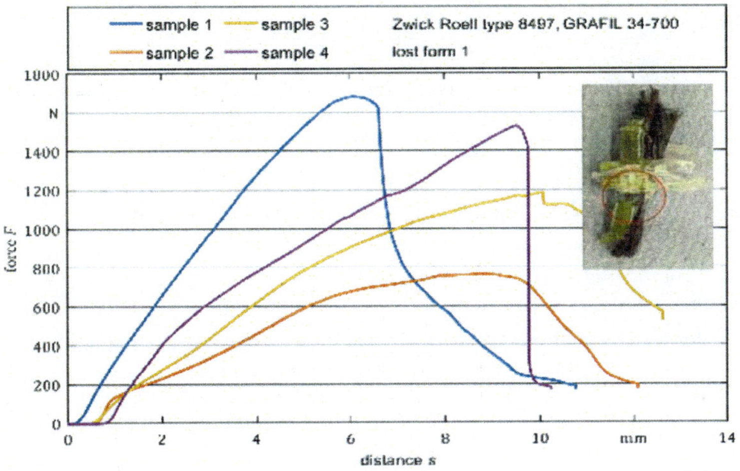

Fig. 10. Broken resin from Lost Form 1 in the experiment, force-displacement diagram over four selected measurement runs when clamped in Lost Form 1.

With Lost Form 2, the highest forces can be achieved on average with 5354.1 N. The lowest measured (3976.7 N) value is also above the required 2800 N (see Fig. 11). The reason for failure in all cases was fiber breakage.

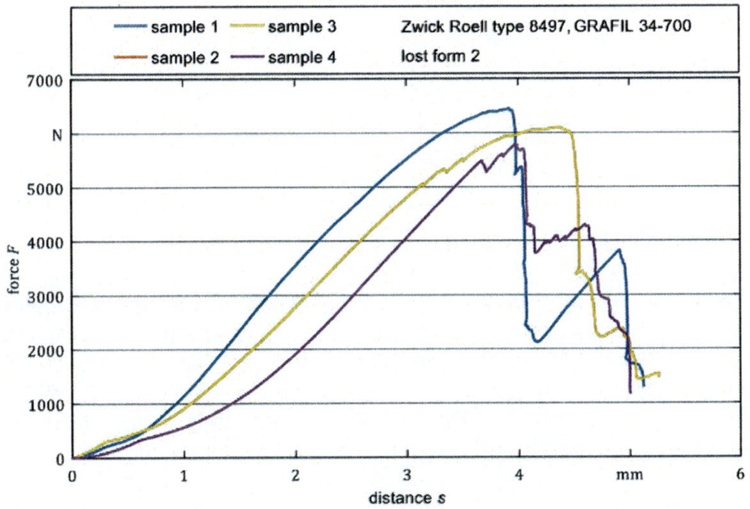

Fig. 11. Force-displacement diagram over four selected measurement runs with clamping in the Lost Form 2

4 Conclusion

In this research project, mechanisms for prestressing carbon fibers that are embedded in mineral cast were investigated. The focus was on the small available installation space and the high prestressing forces of 2800 N.

A force fitting, a frictional, a bionic firmly-binding and a combined firmly bonded and frictional mechanism were considered. The experimental investigation of these mechanisms showed that only the combined firmly bonded and frictional mechanism was capable of transmitting the high prestressing forces to the carbon fibers.

Acknowledgment. The authors would like to thank the German Research Foundation (DFG) for funding this work as part of the project "Prestressed fiber-reinforced mineral cast" (MO2091/11–1).

References

1. Möhring, H.-C., Brecher, C., Abele, E., Fleischer, J., Bleicher, F.: Materials in machine tool structures. CIRP Ann. Manuf. Technol. **64**, 725–748 (2015)
2. Möhring, H.-C., Brecher, Gessler, W., König, A., Nguyen, L.T., Nguyen, Q.P.: Modular intelligent fixture system for flexible clamping of large parts. J. Mach. Eng. **4,** 29–39 (2017)
3. Möhring, H.-C., Wiederkehr, P., Baumann, J., König, A., Spieker, C., Müller, M.: Intelligent hybrid material slide component for machine tools. J. Mach. Eng. **1**, 17–30 (2017)
4. Brecher, C., Sagmeister, B., Fey, M., Kneer. F.: Maschinenarten und Anwendungsbereiche. In: MM Maschinenmarkt, p. 1–4
5. Diederichs, U., Haroske, G., Krüger, W., Mertzsch, O.: Tragelemente aus Polymerbeton. Bautechnik **5**, 306–315 (2002)
6. Reis, J.: Mechanical characterization of fiber reinforced polymer concrete. Mater. Res. **3**, 357–360 (2005)
7. Reis, J., Ferreira, A.: Fracture behavior of glass fiber reinforced polymer concrete. Polymer Test. **2**, 149–153 (2003)
8. Barbuta, M., Harja, M.: Properties of fiber reinforced polymer concrete. Bulletin of the Polytechnic Institute of Jassy, constructions. Archit. Sect. (2008)
9. Kerrouche, A., Boyle, W., Sun, T., Grattan, K., Schmidt, J.W., Taljsten, B.: Strain Measurement using embedded fiber Bragg grating sensors inside an anchored carbon fiber polymer reinforcement prestressing rod for structural monitoring. IEEE Sens. J. **11**, 1456–1461 (2009)
10. Mitsubishi Chemical Carbon Fiber and Composites, Inc.: GRAFIL™ 34-700 12K & 24K Product Data Sheet
11. Corneslissen, B., Rietman, B., Akkerman, R.: Frictional behaviour of high performance fibrous tows: friction experiments. Compos. Appl. Sci. Manuf. **44**, 92–104 (2013)
12. Hwang, D.-H., Zum Gahr, K.-H.: Transition from static to kinetic friction of unlubricated or oil lubri- cated steel/steel, steel/ceramic and ceramic/ceramic pairs. Wear **255**, 365–375 (2003)

Development of Thin-Film Sensors for Data Acquisition in Cold Forging Tools

A. Schott[1](✉), M. Rekowski[1], K. Grötzinger[2], B. Ehrbrecht[3], and C. Herrmann[1]

[1] Fraunhofer Institute for Surface Engineering and Thin Films IST, Bienroder Weg 54E, 38108 Braunschweig, Germany
anna.schott@ist.fraunhofer.de

[2] Institute for Metal Forming Technology, University of Stuttgart, Holzgartenstraße 17, 70174 Stuttgart, Germany

[3] Hahn-Schickard, Wilhelm-Schickard-Str. 10, 78052 Villingen-Schwenningen, Germany

Abstract. The manufacturing of high-quality cold forged components with respect to environmental and economic requirements requires a high process reliability and long tool life. For this purpose, digitization offers new opportunities to increase process understanding by implementing inline data acquisition. Measurement of process parameters close to high loaded forming zones poses a serious challenge. Furthermore, tool wear cannot be detected inline. Therefore, integrated and wear-resistant thin-film sensors represent an innovative approach to realize in situ data acquisition. Hence, a sensor design will be introduced enabling both temperature measurement with spatial resolution and real-time wear detection directly in the forging zone. First, a thermal simulation is performed to determine the temperature range expected in the forming zone to conceive sensor structures and layers. Subsequently, various standard tool coatings are evaluated regarding their tribological and mechanical characteristics. Finally, the embedded system was conceptualized to read out the sensors and send the preprocessed data via USB-interface or BLE.

Keywords: Thin-film sensor · Cold forging · Wear-resistant · Inline process monitoring · Data acquisition

1 Introduction

Metal forming has a long and established history in production engineering. Today's needs of manufacturing companies are improved product quality, reduced costs and delivery time. At the same time, they have to manage increasing energy and emission costs [1]. Cold forging is one suitable process to satisfy these requirements. This manufacturing process is mainly used in fasteners technology and automotive drive train systems, where the components are usually axisymmetric (near) net-shape parts, which can be used without post-machining. The main advantages of cold forging are high productivity (depending on process up to 3 parts per second), advantageous mechanical properties and low energy consumption [2]. New developments in tooling technology

M. Liewald et al. (Eds.): WGP 2022, LNPE, pp. 61–70, 2023.
https://doi.org/10.1007/978-3-031-18318-8_7

help to enhance established process limits and to manufacture highly sophisticated geometrical shapes. In recent times, digitization and online process monitoring gained more and more importance for process understanding and improvement [3]. Data, such as die temperature, punch load or wear status, can be used to detect tool failure, to reduce downtime and thus to increase productivity. An important goal is to gain a deep understanding of the process parameter product structure product property relationships.

Due to the physics of cold metal forming, tools have to withstand highest contact pressure (up to 2500 MPa) as well as challenging tribological and thermal conditions. Tool wear and component geometry are directly affected by these harsh conditions. Tool coatings are one suitable approach to overcome these challenges [4]. Thin-film coatings like hard coatings or diamond-like carbon layers (DLC) are commonly used for cold forging tools [5]. DLC coatings feature a low friction coefficient and high wear resistance leading to an increase tool life [6]. Typically, such coatings are produced by vacuum coating technology such as physical (PVD) or chemical vapor deposition (CVD).

As tooling has significant interaction with the workpiece, this interaction provides information directly related to process conditions. Gathering these information opens up the opportunity to further optimization potentials by means of process understanding. Generating valuable information to enhance production reliability as well as product quality it is necessary to make the transformation from a passive process component to an active smart tool [7].

Therefore, thin-film sensors represent an innovative approach for in situ data acquisition. Directly applied on the tool surface by vacuum coating technology (PVD, PACVD) and structuring technologies they can be designed in a huge variety regarding layer system and sensor structure [8, 9]. Combined with an adequate measurement technology and data transmission the basis for further data analysis and process control is evolved. Wireless sensor networks (WSNs) integrate sensor, embedded computer, wireless communication and distributed intelligent information processing. Amongst others, machinery health monitoring and predictive maintenance is one possible application scenario of WSNs [10]. Their usefulness has already been demonstrated in metalworking industry [11].

The paper is structured as follows. At first, an overview of the system architecture is given, followed by theoretical background of simulation methods. After explaining the coating techniques and characterization methods in more detail, the development of the thin-film sensor is revealed. The results of thermal simulation are the basis for the development of sensor designs. Subsequently, the approach for measurement technology and data transmission is displayed. Finally, a summary of the developed sensory tool is presented and future potential for digitalization is discussed.

2 Technical Background

2.1 System Architecture

The system architecture for the sensory forging tool is depicted in Fig. 1. It shows the schematic structure of a punch coated with thin-film sensors for cold forging technology with the associated data processing system.

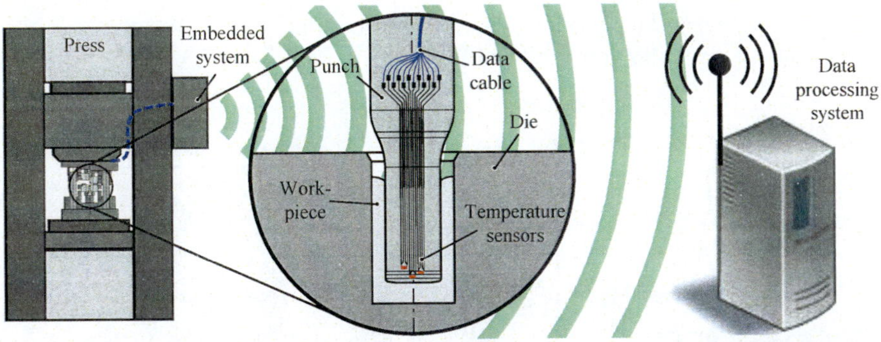

Fig. 1. System architecture from the sensors to the data processing system

The aim of this research is to develop a multifunctional sensor system, directly deposited on tool surfaces, for real-time analysis of process parameters during the pressing process in cold forging. Based on a modified tool concept to integrate the embedded system the thin-film sensor system needs to be designed. The embedded system measures the resistance of the thin-film sensors and calculates the temperature value. This data is then sent to a data processing system by a standard USB interface or wirelessly via Bluetooth Low Energy (BLE). In the data processing system this information can be used to optimize the process. Also, quality data can be gained which can be used to optimize the maintenance time.

2.2 Thermal Simulation of Forging Cycles

The process subjected here is cup backward extrusion, one of the most common cold forging processes. Punch diameter $d = 16$ mm was used to simulate the forging process of cups with an outer diameter $D = 22.5$ mm. Initial billet height was $H0 = 22$ mm and final bottom height $H1 = 6$ mm, where strain $\varepsilon = (D^2 - d^2)/D^2 = 0.49$ was achieved (see Fig. 2a). For the analysis of thermal effects within the punch during the forging process, it is important to consider cyclic heating and instationary ramp-up phase at the beginning of the process, where the tools are still cold. Finite element code DEFORM 2D® was used to create an axisymmetric process model with an elastic-plastic workpiece and elastic tools. Initial workpiece temperature was set to $T = 20$ °C and shear friction model with a friction factor $m = 0.12$ and a heat transfer coefficient $h = 5000$ W/(m^2 K) was used for contact definition between workpiece and tools. Heat transfer to the environment (air, $T = 20$ °C) was defined using heat convection coefficient with a value of 20 W/(m^2 K). Material data for C15 steel was gained from compression tests. The characterization tests were conducted on a thermomechanical testing system Gleeble 3800 C in a temperature range 20 °C $\leq T \leq$ 200 °C with strain rates of 0.1 s^{-1} $\leq \dot{\varphi} \leq$ 10 s^{-1}. The punch temperature was analyzed at several tracking points along the punch surface to characterize the thermal fields for further definition of coating layer design. For the investigation of cyclic heating, the whole process chain, consisting of forging operation, return stroke and transfer time, was considered to gain representative results for thermal analysis. The ejection process was simplified by an extended heat transfer time including also billet transfer, see Fig. 2b).

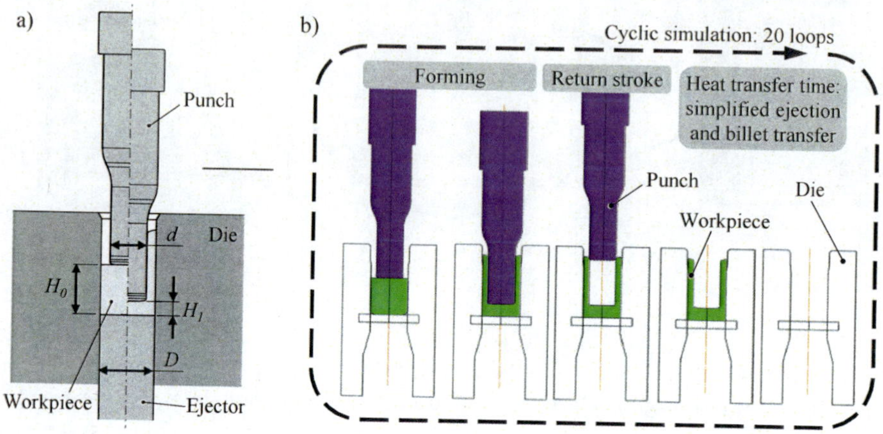

Fig. 2. a) Cup backward extrusion process, b) Cyclic simulation procedure

2.3 Sensor Manufacturing and Coating Characterization Methods

To prepare the thin-film layers of the sensor system different coating techniques were used. The hard coatings aluminum oxide (Al_2O_3), titanium and chromium nitride (TiN, CrN) as well as the metallic layer (Cr) were deposited by physical vapor deposition (PVD) processes. For the diamond-like carbon layer (DLC) and the hydrocarbon layer modified with silicon and oxygen (SICON®) a plasma assisted chemical vapor deposition (PACVD) technique was used. The surfaces of test substrates (1.3505) as well as the tool (ASP30, 1.3343) were polished before coating in a range of about Rz < 0.2 μm. The residual pressure before starting the process was typically < 10^{-3} Pa. Prior to coating, the substrates were cleaned with a water-based cleaning procedure. Instantly after the cleaning procedure samples were mounted into the vacuum chamber. Here, an ion etching step with argon ions for additional surface cleaning was applied. The sensor layer made of chromium was structured in a photolithographic process combined with a wet-chemical etching step. The coating properties were analyzed in terms of coating hardness [12], adhesion [13] and tribological behavior like abrasive wear analysis [14]. For the determination of the friction coefficients μ a ball-on-disk tester under lubricant free conditions at a relative humidity of 50% was used. Balls made of ball bearing steel (100Cr6) with a diameter of 5 mm were used as mating bodies. The normal load was 5 N, the velocity about 30 mm/s and the total sliding distance was 50 m. The equivalent values of the Hertzian pressure were estimated to be in the range of 0.72–1.28 GPa.

3 Results and Discussion

3.1 Thermal Simulation of Cup Backward Extrusion Process

The results of cyclic thermal simulation are shown in Fig. 3. Tracking points were used to visualize temperature rise within each single stroke and to detect stationary areas of tool temperature. Figure 3a) shows the punch temperature at 5 specific tracking points, where the distance to the punch land rises from point 1–5. The cyclic simulation was defined with 5 strokes per minute. An overall process duration of about 4 min (20 strokes) was simulated. Each stroke could clearly be identified with a local temperature maximum at the end of the deformation process, followed by a temperature decrease during ejection and transfer time. Within the amount of forging cycles considered in here, the punch never reached its initial temperature again. A continuous, however decreasing, temperature rise was found. The larger the distance from the punch land, the less intense are the local temperature maxima. Tracking point 1 shows the temperature in the punch land, where most frictional loads are present, while tracking point 5 shows the heating of the punch material due to heat transfer from the front area. This is why the temperature profile in tracking point 5 seems to be smoother than in areas close to the workpiece contact zone. The strongest temperature rise was found during the first 120 s (10 strokes) of the cyclic forging process. Afterwards, the mean temperature in the front punch area reached an almost stationary thermal state, where further heating is still expected.

Figure 3b) visualizes the thermal field in the punch after a specific amount of forming cycles, in each case at the end of forming operation, just before return stroke is initiated. In agreement with the temperature profiles Fig. 3a), the strongest temperature rise could be detected at the beginning of the cyclic heating simulation. Temperature rise of 70 K within 20 strokes in the front punch zone has to be considered in terms of coating design and thermal expansion of the punch.

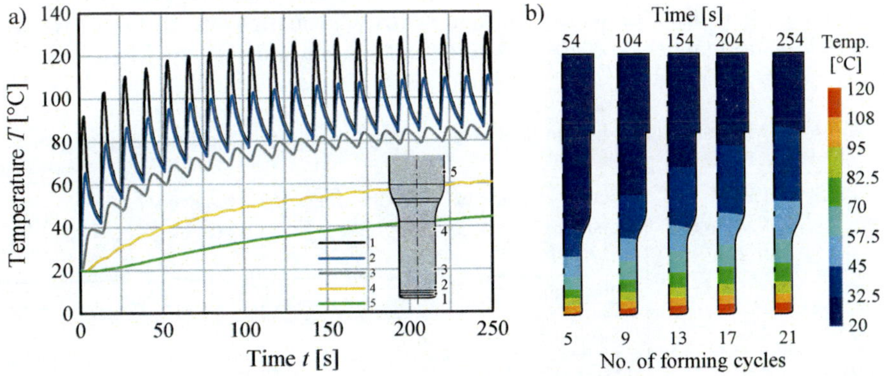

Fig. 3. Results of cyclic simulation a) temperature profiles in different points on punch surface, b) visualization of temperature distribution in punch over time

3.2 Development of the Thin-Film Sensor

Based on the results of the thermal simulation the thin-film layer system was developed. To realize an adequate electrical insulation in the required temperature range between the tool and the sensor structures, an Al_2O_3 coating was chosen and deposited with a thickness of approximately 5 μm. For the sensor layer, a 200 nm metallic layer of chromium was chosen. As this material has a positive temperature coefficient (PTC) it is usable as a temperature sensor by measuring its increasing resistance with rising temperature. This dependence can be described by a linear approximation in the range of ambient temperature up to a few 100 °C. Above the structured sensor layer another layer for electrical insulation and wear protection is applied. The scheme for the thin-film sensor system is shown in Fig. 4a).

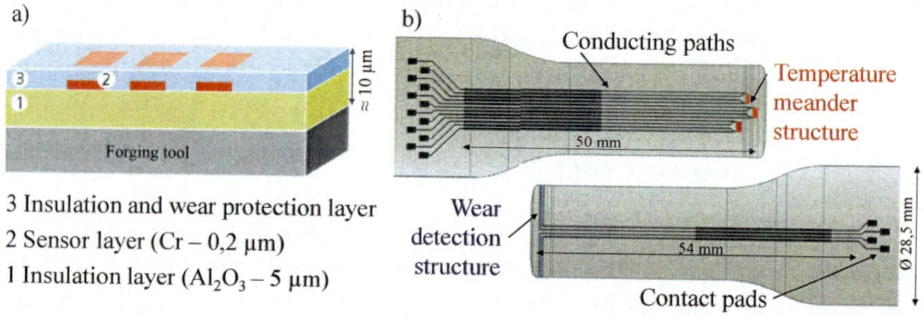

Fig. 4. a) Thin-film sensor system b) Sensor design for temperature and wear detection

Based on the simulation results for expected pressure and temperature during the forging processes, different layer systems were tested and compared regarding their mechanical and tribological properties as wear protection coatings (Table 1). As basic requirements, both the adhesive strength and electrical insulation must be given to prevent an electrical shortcut between the workpiece and the sensor during the forging process.

Table 1. Mechanical and tribological characteristics of different wear protection layer

Top layer system	Layer thickness (μm)	Friction coefficient μ (–)	Microhardness Hupl (GPa)	Abrasive wear rate $w_V (m^3 m^{-1} N^{-1} 10^{-15})$
DLC	2,9	0.22 ± 0.01	10.9	9.7
SICON®	4,0	0.34 ± 0.11	8.1	17.4
Al_2O_3	4,6	0.56 ± 0.20	10.4	20.1

Analysis of the microhardness shows value for the Al_2O_3 layer in the range of 10 GPa, but exhibits very high abrasive wear. Since the friction coefficient is also comparatively high, this layer is less suitable as a wear protection layer despite its excellent electrical insulation properties. The amorphous hydrogenated carbon layer SICON® shows a slightly lower microhardness as well as lower abrasive wear rate compared to Al_2O_3 layer, but at the same time a reduced friction coefficient. Fulfilling the basic requirements, the DLC layer exhibited the best combination of a low friction coefficient, a good microhardness and the lowest abrasive wear rate (Table 1). Compared to the hard coatings, the two layers DLC and SICON® show significantly lower and, above all, especially the DLC a less fluctuating coefficient of friction. Since friction is an important influencing parameter on punch wear in cold forging, these two coatings appear to be suitable as top coating systems.

Beside the approach to use one single top layer for wear protection as well as electric insulation, it was tested to achieve this by a two-layer system. This consists of an electric insulation layer and a standard tool coating. Therefore, industrial established coatings like TiN and CrN were deposited on insulation layers. These standard tool coatings exhibited indeed excellent tribological properties, but it was not possible to achieve a sufficient adhesion to the insulation layer. For this reason, the focus will be on top coatings based on a single layer.

The design of the sensor structures was developed based on the FEM simulation and is depicted in Fig. 4b). The concept developed for temperature measurement consists of three measuring points in the area of the punch tip in order to be able to measure a temperature profile. Therefore, temperature meander structures with an output resistance in the range of 100–250 Ω were designed. Using four-wire sensing each sensor needs four contact pads for wiring. These were placed securely in some distance from the forging zone. The wear measurement will be carried out by means of two conductor loops on the circumference of the punch. These are based on a digital measurement principle and provide a defined resistance when the conductor loop is intact and an infinitely high resistance when the conductors are disrupted. In order to record the high loaded forging zone on the edge of the forging tool in a locally differentiated manner, these were divided into two separate conductor loops. The forging tool with the applied thin-film sensor is depicted in Fig. 5. Each sensor structure will be characterized individually and connected via cable to the embedded system for data transmission.

3.3 Concept for Measurement Technology and Embedded System

The electrical resistance of the thin-film sensor directly correlates with the physical value to be measured. The developed prototype system as shown in Fig. 6 consists of an ADS124 analog to digital converter (ADC) and a PSoC 6 microcontroller unit including a radio module to control the system. The ADS124 contains the analog frontend with current source, reference voltage and a delta sigma ADC with a resolution of 24-Bits.

The software selects the sensor to be measured, sets up the current source for this sensor and starts the analog-to-digital-conversion to measure the voltage drop over the sensor resistor. Knowing the current and the voltage, the system calculates the resistance and from that the physical value – in that case the temperature. This procedure is repeated for all sensors in the system. When all sensor data has been acquired, the information is sent to all connected systems by USB or BLE with a frequency of 100 Hz. The system includes an accumulator, which makes it autonomous for one day. It can be charged

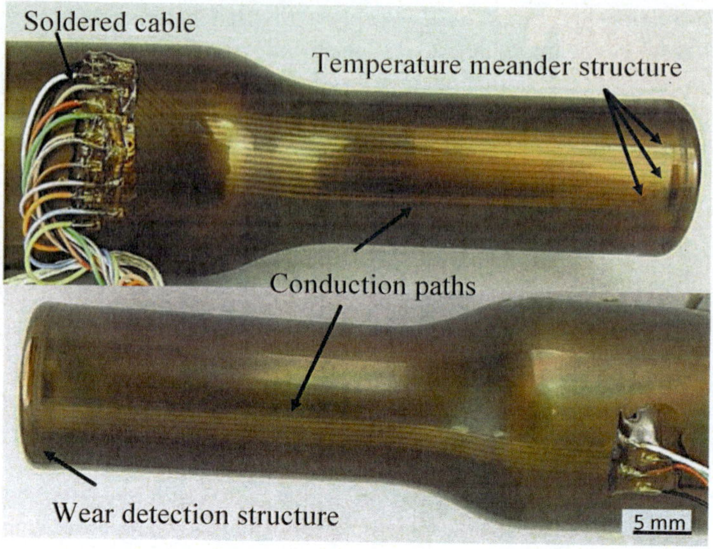

Fig. 5. Forging tool with the thin-film sensor system

over the USB interface. Through a fuel gauge, the external device knows the remaining charge.

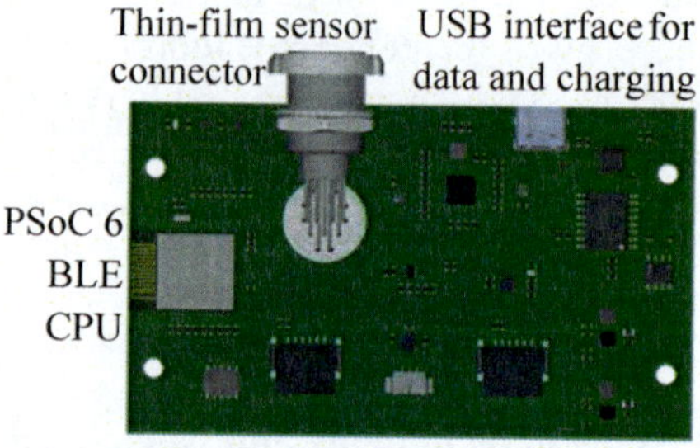

Fig. 6. Embedded main board

For the embedded software design, the PSoC Creator IDE from Cypress was used. To interpret the sensor information, a PC based graphical user interface was programmed with Visual Studio using the programming language C#, where the data is visualized in some charts and stored for later analytics.

4 Summary and Outlook

In this paper, the development steps for thin-film sensors towards intelligent forging tools including the measurement technique and data transfer were presented. A thermal simulation was used to estimate the expected temperature rise during the cold forging process. These results constituted the basis for the selection of the thin-film layer system to withstand the process conditions in terms of temperature. With regard to collected tribological and mechanical test results, it can be stated that the DLC coating shows the best properties for use as protective and electric insulation top coating. The deduced sensor design offers the possibility of measuring the temperature profile at three points in certain distances of the forging tip. Combined with the wear detection structures arranged directly in the high loaded area of the punch the basis for predictive wear detection to prevent part failure was created. Thus, the embedded system was designed to read out the individual thin-film sensors, perform data preprocessing and send this information over a USB-interface or via BLE to a data processing system. The graphical user interface was programmed to visualize and store the sensor data from the embedded system. Furthermore, it will also be used for sensor calibration.

The innovative approach towards the digitalization of cold forging processes shows a high potential to integrate sensors directly in the forging zone to make the change from a passive tool towards an active tool. This sets up the opportunity for an in situ monitoring system and at the same time an increased quality of produced components through an enhanced process understanding.

Further steps will include the implementation of the system in the forging machine to test and evaluate the sensor performance and the measurement system.

Acknowledgment. The research project was carried out in the framework of the industrial collective research program (IGF no. 21520 N). It was supported by the Federal Ministry for Economic Affairs and Climate Action (BMWK) through the AiF (German Federation of Industrial Research Associations eV). The authors would like to thank the IGF for the funding and the FSV and Hahn-Schickard for the support during the project period (Multisensorische Werkzeuge für die Kaltmassivumformung).

References

1. Schulz, K.-P., Riedel, R. (eds.): Nachhaltige Innovationsfähigkeit von produzierenden KMU: Inhalte, Methoden. Rainer Hampp Verlag, Fallbeispiele (2016)
2. Herlan, T.: Optimaler Energieeinsatz bei der Fertigung durch Massivumformung. Springer, Berlin (1989)
3. Liewald, M., Bergs, T., Groche, P., Behrens, B.-A. et al.: Perspectives on data-driven models and its potentials in metal forming and blanking technologies (2022)
4. Pelcastre, L., Hardell, J., Prakash, B.: Galling mechanisms during interaction of tool steel and Al–Si coated ultra-high strength steel at elevated temperature **67**, 263 (2013)
5. Dubar, M., Dubois, A., Dubar, L.: Wear analysis of tools in cold forging: PVD versus CVD TiN coatings **259**, 1109 (2005)
6. Bewilogua, K., Hofmann, D.: History of diamond-like carbon films—from first experiments to worldwide applications **242**, 214 (2014)
7. Cao, J., Brinksmeier, E., Fu, M., Gao, R.X., et al.: Manufacturing of advanced smart tooling for metal forming **68**, 605 (2019)
8. Biehl, S., Rumposch, C., Paetsch, N., Bräuer, G., et al.: Multifunctional thin film sensor system as monitoring system in production **22**, 1757 (2016)

9. Emmrich, S., Plogmeyer, M., Bartel, D., Herrmann, C.: Development of a Thin-film sensor for in situ measurement of the temperature rise in rolling contacts with fluid film and mixed lubrication. Sensors (Basel) **21** (2021)
10. Kandris, D., Nakas, C., Vomvas, D., Koulouras, G.: Applications of wireless sensor networks: an up-to-date survey **3**, 14 (2020)
11. Bal, M.: An industrial Wireless Sensor Networks framework for production monitoring, p. 1442 (2014)
12. DIN EN ISO 14577-1:2015-11, Metallische Werkstoffe_Instrumentierte Eindringprüfung zur Bestimmung der Härte und anderer Werkstoffparameter_Teil_1: Prüfverfahren (ISO_14577–1:2015); Deutsche Fassung EN_ISO_145771:2015, Berlin. Beuth Verlag GmbH
13. DIN EN ISO 6508-1:2016-12, Metallische Werkstoffe_Härteprüfung nach Rockwell_Teil_1: Prüfverfahren (ISO_6508-1:2016); Deutsche Fassung EN_ISO_6508–1:2016, Berlin. Beuth Verlag GmbH
14. Michler, T., Siebert, C.: Abrasive wear testing of DLC coatings deposited on plane and cylindrical parts 163-164, 546 (2003)

Application of Reinforcement Learning for the Design and Optimization of Pass Schedules in Hot Rolling

C. Idzik(✉), J. Gerlach, J. Lohmar, D. Bailly, and G. Hirt

Institute of Metal Forming, RWTH Aachen University, Intzestr. 10, 52072 Aachen, Germany
christian.idzik@ibf.rwth-aachen.de

Abstract. About 95% of all steel products are rolled at least once during their production. Thus, any further improvement of the already highly optimized rolling process, for example reduction of energy consumption, has a significant impact. Currently, most rolling processes are designed by experts based on their knowledge and heuristics using fast analytical rolling models (FRM). However, due to the complex interactions between the processing constraints e.g. machine limits, the process parameters as well as the product properties, these manual process designs often focus on a single optimization objective. Here, novel methods such as reinforcement learning (RL) can detect complex correlations between chosen parameters and achieved objectives by interacting with an environment i.e. FRM. Therefore, this contribution demonstrates the potential of coupling RL and a FRM for the design and multiple objective optimization of rolling processes. Using FRM data e.g. the microstructure evolution, the coupled approach learns to map the current state, such as the height, to process parameters in order to maximize a numerical value and thereby optimize the process. For this, an objective function is presented that satisfies all (technical) constraints, leads to desired material properties including microstructural aspects and reduces the energy consumption. Here, two RL algorithms, DQN and DDPG, are used to design and optimize pass schedules for two use cases (different starting and final heights). The resulting pass schedules achieve the desired goals, for example, the desired grain size is achieved within 4 μm on average. These meaningful solutions can prospectively enable further improvements.

Keywords: Hot Rolling · Process Design · Reinforcement Learning

1 Introduction

In recent years, increased efforts have been made to further optimize energy-intensive processes such as hot rolling and thus make them more sustainable [1]. Hot rolling is one of the most common forming processes for the production of flat metal products with optimized dimensions and properties. About 95% of steel products are rolled at least once during their production as noted by Allwood et al. [1].

J. Lohmar: Deceased

M. Liewald et al. (Eds.): WGP 2022, LNPE, pp. 71–80, 2023.
https://doi.org/10.1007/978-3-031-18318-8_8

Consequently, even small optimizations have a major impact on global energy and material consumption. One major influencing factor regarding the process efficiency in the (hot) rolling process is the process design. Hot rolling processes generally consist of several rolling steps, called passes. During each pass the material is moved through at least one pair of rotating rolls and meanwhile deformed. The pass schedule defines all the parameters for the whole process, e.g. the height-reduction of each pass.

The pass schedule has to guarantee the defined geometric dimensions such as the final height and material properties such as a specific microstructure (grain size). Moreover, it has to consider rolling mill limitations, e.g. maximum allowable rolling force and torque, as well as economic aspects e.g. rolling time and ecological aspects e.g. the energy consumption. Typically, experts design pass schedules based on their knowledge, iterative heuristics and support of predictive models or simulations. However, these approaches often do not optimize for multiple objectives.

Here objective approaches that enable multi-objective optimization can support the experts to find optimized pass schedules. One possible solution is to combine methods of Machine Learning (ML) with physical process models. Scheiderer et al. [2] showed that ML methods, in form of Reinforcement Learning (RL), can design pass schedules while accounting for multiple objectives.

Therefore, in this paper RL algorithms are coupled with a fast rolling model (FRM) to design and optimize pass schedules for the hot rolling process. First, an overview of fast rolling models, approaches for pass schedule design and ML respectively RL is given in the state of the art. In this context, some application of RL for process optimization is shown. Next, the coupling between the RL algorithms and the FRM is detailed. In this paper, the reinforcement learning algorithms Deep Q-Network (DQN) [3] and Deep Deterministic Policy Gradient (DDPG) [4], are used. Additionally, the reward function for adhering to the specified tolerances, machine limits and total energy consumption to enable objective evaluations of pass schedules is presented. Subsequent, the results obtained for two trainings are presented and discussed. Finally, the results are summarized and an outlook regarding the next steps is given.

2 State of the art

This chapter presents the background information and state of the art of the used methods and approaches starting with fast rolling models and an overview of different design approaches of hot rolling processes. Next, ML and more specifically the RL is briefly introduced including current research on using RL for process optimization.

2.1 Fast Rolling Models

Here, conventional FE simulations, which take minutes or hours to run, are not suited because the total computation time of the optimization task depends on the underlying model that should compute as fast as possible. Hence, fast rolling models are used that usually consist of semi-empirical equations derived via mechanical simplifications or physical principles. Several fast rolling models have been developed in the past. Beynon and Sellars [5] present a rolling model called SLIMMER that is able to describe the microstructure evolution and predicts roll force and torque during multi-pass hot rolling. Inspired by their work, Seuren et al. [6] and Lohmar et al. [7] present a model that allows the prediction of force, torque, recrystallized fraction and grain size, among others, including a height resolution and the influence of shear. Another rolling model was developed by Jonsson [8]. It calculates, inter alia, strain, precipitates and dynamic recrystallization in order to predict the ferrite grain size after hot rolling.

2.2 Pass Schedule Design and Optimization

There are several different approaches to design a pass schedule in research and industry [9]. Many researchers and companies developed and proposed approaches dealing with these challenges. For instance, Svietlichnyi and Pietrzyk [10] suggest to set the height reduction in each pass to the maximum value while Schmidtchen and Kawalla [11] propose to set aim for an even distribution of the rolling force. All these approaches are usually used to design a first version of a pass schedule, which then is further optimized with regards to not more than two objectives simultaneously. Typical optimization objectives are grain refinement [5], grain size uniformity [12] as well as maximizing yield strength (YS) and ultimate tensile strength (UTS) [13]. A comprehensive literature review on recent design and optimization approaches for rolling is described by Özgür et al. [14].

2.3 Machine Learning and Its Applications for Process Optimization

Machine learning (ML) describes the ability of algorithms to detect patterns (in data) and learn from their inferences. Rosenblatt [15] proposed a probabilistic model that aims to replicate the functions of biological neurons. These artificial neurons transform an input vector into a scalar output by calculating a weighted sum out of all the input elements and then feeding them through an activation or transfer function. By systematically adjusting the weights, desired results are reproducible. Thus, multiple layers of these artificial neurons, called Artificial Neural Networks (ANN), can learn non-linear interrelations. These ANNs are used in different ML categories like supervised, unsupervised and reinforcement learning. Supervised learning needs a training set of labeled data and is used for classification and regression [16]. The goal of unsupervised learning is to find hidden structures in unlabeled data by clustering and dimension reduction.

The third category, reinforcement learning (RL), is an interacting approach where the algorithm learns by mapping state to actions to maximize a numerical reward [16]. One of the first ideas to use RL in manufacturing was published in 1998 [17]. However, according to Wuest et al. [18], RL is not widely applied in manufacturing, yet.

For hot rolling, Scheiderer et al. [2] published a RL approach that can design pass schedules while accounting for multiple objectives. The authors used a database with simulation data to design and optimize pass schedules. Moreover, Gamal et al. [19] demonstrated that RL in combination with process data identifies model parameters and thus improve predictions for bar and wire hot rolling processes.

2.4 Methodology: Coupling Reinforcement Learning and Fast Rolling Model

In this chapter, the coupling between a FRM and two RL algorithms is shown. For this purpose, the necessary components are described individually. First the used FRM is described, after which the RL algorithms and the optimization objectives are presented.

2.5 Fast Rolling Model

For the coupling with RL an already existing FRM is taken. The FRM, developed at the Institute of Metal Forming (IBF), is described in detail by Seuren et al. [6] and Lohmar et al. [7]. It is based on the slab method and consists of several modules, allowing the prediction of the deformation, the temperature and microstructure evolution as well as rolling forces and torques. In order to describe the material behavior, semi-empirical equations are used. For instance, the microstructure evolution is modelled using static recrystallization and grain growth equations from Beynon and Sellars [5].

The material used in this investigation is S355 since the model parameters are calibrated for this type of steel. S355 is a structure steel according to the European structural steel standard EN 10025–2:2004. The chemical composition is given in Table 1 and was determined using an optical emission spectrometer. The thermal boundary conditions used in the model are in accordance with values typically found in technical literature or were determined experimentally earlier.

Table 1. Chemical composition of S355

	C	Si	Mn	P	S	Cu	Al
Weight in %	0,1	0,3	1,6	0,012	0,001	0,2	0,02

2.6 Reinforcement Learning Algorithms

RL is a trial-and-error interaction approach. It consists of an agent, an environment, which represents the problem, and a set of actions. For each discrete time step, the agent perceives the environmental state and carries out actions, resulting in changes of the environment. Based on the resulting state a numerical value called reward is calculated. The reward assesses the goodness of the performed actions. The value can either be positive or negative, representing a reward or a punishment, respectively. The goal of RL is to choose the actions such that a maximum reward is obtained.

There are numerous RL algorithms available and described in the literature [16]. In this paper, the focus is set on two well-known and established RL algorithms used for process optimizations, Deep Q-Networks (DQN) [3] and Deep Deterministic Policy Gradient (DDPG) [4]. Both algorithms use ANNs to learn the relationships between actions, new states and the resulting rewards. Through interaction, it generates the data samples for training itself. But the samples are correlated, they are stored in a so-called experience buffer from which random samples, called a mini-batch, are taken.

DQN uses one ANN as an approximator for the state-value function which indicates how beneficial it is for an agent to perform a particular action in a state. This state-value function, often referred to as Q function, describes the cumulative long-term reward.

DDPG is very similar to DQN, except for the fact that is uses two ANNs, the critic and the actor network. The critic network has the same task as the DQN, but instead of using just the Q-value to improve action selection, DDPG uses the Q-values to learn a policy using the actor network. This allows to solve more complex problems with even a continuous action space. The general structure of the coupling between the RL and the FRM is exemplarily shown for DDPG in Fig. 1.

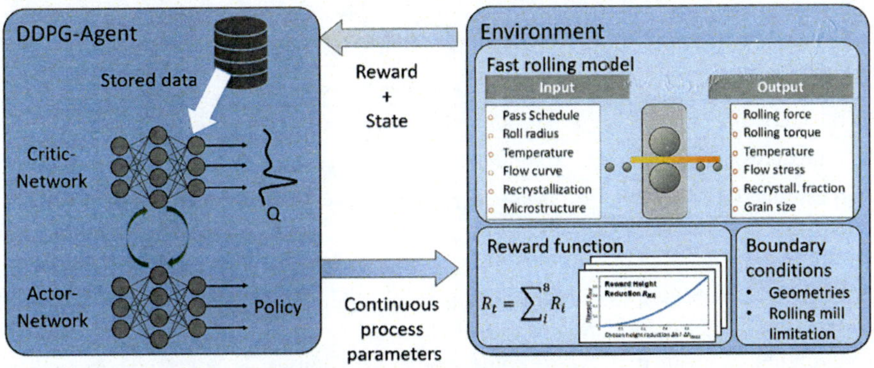

Fig. 1. Coupling of the DDPG algorithm with the FRM

In order to establish a better comparability between the two RL algorithms, the same hyperparameters were used as far as possible. In both DQN and DDPG, the ANNs have a similar structure. The critic networks in DQN and DDPG have four hidden layers with 200 neurons each while the actor network of the DDPG algorithm contains two hidden layers with 200 neurons each. Table 2 shows the most relevant hyperparameters of the two algorithms. They correspond to typical values from the literature and were chosen equally for both algorithms as far as possible.

Table 2. Hyperparameters for the used DQN and DDPG

Hyper-parameter	Discount factor	Learning rate	Experience buffer length	Mini-batch size
	0.95	1e−3	1e5	64

2.7 Optimization Objectives

As mentioned before, the goal is to optimize hot rolling with respect to multiple objectives. For this purpose, the optimization objectives are first defined and converted into a corresponding reward function R. The objectives should be based on real conditions in production and take into account factors such as product properties (height, grain size), limitations (rolling mill limits, final temperature) and process efficiency (energy consumption, process time). Table 3 shows the objectives considered here.

Table 3. Optimization objectives considered in the reward function

Reward component	Objective of optimization
Height R_H	Reward if h_{target} is reached, punish if $h_c < h_{target}$
Grainsize R_d	Reward when the target range of d is reached
Force, torque R_F, R_M	Exploit the rolling mill limit but not exceed it
Temperature R_ϑ	Maximizing final temperature
Energy consumption R_E	Minimizing total energy consumption
Height reduction R_{HR}	Minimizing the pass number
Interpass time R_I	Minimizing the process time

After defining the optimization objectives and thus individual reward components, the actual reward function R is defined, see Eq. 1. An intuitive possibility is to set up a weighted sum of the reward components. This allows easy prioritization so that in the context of this publication, product properties such as the final height R_H and microstructure in terms of the grain size R_d are prioritized and accordingly weighted stronger. The other components are weighted equally, as there is no further preference here. Weighting is necessary otherwise the desired properties like the final height and grain size are achieved within the tolerance. A simple sum of all reward components resulted in pass schedules that did not simultaneously achieve the target height and the target grain size. Through several trials, the weights for R_H and R_d were adjusted so that first the target height and then target grain size was reached.

$$R_0 = 5 * R_H + 3 * R_d + R_E + R_{HR} + R_M + R_F + R_I + R_\vartheta \tag{1}$$

In general, the reward components can be divided into three purposes. The first purpose is to reach a certain target value or area. This applies to the reward components of height R_H, grain size R_d and rolling force R_F as well as torque R_M. In Fig. 2, the definition of the R_H is presented exemplarily. R_d, R_F and R_M are defined similarly.

In addition to achieving certain targets within defined tolerances, there are also objectives that need to be minimized (R_E, R_I) or maximized (R_{HR}, R_ϑ). As in the previous example, the reward components are defined in an uniform way as shown in Fig. 3.

3 Results and Discussion: Designed Pass Schedules

In this chapter, the final pass schedules for two use cases are described, discussed especially in the context of possible novel approaches to design pass schedules. The use cases

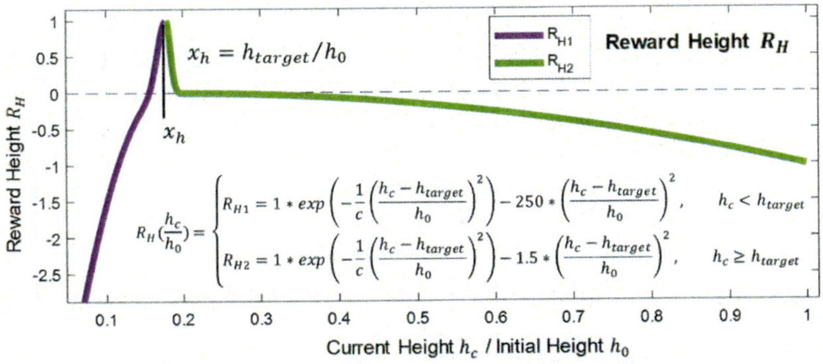

Fig. 2. Reward component for the height

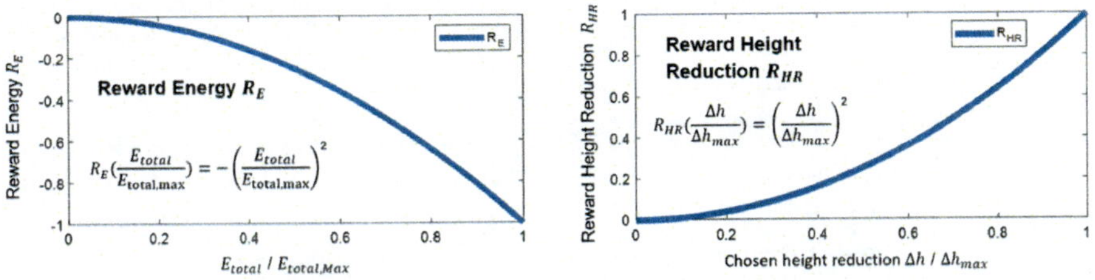

Fig. 3. Reward components for the energy consumption (left) and height reduction (right)

are selected allowing experimental rolling on the laboratory rolling mill at IBF. The initial and final parameters of the two cases can be found in Table 4. Both were trained in 20,000 iterations with the DQN and the DDPG algorithm. These trainings took about 24 h on an average computer (CPU: Intel Xeon E3-1270). If 20,000 FE simulations were calculated instead of the FRM, the training would much longer. The agents could choose the height reduction [3–18.45 mm] and inter-pass time [5–15 s]. The remaining process parameters were held constant (for example the rolling velocity is fixed at 250 mm/s) but can generally be included in the optimization, as well.

Table 4. Initial and target parameters for the two use cases

Use case	Initial parameters				Targets	
	h_0 in mm	b_0 in mm	l_0 in mm	ϑ_0 in °C	h_{target} in mm	d_{target} in µm
1	70	160	320	1100	20 ± 0.4	30 ± 5
2	140	180	500	1200	25 ± 0.5	30 ± 5

First, the resulting pass schedules of the first use case are presented and discussed, see Fig. 4. The pass schedule laid out by DQN consists of a total of four passes, which is the minimum number of passes possible. The most important targets regarding the final properties were achieved only to a limited extent. The final height h_{final} 19.4 mm is slightly outside the desired tolerance, whereas the average grain size d_{final} 30 μm was perfectly achieved. The pass schedule laid out by DDPG consists of a total of four passes and results in targets which are within the desired tolerances (h_{final} 19.8 mm, d_{final} 28 μm). Otherwise, both pass schedules are very similar. In both cases, pass schedules do not show a novel rolling strategy, e.g. the height reduction is reduced with each pass. This practice is very similar to that of experts. Experts would also start with as few passes as possible, starting with large height reductions at the beginning and decreasing the height reduction with each pass. Therefore, the two automatically designed pass schedules do not differ significantly from those designed by experts.

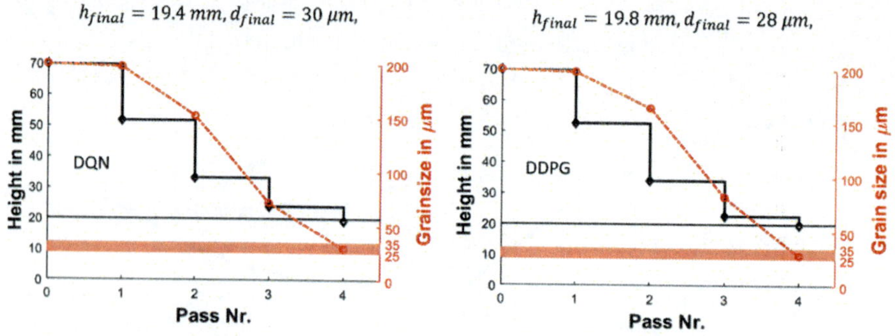

Fig. 4. Final pass schedules for the first use case trained by DQN (left) and DDPG (right)

Comparing the obtained pass schedules for the second use case, see Fig. 5, a lot of similarities can be observed here as well. In both cases, the pass schedules consist of eight passes (minimum number of passes possible), the target value for h_{final} (24.7 and 24.9 mm) is reached while d_{final} (26 and 23 μm) is not, although the results lie very close to the tolerance area. In contrast to the first case, these two pass schedules with the very small height reduction in the last pass show a conspicuousness compared to typical pass schedule designed by experts. Such small height reductions in the last pass would very probably not be suggested by an expert, as it has no added value in process terms. Experts would reduce the height reduction as described above with each pass, but they would distribute it more evenly and not just reduce it by a few mm.

All obtained results show that the concept of coupling a FRM with RL is very promising. Rollable pass schedules were designed and the desired objectives were (almost) reached. However, it is evident that despite stronger weighting in the reward, the final heights are not hit perfectly. Currently, the final height needs to be adjusted slightly to achieve the targets perfectly. Interestingly, there are no noticeable differences between the two RL-algorithms (DQN, DDPG) in the final designed pass schedules for both use cases. Both algorithms can be used for pass plan design.

4 Conclusion

In this paper, two hot rolling processes were designed automatically to demonstrate the promising potential of coupling a process model, FRM and RL. Two different RL

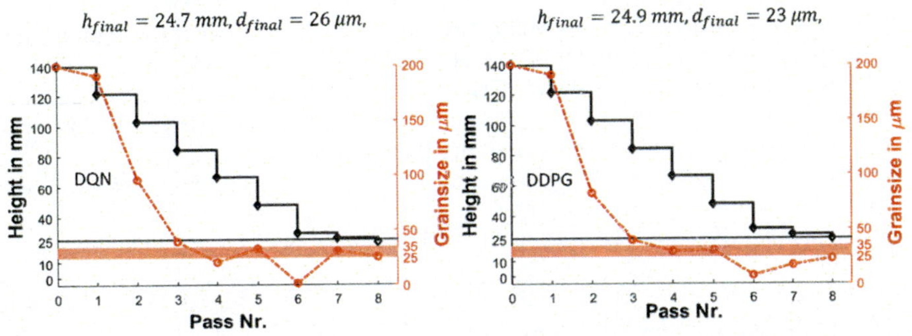

Fig. 5. Final pass schedules for the second use case trained by DQN (left) and DDPG (right)

algorithms, DQN and DDPG, were used for the training to test and compare them. The designed pass schedules after training show that, regardless of the use case and specific RL algorithm used, the coupled approach leads to rollable pass schedules that successfully achieve goals for the most part. In both presented use cases DQN and DDPG lay out similar final pass schedules. It is noticeable that for the first use case ($h_{final} = 20$ mm) presented, both algorithms achieve very good pass schedules in terms of final height and grain size. The average deviation in the final height Δh is 0.4 mm and in the mean grain size Δd is just 1 µm. In the second use case, requiring twice the number of passes compared to the first one, Δh was better (0.2 mm), but Δd was worse (5.5 µm). Comparing the designed pass schedules with typical pass schedules laid out by experts, no novel rolling strategy can be identified. The pass schedules designed by the RL algorithms apply high height reductions at the beginning and reduce them with each pass. In the future, the approach will be extended for online process adaption.

Acknowledgement. Funded by the Deutsche Forschungsgemeinschaft (DFG, German Research Foundation) under Germany´s Excellence Strategy – EXC-2023 Internet of Production – 390621612. We thank Kuan Wang for supporting us in conducting the trainings.

We heartily thank our colleague Dr.-Ing. Johannes Lohmar for his encouragement and for his timely support, guidance and suggestions during this project work.

References

1. Allwood, J.M., Cullen, J.M., Carruth, M.A.: Sustainable materials. With both eyes open; [future buildings, vehicles, products and equipment - made efficiently and made with less new material]. UIT Cambridge, Cambridge (2012)
2. Scheiderer, C., et al.: Simulation-as-a-service for reinforcement learning applications by example of heavy plate rolling processes. Proc. Manuf. **51**, 897–903 (2020). https://doi.org/10.1016/j.promfg.2020.10.126
3. van Hasselt, H., Guez, A., Silver, D.: Deep reinforcement learning with double Q-learning (2015)
4. Silver, D., Lever, G., Heess, N., Degris, T,. Wierstra, D., Riedmiller, M.: Deterministic Policy Gradient Algorithms Proceedings of the 31 st International Conference on Machine Learning, 32. Aufl, Beijing, China, S 387–395 (2014)

5. Beynon, J.H., Sellars, C.M.: Modelling Microstructure and Its Effects during Multipass Hot Rolling. Iron Steel Inst. Jap. **32**(3), 359–367 (1992)
6. Seuren, S., Bambach, M., Hirt, G., Heeg, R., Philipp, M.: Geometric factors for fast calculation of roll force in plate rolling. In: Zhongguo-Jinshu-Xuehui (Hrsg) 10th International Conference on Steel. Metallurgical Industry Press, Beijing (2010)
7. Lohmar, J., Seuren, S., Bambach, M., Hirt, G.: Design and application of an advanced fast rolling model with through thickness resolution for heavy plate rolling. In: Guzzoni, J., Manning, M. (Hrsg) 2nd International Conference on Ingot Casting Rolling Forging. ICRF (2014)
8. Jonsson, M.: An investigation of different strategies for thermo-mechanical rolling of structural steel heavy plates. ISIJ Int. **46**(8), 1192–1199 (2006). https://doi.org/10.2355/isijinternational.46.1192
9. Pandey, V., Rao, P.S., Singh, S., Pandey, M.: A calculation procedure and optimization for pass scheduling in rolling process. A Rew. 126–130 (2020)
10. Svietlichnyj, D.S., Pietrzyk, M.: On-line model for control of hot plate rolling. In: Beynon, J.H. (Hrsg) 3rd International Conference on Modelling of Metal Rolling Processes. IOM Communications, London, S 62–71 (1999)
11. Schmidtchen, M., Kawalla, R.: Fast Numerical simulation of symmetric flat rolling processes for inhomogeneous materials using a layer model—part I. Basic Theory. Steel Res. Int. **87**(8), 1065–1081 (2016). https://doi.org/10.1002/srin.201600047
12. Hong, C., Park, J.: Design of pass schedule for austenite grain refinement in plate rolling of a plain carbon steel. J. Mater. Process. Technol. **143–144**, 758–763 (2003). https://doi.org/10.1016/S0924-0136(03)00363-7
13. Chakraborti, N., Siva Kumar, B., Satish Babu, V., Moitra, S., Mukhopadhyay, A.: A new multi-objective genetic algorithm applied to-hot-rolling process. Appl. Math. Model. **32**(9), 1781–1789 (2008). https://doi.org/10.1016/j.apm.2007.06.011
14. Özgür, A., Uygun, Y., Hütt, M.-T.: A review of planning and scheduling methods for hot rolling mills in steel production. Comput. Ind. Eng. **151**(20), 106606 (2021). https://doi.org/10.1016/j.cie.2020.106606
15. Rosenblatt, F.: The perceptron. A probabilistic model for information storage and organization in the brain. Psychol. Rev. **65**(6), 386–408. (1958). https://doi.org/10.1037/h0042519
16. Sutton, R.S., Barto, A.: Reinforcement Learning. An Introduction. Adaptive Computation and Machine Learning. The MIT Press, Cambridge, MA, London (2018)
17. Mahadevan, S., Theocharous, G.: Optimizing Production Manufacturing Using Reinforcement Learning FLAIRS conference, Bd **372**, S 377 (1998)
18. Wuest, T., Weimer, D., Irgens, C., Thoben, K.-D.: Machine learning in manufacturing. Adv. Chall. Appl. Prod. Manuf. Res. **4**(1), 23–45 (2016). https://doi.org/10.1080/21693277.2016.1192517
19. Gamal, O., Mohamed, M.I.P., Patel, C.G., Roth, H.: Data-driven model-free intelligent roll gap control of bar and wire hot rolling process using reinforcement learning. IJMERR 349–356 (2021). https://doi.org/10.18178/ijmerr.10.7.349-356

Simulation of Hot-Forging Processes with a Temperature–Dependent Viscoplasticity Model

J. Siring[1(✉)], M. Schlayer[2], H. Wester[1], T. Seifert[2], D. Rosenbusch[1], and B.-A. Behrens[1]

[1] Institute of Forming Technology and Machines, Hemnitz, Germany
siring@ifum.uni-hannover.de

[2] University of Applied Sciences Offenburg, Offenburg, Germany

Abstract. Hot forging dies are subjected to high cyclic thermo-mechanical loads. In critical areas, the occurring stresses can exceed the material's yield limit. Additionally, loading at high temperatures leads to thermal softening of the used martensitic materials. These effects can result in an early crack initiation and unexpected failure of the dies, usually described as thermo-mechanical fatigue (TMF). In previous works, a temperature-dependent cyclic plasticity model for the martensitic hot forging tool steel 1.2367 (X38CrMoV5-3) was developed and implemented in the finite element (FE)-software Abaqus. However, in the forging industry, application-specific software is usually used to ensure cost-efficient numerical process design. Therefore, a new implementation for the FE-software Simufact Forming 16.0 is presented in this work. The results are compared and validated with the original implementation by means of a numerical compression test and a cyclic simulation is calculated with Simufact Forming.

Keywords: Martensitic die steel · Plasticity model · Hot forging

1 Introduction

In hot forging processes, the dies are subjected to high loads. These loads can be divided into mechanical, thermal, tribological and chemical loads [1]. The high cyclic thermo-mechanical loads can result in crack initiation and propagation due to material softening. Failure of the forging dies occurs when the crack finally exceeds a critical length [2]. This fatigue behaviour is also known as thermo-mechanical fatigue (TMF) and can cause unexpected failure of the dies [3].

The simulation-based design of forging dies and processes with the help of the finite element method (FE) is increasingly used nowadays [4]. In many cases, simulations are only carried out in relation to the semi-finished products and their forming process using the FE method. This simulation method ensures a cost-efficient development process as the simulation time is low compared to elastic-plastic or thermally coupled simulations considering the dies [5]. The simultaneous or additional consideration of the dies in the simulation process enables an estimation of the stresses on the forging dies under

M. Liewald et al. (Eds.): WGP 2022, LNPE, pp. 81–90, 2023.
https://doi.org/10.1007/978-3-031-18318-8_9

the stresses caused by the forming process and can thus be used to predict the die stability and life time. In addition, numerical simulations can be used to optimise the dies in a cost-effective manner and to increase the service life time by optimisation [6]. Usually isothermal linear-elastic simulations are used, where the focus is on the load in the first cycle. However, the effect of plasticity due to stresses that locally exceed the temperature-dependent yield strength at high operating temperatures is neglected [2]. In addition, cyclic loading on the die results in material softening, which leads to a reduced service life time. This holistic view from the simulation process with the cyclic consideration is the final goal of the project, and this paper presents the necessary material model with the softening behaviour.

The high strength of hot forging steels is a result of the so called Orowan mechanism. Strengthening secondary carbides are precipitated during heat treatment and hinder the slip of dislocations [7, 8]. However, an effect called Ostwald ripening occurs at high temperatures leading to coarsening of the strengthening carbides with time and, thus, reducing the material's strength [9]. Thermal softening of the material was observed in high temperature applications comparable to forging dies [10]. The evolution in the microstructure due to thermo-mechanical loading of hot forging tool steels was for example addressed in [11]. It was found that the microstructure and the bond between the particles depend on the tempering behaviour.

The effects of TMF and thermal softening are well known. Nevertheless, a meaningful implementation in the context of a cyclic viscoplasticity material model for hot forging tool steels is not yet available and therefore not yet accessible in FE-simulations. Zhang et al. [12] developed a material model for the prediction of thermal fatigue for the martensitic steel 55NiCrMoV7, which was commonly used as material for hot forging tool steels. The development of the cyclic anisothermal plasticity model including ageing effects was published in [13].

Jilg et al. developed an approach to assess plastic deformation in FE-simulations for the hot forging tool steel 1.2367 (X38CrMoV5-3) including thermal softening [14]. A kinetic model describes the strengthening effect and the temperature dependent coarsening of the carbides [15]. This model was implemented in the FE-software Abaqus. However, in the forging industry application-specific software is usually used to ensure cost-efficient numerical process design. In addition, the use of application-specific software makes it possible to calculate processes with strong non-linear deformations due to efficient and robust remeshing algorithms.

Within this study it is the aim to implement the plasticity model developed by Jilg et al. [14] in a time-dependent (viscoplastic) formulation for Simufact Forming 16.0 that is commonly used for the design of hot forging processes by means of a user subroutine. Thus, it can be used in future for efficient service life time calculation of complex hot forging processes considering cyclic thermal-mechanical loadings.

2 Model Description and Implementation

2.1 Viscoplasticity Model

The extension of the plasticity model of Jilg et al. [14] to non-isothermal viscoplastic behaviour is based on the works of Chaboche [16]. For simplicity, the viscoplasticity model is presented in the uniaxial formulation in this section whereas the FE-implementation is based on the multiaxial formulation using the von Mises yield function. To describe thermal softening, the model considers a kinetic model, which was originally presented in [14].

The nominal stress σ is calculated using Hooke's Law with young's modulus E and the total strain ε^{tot} by

$$\sigma = E \cdot \left(\varepsilon^{\text{tot}} - \varepsilon^{\text{th}} - \varepsilon^{\text{vp}}\right) \tag{1}$$

considering the thermal strain εth and the viscoplastic strain ε^{vp}. The thermal strain is calculated with the coefficient of thermal expansion αth. The viscoplastic strain ε^{vp} is calculated by integration of the equation

$$\dot{\varepsilon}^{\text{vp}} = \dot{\overline{\varepsilon}}^{\text{vp}} \frac{\sigma - \alpha}{|\sigma - \alpha|} \text{ with } \dot{\overline{\varepsilon}}^{\text{vp}} = \frac{|\sigma - \alpha| - R_p{}^n}{K}. \tag{2}$$

The viscoplasticity model takes time-dependent effects like strain rate dependency into account using the material properties K and n. This model contains kinematic and isotropic hardening. Kinematic hardening is implemented through the variable α named backstress (see Eq. 3) so that the Bauschinger effect occurring under cyclic loading can be described with the model. Isotropic hardening is considered through the yield strength R_p (see Eq. 4). Macaulay brackets ensure that viscoplasticity only occurs if there is an overstress, i. e. expression in the brackets is greater than zero.

The backstress is computed via the following evolution equation representing an extension of the kinematic hardening law proposed by Frederick and Armstrong [17]:

$$\dot{\alpha} = C \cdot \dot{\varepsilon}^{\text{vp}} - \gamma \cdot \dot{\overline{\varepsilon}}^{\text{vp}} \cdot \alpha - R \cdot \alpha + \frac{\partial C}{\partial T}\frac{\dot{T}}{C} \cdot \alpha + \frac{\partial C}{\partial r}\frac{\dot{r}}{C} \cdot \alpha \tag{3}$$

C and γ are the kinematic hardening variables, R is a material property for static recovery which gives a recovery of hardening with time. The last two terms of Eq. 3 are temperature and particle expressions that can be derived from thermodynamic principles presented in [18]. Isotropic softening is implemented through

$$R_p = R_e + Q_\infty \cdot \left(1 - e^{-b \cdot \overline{\varepsilon}\text{vp}}\right). \tag{4}$$

Initially, R_p corresponds to the initial yield strength R_{e} but changes depending on the accumulated plastic strain $\overline{\varepsilon}^{\text{vp}}$. For initiation of softening Q_∞ must be a negative value, $R_{\text{e}}+Q_\infty>0$. The proportionality constant b defines how fast softening occurs with increasing accumulated plastic strain.

The strengthening effect of secondary carbides on hot forging tool steels due to the Orowan mechanism was for example described in Eser et al. [7] and Caliskanoglu et al. [8]. This effect is considered in the viscoplasticity model using the radius r of a representative strengthening particle that is assumed to control the strength of the material. For hot forging tool steels, coarsening of secondary carbides was observed and explained as Ostwald ripening [9]:

$$\dot{r} = \frac{k_c}{r^2} \tag{5}$$

The coarsening leads to thermal softening of the material and is implemented into the viscoplasticity model through the temperature dependent coarsening constant k_c. The particle radius has impact on material properties through formulation of expressions depending on the particle radius [3]. Overall, the material properties are determined based on low-cycle-fatigue (LCF) tests at different temperatures as presented in [4].

2.2 Implementation

It is the aim of this work to implement the viscoplasticity model for the tool steel 1.2367 (X38CrMoV5-3) described in Sect. 2.1 in Simufact Forming. Therefore, some basic definitions that are important for the implementation of the model in Simufact Forming are presented. The specifications from Abaqus are taken from the Abaqus documentation [19] and the Simufact Forming specifications from the MSC Marc Volume D [20].

In previous works the viscoplasticity model was implemented and validated in Abaqus as an external subroutine called User Material (UMAT) using the algorithms described in [21]. Abaqus brings its own implicit FE solver while Simufact Forming uses a modified solver from MSC Marc. For Simufact Forming, a similar environment to UMAT called Hypela-2 is available [20]. This model was used for example in the research of Schmaltz to describe the material behaviour of sheet materials at large shape changes [22]. Hypela-2 is used to compute update of stresses and the internal variables (e. g. accumulated plastic strain or particle radius) at a time point t_{n+1} based on the values of stress and internal variables at time t_n. To this end, the increments of total strain and temperature are provided by the FE-software. Moreover, the consistent material tangent needs to be defined within Hypela-2. In the used configuration strains are provided as logarithmic strains which allow depiction of large deformations. Stresses have to be defined in a rotated coordinate system considering rigid body rotations.

The subroutine uses a predictor-corrector method [23]. In every time increment a check is performed if the stresses are within or outside of the elastic range. Plastic correction of the results is only carried out if the stresses are outside the elastic range. The standard MARC solver uses implicit integration scheme, so the material model is also implemented by means of implicit calculation using the Euler backward method. A schematic representation of the simulation workflow in Simufact Forming is presented in Fig. 1.

At the beginning, Simufact Forming provides several input data of the simulation to the subroutine during initialisation. The material properties are defined in Hypela-2 before the subroutine is called. Communication between the subroutine and Simufact Forming takes place, whereby Simufact Forming provides the forming and the individual parameters to the subroutine. Additional routines of MSC Marc are implemented in Hypela-2 for visualisation of the internal variables in the post processing.

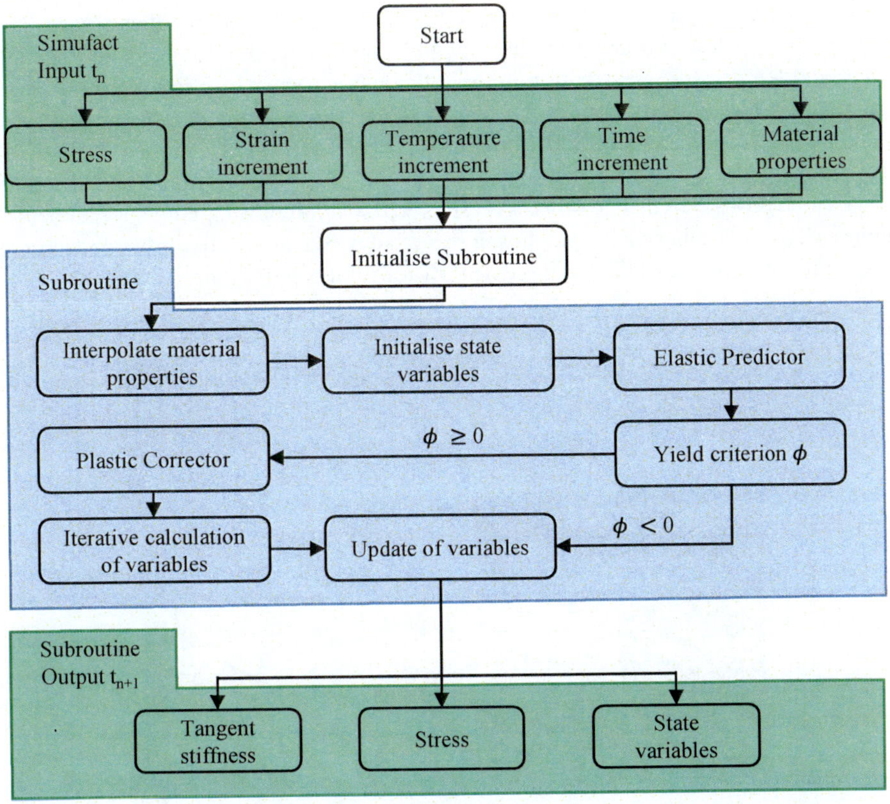

Fig. 1. Schematic representation of the simulation workflow in Simufact forming.

3 Simulation Model in Simufact Forming and Abaqus

For the comparison of the viscoplasticity model in Simufact Forming, simulations are also carried out in Abaqus. For this, a static compression and cyclic tests at various elevated temperatures (between 20 and 650 °C) are simulated, whereby the billet corresponds to the later deformable die in the planned forming process, already initially designed in the work of Behrens et al. [6]. The billet part is computed with the viscoplasticity model and the available material properties such as young's modulus are provided for the material 1.2367 within the material model. The focus is primarily on small deformations, since the later die will also only undergo small deformations.

In Abaqus boundary conditions as the displacement can be specified direct on the geometry. However, in Simufact Forming a press kinematic in combination with dies has to be modelled for a displacement-controlled compression test. To model an isothermal case the upper and lower die are defined as a rigid body without heat conduction. The billet is located between the analytical dies in Simufact Forming (see Fig. 2).

In the first part of the investigation, the focus is on a static compression test (see Fig. 2, left). During these simulations, only one cycle of compression is calculated in order to check the general runability of the material model in Simufact Forming. In this simulation no friction is defined between the billet (purple) and the dies (grey). These parameters are chosen in regard to the Abaqus reference model. Hence, a homogeneous stress and temperature distribution in the billet should occur. A rotationally symmetric cube with a radius of 10 mm and a height of 10 mm is defined as geometry for the simulations. A displacement of 0.1 mm with a forming speed of 1 mm/s is investigated. By choosing these forming parameters, both the linear-elastic range and the plastic range can be considered. The billet is meshed with overall nine equally spaced rectangle elements. This number has been chosen in such a way that exactly only one element is in the middle. The remeshing is switched off for these simulations.

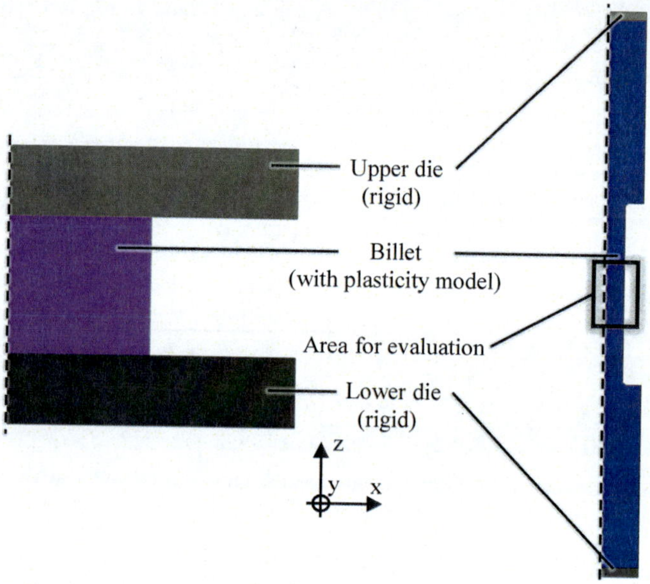

Fig. 2. Exemplary simulation models (left: static; right: cyclic) in simufact forming 16.0.

After the static compression tests, cyclic tests are simulated in the second part of this investigation (see Fig. 2, right). These calculations are carried out for comparison of the cyclic behaviour of the viscoplasticity model. Thus, the runnability of the subroutine must also be guaranteed for this type of simulation. A round tensile specimen is chosen as the geometry. The area used for the evaluation is in the middle of the smaller area of the specimen (area in the black box) and has the same radial dimension as the specimen in the static compression test. By choosing a cyclic loading and unloading, an adherent contact between the dies and the billet must be selected. A larger distance between the contact area and the evaluation area is chosen so that the effects on the stresses do not influence the results from the subroutine. As in the previous simulations, the heat transfer between all components is exhibited. The set mesh has the same edge lengths as in the previous model, so that again only one element is in the middle of the considered area. As before the remeshing in the billet is switched off. The choice of settings aims to achieve a homogeneous stress and strain condition in the evaluation area. A total of twelve cycles over a time of 500 s is examined. The displacement of the upper edge is controlled by the forming speed. The forming speed is linearly reduced / increased to a speed of 1.2 mm/s within 20 s. This specification results in a triangular profile of the forming speed with time.

4 Results and Discussion

After successful implementation of the subroutine in the simulation software Simufact Forming, the results are calculated for the static compression and cyclic tests described in Sect. 3 and the static simulations are compared with the results of Abaqus. Comparison of the results is important for the verification of the viscoplasticity model calculated using Simufact Forming (orange) and Abaqus (blue) as shown in Fig. 3. The evolution of the stress component in forming direction and the accumulated plastic strain with increasing forming time at different temperatures is shown. Up to a forming time of 0.03 s, a linear stress curve is shown, as expected for linear-elastic material behaviour. From 0.03 s to the end of the simulation, a non-linear stress curve is observed. The lines of the accumulated plastic strain follow a straight course up to a time of 0.03 s. After that, an increase of the values can be seen. In both results, the accumulated plastic strain as well as the stress, an increase respectively decrease of the value with increasing temperature can be seen. The influence of isotropic hardening is small in this simulation due to the low plastic strains.

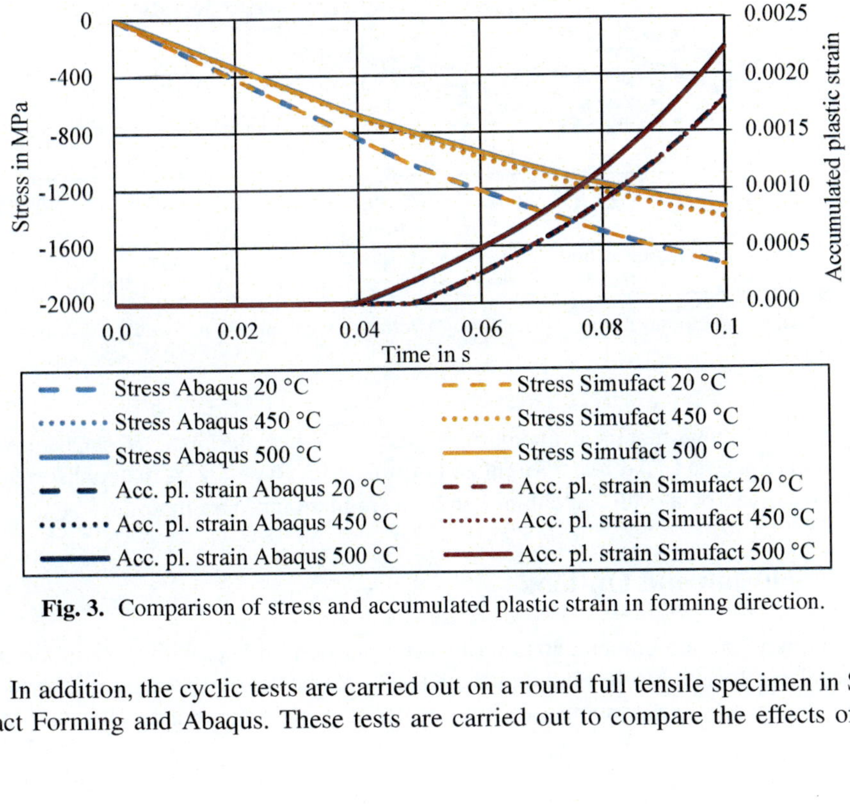

Fig. 3. Comparison of stress and accumulated plastic strain in forming direction.

In addition, the cyclic tests are carried out on a round full tensile specimen in Simufact Forming and Abaqus. These tests are carried out to compare the effects of the viscoplasticity model and the cyclic softening behaviour of the model. The cyclic tests are carried out displacement controlled by specification of the forming speed as mentioned in Sect. 3. Figure 4 shows the results of the cyclic tests as stress-strain hysteresis which is commonly used for the representation of cyclic tests. The figure includes the simulation results at 500 and 600 °C from Simufact Forming and Abaqus. Additionally, results from the first isothermal fatigue tests is presented for 500 and 600 °C marked by triangle symbols for quantitative comparison of the model with the experimental data. The stress values of the fatigue tests are on the same level as the model results for both temperatures. Figure 4 shows that the results of both simulation programmes agree. For 500 °C higher stresses are present and only small stress relaxation appears. The results reach a maximum stress value of about 1150 MPa and a maximum strain value of about 0.008. The initial stresses at 600 °C (900 MPa in the first cycle) are smaller than at 500 °C as indicated by the material properties. The cyclic tests show stress relaxation and thus the impact of the viscoplasticity model. Additionally, material softening occurs which can be evaluated by comparing the stress range in the first cycle to the last cycle at 600 °C. The softening in stress range is 30 MPa from the first to the last cycle of the test. Furthermore, an increase in the maximum strain from 0.008 to 0.01 can be observed.

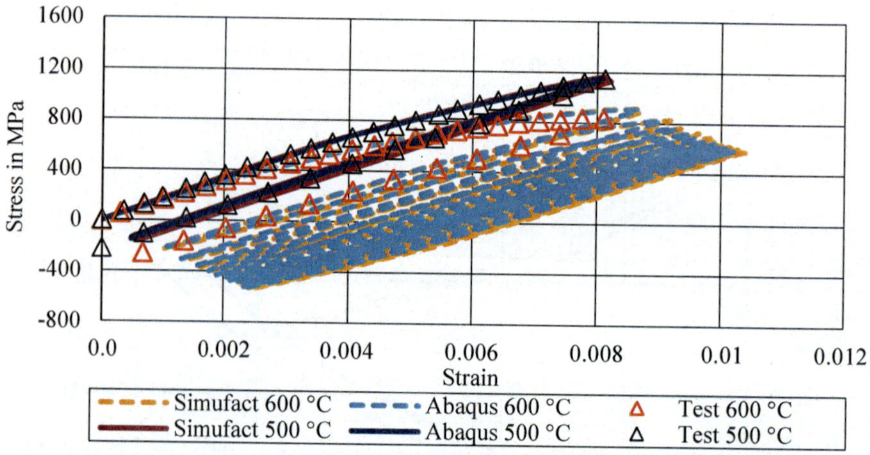

Fig. 4. Experimental and numerical results of cyclic tests calculated in Simufact Forming and Abaqus.

This work shows that the mathematical viscoplasticity material model can be used in the forging-specific FE software Simufact Forming with Hypela-2. A material softening could be proven by a cyclic calculation in both simulation programmes.

5 Conclusion and Outlook

Within this work, the implementation of a viscoplasticity model with material properties available for the hot forging tool steel 1.2367 in the forging specific FE-Software Simufact Forming was demonstrated by means of using a subroutine Hypela-2. The implemented model allows for considering material softening in the common industrial software for development of hot forging dies.

The simulation in this work shows that the integration of the subroutines was possible in Simufact Forming. Furthermore, it was shown that the calculations in Simufact Forming agree with the results from Abaqus for the static compression and cyclic tests. In addition, the cyclic tests demonstrated a softening of the material over the cycles at higher temperature.

In future works the viscoplasticity model has to be used in FE-simulations of forging dies under cyclic loading conditions and validated with experimental tests. Effects such as transient temperature fields and inhomogeneous loads will be considered. The developed and implemented model allows the realistic calculation of the development of plastic strains and stresses within the forging die under consideration of cyclic thermo-mechanical loads as well as their influence on material hardening and softening. This data should be coupled with a life time model. Finally, a realistic prediction of the die's life time can be made which allows cost reduction and resource saving.

Acknowledgement. The results presented were obtained in the research project "Development of a methodology for evaluating the fatigue life of highly loaded hot forming tools based on advanced material models" financed under project number 244928365 by the German Research Foundation (DFG). The authors would like to thank the German Research Foundation for the financial support.

References

1. Behrens, B.-A., Bouguecha, A., Hadifi, T., Klassen, A.: Numerical and experimental investigation on the service life estimation for hot-forging dies. In: Key Engineering Materials, vol. 504–506, pp. 163–168. (2012)
2. Ebara, R.: Fatigue crack initiation and propagation behavior of forging die steels. Int. J. Fatigue **32**(5), 830–40 (2010)
3. Oudin, A., Penazzi, L., Rézaï-Aria, F.: Life prediction of hot work tool steels subjected to thermomechanical fatigue. In: Matériaux and Techniques, vol. 88, pp. 67–72. (2000)
4. Bouguecha, A., Behrens, B.-A., Bonk, C., Rosenbusch, D., Kazhai, M.: Numerical die life estimation of a crack susceptible industrial hot forging process. In: AIP Conference Proceedings, vol. 1896, pp. 190012. (2017)
5. Haselbach, P., Natarajan, A., Jiwinangun, R.G., Branner, K.: Comparison of coupled and uncoupled load simulations on a jacket support structure. Energy Proc. **35**, 244–252 (2013)
6. Behrens, B.-A., Rosenbusch, D., Wester, H., Siring, J.: Numerical investigations on stresses and temperature development of tool dies during hot forging. In: 25th International Esaform Conference on material Forming, Portugal (2022)
7. Eser, A., Broeckmann, C., Simsir, C: Multiscale modeling of tempering of AISI H13 hot-work tool steel—Part 1: prediction of microstructure evolution and coupling with mechanical properties. Comput. Mater. Sci. **113**, 280–291 (2016)
8. Caliskanoglu, D., Siller, I., Ebner, R., Leitner, H., Jeglitsch, F., Waldhauser, W.: Thermal fatigue and softening behavior of hot work tool steels. In: Bergstrom, J. (eds.) Proceedings of the 6th International Tooling Conference, Karlstad University Karlstad, pp. 709–719. (2002)
9. Oriani, R.A.: Ostwald ripening of precipitates in solid matrices. Acta Metallurgica **12**(12), 1399–1409 (1964)

10. Sjostrom, J., Bergström, J.: Thermal fatigue in hot-working tools. Scandinavian J. Metallurgy **34**(4), 221–231 (2005)
11. Zhang, Z., Delagnes, D., Bernhart, G.: Microstructure evolution of hot-work tool steels during tempering and definition of a kinetic law based on hardness measurements. In: Material Science Engineering A, vol. 380(1–2), pp. 222–230. (2004)
12. Zhang, Z., Delagnes, D., Bernhart, G.: Anisothermal cyclic plasticity modelling of martensitic steels. Int. J. Fatigue **24**(6), 635–648 (2002)
13. Zhang, Z., Bernhart, G., Delagnes, D.: Cyclic behaviour constitutive modelling of a tempered martensitic steel including ageing effect. Int. J. Fatigue **30**(4), 706–716 (2008)
14. Jilg, A., Seifert, T.: Temperature dependent cyclic mechanical properties of a hot work steel after time and temperature dependent softening. In: Materials Science and Engineering: A, vol. 721, pp. 96–102. (2018)
15. Jilg, A., Seifert, T.: A temperature dependent cyclic plasticity model for hot work tool steel including particle coarsening. In: AIP Conference Proceedings 1960, pp. 170007. (2018)
16. Chaboche, J.L.: Constitutive equations for cyclic plasticity and cyclic viscoplasticity. Int. J. Plasticity **5**, 247–302 (1989)
17. Frederick, C.O., Armstrong, P.J.: A mathematical representation of the multiaxial bauschinger effect. In: Materials at High Temperatures, vol. 24, pp. 1–26. (2007)
18. Chaboche, J.: Cyclic viscoplastic constitutive equations, Part I: A thermodynamically consistent formulation. J. Appl. Mechan. (ASME) **60**(4), 813–821 (1993)
19. Systèmes, D.: ABAQUS 6.16: Analysis User's Guide (2016)
20. MSC Software GmbH: Volume D: User Subroutines and Special Routines. pp. 312–317. (2018)
21. Seifert, T., Schmidt, I.: Line-search methods in general return mapping algorithms with application to porous plasticity. Int. J. Numer. Methods Eng. **73**, 1468–1495 (2008)
22. Schmaltz, S.: Inverse Materialparameteridentifikation von Blechwerkstoffen für ein anisotropes elasto-plastisches Materialmodell bei finiten Deformationen. In: Schriftenreihe Technische Mechanik, Band 14, Thesis for Doctoral, Erlangen (2015)
23. Wriggers, P.: Nonlinear Finite Element Methods. Springer, Berlin, Heidelberg (2008)

Investigation on Adhesion-Promoting Process Parameters in Steel Bulk Metal Forming

U. Lorenz(✉), K. Brunotte, J. Peddinghaus, and B.-A. Behrens

Leibniz University Hannover, Institute of Forming Technology and Machines, An der Universität 2, 30823 Garbsen, Germany
ulorenz@ifum.uni-hannover.de

Abstract. Surface layers of forging dies are subject to thermal, mechanical and tribological influences during forging. These loads occur combined and result in a variety of tool damages, which shorten tool life. The predominant cause for tool failure is wear. While abrasive wear and crack formation directly cause tool failure, adhesive wear can be equally disruptive as it results in a geometrical deviation of the tool and the formed work piece. However, adhesive wear can also be beneficial in acting as a regenerating, protective layer to the surface of forging dies. This paper deals with the influence of process parameters and billet material on the formation of adhesive wear on forging dies. As adhesive wear is facilitated at elevated temperatures, high thermally loaded dies with a mandrel geometry are investigated in forging tests. During forging, thermal, mechanical and tribological loads on the tools are varied by changing cooling parameters, steel billet material and lubrication strategies. The study presents adhesion-promoting process parameters and tool areas of increased adhesive wear. The results show, that the formation of adhesive wear occurs predominantly at high tool temperatures and in areas with increased material flow, while lubrication and the billet material show little to no impact.

Keywords: Die forging · Adhesive wear · Tool life

1 Introduction

During hot forming of steel, forging dies are exposed to high process related cyclic thermal, mechanical, tribological and chemical loads. Thermal loads are caused by frequent alternations between rapid heating of the tool surface area during forming and the application of cooling lubricants. Mechanical loads are induced by the forces on the tool surface during forging. Tribological stresses are a result of the relative movement between the tool and the deforming work piece [1]. All of these loads individually and combined can cause permanent damage to the tool surface area and ultimately lead to tool failure. However, the main reason for tool failure is wear [2]. Wear is defined as the progressive loss of material from the surface of a solid body. It occurs as detached particles and geometrical deviations in tribologically stressed surface areas [3]. The four wear mechanisms are adhesion, abrasion, tribo-chemical reaction and surface deformation.

M. Liewald et al. (Eds.): WGP 2022, LNPE, pp. 91–99, 2023.
https://doi.org/10.1007/978-3-031-18318-8_10

Adhesion occurs at direct contact between two friction bodies due to atomic interactions [4]. During forging, temperatures in the contact zone between tool and billet are in the range of 600–900 °C, which reduces the wear resistance of the tool surface layer by thermal softening and strongly promotes adhesive processes. During plastic deformation of the billet, the surface layers are in contact with each other and cohesion causes a chemical bond between the two solids. Due to relative movement, the chemical bond leads to the shearing of micro-welded areas, with the softer material adhering to the harder. According to [1], cold welding can be reduced through the selection of suitable tool materials and cooling lubricants. Furthermore, the tool materials for forging dies should feature a high thermal resistance, carbide content and optimal working hardness.

To increase tool life and to reduce unwanted effects, e.g. cyclic thermal shock, forging dies are kept at optimal thermal levels by balancing the effects of indirect heating through contact with the warm billets and cooling through the application of the cooling lubricant. The amount of cooling lubricant has major influence on the die temperature and the tribological conditions on the surface area. In addition, the billet material can further influence the tribological conditions as well as mechanical loads during the forming process. This paper shows the investigation of the influence of these parameters on the promotion of adhesive wear on the surface of hot working tools. Tools are selected and equipped with thermocouples to measure their base temperature (i.e., minimum temperature per cycle) during forging. The amount of coolant and lubricant, as well as the billet material are varied, while billet temperature, tool material and forging cycle times are constant. After forging, the tools are examined optically and metallographically. Thus, the influence of the thermal, tribological and mechanical stresses on the structural changes in the tool surface layer during die forging is investigated.

2 Materials and Methods

2.1 Tools

Forging tests are carried out on a rotationally symmetrical tool (see Fig. 1) with a mandrel surface area being subject to high thermal loads during. These loads occur especially at the convex radius due to the large contact area of the forged part. The tools made of AISI H11 hot working tool steel are heat treated to 48+2 HRC. The tools can display essential structural changes and associated wear, which mainly occurs on the convex tool radius [5]. High surface pressures and thermal loads occur in this area due to the relatively long time of contact and deep penetration of the tool into the work piece. Selected tools with inserted thermocouples are used to characterize the cyclic thermal loads. The temperatures are recorded continuously over the entire forging process during the heating and cooling phases. Therefore, encapsulated Type K thermocouples with a diameter of ø1.5 mm are inserted in machined tool cavity holes with a diameter of ø1.6 mm at a distance of 3 mm to the tool surface, with the thermocouple wiring led out from the tools laterally. Thermal paste is applied for increased heat transfer and decrease reaction time of the thermocouples.

The temperature development of the tools in the forging process is analysed and the tool cooling behaviour characterized. Cooling parameters for different tool base temperatures are determined.

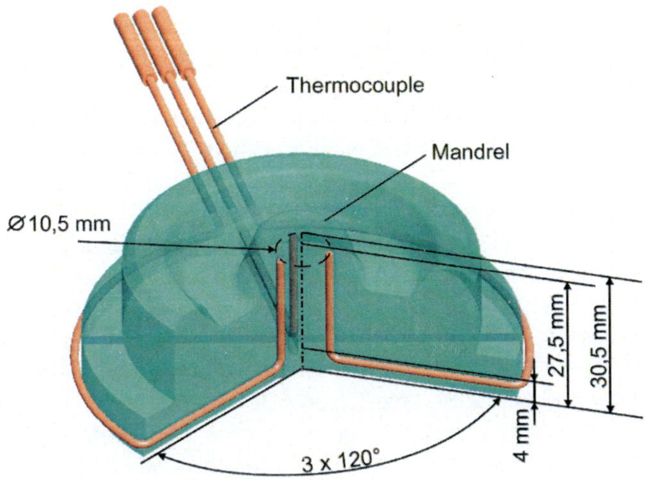

Fig. 1. Thermally loaded tool with channels for three mantled thermocouples.

2.2 Load Variation

Hot working tools are usually cooled and lubricated in a combined process. A suspension of water and lubricant (e.g. Graphite) is applied to the tool surface. The water evaporates and thereby dissipates heat from the tool surface area while the lubricant remains. To carry out forging tests with different cooling gradients while maintaining constant amounts of lubricant, cooling and lubrication are separated. The cooling is carried out by an air-water mixture, which is applied to the tool surfaces using a spraying system (co. Gerlieva). Spraying parameters such as pressure and duration can be specifically controlled during the spray cooling. The spray parameters are varied in such a way that four different tool temperature profiles (100, 200, 300 and 400 °C) are realized. Therefore, tools equipped with thermocouples are used to determine optimal spraying parameters for each base temperature. Coolant spray duration is set to 1000 ms for 100 °C tool base temperature, 430 ms for 200 °C and 250 ms for 300 °C. No water is applied to reach temperature values close to 400 °C. All tools are preheated to a base temperature of 200 °C using heating cartridges.

The same spray nozzle (co. Gerlieva, type 300–55365) and constant spray pressures are used throughout all tests, while the spray duration is varied. Forging cycle times are constant and adjusted to 8.4 s, which is the minimum possible time at maximum spray cooling. Before and after water spraying, air is sprayed on the dies to remove scale and excess water.

In addition, electrostatic powder application technology is used for lubricant application. Powdered boron nitride (co. Henze BNP, type HeBoFill LL-SP 120) is applied as a lubricant. Lubrication is carried out with a powder coating system (co. ITW/GEMA Surface Technology). Studies have shown that this type of lubricant application can significantly reduce tool wear [6]. The feed movement is functionally separated and carried out in two stages, with stage I for cooling and stage II for lubricating the tools. The lubricant application is regulated by the voltage potential, the spray duration and the spray pressure in order to generate a homogeneous lubricating film on the tool surface.

To determine the influence of tribological stress, friction conditions during forging are changed. For this purpose, another test series is carried out without lubricatant, which should lead to a significant increase in friction while increasing the tool heat input. For tool cooling, the strategy for 200 °C base tool temperature is selected.

To determine the influence of superimposed mechanical stress on the structural changes, different billet materials are investigated. The material properties significantly affect the mechanical stress on the tool, resulting in a decisive influence on the wear behaviour. For this study, the steel AISI 4140 is selected which has lower flow stresses and thus lower resistance to deformation during hot forming than steel AISI 1045, which lowers the mechanical stress on the tools. The flow curve of AISI 1045 was measured in a forming simulator (co. DSi, type Gleeble 3800 GTC) through uni-axial cylinder compression tests, which are carried out at constant deformation rates of 10 s^{-1} at process-relevant temperatures of 1200 °C. The flow curve of AISI 4140 was taken from [7] at equal conditions. An overview of the flow curves is shown in Fig. 2. As displayed, the flow resistance of AISI 1045 steel is higher and a more pronounced softening of AISI 4140 at higher strain rates is noticeable.

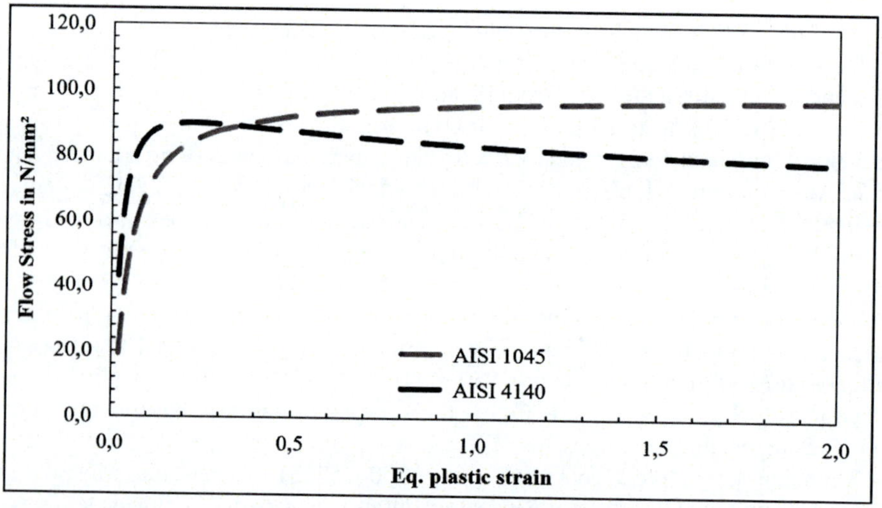

Fig. 2. Flow curves of steel AISI 4140 [7] and AISI 1045 at 1200 °C.

2.3 Forging

To attain repeatable results, serial forging tests are carried out on a fully automated eccentric press (co. Eumuco, type Maxima SP 30d), featuring a maximum nominal force of 3150 kN. Sawn blanks made of steel AISI 1045 and AISI 4140 are used as billets. Before forging, the billets are heated to a temperature of 1200 °C in a continuous induction furnace. The billet temperature is constant throughout all test series and the heated billets are transported automatically into the press by a robot. After each forging cycle, the formed billets are ejected and the scale remaining on the tools is removed with compressed air during pre-blowing. The tools are cooled and then lubricated before the initiation of the next forging cycle. To ensure the test's comparability, the time for each forging cycle is adjusted to 8.4 s and the dies are each loaded with 50 forging cycles.

2.4 Wear Evaluation

The tool surface area is examined optically and high resolution photos are taken from different angles. 3D surface measurements are carried out with a Wide-Area 3D Measurement System (co. KEYENCE, type VR-3200) to determine geometric deviations on the tool contours. To highlight changes, the surface of the forging dies are recorded three-dimensionally before and after forging. All tool mandrels are then separated mechanically by a wet cut-off grinder to characterize the cross-section of the surface layer metallographically. Samples are embedded, polished, etched with 10% alcoholic nitric acid (10% $HNO3$) and the sample microstructures analysed with a light microscope (co. Reichert-Jung, type Polyvar Met).

3 Results

3.1 Thermal Parameters

Figure 3 shows the temperature measurements of the thermocouples for the four base temperatures, each starting at 200 °C. To reach temperatures close to 100 and 300 °C it takes about 20 forging cylces. A temperature of 400 °C cannot be achieved even without coolant application, as much of the thermal energy dissipates during each forging cycle. Therefore, the highest measureable base temperature at a depth of 3 mm beneath the tool surface area with a forging cycle time of 8.4 s was 350 °C, while peak temperatures of about 380 °C are measured. It can be assumed that the actual surface temperature was much higher, due to the thermocouples surface distance. Despite the complete lack of coolant, air was sprayed between all forging cycles for a duration of 0.1 s, which also cools down the dies.

3.2 Optical Evaluation

After forging, the tools are cleaned from lubricant residues and optically evaluated. Due to high thermal loads, the mandrel of each tool was inspected (see Fig. 4). While the tool with a base temperature of 100 °C showed only little adhesive wear, it became increasingly noticeable at higher temperatures. Thus, the mandrels used at 300 and 400 °C show the most adhesive wear. The tool with no lubrication, as well as the one with AISI 4140 as billet material are both forged at 200 °C base temperature. When evaluated optically, both showed similar wear behaviour to the tool that was used to forge AISI 1045 billets at 200 °C base tool temperature. Turning grooves from manufacturing are still visible with all tools after 50 forging cycles.

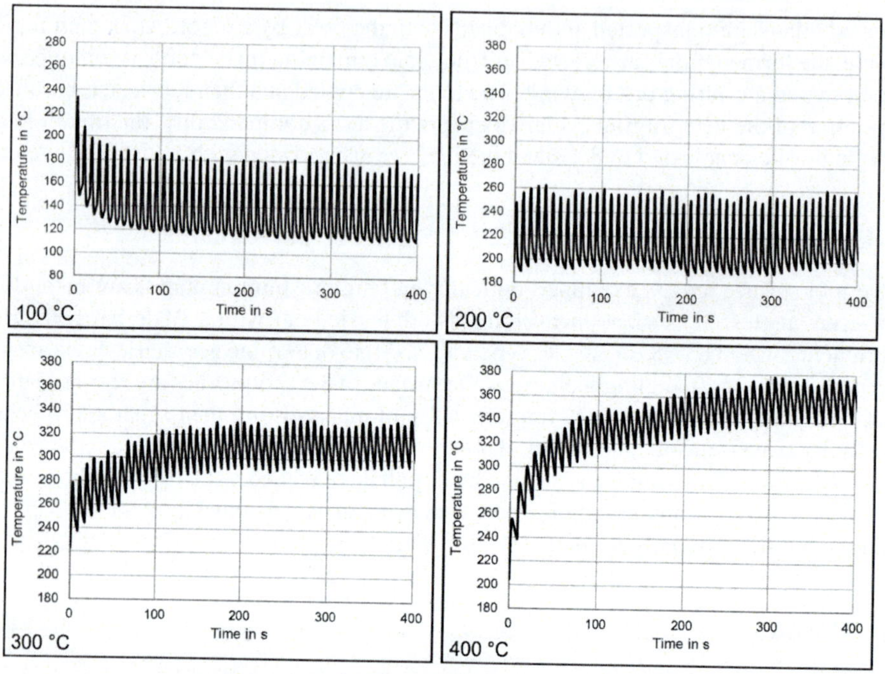

Fig. 3. Base temperature for each parameters set.

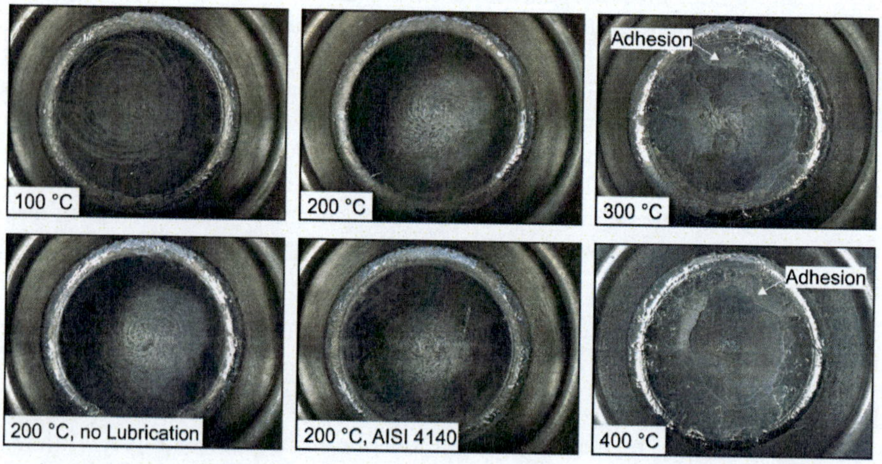

Fig. 4. Wear at the mandrels after forging depending on the tool base temperature.

3.3 3D Surface Measurement

Figure 5 shows the comparison of the tool scans taken before and after forging. Elevated areas are shown as red, unchanged areas green and lowered areas as blue. While even 50 forging cycles can lead to abrasive wear, the deep blue edge areas can be explained by inaccurate overlap between both scans. However, the scans show a correlation between adhesive wear and increased temperatures at the thermomechanically highly stressed mandrel. Due to the highest thermal load at the mandrel radius, most adhesion is localized in nearby surface areas. All tools show some amount of adhesion in the cross areas of the mandrel. In this area there is a slow material flow that supports the formation of adhesive wear.

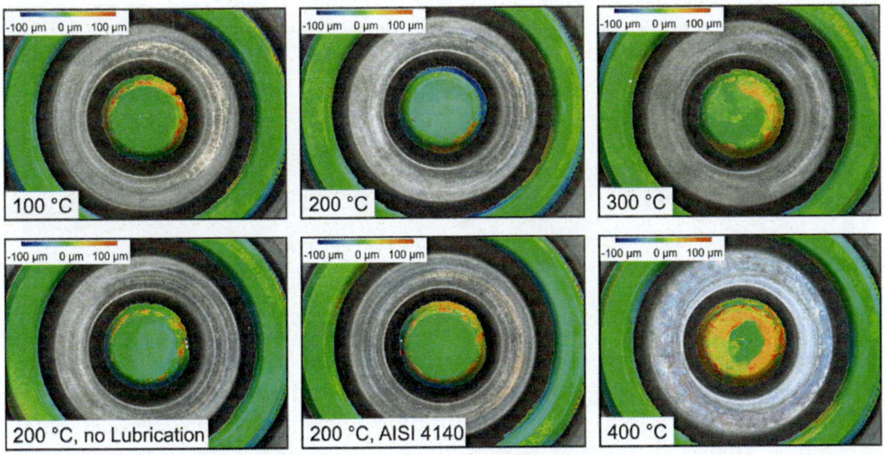

Fig. 5. 3D-Scans of the surface layer changes of the tool thrones after forging.

Meanwhile, the mandrel face is subject to almost no material flow. Here, adhesion occurs at high tool temperatures. A total lack of lubricant seems to have only little impact on adhesion while the change of billet material from the firm AISI 1045 to the softer AISI 4140 does show a slight increase.

3.4 Metallographic Examination

The tools are separated and cross sections of the mandrel areas taken to examine the thickness of the adhesion. The selected cross sections show an increased amount of attached material at higher tool temperatures. The maximum values for visible adhesive thickness are at about 50 μm for the 300 °C tool and 80 μm for the 400 °C tool (see Fig. 6). The adhesion appears to be brittle in all cases and not firmly bonded with the tool material.

4 Summary

To understand the formation of adhesive wear on the surface of forging dies, an examination of the adhesion-promoting parameters is essential. For this purpose, a variety of cooling strategies were created. Forging tools were submitted to different thermal, mechanical and tribological stresses for 50 forging cycles each. It was shown, that high

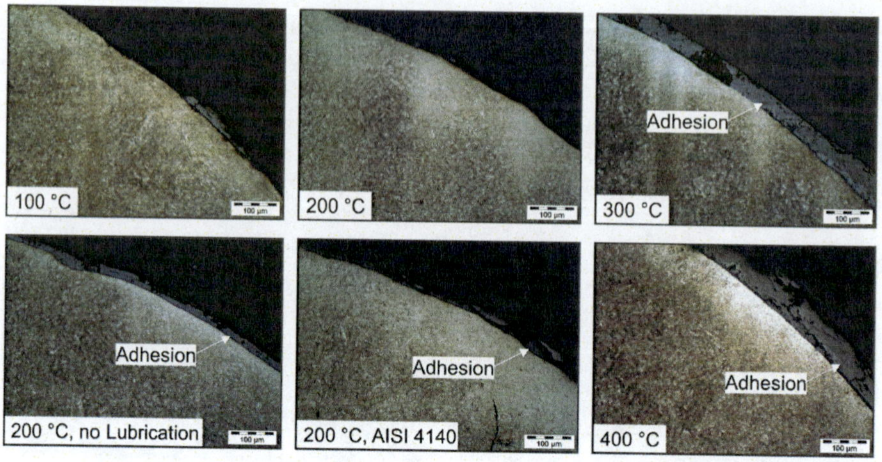

Fig. 6. Cross sections of the tool mandrel surface layers after forging.

tool temperatures promote the formation of adhesive wear even in areas with little material flow, as long as the mechanical pressure is sufficient. This was shown on the mandrel face of a rotationally symmetrical tool. Lubrication strategies have shown a small impact on the formation on adhesive wear. Meanwhile, the billet material can have a larger influence on the formation of adhesive wear. This can be due to both the difference in flow stress, as well as the material specific adhesive properties.

Acknowledgement. The presented investigations are carried out within the project ID 349885770 "Influence of cooling of forging dies on the process-related microstructural changes in the surface zone and their effect on wear behaviour" of the German Research Foundation (DFG). We are thankful for the assistance provided.

References

1. Emamverdiana, A.A., Sun, Y., Wanga, Y.: Current failure mechanisms and treatment methods of hot forging tools (dies)—a review. Eng. Failure Anal. **129**, (2021). https://doi.org/10.1016/j.engfailanal.2021.105678
2. Lange, K., Cser, L., Geiger, M.: Tool life and tool quality in bulk metal forming. CIRP Ann. Manuf. Technol. **41**, 667–675 (1992)
3. Fleischer, G., Gröger, H., Thum, H.: Verschleiß und Zuverlässigkeit. Vieweg, Braunschweig (1992)
4. Chung, S., Swift, H.: Cup-drawing from a flat blank: part I experimental investigations, part II analytical investigations. In: Proceedings of the Institution of Mechanical Engineers, pp. 165. (1951)
5. Behrens, B.-A., Puppa, J., Lorenz, U.: Development of an intelligent hot-working steel to increase the tool wear resistance. In: The 11th Tooling 2019 Conference and Exhibition (2019)
6. Puppa, J., Behrens, B.-A.: Optimization of cooling and lubrication for nitrided and ceramic-coated hot forging dies. Appl. Mechan. Mater. **794** (2015)

7. Behrens, B.-A., Volk, W., Büdenbender, C.: Numerical investigation of thermal and mechanical deviations in a hot forging process of 16mncr5 and 42crmo4 steel. In: 28th International Conference on Metallurgy and Materials (2019)

Finite Element Analysis of a Combined Collar Drawing and Thread Forming Process

E. Stockburger(✉), H. Wester, D. Rosenbusch, and B.-A. Behrens

Institut für Umformtechnik und Umformmaschinen (IFUM), Leibniz Universität Hannover, An der Universität 2, 30823 Garbsen, Germany

stockburger@ifum.uni-hannover.de

Abstract. Collars with internal threads are used in a wide range of products. For the production of threads in sheet metal, collar drawing with subsequent thread cutting or forming is used in progressive tools. For a reduction of the manufacturing costs, collar drawing and thread forming can be combined into a single process step. Since the punch for collar drawing must be retracted after thread forming, an undersized collar must be drawn. The thread forming must be adapted to the collar as well as the process kinematics. Thus, FE simulation is used to analyse the process interactions. Parameters such as feed rate, rotational speed, pre-hole diameter, die diameter, sheet thickness are varied, and their influence on the process are investigated. Based on the findings the best combination of the process parameters is determined. The results will be used to manufacture a tooling system for collar drawing with integrated thread forming in the future.

Keywords: Connection technology · 3D FE modelling · Process analysis

1 Introduction

For a wide range of products, a sheet metal collar with internal thread is required to mount the product to a counterpart firmly, yet removable [1]. The areas of application thus range from automotive and electrical industries to household goods production. In the conventional process chain for producing a sheet metal collar with an internal thread, the first step is to punch the blank and subsequently draw the collar. There are several options for creating the thread with chip formation, such as thread cutting or milling, and chipless, such as thread forming or grooving. The production of internal threads by means of chip generating processes is accompanied by a reduction in sheet thickness, thus weakening the collar. Furthermore, the chips can damage both the thread and the tool [2]. Therefore, in this respect chipless processes offer advantages. Thread forming does not reduce the sheet thickness of the collar but instead hardens the material, which allows joint connections to be more stable compared to cutted threads [3]. This enables collars to be produced deeper for longer threads than with cutting tools. The collars are primarily produced with progressive tools, which consist of several individual stages, and thus form the basis for a commercial production [4]. Each stage usually performs a specific operation on the component. The combination of several operations into one

M. Liewald et al. (Eds.): WGP 2022, LNPE, pp. 100–110, 2023.
https://doi.org/10.1007/978-3-031-18318-8_11

process step thus results in less manufacturing time and increases the potential for cost reduction. Furthermore, a tool combining two work stages results in smaller dimensions, which saves space. Therefore, a combination of the two forming operations into a single step is a reasonable option. Within the scope of this research, a new tool concept is developed, which combines collar drawing with thread forming in order to consequently reduce the number of manufacturing steps for a sheet metal collar with internal thread. In the considered process, the punch for collar drawing must be retracted after thread forming. To realize the retraction of the collar drawing punch, the punch itself and the drawn collar must be undersized compared to usual applications [5]. The finite element (FE) simulation is applied to analyse the combined process, as it is widely used to design processes [6]. Within a sensitivity study various parameters such as the feed rate, rotational speed, pre-hole diameter, die diameter as well as sheet thickness are investigated regarding their interaction and influence on the process. The objective is to investigate which parameter combinations lead to a good filling level of the thread while keeping the required forces and torques at a low level.

2 Materials and Methods

2.1 Numerical Process Modelling

For a numerical analysis, a 3D FE model of collar drawing and thread forming was created in Simufact Forming 16. The numerical model was divided into two parts, whereby the results of the collar drawing served as input for the thread forming. The simulation model consisted of a punch, a down holder, a die and a specimen. The model with the applied boundary conditions is shown in Fig. 1(A). To ensure fine meshing of the thread and thus an accurate modelling of the geometry with an economical calculation time, the process had to be simplified. The blank geometry was varied as full geometry during preliminary investigations and finally modelled as 1/64 of a circular blank as shown in Fig. 1(B). The global thread geometry could not be formed with a helix as in a full model, but the deviations in the local thread geometry were small and therefore the 1/64 model was used. For reference, a blank with a sheet thickness t of 1.5 mm as well as a pre-hole diameter d_1 of 4.4 mm and a die with a diameter d_2 of 10.6 mm were considered for forming the thread size M10. Solid elements of the hexahedron type were used to mesh the blank. In preliminary investigations, the element edge length was varied in order to simulate the shaping of the thread flanks properly. The final element edge length was 0.2 mm with a local refinement of 0.05 mm in the area of the thread flanks. In order to model the high deformations, a remeshing algorithm was implemented, which was activated every time the accumulated plastic strain increased by 0.2. It was assumed that there was no slipping of the blank in the experiment. Therefore the contact of the blank and the die as well as the blank and the blank holder were adhesive. Within the reference case, the blank was formed by the punch with a feed rate f of 75 mm/s and a rotational speed n of 3000 rpm for the thread forming. Collar forming itself was performed without any rotation. The feed rate was selected based on an existing forming press and the rotational speed was correspondingly calculated. In order to model the complex tribological conditions, the combined friction model was used as a simplification with a coefficient of friction of 0.15 and a friction factor of 0.3 [7]. The heat released during the forming process was taken into account by means of a Taylor-Quinney coefficient of 0.9. For the blank material, the high-strength low-alloy steel HX340LAD was considered. To map the flow behaviour, the von Mises yield criterion was used and temperature as well as strain rate dependent flow curves were calculated with the commercial software JMatPro. Based on the data the Johnson-Cook (J-C) hardening model [8] was parametrised, which is widely used for different forming processes [9]. The coefficients are shown in Fig. 1(C). According to the formula, the flow stress k_f is a function of the plastic strain ε_{pl} and the five material parameters A, B, C, m as well as p. $\dot{\varepsilon}_{norm}$ and T_{norm}^{m} are the normalised plastic strain rate and the normalised temperature. The material parameters were determined by fitting the J-C hardening model to the JMatPro data using the least squares error method. The tools were modelled as rigid bodies and the calculation was carried out thermo-mechanically coupled with an implicit solver. To analyse the influence of process and geometrical parameters on the combined collar drawing and thread forming process, the feed rate f, rotational speed n, pre-hole diameter d_1, die diameter d_2 and sheet thickness t were varied according to Table 1.

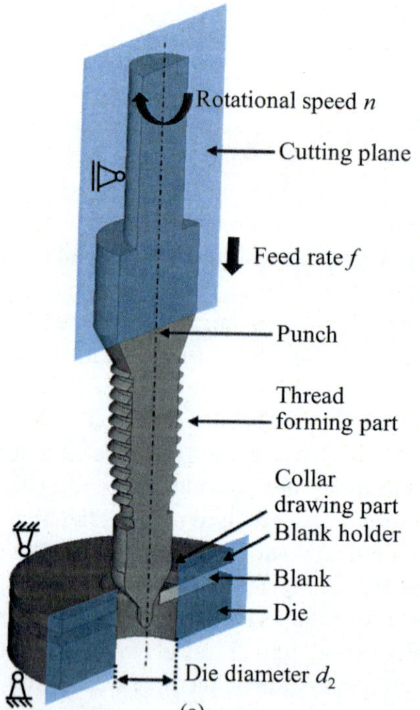

(a)

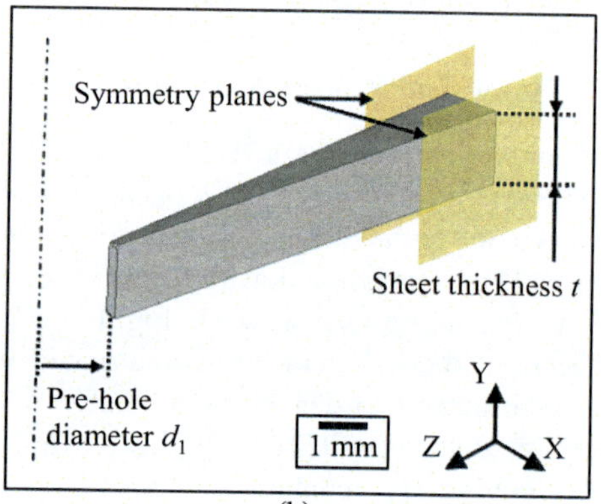

(b)

Parameter	Unit	Value
A	MPa	258
B	MPa	670
C	-	0.015
m	-	0.845
p	-	0.210

$$k_f = \left(A + B\,\varepsilon_{pl}^{p}\right) \times (1 + C \ln \dot{\varepsilon}_{norm}) \times (1 - T_{norm}^{m})$$

(c)

Fig. 1. Simulation model of collar drawing with subsequent thread forming (A), blank geometry (B) and coefficients of the J-C hardening model for HX340LAD (C)

Table 1 Varied process and geometrical parameters

Parameter study	Feed rate in mm/s	Rotational speed in rpm	Pre-hole diameter in mm	Die diameter in mm	Sheet thickness in mm
f	70; 75; 80	3000	4.4	10.6	1.5
n	75	2750; 3000; 3250			
d_1		3000	2.2; 4.4; 6.6		
d_2			4.4	10.6; 10.85; 11.1	
t				10.85	1.5; 1; 0.8

3 Results and Discussion

3.1 Influence of the Parameters on the Filling Level of the Thread

3.1.1 Feed Rate

The specified feed rate f for the corresponding rotational speed n of 3000 rpm is 75 mm/s. In order to investigate the influence of slight variations of the feed rate by the machine control on the collar drawing and thread forming process, the feed rate was varied between 70, 75 and 80 mm/s while maintaining a constant rotational speed, pre-hole diameter, die diameter and sheet thickness. In Fig. 2, the plastic strain distribution of the parameter variation is shown.

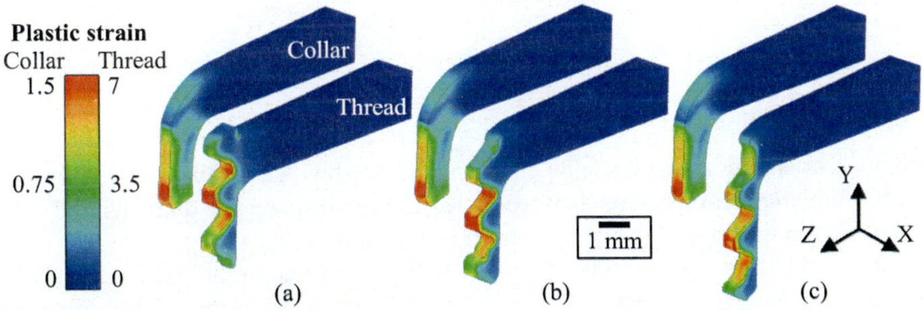

Fig. 2. Plastic strain distribution after collar drawing as well as thread forming for a feed rate f of 70 mm/s (A), 75 mm/s (B) and 80 mm/s (C)

Regarding the collar drawing, the feed rate has no significant influence on the material flow in the investigated process window. Regarding the thread forming, a feed rate of 70 mm/s is too low in relation to the used rotational speed, so that the material is moved against the feed direction. Evidently, 75 mm/s results in a correct thread. 80 mm/s is an excessive feed rate, resulting in a greater amount of material being moved in the feed direction. To conclude, although all investigated feed rates result in a similar filling level of the thread, merely the feed rate of 75 mm/s results in a proper thread.

3.1.2 Rotational Speed

Respectively, the rotational speed n was varied between 2750, 3000 and 3250 rpm, not changing the feed rate, pre-hole diameter, die diameter and sheet thickness. The plastic strain distribution for collar drawing and thread forming is shown in Fig. 3. Since the rotation of the tool only begins during thread forming, no difference between the collars during the variation of rotational speed is noticed. Similarly to the variation of the feed rate, a rotational speed of 2750 rpm is too slow compared to the used feed rate, which moves more material in the feed direction. 3000 rpm results in a proper thread and 3250 rpm is too fast, moving more material against the feed direction. The variation of feed rate and rotational speed points out, that rotational speed needs to be selected and controlled according to the feed rate to ensure a stable process.

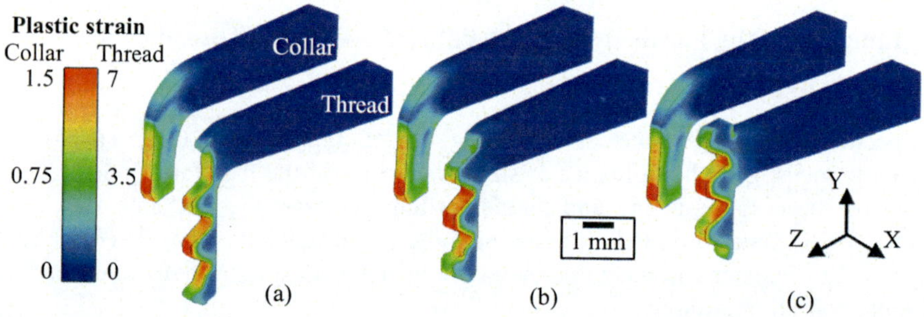

Fig. 3 Plastic strain distribution collar drawing as well as thread forming for a rotational speed n of 2750 rpm (A), 3000 rpm (B) and 3250 rpm (C)

3.1.3 Pre-Hole Diameter

A geometrical variation was made by changing the pre-hole diameter d_1 to 2.2, 4.4 and 6.6 mm, while maintaining a constant feed rate, rotational speed, die diameter and sheet thickness. A comparison of the plastic strain is depicted in Fig. 4.

As expected, the collar becomes shorter with an increasing pre-hole diameter. Correspondingly, fewer teeth are formed during the thread forming when the pre-hole diameter is increased. This is due to the simple fact that as the pre-hole diameter is reduced, more material is available for drawing the collar and hence to form the thread.

3.1.4 Die Diameter

Figure 5 shows the distribution of the plastic strain for the variation of the die diameter d_2. The feed rate, rotational speed, pre-hole diameter and sheet thickness were kept constant. A pre-hole diameter of 4.4 mm was used for the variation of the die diameter d_2, as it was expected that due to the high plastic strains the collar could not be drawn error free

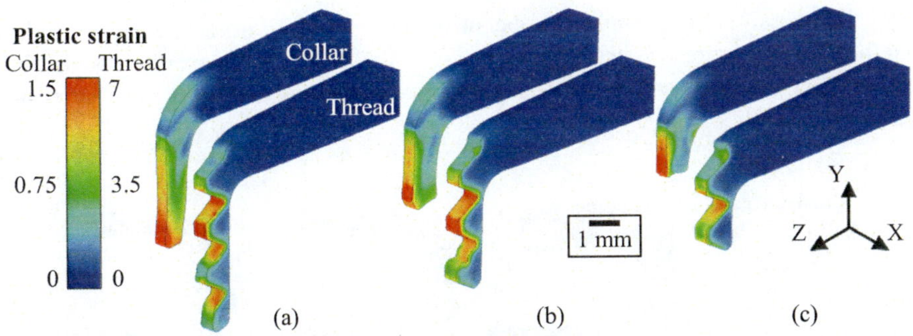

Fig. 4 Plastic strain distribution after collar drawing as well as thread forming for a pre-hole diameter d_1 of 2.2 mm (A), 4.4 mm (B) and 6.6 mm (C)

for the investigated HX340LAD using larger pre-hole diameters. For collar drawing, no significant influence of the die diameter on the collar formation can be detected within the investigated process window. The influence of the die diameter on thread forming can be clearly seen. At a die diameter of 11.1 mm, the thread is not completely filled. When the diameter is reduced to 10.85 mm, an almost complete filling of the thread is achieved. If the diameter is reduced further to 10.6 mm, the thread is fully filled, which results in a high plastic strain along the thread. Therefore, a die diameter of 10.85 mm was used in the further variation of the sheet thickness t.

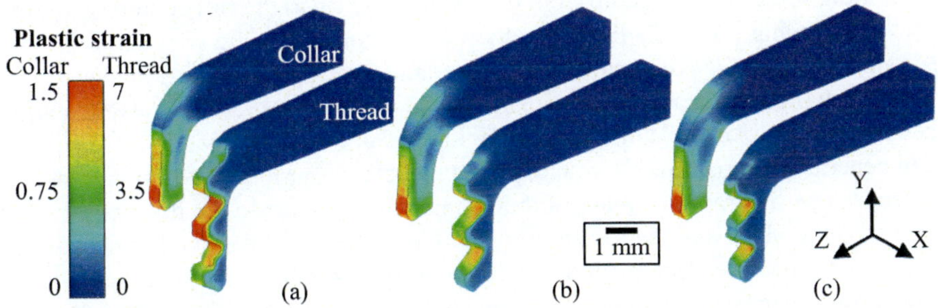

Fig. 5 Plastic strain distribution after collar drawing as well as thread forming for a die diameter d_2 of 10.6 mm (A), 10.85 mm (B) and 11.1 mm (C)

3.1.5 Sheet Thickness

When varying the sheet thickness t between 1.5, 1 and 0.8 mm, the pre-hole diameter, feed rate, rotational speed and die diameter were fixed. The resulting geometries and the distributions of the plastic strain are shown in Fig. 6. During collar drawing, a conical shape of the collar is formed for smaller sheet thicknesses. While thread forming, however, the collar is formed almost cylindrically. As to be expected, the smaller the sheet thickness, the lower the filling level of the thread, since less material is available for forming. Except for the sheet thickness of 1.5 mm, the thread resulting from the sheet thickness of 1 mm could also be sufficient enough to hold a screw, considering that the simulated thread is actually not rotationally symmetric. Nevertheless, the thread resulting by the proposed geometrical parameters for the sheet thickness 0.8 mm is clearly too small.

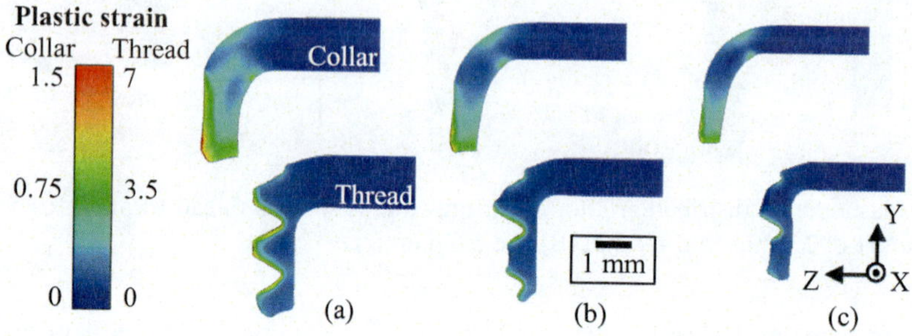

Fig. 6 Plastic strain distribution after collar drawing as well as thread forming for a sheet thickness t of 1.5 mm (A), 1 mm (B) and 0.8 mm (C)

3.2 Influence of the Parameters on the Forming Force and Torque

3.2.1 Collar Forming

The forming force was scaled up from the 1/64 geometry to the full geometry. Figure 7 shows the forming force as a function of stroke for collar drawing with variation of the feed rate f (A), pre-hole diameter d_1 (B), die diameter d_2 (C) and sheet thickness t (D). Since the rotation of the tool begins with the thread forming, the variation of the rotational speed has no effect on the collar formation due to a constant feed rate.

In general, all variations show the same tendencies regarding the forming force over the stroke. The first rapid increase of the forming force is due to the first contact of the punch with the blank and the resulting forming of the edge of the blank. Subsequently, the forming force rises further until the blank is fully bended, at which the maximum forming force is present. Afterwards, the bended blank is drawn creating the collar while the forming force decreases. When varying the feed rate, no significant influence on the forming force is noticeable. An increase of the maximum forming force is noticeable with reduction of the pre-hole diameter and with reduction of the die diameter. With a smaller pre-hole diameter, more material is available for forming and the duration of the process is increased, therefore the maximum forming force increases significantly. For a smaller die diameter, the maximum forming force increases slightly due to a higher thinning of the sheet while collar drawing. When reducing the sheet thickness, a smaller maximum forming force is needed for collar drawing due to less material being formed.

3.2.2 Thread Forming

Similarly to the forming force, the torque was also scaled up from the 1/64 geometry to the full geometry for thread forming. In Fig. 8 the torque is depicted as a function of time

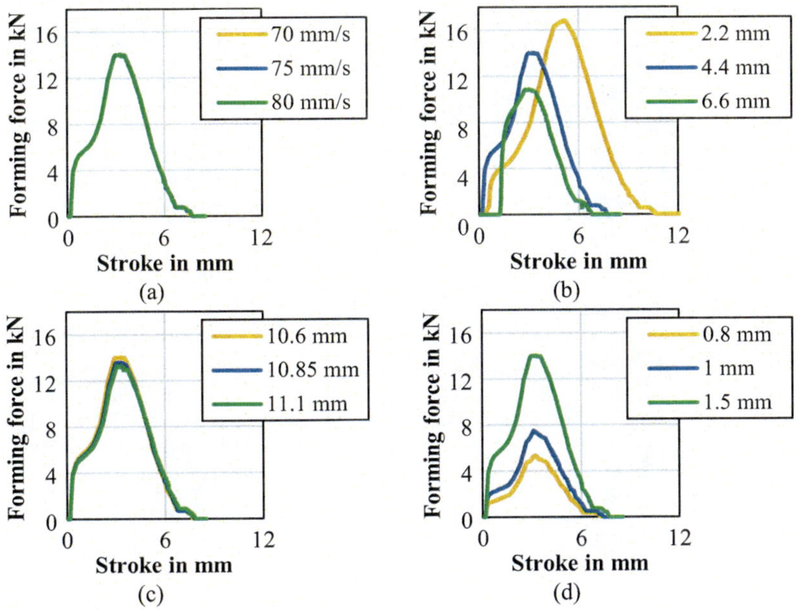

Fig. 7 Effect of the feed rate f (A), pre-hole diameter d_1 (B), die diameter d_2 (C) and sheet thickness t (D) on the forming force while collar drawing

while thread forming with a variation of the rotational speed n (A), pre-hole diameter d_1 (B), die diameter d_2 (C) and sheet thickness t (D).

The constant values of zero in the torque curve result from forming phases in which there was no contact between the blank and the former, since a 1/64 geometry was considered and the former is polygonal. Due to the antiproportional behaviour of the variation of the feed rate and the rotational speed on the filling level of the thread in Figs. 2 and 3, only the effect of the rotational speed on the torque while thread forming is analysed. Generally, all variations show the same tendencies regarding the torque over time. The progressive increase of the torque is due to a slow beginning of the thread forming while the first part of the thread is formed. As soon as the tool is in full contact with the collar, the torque reaches the maximum and remains approximately at this level until the end of the forming. The maximum torque is the highest at a rotational speed of 3000 rpm, lower at 2750 rpm and the lowest at 3250 rpm. This antiproportional behaviour is due to the fact that the filling level of the thread is the highest at 3000 rpm, lower at 2750 rpm and the lowest at 3250 rpm. For the variation of pre-hole diameter, 4.4 mm has the highest maximum torque. Although the collar is longer for a pre-hole diameter of 2.2 mm than for 4.4 mm, the filling level of the thread is higher for 4.4 mm compared to 2.2 mm during thread forming. When varying the die diameter, the maximum torque increases with decreasing diameter. This is likewise due to the filling level of the thread. The diameter 10.6 mm has the highest filling level and therefore the highest maximum torque. When analysing the torque for the variation of the sheet thickness, a decrease of the maximum torque with the reduction of the sheet thickness can be observed. Due to

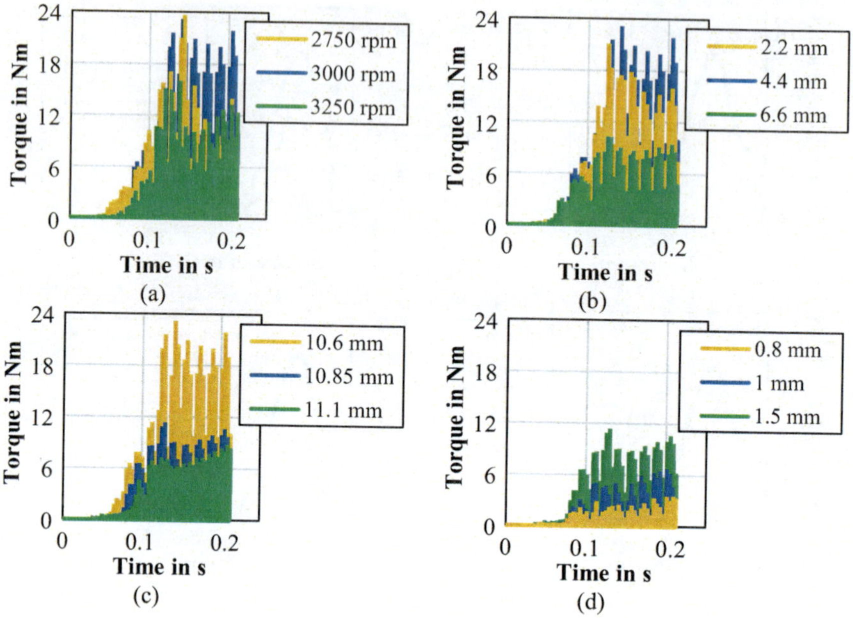

Fig. 8 Effect of the rotational speed n (A), pre-hole diameter d_1 (B), die diameter d_2 (C) and sheet thickness t (D) on the torque while thread forming

the lower sheet thickness, there is a fewer filling level of the thread, as to be expected, and thus a lower maximum torque.

To clarify the interactions between the filling level of the thread and the torque, the contact pressure during thread forming is shown exemplarily for the variation of the pre-hole diameter in Fig. 9. The areas showing high contact pressure are less for the diameters 2.2 and 6.6 mm than for 4.4 mm, resulting in a higher maximum torque for 4.4 mm as a pre-hole diameter.

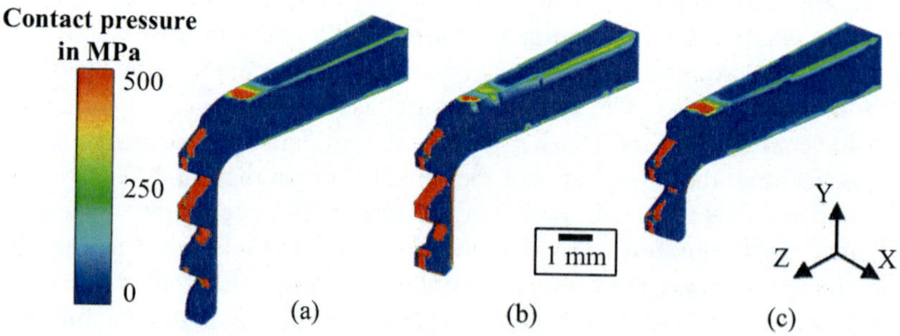

Fig. 9 Contact pressure distribution while thread forming for the variation of the pre-hole diameter d_1 between 2.2 mm (A), 4.4 mm (B) and 6.6 mm (C)

4 Summary and Outlook

This paper introduces the concept of a combined process of collar drawing with thread forming and presents a finite element based sensitivity analysis by varying process as well as geometrical parameters. The effect of the feed rate, rotational speed, pre-hole diameter, die diameter and sheet thickness on the filling level of the thread as well as on process values such as forming force and torque is studied. Even a small change of the feed rate and rotational speed of about 5% has proven to show a high influence on the resulting thread. Therefore, a process control with high precision must be implemented. A feed rate of 75 mm/s as well as a rotational speed of 3000 rpm were shown to fit well together, thus allowing a fast manufacturing of collars with threads. The best fitting pre-hole and die diameters are 4.4 mm as well as 10.85 mm, respectively, which will be verified in a full simulation model and further used for the application. The geometrical parameters pre-hole and die diameters were analysed for a sheet thickness of 1.5 mm, whereas the minimal usable sheet thickness was estimated to be 1 mm. Based on the results, a tooling system for collar drawing with integrated thread forming will be produced in the future and inserted in a forming press with appropriate equipment. Collar drawing and subsequent thread forming experiments with HX340LAD will be performed to evaluate the presented tool design and validate the FE model.

Acknowledgement. This research was supported by the Federal Ministry for Economic Affairs and Climate Action on the basis of a decision of the German Bundestag. It was organized by the German Federation of Industrial Research Associations (Arbeitsgemeinschaft industrieller Forschungsvereinigungen, AiF) as part of the program for Industrial Collective Research (Industrielle Gemeinschaftsforschung, IGF) under grant number 20958N. The authors gratefully acknowledge the support of the AiF.

References

1. Bickford, J.H.: Introduction to the Design and Behavior of Bolted Joints, 3rd edn. Routledge, New York (2017). ISSBN: 978-0824792978
2. Dinger, G.: Dynamic modeling and simulation of the screwing behavior of thread forming screws. J. Manuf. Proc. **20**(1), 374–379 (2015). https://doi.org/10.1016/j.jmapro.2015.06.012
3. Warrington, C., Kapoor, S., DeVor, R.: Experimental investigation of thread formation in form tapping. ASME. J. Manuf. Sci. Eng. **127**(4), 829–836 (2005). https://doi.org/10.1115/1.1951784
4. Balakrishnan, M., Issac, J.C.: Design of the multi-stage progressive tool for blanking a sheet metal component. Int. J. Precis. Eng. Manuf. **15**(5), 875–881 (2014). https://doi.org/10.1007/s12541-014-0411-0
5. DIN 7952–4: Sheet metal anchorage with threads—part 4: dimensions for tools and their design (2020). https://dx.doi.org/10.31030/3141779
6. Behrens, B.A., Rosenbusch, D., Wester, H., Stockburger, E.: Material characterization and modeling for finite element simulation of press hardening with AISI 420C. J. Mat. Eng. Perform. **31**, 825–832 (2022). https://doi.org/10.1007/s11665-021-06216-y
7. Betten, J.: Bemerkungen zum Versuch von Hohenemser. ZAMM **55**(3), 149–157 (1975). https://doi.org/10.1002/zamm.19750550304

8. Johnson, G.R., Cook, W.H.: Fracture characteristics of three metals subjected to various strains, strain rates, temperatures, and pressures. Eng. Fract. Mech. **21**(1), 31–48 (1985). https://doi.org/10.1016/0013-7944(85)90052-9
9. Behrens, B.-A., Dröder, K., Hürkamp, A., Droß, M., Wester, H., Stockburger, E.: Finite element and finite volume modelling of friction drilling HSLA steel under experimental comparison. Materials **14**(20), 5997 (2021). https://doi.org/10.3390/ma14205997

Monitoring of the Flange Draw-In During Deep Drawing Processes Using a Thin-Film Inductive Sensor

T. Fünfkirchler[1](✉), M. Arndt[2], S. Hübner[1], F. Dencker[2], M. C. Wurz[2,3], and B.-A. Behrens[1]

[1] Institute of Forming Technology and Machines (IFUM), Leibniz University Hannover, An Der Universität 2, 30823 Garbsen, Germany
fuenfkirchler@ifum.uni-hannover.de

[2] Institute of Micro Production Technology (IMPT), Leibniz University Hannover, An Der Universität 2, 30823 Garbsen, Germany

[3] DLR Institute for Quantum Technologies, Ulm University, Wilhelm-Runge-Straße 10, 89081 Ulm, Germany

Abstract. The quality of deep-drawn parts is subject to uncontrollable fluctuations, triggered by material property variations and process deviations, which occur despite extensive quality controls along the entire process chain. Monitoring and controlling the draw-in of the sheet material—which is an indicator of a faultless deep drawing process—would allow for a significant increase in process robustness. However, this requires sensor systems suitable for the industrial environment, which so far do not exist. This paper presents a newly developed inductive sensor in thin-film technology for measuring thc flange draw-in. The sensor was designed with the aid of finite-element-analysis and then manufactured using thin-film processes. After integration into a deep-drawing tool, the system was tested and validated. Afterwards, the detection of typical deep-drawing defects was investigated. It was demonstrated that the sensor system can reliably detect both cracks and wrinkles as well as the time at which they occur.

Keywords: Deep drawing · Draw-in sensor · Inductive sensor · Process monitoring · Thin-film sensor

1 Introduction

Deep drawing is one of the most widely used processes in the field of sheet metal forming. The process is used to manufacture complex sheet metal components, which are found in a wide variety of industrial sectors. The quality and reproducibility of deep-drawn parts is of significant importance today, especially in series production. However, due to various factors, such as batch variations, different lubrication conditions, temperature and tool wear, a deep drawing process in series production is subject to certain fluctuations that can have a negative impact on component quality [1, 2]. In order to control the quality and minimize the reject rate, there are various methods and sensors available today that

M. Liewald et al. (Eds.): WGP 2022, LNPE, pp. 111–121, 2023.
https://doi.org/10.1007/978-3-031-18318-8_12

allow process monitoring [3]. The apparent flange draw-in during the deep drawing process is a relevant criterion and can be used to evaluate a deep drawing process [1]. During and after the process the flange draw-in can be analysed to detect the occurrence of typical deep drawing defects such as cracks and wrinkles [4]. Various approaches have been pursued and researched in the past to reliably measure the flange draw-in. Due to various limitations, however, no system has yet qualified for monitoring an industrial series production process.

This paper therefore describes the concept and testing of an inductive measuring system which aims to reliably monitor the flange draw-in in series production.

2 State of the Art in the Field of Flange Draw-In Sensors

In the past, various sensor systems were developed and tested to measure the flange draw-in. Their functional principle was mainly based on tactile, optical and inductive measuring methods.

2.1 Tactile and Optical Sensors

Tactile sensors are based on the detection of the draw-in by touching the sheet metal or the outer edge of the sheet metal. One possibility for tactile flange draw-in measurement is the detection of the sheet outer edge with the aid of a tracked tactile tongue. The distance covered by the tongue is recorded by a displacement transducer and represents the apparent flange draw-in [5]. Another possibility for tactile detection of the apparent flange draw-in was developed by the IFUM. It is based on the detection of the draw-in by a rolling ball sensor, which was installed in the tool [6]. Optical sensors enable non-contact detection of the flange draw. One possibility is process monitoring with the aid of high-speed camera systems, as Yun used in his dissertation [7]. Another way of optically monitoring the flange draw is based on the principle of laser triangulation [8]. An optical receiver picks up the beam reflected from the moving edge of the blank. By detecting the angle, which depends on the distance, the instantaneous location of the scanned blank edge is determined [9].

Despite various investigations, neither of these functional principles have yet been sustainably qualified for use in industrial series processes. The reasons are found in the respective disadvantages that functional principles provide. Tactile tongues offer the danger of buckling or jamming in the tool, which can lead to an undesirable production stop in industry. In addition, this measuring system can only be used with forming tools that do not have a drawing bead [5]. Rolling balls reach their limits due to the required frictional contact when heavily lubricated and are susceptible to contamination. The same applies for optical systems. Furthermore, both measuring systems have to be integrated into the tool, which leads to a local weakening of the tool making it vulnerable to load and media.

Laser triangulation systems can only be used with flat blankholder geometries without drawing beads as a consequence of the straight beam paths. In addition, they have a high susceptibility to mechanical damage and require space outside the tool, which is often not available in series processes [9].

2.2 Inductive Sensors

Due to the great potential of inductive measuring systems, they have been the focus of research for some time [10]. The measurement principle can be divided into the transformational and the parametric principle. Flange draw-in measurements based on the transformational principle require two coils integrated in the forming tool (exciter and receiver coil). The measurement is based on the change of the induced voltage of the receiver coil due to the change of the sensor coverage with the sheet metal during deep drawing (Fig. 1). The parametric measurement principle, on the opposite, requires only the use of one coil in the deep-drawing die. In this case, the coil's inductance changes during sheet drawing [11] based on the electromagnetic coupling of the coil and the sheet. Until now, the sensor coil has been embedded in a non-metallic material, which does not affect the magnetic field of the coil, to protect it from the frictional contact with the heavily loaded sheet material [11, 12]. Special plastics or non-sintered ceramics, which are able to sustain the mechanical load, were used.

A significant advantage of inductive sensors is the low sensitivity to interference compared to optical systems. This applies in particular to vibrations and contamination. Further advantages are the high measuring resolution and the continuous measuring signal [12].

Based on its principle, however, the inductive sensor must be integrated into the deep-drawing tool. Thin-film technology offers particular potential here, making it possible to manufacture thin and at the same time robust sensors for use in forming tools. A sensor of this type has already been developed at the IMPT and tested at the IFUM for tool-integrated temperature measurement for hot stamping processes [13]. Furthermore, a pressure sensitive draw-in sensor has been invented [14]. A thin-film inductive sensor combines the advantages of thin-film sensor and the contactless inductive measurement.

3 Sensor Design and Experimental Conditions

In this paper, the design and testing of a novel flange draw-in sensor for deep drawing, which has been invented at the IMPT, is described. This sensor combines the advantages of inductive flange draw-in sensors described in the state of the art with the advantages of thin-film technology.

3.1 Design of Thin-Film Inductive Sensor for Flange Draw-In Measurement

The newly developed flange draw-in sensor is based on the transformational measuring principle demanding two coils (exciter and receiver coil). However, these are not placed opposite each other in die and blankholder, respectively. Instead, the two coils are positioned closely upon the other in the same substrate. Due to the small distance (10 μm) between the coils, their coupling is increased even without using a magnetic core. Via appropriate measuring electronics, an AC-current with constant amplitude and frequency is applied to the exciter coil, while the secondary voltage on the receiver coil is recorded and further processed with downstream filter and amplifier. Based on the sensor principle, the flange draw-in during the process will result in an increasing or decreasing voltage depending on the frequency and the sheet material. The sensor design especially the coils' dimensions have been designed with the aid of FEA simulation. The manufactured sensor can be seen in Fig. 1.

Fig. 1. Transformational sensor on stainless steel carrier with one turn exciter coil (1) above 25-turn receiver coil (2)

The sensor was placed directly in the blankholder via a tool insert. To meet the demands of a deep drawing process, the sensor has to be protected in the best possible way. An innovative approach was therefore adopted in which the sensor is installed upside down (Fig. 2). As a result, the sensor's surface is not in direct frictional contact with the sheet metal, but only the sensor carrier, on which the sensor is fabricated in thin-film technology. Copper and polyimide are used as conductor and embedding material. The carrier is made of 1.4301 stainless steel to limit the influence on the magnetic field of the coils to a minimum, while providing high wear and pressure resistance in the process.

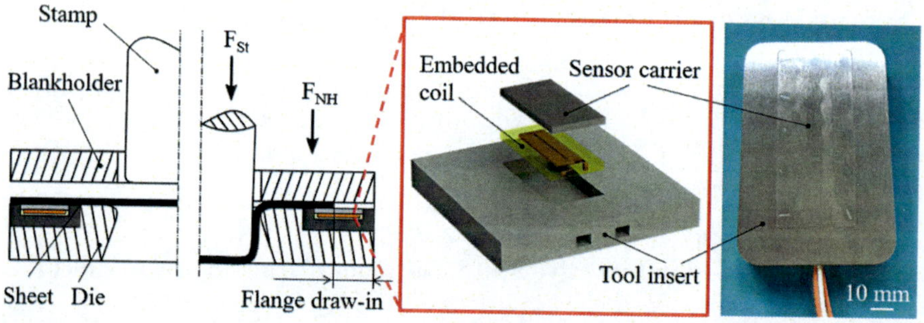

Fig. 2. Sensor concept with optimized installation situation in the forming tool and real sensor

3.2 Design of the Deep-Drawing Tool

For testing the newly developed sensor system, it was integrated into a deep-drawing tool. Its basis consists of a deep-drawing tool for the production of a rectangular cup with dimensions of 220 × 110 mm (Fig. 3). This geometry allows various stress conditions and flange draw-ins to be displayed in the process. By controlling the blankholder force, it is also possible to create typical deep-drawing defects such as cracks and wrinkles.

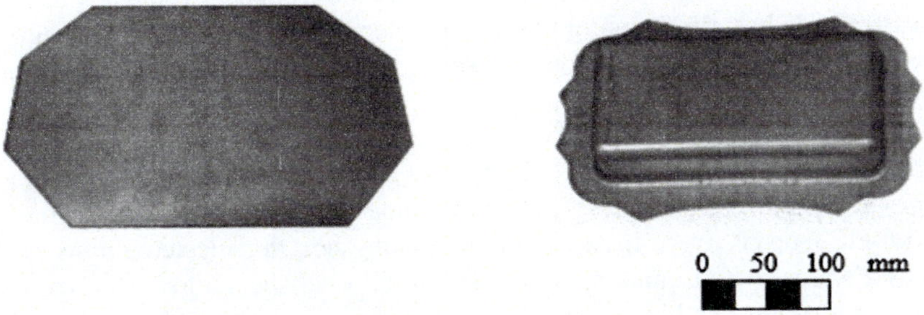

Fig. 3. Sheet metal blank (left) and rectangular cup after deep drawing (right)

The deep-drawing tool was redesigned to accommodate the sensor. Pockets were machined into the tool's blankholder at two positions (corner area and side area), into which the sensor insert can be placed (Fig. 4). The sensor insert itself consists of a tool insert on which the inductive sensor applied on the sensor carrier is screwed. The sensor is inserted with the sensor carrier facing upwards so that the coil is protected by the sensor carrier. After installation, the blank holder was surface milled together with the tool inserts. This minimized gaps between blankholder and tool insert and ensured a flat surface, so that no surface defects were produced on the deep-drawn component. Furthermore, annular grooves were provided to allow the installation of heating wires for temperature control of the tool. The tool was operated in a Dunkes HDZ 400 hydraulic forming press at IFUM.

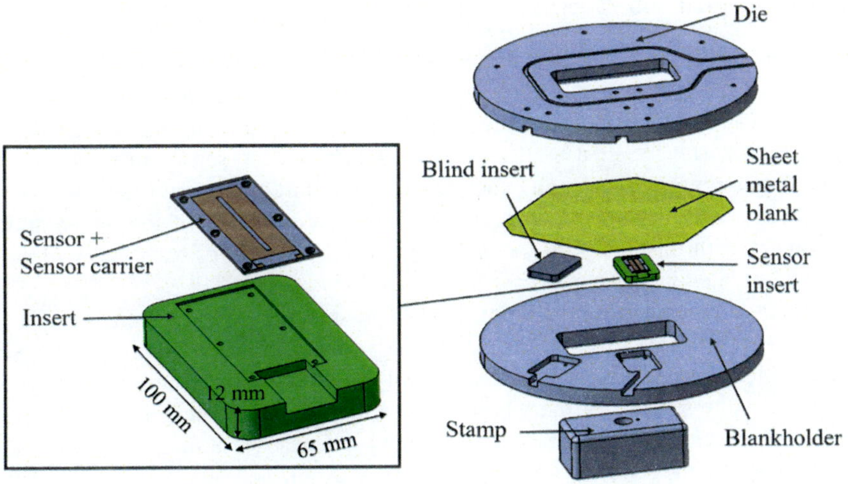

Fig. 4. CAD drawing of the deep-drawing tool with integrated sensor for flange draw-in

3.3 Experimental Conditions

For sensor validation a total of three different test series were carried out with the newly designed tool to validate the sensor. In all test series, the secondary voltage of the receiver coil was recorded as a function of the drawing depth. This serves as a measure of the flange draw-in. As mentioned before, the secondary voltage depends on the measuring frequency and the sheet material and has to be evaluated beforehand.

The first series of tests was used to record an example of a flange draw-in for various materials. The materials used were DP600 steel and AL6-OUT aluminum with a sheet thickness of 1.0 mm. As measuring frequency 1 and 5 kHz have been identified for DP600 and AL6-OUT, respectively. In case of the aluminum sheet the selected frequency causes the sensor's sensitivity to maximize. In case of the steel sheet a lower frequency than 1 kHz results in an even higher sensitivity but this would come at the cost of reducing the number of measuring points.

Rectangular cups with a drawing depth of 40 mm were deep-drawn. In the second series of tests, the more brittle material AL6-OUT was used to provoke cracks and the flange draw-ins were compared with those of good parts. The third series of tests aimed to generate wrinkles. Here, again, the inductively measured apparent flange draw-ins were subsequently compared with those of good parts. Due to the better ability to generate wrinkles, the material DP600 was used with a drawing depth of 50 mm. In all test series, factory-lubricated sheet metal was used, which was not additionally lubricated. However, preliminary tests with lubrication have shown that an oil layer in the volume of a usual application volume for a deep drawing process has no effect on the sensor signal.

4 Test Results

4.1 Sensor Signal and Process Characterization

In order to record an initial sensor signal and to show an example of the apparent flange draw-in, the test setup was put into operation and two rectangular cups (one with the material AL6-OUT and one with the DP600) were deep-drawn. The flange draw-in can be indicated by the measured secondary voltage of the sensor in relation to the drawing depth. Figure 5 shows an example of the measurements recorded with the new sensor for the two different sheet materials. The reproducibility of the flange draw-in measurement with this setup has been demonstrated in previous tests and will not be further investigated in this paper [15].

In the case of the aluminum sheet, it can be found that the voltage increases from the original 3.5 V to approx. 5.7 V. This corresponds to an increase of approx. 63%. Since the sheet material is non-magnetic but electrically conductive, eddy-currents are induced in the sheet by the sensor's primary coil. These eddy-currents reduce the voltage output of the secondary coil. During the process, the draw-in causes less sheet material to cover the sensor, so that the secondary voltage increases. After the voltage initially increases only minimally, a linear course of the flange draw is established from a drawing depth of approx. 15 mm. This course coincides with the expected flange draw-in, which is linear after the initial drawing depth. When observing the course of the DP600 steel, it is noticeable that the secondary voltage drops in contrast to the aluminium sheet. This can be attributed to the operating principle of the sensor. The high permeability of the steel sheet compared to air leads to a decrease in coil inductance with decreasing coverage during the process. Therefore, the output voltage induced in the secondary coil also decreases. In the case of low frequencies, this magnetostatic effect dominates and overlaps the contrary impact of the eddy-currents.

Furthermore, an undefinable fluctuating signal curve can be observed here for the first 5 mm. This phenomenon, which only occurred with the DP600, is currently the subject of further investigations. However, both materials have in common the change in secondary voltage with increasing flange draw-in. As with the aluminium sheet, a linear signal curve is obtained after a drawing depth of approx. 15 mm, up to which the bottom of the drawn part is pronounced. It is noticeable, however, that the change in the voltage is significantly smaller with the DP600. In this case, the voltage drops from an initial 8.3 V to approx. 7.6 V, which corresponds to a decrease of about 9%.

The first series of tests demonstrated the basic suitability of the sensor for monitoring the flange draw-in. Furthermore, it was found that the flange draw-in of ferromagnetic as well as non-magnetic metals can be measured.

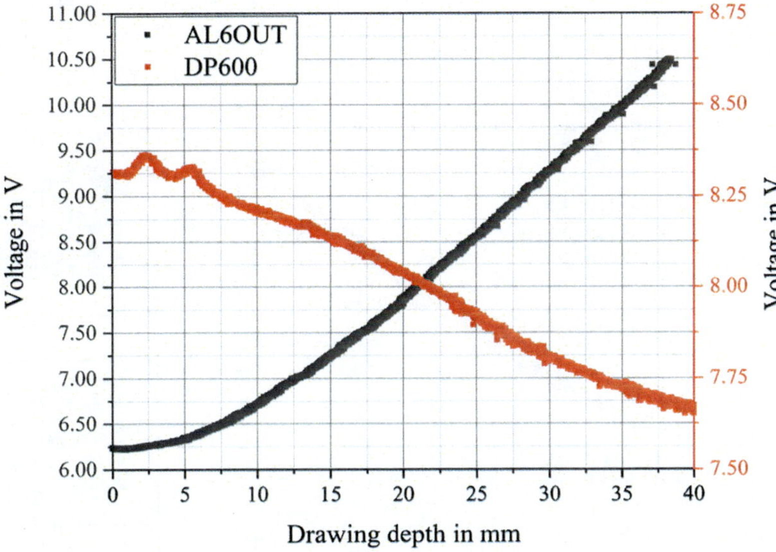

Fig. 5. Inductive measured flange draw-in for different materials

4.2 Detection of Defects Using Thin-Film Inductive Flange Draw-In Sensor

Typical defects that occur during deep drawing are cracks and wrinkles in the component. These can be attributed to various factors. High friction in the process, e.g. due to high blankholder force or low lubrication, leads to cracking. If, on the opposite, the blankholder force is too low, wrinkling can occur. In the series of tests described below, these types of defects were deliberately provoked by varying the blankholder force. The sensor signal was then evaluated and compared with that of good parts.

4.2.1 Crack Detection Through Inductive Flange Draw-In Sensor

Figure 6 shows different flange draw-ins recorded with the inductive thin-film sensor in second test series. The focus here was on the deliberate generation of cracks on the deep-drawn part. This was controlled by varying the blankholder force in the value range between 100 and 200 kN. The material used for this series was aluminum AL6-OUT. While a blankholder force of 100 and 130 kN still produced good parts, blankholder forces of 145 kN and more resulted in the formation of cracking. The cracks could be monitored via the detection of the flange draw-in. Since the cracking of the sheet slows down or completely interrupts the draw-in of the sheet, the sensor can detect this. This was noticeable by a constant secondary voltage instead of the linear drop. Furthermore, Fig. 6 also shows the drawing depth at which the cracks begin to form. It can clearly be seen that with a blankholder force of 200 kN, cracks begin to form at a drawing depth of approx. 25 mm, while with lower blankholder forces this only occurs later at approx. 30 or 35 mm. The fact that in this test series the cracks at a blankholder force of 160 kN formed minimally earlier than at a blankholder force of 145 kN might be due to process-related deviations, in particular cutting and positioning of the sheet, as well as tribological fluctuations.

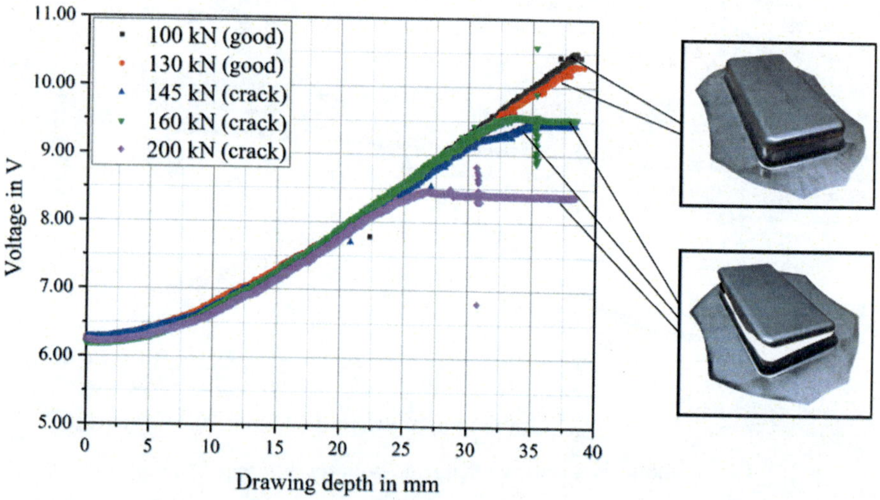

Fig. 6. Course of the secondary voltage as a function of the drawing depth for various blankholder forces from 100 to 200 kN with AL6-OUT as blank material

4.2.2 Detection of Wrinkles Through Inductive Flange Draw-In Sensor

In addition to cracks, wrinkles are also a common defect in deep-drawing. These occur primarily when the selected blankholder force is too low. In the third test series, starting from a good part produced at a blankholder force of 300 kN, the blankholder force was reduced gradually to 70 kN in order to deliberately produce a wrinkled part. As blank material a dual-phase steel DP600 was used, which is in comparison with the aluminum material, more suitable to produce wrinkles because of its higher tensile strength. Figure 7 shows a comparison of the flange draw-in of the two components. In the case of the part deep-drawn at a blankholder force of 300 kN, a linear curve similar to that shown in Fig. 5 can be seen. The component drawn with a blankholder force of 70 kN shows a clear drop in voltage. This can be attributed to the formation of the wrinkles. The sheet material lifts from the blank holder and thus provokes the drop in voltage.

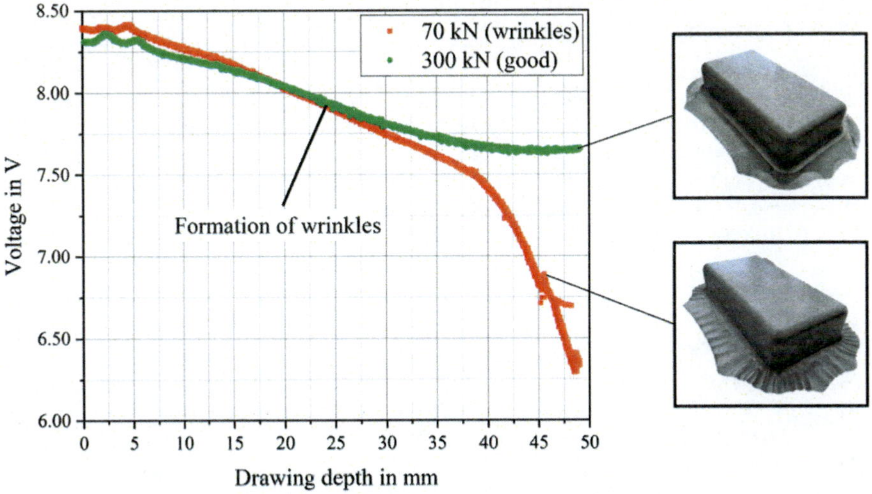

Fig. 7. Course of secondary voltage as a function of drawing depth for blankholder forces 70 and 300 kN with DP600 as blank material

5 Conclusion and Outlook

In this paper, a novel sensor system for measuring flange draw-in during deep drawing was presented. The system is based on an inductive measuring principle and was manufactured using thin-film technology. Its properties in terms of robustness and measurement quality could enable its use in industrial series processes and thus extend the state of the art in the field of monitoring forming processes. The measurement of the flange draw-in is realized by means of a simple tool insert which is surface milled together with the blankholder to ensure a plane surface and only minimal gaps. Further approaches to place the sensor from the other side below the tool surface so that no gap at all exists are currently being pursued. Since the sensor is based on a flexible polyimide substrate, future applications in curved blank holders are plausible.

The test results show that the flange draw-in can be detected based on the sensor system's output voltage. Furthermore, common defects like cracking and wrinkles can be identified. The superposition of the above defects can also possibly be detected in this way. However, this is currently the subject of further investigations. Nevertheless the presented sensor shows promising results with respect to a decision support in the in situ application of deep drawing parts (good part/defective part).

Since common defects induce a different sensor signal than a good part at the exact moment they occur, the sensor also offers the potential to design an intelligent forming tool in the future via a control loop. In this way, the press parameters can be adjusted directly in the process to prevent the formation of rejects. The basis for this is simply calibration before use in the process, depending on the sheet material used. Overall, the rejection rate during deep drawing could be reduced in this way and the quality of the parts could be improved. However, the design and implementation of the control loop in the forming press requires further research. This will allow to identify defective parts reliably, quickly and thus cost-effectively.

Acknowledgements. The research project "Deep drawing with a thin-film inductive sensor for monitoring flange draw-in" from the Research Association for Steel Application (Forschungsvereinigung Stahlanwendung e. V., FOSTA) was supported under grant number 20468N by the Federal Ministry of Economic Affairs and Climate Action (BMWK) through the German Federation of Industrial Research Associations (AiF) as part of the program for promoting industrial cooperative research (IGF) on the basis of a decision by the German Bundestag.

References

1. Brun, M., Ghiotti, A., et al.: Active control of blankholder in sheet metal stamping. In: Procedia CIRP. 31st CIRP Design Conference 2021 (CIRP Design 2021). Twente (2021)
2. Behrens, B.-A., Hübner, S., Wölki, K.: Acoustic emission—a promising and challenging technique for process monitoring in sheet metal forming. J. Manuf. Process. **29**, 281–288 (2017)
3. Bäume, T., Zorn, W., Drossel, W.-G., Rupp, G.: Iterative process control and sensor evaluation for deep drawing tools with integrated piezoelectric actuators. Manufact. Revis. **3**, 1–8 (2016)
4. Mahayotsanun, N., Cao, J., Peshkin, M.: A draw-in sensor for process control and optimization. J. Central South Univer. Technol. **15**, 273–277, USA (2005)
5. Rittmeier, S.: Systemunterstütze Umformung. Dissertation, TU Dresden (2007)
6. Griesbach, B.: In-Prozeß Stofffflußmessung zur Analyse und Führung von Tiefziehvorgängen. Dissertation, Leibniz Universität Hannover, Düsseldorf: VDI Verlag GmbH (Fortschritt-Berichte VDI, 547) (1999)
7. Yun, J.-W.: Stoffflussregelung beim Tiefziehen mittels eines optischen sensors und eines fuzzy-reglers. Dissertation: Leibniz Universität Hannover. Berichte aus dem IFUM (2005)
8. Spindler, J., Breme, M., Hein, C., Struck, R.: The audi toolshop—taking the next step into digital dimension. In: Proceedings of the 5th International Conference on Accuracy in Forming Technology (ICAFT), pp. S. 15–31. (2014)
9. Munser, R., Jacubasch, A., Wagner, U.: Messen beim Pressen: Werkstattstechnik online. Jahrgang 94. Heft 10, pp. S. 544–545. (2004)
10. Mahayotsanun, N., Sah, S., Cao, J., et al.: Tooling-integrated sensing systems for stamping process monitoring. Int. J. Mach. Tools Manuf **49**(7–8), 634–644 (2009)
11. Forstmann, U.: Induktive Wegsensoren zur Überwachung und Regelung des Blecheinzugs beim Tiefziehen. Berichte aus dem Produktionstechnischen Zentrum Berlin, IPK, Berlin (2000)

12. Neumann, A., Hortig, D., Merklein, M.: Measurement of material flow in series production. Key Eng. Mat. **473**, S. 137–144 (2011)
13. Behrens, B.-A., Hübner, S., Niemeyer H., et al.: Hot stamping process reliability—Innovative sensor technology for inline quality assurance. In: Conference Strategies in Car Body Engineering, 22.03.-23.03.2017, Bad Nauheim
14. Biehl, S., et al.: Multifunctional thin film sensor system as monitoring system in production. Microsyst. Technol. **22**(7), 1757–1765 (2016). https://doi.org/10.1007/s00542-016-2831-5
15. Behrens, B.-A., Maier, H.J., Fünfkirchler, T., Arndt, M.: Induktive Flanscheinzugssensorik für das Tiefziehen, Fosta-Forschungsbericht zum Projekt P 1217, Forschungsvereinigung Stahlanwendung e. V (2022)

Parameter Investigation for the In-Situ Hybridization Process by Deep Drawing of Dry Fiber-Metal-Laminates

M. Kruse[1(✉)], J. Lehmann[1], and N. Ben Khalifa[1,2]

[1] Institute of Product and Process Innovation, Leuphana University of Lüneburg, Universitätsallee 1, 21335 Lüneburg, Germany
moritz.kruse@leuphana.de

[2] Institute of Materials and Process Design, Helmholtz-Zentrum Hereon, Max-Planck-Straße 1, 21502 Geesthacht, Germany

Abstract. A newly developed in-situ-hybridization single-step process for the manufacturing of formed fiber-metal-laminates (FML) was introduced in previous works. During the deep drawing process, the fabric layer is infiltrated with a low-viscous thermoplastic matrix in a resin transfer molding process. The matrix polymerizes after the forming is completed. First parts could be manufactured successfully, but the influence of many process parameters continues to be unknown. The interaction of fiber and metal layer (DC04) on the formability of the FML is experimentally investigated by the deep drawing of FML parts without matrix injection. Parameters tested were the blank holding force, tool lubrication as well as different surface treatments of the metal sheet. Fiber breakage was observed after deep drawing of the dry FML. The deep drawn metal sheets were analyzed by surface strain measurements. The formability was then assessed by comparing the measured surface strains to a forming limit curve obtained by Nakajima-tests of the metal-fiber-metal stack. The results of the parameter investigation during dry deep drawing are analyzed to understand the influence of the process parameters on the in-situ hybridization process containing matrix injection.

Keywords: Deep drawing · Fiber-metal-laminates · Formability

1 Introduction

In recent years, increasing concerns about climate change have led to a change in the awareness of sustainability and energy efficiency. The majority of energy use in vehicles is attributed to their fuel consumption [1]. Reducing the weight and therefore the fuel consumption is key to reduce emissions over their entire lifecycle [2]. Fiber-metal-laminates (FML) are one approach for reducing the weight and improving mechanical properties [3]. They consist of several interstacked layers of metal and fiber reinforced plastics. The first and most commonly known type is GLARE (GLAss REinforced aluminum), which was developed by the aircraft industry to improve fatigue resistance [4]. Initially, expensive and time-consuming autoclave processes, which only allowed

M. Liewald et al. (Eds.): WGP 2022, LNPE, pp. 122–130, 2023.
https://doi.org/10.1007/978-3-031-18318-8_13

small curvatures, were used for the production of FML [5]. To increase the geometrical complexity, deep drawing processes were investigated for different types of FMLs. Most are using semi-finished products like prepregs and therefore require several steps [6, 7]. Commonly encountered defects include matrix accumulations [8] or buckling [9] and tearing [10] of fibers and metal sheets. Temperature and blank holder force were identified as parameters with a major influence on the formability of thermoplastic FML cups [10].

A newly developed single-step manufacturing process, combining deep drawing and a thermoplastic resin transfer moulding (T-RTM) process, was introduced by Mennecart et al. [11]. Inexpensive materials such as glass fiber fabrics and steel or aluminum sheets are used. The stack, consisting of two metal sheets with several fiber textile plies in between, is positioned in the tool with a double-dome geometry (Fig. 1). The inner side of the metal sheets is grinded and pretreated with a chemical bonding agent for better bonding between matrix/fibers and metal. The bottom metal sheet has a centered hole to allow for matrix injection from the injection channel inside the punch. The dry stack is deep drawn until plastification in the bottom metal sheet occurs to seal the injection channel between punch and metal sheet. The fabric layer is then infiltrated by a reactive thermoplastic matrix system during continuous deep drawing. After deep drawing, the tool is held closed until polymerization of the matrix is completed.

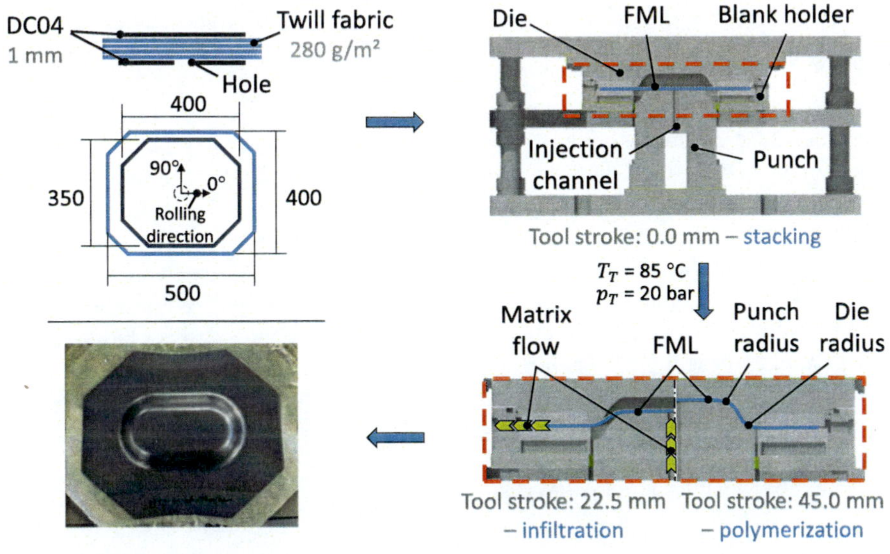

Fig. 1. Schematic illustration of the in-situ hybridization process (adopted from [12]).

Contrary to other processes applying prepregs or liquid compression moulding techniques, where fibers are infiltrated with matrix during the whole forming process, part of the in-situ hybridization process takes place with dry fabric. Therefore, the interaction between dry fibers and metal sheet during the deep drawing process has to be investigated. In previous works, the influence of dry and saturated textile on the formability of metal blanks were investigated by modified Nakajima tests [13]. It was shown, that the friction between metal sheet and fibers significantly reduce the formability of the metal blank. Also, imprints of the fabric structure were observed on the metal sheet under high compressions over 100 MPa.

In this paper, different parameters are investigated by deep drawing of dry FML stacks with the actual part geometry. The goal is to obtain a better understanding of the parameter's influence on the formability in the in-situ hybridization process.

2 Materials and Methods

The same materials as used by Mennecart et al. [11] are chosen. For the glass fiber layer, four plies of twill woven E-glass fabric (280 g/m^2, Interglas 92125 FK800) are selected. The metal sheets are steel DC04 (1.0338) with 1 mm thickness. The lower sheet has a centered hole (19 mm diameter) that would be used for matrix injection. Dimensions and orientations are shown in Fig. 1. The tool is heated to 85 °C and the experiments are conducted on a hydraulic press from Röcher with a maximum force of 2700 kN. The minimum speed of 1 mm/s is used for deep drawing.

After performing the experiments, the produced parts are evaluated with a GOM Argus optical forming analysis system. For that, the outer side of the metal sheets has to be marked with a regular point pattern (diameter 1 mm, distance 1.5 mm) before the experiments. From the displacement of the points, surface strains can be calculated. Marking is performed with an electrolytic marking system EU pulse and electrolyte 701/9.

Three parameters are investigated as presented in Table 1. Each part is produced once. Four different blank holder forces are applied from 140 to 190 kN, as the blank holder force used for the in-situ hybridization process is 160 kN. The influence of the metal sheet pre-treatment is investigated for two different blank holder forces (160 and 190 kN). For that, the inner side of the metal sheets is grinded in 0°-, 90°- direction as well as in small circles and a bonding agent (Dynasalan® Glymo by Evonik) is applied. So far, no lubrication was used in the in-situ hybridization process. Without lubrication, the point pattern is rubbed off in the die radius of the upper metal sheet by the friction between tool and sheet [13]. Therefore, three tool lubricants are chosen to improve the preservation of the point pattern and investigate their influence on the forming process. Deep drawing oil with a viscosity of 260 mPas (Raziol CLF-260 E), Boron nitride extrusion spray (3M) and PTFE foil with a thickness of 0.025 mm are used to reduce the friction between tool and metal sheet.

3 Results

Figure 2 illustrates the strain measurements for top and bottom blank of the reference part with 160 kN blank holder force, no lubrication and without any surface treatment. The forming limit curve (FLC) was obtained as described by Mennecart et al. [13] for DC04 with a dry glass fiber twill fabric interlayer.

Top and bottom blank show a similar strain behavior. While maximum forming is observed in the die radius, the most critical area is the punch radius because of

Table 1. Process parameters (varying parameter italic).

Blank holder force in kN	Metal sheet surface treatment	Lubricant
140, 160, 175, 190	None	None
160	*None, Grinded + Glymo*	None
190	*None, Grinded + Glymo*	None
160	None	*None, Boron Nitride spray, Deep drawing oil, PTFE foil 0.025 mm*

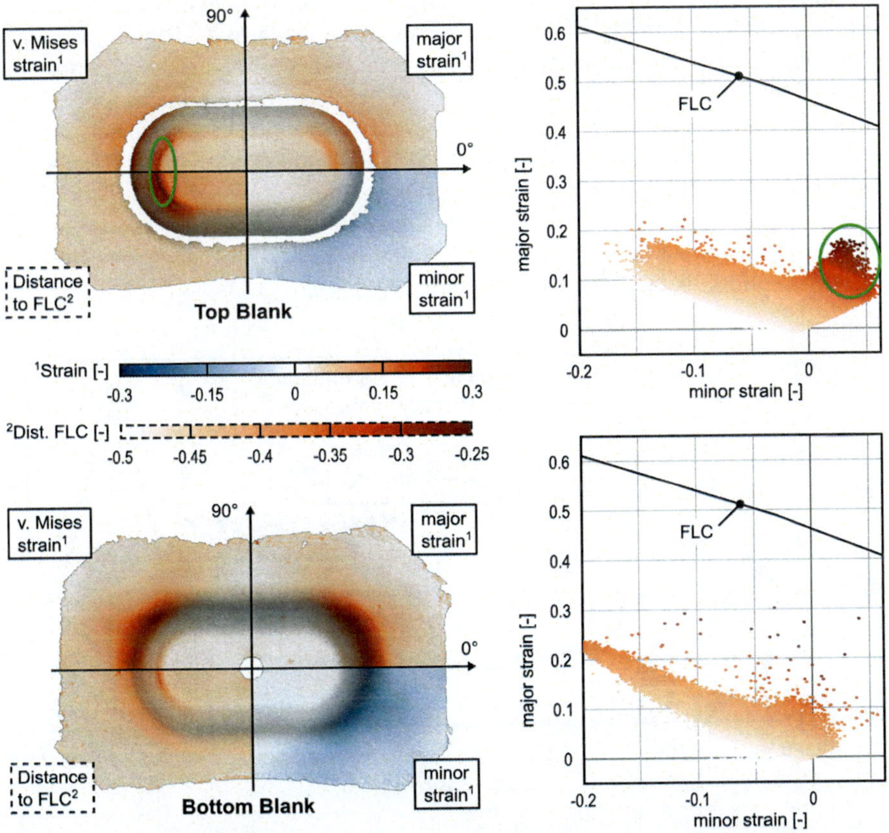

Fig. 2. Surface strain measurements for top and bottom blank (160 kN blank holder force, no lubrication, no surface treatment); Left: Heatmaps v. Mises strain, major strain, minor strain and vertical distance to FLC; Right: Forming limit diagrams.

biaxial tension, as typical for deep drawn parts. For the top blank, the strains in the die radius cannot be measured because of the missing point pattern. From the surrounding measurements, it can be expected, that the strains in the die radius are slightly lower in the top blank. For the more critical punch radius, strains are higher in the top blank.

Failure due to tearing is therefore expected to occur first in the top blank which can also be observed in the forming limit diagram where higher positive minor strains appear together with high major positive strains. No wrinkling was observed in any of the produced parts.

3.1 Influence of the Blank Holder Force

The measured major strains along the 0°-axis for 140, 175 and 190 kN blank holder force are shown in Fig. 3. The results are presented for the top blank on the left and the bottom blank on the right. The strain curves for 160 kN are similar to 140 kN for the top blank and to 175 kN for the bottom blank. Therefore, 160 kN is excluded from the chart for improved visibility. Gaps in the curves from −125 to −110 mm indicate the missing point pattern in the die radius. Gaps from 0 to 10 mm are the injection holes of the bottom blanks. Top and bottom blank as well as different blank holder forces all produce a similar tensile strain pattern. Peaks are visible in each radius. In the part bottom, almost no strain occurs in the bottom blank, while strains around 0.05 occur in the top blank. As already explained, punch radius strains are higher in the top blanks. Strains in the die radius of the top blanks can only be estimated from the curve but are probably lower than for the bottom blanks. Strains in the wall and flange area are similar for top and bottom blank.

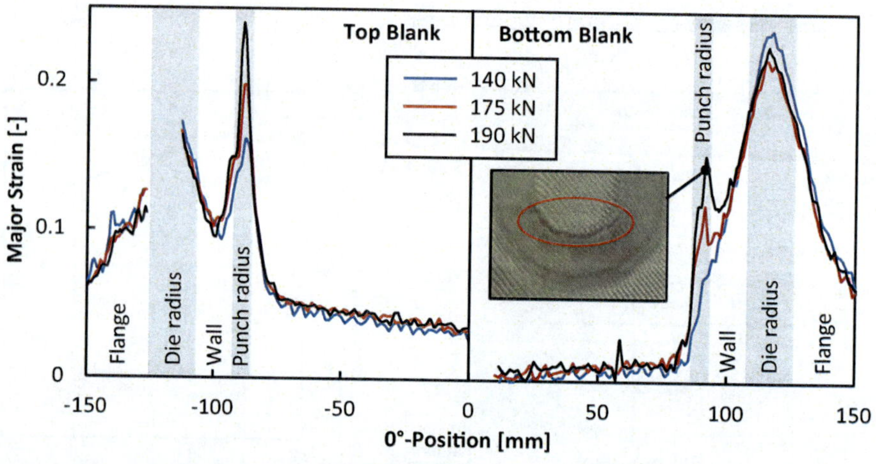

Fig. 3. Major strains along the 0°-axis in the top (left) and bottom (right) blank for different blank holder forces. Image: Fiber tearing in the punch radius.

The major difference in strain for different blank holder forces occurs in the punch radius which is also the most critical because tearing can occur with biaxial tension. Almost no differences are observed in the other areas including the die radius. Therefore, strain values in the punch radius can be used for comparison. As reasonable, the major strain in the punch radius increases with increasing blank holder force for both top and bottom blank. For 140 kN, no peak in strain is observed in the bottom blank. Imprints of the fabric structure on the metal sheet could be observed in the radiuses of top and bottom blank for all blank holder forces. Tearing of the fibers occurred in the punch radius for all blank holder forces. The fibers have a much lower elongation at break than the metal sheets. Therefore, failure of the fabric happens much earlier. No or little fiber breakage could be observed in the in-situ hybridization experiments at similar blank holder forces. The matrix reduces the friction and allows for more relative movement between fibers and metal, which improves the formability of the part as a whole [14].

3.2 Influence of the Metal Sheet Surface Treatment

Maximum major strains in the punch radius are displayed in Fig. 4, for comparison between parts manufactured with and without pre-treatment of the metal sheets. On the bottom blank, the surface treatment has only little influence. No difference is measured for a blank holder force of 160 kN, while a small increase from 0.15 to 0.2 can be observed for 190 kN. However, the pre-treatment has a strong influence on the top sheet. Maximum major strain increases from 0.17 to 0.24 at 160 kN due to grinding and applying the bonding agent. The surface treatment has a similar effect as 30 kN of additional blank holder force. For 190 kN, the metal sheet is tearing during deep drawing when the surface was pre-treated. Hence, the results from the surface strain measurements are in line with the observed tearing as the strain measurements in that area lie above the FLC.

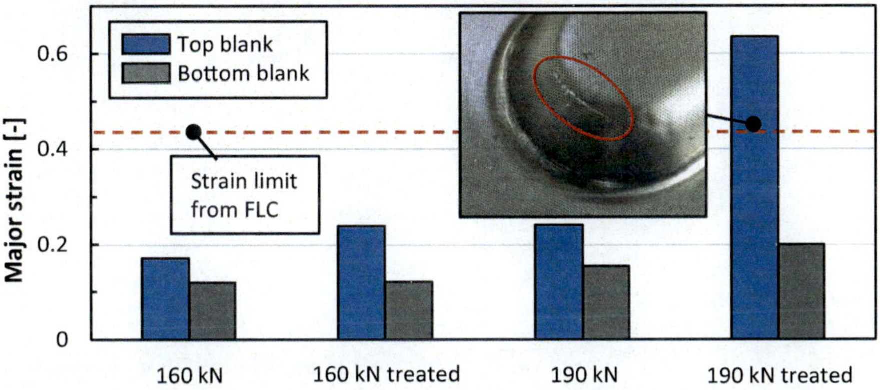

Fig. 4. Influence of the metal sheet pre-treatment on the maximum major strain in the punch radius (for blank holder forces of 160 and 190 kN). Image: Tearing in the punch radius.

For the in-situ hybridization, it can be concluded, that surface treatment possibly has an influence on the forming behavior. In the performed dry deep drawing experiments, the surface treatment has a strong influence on the strains of the metal sheets. However, as was already mentioned in the previous chapter, the matrix reduces the friction between metal and fabric. A smaller influence of the surface treatment can therefore be expected when injecting the low viscous matrix.

3.3 Influence of the Tool Lubrication

Different tool lubrications were tested for a blank holder force of 160 kN. Results are illustrated in Fig. 5. The deep drawing oil has no influence on the forming. Possibly a higher viscosity oil is necessary. Furthermore, oil is infiltrating the fabric on the edges of the sheet. When injecting matrix, it would mix with the oil. The use of boron nitride reduces the strains significantly, in particular in the top sheet. In addition, the point pattern is preserved almost entirely. The PTFE-foil yields the best results with a completely preserved point pattern, strains reduced by approximately factor two and a more even strain distribution, especially in the punch radius. Also, no tearing of the fibers was observed when using the foil. The reduced friction between tool and metal sheet allows for more gliding and therefore less relative movement between fibers and metal. This leads to improved part properties and a preserved Argus point pattern.

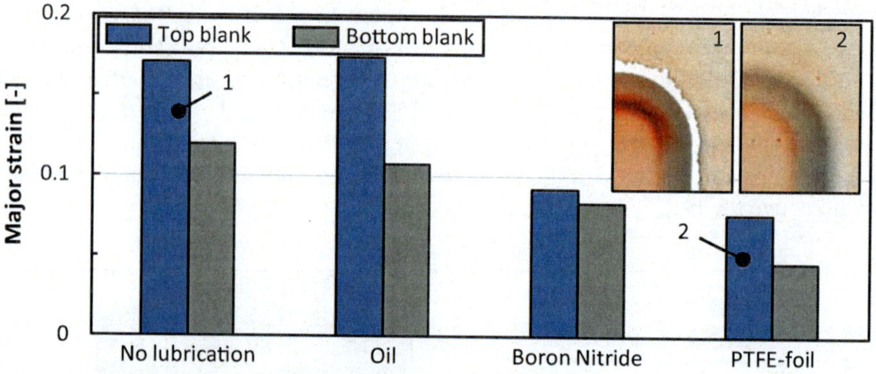

Fig. 5. Influence of different tool lubricants on the maximum major strain in the punch radius (blank holder force 160 kN). Images: Distance to FLC for no lubricant (1) and PTFE foil (2).

4 Conclusion

Three parameters for the deep drawing of fiber metal laminates without matrix injection were investigated to improve and further understand the forming behavior in the presented in-situ hybridization process. The following conclusions are drawn.

- In the punch radius, where tearing can occur, the top blank exhibits higher strains than the bottom blank. The punch radius is also the only area, where the investigated parameters have a significant influence on the measured strains.
- As reasonable, higher blank holder forces lead to higher strains.
- Forming is inhibited by grinding and applying a bonding agent on the metal sheets, probably due to increased friction.
- Tool lubrication preserves the Argus point pattern for strain evaluation and improves the forming by reducing strains. PTFE-foil performs best.

In the in-situ hybridization process, the injected matrix changes the forming behavior. Thus, some parameters might have a slightly different influence in the actual in-situ hybridization process. Therefore, the here tested parameters should be investigated again with matrix injection to quantify the influence of the matrix on the process. Furthermore, matrix injection introduces further parameters and parameter interactions from the RTM process. Further research has to be conducted regarding the influence of these parameters on the in-situ hybridization process.

Acknowledgements. The authors would like to thank the German Research Foundation (DFG) for funding the projects BE 5196/4-1 and BE 5196/4-2. The deep drawing oil was kindly provided by Raziol Zibulla and Sohn GmbH. The authors would like to thank Mr. Marvin Gerdes for the help in performing experiments.

References

1. Mayyas, A., Qattawi, A., Omar, M., Shan, D.: Design for sustainability in automotive industry: a comprehensive review. Renew. Sustain. Energy Rev. **16**(4), 1845–1862 (2012). https://doi.org/10.1016/j.rser.2012.01.012
2. Tisza, M., Czinege, I.: Comparative study of the application of steels and aluminium in lightweight production of automotive parts. Int. J. Lightweight Mater. Manuf. **1**(4), 229–238 (2018). https://doi.org/10.1016/j.ijlmm.2018.09.001
3. Asundi, A., Choi, A.Y.: Fiber metal laminates: an advanced material for future aircraft. J. Mater. Process. Technol. **63**(1–3), 384–394 (1997). https://doi.org/10.1016/S0924-0136(96)02652-0
4. Vlot, A.: Glare: history of the development of a new aircraft material. Kluwer Academy Publication, Dordrecht (2001)
5. Sinke, J.: Manufacturing of GLARE parts and structures. Appl. Compos. Mater. **10**(4/5), 293–305 (2003). https://doi.org/10.1023/A:1025589230710
6. Heggemann, T., Homberg, W.: Deep drawing of fiber metal laminates for automotive lightweight structures. Compos. Struct. **216**, 53–57 (2019). https://doi.org/10.1016/j.compstruct.2019.02.047
7. Blala, H., Lang, L., Khan, S., Alexandrov, S.: Experimental and numerical investigation of fiber metal laminate forming behavior using a variable blank holder force. Prod. Eng. Res. Devel. **14**(4), 509–522 (2020). https://doi.org/10.1007/s11740-020-00974-9
8. Dau, J., Lauter, C., Damerow, U., Homberg, W., Tröster, T.: Multi-material systems for tailored automotive structural components. In: Proceedings 18th International Conference on Composite Materials, Jeju, Korea (2011)
9. Behrens, B.-A., Hübner, S., Neumann, A.: Forming Sheets of metal and fibre-reinforced plastics to hybrid parts in one deep drawing process. Procedia Eng. **81**(7), 1608–1613 (2014). https://doi.org/10.1016/j.proeng.2014.10.198
10. Rajabi, A., Kadkhodayan, M., Manoochehri, M., Farjadfar, R.: Deep-drawing of thermoplastic metal-composite structures: experimental investigations, statistical analyses and finite element modeling. J. Mater. Process. Technol. **215**, 159–170 (2015). https://doi.org/10.1016/j.jmatprotec.2014.08.012
11. Mennecart, T., Werner, H., Ben Khalifa, N., Weidenmann, K.A.: Developments and analyses of alternative processes for the manufacturing of fiber metal laminates. In: Volume 2: Materials; Joint MSEC-NAMRC-Manufacturing USA. American Society of Mechanical Engineers (2018). https://doi.org/10.1115/MSEC2018-6447

12. Werner, H.O., Schäfer, F., Henning, F., Kärger, L.: Material modelling of fabric deformation in forming simulation of fiber-metal laminates—a review on modelling fabric coupling mechanisms. In: 24th International Conference on Material Forming (2021)
13. Mennecart, T., Gies, S., Ben Khalifa, N., Tekkaya, A.E.: Analysis of the influence of fibers on the formability of metal blanks in manufacturing processes for fiber metal laminates. J. Manuf. Mater. Process. **3**(1), 2 (2019). https://doi.org/10.3390/jmmp3010002
14. Kruse, M., Werner, H.O., Chen, H., Mennecart, T., Liebig, W.V., Weidenmann, K.A., Ben Khalifa, N.: Investigation of the friction behavior between dry/infiltrated glass fiber fabric and metal sheet during deep drawing of fiber metal laminates. Prod. Eng. Res. Devel. (2022). https://doi.org/10.1007/s11740-022-01141-y

Numerical Analysis of the Deep Drawing Process of Paper Boards at Different Humidities

N. Jessen(✉), M. Schetle, and P. Groche

Institut Für Produktionstechnik Und Umformmaschinen, Technische Universität Darmstadt, Otto-Berndt-Str. 2, 64287 Darmstadt, Germany
nicola.jessen@ptu.tu-darmstadt.de

Abstract. The socio-political demand for more sustainability is putting a lot of pressure on the packaging industry. In addition to logistics and marketing, packaging today serves preparation, resealability and more purposes. Though more sustainable, paper packaging produced by folding, for example, falls far behind plastic packaging in terms of geometric variety, among other things. These deficits are to be remedied by sustainable formed packaging. The determination of process parameters and material settings in the forming process for paper is complex due to the inhomogeneous natural material and limited formability. Consequently, process layout is mainly based on empirical knowledge. Moisture affects the forming behavior of paperboard significantly. Therefore, process simulations at different moisture levels and distributions within a sample allow more targeted selection of process parameters. This paper paves the way for simulations of cardboard behavior at different moisture levels and reveals the influence of moisture distributions on properties of deep-drawn products.

Keywords: Paper forming · Humidity influence · Numerical simulation of paper forming · Numerical simulation at different moisture contents · Wrinkling of paper

1 Introduction

Sustainability is not only one of the everyday challenges facing every person, but also one of the most important tasks industry and society are faced with. Packaging has a representative role to play and offers every individual the opportunity to contribute to a more sustainable lifestyle. In Germany alone, around 3.2 million tons of short-lived plastic packaging are disposed each year [1] and more than half of this is recycled energetically [2]. Replacing plastic packaging with paper packaging is not a new idea. Nevertheless, its importance is growing steadily, as evidenced by a growing public interest. A study in 2021 revealed that 70% of Germans consider paper packaging to be the most sustainable packaging solution [3]. Due to the relevance of sustainability, they are willing to pay up to 6.5% more for a product when sustainably packaged [3]. Against this background, it is clear that the industry needs better solutions to meet customer demands. Paper packaging does not necessarily have to be more economical

M. Liewald et al. (Eds.): WGP 2022, LNPE, pp. 131–141, 2023.
https://doi.org/10.1007/978-3-031-18318-8_14

in terms of raw material and production costs than existing packaging, but it does have to be more sustainable without compromising the protection or appearance quality to enable the marketing of the packaged goods.

Various types of paper packaging are already used in the food industry. These are mostly folded, wrapped or molded. Each of these packaging production methods has its own disadvantages in terms of geometrical freedom, surface quality and production times. However, for marketing and manufacturing purposes, these properties play a primary role in replacing plastic packaging. For this reason, it is reasonable to produce paper packaging using three-dimensional forming techniques, such as the most common deep-drawing process. Deep drawing is an established method of sheet metal processing and is characterized by a high productivity and at the same time great freedom of geometry and economical implementation.

Establishing the deep drawing process with paper, nevertheless, faces the industry with some material-specific challenges. The natural material requires both a preparation of the material to adjust the production-relevant properties and the specific setting of the process parameters. These settings are usually based on empirical trials. Experimentally determining the ideal parameters is expensive and time-consuming, which is the reason for the use of feasible parameters once they are found. Often a further optimization is avoided. In order to improve the deep drawing results, a wide variety of parameters can be adjusted. In addition to the already mentioned humidity, temperature and blankholder force are also important influencing factors. With regard to temperature, it has been shown that the positive influence of increased temperature on formability is cancelled out by moisture removal due to heat [4]. For this reason, an alternative approach is to introduce both, moisture and temperature into the specimen by means of steam, thereby increasing the deep-drawing quality. In this work, the steam is introduced into the specimen through the blankholder. Thus, the steam can influence the product quality during the forming process. However, it is not clear how the locally adjusted insertion of steam can influence the material anisotropy.

Numerical simulations are established tools to determine production parameters of deep drawing processes in order to achieve higher economic performance and sustainability. This has proven difficult for paper in the past, as the material is dependent on natural imperfections and heterogeneous properties. By considering different pattern of steam introduction during the deep drawing process the complexity of the process design is even higher. Therefore, the process simulation with different moisture settings is of high relevance and discussed in the following.

2 Approach

2.1 Paper as a Material in Forming Technology

Paper is a natural fiber material. It entails many specific and difficult-to-simulate properties, such as inhomogeneities and pores. The production of the material, in which the plant or recycled fibers are dried in a liquid solution on a moving web, causes the fibers to align. A large part of the fibers orient themselves in the direction of the web's movement, known as machine direction (MD), while a smaller part orient themselves perpendicularly to it in cross direction (CD). This fiber alignment results in anisotropic material properties. For example, paper is more stretchable in CD, while in MD it has a higher tensile strength and higher springback after forming [5].

Hydrogen bonds between the fibers are responsible for the fiber bonding, which is why paper is a hygroscopic material. These hydrogen bonds can be dissolved by water molecules, which results in higher extensibility and lower tensile strength of the material in all directions. The properties of the paper can therefore be controlled by the targeted introduction of moisture [6]. The described structure of paper is shown schematically on the left side in Fig. 1.

Since elongation plays a particularly important role in the forming of a material, the papers are examined in tensile tests at different moisture contents. The relative humidity of the respective paper at which it exhibits the maximum elongation is sought. Tensile tests are usually performed in CD, MD and 45° directions. Figure 1 shows the influence of moisture by stress-strain diagrams of a paper in MD and CD at moisture contents of 5 and 15%.

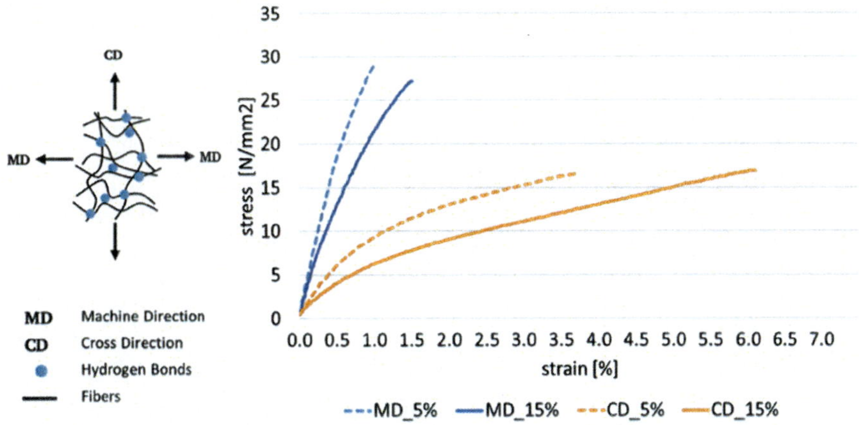

Fig. 1. Left: Paper structure [10, modified]; Right: moisture and direction-dependent stress-strain diagram for the CKB carton by Stora Enso

For most virgin fiber papers, the optimum, represented by the highest elongation, is about 15% moisture content of the material [5]. The moisture content of the paper is defined as follows [7]:

$$mc = \frac{M_{\text{moist}}}{M_{\text{tot}}} = \frac{M_{\text{tot}} - M_{\text{dry}}}{M_{\text{tot}}}, \quad (1)$$

with M_{moist} being the mass of humidity in the paper, M_{tot} being the total mass of the humid paper and M_{dry} being the mass of the paper with 0% humidity [7]. The anisotropic effects can be reduced by using different moistures in different parts of the sample. For example, a moisture that allows the maximum strain to occur in MD and a lower moisture in CD can cause the maximum strains of the two directions to converge. However, this simultaneously leads to a reduction of the maximum strain in CD compared to the possible optimum. For this reason, the geometry of steam injection in the blank holder has to be optimized. Various factors such as anisotropy, specimen geometry, material orientation and targeted drawing depth must be taken into account.

2.2 Deep Drawing of Paper

The deep drawing process of paper is in principal identical to the deep drawing process of sheet metal forming. Differences are caused by additional, quality ensuring measures, like the vapor deposition, which support appearance quality and achievable drawing depth.

However, unlike metals, wrinkles in paper are often not seen as a process-limiting defect and are widely accepted till a certain level. This is due to the fact, that with the current state of research, the forming of paper into three-dimensional geometries with packaging-relevant volumes is only possible with the formation of wrinkles. Nevertheless, wrinkles should be as flat as possible. Wrinkle pressing in the drawing gap or in an optional subsequent calibration step can be applied to counteract this process limit. The more the wrinkles are compressed, the better the surface appearance and the perception of the product quality [8].

Increased moisture leads to increased elongation in the deep drawing process and thus to a higher maximum drawing depth. By introducing moisture in form of steam through the blank holder while the material is being formed, springback can also be reduced and the crease pattern is improved [4].

The quality of a deep-drawn paper product can be evaluated based on various aspects. In the context of this contribution, two points will be considered:

1. Achievable drawing depth before fracture [9].
2. Wrinkle formation: the more wrinkles are formed, the smaller they are in volume, which leads to better compression of the wrinkles and thus better surface quality [9].

Apart from these aspects, the angle between wall and bottom is also an important parameter. It should be as close to 90° as possible. Increased springback leads to higher wall angles, which has a negative effect on product geometry and is unevenly distributed over the circumference due to the anisotropy [9]. The deep drawing geometry considered in this publication is rectangular with dimensions of 175 mm length and 110 mm width. The corners are rounded off with 20 mm radii (cf. Fig. 2).

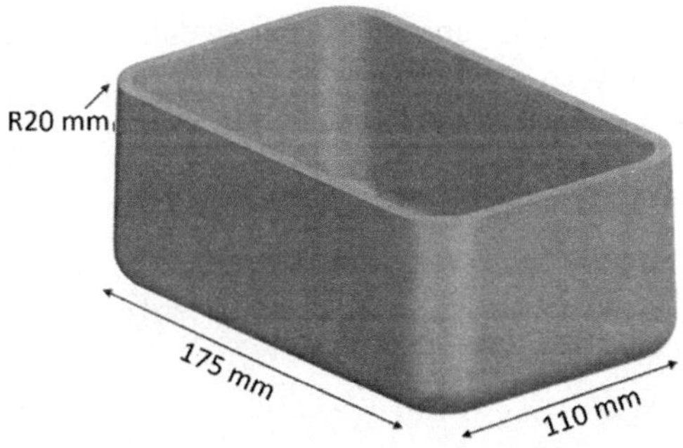

Fig. 2. Deep drawing geometry

Moisture distribution in the deep drawing process

Knowing that moisture can have a positive influence on the deep drawing process, the question arises, whether the introduction of steam can counteract the anisotropy of the material. In addition, the wrinkle pattern, as already explained, is an important issue regarding the product appearance and thus the question is, whether a moisture distribution tailored to the geometry can improve it. In order to investigate this, some experiments with a simple, rotationally symmetrical geometry have been carried out. It was found that the achievable drawing depth is reduced, while an increase in the number and compression of the wrinkles is visible. This results in a visually better appearance as can be seen in Fig. 3. The left side of the Figure shows the deep drawing result at an overall homogeneous15% humidity. It shows higher spring back, as well as higher drawing depth and deeper wrinkles. The right side shows a deep drawn product with a higher moisture in MD (~15%) and a lower moisture in CD (5%). Overall, lower spring back, smaller wrinkles as well as a reduced drawing depth can be detected.

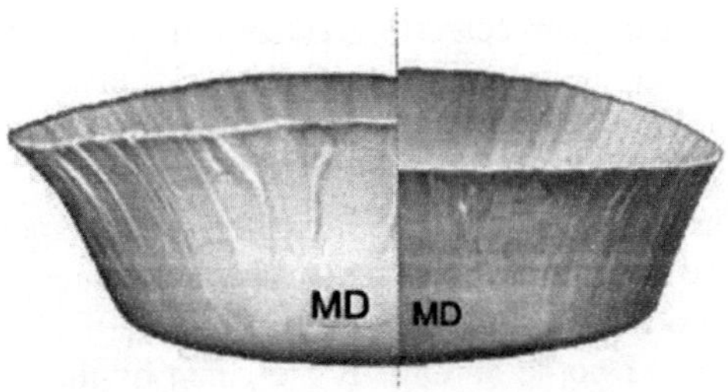

Fig. 3. Deep drawn product at 15% humidity (left) and at ~15% in MD, ~5% in CD (right)

In the next step, the approach of tailored steam distributions with its effects on geometry and wrinkle development and compression will be transferred to a more complex geometry. In order to do this, tools with optimized steam injection geometries are necessary. Time-consuming and expensive experimental test series should be substituted by numerical simulations. This allows not only for an efficient design process for the tools and the process, but also insights into the development of moisture, stress and strain states during the process.

3 Simulations

To analyze the moisture distribution, numerical material models are created at different moisture contents using data from tensile tests. These are then used in differently partitioned samples to simulate the deep drawing process with different moisture distributions.

3.1 Numerical Model of the Deep Drawing Process with Paper

The explicit simulations are performed with 3D elements of type C3D8R in Abaqus 2021. After examining different meshes of the sample with 11 different element numbers between 3620 and 139,780, the variant with 37,956 elements was selected. This resulted in a good compromise between accuracy of the wrinkle mapping, local element distortions and simulation times. The elements have a size of 1 mm × 1 mm × 0.2225 mm. A mass scaling of 200 and the exploitation of symmetry and thus the simulation of a quarter model, have significantly reduced the computational time. The potential values are prefactors of the plastic strain and allow the material-specific anisotropy to be modeled by directional dependence. They are formed via the yield strength ratios or via the Lankford approach. Tension and cupping test material parameters were recorded in experiments of our own at different moisture contents. The other necessary data were taken from Huttel [13]. All tool components - punch, blank holder and die- are modelled as rigid shells. The simulation process starts with the direct contact between punch and sample. A punch path of a maximum of 40 mm is simulated. The specimen lies between the die and the blank holder and is drawn into the drawing gap by the punch movement. Due to the anisotropy of paper, the sample can be deep-drawn in different orientations. In this model, the orientation of the CD fibers is chosen to be parallel to the long edge, since the fibers are subjected to greater strain in the longitudinal direction. The anisotropic plastic material behavior is defined by yield stress and plastic stain, as well as the Hill criterion. Elasticity is characterized by values for Poisson ratio (v_{12} = 0.45; $v_{23} = v_{13}$ = 0.33), elasticity module (E_1 = 1236 N/mm^2; E_2 = 3582 N/mm^2; E_3 = 358.2 N/mm^2) and the shear modulus [G_{12} = 688 N/mm^2; $G_{23} = G_{13}$ = 60 N/mm^2] respectively according to literature [9, 11] as well as experimental findings. The mass density is calculated according to DIN EN ISO 534 using the material thickness and the area related mass [12]. Ductile damage is modelled by the fracture strain and the strain rate defined in the tensile test. The necking is set to 0, as paper does not undergo any necking [13]. Failure is determined by the "displacement at failure" criterion set to 0.6 according to Huttel [13]. The displacement at failure defines the strain deviation between the maximum bearable stress and the stress at fracture. The numerical simulation has been validated for a simplified geometry with respect to wrinkling and achievable drawing depths at 15% humidity. The comparison between experimental and numerical results can be seen on the left side of Fig. 4. The reduced number of wrinkles is a consequence of the number of elements, which is reduced for the benefit of the computation time.

Tailored moisture distribution

Appropriate moisture distributions within the specimen were identified by a partitioning of the specimens. Different variants were considered. Each part of the division was then subjected to the material properties characteristic for specific moisture contents. Four examples are shown and labeled in Table 1. Those examples will be used further on for explanatory purposes in this paper as they show two different ways to partition as well as different moisture gradients within the partitions. Although other gradients were tested those are the most conclusive.

Table 1. Four different moisture distributions and their partitions

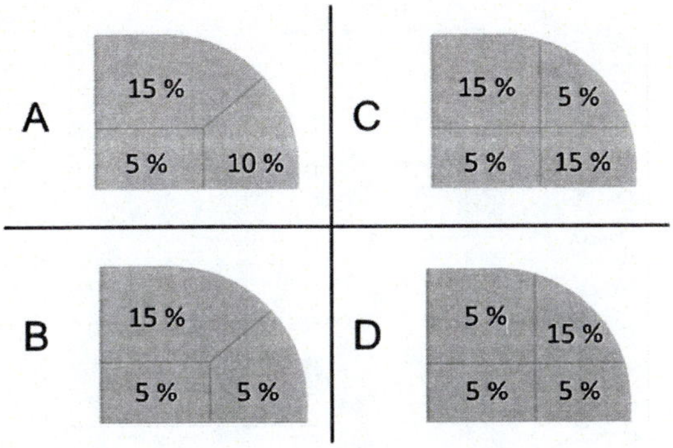

3.2 Results

The geometrically more complex process is modelled with the assumptions described above. The resulting strains are shown on the right side of Fig. 4. The maximum drawing depth was 38 mm. The wrinkle formation can be seen in more details in Fig. 6.

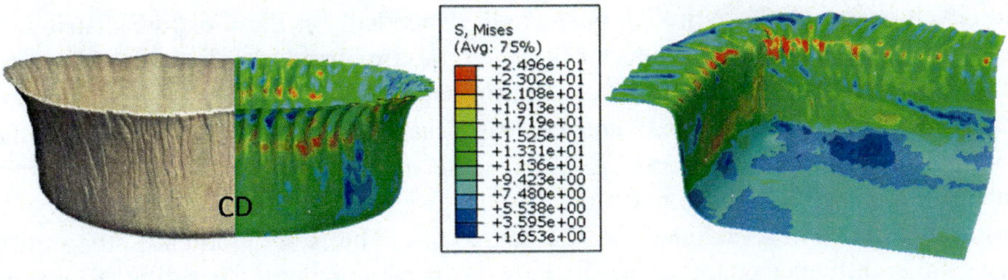

Fig. 4. Verification of numerical simulation (left) and rectangular geometry simulation of paper at 15% humidity (right)

3.3 Influence on the Drawing Depth

Observation of the drawing depth shows that moisture distribution leads to earlier failure based on the "displacement at failure" criterion compared to homogenous moistures of the material specific optimum in all cases. The drawing depths achieved in each case are shown in Fig. 5. The failure took place in the frame area due to rupture. The significant decrease in the achievable drawing depth compared to uniform 15% moisture is attributed to the fact that the maximum elongation can be achieved independently of direction at 15% moisture. A change of this value leads to a lower achievable maximum elongation in the respective direction and thus to earlier failure.

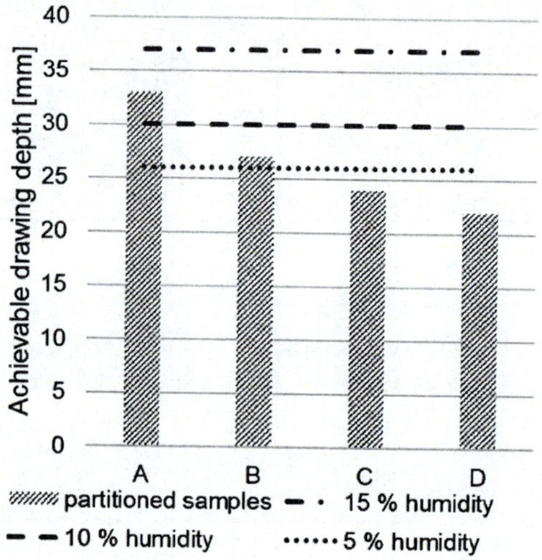

Fig. 5. Achievable drawing depths at different humidities

3.4 Influence on the Wrinkle Formation

The development of the wrinkles is strongly dependent on the moisture distribution and can be controlled and influenced by it. On the one hand, this is evident from the number of wrinkles, but also from their prominence. The wrinkling between variants A and B and between variants C and D are similar in direction, location and number of wrinkles. While A and B resp. C and D differ only by the height of the moisture content in the sections of the partition, A resp. B and C resp. D differ by the way of partitioning. It can be concluded that the degree of the moisture gradient in the sample has a smaller influence on the wrinkling than the distribution of the partition. Therefore, in the following, samples B and D will be considered and compared with each other in order to be able to draw a conclusion regarding the partition distribution. Samples A and C are not discussed further for redundancy reasons. As shown in Fig. 6, B forms more wrinkles than D. Within the same drawing depth of 25 mm and the same number of elements, the specimen with partition B formed 27 wrinkles in the quarter model, the specimen with partition D formed 20 wrinkles and the specimen with 15% homogenous humidity formed 23 wrinkles. Hence, Partition B has more, smaller wrinkles which results in a better appearance. Partition D hat not only the fewest wrinkles but also higher ones, especially in the rounded corner area. In addition, the wrinkles in D run diagonally and have a higher height, which leads to a poorer visual impression.

The wrinkling resulting from B, on the other hand, leads to uniform wrinkles, also in the corner radius, and thus forms an improvement compared to the homogeneous moisture distribution. The better wrinkle appearance by B is due to the fact that the moisture gradient is parallel to the wrinkle formation direction. When partitioning according to C/D, the gradient is diagonal to the direction in which the wrinkles form in the drawing gap. This results in different moisture characteristics and thus different material properties within one wrinkle. The different moisture contents result in diagonal wrinkles, which also affect wrinkles outside the gradient area. As a result, the wrinkles shift, become thicker and lie at an angle, resulting in a poorer surface quality and a deteriorated visual perception of the product.

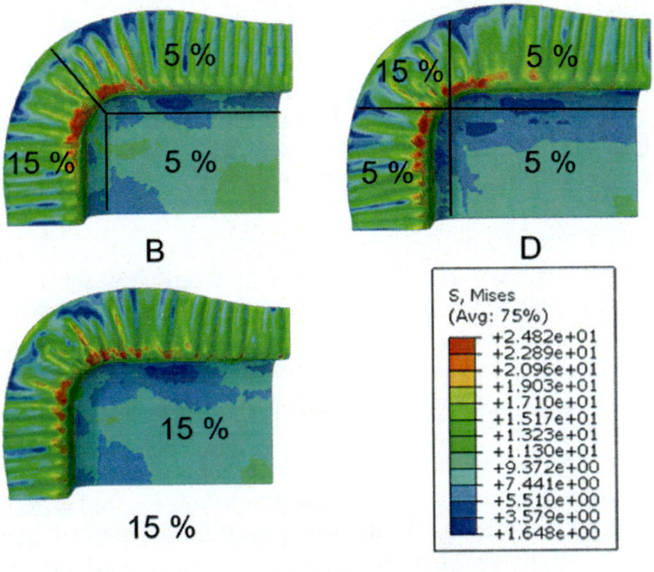

Fig. 6. Wrinkle formation depending on humidity

In variant A/B, however, the wrinkles are located within a moisture range and thus no negative effect is observed on surrounding wrinkles. The gradient probably enhances wrinkling by having areas of varying stress in specimens. The greatest excess material exists in the corner radius, which is why the wrinkling is most pronounced there. Loading of the material to tangential compression in this area causes increased wrinkling. The second most stressed area for compression is the short edge, since the corner radius effect is distributed over a smaller area than in the long edge. At lower moisture levels, the strength of the material is higher, which is why the resistance to wrinkling is higher.

In combination with the hold-down force, a higher resistance probably leads to smaller wrinkles, while a lower resistance leads to fewer but larger wrinkles.

The moisture gradient with decreasing moisture from the long to the short edge thus results in better wrinkle formation, in the sense of more, but smaller wrinkles.

4 Conclusion

It has been shown that moisture distribution within a sample has an effect on the final product in paper board deep drawing. Nevertheless, it must be differentiated whether the more important aim is to achieve the maximum possible drawing depth or to achieve a better surface or appearance. If one works in the limit range of the elongation of a material and thus at the limit of the drawing depth, it is recommended to work with a homogeneous moisture in the range of the material specific optimum, usually 15%. Yet, this is not recommended industrially, since the natural material has a higher degree of inhomogeneity and variations in between as well as within a batch than industrially produced metals, and working in the limit range results in a lot of scrap. On the other hand, moisture distribution can be set on a moisture gradient for packaging products that have a reduced drawing depth. The exact achievable depth has to be validated for each paper using tensile tests and numerical simulations. The resulting moisture distribution, which is adapted to the geometry, makes it possible to obtain the better product. For this purpose, the production times do not have to be extended by the introduction of steam in the process, but the steam introduction of the mold must be suitably designed from the beginning.

By numerically simulating the process prior to die design, significant savings in labor and material resources can be achieved and it can be weighed up whether the better surface justifies the lower drawing depth for the specific product. When designing the moisture distribution, it is important to ensure that the gradient is parallel to the height of the product and thus to the direction of wrinkle formation. Only then, an improved product can be expected. In addition, the gradient should be designed to decrease in the direction of higher compression.

The results presented allow initial approaches for improving the wrinkle appearance of deep-drawn paper products, but there is still potential for further refinement. With regard to anisotropy and drawing depth, new approaches have to be found and tested, since an adjusted moisture distribution did not produce the desired result for this aspect.

References

1. Consultic Marketing und Industrieberatung GmbH Hompeage. https://www.bvse.de/images/pdf/kunststoff/2016/161020_Consultic_Endbericht_2015_19_09_2016_Kurzfassung.pdf. Last accessed 22 July 2022
2. Conversio Market & Strategy Homepage. https://www.vci.de/ergaenzende-downloads/kurzfassung-stoffstrombild-kunststoffe-2019.pdf. Last accessed 22 July 2022
3. Study by Simon-Kucher & Partners Homepage. https://www.simon-kucher.com/de/about/media-center/sustainability-study-2021-fast-ein-drittel-der-deutschen-wuerde-fuer-nachhaltige-produkte-mehr-geld-ausgeben. Last accessed 22 July 2022

4. Franke, W.: Umformung naturbasierter Faserwerkstoffe unter Einflussnahme von Wasserdampf, 1st edn. Shaker Verlag, Düren (2021)
5. Jessen, N., Mushövel, J., Groche, P.: Papier als nachhaltige Alternative in der Verpackungsindustrie. In: 4/2020 VDI Technik und Mensch, pp. 13–14, VDI Bezirksverein Frankfurt-Darmstadt e.V., Frankfurt (2021)
6. Hauptmann, M., Wallmeier, M. et al.: The role of material composition, fiber properties and deformation mechanisms in the deep drawing of paperboard. Cellulose **22**(5), 3377–3395 (2015)
7. Linvill, E., Östsund, S.: The combined effects of moisture and temperature on the mechanical response of paper. Exp. Mech. **54**(8), 1329–1341 (2014)
8. Stein, P.: Individualisierte Formgebung papierbasierter Halbzeuge, 1st edn. Shaker Verlag, Düren (2019)
9. Wallmeier, M., Hauptman, M., Majschak, J.-P.: New methods for quality analysis of deep-drawn packaging components from paperboard. Packag. Technol. Sci. **28**, 91–100 (2014)
10. Stein, P., Franke, W. et al.: Forming behavior of paperboard in single point incremental forming. BioRes **14**(1), 1731–1764 (2019)
11. Wallmeier, M.: Experimental and simulative process analysis of deep drawing of paperboard. Technische Universität Dresden, Dresden (2018)
12. DIN EN ISO 534: Papier und Pappe—Bestimmung der Dicke, der Dichte und des spezifischen Volumens; Beuth Verlag (2012)
13. Huttel, D.: Wirkmedienbasiertes Umformen von Papier, 1st edn. Shaker Verlag, Düren (2015)

Numerical and Experimental Failure Analysis of Deep Drawing with Additional Force Transmission

P. Althaus(✉), J. Weichenhain, S. Hübner, H. Wester, D. Rosenbusch, and B.-A. Behrens

Institute of Forming Technology and Machines (IFUM), Leibniz Universität Hannover, An der Universität 2, 30823 Garbsen, Germany
althaus@ifum.uni-hannover.de

Abstract. Deep drawing is a common forming method, where a sheet metal blank is drawn into a forming die by a punch. In previous research, conventional deep drawing was extended by the introduction of an additional force in the bottom of the cup. The force transmission initiates a pressure superposition in critical areas resulting in a delayed crack initiation. For numerical investigation of the considered process, an accurate modelling of the material failure is essential. Therefore, the parameters of the modified Mohr-Coulomb criterion were identified for the two high-strength steels HX340LAD and HCT600X by means of tensile tests with butterfly specimens. In this research, the fracture modelling is applied in the simulation of deep drawing with and without additional force transmission to enhance the failure prediction. The fracture criterion is validated by experimental deep drawing tests. Finally, the influence of the additional force on the prevailing stress state is evaluated.

Keywords: Sheet metal forming · Deep drawing · Stress-based failure

1 Introduction

Deep drawing is one of the most important manufacturing processes for sheet metal components in the automotive, aerospace and packaging industry. During deep drawing, a sheet metal blank is clamped between a blank holder and a die and formed by a punch. The improvement of deep drawing processes is still the main focus of numerous researches to increase productivity, material utilisation and dimensional accuracy. Furthermore, various possibilities to extend the process limits are being investigated [1]. An extension of the forming limit can be achieved either by strengthening or relieving the force transfer zone in the sheet metal. For a strengthening of the force transfer zone, higher sheet thicknesses or materials with higher tensile strengths can be used [2]. Moreover, locally adapted semi-finished products such as tailored blanks [3] or tailored rolled blanks [4] can be applied to increase the drawing depth.

Relief of the force transmission zone can be achieved by superimposing compressive stresses as this leads to an increased forming capacity of the material. This is already used

M. Liewald et al. (Eds.): WGP 2022, LNPE, pp. 142–151, 2023.
https://doi.org/10.1007/978-3-031-18318-8_15

in forming processes with active media, such as hydro-mechanical deep drawing [5]. Another possibility is the use of a counter punch, which Morishita et al. applied in a deep drawing process with tailored blanks to extend the forming limits [6]. Behrens et al. extended a conventional deep drawing process for the forming of a rectangular cup by using a counter punch, which enables the application of an additional force in the bottom of the cup [7]. Due to the additional force counteracting the forming direction, a pressure-superimposed stress state is created in the bottom and in the transition area from the bottom to the sidewall. The pressure superimposed stress state counteracts the formation of fractures during forming and thus enables an extension of the process window. This was proven in experimental deep drawing tests with two high-strength steels HX340LAD and HCT600 [8]. In a previous study, the forming limit curves of the materials were determined to consider the forming limits in a process simulation. However, necking, which occurs at the radius between the bottom and the sidewall, could not be reliably predicted [9].

In this research, the ductile fracture criterion modified Mohr-Coloumb (MMC) proposed by Bai and Wierzbicki [10] is applied in the process simulation to improve the failure prediction. Peshekodhov et al. proved the suitability of the criterion for a conventional deep drawing process, where cracks occurred in the transition area from the flange to the sidewall [11]. In this work, the criterion is applied to predict the fracture initiation in the bottom radius of the cup during deep drawing with and without additional force transmission. The goal is to improve the fracture modelling in order to fully exploit the potential of this technology for an extension of the process limits. For a validation, experimental deep drawing test are carried out with a special tool setup, which enables the introduction of the additional force in the bottom of the cup.

2 Materials and Methods

2.1 Experimental Setup

For the experimental investigation of deep drawing with additional force transmission and for the validation of the ductile fracture model, experimental tests were carried out. The forming tool shown in Fig. 1 essentially corresponds to the design of a conventional deep drawing tool. This includes the punch, the bottom die and the blank holder.

Additional force was applied to the bottom of the cup by the upper die, which is preloaded by means of a gas pressure spring. A load cell, which is located at the lower end of the punch, allows the punch force to be recorded during the forming process. With the use of a position sensor, force-displacement curves can be evaluated. The tool enables the deep-drawing of rectangular cups in the dimensions of 160 mm x 80 mm and a drawing depth of 55 mm.

The experimental setup was used to determine the process limits regarding the formation of fractures in the deep drawing process with and without additional force transmission as a function of the applied blank holder force. The sheet materials used are HX340LAD+Z (1.0933, short: HX340) and HCT600X-Z100MBO (1.0941, short: HCT600). Both sheet materials are galvanized and have a thickness of 1 mm. First, the sheet metals are cut to size as octagonal blanks. The outer dimensions were 281 mm × 213 mm, corresponding to a deep-drawing ratio of 1.9. The drawing depth of the cups

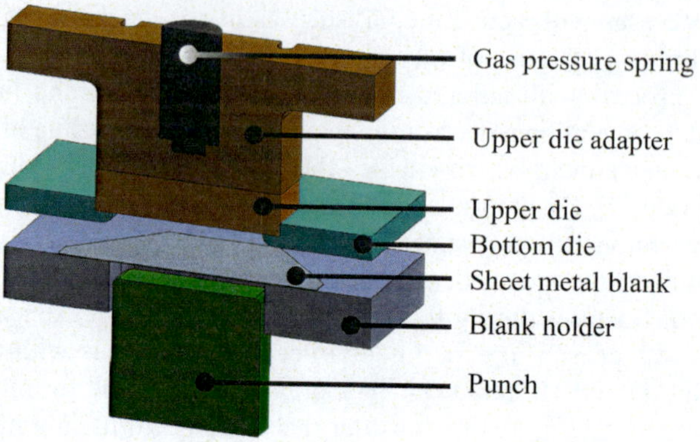

Fig. 1. Tool setup for deep drawing with additional force transmission

was 55 mm and the forming speed was 10 mm/s. In order to ensure uniform tribological conditions, the blanks were lubricated before the deep drawing process with the aid of a spray lubrication system from RAZIOL Zibulla and Sohn GmbH. Raziol CLF 180 was used as the lubricant. The maximum blankholder force at which the components do not fracture was determined both without and with additional force transmission. Therefore, the blank holder force was successively increased in 5 kN steps until a fracture occurred during forming. When a crack appeared, the blank holder force was decreased by 5 kN and the deep drawing was repeated several times. If no crack appeared in five repetitions, the maximum blank holder force was identified. If a crack appeared in the repeated attempts, the blank holder force was again decreased by 5 kN.

2.2 Numerical Validation

For the numerical investigation of the deep drawing process, a simulation model was set up in Abaqus/Explicit, which is shown in Fig. 2 (a). In accordance to the experimental setup, the model consists of the sheet metal, which is clamped between the bottom die and the blank holder, and the forming punch.

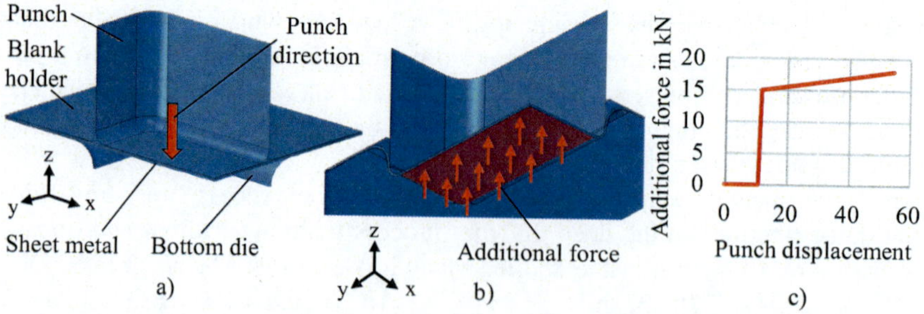

Fig. 2. Numerical simulation model (a) with additional force transmission in form of a pressure field in the bottom of the cup (b) depending on the punch displacement (c)

The simulation was divided into two steps. At first, the force of the blank holder was applied on the sheet metal. Subsequently, deep drawing was carried out by moving the punch with a velocity of 10 mm/s until the final drawing depth of 55 mm is reached. To reduce the computational effort, the symmetry of the process was used and only a quarter of the geometry was depicted.

Instead of the upper die for the additional force transmission, a pressure field is defined at the bottom of the cup. The pressure is activated when the drawing depth of 12 mm is reached. In the real experiment, the force is applied by a gas spring, which is preloaded with 150 bar. This corresponds to an initial force of 15 kN. Due to the increase of pressure in the spring, the force increases to approx. 18 kN at the end of forming (55 mm). The defined pressure field is shown in Fig. 2 (b).

The sheet metal was meshed with solids (C3D8) with an average element edge length of 0.4 mm and three elements across the thickness. The tools were modelled as rigid shells and were also meshed with an element edge length of 0.4 mm. For the sheet materials, elastic-plastic models were defined based on the results of a previous study, where tensile tests were carried out at room temperature according to DIN EN ISO 10002 with the specimens aligned at 0°, 45° and 90° to the rolling direction [7]. The flow curves were extended by means of hydraulic bulge tests for an enhancement of the extrapolation accuracy. For HX340, the approach according to Gosh and for HCT600 a combined approach of Swift and Hocket-Sherby was chosen, because they provided the best approximation to the experimental data. Furthermore, the yield criteria according to Hill was defined for both materials [12]. The flow curves and the yield surfaces with the corresponding parameters are given in Fig. 3.

The failure behaviour of the considered materials was modelled by means of the MMC criterion, which defines the equivalent plastic strain at fracture according to the following equation:

$$\varepsilon_{\mathrm{pl}}^{\mathrm{f}} = \left\{ \frac{\frac{A}{c_2}\left[c_{\Theta}^{\mathrm{s}} + \frac{\sqrt{3}}{2-\sqrt{3}}\left(c_{\Theta}^{\mathrm{ax}} - c_{\Theta}^{\mathrm{s}}\right)\left(\sec\left(\frac{\overline{\theta}\pi}{6}\right) - 1\right)\right]}{\left[\sqrt{\frac{1+c_1^2}{3}}\cos\left(\frac{\overline{\theta}\pi}{6}\right) + c_1\left(\eta + \frac{1}{3}\sin\left(\frac{\overline{\theta}\pi}{6}\right)\right)\right]} \right\}^{-1/n} \tag{1}$$

Here, the fracture strain is depending on the stress triaxiality η, the normalised Lode angle $\overline{\theta}$ and six material-specific parameters. A and n were identified by fitting the flow curves of the materials to the Swift hardening model. The parameters c_{Θ}^{ax} and c_{Θ}^{s} were left at their default value of 1 analogous to [10]. For the remaining parameters c_1 and c_2, tensile test with butterfly specimens were carried out and simulated in a previous study [13]. The parameters were fitted to the experimental data by reducing the sum of the least squares between experimental plastic strain and predicted plastic strain by the criterion. The resulting parameters are given in Table 1. The corresponding fracture curves are shown in Fig. 4.

For contact between the tools and the sheet metal, "surface to surface" contact was used and coulomb´s law with a friction coefficient of 0.1 was defined for the punch and the blank holder. Since the sheet metal was lubricated on the side to the die, the

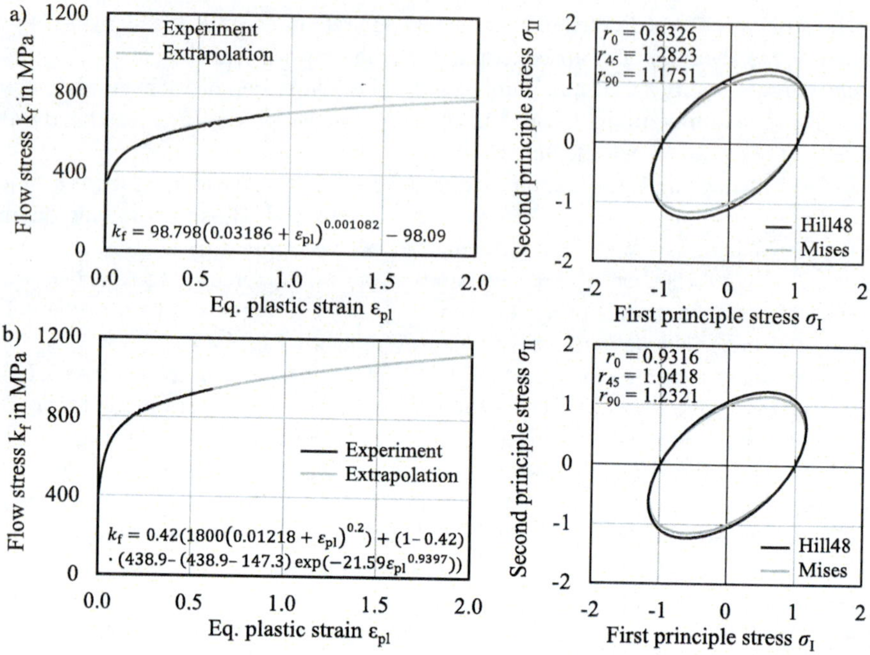

Fig. 3. Flow curves and yield surfaces for HX340 (a) and HCT600 (b) [7]

Table 1. Parameter of the used MMC criterions

	A	n	c_1	c_2
HX340	713.0	0.165	0.208	401.934
HCT600	1018.6	0.143	0.166	560.033

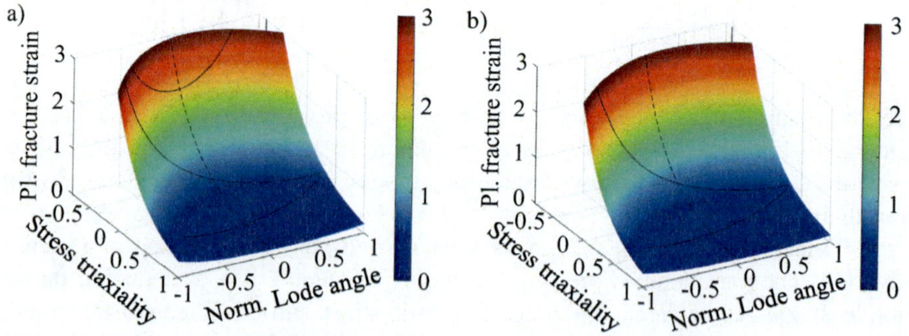

Fig. 4. Plastic fracture strain in the space of stress triaxiality and normalized Lode angle for HX340 (a) and HCT600 (b) based on the MMC criterion

coefficient was reduced to 0.05. To keep the computational effort reasonable, a mass scaling factor of 100 and a time scaling factor of 10 was applied. The application of the scaling factors was tested in an exemplary simulation and showed no significant effect on the results.

3 Results

3.1 Experimental Process Limits

The diagram in Fig. 5 (a) shows the maximum determined blank holder force for deep drawing of HX340 and HCT600. The blank holder force and thus the process window for flawless cups could be increased from 280 to 320 kN for HX340 and from 195 to 230 kN for HCT600 by using additional force transmission. To show the effect of the additional force, exemplary cups made of HX340 are shown in Fig. 5 (b), which were formed with a blank holder force of 320 kN with and without additional force.

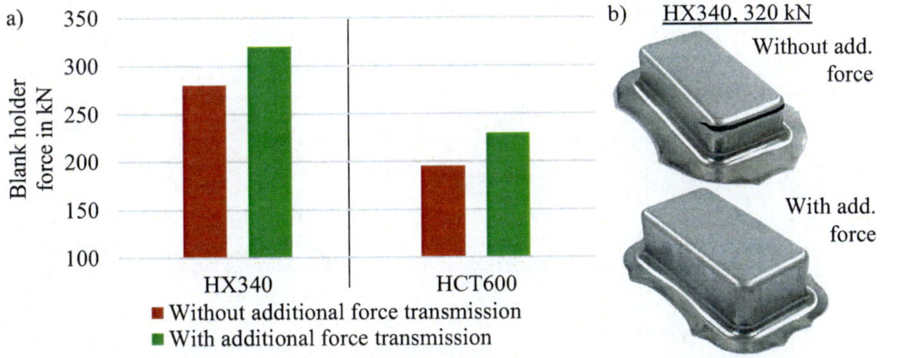

Fig. 5. a) Maximum blank holder force for HX340 and HCT600 and b) cups made of HX340 with a blank holder force of 320 kN with and without additional force transmission.

3.2 Numerical Validation

In Fig. 6, the numerical and experimental force displacement curves are shown for HX340 with Fig. 6 (a) and without Fig. 6 (c) additional force transmission for blank holder forces of 280 and 320 kN. The results of HCT600 are shown in Fig. 6 (b) and Fig. 6 (d) for blank holder forces of 195 and 230 kN. At the beginning of forming, the punch force increases linearly until a drawing depth of approx. 16 mm is reached. Good agreement was achieved by the simulations in the first forming stage. Subsequently, the force measured in the experiments reaches its maximum, followed by a slight decrease until the maximum drawing depth is achieved. In this stage, the force is first overestimated by the simulation, whereas it is underestimated at the end of the forming process. This can presumably be attributed to the fact that a constant friction coefficient is assumed in the simulation, while the friction coefficient is influenced by various factors, such as contact pressure or the thickness of the lubrication film [14]. Another reason for the deviations could be fluctuations between material batches.

The best agreement with a maximum deviation of 5.70% was achieved by the simulation with HX340 and without the additional force transmission, while the largest maximum deviation of 11.38% occurred in the simulation with HCT600 and without additional force in the bottom of the cup. The activation of the additional force is visible by a sudden increase of the punch force at a drawing depth of 12 mm.

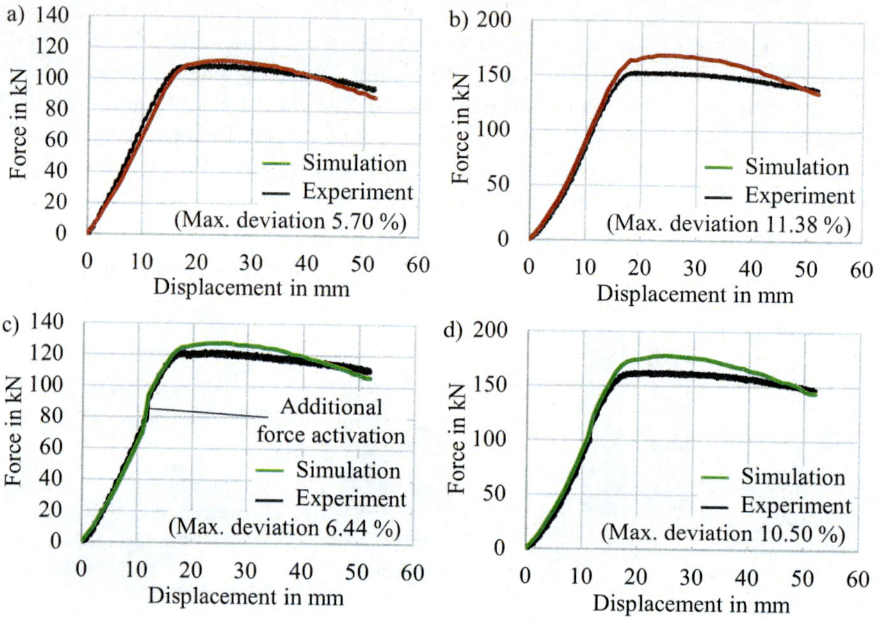

Fig. 6. Force-displacement curves of HX340 with (a) and without (c) additional force as well as HCT600 with (b) and without (d) additional force transmission

In the next step, the prediction accuracy by the MMC criterion was evaluated by comparing the material damage predicted by the simulation to the experimental results without additional force transmission. For HX340, a blank holder force of 320 kN was chosen, because fractures occurred in every experimental test. For HCT600, a corresponding blank holder force of 230 kN was simulated. Figure 7 compares the fractures of the experimental deep drawn cups to the distribution of the damage variable D calculated in the simulation. D is defined according to the following equation:

$$D = \int \frac{d\varepsilon^{\mathrm{pl}}}{\varepsilon_{\mathrm{D}}^{\mathrm{pl}}\left(\eta, \xi(\theta), \dot{\varepsilon}^{\mathrm{pl}}\right)} \tag{2}$$

Here, $\varepsilon_{\mathrm{D}}^{\mathrm{pl}}$ is the equivalent plastic strain at the onset of fracture according to the MMC criterion. It depends on the stress triaxialty η, the plastic strain rate $\dot{\varepsilon}^{\mathrm{pl}}$ and the third stress invariant ξ, which is depended on the Lode angle. Material damage is present, when D reaches or exceeds the value of one. As shown in Fig. 7 (b) for HX340 and Fig. 7 (d) for HCT600, the MMC criterion predicts the initiation of a fracture in the radius at the bottom of the cup. This shows a good agreement regarding the location of the fractures in the experiments. In comparison to the simulation, the experimental deep drawn cups show more progressed cracks, which are present along the entire width of the cups. This crack propagation is not represented by the simulation, since no damage evolution or element removal is taken into account. A comparison of the exact time of fracture initiation difficult, because the experiments had to be stopped manually, when a fracture occurred. Due to the fast processing speed, it was not possible to stop the punch right in time of the fracture initiation. Furthermore, it can be noticed that in the simulations material damage also accumulates in the corner of the sidewall. However no fractures develop in this section, which is in accordance to the experiments. Therefore, the MMC criterion is considered to be well suited for the prediction of fractures in the bottom of the cups during deep drawing.

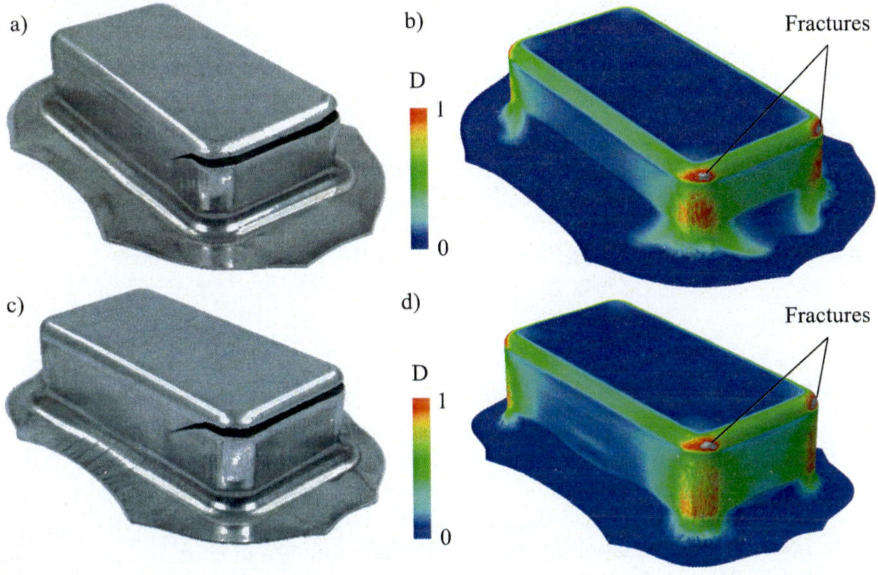

Fig. 7. Experimental and numerical fracture initiation for HX340 (a, b) and HCT600 (c, d)

Subsequently, the simulations shown in Fig. 7 were carried out with the additional force by activating the pressure field in the bottom of the cup. The distribution of the damage variable D is shown in Fig. 8 (a) for HX340. It can be seen, that no fractures are predicted in the simulation with additional force transmission, which is in accordance to the experimental results. For a better understanding of the effect of the additional force, the equivalent plastic strain and the stress triaxiality are evaluated at five element in the radius of the cup, where fractures occurred without the additional force. The averaged values of the elements are shown in Fig. 8 (b) for HX340 and in Fig. 8 (c) for HCT600.

Due to the pressure superposition, the stress state shifts to lower stress triaxialities. Thus allowing higher plastic strains to be reached before material failure occurs.

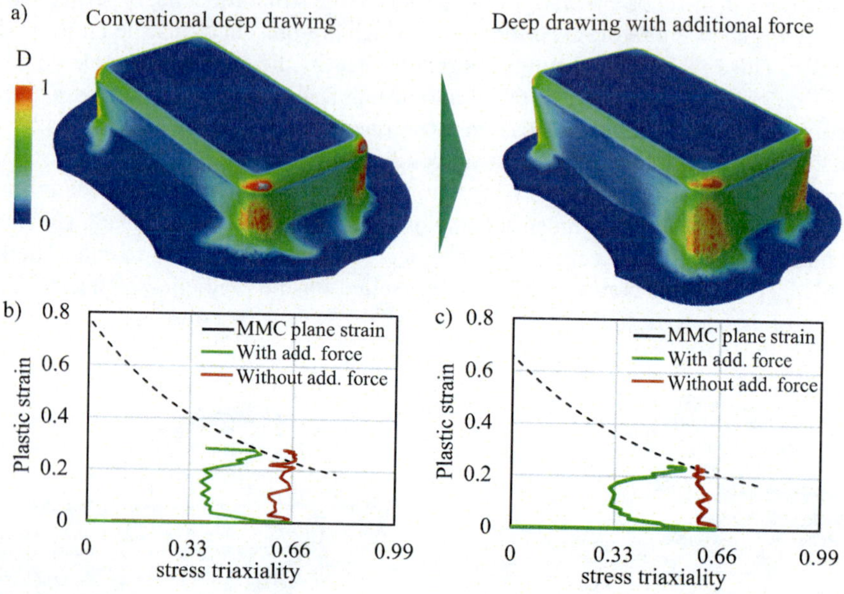

Fig. 8. Simulation results for HX340 without and with additional force transmission (a) and evolution of the plastic strain over stress triaxiality for HX340 (b) and HCT600 (c)

4 Conclusion and Outlook

This contribution deals with the investigation of an extended deep drawing process, where an additional force is applied in the bottom of the cup by a preloaded upper die. Experimental deep drawing test were carried out with the high-strengths steels HX340 and HCT600 under a variation of the blank holder force to identify the process limits regarding fractures in the bottom of the cup. It was found, that the maximum blank holder forces of 280 kN for HX340 and 195 kN for HCT600 could be extended by the additional force transmission to 320 kN and 230 kN, which corresponds to an increase of 14.3% and 17.9%, respectively. Due to the increase of the maximum blank holder force, larger deep drawing ratios can be realised, thus an extension of the process window has been achieved.

Numerical simulations were carried out to predict the material failure by the application of the MMC criterion. The fracture criterion showed good agreement to the experimental results regarding the location of the fracture initiation. Furthermore, the extension of the process limits due to the additional force transmission was successfully considered. The delayed fracture initiation could be attributed to lower stress triaxialities in the bottom radius of the cup, resulting in higher tolerable fractures strains.

Acknowledgements. The results presented were obtained in the project "Extension of the forming limits during deep drawing by additional force transmission" – 212270168. The authors thank the German Research Foundation (Deutsche Forschungsgemeinschaft, DFG) for their financial support.

References

1. Wifi, A., Abdelmaguid, T., El-Ghandour, A.: A review of the optimization techniques applied to the deep drawing process. In: 37th International Conference on Computers and Industrial Engineering (2007)
2. Kumar, D.R.: Formability analysis of extra-deep drawing steel. J. Mater. Process. Technol. **30–131**, 31–41 (2002)
3. Merklein, M., Maren, J., Lechner, M., Kuppert, A.: A review on tailored blanks—production, applications and evaluation. J. Mater. Process. Technol. **214**(2), 151–164 (2014)
4. Meyer, A., Wietbrock, B., Hirt, G.: Increasing of the drawing depth using tailored rolled blanks—numerical and experimental analysis. Int. J. Mach. Tools Manuf. **48**(5), 522–531 (2007)
5. Zhang, S.H., Danckert, J.: Development of hydro-mechanical deep drawing. J. Mater. Process. Technol. **83**(1–3), 14–25 (1998)
6. Morishita, Y., Kado, T., Abe, S., Sakamoto, Y., Yoshida, F.: Role of counterpunch for square-cup drawing of tailored blank composed of thick/thin sheets. J. Mater. Process. Technol. **212**(10), 2102–2108 (2012)
7. Behrens, B.-A., Bonk, C., Grbic, N.,Vucetic, M.: Numerical analysis of a deep drawing process with additional force transmission for an extension of the process limits. IOP Conf. Ser. Mater. Sci. Eng. **179**(1), 012006 (2017)
8. Behrens, B.-A., Bouguecha, A., Bonk, C., Grbic, N., Vucetic, M.: Validation of the FEA of a deep drawing process with additional force transmission. AIP Conf. Proc. **1896**(1), 080024 (2017)
9. Behrens, B.-A., Bouguecha, A., Bonk, C., Rosenbusch, D., Grbic, N., Vucetic, M.: Influence of the determination of FLC's and FLSC's and their application for deep drawing process with additional force transmission. In: Proceedings of 5th International Conference on Advanced Manufacturing Engineering and Technologies, pp. 405–417 (2017)
10. Bai, Y., Wierzbicki, T.: Application of the extended Coulomb-Mohr model to ductile fracture. Int. J. Fract. **161**(1), 1–20 (2010)
11. Gladkov, Y., Peshekhodov, I. A., Vucetic, M., Bouguecha, A., Behrens, B.-A.: Implementation of the Bai & Wierzbicki fracture criterion in QForm and its application for cold metal forming and deep drawing technology. In: MATEC Web of Conferences, vol. 21, pp. 12009 (2015)
12. Hill, R.: A theory of the yielding and plastic flow of anisotropic metals. Proc. Roy. Soc. Lond. **193**(1033), 281–297 (1984)
13. Behrens, B.-A., Rosenbusch, D., Wester, H., Althaus, P.: Comparison of three different ductile damage models for deep drawing simulation of high-strength steels. In: IOP Conference series. Materials Science and Engineering. 1238 012021 (2022)
14. Merklein, M., Zöller, F., Sturm, V.: Experimental and numerical investigations on frictional behaviour under consideration of varying tribological conditions. In: Advanced Materials Research, vol. 966–967, pp. 270–278 (2014)

Efficient Digital Product Development Exemplified by a Novel Process-Integrated Joining Technology Based on Hole-Flanging

D. Griesel, T. Germann(✉), T. Drogies, and P. Groche

Institut Für Produktionstechnik Und Umformmaschine, Technische Universität Darmstadt, Otto-Berndt-Str. 2, 64287 Darmstadt, Germany
germann@ptu.tu-darmstadt.de

Abstract. Increasing weight and energy efficiency requirements drive the use of novel composite materials. For this, metal-polymer sandwich plates are particularly promising. However, their widespread application is hindered by limited formability and a lack of efficient joining technology due to the combination of materials with vastly different mechanical properties. This paper presents an innovative joining element, addressing the special characteristics of sandwich composites by locally compensating the influence of the polymer core. The presented joining elements act as "lost punches" inserted into the sandwich material, opening up the possibility of manifold connection possibilities. In a two-stage hole-flanging process, which can be realized with conventional forming machines, a form and force fit between the punch and the sandwich composite is established. This challenging forming process is discussed, extensively numerically analyzed and the joint geometry is optimized with respect to the resulting joint strength. Furthermore, the achievable limit loads are discussed. Concluding, first prototypes offer an outlook on the industrial application.

Keywords: Integrated functional structures · Collar forming · Sandwich sheets

1 Introduction

Due to increasing demands on structural components in terms of weight and energy savings, lightweight construction with novel composite materials is increasingly gaining attention. By a load-appropriate design, the composite's specific stiffness can be significantly increased compared to monolithic materials. For example, sandwich sheets consisting of stiff cover sheets and a shear-soft core are particularly suitable for bending, as this stress profile allows high material utilization [1]. Sandwich sheets have structured cores (e.g. honeycomb or corrugated structures), solid or foam cores made of metal, plastic or even wood or paper [2]. For many engineering applications, sheets with metallic covers and polymeric cores are particularly suitable [3]. In addition to the benefits of lightweight construction, depending on their composition, sandwich structures offer attractive NVH-properties [4] or have thermal insulation effects [5].

M. Liewald et al. (Eds.): WGP 2022, LNPE, pp. 152–161, 2023.
https://doi.org/10.1007/978-3-031-18318-8_16

Despite all these benefits, sandwich sheets are lacking widespread industrial adoption. The vastly different material properties combined in a single part present a challenge for the forming, compare [6, 7], and joining of sandwich sheets. The design of the manufacturing process is always very close to the process limits [8]. For joining, conventional techniques are only applicable to a limited extent and are presented with additional challenges [5]. Usually, only the cover layers are joined together, either directly or by means of a surrounding structure. Adhesive bonding is performed analogous to conventional sheets [9]. However, this requires expensive and laborious surface pretreatments [10]. Thermal processes to join the cover are mainly used for sandwich sheets with metallic cores [9]. Bolted connections can only achieve limited strength in sandwich composites, because soft cores can barely be pre-stressed without being damaged. Only for suitable core materials, e.g. metallic honeycomb cores, and if local degradation of the core is tolerable pre-stressed connections can be created [11]. Whereas polymer cores have a creep tendency that can lead to a loss of the preload [12]. The integration of local inserts into the core material is associated with considerable additional work [9] and is not possible when processing semi-finished sandwich products. Concluding, these challenges require a complex design process with extensive iterative loops, which must be completed anew for each application [8].

An efficient solution is a simplified, local joining connection which can be used universally for flexible workpiece geometries and thus solves the global problem of design. The targeted form-fit connection enables this efficient joining process. The final joining elements may be manufactured with low unit costs as a mass-produced component similar to bolts. However, their design is a complex process where the constraints are defined by the sandwich structure. In addition, the equipment and tools required to manufacture joining elements are highly cost-intensive. As a result, an iterative empirical design of the joining elements [8] is hardly feasible.

A more economical approach both in terms of time and cost is the usage of digital design processes. Based on experimental material data, a digital twin of the sandwich material is generated and validated by suitable trials. Then, the required iterative loops of the design process for the joining element are carried out on the digital twin [13]. Three major advantages can be achieved this way. Firstly, only by using parameterized models, a full variation of all joining elements characteristic values, as well as its material, is enabled. Restrictions, as in a real manufacturing process, do not need to be considered. Secondly, due to the control over all parameters, individual influences can be analyzed even in the case of strong coupling of individual variables in reality. Thirdly, the complexity of sandwich composite material, joining element geometry and necessary production steps results in a complexity and multitude of process influencing variables, which makes a reproducible investigation of selected parameters almost impossible. Thus, the digital twin represents the ideal solution for an efficient design. Conducting selected validation tests is then sufficient to conclude the design process.

In this paper, this approach is carried out for the development of a process-integrated joining connection in sandwich sheets. In addition to the geometry of the joining element to be inserted, the process feasibility and the performance of the resulting joint must also be optimized. All three parameters are strongly interdependent. The central question is whether the digital development of a complex joining element is possible.

2 Approach

2.1 Process-Integrated Hole-Flanging for Sandwich Structures

As discussed earlier, the key to joining sandwich sheets is to introduce the necessary loads into both sandwich cover layers in a material-appropriate way. Furthermore, the influence of the polymer core must be compensated to avoid a loss of preload.

This can be achieved in a process-integrated matter by combining a rivet-like joining element with hole-flanging of a pre-pierced sandwich sheet, as shown in Fig. 1.

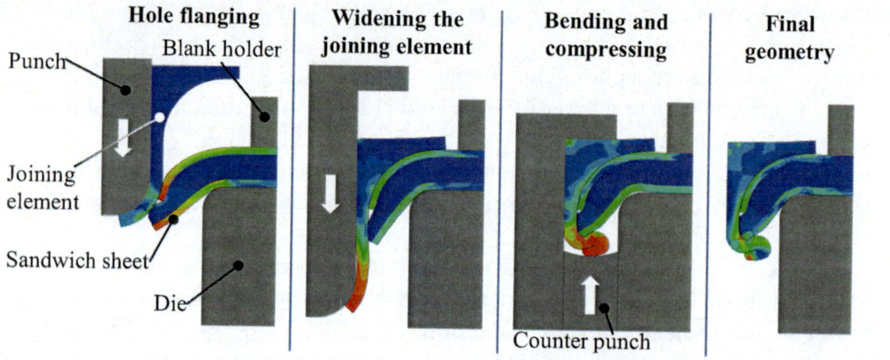

Fig. 1. Sequence of the joining process under investigation (qualitative, for illustration purposes only)

Hole-flanging changes the geometry of the connection point so that the forces exerted by the joining element are redirected as longitudinal forces into the sandwich plane, effectively stiffening the joint and reducing the influence of the shear soft core. By bending the joining element around the flange edge and compressing it, the cover sheets are pressed together and prevented from moving relatively to each other when loaded, further increasing the joint stiffness. In addition, clamping the flange with the joining element ensures that loads are introduced into both cover layers.

With regard to the detailed design of the composite, a multitude of boundary conditions need to be taken into account. In principle, the joining elements and their material thicknesses and stiffnesses must be adjusted to the sandwich. Thus, sufficient ductility for the required formability must be ensured in the joining element. In preliminary tests with elements made of aluminum or brass, for example, showed insufficient formability. Tough steel grades, such as 1.4301, is recommended. With regard to the sandwich composite, a design is required that is capable of withstanding the loads during collar forming. Comprehensive investigations were carried out for this purpose [14]. The desired failure criterion is a sandwich structure that is less stiff than the joining element. If the failure limit is exceeded, the sandwich composite deforms before the joining element pulled out. Also, the joining elements collar height and length have to be coordinated to ensure an optimal form and force fit. Aim of the current investigation is a proof of concept. Detailed investigations of process limits are carried out at a later stage.

2.2 Numerical Process Simulation

The axisymmetric FE model for this process is developed in Abaqus/CAE, utilizing the axial symmetry of the hole-flanging and the joining processes. The model includes a punch, die, blank holder and counter punch as well as the joining element and sandwich sheet. As the process forces and therefore the tool deformations are relatively low and the counter punch surface is hardened for wear reduction, the tools are modeled as rigid parts. Since the large deformations of the polymer core are challenging to simulate, an explicit solver was used. The forming steps are implemented as shown in Fig. 1, where all tool movements are displacement-controlled. The sandwich sheet is implemented as a deformable part with elastic-plastic material properties. It is discretized as a single part with volume elements (CAX4R), which is partitioned into three layers with different material properties. This is equivalent to a perfect adhesive joint and disregards delamination effects. Since previous investigations have confirmed that either the steel sheets or the polymer core fails before delamination can occur at the interface, this ideal joint is a valid simplification. A constant mesh size of 0.06 mm was chosen, which corresponds to 5 elements over the cover sheet thickness [14]. The investigated sandwich material consists of two steel (CR300IF) cover sheets with a thickness of 0.3 mm and a PE/PA core with a thickness of 1 mm, resulting in a total thickness of 1.6 mm.

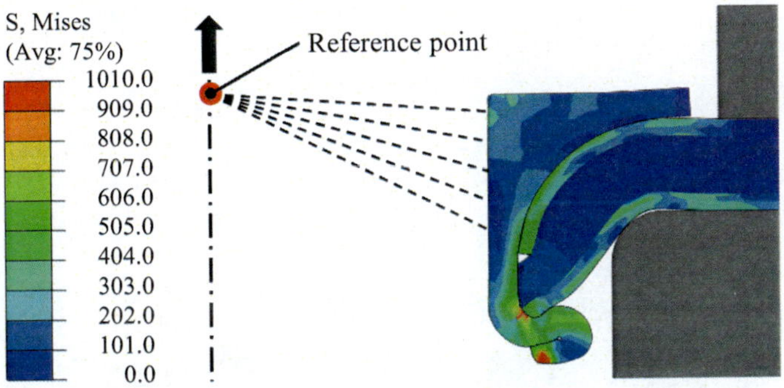

Fig. 2. Simulative test of the joint strength, dotted lines represent a rigid connection as a coupling boundary condition.

The joining element is also implemented as a deformable part (CAX4R) with elastic-plastic material properties. A global mesh size of 0.1 mm has been chosen to guarantee at least 5 elements in the thickness direction of the deformed section. After forming, a separate testing step is conducted numerically to determine the performance of the resulting joint. In this step, a reference point is coupled to the inner surface of the joining element, simulating a thread, and moved by 2 mm in the vertical direction, as shown in Fig. 2. The force required for this motion is then recorded. The entire model was parameterized using the Python interface in Abaqus, which allows varying all process parameters quickly, e.g. including material properties, tool diameters, corner radii, shape of the joining element, travel distances.

In the present study, the joining elements height, counter punch travel and shape were varied and resulting joints were assessed in terms of joint strength and visual impression of the forming result. The evaluated joining elements have an inner diameter of 8 mm, an outer diameter of 15 mm, a thickness of 0.5 mm and a total height of 7…10 mm.

3 Results and Validation

Depending on the shape of the tool and joining element, two different forming phenomena are observed, see Fig. 3.

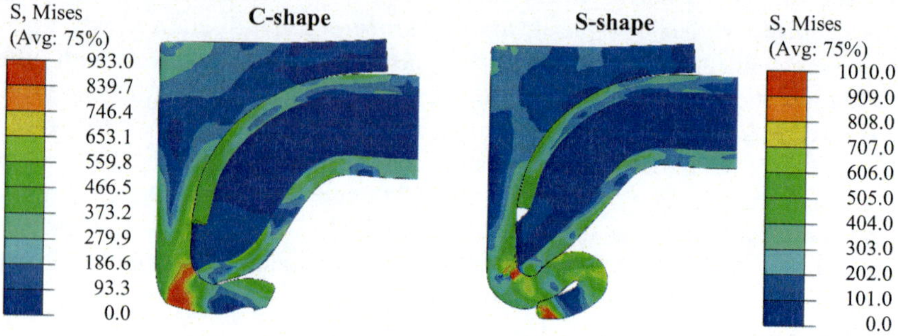

Fig. 3. Different forming results of the joining element.

This is primarily influenced by the free length and therefore the bulge stiffness of the joining element prior to its compression. The C-shape is observed for shorter joining elements (total height 7 mm), while the S-shape occurs for longer ones (total height 8…10 mm). No influence of the final height after compressing is observed.

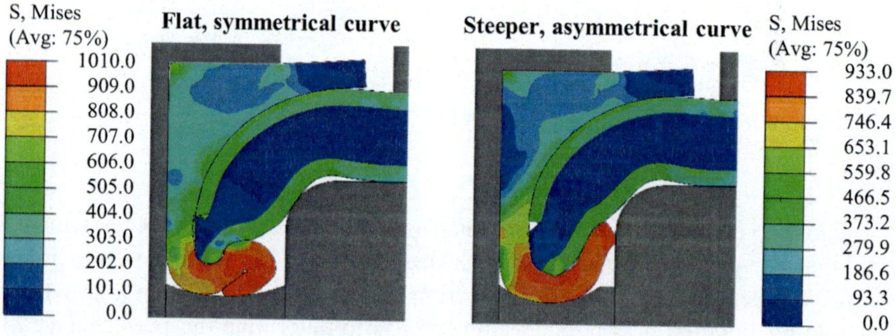

Fig. 4. Influence of counter punch shape for a joining elements height of 8 mm.

The forming results can be further controlled by the counter punch shape, see Fig. 4. Flat curves tend to cause bulging which leads to an S-shape while steeper curves guide the material outwards causing a C-shape. Asymmetrical curves can be used to create a steeper curve because only the inner part of the counter punch takes part in forming the joining element.

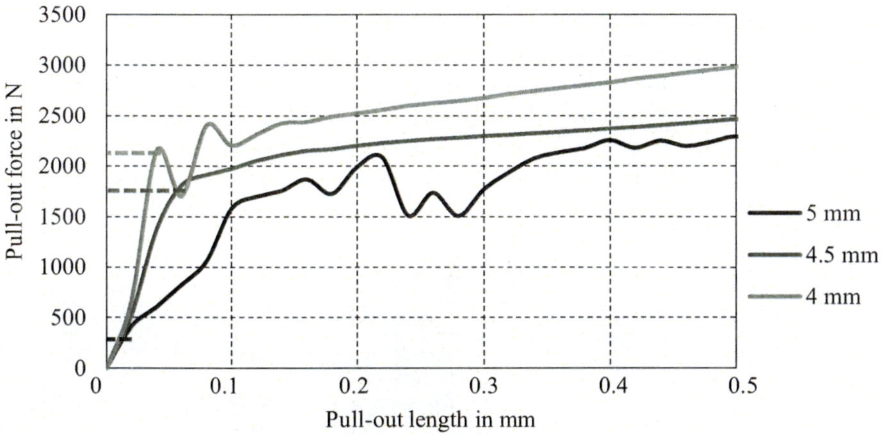

Fig. 5. Simulative comparison of the pull-out force for a joining element with collar height of 10 mm and different final heights after compressing

Figure 5 shows the test results of simulated pull-out tests for joining elements that have been compressed a different amount. The horizontal lines mark the point where the preloads caused by the compression are overcome. Since the force fit is lost at this point, this is considered the maximum achievable joint strength. Further increasing the pull-out force eventually results in plastic deformation of the sandwich sheet and joining element. The joining element with a final height of 5 mm was not compressed completely, so no significant force fit was achieved. While a higher compression and thus a higher preload leads to higher maximum forces, this effect is limited by the flange geometry. Compressing the joint too much leads to unwanted deformation of the top of the joining element, mechanical failure of the sandwich sheet and flattens the desirable flange in the sandwich sheet too much (Fig. 6).

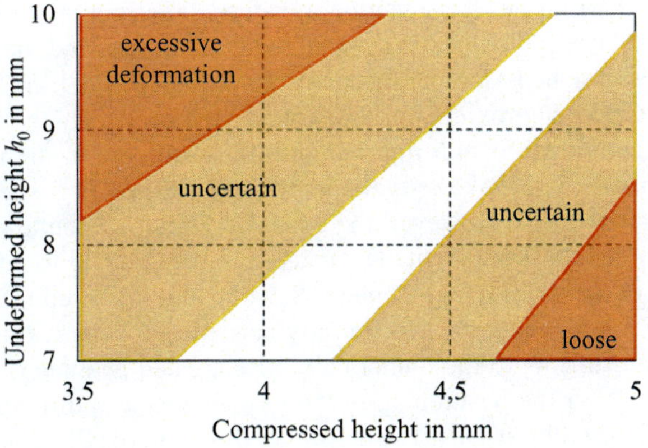

Fig. 6. Process window for the compression of the joining element. The situation at the grid points was simulated, the colored areas interpolated from the simulation results.

For the experimental validation of the gained results from the digital twin, the discussed case of pulling the joining element will be considered. For this purpose, prototypes corresponding to the process step shown in Fig. 1 are produced. Joining elements with a total height of 10 mm are used, which are made of 1.4301. Subsequently, a tension bolt is inserted into the central opening and welded with three spot welds around the circumference between the bolt and the joining elements, as shown in Fig. 7.

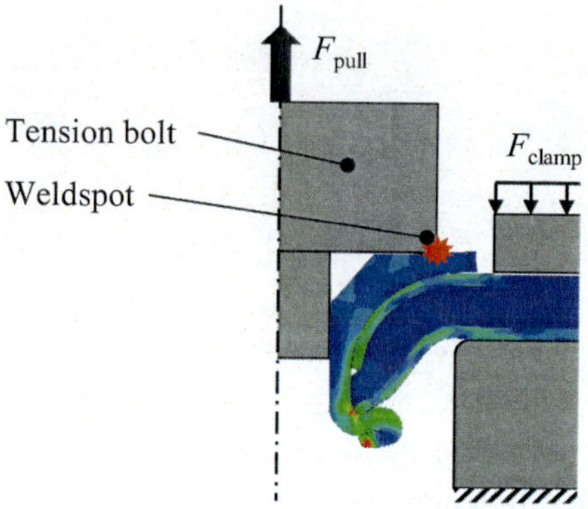

Fig. 7. Prototype test setup: Tension bolt is fixed with three spot welds.

The sandwich composite is fixed in a tensile-compression test bench and the tensile bolt is fixed in the traverse of the tension-compression test bench. Then, the bolt is pulled out at a controlled speed of 5 mm/min and the required tensile force F_{pull} is measured. The longitudinal measurement is conducted by means of a video extensometer. Initially, an elastic deformation of the sandwich composite is visible in the tensile direction. After exceeding a buckling point, failure occurs. Thus, the sandwich sheets collar flips in the tensile direction and a short-term static force level is formed. Failure is indicated by the red line in Fig. 8. In the subsequent phase, the tensile force increases further until the spot welds fail and no further force can be transferred. During this phase, the joining element is continuously pulled out of the sandwich composite and the C- or S-shaped joining elements is stretched. Comparing the results in Figs. 5 and 9, the failure threshold is consistent with close approximation. For four tested specimens with 10 mm initial height and a formed height of 4.5 mm height, the results are in the range of ± 55 N around the numerical value. Likewise, the numerically expected S-shape of the collar is achieved in all specimens. However, the considerably slower build-up of the forces over the extension length is apparent. The cause can be assumed to be a considerably reduced stiffness of the real joining element. Several superimposed influences suggest this conclusion. Preliminary tests are currently carried out using a simplified test rig, which requires adjustments to the numerically investigated geometry. This includes a chamfer ($1.0 \times 45°$) for the positioning in the tool as well as a different point of load application during the pull-out tests. The reason for this was the failure of the epoxy resin adhesive originally planned for the tensile tests. Therefore, the intended and numerically modeled force application (see Fig. 2) on the inner surface of the joining element had to be replaced by a connection via spot welds on the edge of the top surface of the joining element (see Fig. 7). Due to the very low thickness of the element top of only 0.5 mm, there is a low bending stiffness here.

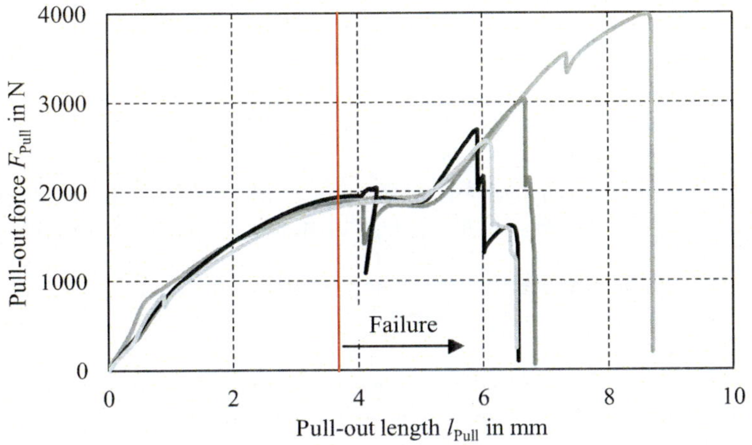

Fig. 8. Comparison of 4 tensile specimens with 10 mm collar height. Shortly below 4 mm, pull-out failure by plastic deformation begins.

The individual steps of the test are shown in Fig. 9. Here, in addition to the metallographic micro section of the joined collar element, a representation of the inserted tension bolt can be seen as well as the state after failure of the collar element.

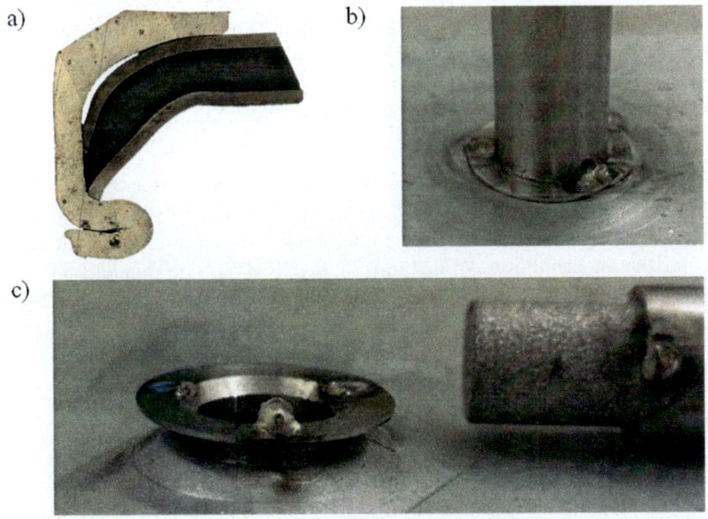

Fig. 9. Single steps during experiments for validation: Finished joining element as a micro section a), joining element with point welded tension bolt b), joining element after failure c).

Figure 9 a) suggests influences of notch effects within the contact area between the sandwich sheet and the joining element, as well as in the area of the bending ground of the s-shaped joining element. Further investigations are required to consider these effects in a reliable process window.

4 Conclusion

The results presented in this study show the design and implementation of a load-appropriate joining element for sandwich composite structures. A digital twin was used for this purpose, which enables the design of the joining elements geometry and the layout of the joint by means of numerical simulation. Thus, the complex and often overlapping influencing values in the design of the joining element as well as their interaction with the complex behavior of the sandwich composite were represented. An individual variation of the different variables could be realized. This allows an optimum geometry to be derived.

Validation tests of the deduced geometry have been carried out using the collar height as an example. The forming behavior of the joint was reproduced with a high degree of agreement. The characteristic bulge of the collar as a function of collar height and counter punch geometry occurred according to the representation in the digital twin. Furthermore, pull-out tests after completed forming reproducibly showed pull-out forces in the anticipated range. One difference between the simulation and the experiments was the significantly longer buildup of the experimental forces until failure. A possible cause can be found in the connection to the load application, which significantly changes the stiffness of the joining element in comparison to the digital twin.

In conclusion, the design of a complex joint connection of a sandwich structure using a digital twin can successfully map the many superimposed influencing variables and considerably simplifies the design process by using comprehensive numerical simulation data. Once the design was purely data-based, it is possible to validate the design on the basis of selected validation tests. Thus, it is shown that the design of a complex joining element via digital twin is successfully feasible.

Acknowledgement. The research presented here is taking place within the IGF project 21405 N of the European Research Association for Sheet Metal Working (EFB). It is supported via the AiF within the funding program "Industrielle Gemeinschaftsforschung und -entwicklung (IGF)" by the Federal Ministry of Economic Affairs and Energy (BMWi) due to a decision of the German Parliament. Furthermore, we would like to thank all industry partners supporting the research project "Fügeelemente-Sandwichkragen".

References

1. Allen, H.G., Neal, B.G.: Analysis and design of structural sandwich panels: the commonwealth and international library: structures and solid body mechanics division. Elsevier Science, Kent (2014)
2. Nutzmann, M.: Umformung von Mehrschichtverbundblechen für Leichtbauteile im Fahrzeugbau. Zugl.: Aachen, Techn. Hochsch., Diss., 2007. Umformtechnische Schriften, vol. 138. Shaker, Aachen (2008)

3. Vijaya Ramnath, B., Alagarraja, K., Elanchezhian, C.: Review on sandwich composite and their applications. Mater. Today Proc. **16**, 859–864 (2019). https://doi.org/10.1016/j.matpr.2019.05.169
4. Harhash, M., Abd, E., Hamid, M.: Forming behaviour of multilayer metal/polymer/metal systems. GDMB Verlag (2017)
5. Palkowski, H., Sokolova, O.A., Carradò, A.: Sandwich materials. In: Crolla, D., Foster, D.E., Kobayashi, T., et al. (eds.) Encyclopedia of Automotive Engineering, pp. 1–17. Wiley, Chichester, UK (2014)
6. Takiguchi, M., Yoshida, F.: Analysis of plastic bending of adhesive-bonded sheet metals taking account of viscoplasticity of adhesive. J. Mater. Process. Technol. **140**, 441–446 (2003). https://doi.org/10.1016/S0924-0136(03)00744-1
7. Engel, B., Buhl, J., Heftrich, C.: Modelling and optimization of lightweight-sandwich-sheets with an adhesive interlayer for the forming process die bending. Procedia CIRP **18**, 168–173 (2014). https://doi.org/10.1016/j.procir.2014.06.126
8. Bruschi, S., Cao, J., Merklein, M., et al.: Forming of metal-based composite parts. CIRP Ann. **70**, 567–588 (2021). https://doi.org/10.1016/j.cirp.2021.05.009
9. Kempf, A.: Entwicklung einer mechanischen Verbindungstechnik für Sandwichwerkstoffe. Aachen, Techn. Hochsch., Diss., 2004, Technische Hochschule Aachen (2004)
10. Ebnesajjad, S.: Handbook of Adhesives and Surface Preparation: Technology, Applications and Manufacturing. Plastics Design Library. William Andrew, Binghamton (2010)
11. Hertel, H.: Leichtbau: Bauelemente, Bemessungen und Konstruktionen von Flugzeugen und anderen Leichtbauwerken, 1st edn. Springer Berlin, Heidelberg (1960)
12. Liebl, J.: InCar plus: Lösungen für automobile Effizienz. ATZ Extra **19**, 8–9 (2014). https://doi.org/10.1365/s35778-014-1302-9
13. Tao, F., Liu, A., Hu, T., et al. (eds.): Digital Twin Driven Smart Design. Academic, London (2020)
14. Griesel, D., Keller, M.C., Groche, P.: Numerical simulation of the hole-flanging process for steel-polymer sandwich sheets. In: Proceedings of the 21st International ESAFORM Conference on Material Forming, p. 160010 (2018)

A Force-Sensitive Mechanical Deep Rolling Tool for Process Monitoring

J. Berlin[1](✉), B. Denkena[1], H. Klemme[1], O. Maiss[2], and M. Dowe[3]

[1] Institute of Production Engineering and Machine Tools (IFW), An der Universität 2, 30823 Garbsen, Germany
berlin@ifw.uni-hannover.de
[2] ECOROLL AG, Celle, Germany
[3] MCU GmbH & Co. KG, Maierhöfen, Germany

Abstract. Deep rolling is an efficient process to increase the service life of highly stressed components such as crankshafts or roller bearings by inducing compressive residual stresses. The residual stresses correspond to the deep rolling force applied. Monitoring the deep rolling force enables the processing result to be assessed. The rolling force is a two-dimensional vector. However, current approaches only allow the measurement of one dimension. Thus, this article presents a force-sensitive deep rolling tool that can measure the applied deep rolling force in two axes. This article, describes the principle of the sensory deep rolling tool and its calibration process. Finally, the sensory properties are evaluated.

Keywords: Mechanical deep rolling · Surface modification · Sensors

1 Introduction

In aerospace or automotive industries, cyclic loads are often the limiting factor for the service life of components, since they lead to fatigue failures [1, 6]. Material fatigue causes more than 80% of all failures occurring during service [4]. Thus, increasing the fatigue strength of the workpiece is important. Surface treatments are an efficient way to achieve this. Surface treatments can be classified in thermochemical, thermal and mechanical treatments [3]. Most commonly used processes for mechanical surface treatment are shot peening, laser shock peening and deep rolling [2]. The advantages of deep rolling are a simple integration in production lines, a high process speed, cost-effectiveness and a large effective processing depth [5, 7]. Deep rolling tools (DRT) use balls or rollers that are pressed against the surface of the workpiece to plastically deform it. Thereby, three positive effects improving fatigue life are achieved. First, the roughness of the surface is reduced, minimizing the notch effect. Secondly, strain hardening is caused by the plastic deformation of the surface, and thirdly, the induction of compressive residual stresses reduces the tensile stresses under cyclic loads [3]. The combination of these effects reduces the crack growth, which usually leads to fatigue failure [6, 7]. Nevertheless, to achieve the desired effect, the rolling force need to be set properly [7, 9]. The force can be applied hydraulically or mechanically [1]. Mechanical

M. Liewald et al. (Eds.): WGP 2022, LNPE, pp. 162–169, 2023.
https://doi.org/10.1007/978-3-031-18318-8_17

deep rolling has the advantage that no hydraulic unit is needed. In mechanical deep rolling, the deep rolling force results from the deflection of a spring that is integrated in the DRT. A changing deflection thus results in a changing force [8]. Geometric errors of the workpiece cause a changing force, as shown in Fig. 1. To reduce this effect, a high compliance of the spring is desirable. Nevertheless, a force error always remains.

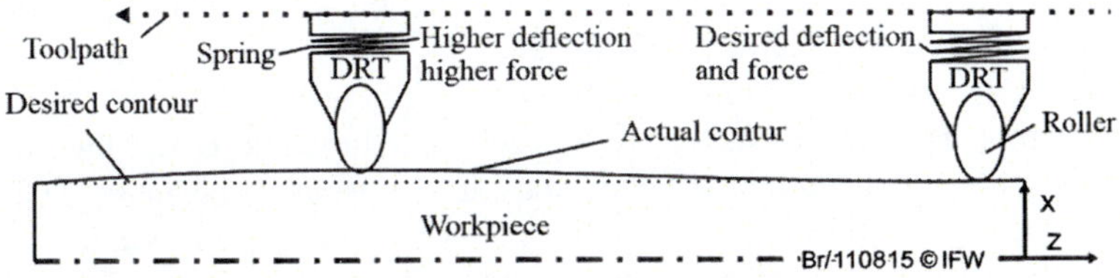

Fig. 1. Basics of mechanical deep rolling

The ability to monitor the rolling force during mechanical deep rolling enables a detection of the force errors and is consequently a requirement for process monitoring. To measure the two-dimensional force vector, a force-sensitive DRT is needed. To equip the DRT with force-sensitive properties, strain gauges (SG) can be used. Using SG is a common way to implement force measurement capabilities into tools or machine components. E.g., in the past, axis-slides [10], spindles [11], cutting tools [14] or clamping systems [12] were fitted with SG. However, there is only one approach known from the state of the art dealing with SG implementation into a DRT. This tool determines one dimension of the rolling force vector by measuring the deflection of the spring that is integrated in the tool [13]. However, SG do not measure the forces directly, but only measure the resulting strain on the material surface. The force is calculated based on the strain using a transfer matrix (M). That matrix transfers the strain signal vector into a force vector. However, machine tools have a high stiffness. Accordingly, the measurable strain is low. Usually, the main challenge is to achieve sufficient strain and a high force-sensitivity, but to avoid a decrease in the tool's stiffness [10–12]. Mechanical DRT, however, need to be compliant to compensate for geometric errors of the workpiece. For the design of a force-sensitive DRT, this means that the stiffness of the DRT components can be reduced to achieve higher strains and thus a higher force sensitivity. The adequate placement of the SG is another challenge as the location of the SG affects the multi-axis force measurement capability. The matrix M can be used to evaluate the placement of the SG by analyzing the column vectors of M. Column vectors that are perpendicular to one another indicate that the matrix M is well conditioned and the placement of the SG is thus favorable [12].

In the first chapter of this work a mechanical model of a DRT for multi-axis force measurement is presented. The mechanical model is then turned into a rough design based on an existing DRT. The rough design is optimized using a simulation and the placement of the SG is evaluated. For the evaluation, the matrix M is examined. On the basis of the simulation results the tool was realized. Finally, the tool was calibrated and evaluated based on measurements.

2 Implementation of the Sensory Capabilities

The minor stiffness requirements of the DRT allow an unusual approach to measure the force with SG. In order to achieve a high sensitivity, a flexible tool holder was designed. This is intended to achieve nominal strains of 1000 μm/m. A high level of sensitivity is therefore to be expected. Figure 2 illustrates a mechanical model of a DRT with a flexible tool holder. The model consists of two perpendicularly aligned bending beams for each direction of force. One SG is attached to each beam to measure the strain. By decoupling the bending beams with the help of a floating joint and a flexure joint, only one beam is deformed significantly for each load direction. Under the load of F_x', the force is transmitted through the tool body to the bending beam x′. The floating joint ensures that the bending beam y′ is loaded with minimal force. Under the load of F_y', the bending beam y′ is subjected to bending. The bending beam x′, on the other hand, only experiences a low tensile load, which leads to significantly lower strains. This distribution of the load is intended to ensure that each SG is sensitive for one direction of force and thereby an adequate differentiation between the directions of force is achieved.

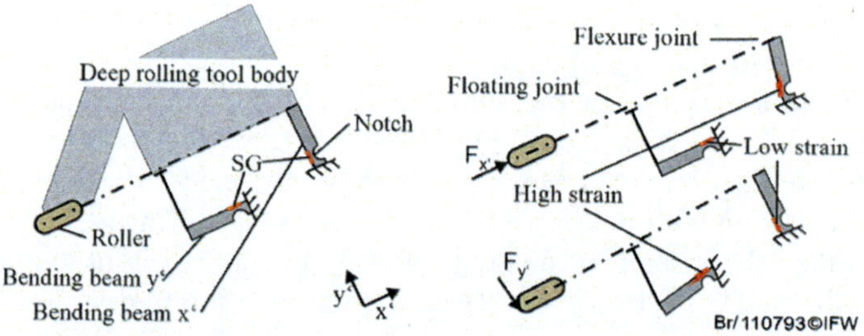

Fig. 2. Mechanical model of the deep rolling tool

In a next step the mechanical model was transferred to a rough design of the DRT. A DRT ECOROLL EG45 was used as a basis for the sensory DRT. The EG45 is designed for a maximum load of 4 kN. The adopted components of EG45 provide a spring travel of 2 mm. The bending beams were integrated into the tool holder of EG45 according to Fig. 3. A thin beam with a thickness of 3 mm was used as a floating joint. The beam can transmit tensile and compressive forces, but offers little resistance to forces transverse to it. The flexure joint was realized by a deep notch in the bending beam x′. Shear forces as well as tensile and compressive forces can still be transmitted. However, the flexure joint is compliant to bending moments. Placing the SGs in the space between the beams ensures protection from external influences like chips, coolant or mechanical damage. Prefabricated SG full Wheatstone bridges (ME type N2A-06) were used. Two electrical resistors of the measuring bridge are aligned to the direction of strain and two are aligned vertically. As a result, the sensitivity is equal to a half-bridge. However, the two additional resistors compensate temperature influences.

A simulation in ANSYS was used to optimize the dimensions of the rough design in Fig. 3. For the simulation model, the tool body was geometrically simplified. The maximum nominal force of 4 kN was applied to the model in x′ and y′ directions and the

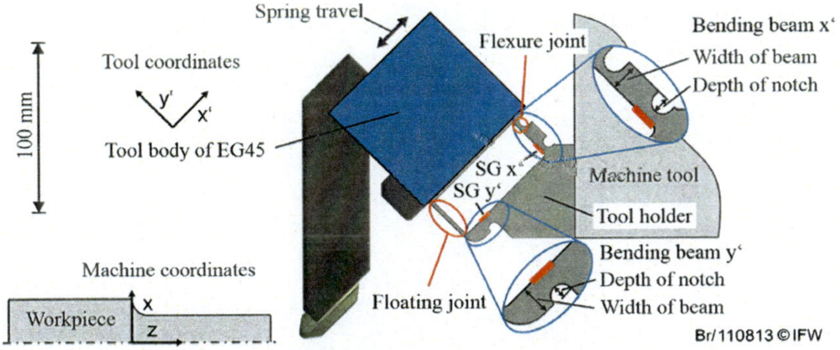

Fig. 3. Rough design of the deep rolling tool

resulting strains were analyzed. In a parameter study, the width of the bending beams and the depth of the notches were varied in order to achieve a maximum nominal strain of 1000 μm/m at the positions of the SG. A width of 8.5 mm and a depth of the notch of 4.5 mm was determined for the bending beam x′ and a width of 10 mm and a depth of 3 mm for the bending beam y′. Figure 4 shows the results of the parameter study. The desired strain at the positions of the SG is almost achieved. That means that, 950 μm/m are achieved for SG x′ and 910 μm/m for SG y′. Thus, the sensitivity is sufficient.

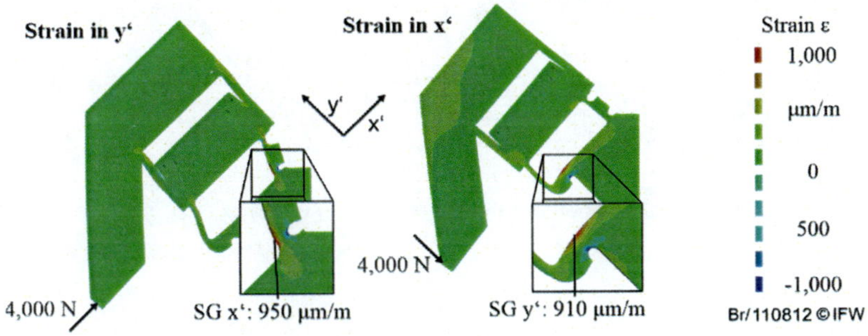

Fig. 4. Results of the Simulation

In order to be able to evaluate the position of the SG on the bending beams, the transfer matrix M was calculated using the simulation results. For this purpose, forces were applied in the machine coordinate system (x/z, Fig. 3) to the DRT. Then, the resulting mean strains at the location of the DMS were used to predict the voltage signal $V_{x/y}$. The k-factor of the SG, the amplification factor vf of the electronics and the reference voltage U_{ref} of the Wheatstone bridge which were used for the calculation are shown in Fig. 5 on the left. M is a 2 × 2 matrix. Thus, to calculate M, a 2 × 2 matrix of F and V is needed. For this purpose, it is necessary to carry out the simulation for two direction of forces (F1/F2). This results in the matrices F and V (Eq. 1). The transfer matrix M is determined by solving (Eq. 2).

$$F = \left[F_1\ F_2\right] = \begin{bmatrix} F_{x,1} & F_{x,2} \\ F_{z,1} & F_{z,2} \end{bmatrix} = M \cdot \begin{bmatrix} V_{x',1} & V_{x',2} \\ V_{y',1} & V_{y',2} \end{bmatrix} = M \cdot V \tag{1}$$

$$M = \begin{bmatrix} M_{1,1} & M_{2,1} \\ M_{1,2} & M_{2,2} \end{bmatrix} = \left[M_1\ M_2\right] = F \cdot V^{-1} \tag{2}$$

Figure 5 (right) illustrates the vectors M1 and M2 of the matrix M graphically. It is obvious that the vectors are almost 90° to each other and M is thus well conditioned. Exactly 90° was not reached because the forces are never transmitted to just one bending beam, but both beams always experience a minimum load. According to [12], however, this confirms that the placement of the SG is favorable for a two-axis force measurement.

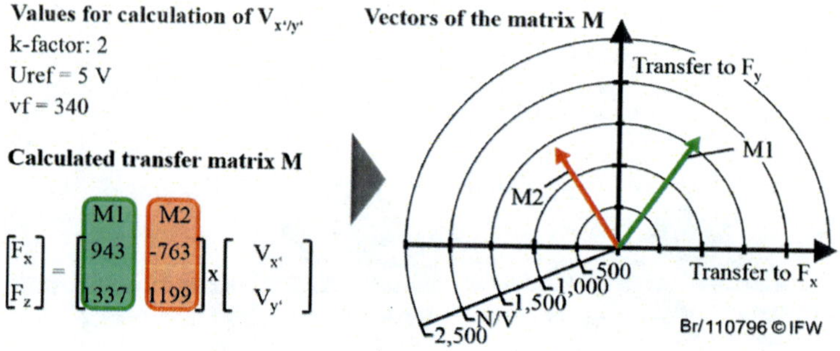

Fig. 5. Calculated transfer matrix from the results of the simulation

3 Calibration and Characterization of the Tool

The DRT was manufactured according to the design shown in Fig. 3 with the optimized parameters found through the simulation. To be able to determine the forces from the strain signals experimentally, the matrix M is needed. This Matrix was already calculated from the simulated results. However, for better accuracy, the matrix M was recalculated based on measured data. For the calculation, the SG signals for 11 different force directions were determined. Therefore, the setup in Fig. 6 was used. A Kistler 9129A multi-coordinate dynamometer supplies the reference force. An adapter with a radius of 15 mm was attached to the dynamometer to enable an application of forces in two dimensions. Forces in x- and z-direction were applied to the DRT while the forces and the SG signal were measured at the same time.

Figure 7 (left) shows the measured values of the SG signals $V_{x'/y'}$ and the forces $F_{x/z}$ which were used to calculate the matrix M. In contrast to the calculation according to

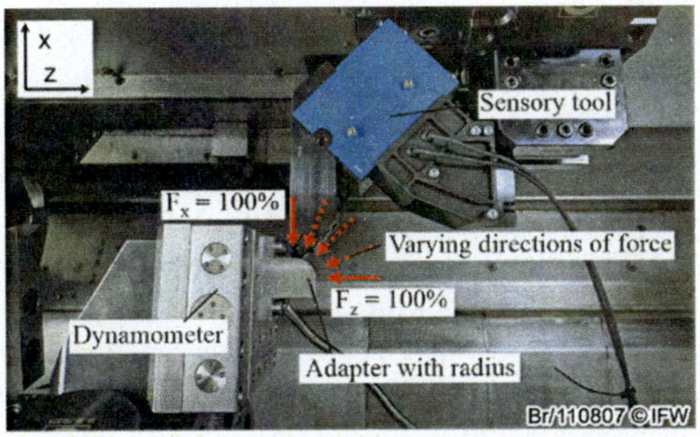

Fig. 6. Calibration of the tool

Eq. 2, the matrices V and F have the dimension 2 × n (n >> 2). The resulting system of equations is therefore overdetermined. The solution was determined using the least squares method according to Eq. 3.

$$M = F \cdot V^T \cdot \left(V \cdot V^T\right)^{-1} \tag{3}$$

The calculated matrix M is shown in Fig. 7 and is compared to the simulated matrix M (middle). When comparing the two transmission vectors M1/2 graphically (right), a parallel alignment can be seen. However, the measured vectors are longer than the simulated ones. The length of the vector is inversely proportional to the sensitivity of the SG. Thus, a longer vector means that the sensitivity is lower. The sensitivity of the real DRT is reduced by a factor of 0.6 compared to the simulated DRT. This can be explained primarily by a reduced transmission of the strain to the SG in practice. In addition, imprecise placement of the SG can lead to deviations. However, the simulated nominal strain of about 1000 µm/m is very high. A slight reduction in sensitivity is thus not a problem. Overall, however, this comparison shows that the tool corresponds to the simulated properties.

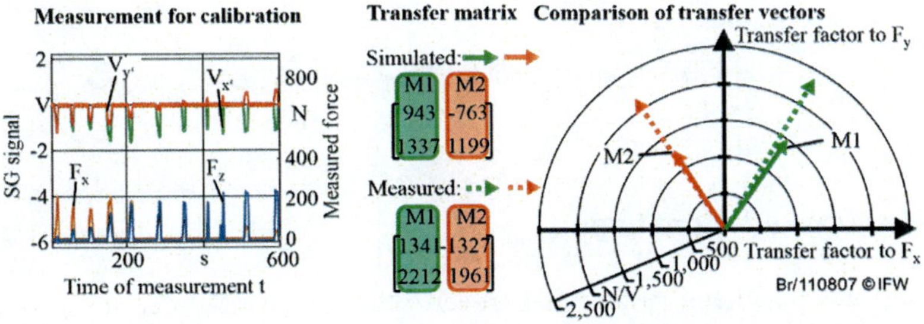

Fig. 7. Comparison of the simulated and measured transfer matrix

For a final evaluation of the force measurement capabilities, the force measured with the DRT was compared to that of the dynamometer. The force was varied from a load in x-direction to a load in z-direction as shown in Fig. 6. Thus, all possible load directions are represented. To avoid overloading, the maximum force was limited to around 1 kN. Figure 8 shows the measured forces. On the left side, the whole measurement is shown for F_x and F_y. At the beginning, the measured Force was applied in x-direction. The cutout A (middle) shows the deviations occurring between the forces measured by the DRT and the dynamometer for a force in x-direction in more detail. It can be seen that a signal noise of about 20 N occurs. On the one hand, the noise comes from the SG itself, but electrical interference from the machine tool adds up. A systematic deviation of 15 N also occurs. This can be caused by an incorrect transmission matrix M. In addition, a non-linear behavior of the tool holder can lead to systematic deviations, because it is not taken into account by the transfer matrix. During the measurement the force was gradually varied from x-direction to z-direction. At the end of the measurement the force in z-direction was dominant. The cutout B shows the force in z-direction in more detail. The systematic deviation decreases to almost zero, but the signal noise is still about 20 N. Over all, a maximum total error of ± 20 N can be assumed. Based on the maximum force measured, this results in a relative error of 2%. Relative to the maximum load of 4000 N the error is 0.5%. Tests with existing tools had shown that a minimum resolution of ± 100 N is required for process monitoring of the deep rolling process. The force-sensitive DRT exceeds this resolution by factor five and is therefore suitable for process monitoring.

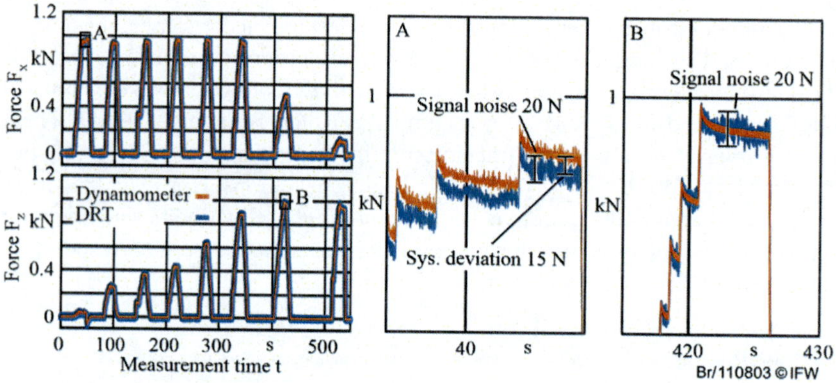

Fig. 8. Validation of the force measurement

4 Summary and Conclusion

The rolling force is a crucial variable for the deep rolling process. A force-sensitive rolling tool thus enables process monitoring. The development, commissioning and characterization of the tool was described in this paper. After introducing the basic concept for the tool, a simulation was used to optimize and evaluate it. Afterwards, the calibration process was described. Finally, experiments revealed that a force measurement with a maximum error of ± 20 N was achieved. In future works, the application of this tool for process monitoring is examined.

Acknowledgements. The cooperation project "Process-monitored and controlled mechanical deep rolling" (ZF4810001LP9/ZF4070523LP9) is funded by the Federal Ministry of Economics and Climate Protection (BMWK) as part of the Central Innovation Program for SMEs (ZIM) and is supervised by the Working Group of Industrial Research Associations (AiF). The ECOROLL AG thanks for the productive cooperation in this project. The MCU GmbH & Co. KG and the IFW thanks for the financial support.

References

1. Delgado, P., Cuesta, I.-I., Alegre, J.-M., Díaz, A.: State of the art of deep rolling. Precis. Eng. **46**, 1–10 (2016)
2. Stichi, M., Schnubel, D., Kashaev, N., Huber, N.: Review of residual stress modification techniques for extending the fatigue life of metallic aircraft components. Appl. Mech. Rev. **67**(1), 010801(pp. 1–9) (2015)
3. Altenberger, I.: Deep rolling—the past, the present and the future. In: 9th International Conference on Shot Peening, pp. 144–155 (2005)
4. Milne, I., Ritchie, R.-O., Karihaloo, B.: Comprehensive Structural Integrity: Cyclic Loading and Fatigue. Elseveir Science Ltd. (2003)
5. Rodríguez, A., López de Lacalle, L.-N., Celaya, A., Lamikiz, A., Albizuri, J.: Surface improvement of Shafts by the deep ball-burnishing technique. Surf. Coat. Technol. **206**(11–12), 2817–2824 (2012)
6. Novovic, D., Dewes, R.-C., Aspinwall, D.-K., Voice, W., Bowen, P.: The effect of machined topography and integrity on fatigue life. Int. J. Mach. Tools Manuf. **44**(2–3), 125–134 (2003)
7. Manouchehrifar, A., Alasvand, K.: Finite element simulation of deep rolling and evaluate the influence of parameters on residual stress. In: 5th WSEAS, pp. 121–127 (2012)
8. Klocke, F., Mader, S.: Fundamentals of the deep rolling of compressor blades for turbo aircraft engines. Steel Res. Int. Surf. Treat. **76**(2–3), 229–235 (2005)
9. Denkena, B., Grove, T., Breidenstein, B., Abrão, A., Meyer, K.: Correlation between process load and deep rolling induced residual stress profile. In: 6th CIRP Global Web Conference, vol. 78, pp. 161–165 (2018)
10. Denkena, B., Litwinski, K.-M., Boujnah, H.: Process monitoring with a force sensitive axis-slide for machine tools. In: 2nd International Conference on System-Integrated Intelligence, vol. 15, pp. 416–423 (2014)
11. Boujnah, H., Denkena, B.: Kraftsensitiver Spindelschlitten zur online Detektion und Kompensation der Werkzeugabdrängung in der Fräsbearbeitung. Tewiss, Hannover (2019)
12. Litwinski, M., Denkena, B.: Sensorisches Spannsystem zur Überwachung von Zerspanprozessen in der Einzelteilfertigung. PZH GmbH, Hannover (2011)
13. Maschinenmarkt.: Werkzeug-Spezialist Ecoroll treibt Entwicklung beim Glatt- und Festwalzen voran. https://www.maschinenmarkt.vogel.de/werkzeug-spezialist-ecoroll-treibt-entwicklung-beim-glatt-und-fest-walzen-voran-a-93256/. Last accessed 7 Apr 2022
14. Zhao, L., Zhao, Y.-L., Shao, Y.-W., Hu, T.-J., Zhang, Q., Ge, X.-H.: Research of a smart cutting tool based on MEMS strain gauge. J. Phys. Conf. Ser. **986**, 012016 (2018)

Optimization of the Calibration Process in Freeform Bending Regarding Robustness and Experimental Effort

L. Scandola(✉), M. K. Werner, D. Maier, and W. Volk

TUM School of Engineering and Design, Department of Mechanical Engineering, Chair of Metal Forming and Casting, Technical University of Munich, Munich, Germany
lorenzo.scandola@tum.de

Abstract. In the freeform bending process, the obtained geometry is determined by the kinematics of the bending head. In order to derive the motion profiles leading to a part within tolerances, the correlations between bending results and machine settings are investigated. This inverse problem is solved by generating calibration curves, whose aim is to correlate the bending radii with the head movement from experimental tests results. Nevertheless, slight adjustments in the shape of the calibration curves show a significant impact on the bending results. In this paper, the robustness of the calibration process is investigated. First, the effect of different interpolating methods is considered. In addition, the influence of the experimental points is examined by comparing the performance of global and local data for the interpolation. Finally, calibration curves obtained with different ratios between the translation and the rotation degrees of freedom are compared. In this way, the interactions between the given parameters are investigated and a more efficient process for calibrating the freeform bending machine can be determined. This allows the reduction of the experimental effort for determining the relation between the machine parameters and the bent result as well as optimizing the process with respect to the geometrical deviations and dimensional stability.

Keywords: Freeform bending · Process optimization · 2D-interpolation · Dimensional accuracy

1 Introduction

The process of free-form bending allows to enhance the freedom of design for bent components in a significant way. With the use of a single tool setup, a virtually unlimited range of bending radii and angles can be achieved in a single component. This is particular attractive for the automotive industry, which requires robust process routes, able to deal with different component designs in a flexible and cost efficient way [1]. In addition, the typical limitation of construction following an arc-line strategy can be overcome, and continuous 3D-geometry developing without constraints in space can be manufactured. This wide range of possibilities comes at the expense of the process configuration, which is still currently performed with a time consuming trial-and-error approach. In contrast to

M. Liewald et al. (Eds.): WGP 2022, LNPE, pp. 170–178, 2023.
https://doi.org/10.1007/978-3-031-18318-8_18

other common form-bound-processes, such as rotary draw bending [2], where the target geometry is determined by the forming tool [3], this kinematic-based process requires the solution of an inverse problem with possibly more than one solution. In order to identify the kinematics of the bending head leading to the target geometry, a reliable calibration must be carried out. In this contribution the robustness and sensitivity of the calibration process is investigated. After the description of the theoretical kinematic model of the process, different calibrations sets are compared regarding their performance and the experimental effort they require. First, the impact of different interpolation methods is analyzed. Subsequently, the choice of the fitting data points is addressed, and local as well global calibrations are considered. Finally, the effect of different ratios between translational and rotational degrees of freedom is investigated. In such a way, an optimal procedure for the determination of the calibration curves in free-form bending can be obtained, reducing the experimental effort as well as time and material use.

2 The Calibration Process

2.1 Machine Degrees of Freedom and Theoretical Parameters

The free-form bending process can be realized through different configurations and setups. The first process concept was introduced by Murata [4], who proposed a 3-axis machine realized with a spherical bearing head [5]. Other designs differ in the realization of the bending head and the number of degrees of freedom [6]. The plant available at the Chair of Metal Forming and Casting is a 6-axis free-form bending machine by the Company J. Neu GmbH [7]. Independently on the practical realization of the head movement, there are always only two active degrees of freedom on the bending plane, namely the translation and the rotation of the bending head as shown in Fig. 1.

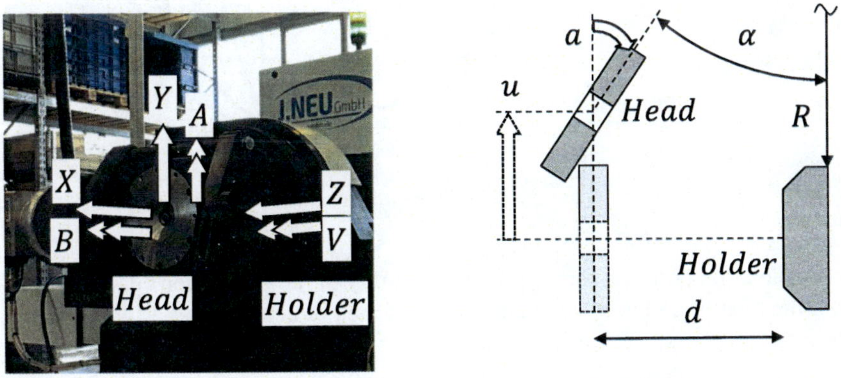

Fig. 1. Degrees of freedom of the machine and on the bending plane

The theoretical values of the translation u and rotation a of the bending head required to generate a target bending radius R can be retrieved with the following equations [8]:

$$u = R(1 - \cos(a)) \quad a = \arcsin(R/)$$

where represents the distance between the bending head and the fixed tool. These calculated values result nevertheless in bents with relevant deviations from the target geometry, since they are computed with a pure geometrical analysis neglecting the material plastic behavior, the geometric inertia of the cross section as well as the forming velocity, tool stiffness and the springback effect. For this reason, the values for translation and rotation are chosen arbitrarily after performing preliminary bending tests and measuring the resulting radii, with the aid of the scanning system T-Scan by Zeiss.

2.2 Robustness of the Calibration Quality

The calibration set for the free-form bending process comprehends two calibration curves, namely the bending radius over the translation and over the rotation:

$$R = f(u) \quad R = f(a)$$

This is because the two degrees of freedom are mechanically uncoupled and different combinations of them can result in the same bending radius featuring different mechanical properties [9]. The most important factors influencing the performance of the calibration curves are the data to fit P_i and their number, the fitting method f and the ratio between the translation and rotation, as represented in Fig. 2.

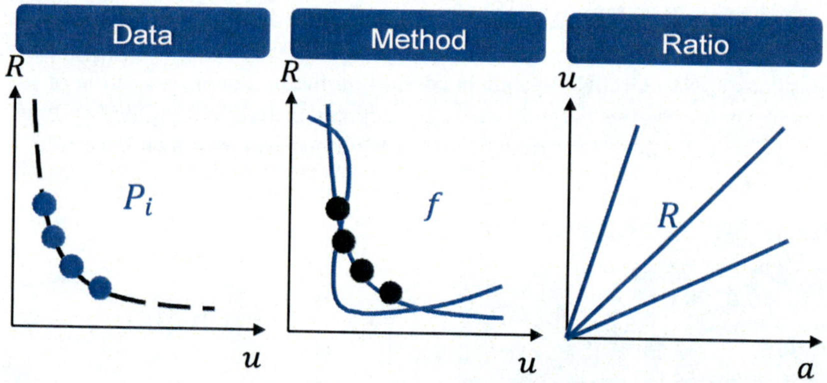

Fig. 2. Investigated factors influencing the calibration performance

As evaluation criterion for assessing the quality of the calibration curve, the smoothness of the resulting fit is considered, as well as the prediction capability for other target bents.

3 Investigation of the Impact of Calibration Parameters

3.1 Effect of Interpolation Approach

First, the effect of five different interpolation methods to fit the experimental data points is investigated. For polynomial interpolation, the following two distinct hyperbolic functions are defined:

$$y_1 = c_1/x \quad y_2 = c_1/x^2 + c_2/x + c_3 \cdot x^2$$

In addition, the performance of cubic spline interpolators [10], Piecewise Cubic Hermite Interpolating Polynomial (PCHIP) approach [11] and the Akima method [12] are compared. The choice of the given methods is justified as follows. The hyperbolic function seems to depict correctly the physics of the problem, as the relation between radius and translation as well rotation should be inversely proportional. Nevertheless, hyperbolic polynomial interpolators can result in smooth curves forcing a non-realistic behavior for some radius range. Cubic splines should then allow to depict a more complex local trend and be more truthful to the input data, but may result in uncontrolled oscillating behavior. The use of PCHIP was proposed and proved successful for the free-form bending of rectangular profiles [13]. In this context, the PCHIP approach as well as the Akima method are employed aiming to reduce the oscillating phenomenon while maintaining the advantages of the spline interpolation, the former flattening more strongly the resulting curve. All tests for generating the interpolating input data are performed on welded tubes of steel P235 with diameter 33.7 mm and thickness 2 mm. To generate the data the radius domain is subdivided into three macro areas, namely mild, moderate and severe bending. Ten tests are performed for every region, resulting in the experimental plan shown in Table 1. The correspondent values of translation and rotation are automatically determined by the machine preliminary setup with a variable ratio.

Table 1. Selection of 30 data points depending on the radius span

	Mild bending	Moderate bending	Severe bending
R [mm]	2000–1100 (step 100)	1100–550 (step 50)	550–80 (step 42)
u [mm]	2.2–3.3	3.5–5.2	5.6–22.6
a [mm]	5.1–7.3	7.8–11.2	11.9–35.1

The response of different interpolating approaches is depicted in Fig. 3, where the experimental data and the resulting interpolations are shown. The first observation concerns the capability of the hyperbolic functions to describe the relation between radius and translation. As it can be seen in Fig. 3a, the first order hyperbolic function y_1 is not reliable for any radius span. The second order function y_2, obtained by fitting the data points corresponding to $R1000$, $R500$ and $R80$ shows a better behavior, but still deviates from the data in the mild bending zone as well as in the severe bending area, where the predicted translation values differ from the target ones of up to 1 mm. A higher consistency with the experimental data can be achieved by employing the proposed spline-based interpolation approaches, as expected. The cubic spline, Akima and PCHIP curves obtained interpolating the whole data set are over imposed in Fig. 3b. Neglecting translation values lower than 2 mm and higher than 30 mm which are not realistic since over the machine limits, it can be seen that the three approaches perform similarly away from the extremes. The cubic spline is affected by the oscillation problem in a more marked way (especially around $u = 22$ mm), while the Akima method returns a curve which is not defined outside the extremes of the interpolating domain. For this reason, in accordance with [13], the PCHIP method is identified as the most stable, robust and preferable method for the retrievement of the calibration curves. Nevertheless, the employment of cubic splines or Akima method can also be recommended over polynomial fitting functions.

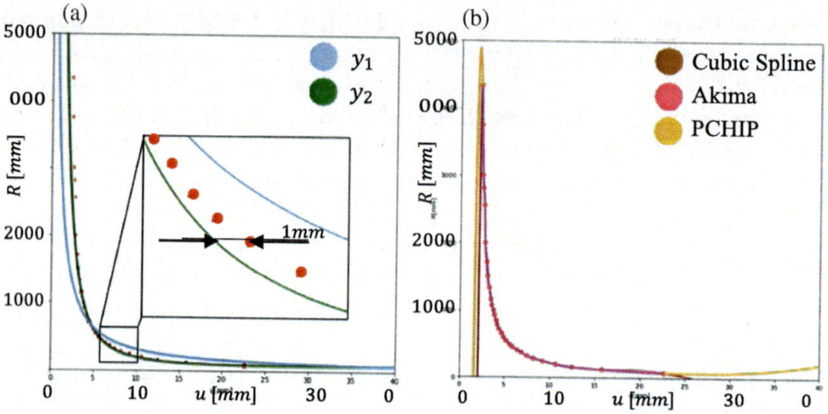

Fig. 3. Influence of interpolating method: hyperbolic functions (a) and spline-based approaches comparison (b)

3.2 Effect of the Choice of Data Points

Subsequently, the calibration curves generated using data points from different radii spans are compared. The influence of the chosen interpolating data range is investigated by comparing the hyperbolic functions and the PCHIP method. Every interpolation approach is used to fit separately the data of the mild (*mild*), moderate (*mod*) and severe (*sev*) radius ranges defined in Table 1, which are compared in Figs. 4 and 5. The results using the polynomial interpolators are shown in Fig. 4.

It can be seen, that the radius span has got a marked influence on the resulting calibrating curves. While the simple hyperbolic function y_1 fails to predict the required translation for every radius span, the second order equation y_2 works well in the corresponding originating span but cannot be used outside this range. This is particularly clear in Fig. 4b, where the mild curve (blue) misses the severe radius area, and the severe curve (green) predicts lower translations values for the mild radius zone. The comparison between the interpolations of the mild, moderate, severe as well as whole bending radius spans with the PCHIP approach are shown in Fig. 5.

The robustness of this interpolating approach with respect to the input data is shown in Fig. 5a, where it is demonstrated that the curve obtained interpolating the whole radius span (black, dashed) corresponds always with the local curves of the mild, moderate and severe spans in the corresponding area. This result allows to define an optimal set of

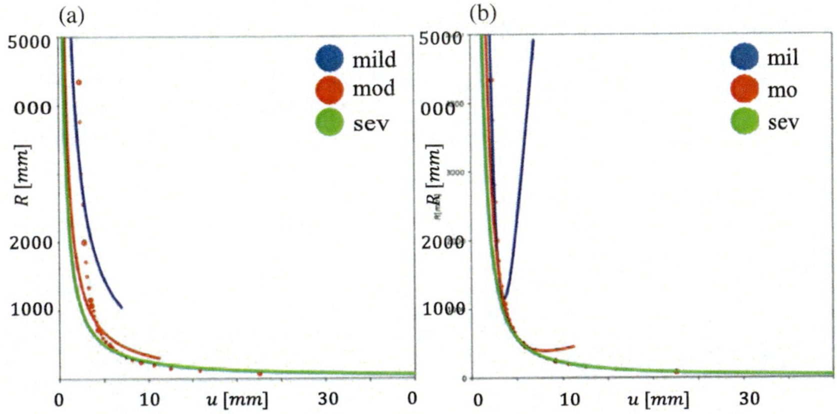

Fig. 4. Influence of interpolating points on hyperbolic fitting functions: y_1 (a) and y_2 (b)

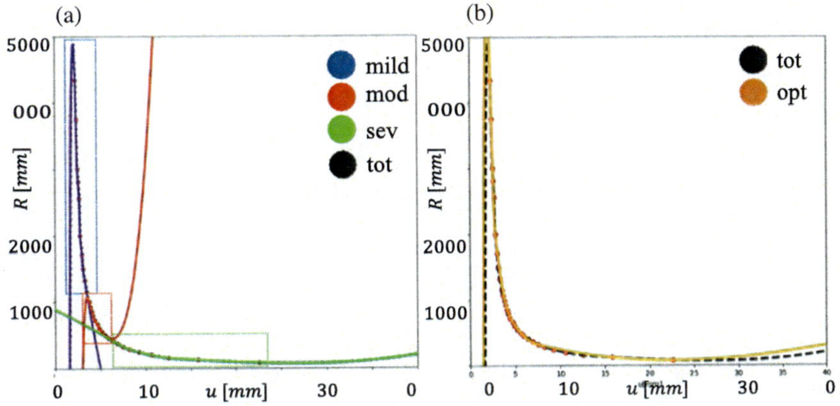

Fig. 5. Influence of interpolating points on PCHIP approach: comparison between mild, moderate, severe and whole span interpolation (a) and identification of optimal data (b)

interpolating data returning a calibration curve valid for the whole radius span with the minimal number of data test. In Fig. 5b the comparison between the reference curve obtained interpolating the whole data set and the reduced curve using 5 data points, corresponding to bending radii of *R*60, *R*0, *R*20, *R*10 and *R*2, is shown. As the difference between the two is negligible, it can be concluded that using the PCHIP approach the influence of fitting data point is less impacting, and that the suggested optimal data set for building the calibration curve requires the data for *R*60, *R*0, *R*20, *R*10 and *R*2 only. On the other hand, polynomial interpolations are strongly affected by the choice of interpolating points and their accuracy outside the considered radius span can be questioned.

3.3 Effect of Translation/Rotation Ratio

Finally, the impact of different combination of translation and rotation values is investigated. To this extent, three fixed ratios, defined as the translation over the rotation, are chosen, namely 0.6, 0.8 and 1.2. The corresponding values of bending head's degrees of freedom and radius are derived by taking a reference sampling of translation values, as shown in Table 2.

Table 2. Input data points for the investigation of different translation/rotation values

	u_{ref} [mm]	6.0	8.0	10.0	12.0	14.0	16.0	18.0
*r*0.6	*a* [g]	10.0	13.3	16.7	20.0	23.3	26.7	30.0
	R [mm]	485	321	230	182	145	124	109
*r*0.8	*a* [g]	7.5	10	12.5	15	17.5	22.5	25
	R [mm]	533	336	250	197	160	116	96
*r*1.2	*a* [g]	5.0	6.7	8.3	10.0	11.7	13.3	15.0
	R [mm]	593	373	286	211	171	142	125

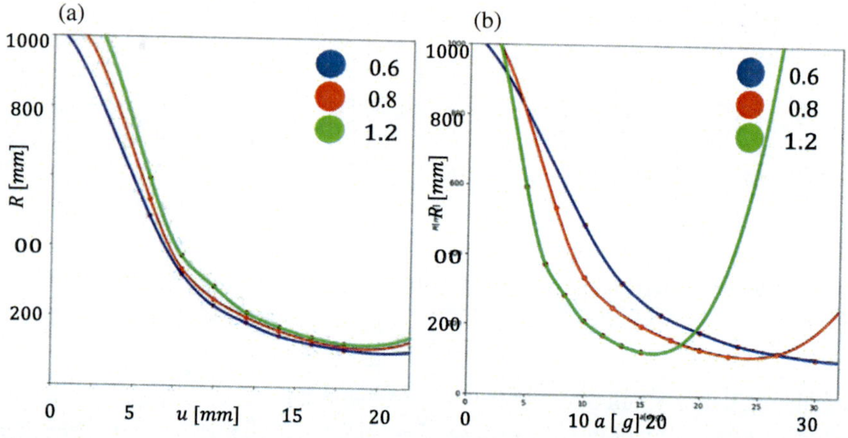

Fig. 6. Influence of ratio between translation and rotation of the bending head on the calibration curves for the translation (a) and rotation (b)

The influence of the degrees of freedom ratio is analysed showing the resulting calibration curves using the PCHIP approach for the translation *u* and the rotation *a*, respectively in Fig. 6a, b. It can be observed that maintaining a constant set of translation values and varying the rotation with a fixed ratio results in different final radii. The calibration curves obtained with the PCHIP approach for the translation are significantly displaced as shown in Fig. 6a, and deviations up to 2.5 mm occur (e.g. in the radius zone between 300 and 350 mm for ratios 0.6 compared to 1.2). Even more evident is the difference arising in the calibration curve for the head's rotation in Fig. 6b, with deviations up to 10 g. This confirms previous investigations related to the properties of the bent components [9] and demonstrates that deriving the kinematic for a target geometry in the free-form bending process is a multi-optimum inverse problem. An interesting further investigation should involve the construction of 3D-surfacebased calibration methods, allowing to distinguish between the resulting multiple optima.

For the scope of this paper, this investigation shows that the translation and rotation degrees of freedom cannot be uncoupled, and that their corresponding calibration curves are only valid for the resulting fixed or variable ratio between them.

4 Summary and Conclusion

In this contribution the calibration process for the free-form bending machine is thoroughly analysed and an optimal strategy for performing the calibration is proposed. The most important parameters affecting the performance of the calibration curves are identified as the starting fitting data, the fitting method and the relation between the degrees of freedom of the process. A comprehensive experimental data set is obtained by discretizing the radius domain in three spans, namely mild, moderate and severe bending. Ten tests are experimentally performed for each range, so to dispose of a reliable basis of 30 data points. First, the impact of different interpolation methods is evaluated, through the comparison of two hyperbolic polynomial functions, the cubic spline approach, the PCHIP method and the Akima strategy. It is concluded that polynomial interpolation returns smooth curves which nevertheless are not truthful to the fitting data in relevant radii zones, hence resulting in reduced accuracy solutions. On the other hand, the spline-based approaches perform very well regarding the adherence to the data, but suffer of oscillations between the extremes of the interpolated domain. The PCHIP approach results as the best compromise and is the strategy of choice, as it allows to control better the over-oscillation arising in the cubic spline and it is defined over the whole domain, in contrast to the Akima interpolator. Successively, the difference between local and global interpolation is analysed. It is shown that while polynomial interpolation responds very differently depending on the choice of the input interpolating data, the PCHIP strategy is much more robust and returns very similar curves when using local as well as global data points. Moreover, it is shown that negligible differences arise when building the calibration curves with this strategy on basis of a wide data-set of 30 points or a more efficient set of 5 points over the whole domain. With this regard it is suggested to employ the PCHIP method and undergo 5 tests close to the values of $R60$, $R0$, $R20$, $R10$ and $R2$. Finally, the impact of different combination of translation and rotation values is investigated by establishing a fixed translation vector and deriving the rotation on basis of a fixed ratio, namely 0.6, 0.8 and 1.2. The results indicate that the two degrees of freedom cannot be uncoupled, and that the resulting calibration curves deviate significantly depending on the ratio of choice, whether fixed or variable. The performed investigations allow to optimize the process of calibration for free-form bending and result in guidelines for the generation of calibration curves for different cross-section and materials. In addition, the results of this research suggests that the problem under investigation belongs to the class of multi-optima inverse problems, which could be tackled increasing the dimensionality of the calibration parameter space. With this respect, further investigations are planned aiming to develop a 3D-surface-based calibration strategy, which could efficiently distinguish between the multi-optimal solutions on basis of a quality criterion. In addition, it is planned to extend this analysis also to different materials and cross-sections, in order to widen the validity of the demonstrated correlations.

List of References

1. Beulich, N., Craighero, P., Volk, W.: FEA simulation of free-bending—a preforming step in the hydroforming process chain. J. Phys. Conf. Ser. **896**, 12063 (2017). https://doi.org/10.1088/1742-6596/896/1/0
2. Engel, B., Hinkel, M.: Analytisch unterstützte Vorauslegung des Rotationszugbiegeprozesses. In: Tagungsband/XXX. Verformungskundliches Kolloquium: Leoben, 14.2.2011; [26.2. bis 1.3.2011, Planneralm, Steiermark, B. Buchmayr, Ed., Leoben: Umformtechnik, pp. 97–102 (2011)
3. VDI Verein Deutscher Ingenieure, VDI 3430:2014-06: Rotationszugbiegen von Profilen. Beuth, Berlin
4. Murata, M., Aoki, Y.: Analysis of circular tube bending by MOS bending method. In: Advanced Technology of Plasticity, pp. 505–508 (1996)
5. Murata, M.: Tube bending of new generation by MOS bending machine. In: 1st International Tube and Profile Bending Conference, pp. 9–18 (2011). Accessed: 29 June 2018
6. Gantner, P.: The Characterisation of the Free-Bending Technique. Thesis (Ph.D.), Glasgow Caledonian University (2008)
7. Neu, J.: 6-Achs Freiformbiegemaschine NSB. [Online]. Available https://www.neugmbh.de/de/freiformbiegen-nsb
8. Gantner, P., Harrison, D.K., de Silva, A.K., Bauer, H.: The development of a simulation model and the determination of the die control data for the free-bending technique. Proc. Inst. Mech. Eng. Part B J. Eng. Manuf. **221**(2), 163–171 (2007). https://doi.org/10.1243/09544054JEM642
9. Maier, D., et al.: The influence of freeform bending process parameters on residual stresses for steel tubes. Adv. Ind. Manuf. Eng. **2**, 100047 (2021). https://doi.org/10.1016/j.aime.2021.100047
10. scipy documentation, Cubic Spline Interpolation. [Online]. Available https://docs.scipy.org/doc/scipy/reference/generated/scipy.interpolate.CubicSpline.html
11. Fritsch, F.N., Butland, J.: A method for constructing local monotone piecewise cubic interpolants. SIAM J. Sci. Stat. Comput. **5**(2), 300–304 (1984). https://doi.org/10.1137/0905021
12. Akima, H.: A method of bivariate interpolation and smooth surface fitting based on local procedures. Commun. ACM **17**(1), 18–20 (1974). https://doi.org/10.1145/360767.360779
13. Werner, M.K., Maier, D., Scandola, L., Volk, W.: Motion profile calculation for freeform bending with moveable die based on tool parameters. In: Proceedings of the 24th International Conference on Material Forming (2021)

Numerical and Experimental Investigations to Increase Cutting Surface Quality by an Optimized Punch Design

A. Schenek(✉), S. Senn, and M. Liewald

Institute for Metal Forming Technology, Holzgartenstraße 17, 70174 Stuttgart, Germany
adrian.schenek@ifu.uni-stuttgart.de

Abstract. Punching is one of the most commonly used production processes in sheet metal working industry. Here, major criterion for the quality of cutting surfaces is a high clean cut proportion. However, the disadvantage of conventional punching processes is that they can only produce clean cut proportions up to 20–50% of the sheet thickness. Until today, more complex processes such as fine blanking are therefore required for a higher cutting surface quality. The content of this paper is a numerical and experimental investigation for a new tool design called "concave punch nose design". The idea of the concave punch nose design is to optimize the cutting edge geometry of conventional punches in order to enlarge clean-cut proportion along the cutting surface despite a process sequence similar to conventional shear cutting. The numerical and experimental investigations presented in this contribution show, that the concave punch nose design increases compressive stresses in the shear-affected zone and therefore significantly raises the cutting surface quality. Compared to conventional punching, concave punch nose design increases clean cut proportions by more than 100%.

Keywords: Punch design · Shear cutting · Cutting surface quality

1 Introduction and State of the Art

Shear cutting is one of the most frequently used manufacturing processes in the sheet metal working industry [1]. Almost every sheet metal component needs to be trimmed or punched during its production chain [2]. Thereby, the component edges resulting from these shear cutting processes must meet increasing quality requirements for components´ functional surfaces [3]. In industrial production processes, high-quality grades for punched and trimmed edges and surfaces are characterized by a small edge draw-in height, a high proportion of clean cut, absence of burrs, low fracture surface heights and narrow manufacturing tolerances (see Fig. 1).

In addition to a high cutting surface quality, the productivity of shear cutting processes is also of importance. The productivity of trimming and punching processes is characterized by high output rates (components per minute), low die costs and low die maintenance costs. These criteria are met in particular with conventional shear cutting or punching processes using single-acting presses. A disadvantage of such conventional

M. Liewald et al. (Eds.): WGP 2022, LNPE, pp. 179–188, 2023.
https://doi.org/10.1007/978-3-031-18318-8_19

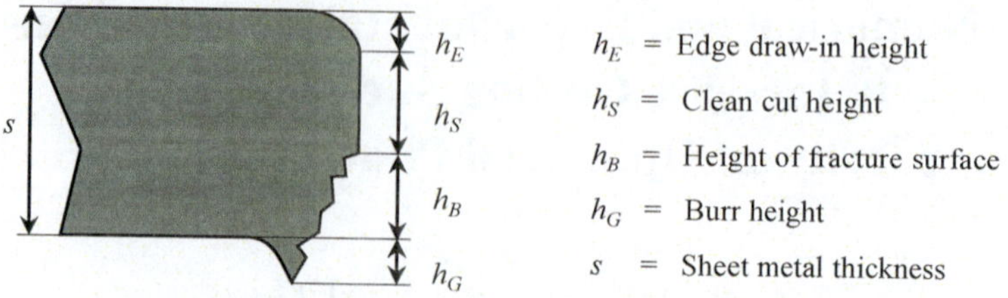

Fig. 1. Characteristic shear cutting surface appearance

shear cutting processes (= normal cutting), however, is that only component edges with comparatively coarse tolerances (IT11) and maximum clean cut proportions (CCP) of up to 50% of the sheet thickness can be produced. If production and method planners are aiming for higher component qualities, more complex cutting processes such as fine blanking, precision blanking or recutting need to be used today. By fine blanking, for example, CCPs (= hs/s) up to 100% and component qualities of tolerance class IT7 can be achieved. Compared to conventional trimming and punching processes, however, the higher tool and process complexity of such precision cutting processes lead to significantly lower output quantities as well as higher tooling and component costs.

In order to achieve high CCPs without the need for complex tool kinematics, Senn and Liewald [4] proposed a new punch design using a concave front surface in 2018. The idea of the so-called concave punch nose design is to optimize the geometry of conventional punches in order to enlarge clean cut proportion along the produced cutting surface (see Fig. 2). In analogy to fine blanking, the basic physical principle behind this process is to induce compressive stresses in the shear-affected zone [5].

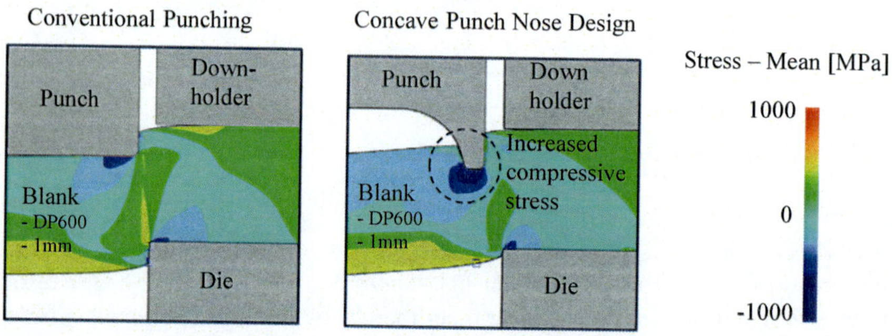

Fig. 2. Conventional punching process (left) and concave punch nose design (right)

The major advantage of using a concave punch nose design as opposed to other precision punching processes is that cutting surface quality can be improved by minor changes in existing punching tools. Thus, punches of conventional shear cutting processes only need to be replaced by the optimized concave punch nose design. However, according to the current state of the art, there is no design method for cutting processes with such filigree punches or blades. Therefore, this paper deals with a numerical analysis of the concave punch nose design considering a wide range of cutting parameters.

2 Investigated Sheet Metal Materials and Simulation Model Setup

Three different sheet metal materials (DC03, DP600 and DP800) were considered for the numerical analysis of the shear cutting process with concave punch nose design in order to investigate a wide range of potential use cases. The yield curves of these sheet materials were determined from experimental tensile tests and subsequently extrapolated using the Hockett-Sherby approach. Figure 3 summarizes the Hockett-Sherby fitting parameters and the fitting result for the investigated sheet metal materials.

Flow curves and extrapolation

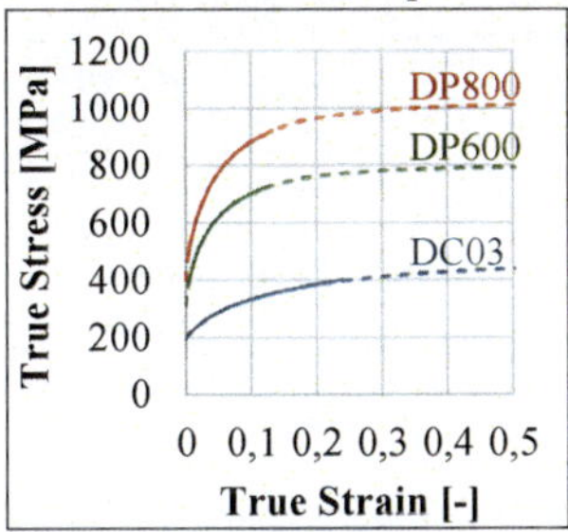

Hockett Sherby extrapolation parameters

$$k_f(\varphi) = k_{f,s} - \left(k_{f,s} - k_{f,0}\right) \cdot e^{-m\varphi^P}$$

	DP600	DP800	DC03
kf_s	794.013	1015.087	457.886
kf_0	311.541	398.240	197.799
m	8.076	7.726	4.508
p	0,692	0,706	0,784

Fig. 3. Experimentally determined flow curves and Hockett Sherby extrapolation parameters

Simulation set up was performed using DEFORM 2D, which is suitable for precise numerical calculation of cutting and punching processes [6, 7]. The damage modelling necessary for a cutting simulation was realized by the approach "Normalized Cockroft Latham", which is commonly used for scientific research [8] and industrial investigations on shear cutting processes [9]. This damage model is defined as a function of maximum principal stress (σ^*) normalized with effective stress (σ_V). The material constant C (see Eq. 1) represents the amount of ductile energy that can be applied to the sheet metal material until fracture occurs. As soon as the parameter C exceeds a material-specific limit value, finite elements are deleted in order to simulate fracture formation.

$$C_{normC\&L} = \int_0^{\overline{\epsilon_f}} \frac{\sigma^*}{\sigma_V} d\overline{\epsilon} \tag{1}$$

Figure 4 shows the geometric structure of the simulation model with the correspondingly varied cutting parameters. A constant punch diameter of 10 mm was chosen for all numerical and experimental investigations. The web width, the web angle, the clearance, the sheet metal thickness and the sheet metal material were varied. The cutting edge radii were kept constant at 50 µm in all simualtions and experiments. The chosen size of the cutting edge radii corresponds to current industrial recommendations for punching high-strength sheet materials [10].

A full-factorial test plan was selected for the numerical investigations, so that in total 540 simulation runs were carried out. The die and the down holder were simulated as rigid bodies. In order to keep the calculation times short, the punch was modelled as a

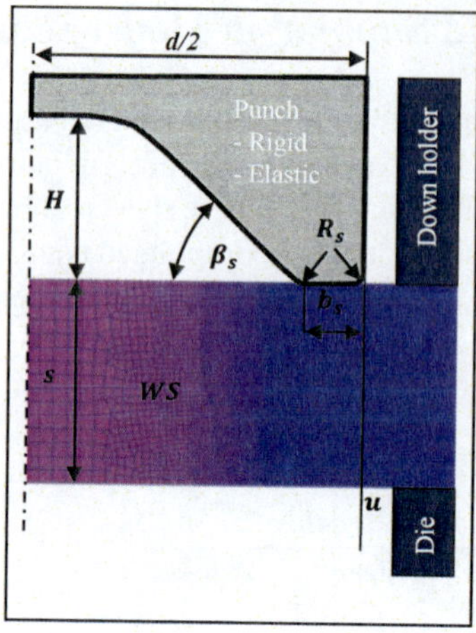

Parameter		Variation
Punch diameter	d	10mm*
Web width	b_s	0.1 – 0.5mm (0.1mm steps)
Web angle	β_s	45° bis 75° (10° steps)
Punch radius	R_s	50µm*
Die radius	R_M	20µm*
Clearance	u	5%, 10%, 15%
Sheet thickness	s	0.5mm, 1.0mm, 1.5mm
Sheet metal material	WS	DP600, DP800, DC03
Height	H	0.8mm*

* Constant for all simulations and experiments

5*4*3*3*3 = 540 Punching simulations

Fig. 4. Numerically investigated parameters

rigid body in a first calculation run. Subsequently, an elastic material model was used for the punch in a second simulation run (E = 210 GPa) in order to carry out a stress analysis by means of force interpolation. For the mesh generation of the blank, mesh windows were used. In the area of the shear affected zone, the smallest element edge length was 6 µm.

3 Results of the Numerical Sensitivity Analysis

In order to evaluate the results of the numerical sensitivity analysis, an image recognition algorithm was developed. This image recognition algorithm was written in the Python programming language and enables a fully automatic analysis of cutting surface characteristics. The algorithm uses a contour detection method to detect the transition areas between the edge draw-in height, clean cut height, fracture surface height and burr height. Figure 5 shows three exemplarily evaluated cutting surface analyses for the sheet metal material DP600.

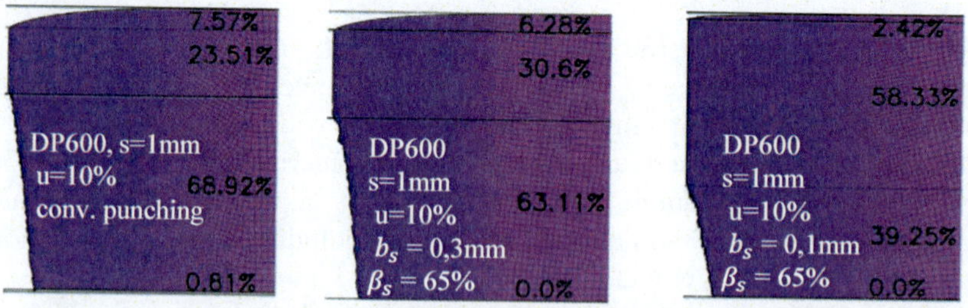

Fig. 5. Comparison between conventionally punched cutting surface and concave punch nose design (DP600, s = 1 mm, u = 10%)

The 3D plots in Fig. 5 give an overview of the overall result of the 540 shear cutting simulations. The cutting parameters web width and web angle were chosen as axes for the 3D plots, since these parameters showed the major influence on the CCP. The more filigree the cutting edge on the cutting punch, the higher is the calculated CCP. This CCP increase can be attributed to two effects. On the one hand, the (local) compressive stresses induced under the punch tip favour the material flow and thus lead to a later crack formation. On the other hand, the small web width reduces the edge draw-in height due to reduced tensile stresses near the blank surface (Fig. 6).

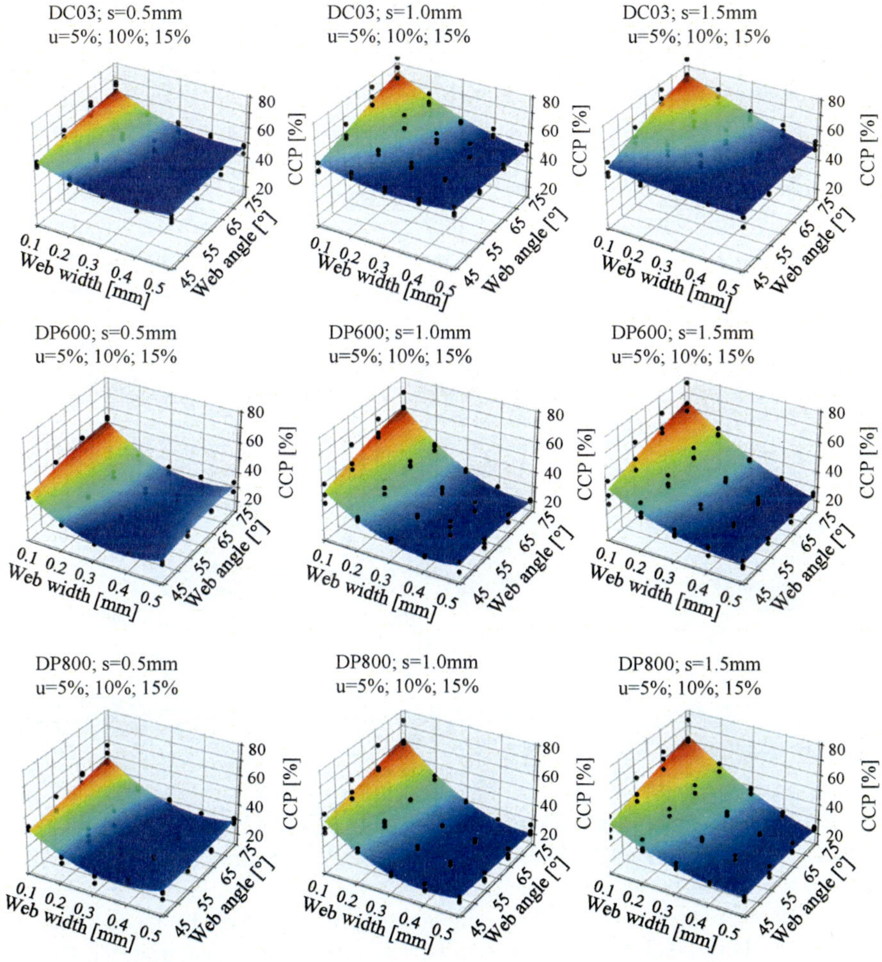

Fig. 6. Results of the numerical sensitivity analysis regarding CCP

In summary, the 3D plots show the following consistent trends:

- The smaller the web width and the steeper the web angle, the higher is the CCP. The punches should therefore be designed as filigree as possible.
- Regarding the CCP, the web width shows the major effect.
- The surface gradients of the 3D plots are steeper for sheet metal materials with a higher tensile strength. Therefore, concave punch nose design seems to be more efficient for high strength sheet metal materials (DP600 and DP800) and less efficient for low strength materials (DC03)
- The cutting parameter "clearance" is often considered as an optimisation parameter in the design of standard punching processes. Regarding shear cutting with punch nose design, however, the numerical investigations show that the web geometry has a comparatively greater influence on the CCP.

The numerical process analysis thus suggests that the punch geometry should be designed as filigree as possible. However, mechanical loads on such fine geometries during shear cutting must be considered in order to prevent immediate break-out of too filigree punch designs. Therefore, obtained punch geometries were meshed in a second calculation step and modelled with a purely elastic material model. By transferring nodal forces from the rigid body simulation (force interpolation), the maximum v. Mises equivalent stresses in the punch tips could be analysed afterwards. Figure 7 summarises the result of this stress analysis.

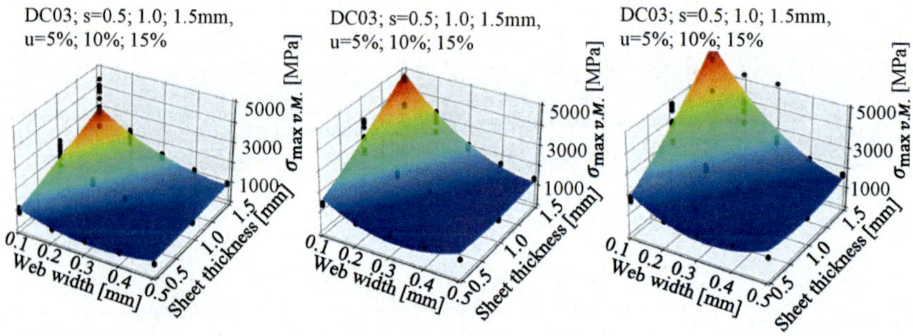

Fig. 7. Results of the numerical sensitivity analysis regarding maximum von Mises stress at the punch tip

Figure 7 shows that the maximum v. Mises equivalent stress strongly depends on the web width and the sheet thickness. The punch stresses mainly result from vertical compressive stresses (punch movement direction). A smaller proportion of the punch stresses results from transverse forces acting laterally on the punch tip. Due to its material properties, HSS steel 1.3343 seems to be a suitable punch material. This punch material has a compressive strength of 3000 MPa and a bending strength of 4000 MPa. Considering these material characteristics and the stresses shown in Fig. 7 the following process limits for a concave punch nose design can be defined:

- 0 mm $\leq$ sheet metal thickness s $\leq$ 1.0 mm
- 0.15 mm $\leq$ web width b_s $\leq$ 0.3 mm

• Tensile strength of sheet metal material ≤ 800 MPa.

4 Experimental Validation

The simulation results described above were validated by experimental investigations. Figure 8 contains an overview of the cutting parameters investigated in these experiments. Punches with a web width of 0.15 and 0.25 mm as well as conventional punches as a reference were produced for the performed punching experiments. The web angle was kept constant at 75°, since comparatively high CCPs are to be expected for the steepest numerically investigated web angle. Furthermore, two different clearances (u = 10%, u = 15%) were investigated. For each parameter constellation, 3 test repetitions were carried out, so that a total of 45 punching tests were performed. A KEYENCE digital microscope was used to analyse the experimentally achieved CCPs.

Parameter		Variation (Experiments)
Web width	b_s	0.15mm, 0.25mm
Web angle	β_s	75°
Punch radius	R_s	50µm
Die radius	R_M	20µm
Clearance	u	10%, 15%
Sheet thickness	s	1.0mm
Sheet metal material	WS	DP600, DP800, DC03
Height	H	0.8mm

Fig. 8. Manufactured punches (left) and experimentally investigated parameters (right)

Figures 9, 10 and 11 show the results of the experimental cutting surface analysis for a clearance with u = 15%. The correlation between the web width and the CCP, which was already predicted by the punching simulations, can be seen very clearly. For all three sheet materials, the CCP increases significantly with decreasing punch web width. Furthermore, the experimental punching investigations prove that a concave punch nose design is particularly efficient for high-strength sheet materials. Compared to normal cutting, a smooth cut increase of up to 30.8% is possible for the lower-strength sheet material. For the sheet materials DP600 and DP800, the CCP increase is 65.3% (DP600) and 109.4% (DP800). No immediate breakout of the filigree cutting edge could be detected on any of the examined punches. Therefore, it can be assumed that the stress analysis carried out in Sect. 3 is a suitable method for the geometrical design of punches with a concave nose design.

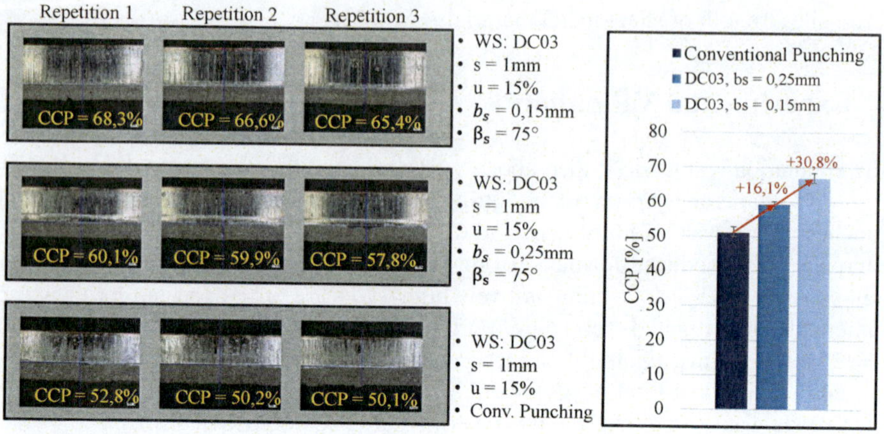

Fig. 9. Experimentally determined CCPs for the sheet metal material DC03 (u = 15%)

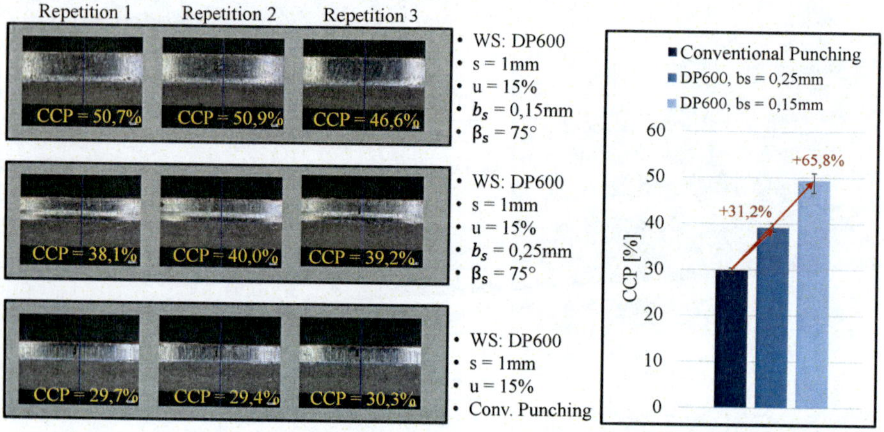

Fig. 10. Experimentally determined CCPs for the sheet metal material DP600 (u = 15%)

In order to validate the punching simulations, the numerically calculated CCPs were compared to the experimentally determined CCPs. The red line in Fig. 12 represents an ideal match between simulation and experiment (CCP Exp = CCP Sim). Furthermore, the mean absolute error (MAE) for the comparison between the numerically calculated CCPs and the experimentally determined CCPs is 3.69%. Due to this small error, the numerically determinded hypersurface models (Sect. 3) can be considered as validated.

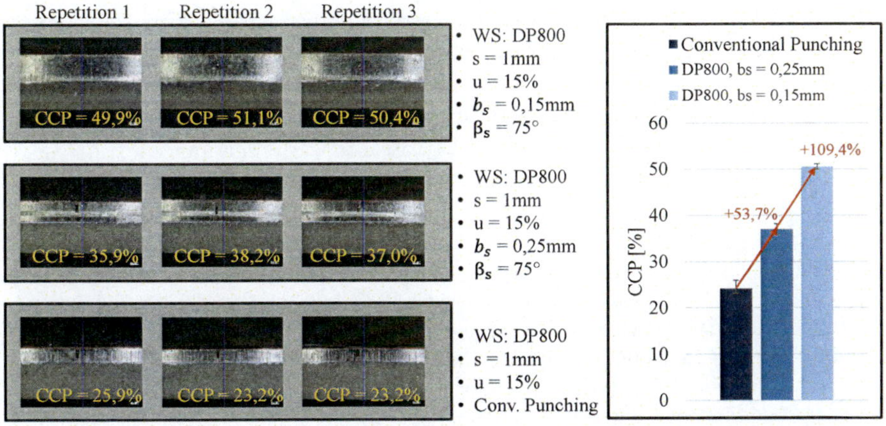

Fig. 11. Experimentally determined CCPs for the sheet metal material DP800 (u = 15%)

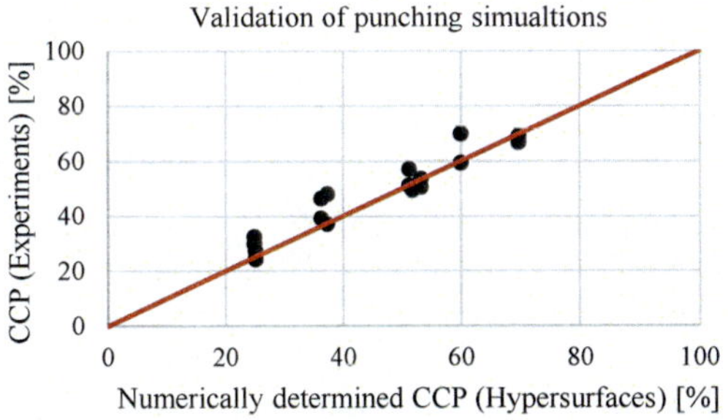

$$MAE = \frac{1}{n}\sum_{i=1}^{n} abs\,(y_i - \hat{y})$$

$$MAE = 3{,}69\%$$

Fig. 12. Validation of the punching simulations (left) and MAE (right)

5 Summary, Conclusions and Further Research

The content of this paper includes numerical and experimental investigations on a new punch design for conventional shear cutting processes called "concave punch nose design". The idea of the concave punch nose design is to optimize the geometry of conventional punches in order to enlarge clean-cut proportion along the cutting surface. In analogy to fine blanking, the basic physical principle behind this process is to induce compressive stresses in the shear-affected zone. The increase in compressive stresses is achieved by a small web attached to the cutting punch as well as a corresponding web angle. Since the existing state of the art contains no design method for cutting processes with such filigree punches or blades, the objective of this paper was a numerical and experimental analysis of the concave punch nose design considering a wide range of cutting parameters. Based on a numerical parameter study, hypersurface models could be obtained that quantify the relationship between the investigated cutting parameters (web width, web angle, sheet metal material, sheet thickness, etc.) and the achievable clean cut proportions (CCP) as well as the maximum v. Mises stresses in the punch tip. Overall, a good agreement was found between the numerically determined results (clean-cut proportions) and the experimental punching tests. Compared to a conventional punching process, a CCP increase of 30.8% was observed for the DC03 sheet material. For the high-strength sheet material DP600, a CCP increase of 65.8% could be achieved. The best results were obtained for the sheet material DP800 with a CCP increase of 109.4%. Future research (IGF research project no. 21053N) concerns endurance testings in order to determine the durability and wear mechanisms of the concave punch nose design in series production.

Acknowledgements. The research project "Erhöhung der Schnittflächenqualität mittels Hohlschneiden" of the Europäische Forschungsvereinigung EFB e.V. is carried out in the framework of the industrial collective research programme (IGF no. 21053N). It is supported by the Federal Ministry for Economic Affairs and Climate Action (BMWK) through the AiF (German Federation of Industrial Research Associations eV) based on a decision taken by the German Bundestag. Additionally, we would like to thank our project partners Stueken GmbH & Co. KG, SCHEUERMANN + HEILIG GmbH, Novelis Switzerland SA, Werkzeugbau Ammer Quick & Partner GmbH, Salzgitter Mannesmann Forschung GmbH, DYNAmore GmbH, Mercedes-Benz Group AG, Hans Berg GmbH & Co. KG, ZF Friedrichshafen AG, Eckold GmbH & Co. KG.

References

1. Hoffmann, H., Neugebauer, R., Spur, G.: Handbuch Umformen – Handbuch der Fertigungstechnik. Carl Hanser Verlag, München (2012)
2. Siegert, K.: Blechumformung – Verfahren, Werkzeuge und Maschinen. Springer, Berlin Heidelberg (2015)
3. Sachnik, P.: Methodik für gratfreie Schnittflächen beim Scherschneiden. Dissertation, TU München (2017)
4. Senn, S., Liewald, M.: Numerical investigation of a new sheet metal shear cutting tool design to increase the part quality by superposed compression stress. J. Phys. Conf. Ser. **1063** (2018)
5. Senn, S., Liewald, M.: Investigation of a new sheet metal shear cutting tool design to increase the part quality by superposed compression stress. IOP Conf. Ser. Mater. Sci. Eng. **651** (2019)
6. Nan.: Deform-2D V12.1 User Manual, SFTC-Deform, Columbus, USA (2022)
7. Uhlmann, E., Von der Scheulenburg, M., Zettier, R.: Finite element modeling and cutting simulation of Inconel 718. CIRP Ann. **56**(1), 61–64 (2007)
8. Han, D., et al.: Investigation of the influence of tool-sided parameters on deformation and occurring tool. Procedia Manuf. **15**, 1346–1353 (2018)
9. Neugebauer, R., et al.: Velocity effects in metal forming and machining processes. CIRP Ann. Manuf. Technol. **60**, 627–650 (2011)
10. Thyssenkrupp Homepage. https://www.thyssenkrupp-steel.com/de/produkte/feinblech-oberflaechenveredelte-produkte/mehrphasenstahl/dualphasenstahl/. Last accessed 2022/05/05

Process Design Optimization for Face Hobbing Plunging of Bevel Gears

M. Kamratowski[1](✉), C. Alexopoulos[1], J. Brimmers[1], and T. Bergs[1,2]

[1] WZL of RWTH Aachen University, Campus-Boulevard 30, 52074 Aachen, Germany
m.kamratowski@wzl.rwth-aachen.de

[2] Fraunhofer IPT, Steinbachstraße 17, 52074 Aachen, Germany

Abstract. The economic efficiency of cutting processes is generally determined by tool wear. Empirical investigations on tool wear in bevel gear cutting, however, are limited to face milling plunging. Hence, the objective of this paper is to examine the influence of the process design on tool wear for an industrial application of face hobbing plunging. For this purpose, first the influence of cutting speed and feed on process characteristics is analyzed with the help of the manufacturing simulation BEVELCUT. Subsequently, cutting trials using the same process parameter variations are performed and their influence on tool wear and machining time is assessed. Finally, the simulation results are compared to the results of the cutting trials in order to identify correlations between the process characteristics and the tool wear.

Keywords: Face hobbing · Bevel gears · Tool wear

1 Introduction and Motivation

Soft cutting of bevel gears with multi-part tool systems consisting of a cutterhead and carbide stick blades is common throughout the manufacturing industry [1]. Nowadays, bevel gears are manufactured on CNC machines with six independent, directly driven axes [2]. Soft cutting of hypoid bevel gears can, on one hand, be distinguished by the applied indexing method and thus, by their flank generation, in continuous indexing (face hobbing) and discontinuous indexing (face milling). On the other hand, soft cutting processes can be classified according to their type of profile generation into plunging and generating. For manufacturing, face milling or face hobbing is combined with either plunging or generating [3].

Regardless of the manufacturing process, the economic efficiency of the cutting process is determined by tool wear [1]. However, due to the complex process kinematics, wear along the cutting edge of the stick blade varies locally. Thus, an analytical description of the tool wear for bevel gear cutting has been proven to be difficult [4–6]. This poses a significant challenge to the process design. To determine efficient process parameters, either experimental studies are performed or the process is analyzed with the help of simulations [4–6]. So far, empirical investigations on the tool wear behavior in bevel gear cutting are limited to face milling plunging [4–7].

M. Liewald et al. (Eds.): WGP 2022, LNPE, pp. 189–198, 2023.
https://doi.org/10.1007/978-3-031-18318-8_20

2 State of the Art

Experimental as well as simulative investigation on tool wear in bevel gear cutting have been conducted solely for face milling plunging so far. KLEIN varied the process and tool design for single blade group trials. Regarding the process parameter variations, the maximum tool life was achieved for a cutting speed of $v_c = 280$ m/min and feed of $f_z = 0.15$ mm. Additionally, tool life was determined by a local maximum of the wear at the corner of the cutting edge in all cases [8].

HERZHOFF and HARDJOSUWITO developed tool life models to predict the tool wear for face milling plunging of bevel gears. Both tool life models are based on empirical models for single flank chips, which were extended by a multi-flank chip factor to consider the cross-section of bevel gears. In HERZHOFF'S model, the individual factors depend on stress and temperature, which were derived by a three-dimensional finite element model of the plunging process. HARDJOSUWITO'S model was composed by a regression analysis of geometrical input data, which was derived through a three-dimensional penetration calculation. The model parameters for both HERZHOFF'S and HARDJOSUWITO'S models were calibrated through bar turning trials [5, 6].

MAZAK investigated the influence of process and tool design on tool wear for face milling plunging through single blade group trials with restricted chip evacuation and simulations. With respect to the process parameter variation, MAZAK observed that the initial feed $f_{z,1}$ of the feed ramp has the highest influence on the tool wear. Furthermore, the highest tool life was achieved for cutting speeds between $v_c = 220$ m/min and $v_c = 240$ m/min and feeds between $f_z = 0.16$ mm and $f_z = 0.18$ mm. Based on the results of the cutting trials, MAZAK developed a first principle based tool life model. [4].

HOU ET AL. set up a finite element based simulation for face milling plunging. After a machining test, images of the worn tool were compared to simulation results. A correlation between high temperature, high pressure areas and tool wear could be identified [7].

Analysis of tool wear for face hobbing plunging has been performed exclusively by simulation so far. HABIBI proposed a method to reduce the tool wear during face hobbing plunging. By locally adapting the cutting edge geometry in a CAD-based penetration calculation, the change in in effective angles was kept to a minimum, which would lead to reduced tool wear [9]. However, no validation of the method through cutting trials has been performed yet.

3 Objective and Approach

As tool life has a significant impact on the economic efficiency of a manufacturing process, knowledge of the influencing factors is of economic interest. The current state of the art shows the lack of investigations on face hobbing of bevel gears. The objective of this paper is to analyze the influence of the process design on tool wear for face hobbing plunging by means of simulations as well as cutting trials. For this purpose, three cutting speeds v_c and three feeds $f_{z,1}$ are combined in a full-factorial manner.

First, a theoretical analysis is performed using the manufacturing simulation BEVELCUT. Based on a planar penetration algorithm, BEVELCUT calculates process characteristics, such as the maximum uncut chip thickness $h_{cu,max}$ or maximum uncut chip length $l_{cu,max}$. Since the cutting speed v_c does not influence the simulation results, solely the influence of different feeds $f_{z,1}$ is examined. Subsequently, the results of cutting trials are presented, in which the influence of the varied process parameters on tool wear and machining time is evaluated. Finally, the simulation results are compared to the results of the cutting trials to identify correlations between the process characteristics and the tool wear.

4 Conceptual Design

The investigated gear design is part of a vehicle transmission with $n = 45$ teeth and a normal module of $m_n = 3.65$ mm. The tool used for the cutting trials was a left-hand cutter head with $z_0 = 17$ blade groups consisting of an inside and an outside blade. The nominal cutter radius was $R_w = 76$ mm.

4.1 Design of Experiments

The cutting trials were conducted on a KLINGELNBERG C30 bevel gear cutting machine in single blade group trials with restricted chip evacuation, cf. Fig. 1. The cutterhead setup used in the analogy trial is similar to the setup used by KLEIN and MAZAK [4, 8]. However, to emulate face hobbing kinematics, the outside and inside blade were placed within one blade group instead of opposite to each other. This ensures that both blades cut the same tooth slot. To regard the chip formation within limited space, dummy blades were placed in front of and behind the blade group. To assure that the dummy blades did not contribute to the cutting process, they were lowered by 1 mm. In order to achieve comparable tool wear in single blade group trials to series production with a fully complement cutterhead, ten workpieces were cut in each trial.

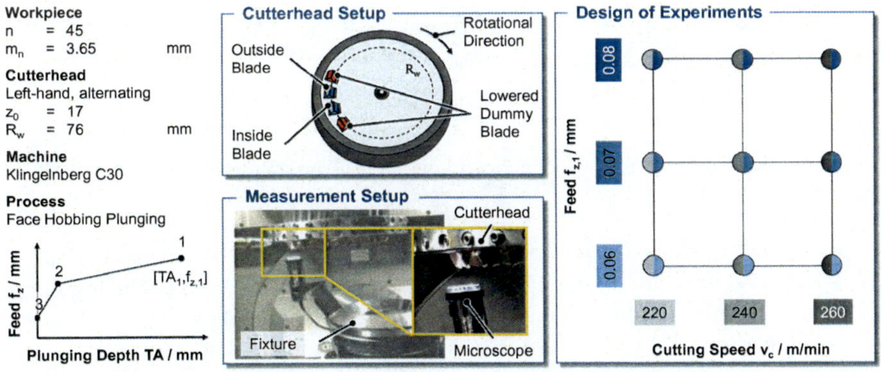

Fig. 1. Trial setup and design of experiments

Tool wear was measured using a DINO- LITE digital microscope inside the machine, cf. Fig. 1. The microscope was mounted on a fixture, which was clamped by the workpiece holder. To obtain comparable images of the worn tools, the microscope was positioned by moving the machine axes with a CNC program. Images of the cutting edge, clearance side and tip of the inside as well as the outside blade were taken. Based on the images of the worn blades, the wear width was measured using a software program.

For the cutting trials, a degressive feed ramp was used, cf. Fig. 1. The cutting process starts at plunging depth TA_3 and ends at full plunging depth TA_1. Solely the initial feed $f_{z,1}$ was varied, because the desired surface finish is achieved during the last cuts, which should not be changed. To investigate the influence of the feed $f_{z,1}$ at different cutting speeds v_c, a full factorial design with three feed levels at $f_{z,1} = 0.06$ mm, $f_{z,1} = 0.07$ mm and $f_{z,1} = 0.08$ mm and three cutting speed levels at $v_c = 220$ m/min, $v_c = 240$ m/min and $v_c = 260$ m/min was applied, cf. Fig. 1.

4.2 Tool and Workpiece Characterization

The stick blades used in the cutting trials were K30 carbide tools coated with AlTiN. The chemical element composition of the cutting material and coating was determined by EDX analysis. The cutting material consisted of 98.4% tungsten carbide and 1.6% cobalt binder. In Fig. 2, the tungsten carbide of the cutting material is depicted in white and enclosed by the black cobalt binder phase. All blades were coated in the same batch with six layers of AlTiN, which resulted in a coating thickness of 6 μm. The coating consisted of 46.1% Ti, 32.9% Al and 21% N.

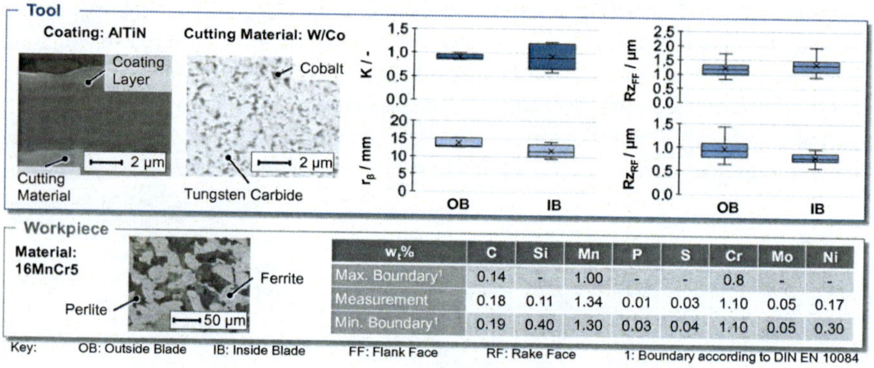

w_t%	C	Si	Mn	P	S	Cr	Mo	Ni
Max. Boundary[1]	0.14	-	1.00	-	-	0.8	-	-
Measurement	0.18	0.11	1.34	0.01	0.03	1.10	0.05	0.17
Min. Boundary[1]	0.19	0.40	1.30	0.03	0.04	1.10	0.05	0.30

Fig. 2. Tool and workpiece characterization

The cutting edge radius r_β and form-factor K of the outside (OB) and inside blade (IB) were determined through optical measurement. The results are depicted using boxplot diagrams, cf. Fig. 2. The mean values of the form-factor for the outside and inside blade are comparable with $K_{OB} = 0.9$ and $K_{IB} = 0.92$ respectively. However, the measured values of the inside blade scatter more than those of the outside blade. The mean value of the cutting edge radius of the outside blade $r_{\beta,OB} = 13.72$ mm is higher than of the inside blade $r_{\beta,IB} = 11.68$ mm.

The maximum height of the roughness profile Rz on the flank face (FF) of the cutting edge and rake face (RF) of the outside and inside blade were obtained through tactile measurement. The maximum height of the roughness profile Rz on the flank face of the inside blade is slightly higher than on the outside blade with $Rz_{FF,IB} = 1.35$ μm compared to $Rz_{FF,OB} = 1.2\ \mu m$. In contrast, the maximum height of the profile Rz on the rake face of the inside blade is with $Rz_{RF,IB} = 0.78\ \mu m$ lower than on the outside blade with $Rz_{RF,OB} = 0.96\ \mu m$.

16MnCr5 case-hardening steel was used as workpiece material, which had a ferrite-perlite microstructure, cf. Fig. 2. The chemical composition was determined by optical emission spectrometry. Except for the mass fraction of manganese, all values were within the tolerance range required by DIN EN 10084 [10].

5 Simulative Analysis

The simulative analysis was performed with the manufacturing simulation BEVELCUT. Based on the workpiece and tool geometry as well as the process parameter, BEVELCUT uses a planar penetration algorithm to determine process characteristics, such as the maximum uncut chip thickness $h_{cu,max}$ or maximum uncut cutting length $l_{cu,max}$. The distribution of the process characteristics along the unrolled profile edge s allows conclusions on tool wear. For example, the maximum uncut chip thickness $h_{cu,max}$ can be used to determine the load at individual points of the cutting edge during the cutting process. The maximum uncut chip length $l_{cu,max}$ represents the contact time between tool and workpiece.

Since BEVELCUT is based on a planar penetration algorithm, the cutting speed v_c does not influence the simulation results. Hence, solely the influence of the feed levels $f_{z,1}$ on the process characteristics was analyzed. For better orientation, the corner radii as well as the cutting edge (CE) and clearance side (CS) are highlighted in the diagrams.

For all investigated feed levels $f_{z,1}$, the maximum uncut chip thickness $h_{cu,max}$ is almost constant along the cutting edge of the outside blade, cf. Fig. 3. At the corner radius of the cutting edge, the maximum uncut chip thickness $h_{cu,max}$ increases significantly and reaches its maximum. The maxima differ depending on the feed $f_{z,1}$. For a feed of $f_{z,1} = 0.06$ mm the maximum uncut chip thickness is $h_{cu,max} = 0.054$ mm, while for a feed of $f_{z,1} = 0.07$ mm a value of $h_{cu,max} = 0.068$ mm and for a feed of $f_{z,1} = 0.08$ mm a value of $h_{cu,max} = 0.087$ mm is calculated. No maximum uncut chip thickness $h_{cu,max}$ occurs at the tip or the corner radius of the clearance side. Along the flank of the clearance side the maximum uncut chip thickness $h_{cu,max}$ is nearly constant.

The distribution of the maximum uncut chip length $l_{cu,max}$ along the unrolled profile edge s of the outside blade is not affected by the change in feed $f_{z,1}$. For all three investigated feed levels $f_{z,1}$ the maximum uncut chip length $l_{cu,max}$ increases slightly along the cutting edge and reaches its maximum at the corner radius of the cutting edge at $l_{cu,max} = 41.45$ mm. While there is no maximum uncut chip length $l_{cu,max}$ at the tip or corner radius of the clearance side, a nearly constant value of $l_{cu,max} = 2.79$ mm is calculated for the flank of the clearance side.

Identically to the outside blade, the maximum uncut chip thickness $h_{cu,max}$ and maximum uncut chip length $l_{cu,max}$ were evaluated along the unrolled profile edge s of the inside blade, cf. Fig. 4. While the distribution of the maximum uncut chip thickness $h_{cu,max}$ is not influenced by the feed $f_{z,1}$, the magnitude changes. Along the cutting edge, the maximum uncut chip thickness $h_{cu,max}$ is nearly constant. Towards the tip of the blade, the maximum uncut chip thickness $h_{cu,max}$ increases significantly until the global maximum of $h_{cu,max} = 0.053$ mm for a feed of $f_{z,1} = 0.06$ mm, $h_{cu,max} = 0.067$ mm for a feed of $f_{z,1} = 0.07$ mm and $h_{cu,max} = 0.084$ mm for a feed $f_{z,1} = 0.08$ mm is reached. Along the clearance side, the maximum uncut chip thickness $h_{cu,max}$ is again nearly constant.

Neither the distribution nor the magnitude of the maximum uncut chip length $l_{cu,max}$ along the unrolled profile edge s of the inside blade change depending on the feed $f_{z,1}$. Along the cutting edge and towards the tip, the maximum uncut chip length $l_{cu,max}$ increases slightly until the maximum of $l_{cu,max} = 58.66$ mm occurs at the tip. At the corner radius of the clearance side, the maximum uncut chip length $l_{cu,max}$ decreases significantly and is then nearly constant at $l_{cu,max} = 15.29$ mm along the clearance side.

Comparing the values of the maximum uncut chip thickness $h_{cu,max}$ of the outside and inside blade for all feed levels $f_{z,1}$, the maxima are approximately equal to the value of the feed $f_{z,1}$. Furthermore, the maxima occur approximately within the same area on the outside and inside blade. In contrast to that, the feed $f_{z,1}$ does not influence the distribution or magnitude of the maximum uncut chip length $l_{cu,max}$. Hence, solely the

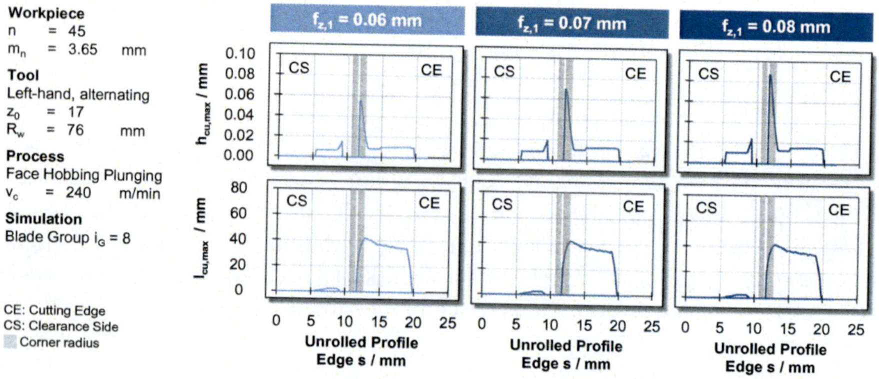

Fig. 3. Process characteristics of the outside blade

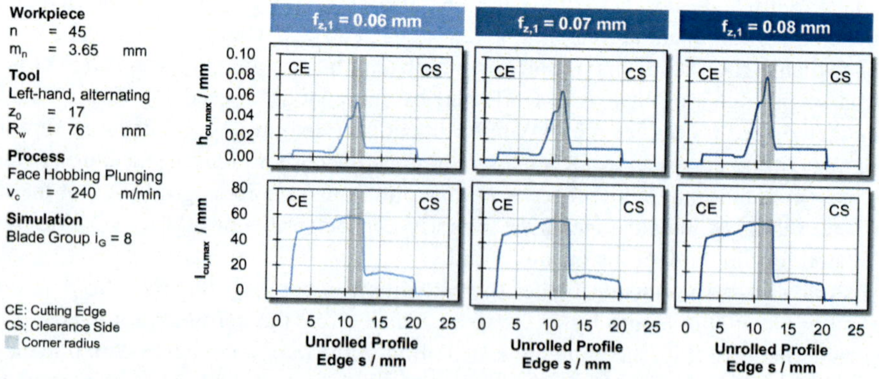

Fig. 4. Process characteristics of the inside blade

load on the blade is influenced by a change in feed $f_{z,1}$ while the contact time between the blade and the workpiece remains unchanged.

6 Cutting Trials

The cutting trials were performed according to the description in Sect. 4.1. The maximum wear width VB_{max} on the cutting edge and tip for the inside and outside blade are depicted in Fig. 5, because no significant wear occurred on the clearance side.

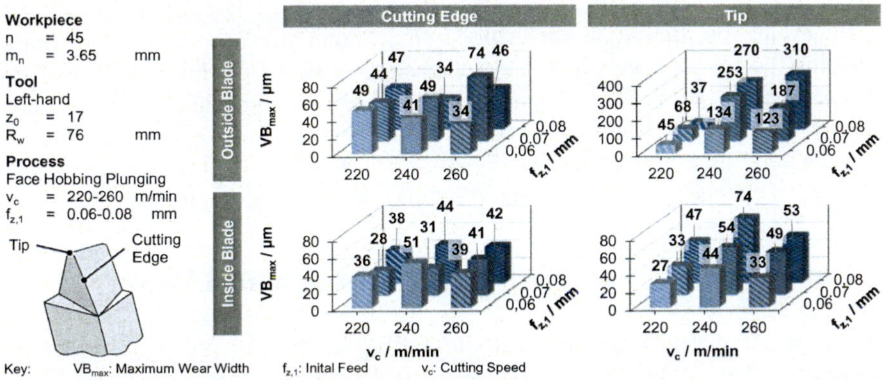

Fig. 5. Maximum wear width on the cutting edge and tip of the outside and inside blade

On the left side of Fig. 5 the maximum wear width VB_{max} on the cutting edge of the outside and inside blade for all cutting speeds v_c and feeds $f_{z,1}$ investigated are displayed. However, no distinct influence of the cutting speed v_c or the feed $f_{z,1}$ on the maximum wear width VB_{max} at the cutting edge can be determined.

The maximum wear width VB_{max} at the tip of the outside and inside blade is depicted on the right side of Fig. 5. At all feed levels $f_{z,1}$, the maximum wear width VB_{max} reaches its maximum at $v_c = 240$ m/min and displays smaller values for higher and lower cutting speeds v_c. An exception is the maximum wear width VB_{max} at the outside blade for a feed of $f_{z,1} = 0.08$ mm, which increases from $VB_{max} = 270$ µm to $VB_{max} = 310$ µm with increasing cutting speed from $v_c = 240$ m/min to $v_c = 260$ m/min. Furthermore, the maximum wear width VB_{max} increases proportional to the feed $f_{z,1}$ with the exception of the maximum wear width VB_{max} at the outside blade for a cutting speed of $v_c = 220$ /min, which decreases from $VB_{max} = 68$ µm at a feed of $f_{z,1} = 0.7$ mm to $VB_{max} = 37$ µm at a feed of $f_{z,1} = 0.08$ mm. In general, the highest maximum wear width VB_{max} was measured at the tip of the outside blade, which hence represents the area to determine the tool life.

The productivity of a process is not only determined by tool wear and thus, by the achieved tool life, but also by the machining time t. Hence, the machining time t has to be taken into account for the evaluation of the cutting trials. Figure 6 summarizes the changes in machining time t in percent, using the trial with a cutting speed of $v_c = 240$ m/min and a feed of $f_{z,1} = 0.07$ mm as reference.

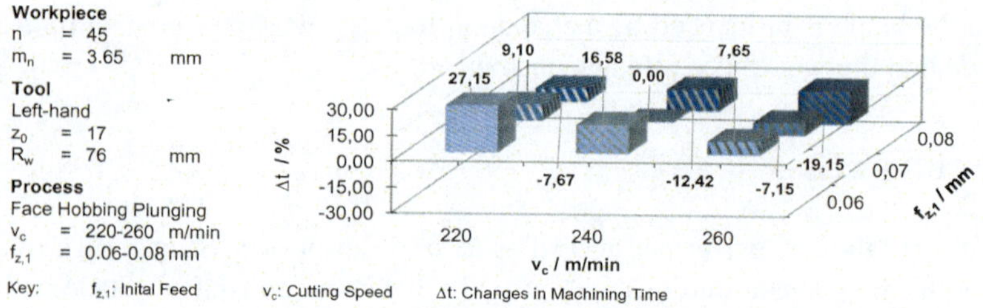

Fig. 6. Influence of the process parameters on the machining time t

The evaluation shows that a reduction in machining time t can be achieved both by increasing the cutting speed v_c and by increasing the feed $f_{z,1}$. However, if the feed is increased by $\Delta f_{z,1} = 0.1$ mm compared to the reference, the machining time is reduced by $\Delta t = 12.42\%$, while an increase of the cutting speed by $\Delta v_c = 20$ m/min results in a reduction of the machining time by $\Delta t = 7.15\%$. Thus, the feed $f_{z,1}$ has a greater influence on the machining time t than the cutting speed v_c in this use case.

7 Comparison of Process Characteristics and Tool Wear

The evaluation of the maximum wear width VB_{max} after the cutting trials in Sect. 6 demonstrates that the highest wear occurs at the tip of the outside blade. To investigate possible causes, the wear at the tip of the outside blade is compared to the distribution of the maximum uncut chip thickness $h_{cu,max}$ along the unrolled profile edge s of the outside blade.

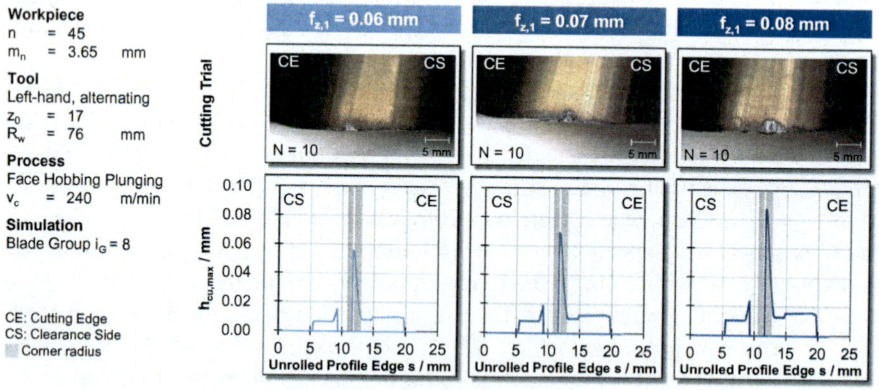

Fig. 7. Comparison of wear and maximum uncut chip thickness $h_{cu,max}$ for the outside blade

The images of the tip of the outside blade after cutting $N = 10$ workpieces with a cutting speed of $v_c = 240$ m/min at all feed levels $f_{z,1}$ are displayed in Fig. 7. For better orientation, the cutting edge (CE) and clearance side (CS) are labeled. All images show chipping, which increases with increasing feed $f_{z,1}$ while the position at which it occurs remains approximately the same. A detailed analysis of course of the maximum uncut chip thickness $h_{cu,max}$ at each feed level $f_{z,1}$ is presented in Sect. 5.

The comparison between wear and maximum uncut chip thickness $h_{cu,max}$, shows that chipping occurs approximately at the same position where the highest maximum uncut chip thickness $h_{cu,max}$ is calculated. Furthermore, the magnitude of the maximum uncut chip thickness $h_{cu,max}$ increases with increasing feed $f_{z,1}$, as does the chipping. However, this possible correlation between maximum uncut chip thickness $h_{cu,max}$ and chipping at the tip of the outside blade requires further investigation for other use cases.

8 Summary and Outlook

The economic efficiency of cutting processes is generally determined by the tool wear. However, investigations on tool wear for bevel gear cutting are limited to face milling plunging. In the presented paper, the influence of the process design on the tool wear for face hobbing plunging of bevel gears was investigated with the help of the manufacturing simulation BEVELCUT and cutting trials. For this purpose, the cutting speed v_c and feed $f_{z,1}$ were varied in a full-factorial manner.

First, the influence of the feed $f_{z,1}$ on the maximum uncut chip thickness $h_{cu,max}$ and maximum uncut chip length $l_{cu,max}$ was investigated through BEVELCUT. The distribution of the maximum uncut chip thickness $h_{cu,max}$ along the unrolled profile edge s of the inside and outside blade was not influenced by the feed $f_{z,1}$. However, the magnitude of the maximum uncut chip thickness $h_{cu,max}$ was proportional to the feed $f_{z,1}$. Regardless of the feed $f_{z,1}$, the maximum chip length $l_{cu,max}$ remained unchanged.

The cutting trials were performed in single blade group trials with restricted chip evacuation. The influence of the process parameters on the tool wear and machining time were evaluated. While no clear influence of the process parameters on the maximum wear width VB_{max} on the cutting edge could be identified, the maximum wear width VB_{max} at the tip of the outside and inside blade increased with an increased feed $f_{z,1}$. Moreover, the maximum wear width VB_{max} at the tip of the inside and outside blade could be reduced by both increasing and decreasing the cutting speed v_c. The machining time was decreased by increasing the cutting speed v_c as well as the feed $f_{z,1}$. The optimum with regards to tool wear and machining time was found at a cutting speed of $v_c = 260$ m/min and feed of $f_{z,1} = 0.07$ mm.

The evaluation of the maximum wear width VB_{max} showed, that in comparison to the inside blade or the flank of the cutting edge in general, significantly higher wear occurred at the tip of the outside blade. Images of the worn tip of the outside blade show chipping along the cutting edge. A comparison between wear and maximum uncut chip thickness $h_{cu,max}$ showed that chipping and the highest maximum chip thickness $h_{cu,max}$ occurred at approximately the same location on the cutting edge. Furthermore, with increasing feed $f_{z,1}$ both chipping and the magnitude of the maximum uncut chip thickness $h_{cu,max}$ increased. However, further investigations are necessary to confirm this correlation.

Further steps include investigating the influence of the tool geometry on tool wear and transferring the optimized process and tool design to series production with a fully complement cutterhead. Moreover, no validated cutting force or chip formation model for bevel gear cutting exists so far. Hence, a development of such models and integration into the manufacturing simulation BEVELCUT could support the simulation-based tool and process design in the future.

Acknowledgments. DFG The authors gratefully acknowledge financial support by the German Research Foundation (DFG) [BE2542/22-1, 389555551] for the achievement of the project results.

References

1. Klocke, F., Brecher, C.: Zahnrad- und Getriebetechnik: Auslegung - Herstellung - Untersuchung - Simulation, 1st edn. Carl Hanser, München (2017)
2. Stadtfeld, H.J.: Gleason Bevel Gear Technology: The Science of Gear Engineering and Modern Manufacturing Methods for Angular Transmissions. The Gleason Works, Rochester, New York (2014)
3. Klingelnberg, J.: Bevel Gear: Fundamentals and Applications. Springer, Berlin (2016)
4. Mazak J.: Method for Optimizing the Tool and Process Design for Bevel Gear Plunging Processes. Diss., RWTH Aachen University (2021)
5. Hardjosuwito, A.F.: Vorhersage des lokalen Werkzeugstandweges und der Werkstückstandmenge beim Kegelradfräsen. Diss., RWTH Aachen University (2013)
6. Herzhoff, S.: Werkzeugverschleiß bei mehrflankiger Spanbildung. Diss., RWTH Aachen University (2013)
7. Hou, F., et al.: Reducing tool wear in spiral bevel gear machining with the finite element method. Gear Solutions, December, pp. 34–40 (2019)
8. Klein, A.: Spiral Bevel and Hypoid Gear Tooth Cutting with Coated Carbide Tools. Diss., RWTH Aachen University (2006)
9. Habibi, M.: Tool Wear Improvement and Machining Parameter Optimization in Non-generated Face-hobbing of Bevel Gears. Diss., Concordia University, Montreal (2016)
10. DIN EN 10084: Einsatzstähle Technische Lieferbedingungen, Berlin Beuth, Juni 2008

Experimental Investigation of Friction-Drilled Bushings for Metal-Plastic In-Mold Assembly

M. Droß[1(✉)], T. Ossowski[1], K. Dröder[1], E. Stockburger[2], H. Wester[2], and B. -A. Behrens[2]

[1] Institute of Machine Tools and Production Technology, Technische Universität Braunschweig, 38106 Braunschweig, Germany
m.dross@tu-braunschweig.de

[2] Institute of Forming Technology and Machines, Leibniz Universität Hannover, 30823 Garbsen, Germany

Abstract. The in-mold assembly process can be used for the production of lightweight hybrid components made of metals and plastics. The connection between the different materials is often realized by a form fit joint. Conventional through-injection points enable the load transfer between the materials. However, through-injection points have disadvantages in the transmission of multiaxial loads. Furthermore, notch effects often occur under load, which can lead to premature failure in the material interface. As a result, the dimensions of the hybrid component or the amount of through-injection points are oversized. In order to increase the bond strength, the use of a friction-drilled bushing was investigated. First, friction drilling tests for varied parameters were performed and analyzed. Second, lap shear tests on hybrid components for appropriate bushings were carried out. The findings obtained have been transferred to the design of a demonstrator. Here, the connection quality between metal and plastic was determined by means of quasi-static and impact load tests. The joint using a friction-drilled bushing thereby confirms the advantages of the enlarged effective area for load transfer compared to conventional through-injection points.

Keywords: Friction drilling · In-mold assembly · Metal-plastic hybrids · Bond strength

1 Introduction

Today's lightweight design concepts for large-scale automotive production are increasingly characterized by the use of hybrid material composite components consisting of fiber-reinforced plastics (FRP) and metal sheet structures. In this context, resource-efficient lightweight design aims at the sensible exploitation of material-specific potentials and enables, ideally, an economic relationship between production costs, increased load-bearing capacity and simultaneous weight reduction. The last point in particular is in the focus of automotive industry due to legal restrictions and socially increasing environmental awareness [1].

M. Liewald et al. (Eds.): WGP 2022, LNPE, pp. 199–208, 2023.
https://doi.org/10.1007/978-3-031-18318-8_21

A key challenge for the large-scale production of material-hybrid composite components is the efficiency of joining processes to overcome the inherent low adhesion between metal and plastic. This is presently carried out through various principles of joining technology and includes adhesive bonding [2], form-fit hybrid injection molding [3], and force-fit bolted joints [4]. Furthermore, there are numerous established mechanical joining processes for FRP-metal composites [5].

A typical process currently used to produce a rigid joint is the in-mold assembly (IMA) process (hybrid injection molding). This encompasses the insertion of metal parts into a mold and their subsequent on-molding or over-molding with mostly short-fiber-reinforced thermoplastics. The connection is achieved by a process-integrated form-fit, which is caused by a number of through-injection points or over-moldings. Parts manufactured in this way use the metal insert as an element for load transfer that passes from the metal joining zone into the plastic reinforcement [6]. However, low transmitting forces by means of the through-injection points can lead to a large number of joints and to material accumulation of FRP and metal, resulting in increased connection areas and oversizing of the component at the same time. Reasons such as these, but also a high demand for safety and structural integrity combined with cost-effective series processing, are responsible for the hesitant use of metal-plastic components in the automotive industry so far [7].

To overcome the challenges in joining metal and plastic within the IMA process, different joining technologies are used. In this paper a new technology is addressed for implementing a form-fit connection between a short-fiber-reinforced plastic and a sheet metal insert structured with a friction-drilled bushing that is capable of bearing loads and meeting the requirements described above. Friction drilling is a forming process in which a bushing is formed in a metal sheet with the use of a rotating carbide mandrel under the action of vertical force. This occurs as a result of the frictional heat generated and the local plasticization in the metal sheet. The process also forms a shoulder on the bushing. The particular advantage of the process is the sufficient one-sided accessibility. In order to increase the interlocking effect of the friction-drilled bushings inside the over-molded plastic structure using the IMA process, the aim is to especially increase the bushing lengths onto the upper side [8].

The bushings are usually used to produce threaded joints with subsequently inserted fasteners. In this process, the bushings formed on the underside of the sheet are used primarily, and the upper bushings are compressed during the forming process. Gies has presented a modification of the classic tool geometry by means of a tool helix angle [9]. Analogous to a fastening screw, the helix angle describes the angle of inclination of the helically rotating forming cleats on the friction drill. Due to its influence on the material flow, the bushing length and the crack development (on the lower bushing), it decisively determines the achievable bushing quality. Positive helix angles up to 37.5° have been proven to increase the transport effect of the displaced material volume in the in-feed direction. The reason for an increased bushing length is explained in the increase of the contact area between tool and workpiece. Therefore, with the aim of producing a high bushing length against the in-feed direction, various negative helix angles (counterclockwise direction of tool rotation) were considered within the scope of this study (see Fig. 1). Once through-molded and over-molded, the bushing acts as a bolt-like connection in the FRP and the collar as an annular undercut in the direction of sheet thickness. As a component of the metallic joining partner, the bushing transfers forces directly into the FRP via an enlarged surface and thus increases the maximum bonding strength.

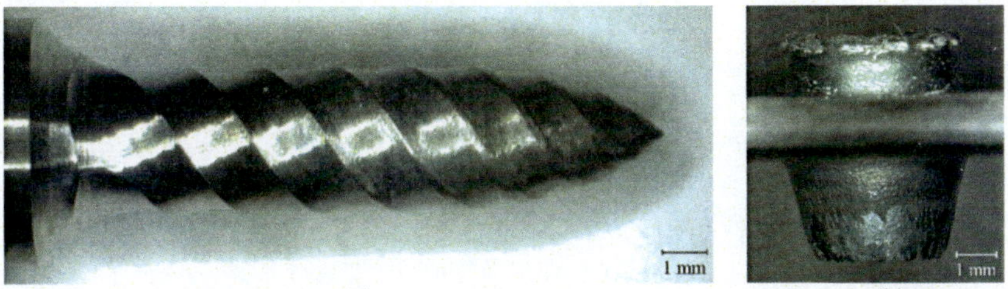

(a) (b)

Fig. 1. Friction drill with 3.7 mm diameter and 60° helix angle (a) and produced bushing (b)

A previous paper was published by the authors describing various modelling techniques and simulation of the friction drilling process parameters as well as their validation [10]. For now, the use of a friction-drilled bushing as an interlocking structure in hybrid structures has not been investigated. The identification of an optimal tool design as well as process parameters for maximizing the upper bushing lengths represent the main research issues of this paper. Therefore, this study presents an analysis of affecting process parameters for friction drilling as well as the investigation of the bonding strength by means of lap shear tests on hybrid components. The results obtained are finally applied in a performance demonstrator, which was tested both quasi-statically and dynamically for its composite strength while being compared with the classic through-injection technology.

2 Experimental Procedures

2.1 Materials

As metal material a high-strength, low-alloy steel (HSLA) HX420 with two different coatings, a zinc coating and a micaceous iron oxide coating, was used. The sheet thickness was 1.5 mm. The material properties, which have been characterized in previous studies [10], are shown in Table 1.

As plastic material, a polyamide 6 with 40 wt.% short glass fiber filling (Ultramid® B3WG8 bk564) was used. The specific properties are summarized in Table 2 and were taken from the CAMPUS® Plastics database [11].

2.2 Friction Drilling Experiments

For the experiments a 5-axis DMU 100 Monoblock milling machine from DMG MORI (DMG MORI AG, 33689 Bielefeld, Germany) was used. The sheet specimens were cut to

Table 1. Material data of the metal component

Description	Units	HX420 LAD
Density	[g/cm^3]	7.85
Elongation at break	[%]	≈ 30
Young's modulus	[GPa]	210
Tensile strength	[MPa]	443
Yield strength	[MPa]	357

Table 2. Material data of the plastic component

Description	Units	Ultramid® B3WG8 bk564 dry/conditioned
Polymer	–	Polyamid 6
Filler	–	40 wt.% glass fiber
Density	[g/cm^3]	1.46
Elongation at break	[%]	3/5.5
Young's modulus	[GPa]	12.7/7.7
Tensile strength	[MPa]	205/125
Melting temperature	[°C]	275
Processing temperature	[°C]	270–290

the dimensions 25 mm × 100 mm and manually deburred. The friction-drilled bushings were inserted using geometrically different tools provided by the company Flowdrill (Flowdrill GmbH, 69469 Weinheim, Germany). The parameters listed in Table 3 were varied in order to determine the influence of the process parameters on the bushing shape. Within the scope of this analysis, their influence on the bushing length on the top and bottom side was compared.

Table 3. Parameter variations

Parameter	Units	Values
Tool diameter	[mm]	2.7 (M3), 3.7 (M4), 4.5 (M5)
Cross-section of shank	[–]	rounded equilateral square
Helix angle	[°]	30, 45, 60
Rotation speed	[1/min]	2500–5800[1]
Feed rate	[mm/min]	50–90[1]

[1]Depending on the diameter

2.3 Metal-Plastic Sample Preparation

All composite specimens were produced on an injection molding machine Engel Victory 330/120 Spex (ENGEL AUSTRIA GmbH, 4311 Schwertberg, Austria). A quick-change fixture from Axxicon (Axxicon Moulds Eindhoven B.V., 5602 BS Eindhoven, The Netherlands) serves for the molds. The molds are kept at a constant temperature by a water temperature control unit. The plastic material was dried at 80 °C for 4 h before each use. For processing, the mass temperature of 270 °C, the injection speed of 25 cm^3/s, a mold temperature of 80 °C and a cooling time of 20 s were set.

To investigate the bonding strength, lap shear tests were carried out in accordance with the testing standard DIN EN 1465 using a universal testing machine of type Zwick 050 (ZwickRoell GmbH & Co. KG, 89079 Ulm, Germany). Prior to the injection molding process, the lower bushings had to be removed manually due to the mold geometry. After production and complete cooling of the test specimens at ambient temperature, the sprue as well as over-molded edge areas and burrs were removed.

Besides the different lap shear tests, a demonstrator based on a stiffened beam structure was selected for testing. The corresponding process chain can be seen in Fig. 2. First, the sheet was press-formed to a u-shaped structure. Afterwards, the zinc-coating was removed at defined locations by removing a surface layer of 100 μm thickness. Subsequently, the bushings (M4) were friction-drilled at these locations. The lower bushings as well as any burr arising were removed. A total of eight bushings were inserted on the beam ground at the intersection points of the rib structure. The diagonal spacing between the friction-drilled bushings is 24.8 mm.

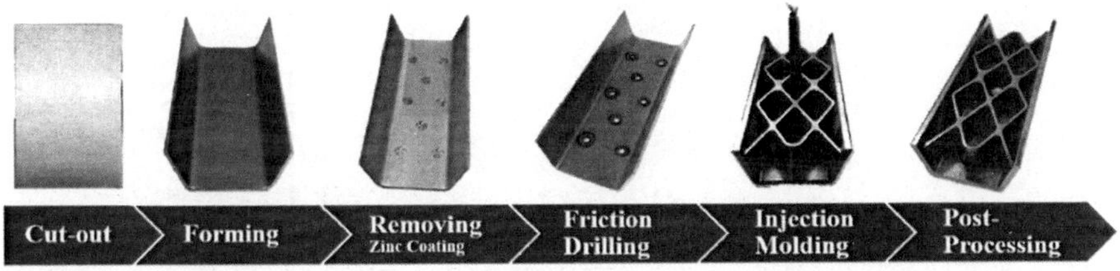

Fig. 2. Process chain for the production of the metal-plastic demonstrator with friction-drilled bushing joining technology

The demonstrator test specimens were subjected to 3-point bending tests according to ISO 178 and ISO 179. The load bearing capacity was investigated in the quasi-static state and under impact conditions. Analogous to the lap shear tests, the quasi-static tests were carried out on a Zwick 050. For the impact tests, an Instron 9250HV (Instron GmbH, 64293 Darmstadt, Germany) drop test rig with a punch diameter of 12 mm and a drop weight of 10.7 kg was used. The test speed was at 5 m/s. To allow a relative comparability between the tests, the same 3-point bending configuration was selected for all tests. The support spacing was set at 120 mm. In each case, a number of 5 specimens were tested.

3 Results

3.1 Effect of Parameter Variation

The increase of the tool diameter itself leads to an increase of the bushing, since there is a trivial correlation between displaced material and tool diameter. It was found that the variation of the tool helix angle has the major geometrical influence on the friction-drilled bushings.

Effect of process parameters onto the bushing length

In [9] the helix angle of 37.5° has been investigated previously. However, the focus was not on increasing the upper bushing length. Therefore, the influence of a helix angle of 45° on the upper and lower bushing length was determined as a function of different speeds and feed rates for a M5 friction drill. In addition, the influence of the direction of rotation was evaluated. For the presentation of the results, an effect diagram is used, which illustrate the percentage influence of the respective parameter set in positive or negative form. In each parameter variation, seven samples were prepared and measured under a microscope. To evaluate the results, the arithmetical averages of the upper and lower bushing lengths were used to calculate the improvement or deterioration in percent. The results obtained for the upper bushing can be seen in Fig. 3.

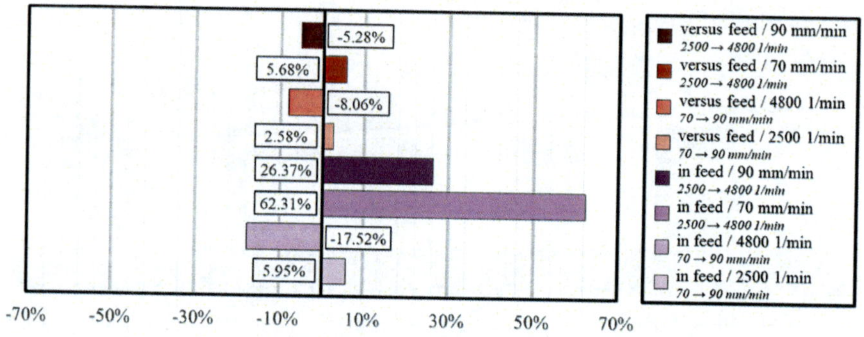

Fig. 3. Comparison of the effect of varying speed or feed for a helix angle of 45° according to the tool rotation direction on the upper bushing length.

There is a significant positive effect of about 62% for a tool rotation in the feed direction, a feed rate of 70 mm/min and a process speed of 4800 1/min compared to a set speed of 2500 1/min. Here, the upper bushing length was increased from 0.86 mm across all parameter variations to an absolute maximum of 1.40 mm. By comparison, a tool rotation versus the feed direction results in a weak effect of 5.68%. Thereby, a maximum bushing length of 1.17 mm is achieved using a rotational speed of 4800 1/min and a feed rate of 70 mm/min. For a tool without helix angle, a maximum bushing length of 1.13 mm was generated with the same set of parameters (4800 1/min and 70 mm/min). The strong effect is assumed to be caused by a longer tool contact time inside the metal sheet, since the bushing was produced at low feed rates and higher rotation speeds.

Analogous to the upper bushing, a low feed rate of 70 mm/min and an increased speed are also advantageous for the lower bushing length. For the length of the lower bushing, a similar behavior but lower effect occurs with a tool rotation direction reverse to the feed direction. A bushing length increase from 2.99 mm (70 mm/min and 2500 1/min) to 3.44 mm (70 mm/min and 4800 1/min) can be achieved.

Effect of the helix angle on the bond strength

The bond strength achieved between the metal sheet and the plastic part depends on the bushing design. As it can be expected, the higher the geometrical resistance of the bushing, the greater the bond strength of the hybrid structure. For this reason, investigations were carried out on the influence of the helix angle respectively the upper bushing length on the achievable bond strength. In each case, a number of seven specimens were tested. The results are shown in Fig. 4.

With an increasing helix angle, the increased bushing length contributes to a locking of the plastic pin within the bushing, so that unbuttoning does not occur anymore. Instead, the pin cracks on its rear side. The crack is clearly visible on the upper side of the plastic sample. At a helix angle of 60° and an associated upper bushing length of about 2 mm, the plastic pin is torn out of the plastic body for all specimens. While the upper bushing length increases with higher helix angles, an increase of the shear stress is not detectable.

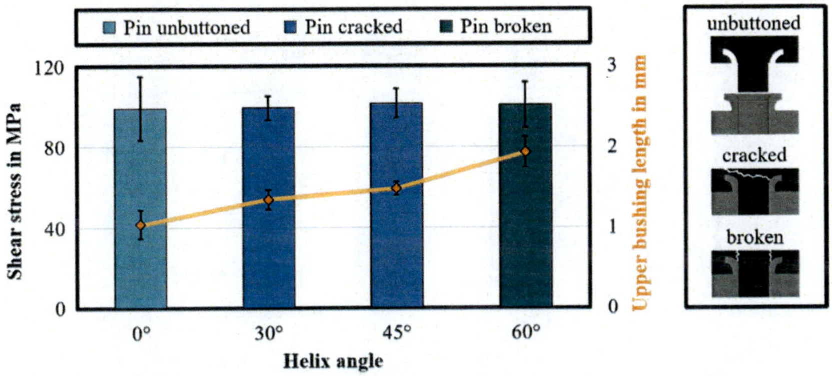

Fig. 4. Comparison of max. Shear stresses and upper bushing lengths for different tool helix angles of a M5 friction drill considering the failure behavior.

3.2 Demonstrator Tests

The results of the quasi-static 3-point bending tests are summarized in Fig. 5. The comparison of the force-displacement curves shows no significant differences between the friction drilling based joining technology and the through-injection point reference joining. This is due to the test setup. The load peaks indicate the failure of particular rib structures. However, the results obtained from the specimens after the test clearly show the positive effect of the friction-drilled bushing (FDB) technology on the bond strength. While the reference specimens undergo a complete loss of bond under load, there is still a rigid bond between metal and plastic for the test specimens with bushings. The performance of additional tests showed that there was no bond between metal and plastic for the reference specimens after the first two load peaks. The results of the impact tests are shown in Fig. 6.

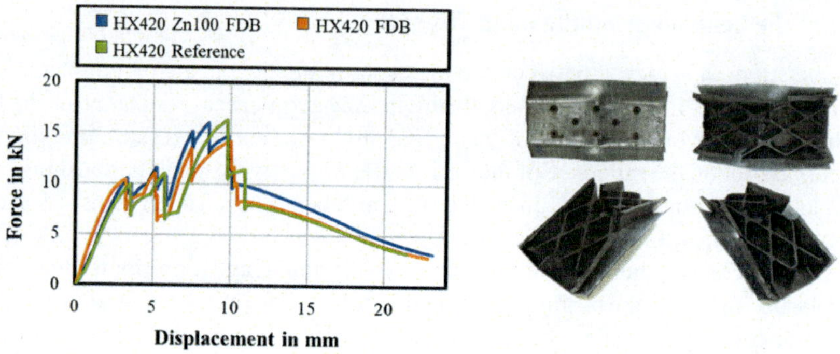

(a) (b)

Fig. 5. Experimental comparison of force-displacement curves for quasi-static tests (a) and one demonstrator specimen with FDB connection after testing (b).

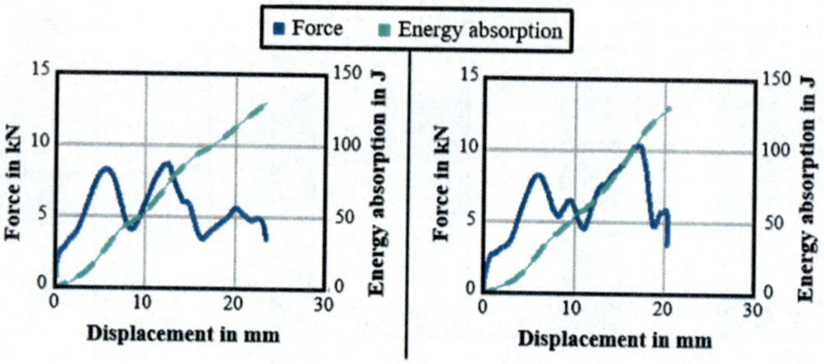

(a) (b)

Fig. 6. Experimental comparison of force-displacement and energy absorption curves for demonstrator specimens with through-injection points (a) and FDB connection (b).

The comparison of the force-displacement curves reveals only minor differences. For the reference specimens, there are two alternating load peaks of equal magnitude. As in the quasi-static tests, these initially represent the rib failure at the supports at the first peak and the subsequent central failure of the plastic body at the second peak. The impact caused a strong spring-back effect, accelerating the test specimens in the direction of the impactor.

Using images of the tests by means of a high-speed camera, a significant loss of connection due to test specimen impact was detected for the reference specimens. In contrast, the test specimens with FDB technology fully maintained a bond between metal and plastic over the entire test. The integrity of the material composite is shown particularly in the energy absorption analysis. When the joint was lost, the reference specimens were able to absorb around 71 J to 12 mm intrusion at 9 kN. The energy absorption of the test specimens with friction-drilled bushings was 112 J to 17 mm intrusion at a maximum load of about 11 kN.

4 Conclusion

This paper presented experimental investigations of a form-fit metal-plastic joint using friction-drilled bushings within the in-mold assembly process. As part of an effect analysis, the influence of the process parameters during friction drilling of the upper and lower side of the bushing was compared. In doing so, the best results were achieved in particular for an increasing helix angle at a low feed rate and high rotation speed.

Finally, the findings were transferred to a demonstrator which was tested quasi-statically and in an impact test. Based on the results of the demonstrator tests, it was possible to show the improved bonding strength of the friction-drilled bushings as well as an increased energy absorption up to 58% compared to classic through-injection points.

Acknowledgements. The authors would like to express gratitude to the companies Flowdrill Fließform-werkzeuge GmbH and Volkswagen AG for providing the friction drills and the materials for the experimental investigations. Furthermore, the authors would like to thank the industrial partners in this research project for the scientific discussion. The Insti-tute of Joining and Welding of the TU Braunschweig is thanked for providing the test equipment for the impact tests.

This research was funded by the Federal Ministry for Economic Affairs and Climate Action on the basis of a decision of the German Bundestag. It was organised by the German Federation of Industrial Research Associations (Arbeitsgemeinschaft industrieller Forschungsvereinigungen, AiF) as part of the program for Industrial Collective Research (Industrielle Gemeinschaftsforschung, IGF) under grant number 20711N.

References

1. Vink, D.: Hybrider Leichtbau wird vielfältiger. Industrie Anzeiger. https://industrieanzeiger.industrie.de/technik/hybrider-leichtbau-wird-vielfaeltiger. Last accessed 2022/02/03
2. Hybride Strukturen mit hoher Festigkeit—Neue Füge-technik für den Leichtbau. In: Kunststoff Magazin. https://www.kunststoff-magazin.de/fvk-werkstoffe/hybride-strukturen-mit-hoher-festigkeit---neue-fuegetechnik-fuer-den-leichtbau.htm. Last accessed 2022/02/07
3. Ridder, H., Schnieders, J.: Hybridspritzgießen. Möglichkeiten und Grenzen. In: Tagung Spritzgießen—Oberflächen von spritzgegossenen Teilen. Hybride Bauteile und Elektromechanik, Düsseldorf. VDI-Verlag, 2007/05
4. Klein, M., Podlesak, F., Höfer, K., Seidlitz, H., Gerstenberger, C., Mayr, P., Kroll, L.: Advanced joining technologies for load and fibre adjusted FRP-metal hybrid structures. J. Mater. Sci. Res. **4**(4) (2015). https://doi.org/10.5539/jmsr.v4n4p26
5. DVS-Reports. https://www.dvs-media.eu/de/neuerscheinungen/. Last accessed 2022/02/03
6. Lanxess Corporation. Plastic/Metal Hybrid Technology. https://techcenter.lanxess.com/scp/americas/en/innoscp/tech/78310/article.jsp?docId=78310. Last accessed 2022/02/06
7. Plastverarbeiter. Verarbeitungsverfahren. Mit hybriden Verfahren zu Hybrid-Bauteilen. https://www.plastverarbeiter.de/verarbeitungsverfahren/mit-hybriden-verfahren-zu-hybrid-bauteilen.html. Last accessed 2022/02/07

8. Eliseev, A., Kolubaev, E.: Friction drilling: a review. Int. J. Adv. Manuf. Technol. **116**(5–6), 1391–1409 (2021). https://doi.org/10.1007/s00170-021-07544-y
9. Gies, C.: Evaluation der Prozesseinflussgrößen beim Fließlochformen mittels DoE. PhD-Thesis, Universität Kassel (2006). ISBN-13 978-3-89958-189-8, Kassel
10. Behrens, B.-A., Dröder, K., Hürkamp, A., Droß, M., Wester, H., Stockburger, E.: Finite element and finite volume modelling of friction drilling HSLA steel under experimental comparison. Materials **14**, 5997 (2021). https://doi.org/10.3390/ma14205997
11. CAMPUS®—a material information system for the plastics industry. https://www.campusplastics.com/ (2021)

Localization of Discharges in Drilling EDM Through Segmented Workpiece Electrodes

K. Thißen[1(✉)], S. Yabroudi[1], and E. Uhlmann[1,2]

[1] Institute for Machine Tools and Factory Management, Technische Universität Berlin, Pascalstraße 8–9, 10587 Berlin, Germany
thissen@iwf.tu-berlin.de

[2] Fraunhofer Institute for Production Systems and Design Technology IPK, Pascalstraße 8–9, 10587 Berlin, Germany

Abstract. In drilling EDM different flushing methods and electrode geometries are applied to improve the process stability and avoid form deviations as a consequence of discharges in the lateral working gap. These methods are usually reviewed indirectly by determining target parameters of the process, the tool electrode or the bore hole. A localization of discharges along the bore hole however provides a quantitative measure for the ratio of discharges in the frontal and lateral working gap. This paper presents a setup for localizing discharges by use of a workpiece electrode consisting of electrically insulated segments and performing signal analyses. Tool electrodes with interior and exterior flushing channels are analyzed with this experimental setup to compare the effectiveness of the resulting flushing conditions. It was found that the use of a helix tool electrode leads to an increased material removal rate by at least 11% and a considerable reduction of lateral discharges.

Keywords: Electric discharge machining · Drilling EDM · Discharge localization · Signal analysis

1 Introduction

The continuous developments in the field of electrical discharge machining (EDM) and especially drilling EDM for applications in the tool and mold making, automotive and aerospace industry are driven by fundamental knowledge about the complex processes during machining operations [1]. The material removal mechanism of EDM by electrical discharges leads to the generation of process residuals like debris particles and gas bubbles. An accumulation of debris in the working gap leads to arcing and short-circuits, related inaccuracies and process instabilities [2]. The effective removal of debris and gas bubbles out of the working gap is crucial to impede the local concentration of consecutive discharges, prevent related process inefficiencies and maintain stable machining. Especially in bottom-closed cavities with large scale during sinking EDM or with high aspect ratios φ during drilling EDM applications, debris and gas bubbles are most likely to accumulate. The sufficient removal of the process residuals is consequently a focus of various research activities in the last decade.

M. Liewald et al. (Eds.): WGP 2022, LNPE, pp. 209–218, 2023.
https://doi.org/10.1007/978-3-031-18318-8_22

Different authors investigate the flow field in the frontal and lateral working gap by means of computational fluid dynamic (CFD) simulations and high-speed camera observations. CETIN ET AL. [3] analyzed various electrode jump configurations for an aspect ratio of $\varphi = 5$ and figured out, that the hole conicity α is considered to be the result of localized secondary discharges caused by debris-rich regions inside the lateral working gap. AYESTA ET AL. [4] stated that during the machining of high aspect ratio slots, there is a sinking depth d_s in which the process destabilizes and the erosion duration t_{ero} starts to grow. The application of side flushing and electrode jumps as high as the workpiece surface are necessary to obtain a total removal of debris out of the cavity. DOMINGOS [5] used a piezo-mechanical system with an optimized frequency of $f = 200$ Hz and amplitude of $A = 4$ μm to introduce vertical vibrations to graphite tool electrodes during the sinking EDM of seal slots with an aspect ratio $\varphi > 11$ in nickel-base casting alloy MAR-M247. A signal analysis showed that the vibration-assisted EDM technology keeps the debris particles in motion inside the lateral working gap and therefore increases the process stability through the reduction of arcing and short-circuit pulses as well as increasing the number of events n per time unit. MEENA ET AL. [6] used copper tool electrodes to investigate different flushing methods for micro-ED drilling with high aspect ratios of $\varphi = 12.5$ in copper workpieces. Using passive immersion of the tool electrodes presented best results for the hole conicity α whereas circularity was assured by pressure flushing only. Any application of side flushing resulted in secondary discharges downstream of the side flushing jet inside the lateral working gap and consequently a loss of circularity. LI ET AL. [7] intensively investigated the gradually deteriorating process conditions and complex events in the working gap during micro-ED drilling using signal analysis and a high-speed camera. Amongst others, the authors found that especially the lateral working gap width s_l provides a space for debris accumulation and electrophoresis, and increases the probabilities of debris concentration, ultimately resulting in nonuniform material removal.

PLAZA ET AL. [8] indirectly confirmed the hypothesis that the volume of helical grooves in the exterior surface of cylindrical tool electrodes would incorporate the debris originating from drilling EDM of the titanium alloy Ti6Al4V and thus improve the flushing conditions. The lower flute angle of $\alpha_1 = 15.5\ °$ as well as flute angles of $\alpha_2 > 46\ °$, led to increased instabilities due to inferior flushing conditions and thus increased the erosion duration t_{ero}. Regardless of the geometry of the helix electrodes, strong process instabilities occurred above aspect ratios of $\varphi = 5.6$. Since the relative frontal wear ϑ_{lF} at the tip of the tool electrode led to the loss of the helix geometry, two process steps of dressing had to be performed, enabling an increase of the aspect ratio to a value of $\varphi = 9$. NASTASI AND KOSHY. [9] used exterior flushing channels for rotation-induced removal of particles from the working gap and optimized these geometries based on CFD simulations. According to the authors, when using a helical electrode, the removal of contaminated dielectric out of blind holes upwards occurs through the helical groove itself, while the inflow of dielectric takes place through the lateral working gap. This was supported by CFD simulations of YABROUDI [10]. In the case of axially fluted electrodes, both the removal and the supply of fluid predominantly occur within the vertical flute. The greatest increase in the material removal rate (MRR) $\dot{V}_W$ by 300% compared to cylindrical electrodes could be achieved with a deep single straight flute.

KUMAR AND SINGH [11] as well as WANG ET AL. [12] applied helical flutes to tungsten carbide tool electrodes for micro-ED drilling to reach aspect ratios of $\varphi < 7.5$ and $\varphi = 2$ respectively. Both works concluded that the additional exterior flushing channels generate rotation-induced debris removal that results in eliminating the occurrence of arcing and short-circuiting, decreasing the electrode wear rate (EWR) $\dot{V}_E$ as well as the probability of secondary discharges compared to cylindrical electrodes.

KOJIMA ET AL. [13] performed a 2D localization of discharges during sinking EDM already in 1992 to observe spark locations under different conditions. The authors used branched wires to supply the pulse current to the graphite tool electrode and four current sensors to measure the four converged current values that vary in dependence of the discharge location. KÜPPER ET AL. [14] also aim at localizing discharges in the future, but over the length of the wire during wire EDM and by use of an online process monitoring system based on a signal analysis that incorporates a field programmable gate array (FPGA). The authors distinguished the discharges in normal pulses and short-circuit pulses by calculating their discharge energy W_e and using thresholds for the discharge current i_e and discharge duration t_e. DI CAMPLI ET AL. [15] used the commercially implemented so-called discharge location tracker (DLT) which is capable of tracking and locating spark positions on the wire in wire EDM in real time. The authors used the discharge locations and discharge energies W_e for real-time computations of a wear model to develop wire breakage prevention strategies for online process control.

In contrast, HUANG ET AL. [16] aimed at the prevention of discharges from the cylindrical side surface of WC tool electrodes during micro-ED milling by coating the tool electrodes with TiN and ZrN thin films using magnetron sputtering. The authors noted a reduced EWR and overcut for the benefit of an improved process stability and microscale component machining quality.

The most common approach to perform signal analyses involves the use of thresholds for characteristics of the voltage and current signals and associated gradients to identify individual discharge types [17, 18]. This is often supplemented by sensitivity analyses regarding the choice of threshold values [19].

The effectiveness of the methods described above to prevent or reduce discharge events in the lateral working gap is usually reviewed indirectly by determining the MRR, the EWR, the surface roughness Ra or hole conicity α of the bore hole. The localization of discharges along the vertical processing direction however is not yet part of the portfolio of process target variables but promises to be a quantitative measure for the efficiency of flushing methods applied or other approaches to prevent discharges in the lateral working gap. This is why this work aims at distinguishing the location and assigning the corresponding voltage and current signals of discharges to the frontal and the lateral regions of the working gap during drilling EDM with tool electrodes with interior and exterior flushing channels.

2 Methods and Materials

2.1 Experimental Setup

Drilling experiments were carried out on the machine tool AGIE compact 1, AGIE CHARMILLES SA, Switzerland. A workpiece electrode was used that consisted of two identical workpieces of ELMAX Superclean (X170CrVMo18–3-1) with a height of $d_{wp} = 5$ mm. These two halves are separated and electrically insulated from each other using a rubber layer with a thickness of $d_{rl} = 1.25$ mm. 3D-printed spacers made of polylactic acid (PLA) allowed to clamp the workpiece electrode to the machine tool bed while maintaining the electrical insulation between both halves as can be seen in Fig. 1. Contacting each workpiece half electrically and measuring the currents i using two TCP303 current probes with TCPA300 current probe amplifiers by TEKTRONIX, INC., Beaverton, USA, enables to distinguish both current flows separately. The gap voltage u was measured using a voltage probe with a scaling factor $r_u = 100$ to attenuate the signal amplitude. All signals were sampled with a USB-oscilloscope of type PicoScope 3405D by PICO TECHNOLOGY, Cambridgeshire, UK. A schematic setup of the data acquisition setup is shown in Fig. 1.

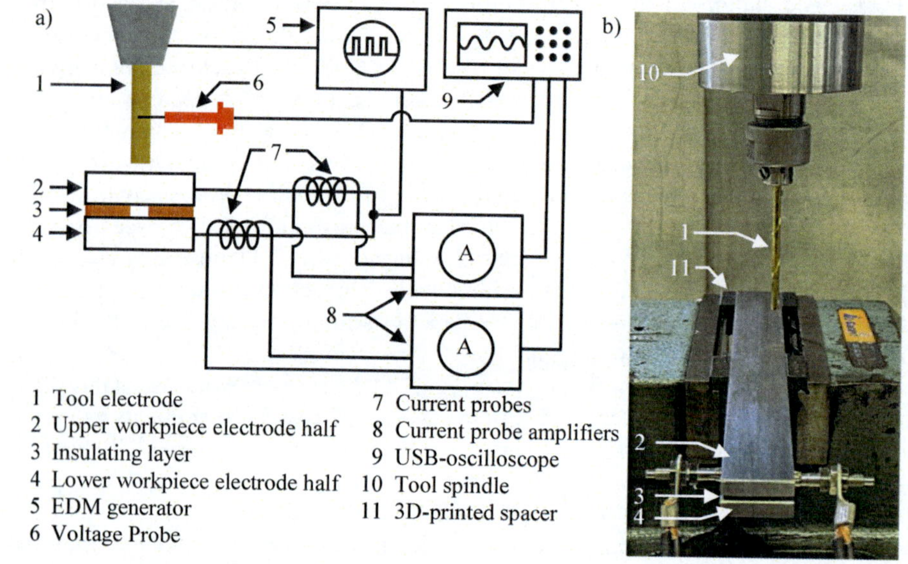

Fig. 1. Experimental setup for drilling EDM experiments and parallel signal data acquisition a) schematic experimental setup b) two-piece workpiece electrode inside the machine tank

Four different tool electrode geometries were chosen to analyze the influence of different flushing configurations performing one drilling experiment each. These supplement prior simulations and experiments using these geometries that revealed varying behavior in the removal of debris and therefore different probabilities of lateral discharges [20]. The four types of tool electrodes exhibit different interior and exterior flushing channels and are accordingly named rod-1-channel (R1C) and rod-4-channel (R4C), Fig. 2 a), as well as helix-1-channel (H1C) and helix-4-channel (H4C), Fig. 2 b). Figure 2 c) and d) visualize the general flushing configuration for the use of tool electrodes.

All experiments were carried out using relaxation type discharges, Table 1.

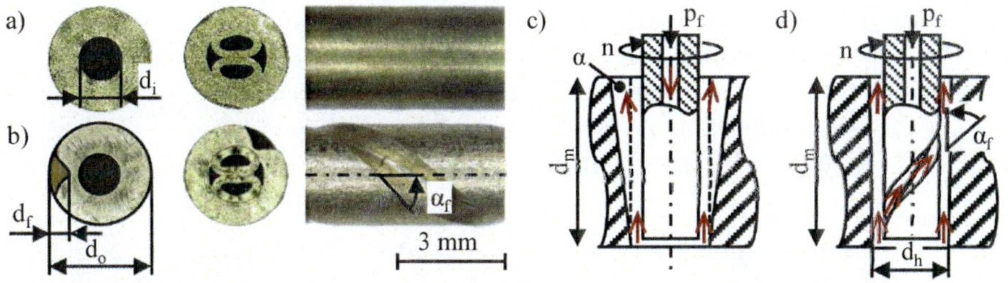

p_f	- Flushing pressure	d_m	- Machining depth	d_o	- Outer diameter
n	- Rotational speed	d_h	- Hole diameter	d_f	- Flute depth
α	- Conicity	d_i	- Inner diameter	α_f	- Flute angle

Fig. 2. Types of tool electrodes used for the drilling EDM experiments

Table 1. Processing parameters for drilling EDM experiments.

Parameter	Variable	Value
Open circuit voltage	$\hat{u}_i$	180.00 V
Discharge capacity	C_e	1.32 μF
Charging current	i_c	4.00 A
Ontime	t_{on}	75.0 μs
Offtime	t_{off}	5.60 μs
Rotational speed	n	5,000.00 1/min
Flushing pressure	p_f	1.00 MPa
Polarity		negative

2.2 Signal Analysis Algorithm

The basis of the signal analysis for the localization of discharges is a semi-automated software tool presented in prior work, enabling data acquisition and analysis [21]. With the implemented means of data acquisition, the sampling frequency of $f_s = 10.4$ MHz results in a time of measurement of $t_{meas} = 24.04$ ms, which directly defines the time interval of uninterrupted signal acquisition. The corresponding ratio of measurement r_{meas} is calculated by excluding the time delay $t_{delay} = 66.00$ ms that results from buffering, transfering and saving of each data file:

$$r_{meas} = \frac{t_{meas}}{t_{meas} + t_{delay}} = 0.26. \quad (1)$$

Thus, r_{meas} is a measure of the percentage of time during which electrical signals are recorded within each measurement cycle.

The classification algorithm used for detection and characterization of discharge events from the signals of gap voltage u and current i is based on the method presented by NIRALA AND SAHA [18] and the classification is carried out subsequent to the experiments. The basic principle of the algorithm is based on edge detection whenever a certain threshold is exceeded (rising edge) or undercut (falling edge). The occurrence of these signal edges stems from the specific physical phenomena inside the working gap, i.e. a falling edge of the gap voltage u with a simultaneously rising edge of the current i typically indicates the begin of a spark. Figure 3 shows an exemplary excerpt from one of the measurements, performed with the setup depicted in Fig. 1.

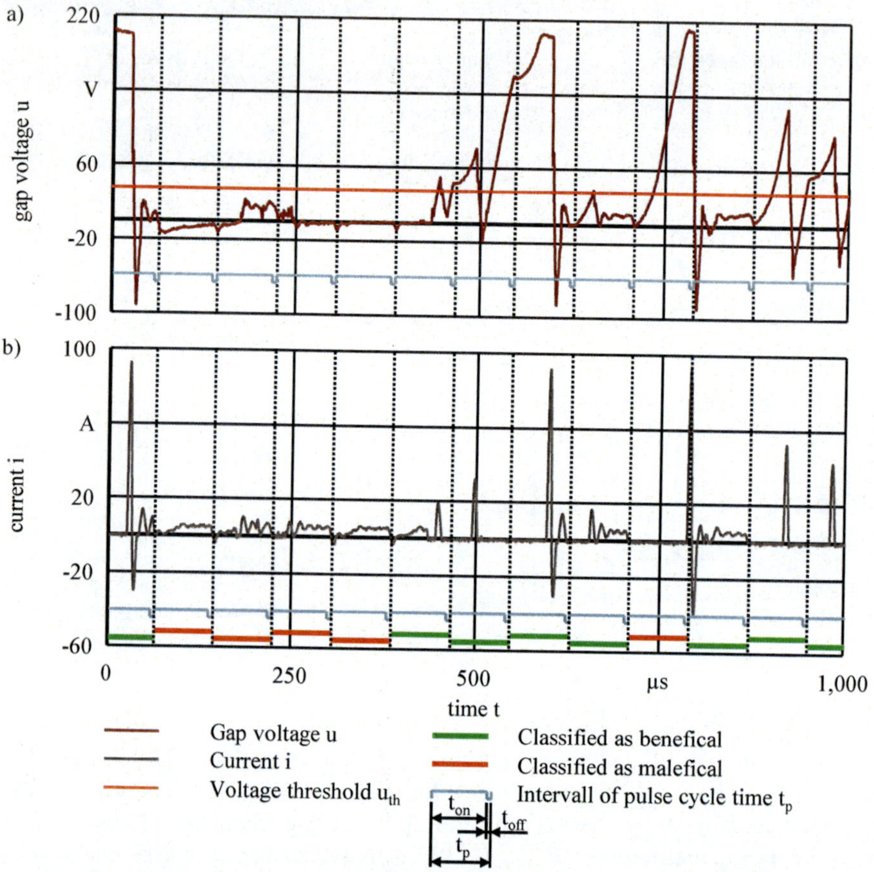

Fig. 3. Exemplary signal waveforms including event classification as benefical and malefical or favoring and detrimental to material removal respectively; a) gap voltage u; b) current i

The blue line marks the interval of the pulse cycle time t_p, that is the sum of the ontime t_{on} and the offtime t_{off} and indicates the charging phases of the EDM generator. This interval is taken as a time reference for assigning individual events. The voltage threshold u_{th} is used to distinguish events as being beneficial or maleficial for the drilling EDM process. If a falling voltage edge is falling below the voltage threshold u_{th} and exists in close temporal proximity to a rising current edge, the event is registered as beneficial. Current edges without a respective voltage edge mark events that are counted as maleficial.

3 Results

The evaluation of the experimental results only considers processing results for the machining depth 5 mm $\leq d_m \leq$ 10 mm, i.e. for drilling in the lower workpiece half. A more detailed insight into the drilling performance of tool electrodes with interior and exterior flushing channels is given by UHLMANN ET AL. [20], where deviations in the process results compared to the present work arise from the modified experimental setup using a split workpiece, varied flushing parameters, the different aspect ratio φ as well as the fact that blind holes have been machined instead of through holes. It is noted that the overall material volume that needs to be removed for drilling through holes is dependent on the frontal area A_e and flushing cross-section A_f of the tool electrodes and therefore varies [20]. One reason for this is the pin that remains in the center of the bore hole, mostly for 1-channel electrodes. This is why the erosion duration t_{ero} for 4-channel tool electrodes is usually higher than for 1-channel tool electrodes.

Figure 4 shows a comparison of the total number of events n plotted over the machining depth d_m. The black bars display all events that were recorded in the upper workpiece half and are highly undesired. The blue bars indicate those events in the lower workpiece half. It can be seen that from H1C over R1C and H4C to R4C a general increase of the number of events occurs. The same tendency is reflected by the process results of the four drilling experiments performed, namely the erosion duration t_{ero} and the MRR of all four types, also given in Fig. 4. The repeatability concerning the process results is attested in direct comparison with the prior results that were based on three drilling experiments for each type of tool electrode [20].

Concerning the discharges that can be localized in the lateral working gap, a gradually decrease with the machining depth d_m can be stated. This decrease converges to zero for the rod types and is accelerated by using the helical exterior flushing channel in case of H1C. For the H1C and H4C types the number of events undercuts n = 500 after a machining depth of d_m = 6.5 mm and d_m = 6.8 mm but is retained on the level of 300 $\leq$ n $\leq$ 500 throughout. The R1C and R4C type reach this decreased level of lateral discharges only after machining depths of d_m = 7.8 mm and d_m = 7.1 mm respectively.

These correlations are also mirrored in Fig. 5, where the total number of events n is split into events that favor the material removal and events that destabilize the process and are therefore detrimental to material removal. While the electrical signals possess similar characteristics and corresponding energy input for each classified event type and each type of tool electrode, the deduction has to be drawn from the graphs that the material removal effectiveness though differs for each type of tool electrode. This is a direct consequence of the different flushing conditions and the fact that the discharge energy W_e is on the one hand distributed in diverse ratios between e.g. the workpiece, the tool electrode, solidification of debris particles and the dielectric fluid. On the other hand, the major part of the discharge energy W_e is transferred and stored into the dielectric fluid whereas actual material removal is represented by less than 1% of the discharge energy W_e [1, 22]. It can be stated that it takes fewer discharge events for the helix type compared to the rod type tool electrodes to remove a similar amount of workpiece material. In the particular case of H1C, this corresponds to a 11% higher MRR or 10% lower erosion duration t_{ero} compared to R1C and favoring as well as detrimental discharges in the upper workpiece are considerably reduced. In case of through holes the improvements in MRR using helix type electrodes amount to 111% without interior flushing channels and to 28% for H1C [20].

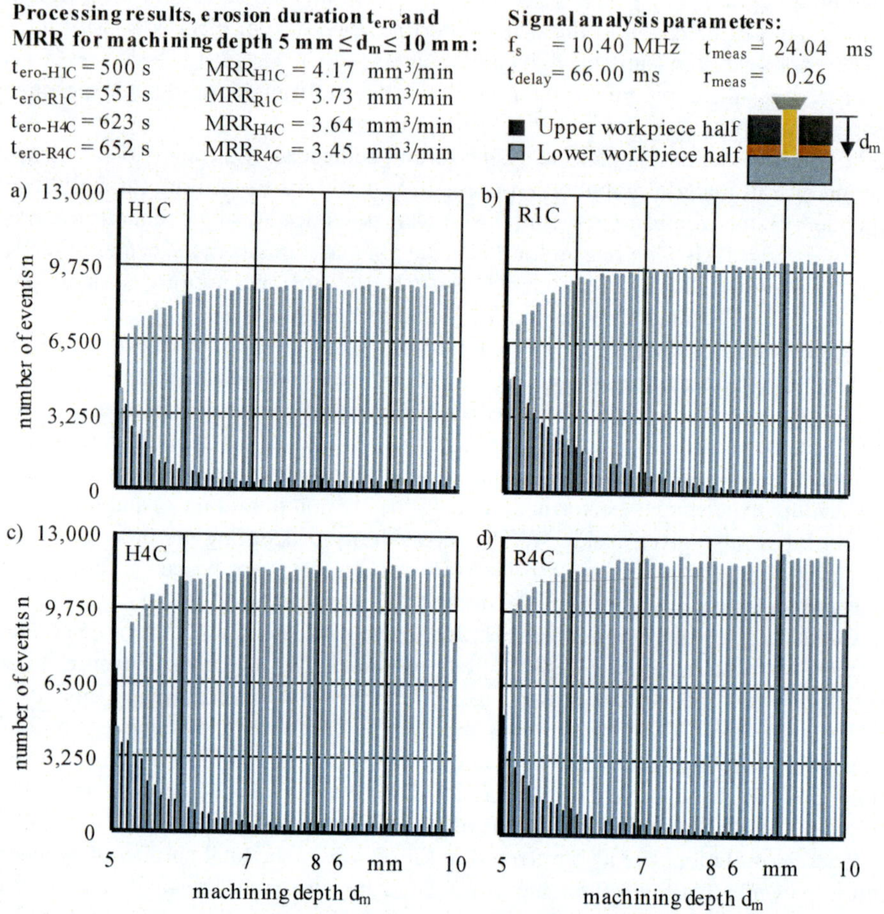

Fig. 4. Comparison of the overall number of events n occurring at the upper and lower workpiece half for the four types of tool electrode a) H1C; b) R1C; c) H4C; d) R4C

With increasing machining depth d_m the detrimental number of events n is gradually reduced with the H1C type in the lower workpiece but is retained for the H4C type, Fig. 5 b). From a machining depth $7\text{ mm} \leq d_m \leq 8\text{ mm}$ on the amount of lateral discharges in the upper workpiece stays on a constant level for the helix type electrodes whereas it approaches zero for the rod types. This leads to the conclusion that the specific geometry of the helical flute does not yet fully incorporate the debris efficiently for the purpose of evacuation aside from the lateral working gap. Reasons for this can be found in the additionally introduced turbulence and the not optimized combination of the fluid mechanic parameters flushing pressure pf and rotational speed n [20].

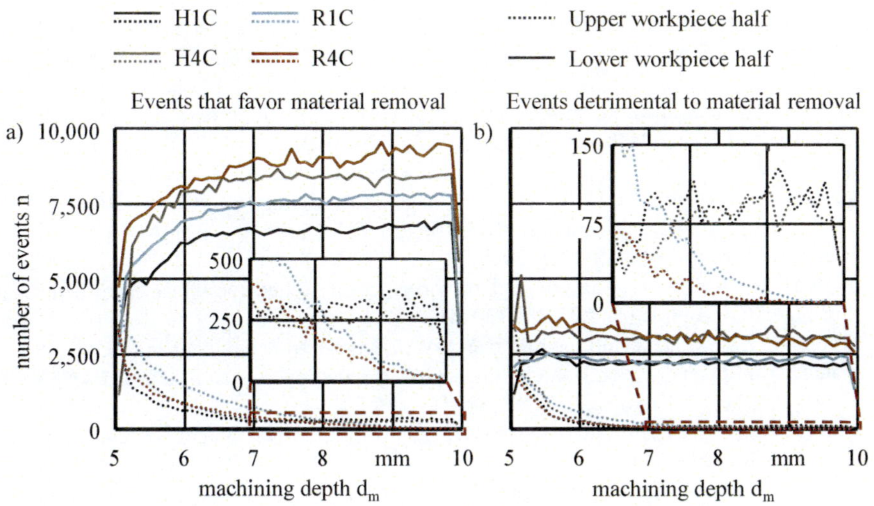

Fig. 5. Comparison of the number of events n occurring at the upper and lower workpiece half for all four types of tool electrode, classified as a) favoring; b) destabilizing

4 Conclusions

The experiments presented in this work demonstrate a quantitative measure for the ratio of discharges in the frontal and lateral working gap in the case of drilling EDM by distinguishing the different types of discharge events and localizing them along the vertical processing direction. This hence promises to be a valuable approach to explain differences in the machining performance and the efficiency of flushing methods applied or even other approaches to prevent discharges in the lateral working gap on a basic level. With the new method, indications could be provided that not only process results can be improved but also the occurrence of arcing and short-circuiting can be vastly decreased using additional exterior flushing channels in the form of a helix compared to cylindrical electrodes. Further experiments using the new method to localize discharge events along the processing direction are necessary for a more generalization of the conclusions drawn.

The presented setup can easily be adapted for other forms of EDM such as sinking or wire EDM. Further splitting and thinning down the workpiece electrode segments as well as the insulation layer would allow to achieve even more detailed localizations.

References

1. Kunieda, M., Lauwers, B., Rajurkar, K.P., Schumacher, B.M.: Advancing EDM through fundamental insight into the process. CIRP Ann. **54**(2), 64–87 (2005)
2. Klocke, F., König, W.: Fertigungsverfahren Band 3: Abtragen, Generieren und Lasermaterialbearbeitung. 4th edn. Springer, Heidelberg, Berlin (2007)
3. Cetin, S., Okada, A., Uno, Y.: Effect of debris distribution on wall concavity in deep-hole EDM. Int J Jap Soc Mech Eng **47**(2), 553–559 (2004)
4. Ayesta, I., Flaño, O., Izquierdo, B., Sanchez, J.A., Plaza, S.: Experimental study on debris evacuation during slot EDMing. Procedia CIRP **42**, 6–11 (2016)

5. Domingos, D.C.: Die-Sinking EDM of High Aspect Ratio Cavities in Nickel-Base Alloy. In: Berichte aus dem Produktionstechnischen Zentrum Berlin. Fraunhofer Verlag, Stuttgart (2015)
6. Meena, V.K., Azad, M.S., Mitra, S.: Effect of flushing condition on deep hole micro-EDM drilling. Int J Mach Machinab Mat **12**(4), 308–320 (2012)
7. Li, G., Natsu, W., Yu, Z.: Elucidation of the mechanism of the deteriorating interelectrode
8. environment in micro EDM drilling: Int J Mach Tools Manuf **167**, 103747 (2021)
9. Plaza, S., et al.: Experimental study on micro EDM-drilling of Ti6Al4V using helical electrode. Prec Eng **38**, 821–827 (2014)
10. Nastasi, R., Koshy, P.: Analysis and performance of slotted tools in electrical discharge drilling. CIRP Ann. **63**(1), 205–208 (2014)
11. Yabroudi, S.: Einsatzbewertung neuartiger Werkzeugelektroden mit außenliegenden Spülkanälen beim funkenerosiven Bohren mittels CFD Simulationen. 37th CADFEM ANSYS Sim Conf (2019)
12. Kumar, R., Singh, I.: Productivity improvement of micro EDM process by improvised tool. Prec Eng **51**, 529–535 (2018)
13. Wang, K., Zhang, Q., Zhu, G., Liu, Q., Huang, Y.: Experimental study on micro electrical discharge machining with helical electrode. The International Journal of Advanced Manufacturing Technology **93**(5–8), 2639–2645 (2017). https://doi.org/10.1007/s00170-017-0747-6
14. Kojima, H., Kunieda, M., Nishiwaki, N.: Understanding discharge location movements during EDM. Proc. ISEM X, 144–149 (1992)
15. Küpper, U., Herrig, T., Klink, A., Bergs, T.: Evaluation of the Process Performance in Wire EDM Based on an Online Process Monitoring System. Procedia CIRP **95**, 360–365 (2020)
16. Di Campli, R., Maradia, U., Boccadoro, M., D'Amario, R., Mazzolini, L.: Real-Time Wire EDM Tool Simulation Enabled by Discharge Location Tracker. Procedia CIRP **95**, 308–312 (2020)
17. Huang, C.H., Yang, A.B., Hsu, C.Y.: The optimization of micro EDM milling of Ti-6Al-4V using a grey Taguchi method and its improvement by electrode coating. Int J Adv Manuf Techn **96**, 3851–3859 (2018)
18. Dauw, D.F., Snoeys, R., Dekeyser, W.: Advanced Pulse Discriminating System for EDM Process Analysis and Control. CIRP Ann. **32**(2), 541–549 (1983)
19. Nirala, C.K., Saha, P.: Evaluation of μEDM-drilling and μEDM-dressing performances based on online monitoring of discharge gap conditions. The International Journal of Advanced Manufacturing Technology **85**(9–12), 1995–2012 (2015). https://doi.org/10.1007/s00170-015-7934-0
20. Belotti, M., Qian, J., Reynaerts, D.: Breakthrough phenomena in drilling micro holes by EDM. Int J Mach Tools Manuf **146**, 103436 (2019)
21. Uhlmann, E., Polte, M., Yabroudi, S.: Novel advances in machine tools, tool electrodes and processes for high-performance and high-precision EDM. Procedia CIRP (2022) https://www.sciencedirect.com/science/article/pii/S2212827122015220, https://doi.org/10.1016/j.procir.2022.10.080
22. Thißen, K., Streckenbach, J., Santibáñez Koref, I., Polte, M., Uhlmann, E.: Signal analysis on a single board computer for process characterisation in sinking electrical discharge machining. In: Proceedings of the11th Congress of the German Academic Association for Production Technology (WGP), pp. 169–176. Springer, Heidelberg (2021)
23. Oßwald, K., Schneider, S., Hensgen, L., Klink, A., Flocke, F.: Experimental investigation of energy distribution in continuous sinking EDM. CIRP J Manuf Sci Techn **17**, 36–43 (2017)

Experimental Studies in Deep Hole Drilling of Ti-6Al-4V with Twist Drills

M. Zimon(✉), G. Brock, and D. Biermann

Institut of Machining Technology, TU Dortmund University, Baroper Straße 303, 44227 Dortmund, Germany
mike.zimon@tu-dortmund.de

Abstract. Due to the attractive material properties such as high specific strength and very good corrosion resistance, titanium and its alloys are frequently used in the aerospace, automotive and chemical industries. However, the low thermal conductivity leads to high thermomechanical tool loads during machining, making an optimal cooling lubricant supply to the cutting edge essential. For an optimized design of the tools a realistic simulation is necessary. Since the viscosity of deep hole drilling oils is strongly temperature dependent, it plays a significant role in the comprehensive simulation of the process. In this study the thermomechanical tool loads on TiAlN-coated helical deep hole drills during machining of the titanium alloy Ti-6Al-4V (Grade 5) are investigated and will serve as input for a three-dimensional finite element method (FEM) chip formation simulation focusing on the temperature distribution. The experimental investigations are carried out with successively varying process parameters of cutting speed, feed rate and cooling lubricant pressure. The knowledge gained in this study is of fundamental importance, as it serves as the basis for the future development of a fluid-structure interaction (FSI) simulation in order to be able to take the temperature influence on the cooling lubricant flow into account.

Keywords: Deep hole drilling · Finite element method (FEM) · Titanium alloy · Ti-6Al-4V

1 Introduction

Due to superior properties such as very high specific strength, good corrosion resistance and biocompatibility, titanium alloys are frequently used in the aerospace, automotive and chemical industries as well as in medical technology [1]. However, the low thermal conductivity leads to extreme temperatures at the tool cutting edge during machining, which can locally rise up to 900 °C under dry conditions [2]. A large proportion of the heat generated must be dissipated by the tool. In combination with the mechanical stresses and the chemical reactivity, the high thermal stresses lead to increased tool wear, making titanium alloys one of the difficult-to-cut materials [3, 4]. In addition to wear phenomena, total tool failure in the form of chipping often occurs during machining. This can be minimized by a wear-resistant coating, composed of materials such as TiAl- or supernitride coatings [5, 6]. To reduce thermal stresses, a supply of cutting fluid is

M. Liewald et al. (Eds.): WGP 2022, LNPE, pp. 219–227, 2023.
https://doi.org/10.1007/978-3-031-18318-8_23

essential, especially at high cutting speeds [7]. On the one hand, the heat is directly dissipated by the cutting fluid and on the other hand, the friction is reduced significantly, resulting in less generated heat in the process. Therefore, cutting fluids with high thermal conductivity and high specific heat capacity are ideal for machining titanium alloys. The application of high-pressure coolant (HPC) additionally improves chip formation, chip breaking and chip evacuation [8]. Chip evacuation is particularly important in deep hole drilling, as the risk of chip jamming increases with the drilling depth [9, 10]. As soon as the drilling depth l_t exceeds ten times the drilling diameter D (l/D-ratio > 10), it is categorized as deep hole drilling [11]. The classic deep hole drilling tool has an asymmetrically arranged cutting edge and is used in horizontal drilling machines. The symmetrical double-edged tools, which are used up to a diameter of $D = 12$ mm and a l/D-ratio up to $l/D = 30$, enable the use in machining centers [12, 13]. Compared to the single-edged tools, the twist drills allow for a higher feed rate and therefore a higher productivity [14]. Due to the length of the tools, a guidance in the form of a drill bush or a pilot hole is required at the beginning of the drilling process. Furthermore, the tools have internal cooling channels that lead the cooling lubricant to the bottom of the hole [15]. In deep hole drilling, the lubrication effect is of particular importance, so that often non-water-miscible cooling lubricants, such as so-called deep hole drilling oils, are used. Some fluid properties such as density and kinematic viscosity are dependent on temperature and pressure [16, 17].

This study investigates the helical-fluted deep hole drilling of Ti-6Al-4V with a three-stage variation of cutting speed, feed, and cooling lubricant pressure. The focus is on determining the thermomechanical tool loads. The obtained results will serve as input for a three-dimensional finite element simulation.

2 Experimental Investigation

The experimental investigations were carried out on the deep hole drilling machine KTE40–1000 manufactured by TIBO Tiefbohrtechnik GmbH. The Samples of titanium alloy Ti-6Al-4V (Grade 5) were used in a cylindrical shape with a diameter of $D_{S1} =$ 50 mm and a length of $l_{S1} = 105$ mm for the analyses of the mechanical loads and with a diameter of $D_{S2} = 20$ mm and a length of $l_{S2} = 55$ mm for the determination of the temperatures at the cutting edges. The chemical composition of the investigated material is listed in Table 1. The deep holes were drilled with a helical-shaped deep hole twist drill made of solid carbide with a TiAlN coating on the tool tip. This coating protects the highly stressed cutting edges from wear and enables high-performance feed rates. The drills have a diameter of $D_1 = 5$mm and are designed for a maximum l/D-ratio of $l/D = 20$. In addition, the tools have internal cooling channels that supply the cutting oil directly to the cutting edge. The used deep drilling oil Blasomill 10 DM by Blaser has a kinematic viscosity $\nu = 10$ mm^2/s at the reference temperature of $T = 40$ °C and a density of $\rho = 0.85$ g/cm^3 at $T = 20$ °C.

The tool properties and the experimental setup for the measurement of the mechanical tool loads and are presented in Fig. 1. The twist drill was inserted on the machine side into the HSK 63 holder and the specimen was placed into the Kistler 4-component dynamometer type 9272, which recorded the feed force and the drilling torque during the

Table 1. Chemical composition of the workpiece material Ti-6Al-4V (wt%)

Ti	Al	V	Fe	O	C	N	H
Bal	5.50–6.75	3.50–4.50	0.40	0.20	0.08	0.05	0.015

experiment. It was connected to the measuring computer via an amplifier and scanned the mechanical load with a sample rate of $f_S = 1000$ Hz. In order to prevent the measurement results from being influenced by the contact of a drill bush with the sample, the tool was guided through pilot holes with a depth of $l_{ph} = 15$ mm in the first drilling phase. Blind holes with a depth of $l_{t1} = 100$ mm were drilled in a drilling pattern in random order, to avoid a possible influence of the previous borehole heat on the subsequent ones. According to the recommendation of the tool manufacturer, the process parameters cutting speed ($v_c = 20; 30; 40$ m/min), feed rate ($f = 0.08; 0.1; 0.12$ mm) and coolant lubricant pressure ($p = 40; 80; 120$ bar) were varied successively. For statistical validation, each test series was repeated four times for each parameter combination. Every time the cutting speed or feed rate was changed, a new tool was used. Tool wear at this small drilling depth is very low and therefore the influence on the measured values is negligible. After each manufactured borehole, the chips were collected and documented with a digital camera.

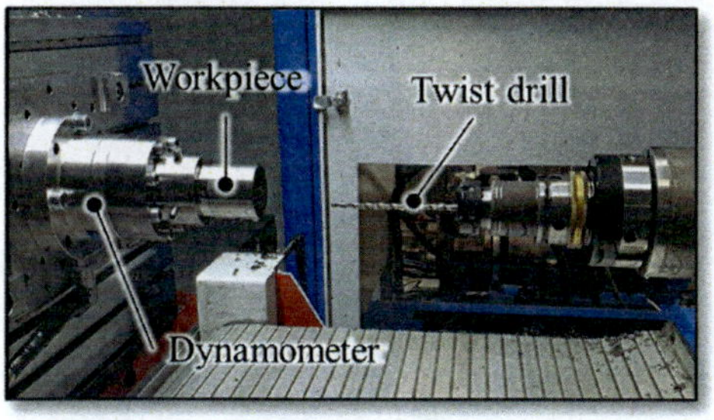

property	value
tool diameter	$D_l = 5$ mm
drill type	deep hole twist drill
tool material	solid carbide
coating	TiAlN (tip coated)
length	$l_l = 158$ mm
drill point angle	$\sigma = 135°$

Fig. 1. Experimental setup for tool load measurement and an overview of tool properties

The tool temperature at the cutting edge was measured with a two color ratio pyrometer FIRE 3 by en2Aix. For this purpose, an optical fiber with a diameter of $D_{Fiber} =$ 330 μm was placed axially into the specimen. In order to ensure that the fiber is not damaged during insertion, a sufficiently large pilot hole with a diameter of D = 0.8 mm at a distance of r = 2.1 mm was selected. This allows a constant measurement of the temperature at the cutting edge. The setup is presented in Fig. 2. In this arrangement, the fiber is machined together with the workpiece. This occurs after a drilling path of l_t = 25 mm, so that a stationary temperature has already set in at the cutting edge, which means that a shorter specimen geometry is sufficient. The thermal radiation emitted in the drilling process is transferred trough the fiber optic to the pyrometer at a sampling rate of $f_S = 50$ kHz. In the pyrometer, the temperature is determined by the quotient of the radiation intensity of two nearby wavelengths. For each parameter variation, 10 measured values were evaluated and averaged.

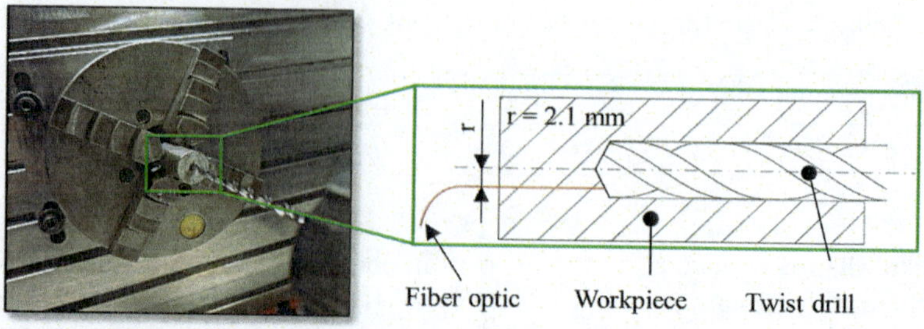

Fig. 2. Experimental setup for the determination of the temperatures at the cutting edges

3 Results and Discussion

The feed forces and drilling torques determined during the experimental investigations are illustrated in Fig. 3. They are shown as a function of cutting speed and feed rate at a constant cooling lubricant pressure of $p = 80$ bar. While the values of the feed force at a cooling lubricant pressure of $p = 40$ bar are slightly lower, increasing the pressure to $p = 120$ bar does not lead to any significant change. The influence of the cooling lubricant pressure on the drilling torque is also negligible.

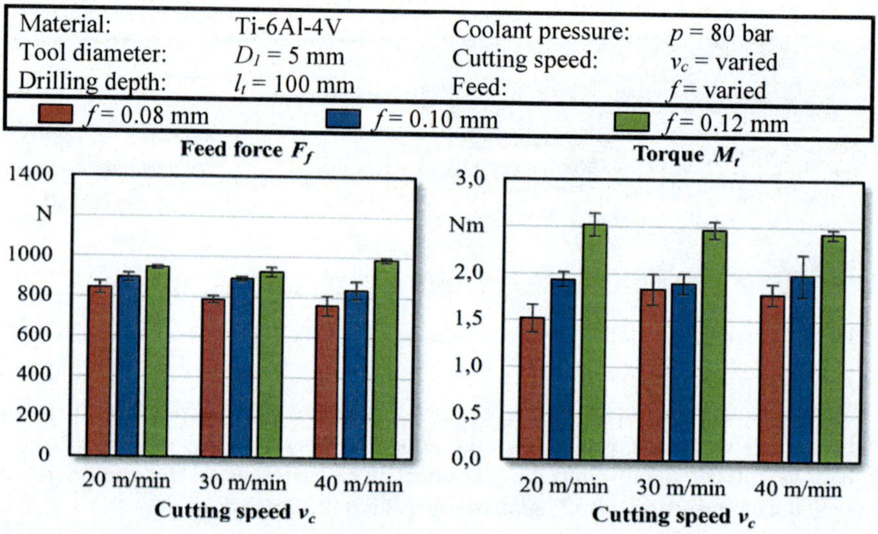

Fig. 3. Analysis of the mechanical load at a cooling lubricant pressure of $p = 80$ bar

The results of all three investigated pressure stages show an identical influence of the cutting speed and feed on the feed force and drilling torque. An increase in the feed rate is directly accompanied by an increase in the feed force and drilling torque. This can be explained by the increase of the cutting cross section at higher feed rates. The cutting speed has no clear influence on the measured variables. An increase in cutting speed tends to result in lower feed forces. One reason for this behavior may be thermal softening of the material at higher temperatures [18]. In the case of the drilling torque, an increase in the cutting speed at the two smaller feed rates leads to higher drilling torques. At a feed rate of $f = 0.12$ mm the drilling torque decreases with increasing cutting speed.

The primary chip shape observed in the present study consists of squeezed spiral chips. Since the influence of the feed rate has the least influence on the size of the chips, the Fig. 4 shows the documented chip shapes at constant feed rate of $f = 0.1$ mm.

Material:	Ti-6Al-4V	Coolant pressure:	p = varied
Tool diameter:	D_l = 5 mm	Cutting speed:	v_c = varied
Drilling depth:	l_t = 100 mm	Feed:	f = 0.1 mm

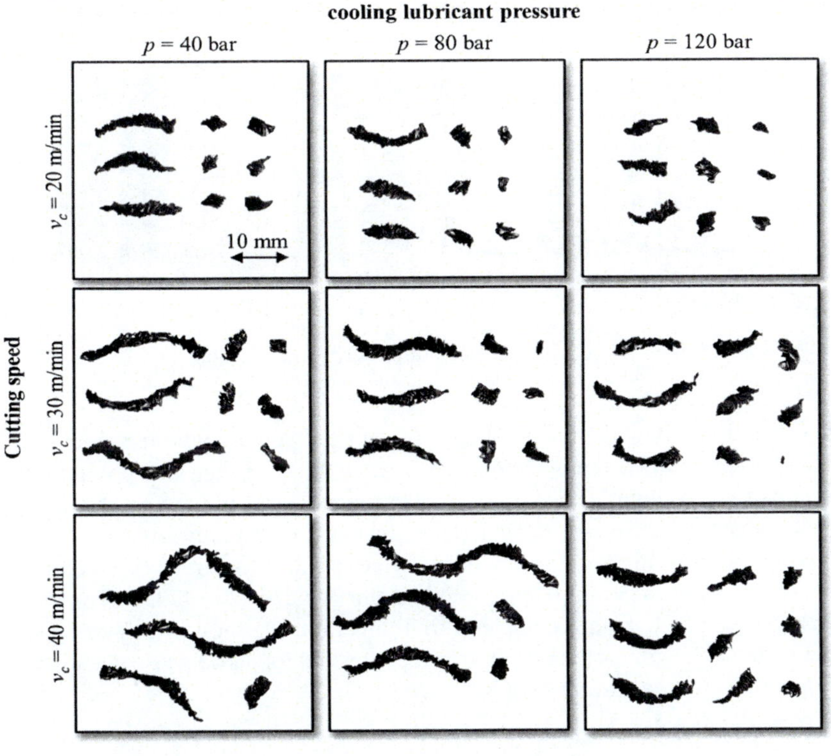

Fig. 4. Chip form analysis for a constant feed rate $f = 0.1$ mm

A clear influence of the cooling lubricant pressure on the size of the chips can be observed. An increase in pressure leads to an earlier chip breaking, resulting in significantly smaller chips. This facilitates their removal from the die, reducing the risk of chip jamming and improving process stability. Considering the cutting speed, small chips are produced at small values. The length of the chips increases with the cutting speed. One reason for this is the changed flow behavior of the material due to the thermal softening at higher cutting speeds.

Furthermore, the temperatures occurring during the machining of Ti-6Al-4V have been analyzed in the experimental investigations. Figure 5 shows the results for the variation of cutting speeds and feed rates at a constant coolant pressure of $p = 80$ bar.

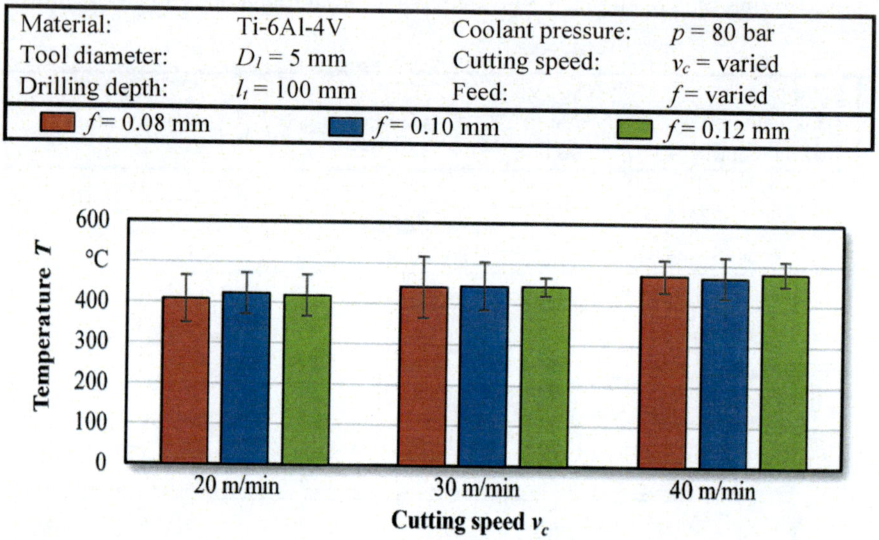

Fig. 5. Analysis of the occurring temperatures during the machining of Ti-6Al-4V

The evaluations show that the cooling lubricant pressure and the feed rate have no significant influence on the temperatures measured in the test setup. One possible reason for this is that the cooling lubricant does not manage to wet the tool in the measured effective zone and thus does not lead to a reduction in the local temperatures. The cutting speed, on the other hand, has a nearly linear effect on the temperatures. According to this, the cutting edge temperature for $v_c = 40$ m/min, which is $T = 473.6$ °C on average, is approx. $\Delta T = 65$ °C higher compared to Temperature at the lowest cutting speed. This observation strengthens the assumption that the feed forces decrease due to thermal softening with increased cutting speeds.

4 Summary and Outlook

This paper presents the experimental study of the deep hole drilling process in titanium alloy Ti-6Al-4V (Grade 5). The coated TiAlN helical-shaped deep hole twist drills made of solid carbide with a TiAlN coating on the tool tip and a diameter of $D_l = 5$ mm were used for manufacturing blind holes with a drilling depth of $l_{t1} = 100$ mm for the determination of the mechanical loads and with a drilling depth of $l_{t2} = 50$ mm for the investigation of the temperature. Based on the recommendations of the tool manufacturer, the process parameter cutting speed ($v_c = 20$; 30; 40 m/min), feed rate ($f = 0.08$; 0.1;

0.12 mm) and coolant lubricant pressure ($p = 40; 80; 120$ bar) were varied successively. The following conclusions can be drawn from this study:

- The cooling lubricant pressure has no significant influence on the drilling torque and on the temperatures generated in the process. Only at a pressure of $p = 40$ bar a slight reduction in the feed forces is detected.
- The mechanical loads are primarily dependent on the feed rate. Both feed forces and drilling torques increase with increasing feed rate.
- Regarding the cutting speed, it was observed that an increase tends to result in decreasing feed forces.
- The primary chip shape consists of squeezed spiral chips. The size is independent of the feed rate. While an increase in coolant pressure leads to smaller chips, increase in cutting speed leads to longer chips.
- The temperatures occurring during the deep hole drilling of Ti-6Al-4V are mainly influenced by the cutting speed. The temperature at a cutting speed of $v_c = 40$ m/min, which is $T = 473.6$ °C on average, is approx. $\Delta T = 65$ °C higher compared to temperature at the lowest cutting speed.

For further chip formation investigation, finite element analyses will be carried out using the software DEFORM v12.1. The drilling process is to be modeled three-dimensionally, containing the drilling tool tip, the bore wall and the plastically deformable hole bottom. With the simulations' high computational power demand, the focus will be placed on an optimized workpiece model and mesh, enabling the timesaving study of varying process parameters. As a realistic modeling of the material behavior is required to reliably predict the chip morphology and process loads, the flow stress will be calculated according to the Johnson-Cook equation, parameterized based on tests executed at the Institute of Machining Technology (ISF) using a Split Hopkinson pressure bar in conjunction with an inductive heating of the specimens. Therefore, the high strain rates' and temperatures' impact occurring during the drilling process can be mapped in the simulation. For the contact condition modeling, results from friction characterization experiments conducted at the ISF will be implemented, considering the relation between the relative velocity and the corresponding friction. The finite element analyses' validation provides for the results' comparison with experimental data. For this purpose, the measured tool torques and feed forces presented in this paper as well as the evaluated chips will be utilized. Additionally, the effect of fracture criteria will be studied in order to further improve the modelling.

The simulative investigations of the cooling lubricant flow will be carried out using the software ANSYS CFX. For this purpose, the model with the filled fluid is generated from the negative of the CAD model of the drill. The results are then compared with experimental high-speed recordings of the cooling lubricant flow.

Acknowledgements. Funded by the Deutsche Forschungsgemeinschaft (DFG, German.

Research Foundation)—Gefördert durch die Deutsche Forschungsgemeinschaft (DFG)—Projektnummer: 317373968.

References

1. Peters, M.: Titan und Titanlegierungen. WILEY-VCH Verlag GmbH & Co. KGaA, Weinheim (2002)
2. Sun, S., Brandt, M., Dargusch, M.S.: Machining Ti-6Al-4V alloy with cryogenic compressed air cooling. Int. J. Mach. Tools Manuf. **50**, 933–942 (2010). https://doi.org/10.1016/j.ijmachtools.2010.08.003
3. Ezugwu, E.O., Wang, Z.M.: Titanium alloys and their machinability—a review. J. Mater. Process. Technol. **68**, 262–274 (1997). https://doi.org/10.1016/S0924-0136(96)00030-1
4. Yuan, C.G., Pramanik, A., Basak, A.K., Prakash, C., Shankar, S.: Drilling of titanium alloy (Ti6Al4V)—a review. Mach. Sci. Technol. **25**, 637–702 (2021). https://doi.org/10.1080/10910344.2021.1925295
5. Rahim, E.A., Sharif, S.: Tool failure modes and wear mechanism of coated carbide tools when drilling Ti-6Al-4V. IJPTECH **1**, 30 (2007). https://doi.org/10.1504/IJPTECH.2007.015342
6. Sharif, S., Rahim, E.A.: Performance of coated- and uncoated-carbide tools when drilling titanium alloy—Ti-6Al4V. J. Mater. Process. Technol. **185**, 72–76 (2007). https://doi.org/10.1016/j.jmatprotec.2006.03.142
7. Ezugwu, E.O., Da Batista Silva, R., Falco Sales, W., Rocha Machado, A.: Overview of the Machining of Titanium Alloys Encyclopedia of Sustainable Technologies, pp. 487–506. Elsevier (2017). https://doi.org/10.1016/B978-0-12-409548-9.10216-7
8. Stolf, P., Paiva, J.M., Ahmed, Y.S., Endrino, J.L., Goel, S., Veldhuis, S.C.: The role of high-pressure coolant in the wear characteristics of WC-Co tools during the cutting of Ti-6Al-4V. Wear **440–441**, 203090 (2019). https://doi.org/10.1016/j.wear.2019.203090
9. Baumann, A., Oezkaya, E., Schnabel, D., Biermann, D., Eberhard, P.: Cutting-fluid flow with chip evacuation during deep-hole drilling with twist drills. Eur. J. Mech. B. Fluids **89**, 473–484 (2021). https://doi.org/10.1016/j.euromechflu.2021.07.003
10. Klocke, F., Keitzel, G., Veselovac, D.: Innovative sensor concept for chip transport monitoring of gun drilling processes. Procedia CIRP **14**, 460–465 (2014). https://doi.org/10.1016/j.procir.2014.03.096
11. Gerken, J.F., Biermann, D.: Concept of a mechatronic system for targeted drill head direction and angular alignment control in BTA deep hole drilling. In: Behrens, B.-A., Brosius, A., Hintze, W., Ihlenfeldt, S., Wulfsberg, J.P. (eds.) Production at the Leading Edge of Technology. Lecture Notes in Production Engineering, pp. 215–224. Springer Berlin Heidelberg, Berlin, Heidelberg (2021). https://doi.org/10.1007/978-3-662-62138-7_22
12. Biermann, D., Bleicher, F., Heisel, U., Klocke, F., Möhring, H.-C., Shih, A.: Deep hole drilling. CIRP Ann. **67**, 673–694 (2018). https://doi.org/10.1016/j.cirp.2018.05.007
13. Biermann, D., et al.: Thermal aspects in deep hole drilling of aluminium cast alloy using twist drills and MQL. Procedia CIRP **3**, 245–250 (2012). https://doi.org/10.1016/j.procir.2012.07.043
14. Biermann, D., Blum, H., Frohne, J., Iovkov, I., Rademacher, A., Rosin, K.: Simulation of MQL deep hole drilling for predicting thermally induced workpiece deformations. Procedia CIRP **31**, 148–153 (2015). https://doi.org/10.1016/j.procir.2015.03.038
15. Verein Deutscher Ingenieure e.V.: Tiefbohrverfahren. Beuth Verlag GmbH, Berlin, vol. 25.080.40 (2006)
16. Al-Maamari, R.S., Houache, O., Abdul-Wahab, S.A.: New correlating parameter for the viscosity of heavy crude oils. Energy Fuels **20**, 2586–2592 (2006). https://doi.org/10.1021/ef0603030
17. Watter, H.: Fluide und Fluideigenschaften. In: Watter, H. (ed.) Hydraulik und Pneumatik, pp. 7–53. Springer Fachmedien Wiesbaden, Wiesbaden (2017). https://doi.org/10.1007/978-3-658-18555-8_2

18. Patil, S., Jadhav, S., Kekade, S., Supare, A., Powar, A., Singh, R.: The influence of cutting heat on the surface integrity during machining of titanium alloy Ti6Al4V. Procedia Manuf. **5**, 857–869 (2016). https://doi.org/10.1016/j.promfg.2016.08.073

Determination of Largest Possible Cutter Diameter of End Mills for Arbitrarily Shaped 3-Axis Milling Features

M. Erler(✉), A. Koch, and A. Brosius

Chair of Forming and Machining Processes, Technische Universität Dresden, 01062 Dresden, Germany
martin.erler@tu-dresden.de

Abstract. Milling is one of the most frequently used manufacturing processes. End milling cutters, which are available in a wide variety of designs and sizes, are used very frequently due to their universal applicability and low manufacturing costs. During NC programming and the associated path planning, the tool to be used for machining a specific area is determined. The choice of tools plays a crucial role here. In particular, their diameter has a significant influence on the efficiency of processing. In order to choose the best tool, it is necessary to know which tool can be used for machining a specific part of the differential volume. The geometric accessibility sets an upper limit for the tool diameter. This paper presents an algorithm for calculating the largest possible tool diameter for end mills for any point within concave polygons. The heuristic determines the largest possible tool diameter and the associated position of the tool guide point of the end mill within an adjustable error corridor. The algorithm allows for the efficient calculation of the distribution of the largest possible tool diameters for any arbitrarily shaped 3-axis milling feature. Possible areas of application are optimization of tool usage, automated tool selection, feature recognition, automated NC programming and automated process planning.

Keywords: Cutter size determination · Tool diameter calculation · CAPP

1 Introduction

Selecting the right milling tool is an important part of the work preparation [1]. Typically, several tools are assigned to different machining areas. The selection of a single tool is subject to various restrictions that must be met by the tool parameters. Besides the tool type itself, the diameter is the most important parameter. On the one hand it has a significant influence on the most important parameters of the cutting process. On the other hand, the diameter is limited by the workpiece geometry, as it must not collide with it, but at the same time it must reach areas with limited accessibility. In order to select the best possible cutter, it is necessary to know the largest possible diameter. However, since accessibility varies, the largest possible diameter must be known for all locations [2].

M. Liewald et al. (Eds.): WGP 2022, LNPE, pp. 228–237, 2023.
https://doi.org/10.1007/978-3-031-18318-8_24

Knowledge of the tool diameter is particularly important for roughing, since it is often necessary to machine with the largest possible tool diameter in order to achieve the largest possible metal removal rate.

In this paper, a method is presented with which it is possible to determine the largest possible tool diameter for an end mill for a given point within an arbitrarily shaped area to be machined.

2 Related Work

When determining the tool diameter in the context of automated tool selection, a distinction must be made between roughing and finishing [3]. In finishing, the surface geometry determines the tool diameter, since it must not produce undercuts when in contact with the surface [4]. In roughing, the tool diameter is chosen to be as large as possible, as this generally allows the greatest metal removal rate to be achieved [5]. The limiting factor is the accessibility. The determination of the tool diameter is often combined with the selection of several tools in order to be able to machine differently accessible areas as effectively as possible [6]. The approaches used in both single and multiple selection to determine the largest possible diameter can be divided into 3 main categories: Direct reading (as a result of simplicity feature), trying with conditions or rule based (with preliminary subdivision if necessary) and calculating.

2.1 Direct Readout

Direct reading is limited to prismatic or very simple features/areas whose geometry has no or hardly any changes in one dimension (for examples see Fig. 1). Thus, the maximum diameter can be extracted directly, if the feature is parameterized or measured. It only has to be known which feature type it is and how it has to be measured.

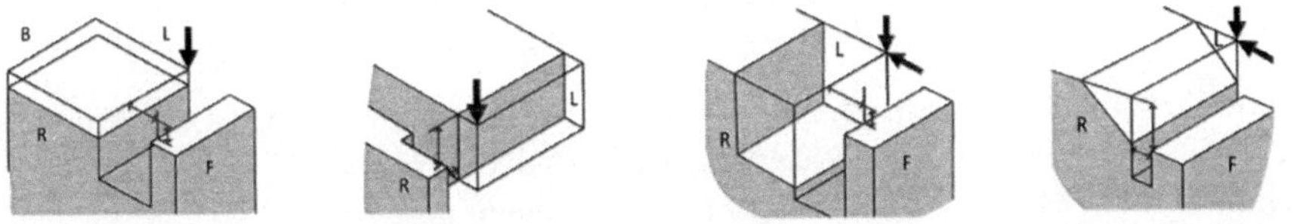

Face milling 4-sided interference Face | Face/Side milling 2-side interference Side | Face/Side milling 3-side interference Step | Chamfer milling 3-side interference Chamfer

Fig. 1. Example of feature types [7]

This approach is not universal, but is commonly used for 2.5-axis systems [8].

2.2 Trying

Trying out a tool is the most common approach and has been studied in numerous variations and extensions. The basic idea is to select a tool and then try out whether and which areas can be processed with it. The approaches differ with respect to the methodology for trying out, additional rules to be observed and, if necessary, subdivision carried out beforehand.

Bala and Chang subdivide the area to be processed by means of pseudo-paths for the tools to be tested [9]. Whether a tool is suitable for a pseudo-path is then determined by geometric cutting tests. A similar approach is taken by Lee and Daftari [10]. However, they divide the volume to be cut into virtual rectangular pockets. For these, the tool knives can then be read directly from the dimensions of the pockets. An extension of this is the subdivision into areas to be machined by the tool by means of accessibility analysis. If the accessible area is larger than the tool diameter, the tool is valid [11]. Also related to trial and error is the generation of offset contours [12] and volumes [13]. If the tool does not cut finished part contours, it is valid (see Fig. 2).

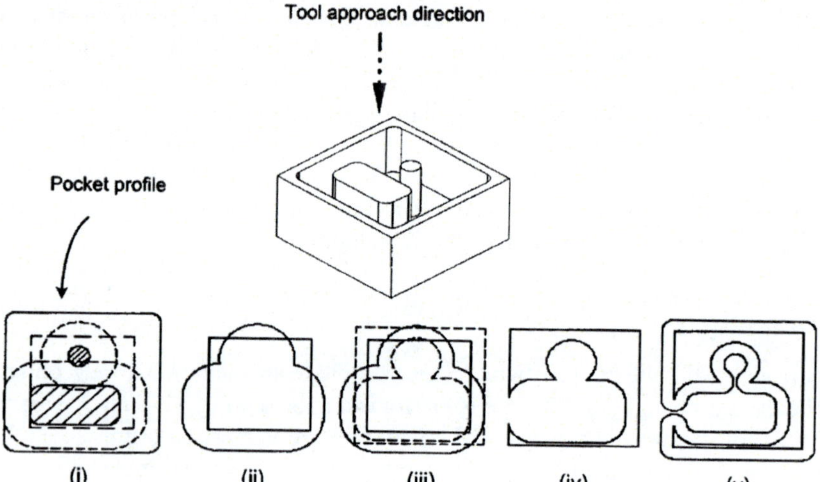

Fig. 2. Generation of offset volume by sweeping tool [13]

However, the most widespread approaches are those using NC/CAM simulation. Here, an attempt is made to generate paths for the selected tool using the path generation function of a CAM program. Valid paths are collected and the best tool is selected by means of a fitting function [14] or several tools are combined to an optimal tool set [3]. These approaches are basically universal, but can be very ineffective for complex parts and many tools. In addition, the results of the fitting function depend on the selected path strategies.

2.3 Calculation

The best known approach for direct calculation of the largest possible tool is the Voronoi diagram [15]. Originally having been restricted to purely convex contours [16] and later on to convex contours with in-islands [17], the algorithm now allows for the calculation of the largest possible tool diameter for arbitrary contours [18] (Fig. 3).

With Voronoi Mountain, this approach was significantly further developed [6]. This algorithm does not only determine one diameter for one specific geometry, but functions describing the maximum diameter at any given location. This information can be interpreted as a height value (Fig. 4). Seth and Stori extended it again so that it can be applied to open contours with islands [19]. Using this information, an optimal tool set can then be determined by means of a fitting function [20].

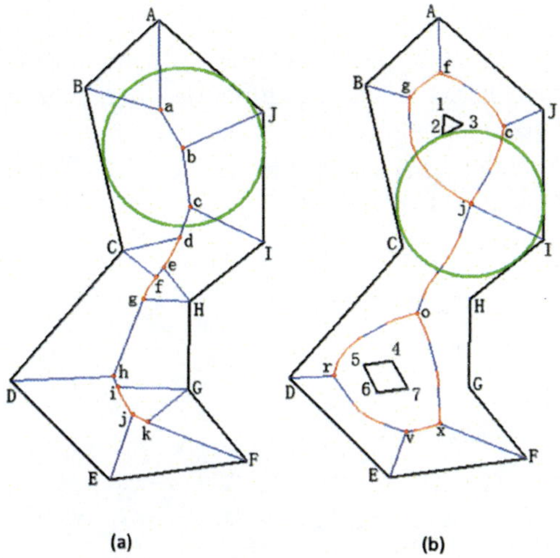

Fig. 3. Voronoi diagram of a simple (a) and complex polygon (b) [21]

Fig. 4. Voronoi mountain for sample geometry [19]

3 Description of the Algorithm

3.1 Model

The surface of the finished part is represented by points. As this paper focuses on 3-axis milling, it is sufficient to consider two-dimensional coordinates. Therefore, the surface points are projected into the plane perpendicular to the tool axis. The presented algorithm interprets these 2D points as obstacles, that must not be intersected. The second constraint for calculating the maximum tool radius requires the point to be removed to be within the range of the end milling tool, which is represented by a circle with center cutter location (CL) (see Fig. 5). Given this setup, the algorithm tries to find the biggest possible circle, which contains r and does not contain any surface point *p*.

To provide a fast search for the closest point, surface points are organized in suitable data structure such as grids or trees.

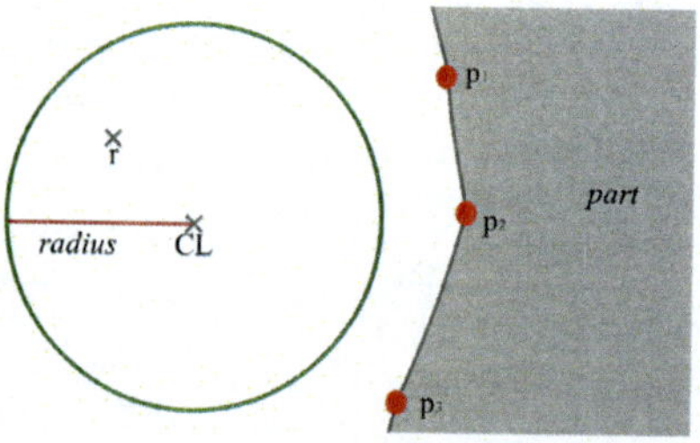

Fig. 5. Schematic representation base model: 1st constraint: part surface points (p_{1-3}) must not be intersected by circle, 2nd constraint: point to be removed (r) must be inside the circle.

3.2 Functionality

In general, the algorithm has an initialization part and a loop part, which itself consists of two parts.

1. moving the cutter location by f and
2. increasing the cutter radius by s

which are executed alternately until the abort condition is fulfilled (Fig. 6).

Initialization. To run properly, the initial values of CL and radius need to be set in a way, that the abort conditions are not fulfilled. This initial value problem is solved by setting the starting value for CL to the point to be removed and the initial radius to the distance between CL and the closest point.

There is also an overall maximum value for the tool radius. This represents a real world limitation to tool sizes and also prevents the algorithm from running endlessly.

Loop. The Loop consist of two steps, which are repeated until one or more abort conditions are fulfilled. The point with the minimum distance to CL is calculated. This is

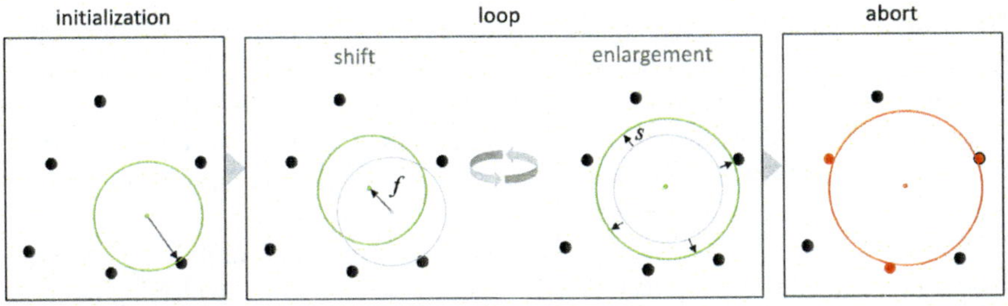

Fig. 6. Working principle of algorithm

equivalent to the largest possible radius. If the radius is larger than the overall maximum tool radius, the algorithm ends and returns the maximum tool radius. If the distance is smaller than the radius, the radius is increased by s with s being an adjustable constant and step one is repeated. Otherwise step two is performed and CL is moved away from the closest point by step size f with f being an adjustable constant.

The full flow chart of the algorithm is shown in Fig. 7.

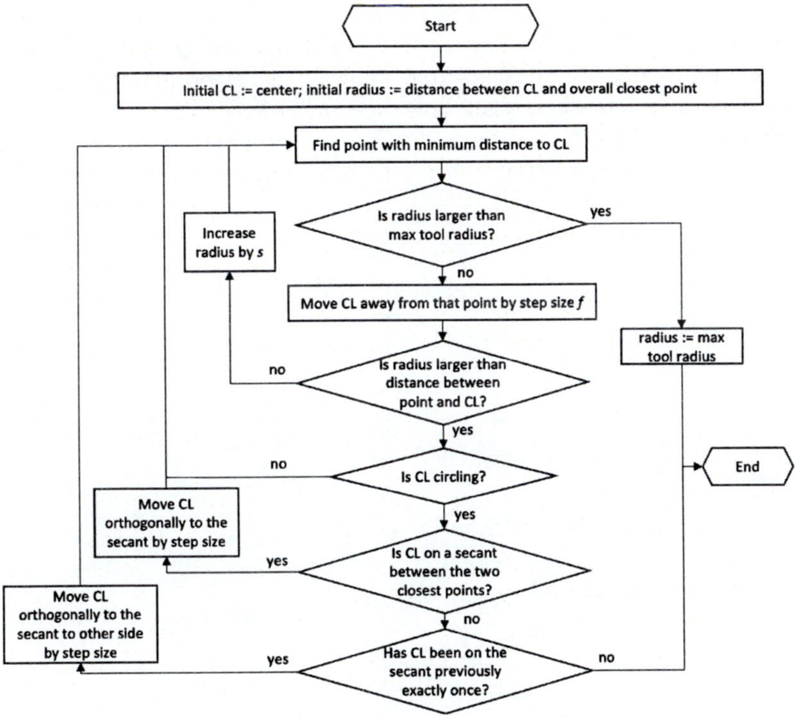

Fig. 7. Flow diagram of the algorithm.

Abort. The loop ends, when no more shifting or enlarging is possible without violating the constraints. However, shifting multiple times in succession can cause CL to oscillate between boundaries. To detect this case, the geometric mean value of CL is additionally monitored. It is evaluated as an additional termination criterion if the geometric mean value of CL has changed by less than c after a certain number of straight and orthogonal shifts without increasing the radius.

The value of c is critical. If it is too small, the circling condition is regarded as false even though the CLs are actually circling. If it is too big, the circling condition is regarded as true even though the centers are not circling.

If CL is circling it might be caught on the secant of two points. Due to the movement direction of step two, CL would be stuck on the secant. To escape that, CL is moved to a point that is located orthogonally to the secant with a distance of the step size s. Then, the search is started again from the start of the loop. This orthogonal search is repeated for the opposite orthogonal direction as well.

4 Results and Discussion

The sequence of the algorithm in the normal case is shown in Fig. 8. First, the circle is moving away from point 1 and increasing size, until CL reaches the line bisecting the plane between point 1 and 2 and zigzags around that line until the circle reaches point 3 where there is no room left to increase the circle and the algorithm stops.

It can be seen well that before reaching the termination criterion CL wanders around the final midpoint. This is the rule because only in special cases the circle reaches exactly the position and size where no magnification or movement is possible at the same time.

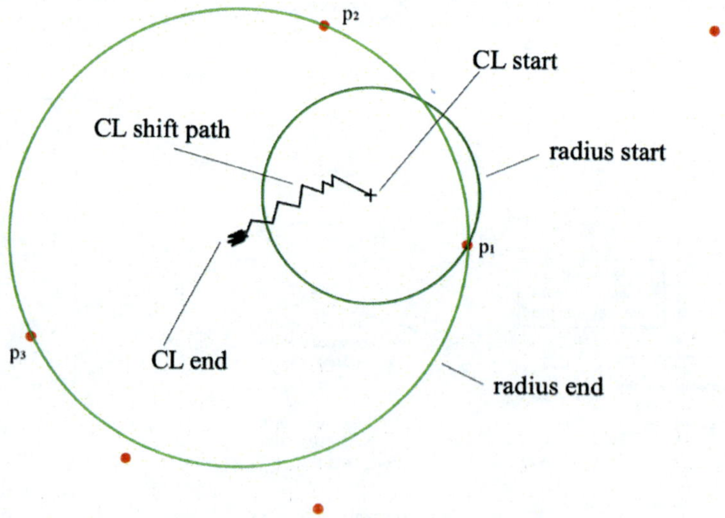

Fig. 8. Increasing tool size. Dark green circle: starting radius. Black zigzag line: iterative CL positions. Light green circle: final tool size.

4.1 Time Complexity

The following findings result from tests based on 10 sample contours (see Fig. 9), which were represented by a varying number of points ranging from 100.000 to 1.000.000.

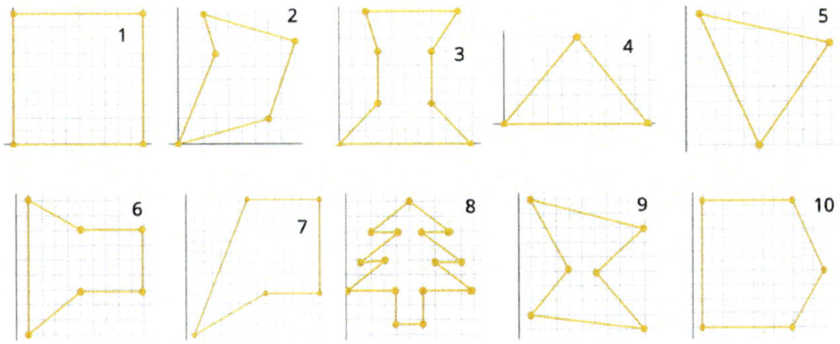

Fig. 9. Used sample contours

The main influencing factor on the runtime is the number of iterations, which is mainly dominated by f and s linearly. The maximum possible tool radius also limits the number of iterations, but only in case of very large contours or areas that are not limited to all sides, such as open-sided pockets.

The main advantage of the algorithm is its linear complexity regarding the number of points with a very low slope of the straight. On an average business-laptop, it is possible to compute the largest tool radius for an average arbitrary contour with over 1Mio points in less than a millisecond (see Fig. 10).

4.2 Conclusion

Fast calculation of largest possible tool diameter on high resolution contours opens a wide variety of applications. In terms of manufacturing planning, the distribution of tool sizes can be calculated for a complete work piece at a very high density (Fig. 11).

It can be applied to any possible geometry and has no special requirements in terms of its definition. Furthermore the presented algorithm provides the opportunity to combine it with additional constraints, which can be taken into account.

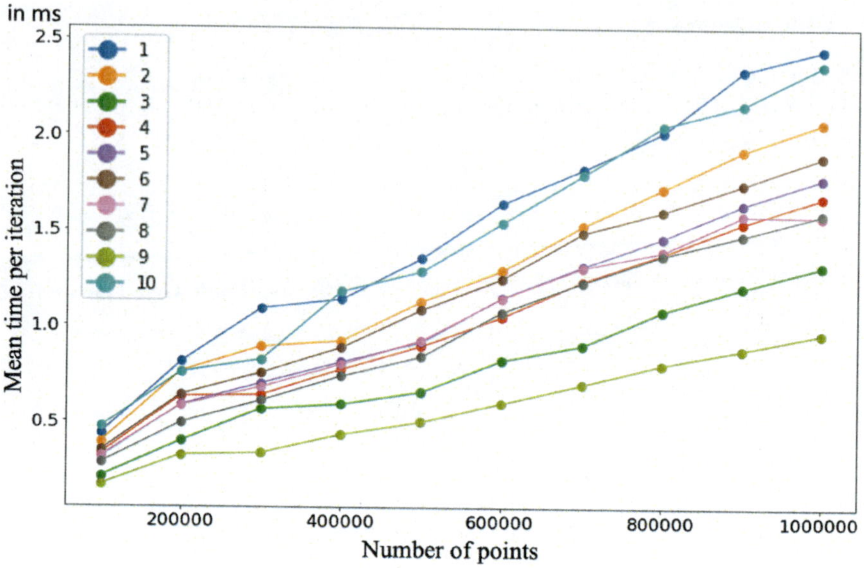

Fig. 10. Linear time complexity.

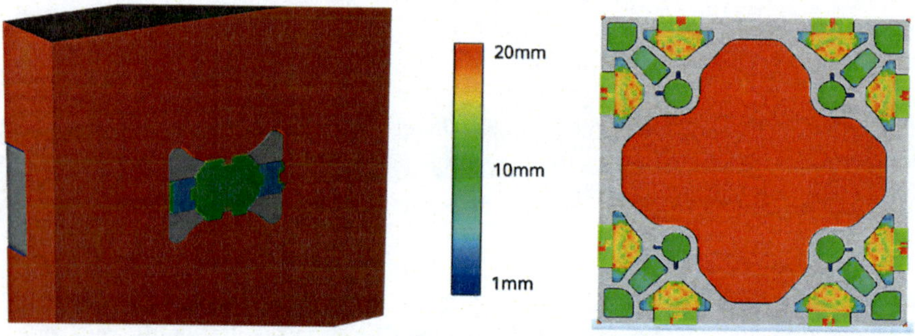

Fig. 11. Distribution of largest possible cutter size as heatmap for two sample parts

References

1. Li, X., Zhang, S., Huang, R., Huang, B., Xu, C., Zhang, Y.: A survey of knowledge representation methods and applications in machining process planning. Int. J. Adv. Manuf. Technol. **98**(9–12), 3041–3059 (2018). https://doi.org/10.1007/s00170-018-2433-8
2. Xu, X., Wang, L., Newman, S.T.: Computer-aided process planning—A critical review of recent developments and future trends. Int. J. Comput. Integr. Manuf. **24**(1), 1–31 (2011). https://doi.org/10.1080/0951192X.2010.518632
3. Mwinuka, T.E., Mgwatu, M.I.: Tool selection for rough and finish CNC milling operations based on tool-path generation and machining optimization. Adv. Prod. Eng. Manag. **10**(1), 18–26 (2015). https://doi.org/10.14743/apem2015.1.189
4. Lee, Y.S., Chang, T.C.: CASCAM-An automated system for sculptured surface cavity machining. Comput. Ind. **16**(4), 321–342 (1991). https://doi.org/10.1016/0166-3615(91)90073-I

5. You, C.F., Sheen, B.T., Lin, T.K.: Selecting optimal tools for arbitrarily shaped pockets. Int. J. Adv. Manuf. Technol. **32**(1–2), 132–138 (2007). https://doi.org/10.1007/s00170-005-0320-6
6. Veeramani, D., Gau, Y.: Selection of an optimal set of cutting-tool sizes for 2D pocket machining. Comput. Des. **29**(12), 869–877 (1997). https://doi.org/10.1016/S0010-4485(97)00042-0
7. Ji, W., Wang, L., Haghighi, A., Givehchi, M., Liu, X.: An enriched machining feature based approach to cutting tool selection. Int. J. Comput. Integr. Manuf. **31**(1), 1 (2018). https://doi.org/10.1080/0951192X.2017.1356472
8. Maropoulos, P.G., Baker, R.P.: Integration of tool selection with design part 1. Feature creation and selection of operations and tools. J. Mater. Process. Technol. **107**(1–3), 127–134 (2000). https://doi.org/10.1016/S0924-0136(00)00686-5
9. Bala, M., Chang, T.-C.: Automatic cutter selection and optimal cutter path generation for prismatic parts. Int. J. Prod. Res. **29**(11), 2163–2176 (1991). https://doi.org/10.1080/00207549108948076
10. Lee, Y.S., Daftari, D.M.: Feature-composition approach to planning and machining of generic virtual pockets. Comput. Ind. **31**(2), 99–128 (1996). https://doi.org/10.1016/0166-3615(96)00027-9
11. D'Souza, R.M.: On setup level tool sequence selection for 2.5-D pocket machining. Robot. Comput. Integr. Manuf. **22**(3), 256–266 (2006). https://doi.org/10.1016/j.rcim.2005.06.001
12. Kyoung, Y.M., Cho, K.K., Jun, C.S.: Optimal tool selection for pocket machining in process planning. Comput. Ind. Eng. **33**(3–4), 505–508 (1997). https://doi.org/10.1016/S0360-8352(97)00179-4
13. Lim, T., Corney, J.R., Clark, D.E.R.: Exact tool sizing for feature accessibility. Int. J. Adv. Manuf. Technol. **16**(11), 791–802 (2000). https://doi.org/10.1007/s001700070013
14. Spanoudakis, P., Tsourveloudis, N., Nikolos, I., Ia, E.N.G.: Optimal selection of tools for rough machining of sculptured surfaces. Imecs 2008 Int. Multiconference Eng. Comput. Sci. Vols I Ii **II**, 1697–1702 (2008)
15. Voronoi, G.: Nouvelles applications des paramètres continus à la théorie des formes quadratiques. Premier mémoire. Sur quelques propriétés des formes quadratiques positives parfaites. J. für die reine und Angew. Math. (Crelles Journal) **1908**(133), 97–102 (1908). https://doi.org/10.1515/crll.1908.133.97
16. Aggarwal, A., Guibas, L.J., Saxe, J., Shor, P.W.: Linear time algorithm for computing the Voronoi diagram of a convex polygon. Conf. Proc. Annu. ACM Symp. Theory Comput. 39–45 (1987). https://doi.org/10.1145/28395.28400
17. Fortune, S.: A sweepline algorithm for Voronoi diagrams. Algorithmica **2**(1–4), 153–174 (1987). https://doi.org/10.1007/BF01840357
18. Chin, F., Snoeyink, J., Wang, C.A.: Finding the medial axis of a simple polygon in linear time. Discrete Comput. Geom. **21**(3), 405–420 (1999). https://doi.org/10.1007/PL00009429
19. Seth, A., Stori, J.A.: Optimal tool selection for 2.5D milling, part 1: A solid-modeling approach for construction of the Voronoi mountain. Int. J. Comput. Integr. Manuf. **18**(4), 294–307 (2005). https://doi.org/10.1080/09511920512331319645
20. Seth, A., Stori, J.A.: Optimal tool selection for 2.5D milling, part 2: a Voronoi mountain approach for generalized pocket geometries. Int. J. Comput. Integr. Manuf. **18**(6), 463–479 (2005). https://doi.org/10.1080/09511920512331319636
21. Shen, Z., Yu, X., Sheng, Y., Li, J., Luo, J.: A fast algorithm to estimate the deepest points of lakes for regional lake registration. PLoS One **10**(12), e0144700 (2015). https://doi.org/10.1371/journal.pone.0144700

Investigation of the Effect of Minimum Quantity Lubrication on the Machining of Wood

A. Jaquemod(✉), K. Güzel, and H.-C. Möhring

Institute for Machine Tools (IfW), University of Stuttgart, Holzgartenstr. 17, 70174 Stuttgart, Germany

andre.jaquemod@ifw.uni-stuttgart.de

Abstract. Climate change, scarcity of resources and sustainability are increasingly becoming the focus of social and political attention. In this context, the importance of timber construction in particular is increasing. The use of wood as a building material offers additional storage capacity for the greenhouse gas CO_2, which is bound during tree growth. In this way, another extremely effective CO_2 reservoir can be created alongside the natural forest reservoir. In order to promote the establishment and further development of timber construction, there is a particular need for action in the machining of construction elements. To this end, it is necessary to investigate and develop solution approaches in terms of machines, processes and tools in order to optimize the manufacturing processes in timber construction and increase productivity. In the industrial environment, wood materials are usually machined dry. Unfavorable process parameters can lead to thermal problems that can have a negative effect on the machining qualities. In contrast to dry machining, no scientific findings are yet available for wood machining when using minimum quantity lubrication (MQL). This paper discusses the results on the influence of the use of minimum quantity lubrication when grooving spruce beams. Within the scope of experimental tests, different process parameters were varied and the effects on process forces and surface characteristics of the workpieces were analyzed.

Keywords: Wood machining · Lubrication · Surface analysis

1 Introduction

On the way to a resource-conserving and climate-neutral future, there will be no getting around the sustainable use of wood as a building material. Sustainability, climate protection and resource conservation are and will increasingly become central issues in politics and society. As a naturally (re)growing and organic building and construction material, wood is one of the most important renewable resources that can replace energy-intensive, finite materials as well as petroleum-based, fossil resources [1]. During the growth period, wood binds the greenhouse gas CO_2 and releases oxygen, which is vital for many living creatures, including humans [2]. However, wood not only stores CO_2 during growth, but also as a building material, wood can contain CO_2 that has already

M. Liewald et al. (Eds.): WGP 2022, LNPE, pp. 238–245, 2023.
https://doi.org/10.1007/978-3-031-18318-8_25

been bound [3]. Among other things, this has led to a steady increase in demand for wood as a building material. In Germany, for example, one in five new houses is already built with wood [4]. In addition, wood is also attractive in terms of production technology, as its processing is associated with a lower energy input compared to other materials [5]. Nevertheless, there is a need for action in manufacturing to make timber construction even more attractive. In particular, an optimization of serial production with regard to machine, process and tool is to be aimed at. Initial investigations and optimizations have already been explored, as [6] illustrates in summary.

Within the framework of the project "Cost-effective, highly insulating, layer-reduced, grade-pure, adhesive-free, digitally manufactured solid wood construction: Manufacturing and Joining Optimization for Multistory Exterior Wall Structures Made of Slotted Edge Solid Timber" from the Future Construction Research Grant, design and manufacturing strategies based purely on solid wood will be developed [7]. One way to optimize the process is to use lubrication. In woodworking, the process is usually dry machining. By using a cooling lubricant, a new tribological condition could be generated, which could have a positive effect on the thermal conditions and the cutting process. In the best case, an increase in cutting and feed rates and therefore an increase in productivity could be achieved without a reduction in the quality of machining. There are several types of lubrication. Machines in woodworking are usually only partially encapsulated, so that flood lubrication can not be implemented easily. Minimum quantity lubrication is a simpler method to implement.

In woodworking, there have been hardly any scientific studies on the use of MQL. On the other hand, numerous publications analyze the influence of MQL in the machining of metal and fiber-reinforced plastic composites [8–11]. Studies on the influence of MQL on surface quality have also been carried out in this field. References [12, 13] have shown that the roughness parameters S_a and R_a decrease when MQL is used.

For workpieces or components made of wood, gluing and painting often follow in the woodworking value chain. Consequently, high demands are placed on the machined surfaces, which can usually be achieved with a follow-up sanding process [14]. In order to reduce the associated effort, the aim should be to achieve the lowest possible roughness for the sawed surfaces. If a coating with powder coating or with foils is to be applied, a "minimal roughness" must be present according to [15]. Another advantage of low roughness is a reduced requirement for paint [16, 17]. In addition to the resource-saving aspect, this could also be of economic interest. In the following, the results of the experimental investigation of the use of MQL in the circular sawing of solid wood are presented.

2 Experimental Approach

The experiments were carried out with reference to those performed in [18]. The cuts were performed using the 5-axis machining center "Maka PE 170". A carbide circular saw blade from Leuco was used. The saw has a diameter of 350 mm with a cutting width of 3.5 mm and a number of teeth of 42. The teeth are positioned as alternate teeth. To investigate various influencing parameters the rotational speed (2500, 4000, 5500 min^{-1}) and feed rate (4 and 16 m/min) were varied. All cuts were made in a synchronous-cutting process and at a cutting depth of 40 mm.

Spruce cuboids (100 × 100 × 200 mm) were used as experimental material. For force measurement, these samples are clamped on a Kistler load cell type 9272. In each case, 100 mm long slots are sawn alongside the fiber. The sampling rate of the Kistler load cell is 1000 Hz and the data is recorded and subsequently evaluated using the institute's own LabView program.

To enable an investigation of the influence of the MQL, compressed air is used to feed the lubricant to the cutting edge. For this purpose, the MQL spray head from HPM Technologies was installed in the machine by using a magnetic arm. The orientation of the spray head was set at an angle between the nozzle and the circular saw blade of approximately 60° and frontal to the circular saw blade. This allows sufficient lubricant supply and good lubrication of the secondary cutting edges to reduce friction. The lubricant used was a "Survos Standard" from HPM Technologies, which is advertised as having an evaporation rate of up to 100%. The ignition point of the lubricant is 62 °C, while the evaporation point lies between 180 and 235 °C. Its density is 0796 g/cm3 and the lubricant is colorless, which means that there should be no visual change to the wooden surface. The entire test setup for force measurement under dry and lubricated conditions is shown in Fig. 1 for illustration.

Not only the passive force, which is decisive for the surface quality, is to be used as a criterion for evaluating the machining variants "dry" and "MQL", but also the cut surface and thus the roughness parameters R_a and R_z are to be considered. For this purpose, three measurements per cutting surface are carried out using the InfiniteFocusG5 from Alicona. The measurements are carried out at the cutting edge entry, in the center of the specimen and at the cutting edge exit.

Fig. 1. Experimental setup for force measurement while circular sawing spruce under dry and MQL conditions (Image: IfW Stuttgart)

3 Results

Figure 2 shows a summary of the results for a rotational speed of 2500 min^{-1}. Looking at the surfaces of the specimens, it can be clearly seen that the use of lubricant reduces the occurrence of post-cutting marks. There is also a reduction in the offsetting at the cutting edge exit. This relationship can be recognized not only visually, but also based on the measured roughness values.

The roughness R_z averaged during dry machining is 84.6 μm and drops to 60.6 μm with the use of the lubricant. Similarly, this reduction can be seen in the average roughness R_a. This suggests a decrease in the force component that is decisive for the quality of the cut surface, the passive force. If the curves of the passive force are plotted over time (Fig. 2), this relationship can be confirmed. The passive forces increase for both machining variants until they drop again to 0 N. When looking at the passive force of dry machining (dry), changes in the force curve can be seen, especially when the cutting edge runs out. This part of the passive force is decisive for the cutting surface. Considering the previously viewed images of the cut surfaces and the post-cut traces visible on them, it is obvious that the final surface finish is only created by the outgoing saw blade. If, on the other hand, the force curve of the passive force is considered under the use of MQL, it runs smoother and steadier, especially at the cutting edge exit. In terms of magnitude, a difference can also be observed between the two machining variants. The passive force reaches its maximum at 18 N during dry machining, whereas it has a maximum of 14.3 N during MQL machining. Such a reduction can be attributed to the reduced friction due to the lubricant.

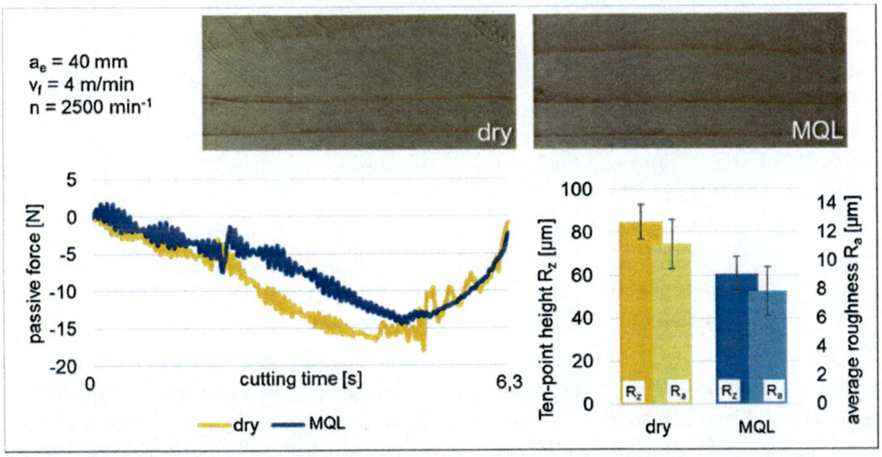

Fig. 2. Results overview of the different machining variants dry and MQL at a rotational speed of 2500 min^{-1} (Image: IfW Stuttgart)

The test sequence with a rotational speed of 4000 min^{-1} shows a similar tendency (Fig. 3). A reduction can be achieved in the cutting pictures and in the roughness values using MQL. Furthermore, it can be stated that there is not only an improvement in the roughness values due to the use of MQL, but also due to the increase in the rotational speed. This can be justified by a smoother and more stable running behavior at higher speeds, as well as a lower chip volume. If the course of the passive force is considered, a reduction in the amount due to the increase in speed can also be seen here. This suggests that the amount is more decisive for the surface quality than the course. Regardless of the rotational speed, the passive force is again lower with minimum quantity lubrication than for dry machining.

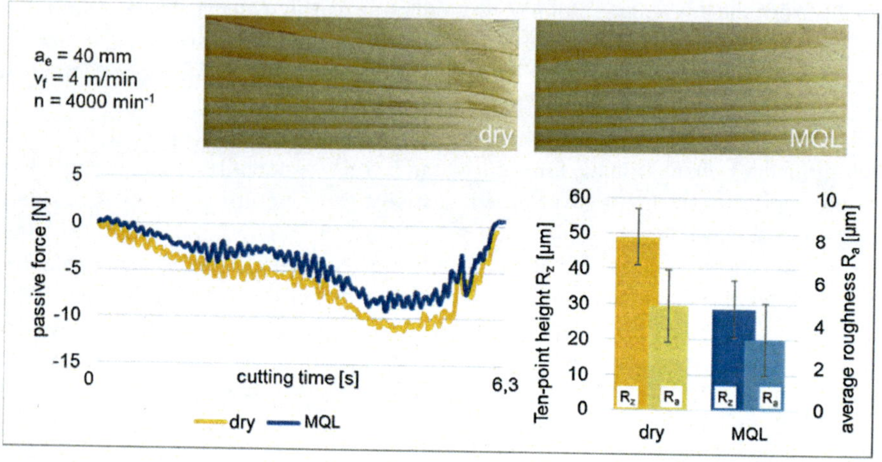

Fig. 3. Results overview of the different machining variants dry and MQL at a rotational speed of 4000 min^{-1} (Image: IfW Stuttgart)

If the maximum rotational speed of 5500 min^{-1} is now examined, a change in the cutting pattern can first be seen (Fig. 4). The quality feature of burn marks dominates the appearance of dry machining. This is where a thermical change of the surface occurs. It can already be seen from visual observation that there is a reduction in these burn marks with the use of MQL.

Looking at the roughness values, the tendency to decrease by increasing the rotational speed during dry machining can be further confirmed. However, when comparing the two machining variants at the maximum rotational speed, no improvement in roughness is evident due to the use of MQL. In this case, the values increase from $R_{z,\ dry}$ 40.47 µm to $R_{z,\ MQL}$ 57.24 µm. If the passive force is once again considered, it supports the results shown here. Again, a reduction in magnitude can be seen due to the increase in rotational speed. However, it is also visible that, contrary to the results previously considered, the maximum passive force for MQL machining is greater at 8.8 N than for dry machining (6.2 N). By increasing the rotational speed, there is also an increase in the circumferential speed. This could lead to a partial quantity of lubricant droplets being detached from the cutting edge due to the centrifugal force, resulting in little or no lubricating effect. However, the reduction in burn marks suggests that the lubricant has at least a partial effect on machining. It is possible, that the lubricant still has a cooling effect to a certain degree by evaporating when it comes into contact with the saw blade, thereby reducing the burn marks, but that there is no lubricating effect. However, it must also be considered that wood is a natural, anisotropic and inhomogeneous material. Various influencing variables such as branch growth, resin galls and torsional growth lead to different properties within tested specimens and to a corresponding variation in machinability.

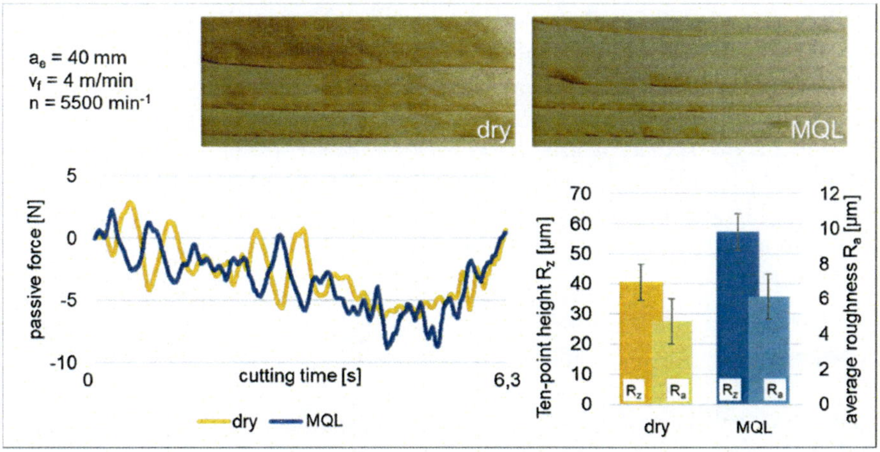

Fig. 4. Results overview of the different machining variants dry and MQL at a rotational speed of 5500 min^{-1} (Image: IfW Stuttgart)

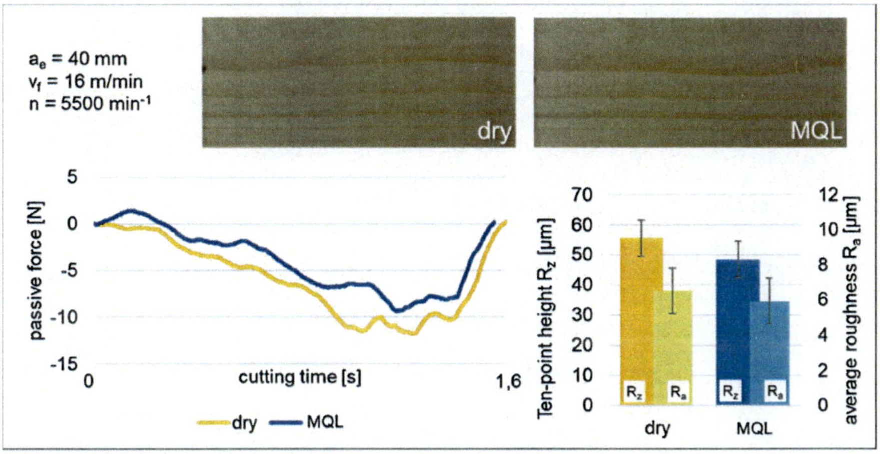

Fig. 5. Results overview of the different machining variants dry and MQL at a rotational speed of 5500 min^{-1} and a higher feed speed of 16 m/min (Image: IfW Stuttgart)

Finally, the influence of MQL on an increase of the feed rate to 16 m/min was investigated to analyze the potential of an increase in productivity. Analogous to the first results, an improvement using MQL can already be recognized from the cutting pattern in Fig. 5. Once again, there is a decrease in the number of post-cut traces. In addition, the roughness values again confirm the associated improvement in surface quality. The roughness R_Z is 55.5 µm for dry machining and drops to 48.6 µm with the use of the lubricant. When the passive force is considered, a direct correlation can again be seen. The passive force is greater with dry machining than with MQL machining.

However, when these results are compared to those at 5500 min^{-1} and a lower feed rate of 4 m/min, an increase in both roughness and passive force can be seen. This can again be justified by the chip volume, which increases due to the higher feed rate.

The significant reduction in passive forces and the associated decrease in roughness underline the thesis that the use of MQL also leads to a reduction in friction in woodworking. The appearance of burn marks indicates that the lubricant no longer reduces roughness. Judging from the cutting pattern, a cooling effect occurs. The burn marks are reduced by the lubricant. However, the lubricant no longer contributes to an improvement of the roughness.

4 Summary

In the course of the investigations, it was shown that the use of MQL leads to an improvement of the surface quality. The passive forces decrease both through higher cutting speeds and using MQL. This is associated with lower roughness. The use of the lubricant can be used to increase productivity while maintaining the quality (roughness).

Only burn marks change the roughness values. Here, too, an improvement in surface quality can be recorded. The area affected by burn marks can be significantly reduced using MQL. However, there is an increase in roughness and passive force.

The results show great potential for the use of MQL in wood machining. This potential has to be worked out in further investigations. In particular, the influence of the feed rate should be investigated further in order to increase productivity and optimize the machining process. The tool life and the thermal influence of the lubricant and lubricating jet should also be investigated.

Acknowledgements. The authors would like to thank the German Federal Ministry of Housing, Urban Development and Construction (BMWSB) for funding the project "Cost-effective, High-insulation, Layer-reduced, Grade-pure, Adhesive-free, Digitally Manufactured Solid Wood Construction: Manufacturing and Joining Optimization for Multi-story Exterior Wall Structures Made of Slotted Edge Solid Timber" (10.08.18.7–20.54) under the "Zukunft Bau" research grant.

References

1. Westkämper, E., Warnecke, H.-J.: Einführung in die Fertigungstechnik, 5. Teubner Verlag, Aufl (2002)
2. Technische Universität München- Holzforschung München: Bauen mit Holz = aktiver Klimaschutz. TU München (2020)
3. Walz, A., Taverna, R., Stöckli, V.: Effektiver Klimaschutz durch den Wald—Holz nutzen ist wirksamer als Vorräte anhäufen. Wald und Holz **4**(10), 37–40
4. Holzbau Deutschland—Bund Deutscher Zimmermeister im Zentralverband des deutschen Gewerbes: Lagebericht Zimmerer/Holzbau 2021 Stand 2021. Internet: https://www.holzbau-deutschland.de/fileadmin/user_upload/eingebundene_downloads/Lagebericht_2021_mit_Statistiken.pdf. Letzter Zugriff 09 Jan 2022
5. Gottlöber, C.: Zerspanung von Holz und Holzwerkstoffen: Grundlagen–Systematik–Modellierung–Prozessgestaltung. Carl Hanser Verlag GmbH Co KG (2014)

6. Möhring, H., Eschelbacher, S., Güzel, K., Kimmelmann, M., Schneider, M., Zizelmann, C. Häusler A., Menze, C.: En route to intelligent wood machining–current situation and future perspectives. J. Mach. Eng. **19** (2019)
7. ZukunftBau Forschungsförderung: Kostengünstige, hochdämmende, schichtenreduzierte, sortenreine, klebstofffreie, digital gefertigte Holzmassivbauweise: Herstellungs- und Fügungsoptimierung für mehrgeschossige Außenwandkonstruktionen aus geschlitzten Kantvollhölzern. https://www.zukunftbau.de/projekte/forschungsfoerderung/1008187-2054. Letzter Zugriff 09 Jan 2022
8. Nagaraj, A., Uysal, A., Jawahir, I.S.: An investigation of process performance when drilling carbon fiber reinforced polymer (CFRP) composite under dry, cryogenic and MQL environments. Procedia Manuf. **43**, 551–558 (2020)
9. Iskandar, Y., Damir, A., Attia, M. H., Hendrick, P.: On the effect of MQL parameters on machining quality of CFRP. In: ICCM International Conferences on Composite Materials, vol. 2013, pp. 3281–3290 (2013)
10. Ekinovic, S., Prcanovic, H., Begovic, E.: Investigation of influence of MQL machining parameters on cutting forces during MQL turning of carbon steel St52-3. Procedia Eng. **132**, 608–614 (2015)
11. Pervaiz, S., Anwar, S., Qureshi, I., Ahmed, N.: Recent advances in the machining of titanium alloys using minimum quantity lubrication (MQL) based techniques. Int. J. Precision Eng. Manuf.-Green Technol. **6**(1), 133–145 (2019)
12. Prarone, P., Robiglio, M., Settineri, L., Tebaldo, V.: Milling and turning of titanium aluminides by using minimum quantity lubrication. In: Elsevier ScienceDirect (2014)
13. Hadad, M., Sadeghi, B.: Minimum quantity lubrication-MQL turning of AISI 4140 steel alloy. J. Clean. Prod. **54**, 332–343 (2013)
14. Hänsel, A., Prieto, J. (eds.): Industrielle Beschichtung von Holz und Holzwerkstoffen im Möbelbau. Carl Hanser Verlag GmbH Co KG (2018)
15. Verein Deutscher Ingenieure e.V. (VDI): VDI-Richtlinie 3414 Blatt 1—Beurteilung von Holz- und Holzwerkstoffoberflächen. Beuth Verlag, Berlin (2019)
16. Heid, H., Reith, J.: Malerfachkunde, 5. Auflage Vieweg+Teubner Verlag, Wiesbaden (2010)
17. Rabehl, E.W.: Die moderne Lackierung und Beschichtung von Holz. Holz als Roh- und Werkstoff (1969)
18. Güzel, K., Jaquemod, A., Stehle, T., Möhring, H.C.: Einsatzpotential von MMS in der Holzbearbeitung. WT Werkstatttechnik **112**, 67–72 (2022)

Fluid Dynamics and Influence of an Internal Coolant Supply in the Sawing Process

C. Menze[1](✉), M. Itterheim[1], H.-C. Möhring[1], J. Stegmann[2], and S. Kabelac[2]

[1] Institute for Machine Tools, University of Stuttgart, Holzgartenstrasse 17, 70174 Stuttgart, Germany
christian.menze@ifw.uni-stuttgart.de

[2] Institute of Thermodynamics, Leibniz University Hannover, An Der Universität 1, 30823 Garbsen, Germany

Abstract. This paper contains the first steps for an analysis of the fluid dynamics in a circular sawing process with an internal cooling as part of an ongoing project. The analysis takes place with a test rig for orthogonal cutting of the IfW of the University of Stuttgart, which is modified to be able to create an artificial narrow closed cutting gap. It is built with sapphire glasses on the sides to make the fluid dynamics and the chip formation in the gap visible and analyzable. The investigation of the fluid dynamics carried out with the PIVLab software, which is a module in Matlab.

Keywords: Sawing · Fluid dynamics · Coolant

1 Introduction

Circular sawing is an important intermediate step in the production of semi-finished products in industrial manufacturing systems. Modern circular sawing tools consist of a disc-shaped base blade and welded-on carbide cutting edges that perform a circular cutting motion [1]. A special aspect of the cutting process is a narrow cutting gap that forms. This means that the cutting process takes place in a chip space and a narrow, almost closed cutting gap. This makes it difficult to deliver coolant to the cutting zone at the tool centre point (TCP).

Due to the low thermal conductivity, high specific strength and wear-promoting chemical reactivity, titanium alloys are among the materials that are difficult to machine [2]. Especially a low thermal conductivity leads to an increased thermal load on the tool during the machining of these materials. This can lead to melting and subsequent breakout of the soldered joints between the base plate and the cutting insert of the tool [3].

The internal coolant supply (ICS) is used, for example, during drilling to reliably transport the coolant into zones that are difficult to access. This allows the coolant to be conditioned and used effectively and leads to improved heat and chip removal from the cutting zone.

M. Liewald et al. (Eds.): WGP 2022, LNPE, pp. 246–255, 2023.
https://doi.org/10.1007/978-3-031-18318-8_26

The influence of a coolant on the machining process is complex and the detailed physical interactions are still insufficiently understood [4]. However, independent studies show that the use of a coolant generally leads to a reduction in cutting forces [5], shortening of the chip-tool contact length [6] and a change in the chip shape or curvature [7]. During sawing, chip curling is of increased importance, as unfavourably shaped chips can cause chip clamping, which exposes the tool to increased loads and can even lead to tool breakage. The mechanisms of chip curling are still unclear. Explanations mainly describe chip curl by [8] (1) the compressive stress in the chip lower layers; (2) the detention in the chip lower layers; (3) the thermal stress in the cooling process. Hongtao et al. [9] showed with experimental tests with tools with restricted contact length that an internal bending moment must probably exist. This bending moment results from the frictional forces in the primary shear plane and the rake face of which the direction vectors are not collinearly aligned. Using an analytical slip line model, the authors calculate the resulting bending moment.

Since the investigation of the chip-fluid interaction is complex and difficult, numerical simulation tools are often used to analyse the interactions [10]. Oetzkaya et al. [11, 12] used polyamide particles to analyse the flow properties during drilling. By using a transparent acrylic workpiece, the particle movement could be visualised with a high-speed camera, and it was possible to prove experimentally and simulatively that parts of the tool cutting edge are not sufficiently supplied with coolant.

The ICS might have a comparable effect in the circular sawing process and lead to increase the tool life and to supply the coolant conditioned to the process in a resource and environmentally friendly way. As part of an ongoing project, this paper shows the experimental investigation of the influence of a coolant with ICS on the sawing process. In addition, the fluid dynamics of the coolant in the chip space and cutting gap are visualised.

2 Preparation of Experiments

The investigation and observation of chip formation during circular sawing is difficult due to the motion of the tool coordinate system (superposed translation and rotation). For this reason, the complex tool movement is reduced to an analogy experiment with a static tool coordinate system. This makes it possible to observe the chip formation process as well as the flow behaviour of an ICS inside the chip space. As shown in [13], in a first step two-tooth segments are cut out of a circular saw blade. These are fixed in a tool holder. A special test machine for linear cutting movements (to investigate orthogonal and oblique cutting) is used as a test rig [14]. The basic set-up consists of a tool holding system, with which the cutting depth can be adjusted via a feed axis. In the cut, the tool is held in a fixed position. The workpiece is fixed in a clamping device on a feed axis with a linear motor and carries out the cutting movement. The cutting forces are measured via a Kistler dynamometer, which is fixed between the workpiece and the feed carriage.

2.1 Modification of the Test Rig

In the real sawing process, the cutting process takes place in a cutting gap. To simulate these conditions and at the same time ensure process monitoring, scratch-resistant sapphire glasses are used as an artificial cutting gap (Fig. 1).

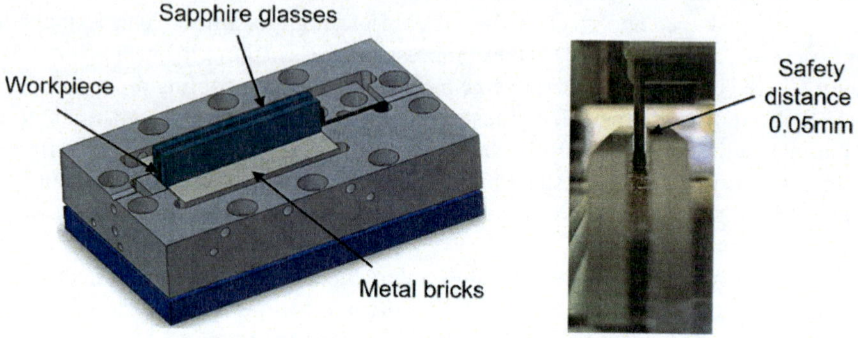

Fig. 1. Clamping device and artificial cutting gap

The gap is designed in such a way that there is a safety distance of 0.05 mm between the tool and the sapphire glass on both sides. This is achieved by making the workpiece sample 0.1 mm thicker than the cutting width of the tool. Because of the cutting forces, it must be ensured that the workpiece is firmly clamped. However, the sapphire glasses must be applied to the workpiece with a gentler pressure. This was realised by means of a special clamping device. On one side there is a slope of 8° to create a high pressure on the material sample, which is applied with two ramped pieces. One of them is underneath the sapphire glass and the other one behind it. The sapphire glasses are slightly pressed against the material sample with two metal bricks. One brick is placed on the small, ramped brick and the other one on a plateau on the other side of the material sample.

The coolant is introduced via the ICS using a pressure accumulator system that is filled with a hand pump. A coaxial valve is triggered via a control signal, whereby the coolant is abruptly directed from the pressure accumulator into the tool. The coolant finally reaches the chip space in the cutting gap via a hole (Ø 1.5 mm) drilled in the tool (twotooth segment).

2.2 Design of Experiments and Measuring Analytics

The aim of the study is to analyse the influence of a coolant with ICS on the cutting process under the conditions of a sawing process in the cutting gap. The material used is Ti6Al4V. The cutting speed in the test is 30 m/min with a depth of cut of 0.05 mm. The pressure of the inflowing fluid is adjusted in different steps and is 2; 4; 6 and 8 bar. As coolant, water is used. For later analysis of the flow characteristics with Particle Image Velocimetry (PIV), polyamide tracer particles (density 1.016 g/cm^3, Ø 90 μm) are added to the liquid.

For the analysis, a high-speed camera (type "Os8 - S3" from Imaging Solutions GmbH) with a focusing objective is used and aligned perpendicular to the cutting edge.

The objective is shown in Fig. 2 on the left and provides the shown flow picture from this angle. The inlet is directed with an angle of 5° towards the engaging tooth.

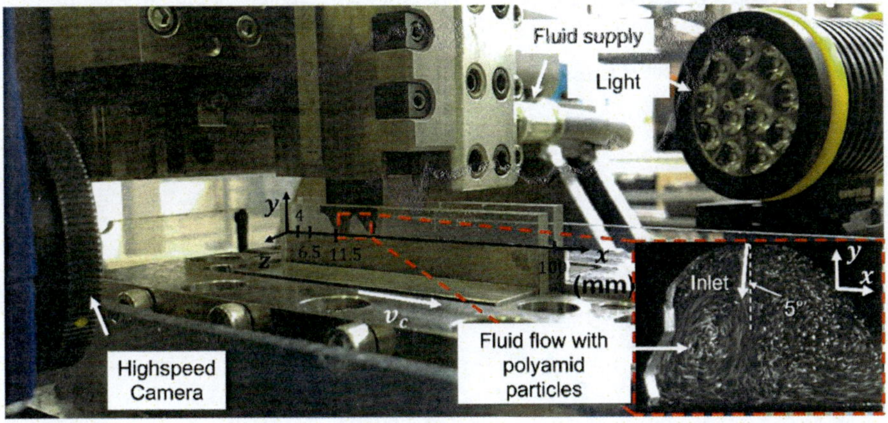

Fig. 2. Setup of the test rig with an example of the shown angle of the camera view

The illumination is provided by two Constellation 120E spotlights from Veritas and one of which is positioned in the uplight and one in the downlight. The pipe connected to the tool holder from the right-hand side is connected to the valve for coolant control. The marked area is the artificially created flow gap (x-y-plane).

For the analysis of the fluid dynamics, the software PIVLab [15] is used. PIVLab is programmed based on Matlab and is made available there as a module. It allows the particle movements within a flow to be observed and followed. As a result, the velocity distribution of the flow is displayed in a vector diagram.

3 Influence of Coolant on the Cutting Forces in the Cutting Gap

Important parameters for the cutting process are the resulting cutting forces. This section examines how the coolant pressure of an ICS affects the cutting forces. In addition, the cutting forces under the conditions dry, external coolant supply and ICS are compared with each other.

3.1 Investigation of ICS with Different Pressures

For the design of an ICS during sawing, knowledge of the effect of the coolant pressure on the cutting process is essential. Figure 3 shows a comparison of the cutting forces under different coolant pressures over the cutting distance in the cutting gap. The cutting forces are influenced by the coolant pressure. While the force curve is similar in the tool engagement and up to a time of approx. 50 ms, a different behaviour becomes visible in the further course of the cut. Especially the comparison towards the end of the cut (at approx. 180 ms) shows that a coolant pressure of 8 bar leads to an increasing and 2 bar to a decreasing cutting force characteristic. This suggests that there is a correlation between the cutting parameters and the pressure of the internal cooling. For the selected cutting parameters, the variant with 2 bar delivers the best results of the tested internal pressures and is therefore compared in the following with dry cutting and external cooling.

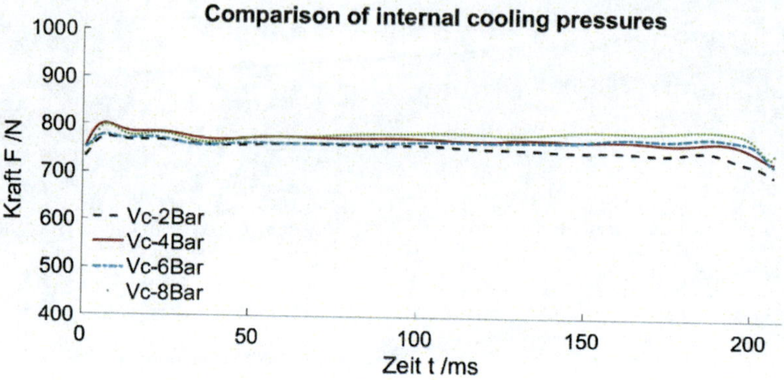

Fig. 3. Analysis of the cutting forces at different internal fluid pressures

3.2 Investigation of Different Cooling Variants

In the next step, different cooling variants are compared with each other. For this purpose, cutting tests are carried out in dry condition, with external coolant supply and ICS (2 bar) under the cutting parameters presented in Sect. 2.2 and compared with each other. The external coolant supply is carried out with a pump and the coolant is fed laterally onto the base plate of the two-tooth segment (similar to the real process conditions).

Figure 4 shows the determined force curves. These start at a similar magnitude, but the cutting forces run differently over the cutting length. While with dry machining and external cooling the forces increase again after a short drop, the forces decrease continuously with ICS. Under the selected cutting parameters and the cooling variants considered, the best results are obtained with internal cooling at a pressure of 2 bar.

4 Analysis of the Fluid Dynamics

The investigation of the fluid dynamics is carried out using the PIV system presented in Sect. 2.2. For this purpose, a static momentary image is taken. This makes it possible to record the flow over a longer period of time under the same geometric conditions with the high-speed camera. The flow can then be analysed and the influence of a chip in the chip space on the flow characteristics can be investigated. Therefore, the cutting process is abruptly stopped at several defined positions (Fig. 2). This allows the different chip characteristics to be analysed regarding their influence on the resulting fluid dynamics. To be able to carry out this analysis in a static state, an additional test (without cutting) is carried out in which the flow structure in a moving (workpiece performs cutting movement) and stationary system is compared.

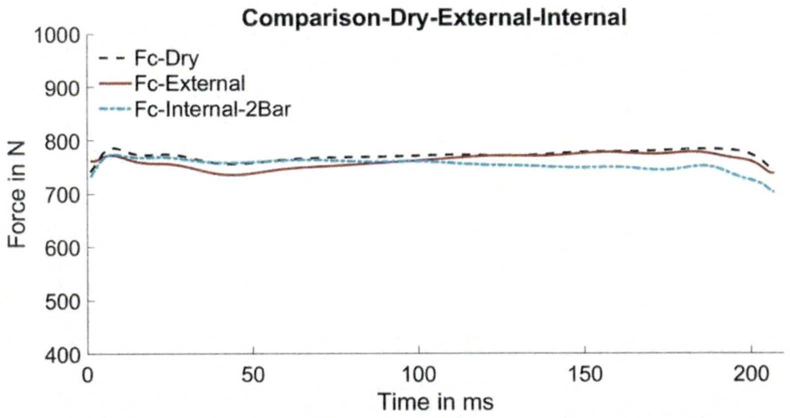

Fig. 4. Comparison of the cutting forces of the machining (dry, internal and external cooling)

4.1 Comparison of the Fluid Dynamics with a Static and Dynamic Tool

Therefore, the flow properties of a static and dynamic test were compared with each other. In the static test, the workpiece is not moved, and the flow propagates without cutting speed. In the dynamic test, a cutting movement takes place without chip removal, which is directly above the workpiece surface. The focus of the experiment is on the propagation of the flow and significant differences in the flow characteristics. The two flows are shown in the following Fig. 5.

The characteristics of the flow are almost identical in both experiments. The inflow can be seen in the centre of the image and the two resulting vortices move at different speeds. The faster vortex is on the left side near the cutting edge. The faster area on the right side can be explained by the gap that exists between the previous tooth and the workpiece (caused by the linear analogy test). This gap is caused by the round geometry of the entire saw blade and cannot be avoided within the framework of this test. Based on these results, the complete test plan, in particular the static recordings with chips, can be carried out with a guarantee of comparability.

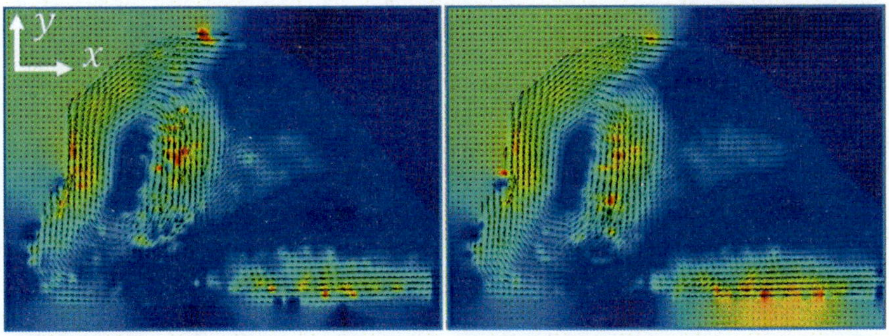

Fig. 5. Results of the static (left) and dynamic (right) fluid dynamics analysis

4.2 Analysis of the Fluid Dynamics

Therefore, three different positions of the chip formation are considered, which are located at the three positions of 4, 6.5 and 11.5 mm (cutting distance Fig. 2).

Position 1: 4mm (Fig. 6). At this point the chip contacts the workpiece surface for the first time and the flow is still similar to the flows without chip (Fig. 5). It is noticeable that the right vortex is much less visible and the left vortex is much more pronounced. This suggests that the stronger left vortex supplies the chip and the cutting edge with sufficient fluid. It can also be assumed that the flow influences the chip by causing it to be rolled up in a smaller size than with external cooling or dry machining.

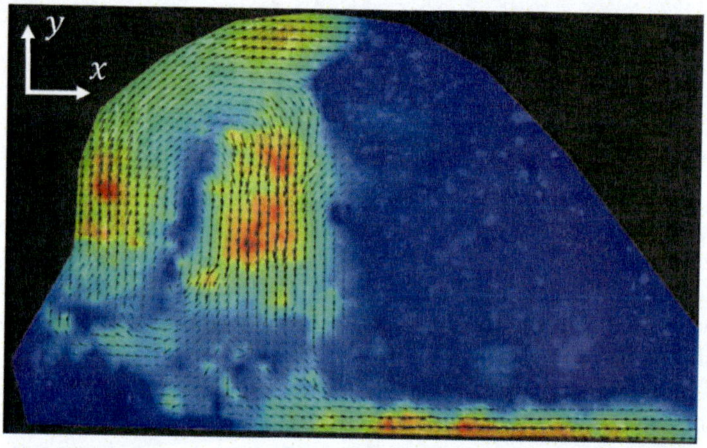

Fig. 6. Analysis of the fluid dynamics on position 1 (4 mm)

Position 2: 6.5mm (Fig. 7). Here the chip starts to roll up and the flow clearly shifts to the left side above the chip and the right vortex has almost completely disappeared.

Position 3: 11.5mm (Fig. 8). At this time the chip is completely curled once and the flow shifts to the right side. This increases the flow speed in the direction of the exit from the chip space and the right vortex. The chip rolls up tighter and is pressed towards the cutting edge by the flow. This is a positive effect for the cutting process, as the chip space can be better utilised with a tightly rolled chip. In addition, the friction at the chip space is minimised and thus the forces during the cutting process are improved. The effect of curling is confirmed by the analysed image where the vectors along the chip surface point in the direction of the cutting edge or the workpiece. At position 3, the right-hand vortex is again clearly visible in the analysed image, while the left-hand vortex is difficult to see currently.

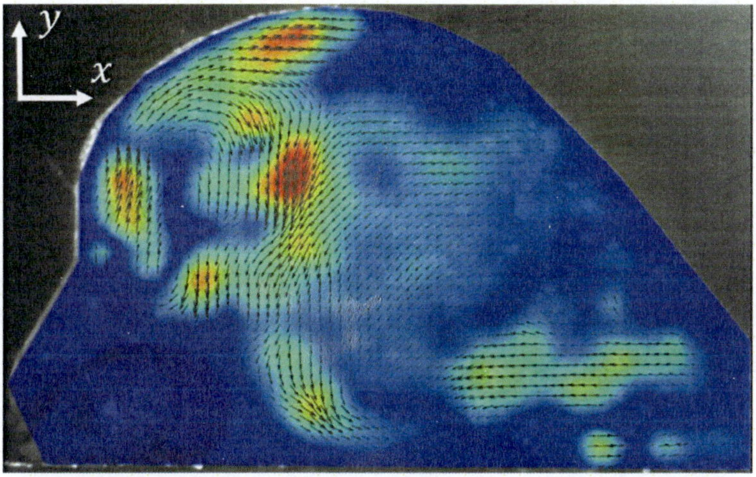

Fig. 7. Analysis of the fluid dynamics on position 2 (6.5 mm)

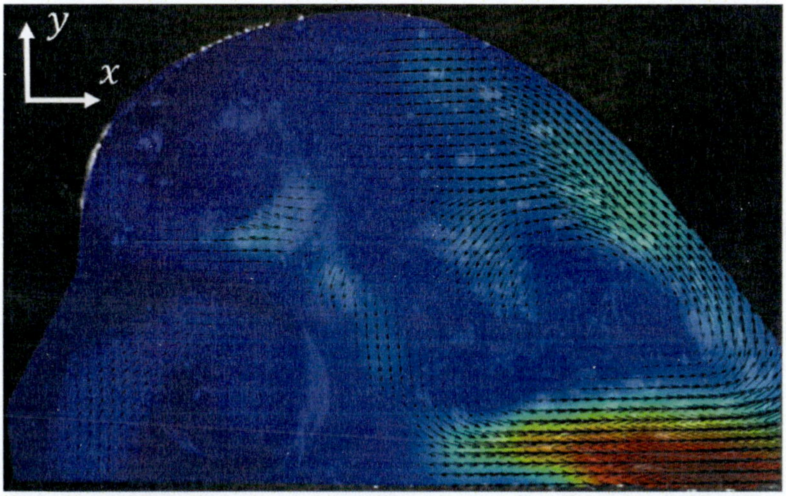

Fig. 8. Analysis of the fluid dynamics on position 3 (11.5 mm)

5 Summery and Future Work

This paper shows the investigation of the influence of a coolant with ICS on the circular sawing process by means of an analogy test. For this purpose, a special test set-up with a transparent artificial cutting gap was presented.

The investigation of the cutting forces with different coolant pressures showed that there is a correlation between the supply pressure and the process parameters. The adjusted pressure of 2 bar showed the lowest cutting forces.

Furthermore, it could be shown that the ICS under the same process boundary conditions had the lowest cutting forces compared to an external coolant supply and a dry cut.

The analysis of the fluid dynamics showed with an empty chip space (without chip) a flow structure with two characteristic vortices. The vortex rotating towards the cutting edge promises a good cooling effect of the TCP due to a constant flow of fluid around it. A chip in the chip space changes the flow. The evaluation of the PIV showed a flow strongly directed towards the chip. It can be assumed that this results in pressure forces acting on the chip, which promote tighter curling.

In future work, the correlation of the process parameters with the ICS setting variables will be investigated in more detail. The aim is to examine which parameter combinations lead to a reduction in the process forces and to chip formation with the closest possible curl diameter.

Acknowledgement. The authors appreciate the funding by the German Research Foundation (DFG) – project number 439925537.

References

1. Tandler, T., Becker, D., Eisseler, R., Stehle, T. Möhring, H.-C.: Effekt der Sägekinematik auf die Prozesseffizienz/Kinematic variation with a circular sawbalde process. wt Werkstattstechnik online **111**(01–02), 2–7 (2021)
2. Ezugwu, E.O., Wang, Z.M.: Titanium alloys and their machinability—a review. J. Mater. Process. Technol. **68**(3), 262–274 (1997)
3. Weiland, S., Drewle, K., Ochs, T., Möhring, H.-C.: A pragmatic approach to high speed circular sawing. In: 14th Conference on High Speed Machining, 17–18 April 2018. San Sebastian, Spanien (2018)
4. DeChiffre, L.: Lubrication in cutting—critical review and experiments with restricted contact tools. ASLE Trans. **24**(3), 340–344 (1981)
5. Courbon, C., Kramar, D., Krajnik, P., et al.: Investigation of machining performance in high-pressure jet assisted turning of Inconel 718: an experimental study. Int. J. Mach. Tools Manuf. **49**(14), 1114–1125 (2009)
6. Ellersiek, L., Menze, C., Sauer, F., Denkena, B., Möhring, H.-C., Schulze, V.: Evaluation of methods for measuring tool-chip contact length in wet machining using different approaches (microtextured tool, in-situ visualization and restricted contact tool). Prod. Eng. **260**(3), 310 (2022)
7. de Chiffre, L.: Mechanics of metal cutting and cutting fluid action. Int. J. Mach. Tool Des. Res. **17**(4), 225–234 (1977)
8. Findley, W.N., Reed, R.M.: The influence of extreme speeds and rake angles in metal cutting. J. Eng. Ind. **85**(1), 49–64 (1963)
9. Hongtao, Z., Peide, L., Rongsheng, H.: The theoretical calculation of naturally curling radius of chip. Int. J. Mach. Tools Manuf. **29**(3), 323–332 (1989)
10. Liu, H., Peng, B., Meurer, M., et al.: Three-dimensional multi-physical modelling of the influence of the cutting fluid on the chip formation process. Procedia CIRP **102**, 216–221 (2021)
11. Oezkaya, E., Baumann, A., Michel, S.D., Eberhard, P., Biermann, D.: Cutting fluid behavior under consideration of chip formation during micro single-lip deep hole drilling of Inconel 718. Int. J. Model. Simul. **1**, 1–15 (2022)

12. Oezkaya, E., Michel, S., Biermann, D.: Experimental and computational analysis of the coolant distribution considering the viscosity of the cutting fluid during machining with helical deep hole drills. Adv. Manuf. **9**, 12484 (2022)
13. Menze, C., Gutsche, D., Eisseler, R., Stehle, T., Möhring, H.-C., Stegmann, J., Kabelac, S.: Visualisierung der Spanbildung beim Sägen mit IKZ; Spanbildung im geschlossenen Schnittspalt beim Sägen mit innerer Kühlschmiermittelzufuhr. WT WerkstattsTechnik **1**(2), 50–54 (2022)
14. Storchak, M., Drewle, K., Menze, C., Stehle, T., Möhring, H.-C.: Determination of the tool–chip contact length for the cutting processes. Materials **15**, 3264 (2022). https://doi.org/10.3390/ma15093264
15. Thielicke, W., Sonntag, R.: Particle image velocimetry for MATLAB: accuracy and enhanced algorithms in PIVlab. J. Open Res. Softw. **9**, 12 (2021). https://doi.org/10.5334/jors.334

Investigation of the Weld Line of Compression Molded GMT and UD Tape

J. Weichenhain(✉), J. Wehmeyer, P. Althaus, S. Hübner, and B. -A. Behrens

Institute of Forming Technology and Machines, Leibniz Universität Hannover, An Der Universität 2, 30823 Garbsen, Germany
weichenhain@ifum.uni-hannover.de

Abstract. The use of fiber-reinforced plastics (FRP) is essential to meet future lightweight construction requirements in the automotive industry. Therefore, resource- and cost-saving as well as innovative manufacturing processes are needed, which enable large-scale series production. One example is the manufacturing of such components from glass mat-reinforced thermoplastics (GMT) and unidirectional fiber tapes (UD tapes) by combined compression molding and thermoforming. It may be necessary to insert multiple GMT pieces into the mold to improve mold filling and extend the process limits. In the forming process, the different material fronts of the GMT blanks collide and are joined by fusing and consolidating the matrix material of the composite. This area is called the weld line. In this paper, the influence of the weld line on the tensile strength of the workpieces was investigated. First, two GMT blanks were placed next to each other and formed in a heated plate mold. On the basis of tensile specimens that were cut across the weld line, it was found that the weld line causes a marked decrease in tensile strength. In further investigations, the UD tapes were added into the process. Due to the reinforcement of the UD tapes, the decrease in tensile strength due to the weld line was significantly reduced.

Keywords: FRP · Compression molding · GMT · UD-Tape · Weld line

1 Introduction

In the context of the energy transition and sustainability, which are steadily increasing driven by social trends, electromobility plays a central role in reducing the consumption of fossil fuels in the future. New manufacturing challenges are constantly arising in this process. On the one hand, there are the requirements for lightweight construction to reduce the overall vehicle weight and to enhance the ranges. On the other hand, the production of new vehicle components, especially battery housing structures are needed in electric vehicles to store the battery units. The lightweight requirement as well as the new components can be met by using composites like fiber-reinforced plastics (FRP). To achieve these goals, a cost- and resource-saving production for FRP components, which are suitable for large-scale production, is required.

M. Liewald et al. (Eds.): WGP 2022, LNPE, pp. 256–264, 2023.
https://doi.org/10.1007/978-3-031-18318-8_27

1.1 Fiber-Reinforced Thermoplastics

Fiber-reinforced thermoplastics (FRP) offer a high potential to overcome the mentioned challenges [1]. In recent years, the use of FRP has been steadily increasing and has thus become very important to achieve future lightweight construction goals. FRP can be differentiated into various semi-finished products. For example, there are long fiber reinforced thermoplastics (LFT), glass mat reinforced thermoplastics (GMT), fiber reinforced unidirectional tapes (UD tapes) and organic sheets [2]. Typical materials are glass or carbon fibers and polyamide 6 (PA6) or polypropylene (PP) as matrix material.

Under elevated thermal conditions—depending on the matrix material—the thermoplastics can be formed in stamp forming processes, which enable a large scale production [3, 4]. In addition to consolidating the matrix materials, they can also be joined by other joining processes, such as clinching, at higher temperatures [5].

In previous works, Behrens et al. has already successfully produced a battery shell by thermoforming glass fiber organic sheets [3, 6]. Problems like wrinkling, fiber buckling and fiber cracking were identified and investigated. However the design complexity of the battery shell was limited due to the fiber structure of the organic sheet. To achieve the manufacturing of parts with higher geometrical complexity, other manufacturing processes like compression molding of GMT are suitable [7].

1.2 Compression Molding and Thermoforming

The compression molding process is widely used in the automotive industry for the production of FRP components from GMT and LFT [8]. The process consists of four essential steps: heating, transfer, forming and consolidation. First the semi-finished product is heated above the melting temperature and below the disintegration temperature of the matrix material. Then, it is transferred to the mold with the shortest possible transfer time to prevent premature cooling [9]. While forming, the fibers flow within the matrix along the cavities until the mold is filled. In the forming process the outer layers of the matrix act as additional lubrication [10]. After the consolidation time—a time of 5 s/mm thickness is recommended [11]—the finished part can be removed and prepared for the following steps.

Weld lines represent an unavoidable defect in the production of plastic components, e.g. by means of compression or injection molding. They are caused by the collision of several material fronts in the flow process [12]. One consequence of these weld lines is a decrease in mechanical tensile strength. Various investigations have shown, that the reduction of the tensile strength in the weld line can be influenced by the process parameters pressure and temperature, but not prevented [13, 14, 15].

In the case of compression molding of GMT it is possibly necessary to insert multiple GMT blanks into the mold, due to the dimensions and the complexity of the geometry. In addition to the general decrease in tensile strength of the matrix materials in the weld line, the fibers used in the compression molding of GMT have an additional influence. GMT achieves higher tensile strength than LFT, due to the long fibers needled together [16]. If two material fronts now collide, there is a lack of interaction between the fibers in the area of the weld line, which means that a further decrease of tensile strength may be expected.

2 Experimental Setup

2.1 Objects of Investigation

To investigate the influence of the weld line on the maximum tensile strength, appropriate specimens must be prepared. A forming tool (mat. 1.0570), which corresponds to a flat square area of 300 × 300 mm^2 (see Fig. 1), is used for this purpose. The lower tool also includes dipping edges to prevent material leaking out. The upper and lower tool are equipped with heating cartridges and sensors for the setting of the temperature.

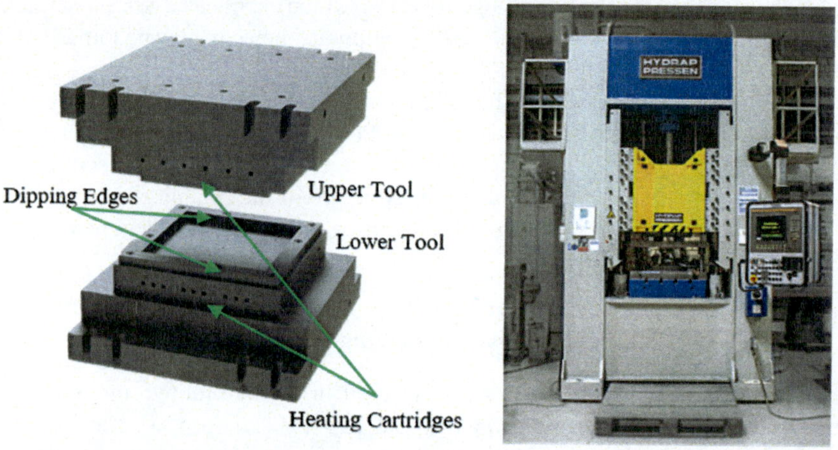

Fig. 1. Forming tool (left) and hydraulic press Hydrap HPDZb 63 at the IFUM (right)

For the investigations, GMT in form of 4.2 mm thick sheets obtained from Mitsubishi Chemical Advanced Materials Composites AG was used. It is based on a PA6 matrix with a glass fiber content of 30 wt%. The Fiber length is up to 50 mm. The UD-Tape was received from BÜFA Thermoplastic Composites GmbH and is based on a PA6 matrix as well. The glass fiber content is 60 wt%.

Since the PA6 has a melting temperature of 220 °C, the semi-finished products should provide a temperature of at least 260 °C to achieve the best suitable formability and consolidation [3, 4, 6]. To compensate the cooling caused by the transfer, the materials are heated up in a convection oven, which was set at 280 °C. This is the maximum temperature to ensure sufficient flowability without thermal degradation effects. The semi-finished products were manually transferred to the forming tool. The forming process was carried out on the hydraulic press Hydrap HPDZb 63.

Two series of tests are carried out. In test series 1 (see Table 1), pure GMT specimens are produced under variation of die speed and tool temperature with and without a weld line. The initial distance in test series 1 is 20 mm which is the distance between the two GMT blanks when they are inserted into the mold. Thereby, the influence of these parameters can be investigated and determined for the second series of tests.

In the second series of tests, specimens with different configurations of the weld line and two specimens with additional UD tape reinforcement are produced and the

Table 1. Forming parameters for the specimen with and without weld line (test series 1)

Sample	1	2	3	4	5	6	7	8
Tool temperature [°C]	90	90	110	110	90	90	110	110
Die speed [m/s]	30	60	30	60	30	60	30	60
Weld line	No	No	No	No	Yes	Yes	Yes	yes

maximum tensile strength is determined. The different configurations in test series 2 are shown in Table 2.

The materials are formed as plates with a thickness of about 1.5 mm. Tensile specimens measuring 30×120 mm^2 are then cut out of the sheets using guillotine shears. In test series 2 the UD tape is positioned between two GMT layers with the fibers in the tensile direction. The maximum tensile strength is determined by pulling the specimens to the crack on a tensile testing machine. Each parameter combination is repeated 5 times.

Table 2. Configuration for the tensile test specimen (test series 2)

Configuration	Weld line	Initial distance	With UD-tape
1	No	–	No
2	Yes	20 mm	No
3	Yes	Contact	No
4	Yes	Overlap (20 mm)	No
5	No	–	Yes
6	Yes	20 mm	Yes

3 Results

3.1 Influence of the Forming Parameters

The diagram in Fig. 2 shows the maximum tensile strength of the GMT samples of test series 1 depending on the die speed and the tool temperature. Overall, only minor influences on the maximum tensile strength of the specimen can be seen due to these forming parameters.

The specimen without a weld line have a decrease of the tensile strength from 63 to 59 MPa with increasing die speed, whereas no significantly decrease of the tensile strength is detectable at 110 °C. For the specimen with a weld line, the tensile strength decreases from 36 to 33 MPa at 90 °C and from 37 to 31 MPa at 110 °C. The decrease of the tensile strength could be due to the fact that the fibers cannot flow fast enough

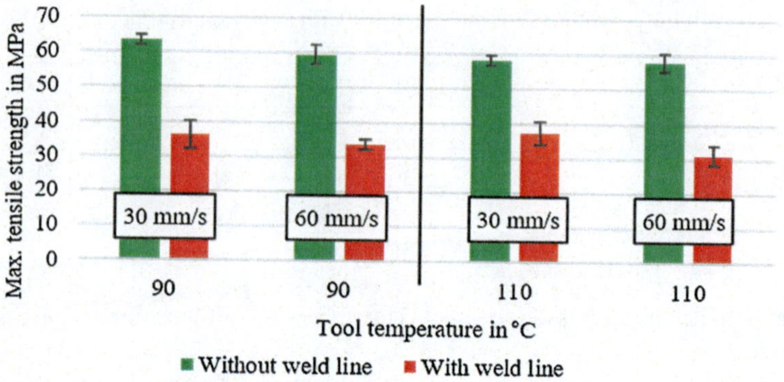

Fig. 2. Influence of the weld line on the tensile strength of GMT specimens at different forming parameters

when the flow velocity is increased. This effect is intensified at higher temperature due to better flowability of the matrix material.

The results of the first series of tests show that the presence of a weld line results in a significant reduction in the maximum tensile strength. This could be due to the fact that the material fronts are no longer sufficiently tempered when they collide, so that the flowing into each other of the fibers and the consolidation of the matrix is hindered.

The specimen of the further series of tests will be produced at 30 mm/s, since the higher die speed results in a decrease of the tensile strength. Furthermore, the tool temperature will be set at 110 °C, as this allows for better mold filling of the component in later applications.

3.2 Weld Line in GMT Samples

The photo in Fig. 3 shows the cracked tensile specimens with and without a weld line. It is clearly visible that the samples are torn exactly in the area of the weld line. This shows that the weld line is the weakest part of the tensile specimen.

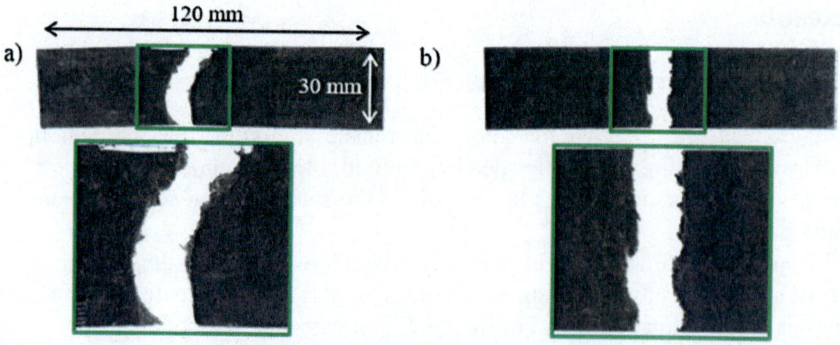

Fig. 3. Cracked tensile specimens a) without weld line (conf.1) and b) with weld line (conf. 2)

The diagram in Fig. 4 shows the maximum tensile strength of the samples. The configuration of the specimens is shown in Table 2. Configuration 1 serves as reference and was produced without weld line and UD tape. A maximum tensile strength of about 73 MPa was achieved. In configuration 2, the two GMT pieces were inserted into the tool with a distance of approximately 20 mm. Due to the weld line that forms as a result during compression molding, a maximum tensile strength of only about 50% (~36 MPa) was observed.

For comparison, further specimens were produced by positioning the GMT pieces in the mold on impact (configuration 3) and overlapping (configuration 4). In both cases, comparable tensile strengths of approximately 75 MPa and 67 MPa were achieved. Including the variation, no significant difference in the maximum tensile strength to the specimen without weld line (configuration 1) is thus recognizable.

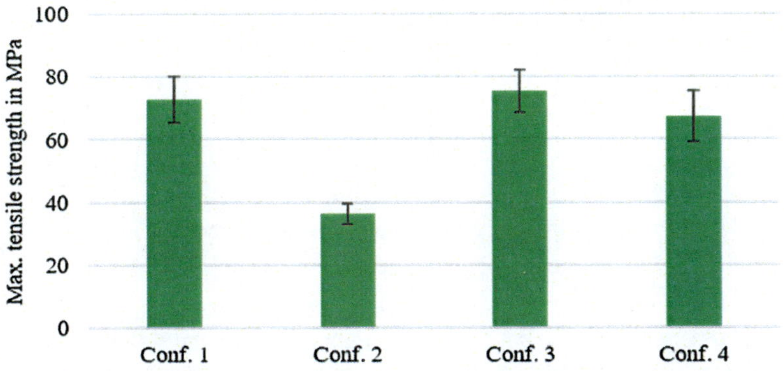

Fig. 4. Maximum tensile strength of the GMT samples with and without weld line

For the samples of configuration 2, the GMT pieces were inserted into the tool with 20 mm spacing. This means that the flow paths until the material fronts meet are longer, which could mean that the matrix material has already cooled down in comparison to the other configurations. In this case both the consolidation and the flowing into each other of the matrix and the fibers could be hindered and thus lead to the formation of a weak point in the material.

3.3 Weld Line in Combined GMT/UD-Tape Samples

The additional influence on the maximum tensile strength due to the use of single-layer UD tape in the tensile direction is shown in Fig. 5. The configurations 1 and 2 serve as reference.

The tensile specimens of configuration 1 and 5 have each been produced without a weld line. Here, the use of UD tape results in an increase in the maximum tensile strength of about 35% from 70 MPa without and 95 MPa with UD tape. This increase is due to the unidirectional designed fibers in the UD tape, which allow a higher load capacity than the twisted fibers in the GMT.

Despite the use of UD tape in configurations 5 and 6, a significant decrease in the maximum tensile strength due to the weld line can be seen. With configuration 6, a tensile strength of 62 MPa was achieved, which is less than the GMT samples without the weld line. The comparison between configurations 2 and 6 shows the influence of the UD tape when a weld line is present. The insertion of the UD tape increases the maximum tensile strength from 36 to 62 MPa. This corresponds to a significantly higher increase than in the samples without a weld line.

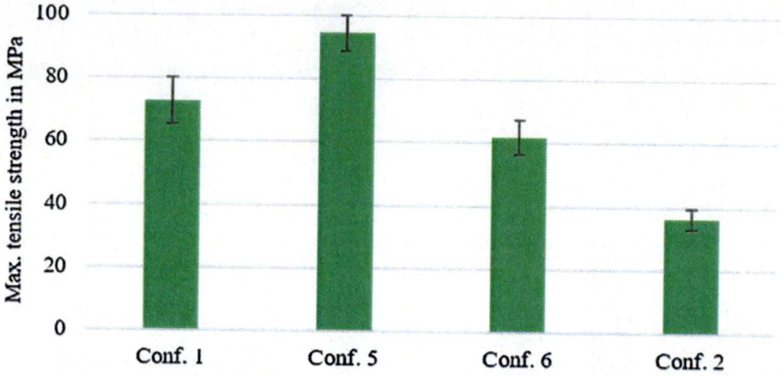

Fig. 5. Maximum tensile strength of the samples with and without UD tape and weld line

The cracking behavior of the samples with UD tape is comparable to the samples without UD tape (see Fig. 3). As soon as a crack occurs in the GMT, the fibers of the UD tape tear at the same point. Accordingly, the samples with a weld line crack in the area of this weak point.

4 Conclusion and Outlook

Overall, a significant decrease in tensile strength was observed due to the appearance of a weld line in the specimen. The differences resulted in a decrease of up to 50%. The investigations showed that the decrease in tensile strength is not solely due to the insertion of several GMT pieces. If the GMT pieces are on impact or overlapping, the tensile strength is comparable to that of samples without a weld line. Only by an increased initial distance the influence of the weld line results in a significant decrease in tensile strength.

Accordingly, it can be deduced that if the part geometry requires the insertion of multiple GMT blanks, they should be positioned as close to each other as possible to minimize the tensile strength reduction due to the weld line.

Furthermore, it was shown that the use of single-layer UD tape allows a significant increase in the tensile strength for both, the sample with and without a weld line. However, the single layer UD tape was not enough to compensate the decrease in tensile strength by the weld line compared to a monolithic GMT part. It can be assumed that the use of several layers of UD tape should lead to a corresponding increase in this reinforcement effect. This will be subjected in further studies.

Acknowledgements. This research and development project is funded by the German Federal Ministry of Education and Research (BMBF) within the funding initiative "Research Campus - Public-Private Partnership for innovation" (funding code: 02P18Q745) and implemented by the Project Management Agency Karlsruhe (PTKA). The author is responsible for the content of this publication.

References

1. Friedrich, K., Almajid, A.A.: Manufacturing of advanced polymer composites for automotive applications. In: Applied Composite Materials vol. 20, pp. 107–128 (2013)
2. Gebai, S., Hallal, A., Hammoud, M.: Composite materials types and applications: a review on composite materials. In: Mechanical Properties of Natural Fiber Reinforced Polymers. Emerging Research and Opportunities, IGI Global, pp. 1–29 (2018)
3. Behrens, B.-A., Hübner, S., Bonk, C., Bohne, F., Micke-Camuz, M.: Development of a combined process of organic sheet forming and GMT compression molding. In: Procedia Engineering, vol. 207, pp. 101–106. Cambridge (2017)
4. Behrens, B.-A., Hübner, S., Neumann, A.: Forming sheets of metal and fibre-reinforced plastics to hybrid parts in one deep drawing process. In: Procedia Engineering, vol. 81, pp. 1608–1613 (2014)
5. Behrens, B.-A., Hübner, S., et al.: Forming and joining of carbon-fiber-reinforced thermoplastics and sheet metal in one step. In: Procedia Engineering, vol. 183, pp. 227–232 (2017)
6. Behrens, B.-A., Raatz, A., Hübner, S., Bonk, C., Bohne, F., Bruns, C., Micke-Camuz, M.: Automated stamp forming of continuous fiber reinforced thermoplastics for complex shell geometries. In: Procedia CIRP, vol. 66, pp. 113–118 (2017)
7. Behrens, B.-A., Weichenhain, J., Althaus, P., et al.: Investigation of a compression molding process for the variant flexible production of a GMT battery shell. In: Production at the Leading Edge of Technology, Proceedings of the 11th Congress of the German Academic Association for Production Technology (WGP), Sept 2021, pp. 20–28. Dresden (2021)
8. Kurcz, M., Baser, B., Dittmar, H., Sengbusch, J., Pfister, H.: A case for replacing steel with glass-mat thermoplastic composites in spare-wheel well applications. In: Society of Automotive Engineers World Congress: Technical Paper (2005)
9. Giles, H., Reinhard, D.: Compressing moulding of polypropylene glass composites. In: 36th International SAMPE Symposium and Exhibition, pp. 556–570. San Diego, California (1991)
10. Guiraud, O., Dumont, P.J.J., Orgéas, L., Favier, D.: Rheometry of compression moulded fibre-reinforced polymer composites: rherology, compressibility, and friction forces with mould surfaces. Compos. Part A: Appl. Sci. Manuf. **43**(11), 2107–2119 (2012)
11. Wakeman, M.D., Cain, T.A., Rud, C.D., Brooks, R., Long, A.C.: Compression moulding of glass and polypropylene composites for optimized macro- and micro-mechanical properties II. Glass-mat-reinforced thermoplastics. Compos. Sci. Technol. **59**, 709–726 (1999)
12. Malloy, R. A.: Plastic Part Design for Injection Molding, An Introduction 2nd edn. Carl Hanser Verlag (2010)
13. Halilovic, J., et al.: Effect of injection molding parameters on weld line tensile strength. J. Trends Develop. Machinery Assoc. Technol. **21**(1), 13–16 (2018)
14. Chookaew, W., et al.: An investigation of weldline strength in injection molded rubber parts. In: 10th Eco-Energy and Materials Science and Engineering. Energy Procedia, vol. 34, pp. 767–774 (2013)
15. Vaxman, A., Narkis, M., et al.: Weld line characteristics in short fiber reinforced thermoplastics. Polymer Compos. 161–168 (1991)

16. AVK – Industrievereinigung verstärkte Kunststoffe e. V.: Handbuch Faserverbundkunststoffe/Composites – Grundlagen, Verarbeitung Anwendung. 4. Auflage (2013)

In-situ Computed Tomography and Transient Dynamic Analysis of a Single-Lap Shear Test with a Composite-Metal Clinch Point

Daniel Köhler[1(✉)], Richard Stephan[2], Robert Kupfer[1], Juliane Troschitz[1], Alexander Brosius[2], and Maik Gude[1]

[1] Institute of Lightweight Engineering and Polymer Technology, Technische Universität Dresden, Dresden, Germany
daniel.koehler3@tu-dresden.de

[2] Chair of Forming and Machining Processes, Technische Universität Dresden, Dresden, Germany

Abstract. Clinching is a well-established joining technology, e.g. in automotive production, because of its cost-efficiency and the ability to join different materials at low cycle times. Nowadays, a detailed quality inspection of clinch points is usually carried out ex-situ, e.g. via macroscopic examination after joining. However, only 2D-snapshots of the complex three-dimensional and time-dependent forming and damaging phenomena can be made. The closing of cracks and the resetting of elastic deformations due to unloading and specimen preparation are also disadvantageous. In contrast, the use of non-destructive in-situ testing methods enables a deeper insight into the joint deformation and failure phenomena under specific load conditions. In this paper, progressive damage is observed during the single-lap shear testing of a clinch point using in-situ computed tomography (CT) and transient dynamic analysis (TDA). The TDA can continuously monitor the characteristic dynamic response of the joint, which is sensitive to damage and process deviations. In-situ CT creates 3D images of the inner structure of the clinch point at specific process steps. In this work, the sensitivity of both testing methods to detect damage in joints with EN AW 6014 and glass fibre reinforced polypropylene (GF-PP) is evaluated. As a reference, joints with both joining partners made of aluminium alloy (EN AW 6014) are analyzed. It is shown, that TDA and in-situ CT has the potential to identify joint quality as well as critical processing times.

Keywords: Computed tomography · Active acoustic testing · Clinching

1 Introduction

Clinching is a cost-efficient joining process that allows different materials to be connected. This is particularly important in lightweight design in order to optimize the use of materials. In most cases, destructive testing methods are used for the characterization of clinch points. The preparation of a macroscopic examination is a standard method in addition to strength tests by shear-, peel- or head tensile testing [1]. A macroscopic

M. Liewald et al. (Eds.): WGP 2022, LNPE, pp. 265–275, 2023.
https://doi.org/10.1007/978-3-031-18318-8_28

examination enables the measurement of the geometric characteristics of the clinch point (undercut, neck thickness and symmetry) in a specific cross-section. Due to the resetting of elastic deformations during specimen preparation, it is possible that cracks close or cracks are not in the plane of the macroscopic examination and thus cannot be detected [2].

Using mechanical testing, maximum loads in a specific direction and the predominant failure mechanism (unbuttoning, neck breakage or a combined failure) can be determined [3]. With these methods, it is hardly possible to draw conclusions about the material flow during loading, internal damage or the failure chronology. The existing characterization possibilities can be complemented by the use of in-situ CT [4]. This imaging analysis method uses the X-ray attenuation of a test object in order to obtain a three dimensional reconstruction of the object in high resolution. For this purpose, two-dimensional radiographs of the object are created from multiple angles.

Since CT is a time-consuming procedure, it can be combined with the TDA, as described in [5]. The TDA investigates the dynamic behaviour of the clinch point by selectively introducing structure-borne sound waves and recording the response behaviour. This method offers the possibility to obtain data about the clinch point rapidly, cost-effectively and non-destructively. The method has already been applied with regard to bolted joints. Bournine et al. attempted to maximize the damping of a bolted connection while maintaining the maximum load capacity of the bolted structure [6]. Wang et al. investigated the relationship between the damping properties of a bolted joint and the preload [7]. Wolf et al. showed a change in the structure-borne sound energy dissipated in the bolted joint with varying tightening torque [8]. However, the interpretation of the data with regard to clinched joints as well as the application limits of the TDA method is still subject of ongoing research [9].

In order to evaluate the application of a combined TDA and in-situ CT, this paper examines clinched specimens consisting of an aluminium (Al) sheet (EN AW 6014) and a glass fibre reinforced polypropylene (GF-PP) sheet. To evaluate the influence of the highly damping polypropylene on the TDA, the tests were also carried out on specimens with both joining partners made of EN AW 6014.

2 Materials and Methods

2.1 Sample Preparation

Lubricated 2 mm thick aluminum sheets made of EN AW 6014-T4 (Advanz™ 6F-e170, Novelis Inc., Atlanta, USA) are used for the test specimens. The GF-PP sheets consist of glass fibre reinforced polypropylene. Both materials GF-PP [10] and EN AW 6014 [11] are typically applied in the automotive industry. For the Al-Al joints, a 0.01 mm thick tin foil is positioned between the sheets in order to enhance the visibility of the sheet-sheet interface in the CT-scan. The sheets and the foil are clinched with a punch A50100 and a die BE8012 (both from TOX PRESSOTECHNIK GmbH & Co.KG, Weingarten, Germany). Then, the specimen is solution heat-treated and artificially aged (T6) at 185 °C for 20 min. The Al-GF-PP joints are manufactured using the hotclinching method described in [12] with a process temperature of 70 °C. Here, the aluminium plate is positioned at the punch-faced side. The punch A58100 with a fillet radius of 0.25 mm

(TOX PRESSOTECHNIK) and a BE-type die with an initial anvil position of 1 mm is used. The procedure is explained in detail in [13].

The shear specimen dimensions according to ISO 12996, see Fig. 1c. In order to ensure a concentric load introduction, two shim plates are adhesively bonded onto the ends of the sheets. In contrast to the mentioned norm, the specimen width is 37 mm.

(cf. Fig. 1c) to fit the clamps. The specimen is pulled along the main axis in the in-situ CT, featuring a tensile testing machine ZwickRoell Z250 (ZwickRoell GmbH & Co. KG, Ulm, Germany) with the lower traverse moving.

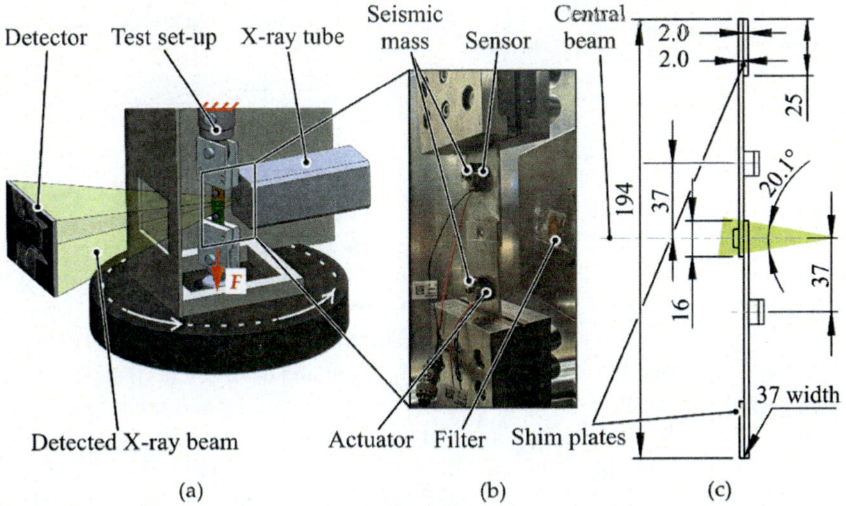

Fig. 1. Experimental setup for in-situ CT and TDA during shear testing of clinched specimens

2.2 Shear Test and CT-Setup

The CT system FCTS 160-IS (FineTec FineFocus Technologies GmbH, Garbsen, Germany) consists of the X-ray source FORE 160.01C TT (160E3 V, 1E-3 A, 80 W) and a flat panel detector (3200 × 2300 pixel, 405 × 290 mm active area).

The clinch point is CT- and TDA-scanned at six crosshead travel steps whereby the crossbeam remains at constant displacement. The crosshead travel (CHT) is selected in a way the sample displacement follows a fixed displacement step size. For the Al-Al specimens a linear relation and for the Al-GF-PP specimens a non-linear relation between crosshead travel and sample displacement is measured. A preload of 60 N is applied. The parameters for the CT scans are summarized in Table 1.

2.3 TDA Setup

To perform the TDA, structure-borne sound waves have to be introduced into the sample. This is realized by a piezoelectric ring stack actuator from the manufacturer PI Ceramic

Table 1. CT-parameter of the CT-system

Parameter	Unit	Value
Acceleration voltage	kV	150
Tube current	μA	30 (Al-Al specimen) 45 (Al-GFRP specimen)
X-ray projections	–	1440 (4 per 1°)
Exposure time	ms	625
Resolution	μm	7.58
Magnification	–	16.8
Filter	mm	0.1 (copper)

GmbH with the type designation P-016.00H. A similar piezoelectric ring stack is used to record the signal on the receiver side. In the following, these piezoelectric ring stacks are referred as actuator and sensor respectively. In order to improve the sensitivity of the sensor and to intensify the coupled sound of the actuator, both are equipped with an effective seismic mass of 9.5 g, (cf. Fig. 1b, c). The application of the seismic mass reduces the resonance frequency of the actuator and sensor from 144 kHz to pprox. 76 kHz. This improves the sensitivity and still maintains a sufficient distance from the maximum signal frequency of 22 kHz used in the test. Actuator and sensor are mounted on the specimen with a tightening torque of 1 Nm. Holes of 3.3 mm are drilled for the bolts at a distance of 37 mm from the clinch point on the respective sheet (cf. Fig. 1c). Besides mounting, the fastening also generates the compressive stress in the actuator that is necessary for dynamic operation. An analogue output of the data acquisition card NI6356 is used to excite the actuator. The alternating signal from the data acquisition card in the form of a sine with 4 V peak is converted via an analogue signal amplifier of the type HV-LE150-100-EBW from the manufacturer piezosystem jena GmbH into an excitation with 67 V peak for the actuator. The analogue output signal of the sensor passes through a first-order high-pass filter with a cut-off frequency of 147 Hz in order to reduce the interspersed influences of the surrounding electrical devices. The filtered signal is amplified by a Kistler 5018 charge amplifier and then sampled by the NI6356 using a sampling rate, which always equals eight times the excited frequency.

During the TDA, a specific frequency is always excited for a duration of two seconds. The measuring program always stores the sampled signal from the final half second of each frequency in order to process only data from the steady state. After these 2 s, the frequency to be stimulated is increased by 50 Hz. The spectrum of the TDA starts at 200 Hz and finishes at 22,000 Hz. For each excited frequency, the amplitude value is taken and added to the frequency spectrum.

3 Results

3.1 CT-Results

Figure 2 shows six CT images of Al-Al sample C04_A_CV_240_ZS_180 at different process steps of the shear test. The interface between the sheets in the clinch point area, which is highlighted by the tin foil, can be clearly seen. It also makes the undercut of the clinch point in Fig. 2a visible. The chronology of the present unbuttoning failure mechanism can be clearly traced (cf. Fig. 2c–f). There is a significant stretching of the right neck area of the punch-side sheet from Fig. 2b–f. Despite the shear between the sheets in this area, the tin foil remains free of macroscopic ruptures. However, in the left half of the images, the punch-side joining partner deforms and the bottom glides along the neck area of the die-sided joining partner. At this are the tin foil breaks. At the same time, the bottom of the punch-sided sheet is bended off the die-sided sheet, reducing the contact area between both sheets.

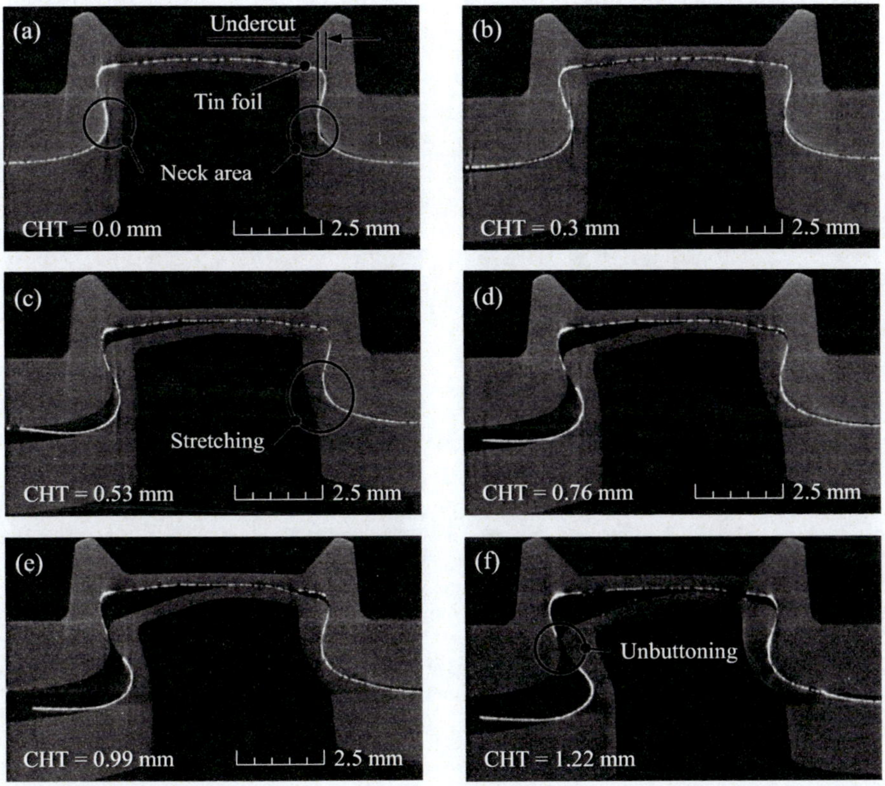

Fig. 2. CT scans of sample C04_A_CV_240_ZS_Z180 at different CHT steps from 0.0 mm (a) to 1.22 mm (f)

Figure 3 shows the CT scans of sample A01_FKVA_CV_03 during the shear tensile test on different CHT. Initially, there is a crack in the Al joining partner in the neck area visible (cf. Fig. 3a). In the FRP joining partner a thin FRP layer with lower fibre volume content remains at the bottom side of the clinching point. Additionally, there is a gap between the FRP and the Al joining partner. Despite the initial neck crack in the Al partner, the failure of the specimen is caused by unbuttoning (cf. Fig. 3f). While the Al-Al specimen was unbuttoned under strong plastic deformation of the punch-side sheet, no deformation of the Al joining partner is visible in the Al-GF-PP specimen. Instead, the bottom and neck area of the punch-side sheet increasingly press into and deforms the right clinch point shoulder of the die-side GF-PP joining partner. Thereby, the two sheets increasingly separate from each other. At the same time, the contact area on the left side of the clinching point decreases until unbuttoning results, see Fig. 3f.

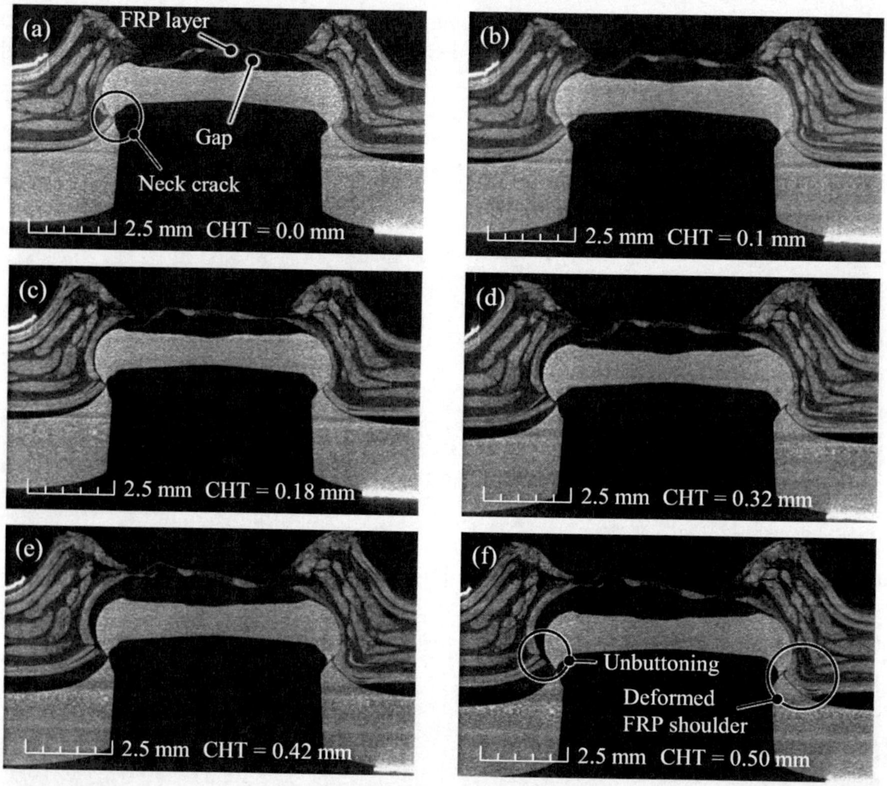

Fig. 3. CT-Scans of sample A01_FKVA_CV_03 at different CHT steps from 0.0 mm (a) to 0.5 mm (f)

3.2 Results of the Combined TDA and Single Lap Shear Test

The TDA result is a frequency spectrum of the amplified sensor signal for each measuring step. In Table 2, the TDA amplitudes are averaged at the time of achieving the mentioned test preload (60 N), i.e. before the actual shear test. In addition, the maximum force achieved in shear test is given.

Table 2. Averaged TDA amplitudes and maximum forces

No	Specimen	Material bottom plate	TDA amplitude in mV	Max. Shear force in N	Crosshead travel in mm
1	A01_FKVA_CV_01	GF-PP	79	499	0.50
2	A01_FKVA_CV_03	GF-PP	103	526	0.10
3	C04_A_CV_240_ZS_Z180	EN AW 6014	192	1972	1.22
4	C04_A_CV_238_ZS_Z180	EN AW 6014	209	1998	0.30

For both types of specimens, increasing maximum shear force will result in larger TDA amplitudes as measurement signals. The average TDA amplitudes of the Al-GF-PP specimens are about half the size of the Al-Al specimens.

In order to investigate whether the signal is sufficient to carry out a measurement on strongly damping materials, it is set in relation to the measured noise. The signal-to-noise ratio (SNR) defines the ratio of the effective value of the useful signal to the effective value of the interfering signal on a logarithmic scale using the unit dB. Four SNR graphs are shown in Fig. 4 as indicators for signal quality. The SNR graphs of all clinched joints are very similar. In order to maintain clarity, only the SNR graphs with the highest and with the lowest average SNR of both sample types are shown. All graphs show a strong increase of the SNR with increasing frequency. In the frequency range between 0 to 5 kHz the SNR of Al-GF-PP samples is mostly smaller than 10 dB. For Al-Al samples, the SNR of 10 dB is already stable at frequencies from 2 kHz. For high frequencies, the SNR increases to 80–100 dB for all samples. The largest average SNR value of Al-GF-PP samples (50.6 dB) is 2.6 dB smaller than the smallest average SNR value of Al-Al samples. For all samples, it can be noted that the SNR decreases as the shear test progresses. The selected four values also show this behaviour, as the highest average SNR values are obtained at the beginning of the test and the lowest at the end of the shear test. Because the noise is only measured once, this observation results from the average decreasing signal power at the sensor as the shear test progresses.

In Fig. 5 it can be seen, that the frequency spectra for different CHT show a very similar course. Particularly obvious is the continuous shrinking and finally disappearance of the large peak at 16.5 kHz with increasing CHT. Also noticeable is the peak shifting to smaller frequencies and shrinking at the same time, which is at 21 kHz at the beginning of the test.

Figure 6 shows four TDA frequency spectra of sample No. 3, Table 2 at different crosshead travel (CHT) steps. Overall, the changes in the frequency spectra for this Al-Al sample are less continuous than for sample No. 2. In order to achieve clarity, fewer spectra are shown than in Fig. 5. Beside the fact, that for both material combination, Al-Al and Al-GF-PP, differences in the frequency spectrum occurs, the amplitudes of both material combinations are getting larger with increasing frequency. The peaks in the spectra of sample No. 3 with a maximum of 2.3 V are significantly larger than the largest peaks of sample No. 2 with about 0.6 V. It is noticeable that both samples have

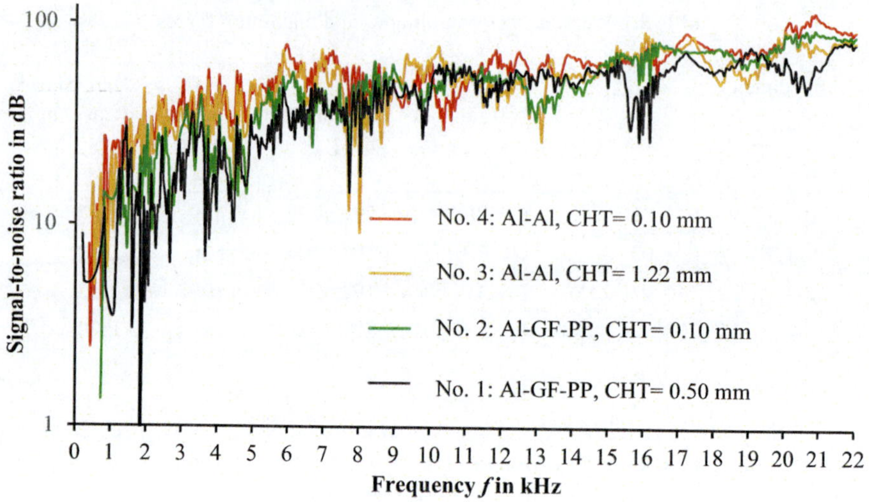

Fig. 4. Signal-to-noise ratio of different specimens

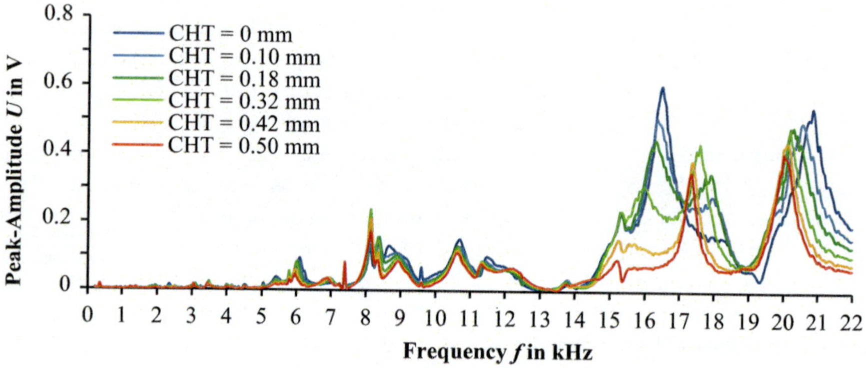

Fig. 5. Frequency spectra of the peak amplitudes of the TDA signal of the sample No. 2

their largest peak at about 16.5 kHz. However, while this peak in sample No. 2 becomes steadily smaller as the shear tensile test progresses, in sample No. 3 it also becomes smaller at first and then increases again to the original height.

4 Summary, Discussion and Conclusions

With conventional characterization methods, it is hardly possible to draw conclusions about the material flow during loading of a clinch point. Additionally, internal damage or the failure chronology remain unknown. The in-situ CT in combination with TDA has the potential to compensate these drawbacks. In this work, the feasibility of the combined measurement is investigated et al. -GF-PP. The results are compared with those from Al-Al joints.

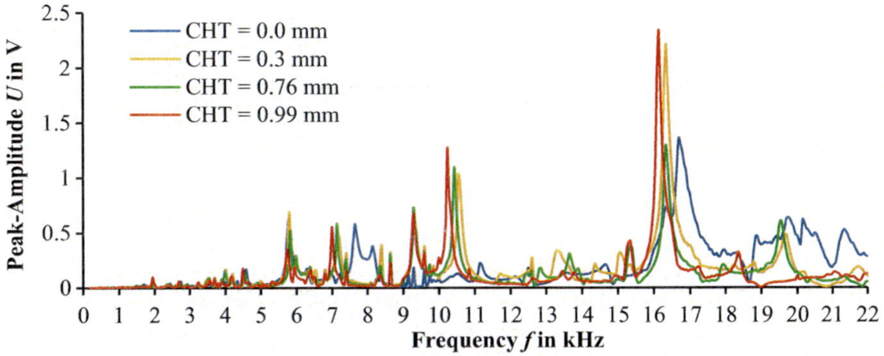

Fig. 6. Frequency spectra of the peak amplitudes of the TDA signal of sample No. 3

Using in-situ CT, the chronology of specimen failure during shear testing can be easily traced. Both types of specimens fail by unbuttoning. However, a clearly different failure development is recognizable. The Al-Al specimens fail under heavy deformation of the neck area of the punch-side sheet with a significant change in geometry. In contrast, the Al-GF-PP specimens do not show any change in the geometry of the punch-side Al sheet. Instead, the neck and head of Al sheet presses increasingly into the shoulder of the GF-PP joining partner and deforms it. This precise characterisation of the failure of the two specimens enables the conclusion of recommendations for improving of the shear strength. For the Al-Al specimens, the maximum shear force could be improved if the top sheet's unbuttoning resistance is increased. For example, this can be achieved by adapting the clinching tools' geometry resulting in an increased undercut. To improve the shear strength of the PF-PP Al composite, the strength of the GF-PP against impression would have to be improved.

In TDA, a significantly greater damping of the fibre-reinforced plastic compared to aluminium is measured. This is particularly evident in the lower amplitudes over the entire frequency range of the TDA, but also in a reduced SNR compared to the Al-Al samples. The averaged SNR for Al-GF-PP samples is 2.5–14.2 dB lower than for Al-Al samples. In principle, the noise that reaches the sensor in the form of structure-borne sound is also damped by the plastics, but other influences via airborne sound, electrical influences via induction and the noise of the measuring amplifier remain of the same magnitude with a significantly reduced signal amplitude at the same time.

TDA on Al-Al samples achieves good signal quality above 2 kHz. In the case of Al-GF-PP samples, the signal quality achieves a sufficient SNR of stable above 10 dB from 5 kHz onwards. One conclusion is that in terms of signal quality, TDA can therefore be implemented above this frequency, even with the highly damping material GF-PP.

The interpretation of the TDA data is still a subject of research. However, by combining the data with the in-situ CT, explanations for some of the observations in the TDA results can be found. So, at both samples the amplitudes decrease with increasing crosshead travel. This observation could be explained by the continuous reduction of the contact area between the joining partners, which is recognizable in both sample types (cf. Figs. 2, 3). TDA results of the Al-Al samples show less continuously and smooth changes of the TDA characteristics than the Al-GF-PP samples. This could be explained by the strong deformation of the clinch point area on the punch side (cf. Fig. 2c–f) and the resulting change in dynamic properties. The Al-GF-PP samples, on the other hand, do not show such macroscopic changes in the geometry of the joining partners. A continuous pressing of the aluminium into the reinforced plastic during joining as well as the continuously increasing sheet distance (cf. Fig. 3) is a possible explanation for the mentioned typical characteristic of the TDA signal. The continuous changes in the TDA signal during the shear test of Al-GF-PP samples could be a utilisable feature for monitoring the clinching process by TDA. It is possible that these features can be used to identify the failure characteristics during a shear test using TDA.

5 Outlook

Beside the possibility for online monitoring of the clinching process by TDA another option is the combined usage of TDA and in-situ CT. The continuous online monitoring by TDA could be used to identify a critical time during clinching, when cracks or other failure mechanisms starts. At these relevant points of interest, the joining process would be stopped and the time intensive scanning process of in-situ CT can start. Moreover, TDA could be used for other applications, such as process monitoring or the inspection of clinch joints located on structures in operation. Both applications will be investigated in future.

Acknowledgements. This research was funded by the German Research Foundation (DFG) within the project Transregional Collaborative Research Centre 285 (TRR 285) (project number 418701707), sub-project C04 (project number 426959879).

References

1. Kupfer, R., et al.: Clinching of aluminum materials—methods for the continuous characterization of process, microstructure and properties. J. Adv. Joining Process. J. Pre-Proof (2022). https://doi.org/10.1016/j.jajp.2022.100108
2. Böhm, R., et al.: A quantitative comparison of the capabilities of in situ computed tomography and conventional computed tomography for damage analysis of composites. Compos. Sci. Technol. **110**(2015), 62–68 (2015). https://doi.org/10.1016/j.compscitech.2015.01.020
3. ISO 12996:2013: Mechanical joining—destructive testing of joints—specimen dimensions and test procedure for tensile shear testing of single joints
4. Köhler, D., Kupfer, R., Troschitz, J., Gude, M.: In situ computed tomography—analysis of a single-lap shear test with clinch points. Materials **14**(8), 1859 (2021). https://doi.org/10.3390/ma14081859
5. Köhler, D., Sadeghian, B., Kupfer, R., Troschitz, J., Gude, M., Brosius, A.: A method for characterization of geometric deviations in clinch points with computed tomography and transient dynamic analysis. Key Eng. Mater. **883**, 89–96 (2021). https://doi.org/10.4028/www.scientific.net/KEM.883.89
6. Bournine, H., Wagg, J., Neild, S.A.: Vibration damping in bolted friction beam-columns. J. Sound Vib. **330**(8), 1665–1679 (2011). https://doi.org/10.1016/j.jsv.2010.10.022

7. Wang, F., Huo, L., Song, G.: A piezoelectric active sensing method for quantitative monitoring of bolt loosening using energy dissipation caused by tangential damping based on the fractal contact theory. Smart Mater. Struct. **27**(1) (2017). https://doi.org/10.1088/1361-665X/aa9a65
8. Wolf, A., Lafarge, R., Kühn, T., Brosius, A.: Experimental analysis of mechanical joints strength by means of energy dissipation. In: Proceedings of the 21st International ESAFORM Conference on Material Forming, AIP Publishing (2018). https://doi.org/10.1063/1.503488
9. Sadeghian, B., Guilleaume, C., Lafarge, R., Brosius, A.: Investigation of clinched joints—a finite element simulation of a non-destructive approach. In: Behrens, B.-A., Brosius, A., Hintze, W., Ihlenfeldt, S., Wulfsberg, J.J. (eds.) WGP 2020. LNPE, pp. 116–124. Springer, Heidelberg (2021). https://doi.org/10.1007/978-3-662-62138-7_12
10. Puri, P., Compston, P., Pantano, V.: Life cycle assessment of Australian automotive door skins. Int. J. Life Cycle Assess. **14**, 420–428 (2009). https://doi.org/10.1007/s11367-009-0103-7
11. Bhattacharya, R., Stanton, M., Dargue, I., Williams, G., Aylmore, R.: Forming limit studies on different thickness aluminium 6xxx series alloys used in automotive applications. Int.J. Mater. Form. **3**, 267–270 (2010). https://doi.org/10.1007/s12289-010-0758-4
12. Gröger, B., et al.: Clinching of thermoplastic composites and metals—a comparison of three novel joining technologies. Materials **14**(9), 2286 (2021). https://doi.org/10.3390/ma14092286
13. Vorderbrüggen, J., Gröger, R., Kupfer, R., Hoog, A., Gude, M., Meschut, G.: Phenomena of forming and failure in joining hybrid structures—experimental and numerical studies of clinching thermoplastic composites and metal. In: AIP Conference Proceedings, vol. 2113, p. 050016 (2019). https://doi.org/10.1063/1.5112580

Development of Pressure Sensors Integration Method to Measure Oil Film Pressure for Hydrodynamic Linear Guides

B. Ibrar[1(✉)], V. Wittstock[1], J. Regel[1], and M. Dix[1,2]

[1] Institute for Machine Tools and Production Processes (IWP), Chemnitz University of Technology, Reichenhainer Straße 70, 09126 Chemnitz, Germany
burhan.ibrar@mb.tu-chemnitz.de

[2] Fraunhofer Institute for Machine Tools and Forming Technology IWU, Reichenhainer Straße 88, 09126 Chemnitz, Germany

Abstract. The accuracy of the machined workpiece depends on the true conditions of the hydrodynamic linear guides. Linear guides based on hydrodynamic lubrication are still in use due to their high damping coefficient and high load carrying capacity. Measuring the true conditions of the hydrodynamic linear guide is important to achieve high accuracy. Until today, pressure measurement has not been established for linear guides. Oil film pressure is one of the important factors explaining the operating conditions of the hydrodynamic linear guides. The main goal of this study is to develop a method in which pressure sensors were installed in the lubrication gap to measure the oil film pressure under realistic condition of a hydrodynamic linear guide. A hydrodynamic linear guide testing rig with varying load capability was used as main test device. Multiple miniature pressure sensors were installed in a stationary rail in different manners to get the realistic oil film pressure distribution along the length of the slide. Additionally, the sensors were also calibrated hydrostatically and hydrodynamically with variable frequency and amplitude. Due to the problem of the enormous influence of air inside the lubrication gap of the testing rig, the measured pressure by the sensors showed that the numerical results have to be adapted to the experimental oil film pressure, which is lower. A new sensor's integration method has shown great improvements in estimating the oil film pressure of hydrodynamic guides experimentally.

Keywords: In-line measurement · Hydrodynamic linear guides · Fluid film lubrication · Coupled CFD simulation · High damping coefficient

1 Introduction

Hydrodynamic guides are of significant importance for high quality manufacturing processes in the machining industry. The application range of hydrodynamic guides is growing due to the increasing size of machines or machine tools. As a result of their good damping coefficient, low cost and high efficiency and therefore the high surface quality. Whereas, the surfaces machined by using machines with roller bearing guidance are of

M. Liewald et al. (Eds.): WGP 2022, LNPE, pp. 276–287, 2023.
https://doi.org/10.1007/978-3-031-18318-8_29

relatively lower quality. Additionally, hydrostatic guides are also used in the machining industry, but due to constant oil supply system, the operating costs increase, which reduces the efficiency of such guides. Hydrodynamic guides, on the other hand, have an unstable floating behaviour of a carriage. This is due to the relative speed-dependent oil pressure within the lubrication gap of the parallel surfaces. Feed rate also has a great influence on the instability of floating slides.

Zhang et al. [1] has compared the floating heights measured experimentally and using CFD model (based on the Reynolds differential equation) of the machine table of hydrodynamic linear guides. In her investigations, eddy current sensors were used to measure floating heights for constant and variable speed. However, oil film pressure could not measure because a suitable pressure measurement method was not available.

One of the key parameters in hydrodynamic guides is the oil film pressure, which influences the operation of guides [2, 3]. Experimental measurement of the oil film pressure is a challenging task. The main goal of this study is to develop a geometrical setup to measure the oil film pressure in linear hydrodynamic guides, which also recognizes the air influence inside the lubrication gap.

2 State of the Art

In the last decade, studies have been conducted to measure the key parameters like oil film pressure, floating heights and oil film temperature for hydrodynamic journal bearings experimentally but for linear guides, no method has been developed.

For instance, thin film pressure sensors consisting of thin material layers (with a total thickness of 6 μm) on the sliding surface of the bearing were used to measure the oil film pressure of journal bearing [4–8]. Mihara and co-workers have carried out oil film pressure measurements as well as temperature and strain measurements of the bearing surface in an engine test. The sensors have been deposited by the physical vapour deposition (PVD) technique onto the bearing surface. Using the pressure sensor e.g. at the bottom of the main bearing of a diesel engine the maximum oil film pressure of 45 MPa at full engine load has been measured [8]. Masuda et al. [9] carried out testing rig experiments in which oil film pressure in a connecting rod big-end bearing was measured by a semi-conductor strain gauge-type transducer embedded in the surface of the crankpin. In 2005, Sinanoglu et al. monitored oil film pressure of journal bearing in 16 tubes of manometer at various rotational speeds. Twelve manometer tubes were placed around the circumference with 30° between the tubes and 4 more tubes were located along the bearing length [10]. Optical pressure sensors were used to measure the oil film pressure in journal bearings by Ronkanien et al. [11].

Krampert et al. [12] has used the approach based on the piezo resistive diamond like carbon (DLC) coating DiaForce. To capture the load on a linear element by measuring the stresses or deformation resulting from the rolling element contact on the side of the runner block. Direct strain measurement method based on DLC sensors to estimate the load distribution in linear guide bearings. He placed a group of three sensors on the back of each of the four raceways of the runner block and measure the strain introduced by loading the rolling element above the sensor elements [13]. But there is no method to install the sensors inside the lubrication gap to measure the oil film pressure for linear guides.

Another functional parameter of interest in journal bearings is the oil film thickness. The eddy current sensors have been used to measure the oil film thickness and shaft trajectories in engine bearings by Moreau et al. [14]. Optical sensors can also be used to measure the oil film thickness in hydrodynamically lubricated bearings. The optical sensor has been typically mounted flush with the bearing surface and the light is transmitted through the lubricant film and reflected from the shaft surface back to the optical sensor [15–17]. The sensors mentioned above are embedded in materials and utilize advanced techniques like Fibre Bragg Grating [18] or Fabry-Perot interferome-try [19], which makes them complicated and expensive to produce.

3 Experimental Setup

Three strain gauge pressure sensors have been used in this study to measure the oil film pressure of hydrodynamic linear guides. Because of their ability to measure both static and dynamic loads. Whereas, piezoelectric sensors can only support vibrations, acceleration, and dynamic pressure measurement. Three similar commercial pressure sensors, namely PS-1, PS-2 and PS-3 by Kyowa [20] were used as shown in Fig. 4. Pressure sensors have a measuring range up to 7 MPa and have a safe temperature range from −20 to 70 °C. Pressure sensors are known by the model name "PS-70KD M2". They have a round sensing surface with 6 mm diameter, length of 6.7 mm and a sensing surface thickness of 0.6 mm. Additional sensor named 'Burster 8210' was also used to see if sensors (PS-1, PS-2 and PS-3) show same behaviour as Burster 8210. Pressure sensor Burster is also a strain gauge precision pressure sensor with the measuring range of 0–1 MPa.

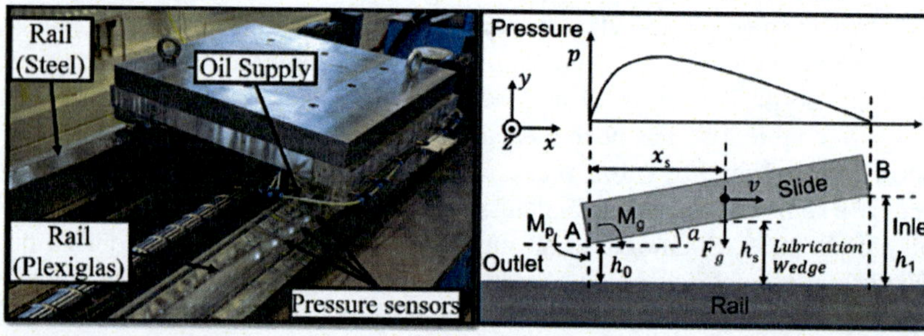

Fig. 1. Left: Experimental setup of hydrodynamic linear guide and Right: Schematic depiction of floating behaviour of guidance [21].

A machine table of a testing rig can move back and forth with a speed of up to 100 m/min and it has a stroke length of about 1.6 m. A ball screw drive is used to drive the table with the help of a motor because of its low friction and high precision. A force sensor is installed in between the machine table and motor to know the exact force applied during the operation. The machine table slides on two different rails, which can be seen in Fig. 1. The high-performance steel rails are fully machined, through hardened, straightened and ground. Four distance sensors are integrated on all 4 ends of the machine table to measure the floating heights and floating angle of the machine table. Extra weights can also be attached on top of the machine table. The machine table has three connecting surfaces with the steel rail and one with Plexiglas rail, where on both sides of the steel rail, roller bearings are used to restrict motion in other directions. Whereas, both rail (steel and Plexiglas) and carriage are separated by hydrodynamic lubrication during application.

To develop the integration method of strain gauge pressure sensors, one rail and machine table is replaced with acrylic glass (levelled and ground) material to have visual ability inside the lubrication gap. It helps to see how air plays a significant role inside the lubrication gap. The presence of air can reduce the sensor's ability to measure an actual pressure inside the lubrication gap. Although the disadvantage of acrylic glass rail is, that the normally used inductive (eddy current) sensors to measure lubrication gap don't work. The idea is to integrate the pressure sensors in steel rails to get realistic oil film pressure and floating heights in future work.

4 Calibration of Pressure Sensors

4.1 Hydrostatic Pressure Calibration

Hydrostatic calibration of sensors was performed using the apparatus shown in Fig. 2. A dead-weight tester is used to test if the sensors show accurate results according to the weights used. Pressure sensors PS-1, PS-2 and PS-3 were calibrated before they can be installed in the lubrication gap of hydrodynamic guide. Additionally, pressure sensor 'Burster 8210' was also connected to see if pressure sensors (PS-1, PS-2 and PS-3) show same behaviour as Burster.

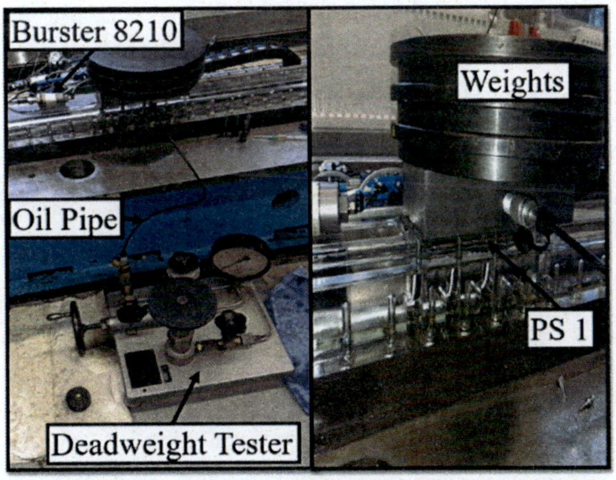

Fig. 2. Apparatus for hydrostatic calibration.

4.2 Hydrodynamic Pressure Calibration

The apparatus used to test the sensors hydrodynamically is shown in Fig. 3. Where, a block from acrylic glass with a through hole in the middle was used. The sensor was being tested, installed on the bottom side and on the top side, the piston oscillates to generate pressure fluctuations. Oscillation of the piston was achieved by direct connection between the piston and shaker. The cylinder has a diameter of 10 mm and height of 60 mm whereas, the sensor's diameter is 6 mm. The piston is directly connected with the shaker and can oscillate inside the cylinder. The oil is filled inside the cylinder up to the height of 40 mm to have enough space for piston movement.

Additional sensor 'Burster 8210' was also connected with an oil cylinder as shown in Fig. 3. The experiments were performed for different frequencies from 5 to 20 Hz with an increment of 5 Hz and different amplitudes used are shown in Table 1.

Table 1. Corresponding amplitudes of shaker in mm for 10 Hz frequency were measured experimentally.

Voltage applied to Shaker (V)	1	3	5
Peak-Peak Amplitudes of Piston (mm)	0.052	0.13	0.21

It was important to know the percentage of losses due to the presence of air and leakages. As when the piston moves in an upward direction the air flows inside an oil cylinder due to moderate sealing between the piston and cylinder's wall. The reason behind using a sealing ring with moderate sealing was to get the complete conversion of movement of the shaker to the piston. The larger ring was also used but the movement of the piston was restricted due to high friction and it had very low-pressure fluctuation. A sealing band was also used in between the screws to have better sealing.

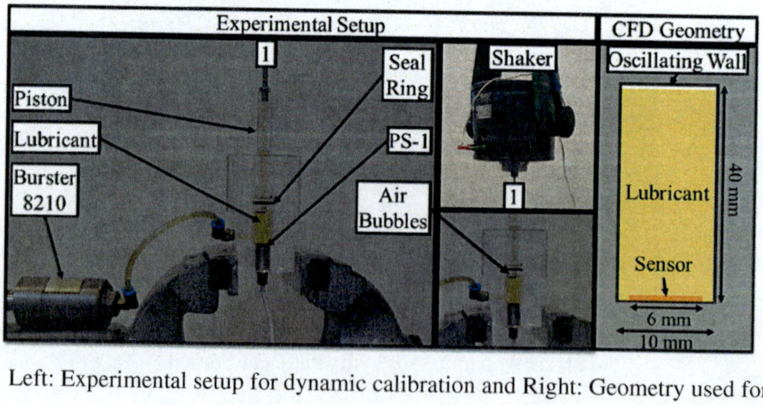

Fig. 3. Left: Experimental setup for dynamic calibration and Right: Geometry used for CFD.

To measure losses, CFD simulations were performed with the same amplitudes of oscillations (see Table 1) as in the experiments. The model was meshed in ANSYS-ICEM and simulated in ANSYS-Fluent software. An oil cylinder with a height of 40 mm and a diameter of 10 mm was used. Where, a top-wall of oil cylinder oscillates to generate pressure fluctuation. Due to the oscillation of the top wall, the layering method was used to have an appropriate mesh for each time step. The oscillation of the top wall has been defined using the user defined function. The simulations have been performed for one second for each case.

Numerical pressure measured were higher than the experimental one if the cylinder is filled completely with oil and there is no leakage. But to compensate losses in experimental setup, parametric study has been performed to estimate air volume percentage. Air volume percentage was measured for each amplitude for only 10 Hz frequency and then used the same values for other cases (15 and 20 Hz) to validate our experimental results.

5 Development of the Integration Method of Pressure Sensors

In this study, the integration method of the pressure sensors in the rail has been developed to improve the oil film pressure measurement. Different pressure sensor's integration methods used in this study are shown in Fig. 4. The actual testing rig has both rails out of steel but for the development of pressure sensors, an acrylic glass rail was used on one side because it is transparent. Another reason of using an acrylic glass rail was that it is easier to change the pressure sensor's integration method. However, one of the problems observed during the development is that the oil film pressure measured by sensors was smaller than the numerical results and the pressure curve was different from our expectations. The reason behind the irregularity and negative peaks was due to the presence of air inside the lubrication gap.

5.1 Geometry Variant I

Initially, two pressure sensors were integrated in the rail with different integration methods as shown in (Geometry Variant I) Fig. 4. Pressure sensor 1 (PS-1) is installed in the middle of the rail in such a way that the sensor's surface is directly connected to the lubrication gap. It was also challenging to install the sensor with precaution so the sensor's surface is not get damaged during operation. PS-2 was installed on the side of the rail in the same way. The sensor's surfaces of PS-2 is not directly connected to the lubrication gap but there is a small oil cavity above the sensor's surface and oil can flow from the lubrication gap to the oil cavity through a very small hole. The oil chamber above the sensor has also another through hole which opens on the side of the rail, which is used to fill the oil or to remove the air from the oil cavity. The set screw was used to keep the side hole closed during the measurements.

The integration method of PS-2 was more feasible and safe but it has been observed that after some measurements the air bubbles were trapped inside the oil chamber which should be removed. Although, the integration method of PS-1 was not safe as compared to the PS-2 but the problem of air entrapment was not an issue. However the presence of air inside the lubrication gap affected the oil film pressure measurements due to the air compressibility. This is the reason that it was a difficult task to perform numerical simulation to measure the volume percentage of air inside the lubrication gap and location of air bubbles inside lubrication gap.

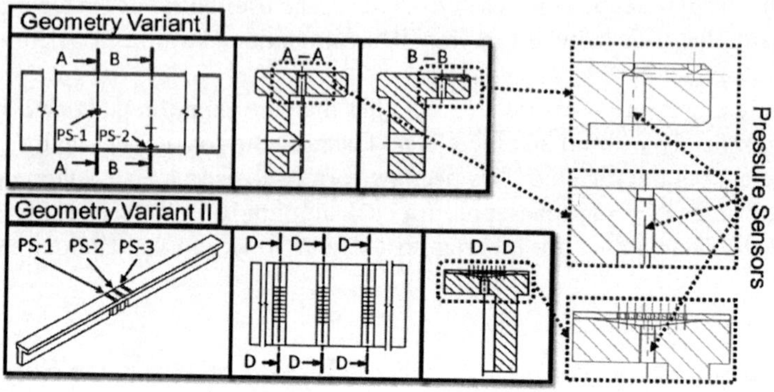

Fig. 4. Integration method of sensors for both geometry variants (Top-Left = Top-view of the rail, Top-Middle and Top-Right = Cut view of sensors PS-1 and PS-2 respectively, Bottom-Left = Isometric view of variant II, Bottom-Middle = Top-view of the rail and Bottom-Right = cut view of pressure sensors PS-1, PS-2 and PS-3).

5.2 Geometry Variant II

It has been assumed that there is no unrealistic variation of the oil film pressure in the width direction of the lubrication gap. And this study is more focused on measuring the pressure along the length of the slide. This is the reason a new geometry has been developed where the groove along the whole width has been introduced which can be covered by the strip made of the same material. In Fig. 4, the geometry variant II can be seen and it is apparent that the lower surface of the oil compartment is tapered so the focus should be on the centre where the sensor sits. The above strip which has several small holes is glued with the rail to avoid leakages. Also, the different sizes of small holes (0.3, 0.7, 1.2, 1.5 and 2 mm diameter) in the strip have been tested numerically and experimentally to reduce the friction losses and more importantly to avoid air entrapment in the oil chamber. In this study, a strip with 1.2 mm diameter of 9 small holes is used (see Fig. 4 Bottom-Middle). For this study, it was important to have similar geometry for all three sensors to make sure that there is no influence on the position of the sensor. To avoid the influence of acceleration and deceleration, the pressure sensors must be installed in constant velocity range.

Another important parameter which can play an important role in estimating the oil film pressure is the number of lubrication grooves. Initially, only two lubrication grooves were introduced. It had been observed that the maximum pressure peaks were only captured when the lubrication groove and the sensor are at the same position. Which means that with 2 lubrication grooves only 2 pressure peaks were measured. To have a better resolution of the pressure curve, the lubrication grooves have been increased to six and it improved the resolution of the pressure curve significantly.

6 Results

6.1 Calibration of Pressure Sensors

Initially, sensors were statically calibrated using a Dead-weight tester. Results of pressure sensors PS-1, PS-2 and PS-3 were compared with Burster 8210. In Fig. 5, it can be seen that for PS-2 and PS-3 the results are very accurate and close to the ideal and pressure sensor Burster. However, in PS-1 case, PS-1 shows accurate results but pressure sensor Burster shows little higher values than the PS-1 and ideal curve. The reason for little deviation of Burster from PS-1 is due to the presence of air, or due to inappropriate setting of zero error. However, pressure sensors PS-1, PS-2 and PS-3 will be used to measure oil film pressure of hydrodynamic guide.

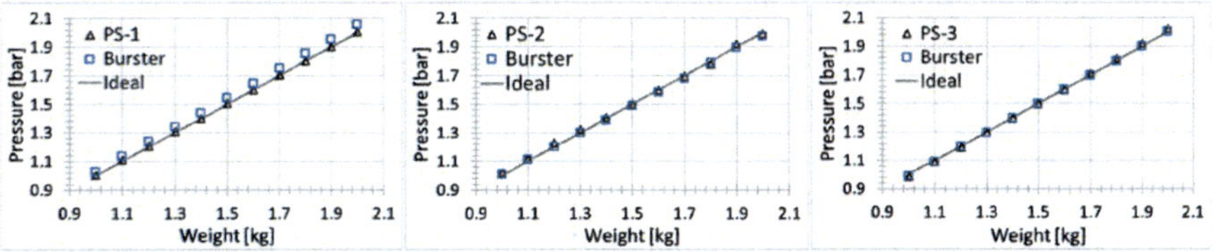

Fig. 5. Validation of all three sensors (PS-1, PS-2 and PS-3) with Burster 8210 with the help of a Dead-Weight tester (Static Calibration using pressure sensor installed in the rail).

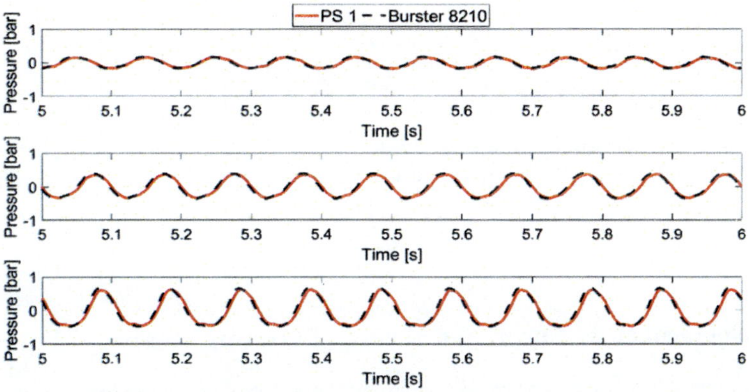

Fig. 6. Dynamic validation of pressure sensor at different oscillation amplitudes (0.03 mm (Top), 0.13 mm (Middle) and 0.21 mm (Bottom)) and for frequency equal to 10 Hz (using apparatus shown in Fig. 3).

Furthermore, the pressure sensors were also tested dynamically with the help of experimental setup shown in Fig. 3. Sensors were tested for various frequencies and amplitudes. In Fig. 6, both pressure sensors (PS-1 and Burster 8201) installed in a cylinder made of acrylic glass in which the piston oscillates, showed similar results. Results from both sensors were compared for different frequencies (5, 10, 15 and 20 Hz) and different oscillation amplitudes (0.05, 0.13, and 0.21 mm).

To validate both pressure sensors (PS-1 and Burster 8210) dynamically, CFD simulations were performed. Pressure fluctuations were estimated due to the oscillation of the piston. In CFD simulations, a small volume percentage air is introduced in an oil cylinder in order to compensate the losses in experimental setup. Air percentage was estimated through parametric study by comparing the numerical results with the experimental results. In Fig. 7, the experimental peak-to-peak pressure is compared with numerical peak-to-peak pressure for various frequencies and amplitudes.

The air volume percentages estimated for 10 Hz are 0.27%, 0.33% and 0.39% for 1V, 3V and 5V respectively. It shows a very good agreement between the experimental and numerical results for the case of 10 Hz frequency. However, it was also interesting to see that for 1V results are also in a very good agreement for frequencies like 15 Hz and 20 Hz. Although, peak-to-peak pressure for higher amplitudes (3V and 5V) and higher frequencies (15 Hz and 20 Hz) are not in good agreement, which is due to many reasons. Firstly, experiments were performed in one run and it has been observed that air bubbles inside the cylinder increases over time. Secondly, for higher frequencies, actual oscillations corresponding to 3V and 5V were little different.

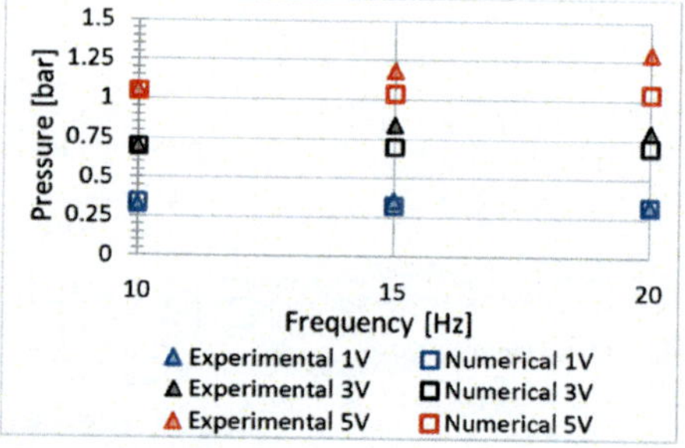

Fig. 7. Numerically validation of pressure sensors PS-1 and Burster 8210 at different frequencies (10, 15, 20 Hz) and oscillation amplitudes results from different voltages applied to shaker (1V, 3V, and 5V).

6.2 Development of the Integration Method of Pressure Sensors

In Fig. 8, the pressure measured by using sensors integrated using the integration method explained in geometry variant 1 section. Oil film pressure measured using PS-1 and PS-2 deviate from the expectations. One of the reasons for not getting expected pressure curves, but only the noise, is because of the measurement range of the sensors. A sensor's surface has a diameter of 6 mm, which means the measuring range is small. Consequently, the volume percentage of air may be too high in between the slide surface and the sensor's surface, which leads to the inaccurate pressure curves. It has been also observed that all sensors integrated the way it is explained in geometry variant I showed different results, moreover the oil film pressure of the sensors itself was not reproducible.

In Fig. 8, the oil film pressure of PS-1 and PS-2 is compared for three velocities (10, 20, and 30 mm/s) and for both forward and backward strokes. Where the distance

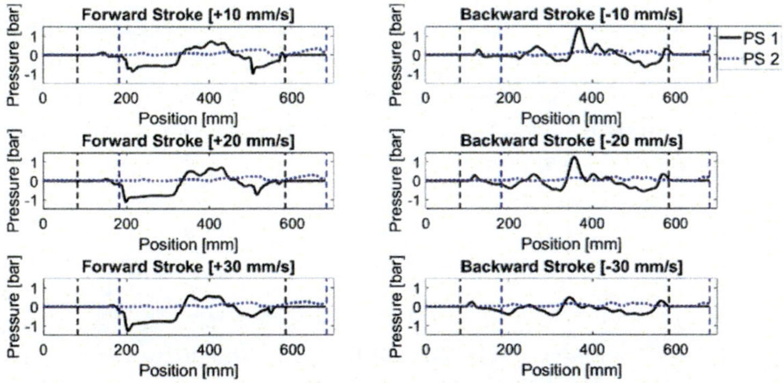

Fig. 8. The pressure measured for forward and backward strokes inside the lubrication gap using the sensors installed or integrated in the rail as explained in geometry variant I.

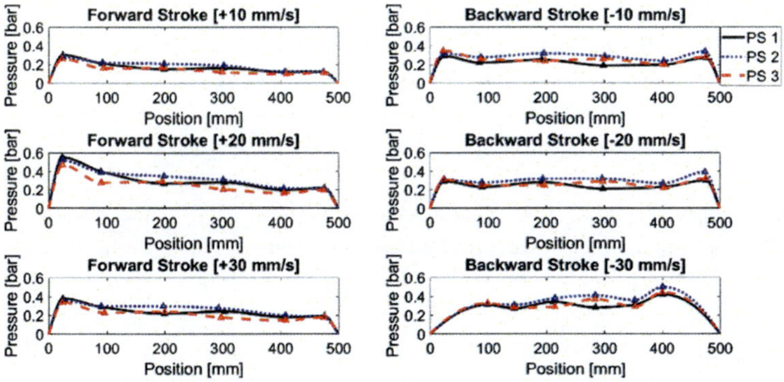

Fig. 9. The pressure peaks (6 peaks for 6 lubrication grooves) measured for forward and backward strokes inside the lubrication gap using geometry variant II.

between black dotted vertical lines show stroke length of the slide for sensor PS-1 and blue one for PS-2. Pressure curve at higher velocities were also acquired, but a constant velocity range at higher velocities could not be achieved. It can be seen in Fig. 8, that it is evident that both sensors do not produce identical results. PS-1 showed some peaks in positive and negative directions, whereas, PS-2 showed almost a constant line. The negative pressure peaks also supports the statement that there is a large percentage of air inside the lubrication gap. To reduce the effects of air on pressure measurement, sensor's integration method was improved as shown in Fig. 4 (geometry variant II).

Oil film pressure has been measured using geometry variant II. Results are compared for different velocities (10, 20 and 30 mm/s) and for both forward and backward strokes as shown in Fig. 9. A pressure curves were acquired by using the six peaks (corresponding to six lubrication grooves) represented by small triangles in Fig. 9. As it can be seen from the results that with the help of geometry variant II the measurement of oil film pressure significantly improved and it shows a maximum pressure in an acceptable range. Whereas, pressure exerted by the machine table at rest position is little above 0.2 bar on one rail. Another noticeable finding is that the variation in between pressure sensors is minimal, and all three sensors show same behaviour. It is also very interesting to see that sensors captured variable pressure along the length of the slide. Additionally, lubrication method also plays a very important role in estimating the oil film pressure inside the lubrication gap. With better lubrication, there is a less chance of air entrapment inside the lubrication gap and oil compartments.

7 Conclusion

Different integration methods of pressure sensors have been tested and compared. By using a novel approach based on the integration of the miniature pressure sensors inside the lubrication gap, an approach to measure the true conditions of hydrodynamic linear guides. Different sensor integration methods have been compared experimentally and significant improvements have been observed in estimating the oil film pressure. However, other influencing parameters (number of lubrication grooves and lubrication method) need to be considered in further investigations.

In the context of further research, the effects of the lubrication method and acceleration profile of the machine table on the pressure measurement and the friction coefficient are to be investigated. Additionally, the pressure sensors will be integrated into the steel rail and the results will be compared with numerical CFD results.

Acknowledgements. This work is funded by the Deutsche Forschungsgemeinschaft (DFG, German Research Foundation)—Project-ID 285064832. Authors thank for the financial support.

References

1. Zhang, Y., Wittstock, V., Putz, M.: Modellierung des Aufschwimmverhaltens hydrodynamischer Linearführungen bei konstanter Geschwindigkeit. Forschung im Ingenieurwesen **83**, 1–6 (2019). https://doi.org/10.1007/s10010-019-00308-x
2. Bhushan, B.: Introduction to Tribology. John Wiley & Sons Inc, USA (2002)
3. Brewe, D.E.: Slider bearings. In: Bhushan, B. (ed.) Modern Tribology Handbook, vol. 2, pp. 969–1039. CRC Press, USA (2001)
4. Ichikawa, S., Mihara, Y., Someya, T.: Study on main bearing load and deformation of multi-cylinder internal combustion engine: Relative inclination between main shaft and bearing. JSAE Rev. **16**, 383–386 (1995)
5. Mihara, Y., Hayashi, T., Nakamura, M., and Someya, T.: Development of measuring method for oil film pressure of engine main bearing by thin film sensor. JSAE Rev. **16**, 125–130 (1995)
6. Mihara, Y., Kajiwara, M., Fukamatsu, T., Someya, T.: Study on the measurement of oil-film pressure of engine main bearing by thin-film sensor—The influence of bearing deformation on pressure sensor output under engine operation. JSAE Rev. **17**, 281–286 (1996)
7. Mihara, Y., Someya, T.: Measurement of oil-film pressure in engine bearings using a thin-film sensor. Tribol. Trans. **45**(1), 11–20 (2002)
8. Someya, T., Mihara, Y.: New thin film sensors for engine bearings. In: Proceedings of the CIMAC Congress 2004, p. 16. Kyoto. Paper No. 91
9. Masuda, T., Ushijima, K., Hamai, K.: A measurement of oil film pressure distribution in connecting rod bearing with test rig. Tribol. Trans. **35**(1), 71–76 (1992)

10. Sinanoğlu, C., Nair, F., Karamış, M.B.: Effects of shaft surface texture on journal bearing pressure distribution. J. Mater. Process. Technol. **168**, 344–353 (2005)
11. Ronkainen, H., Hokkanen, A., Kapulainen, M., Lehto, A., Martikainen, J., Stuns, I., Valkonen, A., Varjus, S., and Virtanen, J.: Optical sensor for oil film pressure measurement in journal bearings. In: Proceedings of Nordtrib 2008, 13th Nordic Symposium of Tribology, p. 12. Tampere, Finland (June 10–13, 2008)
12. SMSI 2021—Sensors and instrumentation. digital, 03/05/2021–06/05/2021: AMA Service GmbH, Von-Münchhausen-Str. 49, 31515 Wunstorf, Germany
13. Krampert, D., Unsleber, S., Janssen, C. and Reindl, L.: Load measurement in linear guides for machine tools. Sensors (Basel, Switzerland) **19**(15) (2019). https://doi.org/10.3390/s19153411
14. Moreau, H., Maspeyrot, P., Chomat-Delalex, A.M., Bonneau, D., Frene, J.: Dynamic behaviour of elastic engine main bearings: theory and measurements. Proc. Instn. Mech. Eng. Part J. Eng. Tribol. **216**, 179–193 (2002)
15. Jaloszynski, T.M., Evers, L.W.: Dynamic film measurements in journal bearings using an optical sensor. SAE Tech. Paper Series **970846**, 95–103 (1997)
16. Höbel, M., Haffner, K.: An on-line monitoring system for oil-film, pressure and temperature distribution in largescale hydro-generator bearings. Meas. Sci. Technol. **10**, 393–402 (1999)
17. Glavatskikh, S., Larsson, R.: Oil film thickness measurement by means of an optical lever technique. Lubr. Sci. **13**, 23–35 (2000)
18. Valkonen, A., Kuosmanen, P., Juhanko, J.: Measurement of oil film pressure in hydrodynamic journal bearings. In: Presented at the 7th International DAAAM Baltic Conference—Industrial Engineering, Tallin Estonia (2010)
19. Binu, K.G., Yathish, K., Mallya, R., Shenoy, B.S., Rao, D.S., Pai, R.: Experimental study of hydrodynamic pressure distribution in oil lubricated two-axial groove journal bearing. Mater. Today: Proc. **2**(4–5), 3453–3462 (2015). https://doi.org/10.1016/j.matpr.2015.07.321
20. Website for Pressure sensors. https://www.kyowa-ei.com/eng/product/category/sensors/ps-d/index.html. Last Accessed 18 Apr 2022
21. Zhang, Y., Wittstock, V., Putz, M.: Simulation for instable floating of hydrodynamic guides during acceleration and at constant velocity. J. Mach. Eng. **18**(3), 5–15 (2018). https://doi.org/10.5604/01.3001.0012.4602

Check for updates

Multivariate Synchronization of NC Process Data Sets Based on Dynamic Time Warping

J. Ochel(✉), M. Fey, and C. Brecher

Laboratory for Machine Tools and Production Engineering (WZL), RWTH Aachen University, 52074 Aachen, Germany
j.ochel@wzl.rwth-aachen.de

Abstract. Various sensors as well as the numeric control serve as sources for the acquisition of process data in operating machine tools. Since the manufacturing industry acknowledges the value immanent to the data, numerous approaches to analyze and exploit large amounts of data have been developed. Generating comparable data sets represents a general challenge when collecting data in real production environments. Multiple external interferences, such as interventions by the operator, alter the manufacturing process and the data set. In order to ensure the transferability of results, a standardized preprocessing and the comparability of data sets, such interferences need to be eliminated by a synchronization algorithm. In this paper, a novel approach is presented, which allows for a synchronization of numeric control process data sets considering multivariate raw data. The approach is able to align data sets of different lengths considering local process modifications. Reliably synchronizing data sets, the presented algorithm aims to support data preprocessing in manufacturing environments and, thus, facilitates the application of data-driven solutions for production optimization.

Keywords: Data mining · Data synchronization · Dynamic Time Warping

1 Introduction

The manufacturing industry faces a demand for economic efficiency and environmental sustainability. In order to meet these demands, the efficiency of production processes is optimized by data analytics methods combined with physical models describing the machine and manufacturing process behavior. In contrast to conventional approaches, they incorporate external and stochastic effects to generate tangible insights [1].

Facing growing data volumes, mining manufacturing data is becoming increasingly important. One essential challenge is generating comparable data sets in heterogeneous production environments. Various external interferences, such as interventions by the operator, or varying conditions for production, e.g. different initial axis positions when the numeric control (NC) program is started or running a process on different machines, alter the manufacturing process and therefore the data set. These interferences obstruct the transferability of analytics results from one data set to another.

Ensuring a standardized data preprocessing, comparable data sets and transferable analytics results drives the need for a locally adaptive synchronization of manufacturing

M. Liewald et al. (Eds.): WGP 2022, LNPE, pp. 288–296, 2023.
https://doi.org/10.1007/978-3-031-18318-8_30

process data sets. To this end, a novel algorithm is proposed, which exploits multiple signals captured from the NC, i.e. axis positions and drive currents, to align data sets of different lengths. Local process modifications and interferences are considered. Reliably synchronizing data sets, the presented algorithm aims to support data preprocessing in manufacturing environments and, thus, facilitates the application of data-driven solutions for production optimization.

2 State of the Art

In production engineering, data sets are typically made comparable by creating identical boundary conditions for the metrological process. A human expert often selects the data of interest by defining time ranges manually. That is why synchronization approaches considering the specificities of production environments are rather rare. They are used to synchronize measured and predicted cutting forces [2], to combine simulation and sensor data [3] and for golden batch monitoring [4]. However, existing strategies from preparation and mining of time series in general can be adapted.

The application of cross-correlation is one way to synchronize data sets. One data set is slid across the other while their correlation is calculated. The location of the peak within the emerging correlation function reveals the time delay for which both data sets exhibit maximum synchrony. The data sets are typically assumed stationary. However, existing approaches of detrended cross-correlation also enable the analysis of nonstationary time series [5]. An exemplary application of cross-correlation in manufacturing is the detection of causalities of production processes by identifying time lags in production state data [6]. Cross-correlation approaches aim to retrieve the initial time delay, e.g. due to different process start times, between two otherwise synchronized data sets. Thus, they are unsuitable for synchronizing processes with local modifications.

A more sophisticated way of synchronizing time series data is the analysis of Cross Recurrence Plots, a visualization of two data sets embedded in the same phase space. Distortions of its main diagonal enable the derivation of a rescaling function. Reliably detecting these distortions poses a challenge depending on the data sets analyzed [7].

The synchronization of time series data with oscillating properties can be achieved by measuring the instantaneous phase synchrony. A necessary step is splitting the data into phase and power, e.g. through a Hilbert transform. Such approaches are applied in medical and neurosciences [8]. However, manufacturing data sets exhibit oscillating properties only partially, e.g. when milling a circular pocket.

Another method to synchronize data sets is Dynamic Time Warping (DTW). The algorithm calculates the minimum Euclidean distance between every data point of one data set and a series of close data points of the other data set. This way, it is not only possible to compare data sets of different lengths but also to assign multiple data points in one data set with a single data point in the other. Hence, the data is non-linearly warped in time. The warping path illustrates how one data set progresses relative to the other [9]. Because of its computational complexity, much effort has been spent on accelerating DTW calculations [10, 11]. Given its desirable features and efficient implementations, DTW is a promising basis for synchronizing NC process data sets.

3 Synchronization of Process Data Sets

The overall approach can be applied to numerous NC processes in manufacturing. For illustration purposes, milling processes serve as an example in this paper.

The synchronization algorithm is based on time series raw data from the machining process captured by the NC. These are the positions and drive currents of all positioning axes (e.g. X, Y, Z, A, B), the drive current of the spindle as well as the active line number of the NC program. In order to synchronize two different data sets representing two instances of a single process adaptively, a four-step approach is applied (Fig. 1).

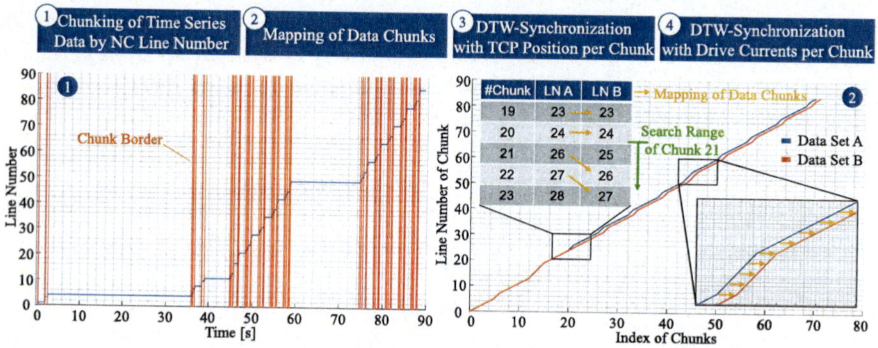

Fig. 1. Synchronization procedure for two sample data sets with data chunking (1) and chunk mapping (2)

Chunking of Time Series Data by NC Line Number (1). As a first step, all signals are split into chunks of constant NC program line numbers. The size of these chunks ranges from a single to multiple thousand data points representing short movements up to whole (cycle-based programmed) features. Separating these chunks breaks down the general challenge of synchronizing whole data sets to synchronizing individual movements. The overall performance of the algorithm is increased by domain-specific knowledge: all signals of two data sets are reliably aligned when both line numbers change accordingly, since a programmed machine movement is completed.

As the NC broadcasts the line number as discrete time series data, chunking is achieved by detecting every change in these integer line number values (Fig. 1, left). This chunking procedure works only if the NC programs of the data sets to be synchronized are identical. Nevertheless, there are other valid ways for chunking, e.g. by discrete events like tool changes, if the NC programs differ.

Mapping of Data Chunks (2). After identifying chunks according to constant line numbers, they need to be mapped across both data sets with respect to their actual value.

Ignoring loops, the NC program is executed from top to bottom. Its line numbers are therefore generally ascending with few exceptions: the line number is reset for subprograms and control-specific cycles. These exceptions are found by detecting multiple occurring line numbers and they are initially skipped in the following steps.

Synchronization is achieved by computing a distance vector, whose elements represent the difference between the first line number of one data set (A) and all line numbers of the other (B). Two chunks match where the difference equals zero (Fig. 1, right). To increase efficiency and avoid inconsecutive matches, the following distance vector represents the difference between the second line number of data set A and all line numbers of data set B succeeding the previous match. Concatenating the distance vectors for every line number in data set A, a sparse distance matrix is generated adaptively.

Individual line numbers might exist in one data set but not in the other, even if both data sets are based on the same NC program. This is because the NC is able to execute single NC program lines faster than the data acquisition sampling rate. Subsequently, a chunk without matching line number is merged with the chunk of the previous match.

Finally, all initially found exceptions are handled. Since their line number occurs multiple times, the corresponding distance vector exhibits more than one value of zero, i.e. more than one match. Respecting the consecutive order of the NC program, the correct match is found between the matches of the previous and following line number.

DTW-Synchronization with TCP Position per Chunk (3). Once this pre-synchronization is achieved, matching data chunks in both data sets cover the same machine movement. This movement, however, can be misaligned or locally scaled over time. Therefore, the locally adaptive synchronization is applied on each chunk independently, simplifying and speeding up the algorithm due to the reduced search space.

Since the time signal cannot be used for synchronization of the chunks, other signals need to be exploited. The overall tool center point (TCP) path is identical for two instances of the same NC process, which is why the positional data is suitable for synchronization. The TCP position is defined as a vector of positions of all axes. The individual synchronization of each positional signal is not applicable as it results in multiple inconsistent data alignments impossible to unify. Reducing this vector prior to synchronization to a one-dimensional signal by calculating the total traveled distance leads to poor data alignment of the individual positional signals, e.g. due to varying process force-induced positional displacements. Additionally, the total traveled distance does not reveal the TCP position during the start of processing and might therefore lead to errors in data alignment. As a result, considering all positional signals simultaneously in a multivariate manner improves the accuracy of data synchronization, even if a single axis does not change its position for an extended period of time.

Current implementations of DTW do not only allow for an exact alignment of multivariate data but are also fast, parallelizable and need linear memory only [12]. Hence, such approaches are exploited to compute a multivariate alignment of the positional data of all linear and rotational axes (Fig. 2, top). While the total DTW-distance is a measure of how well the data sets are aligned, the resulting warping path represents the alignment of both data sets under local scaling (Fig. 3, top).

DTW-Synchronization with Drive Currents per Chunk (4). The synchronization based on the positional signals is applicable, where at least one axis is moving. Where no axis moves at all, e.g. during a tool change, the data alignment deteriorates as the lowest DTW-distance solely depends on random noise of the positional signals. Since

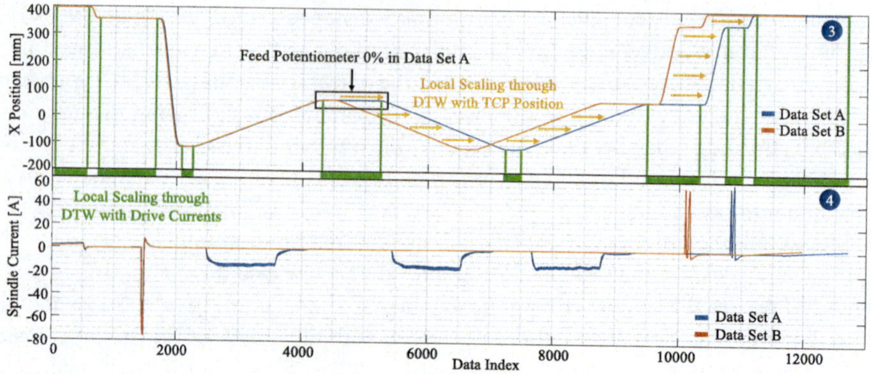

Fig. 2. Concept of DTW-Synchronization with TCP Position (3, orange, represented by X position) and DTW-Synchronization with Drive Currents (4, green, represented by spindle current)

the positional signals are constant, their actual mapping is irrelevant for an accurate synchronization. The spindle and drive currents, though, can change significantly, e.g. due to a spindle speed adjustment, and are therefore inaccurately aligned.

In order to overcome this challenge, areas with no axis movement are identified in the position-synchronized data set, where the velocity of all positioning axes falls below a standstill threshold. The DTW algorithm is applied again on these areas, now considering the spindle and all drive currents to ensure their alignment over the whole data set. The resulting warping path locally replaces the previously calculated path.

Synchronizing two data sets purely based on the spindle and all drive currents is not expedient, since the currents differ under varying loads, e.g. due to inhomogeneous material. Such differences are impossible to occur during positional standstills.

The concept is depicted in a simplified manner for illustration purposes in Fig. 2. The example shows the data corresponding to a single cycle-based programmed NC line, i.e. a single data chunk, for which only the X-axis moves (Fig. 2, top). While data set A corresponds to a real production run with interferences by the operator, data set B represents a production run without interferences and material removal (air cut) for compensation purposes. The spindle current serves as an illustration for the data alignment based on all drive currents (Fig. 2, bottom).

4 Experimental Application

The presented approach is applied on a 3-axis-milling-process on a Heller HF3500 machining center. The internal trace functionality of a Siemens 840D sl NC is used to record the axis positions, the NC program line numbers, the drive currents and the spindle current at 500 Hz (position control cycle). The manufacturing process is executed two times creating two data sets. Data set A represents a ramp up process exhibiting multiple feed rate overrides by the operator, while data set B represents a series production process running unaffectedly. Consequently, data set A is more time-consuming.

The algorithm successfully synchronizes the two data sets overcoming their initial misalignment and local scalings (Fig. 3, bottom). This way, analytics results from data set A can be transferred to data set B and vice versa.

The warping path illustrates the relationship of the temporal progress of the two data sets (Fig. 3, top). Where it exhibits a (nearly) horizontal line, the operator decelerated or stopped process A, while process B ran unaffectedly. A detailed comparison of the warping path before and after the current-based synchronization (cf. Sect. 3, Step 4) reveals to what extent the current-based synchronization alters the initial data alignment. The warping paths deviate to a small degree where all positional axes stand still.

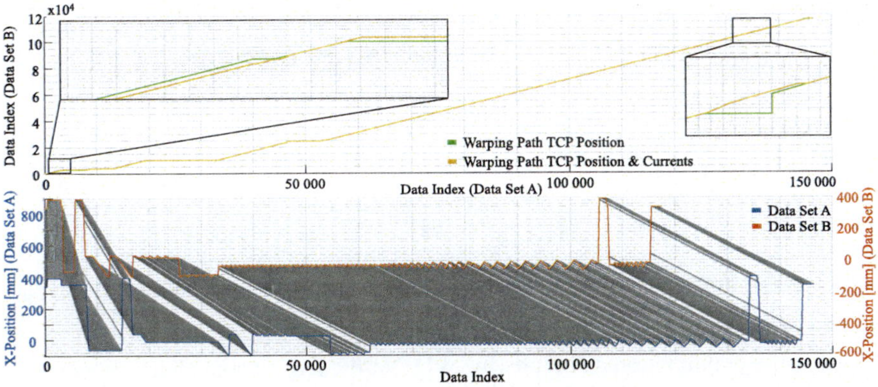

Fig. 3. Synchronized Data Sets with Warping Path (top) and Alignment of X-Position (bottom)

The spindle current of both data sets after synchronization serves as an example of how well all currents are aligned through the presented algorithm (Fig. 4). The green dots indicate where the current-based synchronization replaces the position-based synchronization (cf. Sect. 2, Step 4). While the whole data set consists of 148645 data points, the current-based synchronization is applied to 39258 data points (26.41%). Evidently, the replacements take place where tool changes or spindle speed adjustments occur, which are represented by peaks in the spindle current. Green dots also appear infrequently during machine standstills, e.g. when the direction of machine movement changes significantly (Fig. 4, left). As expected, this is where the positional synchronization is imprecise, since changes in the drive or spindle currents are not covered.

Again, a detailed comparison of the alignment of signals before and after the current-based synchronization (Fig. 4, right) reveals that the small changes through the current-based synchronization improve the alignment of all currents significantly.

There are clear deviations between the spindle current of data set A and B (Fig. 4, orange areas). They can be attributed to a manual feed rate override during cutting effectively stopping the NC program execution. The spindle rotation remains in a stable state without deliberate speed adjustments or any resistance due to material removal. Consequently, the spindle current drops to a baselevel close to zero and rises again as soon as the NC program execution continues (data set A). However, the rise and fall in spindle current are not contained in data set B, such that there is no equivalent value to be matched. Thus, the algorithm keeps the signals of data set B constant until process A continues and both data sets are aligned again. This observation highlights that the proposed approach synchronizes data sets, but does not necessarily make them equal. Additionally, such effects justify why a purely current-based synchronization is not expedient for the alignment of data sets.

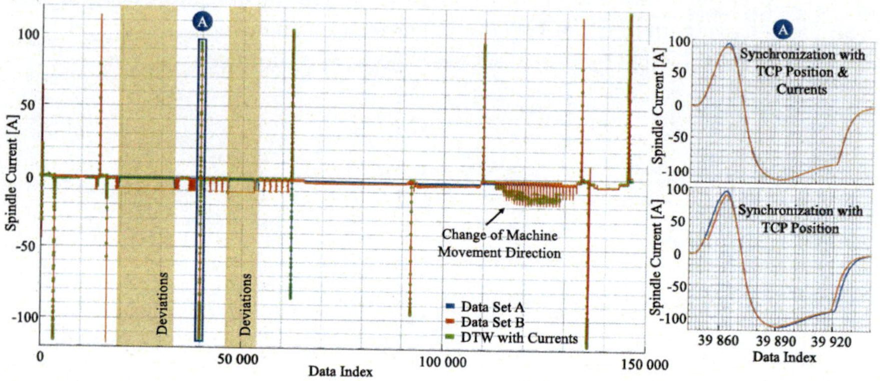

Fig. 4. DTW-Synchronization with and without Drive Currents on Spindle Current Signal

Due to the computational complexity of DTW, the largest data chunk dominates run time and memory usage of the algorithm. While NC programs generated by modern post processors usually contain many lines commanding short movements, large data chunks emerge especially from control-specific cycles. Initially compressing the raw data set before applying the algorithm increases computational efficiency significantly.

5 Conclusions and Future Work

Data analytics methods are exploited to increase efficiency and sustainability of production. The automated mining of increasingly large data volumes demands comparable data sets despite heterogeneous production environments and external interferences.

The algorithm presented in this paper reliably synchronizes NC process data sets exhibiting local scaling based on effortlessly recordable data: the positions and drive currents of all positioning axes, the drive current of the spindle as well as the active line number of the NC program. It subdivides the process into data chunks enabling a parallelizable and fast data alignment based on an exact multivariate DTW approach with linear memory. The presented algorithm is applicable to a wide range of NC processes.

The algorithm facilitates the transferability of analytics results between data sets. Especially the results of elaborate data mining approaches, e.g. the pattern-based segmentation of NC process data [13], become adaptable to other instances of the same process. Moreover, the presented approach can be exploited for data-driven models, e.g. calculating process forces [14], demanding an aligned reference process to compensate unwanted side effects.

In future work, precision, robustness and speed of the algorithm are quantified and tested against other NC process data sets. Moreover, other areas of application in model-based data analytics are investigated. The combination with existing data-driven solutions is a promising approach to accelerate the generation of tangible insights and, thus, further exploit process data for manufacturing optimization.

Acknowledgements. This paper was written as part of the research project "Semantic Data Analysis for Ma-chining Processes (SeDaZ)". The project 22277 N of the Research Association Pro-gramming Languages for Manufacturing Equipment e. V. (FVP) is funded by the German Federation of Industrial Research Associations (AiF) within the program for the promotion of joint industrial research (IGF) by the German Federal Ministry for Economic Affairs and Climate Action (BMWK) based on a resolution of the German Bundestag.

References

1. Brecher, C., Biernat, B., Fey, M. et al.: Data science in production. In: Bergs, T., Brecher, C., Schmitt, R., Schuh, G. (eds.) Internet of Production—Turning Data in Sustainability, AWK 2021, pp. 202–236. Apprimus, Aachen (2021). https://doi.org/10.24406/ipt-n-640534
2. Wan, M., Zhang, W.H., Tan, G., et al.: An in-depth analysis of the synchronization between the measured and predicted cutting forces for developing instantaneous milling force model. Int. J. Mach. Tools Manuf. **47**(12–13), 2018–2030 (2007). https://doi.org/10.1016/j.ijmachtools.2007.01.012
3. Finkeldey, F., Saadallah, A., Wiederkehr, P., et al.: Real-time prediction of process forces in milling operations using synchronized data fusion of simulation and sensor data. Eng. Appl. Artif. Intell. **94**, 103753 (2020). https://doi.org/10.1016/j.engappai.2020.103753
4. Yeh, C. M., Zhu, Y., Dau, H. A. et al.: Online Amnestic DTW to allow real-time golden batch monitoring. In: Proceedings of the 25th ACM SIGKDD International Conference on Knowledge Discovery & Data Mining, pp. 2604–2612. Association for Computing Machinery, New York (2019). https://doi.org/10.1145/3292500.3330650
5. Shen, C.: Analysis of detrended time-lagged cross-correlation between two nonstationary time series. Phys. Lett. A **379**(2), 680–687 (2015). https://doi.org/10.1016/j.physleta.2014.12.036
6. Saller, D., Kumova, B.I., Hennebold, C.: Detecting causalities in production environments using time lag identification with cross-correlation in production state time series. In: Rutkowski, L., Scherer, R., Korytkowski, M., Pedrycz, W., Tadeusiewicz, R., Zurada, J.M. (eds.) ICAISC 2020. LNCS (LNAI), vol. 12416, pp. 243–252. Springer, Cham (2020). https://doi.org/10.1007/978-3-030-61534-5_22
7. Marwan, N., Thiel, M., Nowaczyk, N.: Cross recurrence plot based synchronization of time series. Nonlinear Process. Geophys. **9**, 325–331 (2002). https://doi.org/10.5194/npg-9-325-2002
8. Pedersen, M., Omidvarnia, A., Zalesky, A., et al.: On the relationship between instantaneous phase synchrony and correlation-based sliding windows for time-resolved fMRI connectivity analysis. Neuroimage **181**, 85–94 (2018). https://doi.org/10.1016/j.neuroimage.2018.06.020
9. Shou, Y., Mamoulis, N., Cheung, D.W.: Fast and exact warping of time series using adaptive segmental approximations. Mach. Learn. **58**(2–3), 231–267 (2005). https://doi.org/10.1007/s10994-005-5828-3
10. Salvador, S., Chan, P.K.: FastDTW: toward accurate dynamic time warping in linear time and space. Intell. Data Anal. **11**(5), 561–580 (2007)

11. Geler, Z., Kurbalika, V., Ivanovic, M. et al.: Dynamic time warping: Itakura vs Sakoe-Chiba. In: 2019 IEEE International Symposium on Innovations in Intelligent Systems and Applications, pp. 1–6. IEEE, New York (2019). https://doi.org/10.1109/INISTA.2019.8778300
12. Tralie, C., Dempsey, E.: Parallelizable Dynamic Time Warping Alignment with Linear Memory. In: Proceedings of the 21st International Society for Music Information Retrieval Conference, pp. 462–469. International Society for Music Information Retrieval, Montreal (2020). https://doi.org/10.48550/arXiv.2008.02734
13. Ochel, J., Fey, M., Brecher, C.: Semantically meaningful segmentation of milling process data. In: Behrens, B.-A., Brosius, A., Drossel, W.-G., Hintze, W., Ihlenfeldt, S., Nyhuis, P. (eds.) WGP 2021. LNPE, pp. 319–327. Springer, Cham (2022). https://doi.org/10.1007/978-3-030-78424-9_36
14. Aslan, D., Altintas, Y.: Prediction of cutting forces in five-axis milling using feed drive current measurements. IEEE/ASME Trans. Mechatron. **23**(2), 833–844 (2018). https://doi.org/10.1109/TMECH.2018.2804859

Investigation of the Process Limits for the Design of a Parameter-Based CAD Forming Tool Model

J. Wehmeyer[2](✉), R. Scheffler[1], R. Enseleit[1], S. Kirschbaum[1], C. Pfeffer[2], S. Hübner[2], and B. -A. Behrens[2]

[1] Society for the Advancement of Applied Computer Science (GFaI), Berlin, Germany
[2] Institute of Forming Technology and Machines, Leibniz Universität Hannover, An Der Universität 2, 30823 Garbsen, Germany
wehmeyer@ifum.uni-hannover.de

Abstract. Industrial product development today is faced with the challenge of achieving shorter creation cycles to keep up with international competition. This causes constantly changing requirements for the geometry of the components and thus for the used forming tools. These tools must be designed much faster so that customer requirements are met quickly, which is feasible through a parametric CAD design. As part of a cooperative research project involving the GFaI and the IFUM, a fully parametric CAD model for a sheet-bulk metal forming process was developed. With this tool it is possible to produce cylindrical components with internal and external gearing by combined sheet and bulk forming operations. For this purpose, the CAD model of the tool system is divided into different assemblies. Each assembly consists of various components which relate to each other. Furthermore, the dependencies between the assemblies were built up parametrically via global constrains. An initial structure of the CAD model including constraints is described in this paper. In addition, various process limits are determined by means of experimental tests and calculations. In the first stage of the forming process, blanks are deep-drawn into cups. Due to the geometry of the gears, round cup forming tests were conducted to examine the drawing ratio for different materials (DC04, DP600 and HC260LA). The characteristic values are converted into parameter limits for the new CAD model. Thus, the forming tool can be designed depending on the material used and the required gear size, which can reduce the development time in the future.

Keywords: Model-based systems engineering · Sheet metal forming · Parametric three-dimensional computer-aided models (3D CAD)

1 Introduction

Parameter-based design is widespread in today's product development. Areas of application are, for example, vehicle and aircraft construction, as well as plant and tool construction. The advantage of this design methodology is the fully automatic reconstruction and realignment of all components of a 3D CAD model after the user has changed a parameter, such as the dimensions of a component.

M. Liewald et al. (Eds.): WGP 2022, LNPE, pp. 297–306, 2023.
https://doi.org/10.1007/978-3-031-18318-8_31

3D CAD systems have been used for modelling of deep-drawing tools for years. Increasing product diversity and complexity, and thus growing cost pressure, require companies in the sheet metal processing industry to use parametric 3D CAD systems [1, 2]. These tools allow to integrate the product logic and design knowledge directly into the CAD models. Thereby, new variants and customized designs can be generated quickly by changing parameters [3].

However, the initial construction of a completely parametric CAD model is made considerably more difficult by the precise planning and modelling of parameter interrelations. Those relations can be in and between the individual components as well as assemblies. This brings additional time and cost pressures for designers [4]. Therefore, a parametric model structure is rarely used. As a result, the product logic as well as the design knowledge are subsequently entered into the CAD model, which is very error-prone and can lead to model instability [5, 6]. Currently, there are no suitable methods and support tools for simple and clear modelling of parametric relationships [7]. Consequently, the designer lacks a tool integrated into a 3D CAD system with which the main and secondary functions, structural and parameter interrelationships can be modelled consistently [8]. Due to this, in this paper an approach is shown in which it is possible to design a tool for a multi-stage process on the basis of boundary conditions.

2 Experimental Setup and Procedure

In the following chapter, first the forming tool is presented for which the parametric design is carried out. After that the materials used as well as the experimental setup are presented. Round cupping tests are carried out to determine the forming limits of the material and to determine the constraints for the parametric design. This is done by the investigation of the limiting drawing ratio (LDR).

2.1 Tool Structure

The forming tool considered in this research combines sheet and bulk forming operations [9]. This multi-stage forming process has been developed at the IFUM in recent years [10]. With the developed forming tool, gear wheels are produced. Figure 1a shows the forming tool integrated in a hydraulic press. In a sheet metal forming process, very high contact normal stresses of solid forming prevail in addition to the long sliding paths of the sheet metal forming processes. In order to meet these process requirements, the tool active elements were made of the high-speed steel 1.3343. The required tool strength was achieved by hardening and subsequent polishing of the tool surfaces. The hardened tools are sensitive to the tensile stresses that occur due to the high process forces, especially in the die tooth cavity [11]. Their compensation was aimed by a strip-wound reinforcement of the die. This high-strength pretensioning tool system, shown in Fig. 1b (Die structure), enables a controlled adjustment of the die inner diameter by means of a conical die sleeve. Such a die preload provides a wider strain range during loading. The strength of the strip-wound reinforcement depends on the material of the steel strip wrapped around the carbide winding tube.

The forming stage of this process (Fig. 1b) includes three different manufacturing operations. First, the sheet is deep drawn without a blank holder, upsetted and shear cutted. In the second forming step the flange forming and the wall ironing are performed. The third process step is the sizing of the gear. Process step 2 and 3 are not further considered in the modelling.

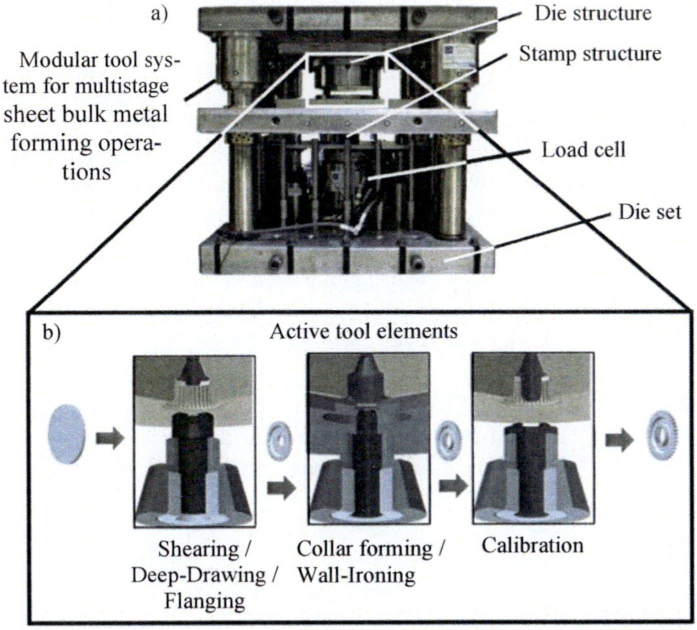

Fig. 1. **a)** Forming tool, **b)** Process sequence and active tool elements.

2.2 Sheet Materials

The sheet materials used are DC04, DP600 and HC260LA. The selection influences the limiting drawing ratio, thus the possible cup height and therefore the possible tooth width of the gear.

The characteristics of the used materials are shown in Table 1.

2.3 Experimental Setup and Test Procedure

The deep-drawing of round cups is primarily conducted to determine the maximum possible limiting drawing ratio for each material. The LDR describes the quotient of the maximum diameter of the blank $D_{0,max}$ to the deep drawing punch diameter d_0 and is calculated by Eq. 1.

$$LDR = \frac{D_{0,max}}{d_0} \tag{1}$$

Table 1. Material properties DC04, DP600 and HC260LA

Material	Material number	Ultimate tensile strength UTS [MPa]	Yield strength YS [MPa]
DC04	1.0338	270–350	210
DP600	1.0936	580–670	470
HC260LA	1.0480	350–430	330

For this purpose, a round cup tool was installed in the hydraulic forming press Hydrap HPDZb 63 of the IFUM, shown in Fig. 2. The diameter of the punch was 48 mm. Circular blanks with a material thickness of 2 mm were used from the respective material and the Beruforge 152D was used lubricantion. The blank holder is not used in this test-setup, as the sheet-bulk metal forming process also runs without a blank holder. Therefore, spacers were used, to keep the gap between the drawing ring and the blank holder constant above the sheet thickness.

To determine the LDR for the different materials the diameter of the blanks was varied. The starting diameter of the sheet was 100 mm. This corresponds to a drawing ratio of 2.08. The diameter was successively increased in 1 mm steps until failures appeared on the cup. Types of failure can be folds of the 1st and 2nd type, as well as cracks at the punch edge run-out or in the flange area, the latter appearing the most.

3 Results

In this chapter the results of the experimental test are shown. Furthermore, the structure of the CAD model of the forming tool is presented and the implementation of the experimentally detected drawing ratio is described.

3.1 Drawing Limit Ratio

Cup tensile tests were carried out to determine the limiting drawing ratio. During the trials, the diameter of the circular blanks was increased until errors occurred. Figs. 2 and 3 shows 2 deep drawn cups of the material DP600. The left cup represents the limit of deep drawing without the occurrence of a defect. A strong unevenness of the cup rim can be detected. The right cup shows an exceedance of the maximum drawing ratio. The bottom of the cup is cracked.

For DP600 and HC260LA a diameter of 101 mm already led to components with cracks. For DC04 the maximum diameter of the blanks could be increased to 105 mm without errors occurring. The tests were repeated five times. The drawing ratio was then determined from the diameters with the Eq. 1 shown in Chapter 2.3. The calculated results for the limiting draw ratio for the materials are shown in Table 2.

The deep drawing steel DC04 has the highest limiting drawing ratio with 2.19. The DP600 and HC260LA have the same LDR with 2.08. This seemed not expected due to the different tensile strengths of DP600 and HC260LA but however HC260LA is a

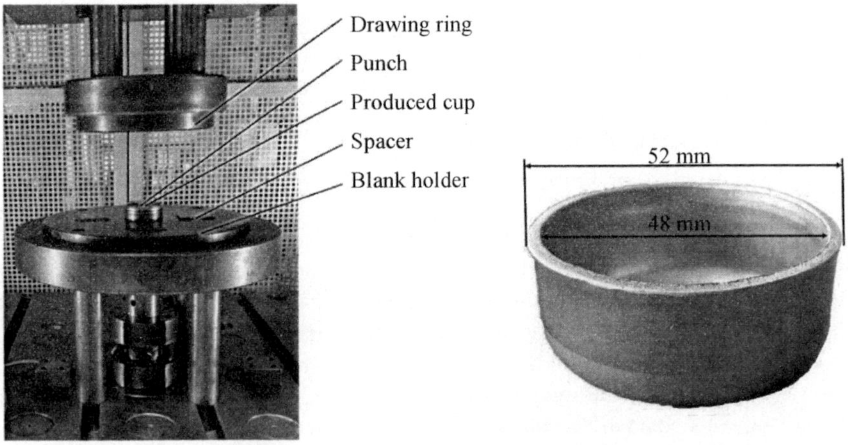

Fig. 2. Experimental set-up cup deep-drawing and deep drawn cup sample

Fig. 3. Failure limit DP600 left maximum possible diameter, right failed part

Table 2. Calculated results for the limiting draw ratio for the materials

DC04	*LDR* = 2.19
DP600	*LDR* = 2.08
HC260LA	*LDR* = 2.08

steel with high yield strength for cold forming and DP600 is a dual-phase steel which is specially developed for deep drawing. The forming limit curves of both materials are almost similar in the deep drawing area.

These determined values are taken for the design of the parameter based CAD-modell with a safety factor of 10%. This results in a LDR of 1.98 for DC04 and a LDR of 1.89 for DP600 and HC260LA.

3.2 Structure of the CAD Model and Import of the Experimental Results

In order to incorporate the experimentally determined different material restrictions into the design, the parametric design of the forming tool can be used. In the final interface of the parametric designed CAD-model, users will be given the opportunity to adapt a tool to their required component without having to acquire comprehensive knowledge about this forming process.

Since the component is a gear, the most important parameters will be those that characterise a gear. In order to select a gear to match another gear, the module must first fit in order to produce the matching tooth form to the gear that may already exist. In addition, a pair of gears is always characterised by a transmission ratio *i* which gives the ratio of the numbers of teeth of the gears. This indicates the ratio of the number of teeth of the two gears. Thus, the number of teeth must also be specified, but the number of teeth must always be a whole number, as no half teeth can be manufactured. Users must therefore select a suitable number of teeth and module for their application in order to be able to realise the corresponding centre distances. Module and number of teeth are the parameters that describe the shape and size of the gear.

In addition, a parameter is required that specifies the tooth thickness. This parameter has a decisive influence on the cup height, as the teeth are compressed from the cup height. Furthermore, the inner diameter is set, which is punched out in the forming process. An internal gearing could be produced on this inner diameter in further production stages. The choice of materials is between DP600, DC04 and HC260LA. This influences the later strength and also the limits in which the component with the dimensions must be located. In summary, the parameters that relate to the final product are number of teeth, module, tooth width, inner diameter and material.

In addition, there is a parameter which simplifies working with the model. This is called press depth and with it the model can be moved in and out to visualise the pressing process. These parameters are limited in order to prevent the user from entering them incorrectly. This is to ensure that the component created can also be finished with the processes and that the maximum dimensions of the final product are defined, as the sheet thickness is set to 2 mm. Thus, pitch circle diameters over 100 mm of a gear with 2 mm thickness are hardly relevant in practice and are not permitted here. In the same way, a smallest diameter of 30 mm is also specified in order not to manufacture the tool with profile thicknesses that are too small. This could complicate the production of the tool components and reduce the fatigue strength.

These before mentioned parameters are called global parameters in the following and with the help of these global parameters local parameters in the individual components are controlled. The basic structure is the same for all components. Components that change their shape by changing the global parameters are linked to the global parameters. With the help of this information, each component calculates the required local parameters. These local parameters are then used to control the dimensions in the sketches. In this way, the structure of the individual parts is very similar and it is possible to familiarise oneself more quickly with the design.

Subsequently, the parameter-based structure of the CAD model was created with the CAD software CATIA V5-6R. The model was divided into various assemblies and is shown in Fig. 4. The assemblies represent components that are rigidly connected to each other in the assembled state.The user can now design the gear by entering parameters such as the number of teeth, module, inner diameter, tooth thickness and material so that the forming tool required to produce the gear is generated automatically. For this purpose, the individual components are related to each other by using rules which can be created with the help of the knowledge adviser function of CATIA. The selection of the material and the associated checking of the limiting drawing ratio is a combination of a test and a rule and is explained as an example below.

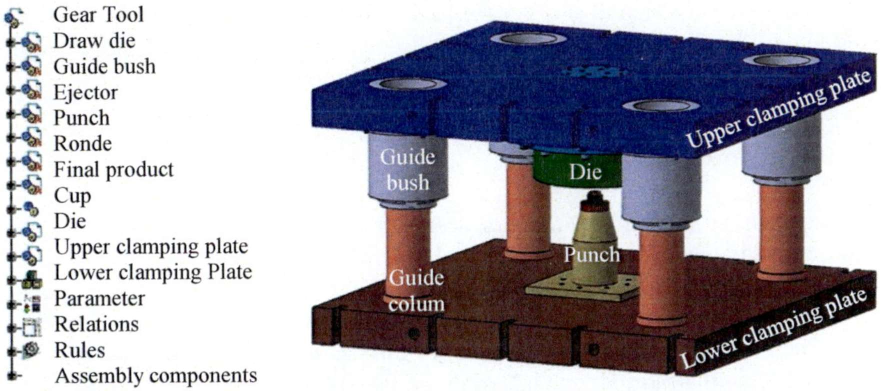

Fig. 4. Structure tree and parameterized CAD model of the forming tool

The materials had to be chosen in numbers for the selection, as a selection of words is not feasible. In the following, the materials are named as follows:

- Material 1 = DP600,
- Material 2 = DC04,
- Material 3 = HC260LA.

First, a rule, which is already implemented in the CAD-model, checks which material is currently selected and activates the appropriate check, at the same time deactivating the other checks. This rule can be implemented for example by an if-loop.

if (Material == 1)	
{	
	'Relation\Material 2\Activity' = False
	'Relation\Material 3\Activity' = False
	'Relation\Material 1\Activity' = True
}	

If, for example, material 1 is selected, only test 1 is activated. In this test, the limiting drawing ratio is measured and compared with the maximum possible value. In addition, it is checked once again whether material 1 is selected. The conditions that must be fulfilled (mentioned in 2.3 formula 1: $LDR = \frac{D_{0,max}}{d_0}$) are described in program syntax as followed:

$$\frac{\text{(blank\textbackslash blank diameter /2)}}{\text{final product\textbackslash cup\_inner\_diameter\_final product}} < 1.98 \text{ and Material} == 1$$

If this is not fulfilled, the warning message that the desired material is not suitable appears. This check ensures whether the limiting drawing ratio has been exceeded and whether the correct material has been selected. Without the rule, a warning message would always be displayed. If the rule is fulfilled, the CAD model of the sheet metal forming tool for the desired gear will be generated automatically.

4 Conclusion and Outlook

In this paper, the structure of a parametric design of a tool for sheet and bulk metal forming was described in which the users of the final interface are given the opportunity to adapt the tool to their required component without having to acquire comprehensive knowledge about the forming process or the construction. The manual reconstruction of a new tool is therefore no longer necessary, which saves time and costs. For the determination of the limiting drawing ratio for the materials DC04, DP600 and HC260LA, cupforming tests were carried out. These values were taken over into the design of the parameter based CAD-model with a safety factor of 10%.

The CAD model was then presented with its assembly subdivision and the parameters entered by the user were presented as well. Furthermore an example of a rule was given to show how the parameter of the limiting drawing ration is integrated in the parameter based CAD-model.

In further investigations the main goal is to develop a better graphical language to facilitate analysis of the model structure. The CAD model itself has poor representation in typical CAD software. While the hierarchical structure of the model is clear and easily understandable through known graphical representation (a so-called tree view), the additional relationships that make up the details of the model often aren't displayed graphically at all. As a first step towards this goal, the CAD model has been modeled in SysML. Figure 5 shows a section of the model setup in SysML.

An evaluation of that graphical representation and the development of a graphical domain-specific language for CAD models is ongoing.

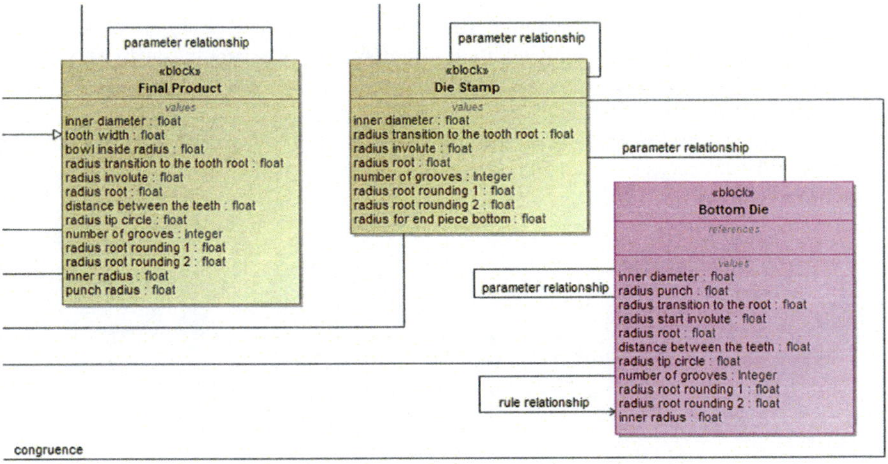

Fig. 5. Section of the CAD model modeled in SysML

Acknowledgements. The authors thank the German Research Foundation (DFG) for the financial support of the research project "Method for the Model-Driven Design of Deep Drawing Tools", project numbers BE 1697/164-3 and BA 6300/1-3.

References

1. Behrens, B.-A., Betancur Escobar, S., Almohallami, A., Weigel, N., Vucetic, M., Stukenborg, E., Colsman, C., Lerch, M., Nolte, I., Wefstaedt, P., Lucas, K., Bouguecha, A.: Production of patient-individual hip cups by sheet metal forming: simulation-based planning and metal forming adapted design method. In: Advanced Material, Research, vol. 907, pp. 253–264 (2014)
2. Voges-Schwieger, K., Behrens, B.-A., Hübner, S.: Enhancing deep drawing processes by using a thermomechanical tool design. In: Key Engineering Materials, pp. 410–411(2009)
3. Bogado, W.H., Dvorkin, E., Idelsohn, S.R., Oate, E., Scheer, S.: A systematic approach to cad systems customization. In: Computatitional Mechanics, pp. 1–9 (1998)
4. Erhan, H., Salmasi, N., Woodbury, R.: ViSA: A parametric design modeling method to enhance visual sensitivity control and analysis. Int. J. Architectural Comput. **08**(04), 463 (2010)
5. Brujic, D., Ristic, M., Mattone, M., Maggiore, P., De Poli, G.P.: CAD based shape optimization for gas turbine component design. In: Structural and Multidisciplinary Optimization, ProQuest, pp. 647–659, Springer-Verlag (2010)
6. Lin, B., Hsu, S.: Automated design system for drawing dies. Expert Syst. Appl. **34**(3), pp.1586–98 (2008)
7. M. Tarkian: Design Automation for Multidisciplinary Optimization, A High Level CAD Template Approach. In: Linköping studies in science and technology. Dissertations, no. 1479, p. 17 (2012)
8. Myung S., Han S.: Knowledge-based parametric design of mechanical products based on configuration design method. In: Expert Systems with Applications, vol. 21, pp. 99–107. Elsevier Science Ltd. (2001)

9. Behrens, B.-A., Hübner, S., Müller, P., Besserer, H.-B., Gerstein, G., Koch, S., Rosenbusch, D.: New multistage sheet-bulk metal forming process using oscillating tools. In: Metals (2020)
10. Behrens, B.-A., Bouguecha, A., Vucetic, M., Hübner, S., Rosenbusch, D., Koch, S.: Numerical and experimental investigations of multistage sheet-bulk metal forming process with compound press tools. In: Key Engineering Materials, pp. 1153–1158 (2015)
11. Lange, K., Kammerer, M., Pöhlandt, K., Schöck, J.: Impact extrusion. Economical production of metallic precision workpieces, pp. 335–337 (2008)

Embossing Nanostructures

D. Schmiele(✉), R. Krimm, and B. -A. Behrens

Leibniz Universität Hannover, Hannover, Germany
schmiele@ifum.uni-hannover.de

Abstract. At present, optical components are costly and complex to manufacture. The costs often are a decisive factor in developing and manufacturing of optical components and sensors. The goal of the cluster of excellence PhoenixD, a major cross-disciplinary initiative, is the time- and cost-efficient production of optical systems. One promising approach is the accurate molding of micro- and nanostructures in a precisely controllable embossing process. Embossing as a manufacturing process for structured functional surfaces enables high output rates at low costs per component. However, embossing of micro- and nanostructures in particular requires high demands concerning the precision of the used machines and tools as well as on the precision of the positioning accuracy of actuated active parts. Machine- and tool-related disturbances are often unavoidable—these include guide inaccuracies, bearing clearances or temperature-related expansions in the powertrain. All these effects can be counteracted by means of an active process control. For this reason an embossing device is being developed which enables the die to be positioned precisely so that micro- and nanostructures can be transferred reproducibly with a high quality. In addition to the high positioning accuracy, this embossing device should also provide high embossing forces. This leads to an expansion of the material spectrum in microembossing and enables a variety of new applications. In this paper various concepts are presented and analyzed concerning their suitability for the precise embossing of fine structures by means of multi-body-simulation with regard to their deformation under load. In addition, a test bench of an electromagnet-spring system is introduced.

Keywords: Embossing · Nanostructures · High accuracy

1 Introduction

Today, the production of highly functional precision optical systems is based on several complex, individual components, which are often assembled costly in serial processes. The associated high costs prevent mass use, for example in life sciences, product engineering and sensor technology.

Future generations of optical systems to be realized in PhoenixD will be based on fewer or even just a single optical element that integrates the required functions in a much more compact and resource-efficient way.

Figure 1 shows an intended demonstrator which allows the application of several optical functions, such as beam path manipulation by means of switchable mirrors or coupling/uncoupling processes with Bragg gratings, in a very small space.

M. Liewald et al. (Eds.): WGP 2022, LNPE, pp. 307–313, 2023.
https://doi.org/10.1007/978-3-031-18318-8_32

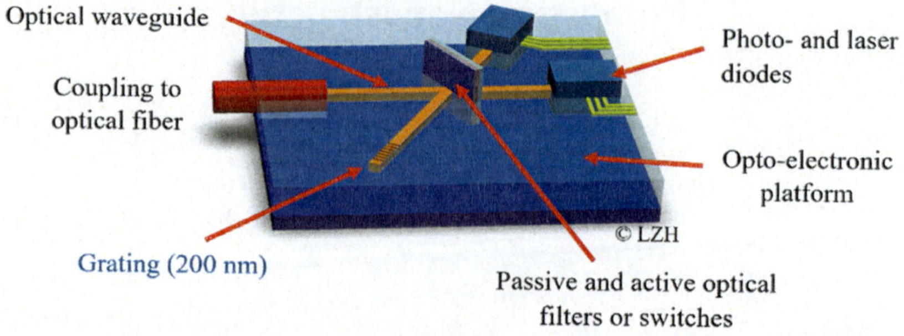

Fig. 1. Demonstrator with several optical functions

There are many challenges and innovative components in the production of the demonstrator. The opto-electronic platform, for example, is manufactured from polyetheretherketone (PEEK) in an injection molding process. The cavities for the optical fibers, filters and photodiodes are already inserted. The design of the diode pocket is intended to ensure self-alignment of the diode with the optical waveguide. After positioning the diodes and filters, the cavities of the optical waveguides are filled with polymethymethacrylate (PMMA) and then a coupling grating with a grating interval in the submicrometer range is going to be embossed into the optical waveguide.

During embossing, the component surface is formed under high pressure [1]. For example, holographic design elements are embossed directly into the surface of sheet metals [1] or machine-readable security features on ID documents are produced by embossing, where even the smallest geometric differences impair the function.

The desired goal of transferring nanostructures with optical quality is associated with very high demands on the accuracy of the embossing process. These high requirements cannot be achieved even with special forming machines for embossing processes, since not all effects of disturbance variables such as bearing play can be compensated.

To avoid these influences, a free-standing embossing device is being designed with which the reproducible embossing of microstructures can be realized cost-effectively and in large quantities.

2 State of the Art

Injection molding, hot stamping or nanoimprinting processes are used to manufacture micro- and nanostructures in large quantities. These are replication processes in which a master structure is molded in the component. These master structures are produced with resolutions in the submicrometer range in an upstream process using micro machining [3] or ion beam writing [4].

In Fig. 2 a sketch of a screw injection molding machine is shown. During the injection molding process, the plastic granules are fed from a hopper to the screw channel and plasticized due to the frictional heat caused by rotation and additional external heating. The plastic melt is then forced through a nozzle into the cavities of the mold under very high pressure. To compensate for material shrinkage, holding pressure is maintained during the cooling phase. After complete solidification of the material, the mold is opened and the part ejected. Injection molding allows complex molded parts to be produced in high quality and in large quantities [5].

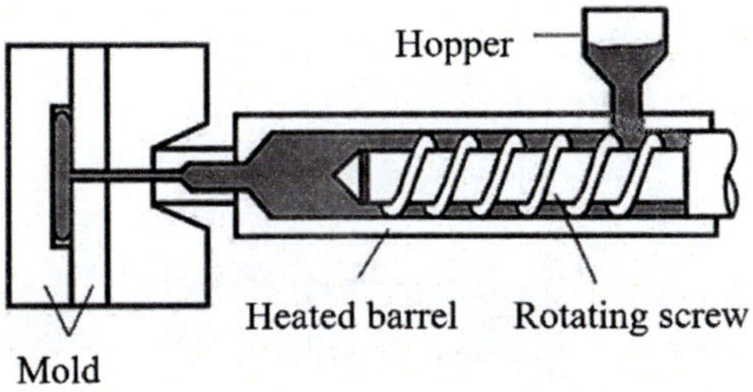

Fig. 2. A sketch of screw injection molding machine [5].

In contrast to the primary shaping of the injection molding process, the polymer films are formed into structured parts during hot embossing. The process sequence is shown in Fig. 3. In this process, the mold insert and a plastic film (Fig. 3a) are heated by an external heating to the desired embossing temperature (Fig. 3c) using a contact force (Fig. 3b). The embossing force is then applied and, after a defined holding time, the mold insert is cooled (Fig. 3d), the mold is opened (Fig. 3e) and the part is demolded (Fig. 3f) [6].

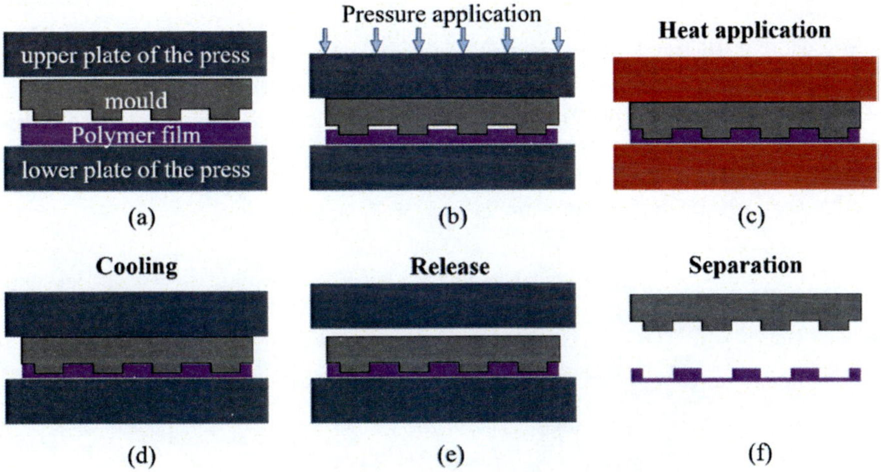

Fig. 3. Schematic of the hot embossing process

The nanoimprint process can be classified into thermal nanoimprint and UV nanoimprint. The process flow in thermal nanoimprinting is very similar to hot embossing. Instead of thick polymer films, as in hot embossing, thin polymer layers are used on a hard substrate base and the structure is formed out under pressure and heat. In UV nanoimprinting, a UV-curable polymer with low viscosity is applied to the substrate

(Fig. 4i), the cavities are filled with low pressure, and UV light is used to crosslink the polymer through the transparent stamp (Fig. 4ii). After structural stabilization, the part is unmolded (Fig. 4iii) [7].

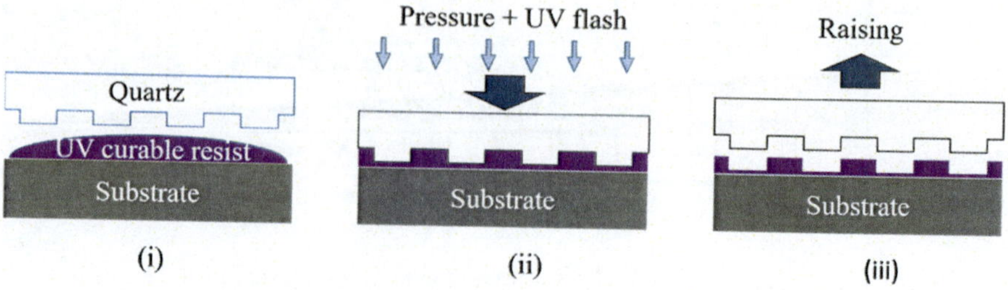

Fig. 4. Schematic of the UV-Nanoimprint process

3 Concepts of Embossing Devices

It is not possible to transfer structures to semi-finished products at a defined position when using the abovementioned processes. This limits the use in the production of complex optical components. The following approaches of embossing devices compensate these limitations and thus establish the embossing process in the mass production of optical components.

To ensure a tilt-free embossing stroke, the stamp position should first be precisely recorded in real time with four distance sensors. The recorded position is compared with the target specification and used as a manipulated variable for the four actuators provided. For this purpose, requirements and specifications have been initially worked out. Based on this, two potential solution concepts were developed and compared simulative with regard to their elastic performance under load in Figs. 5 and 6.

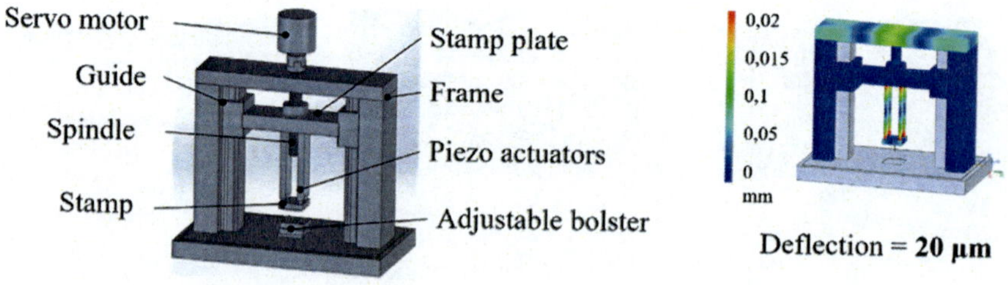

Fig. 5. Investigated concept with piezo actuators and deflection analysis

In one solution (Fig. 5), firstly the punch is pre-positioned using a spindle and then the embossing stroke is executed via four piezo actuators. In another concept, a belt drive is used for pre-positioning and the embossing force is applied using four electromagnets (Fig. 6). In both concepts, capacitive distance sensors with a measuring

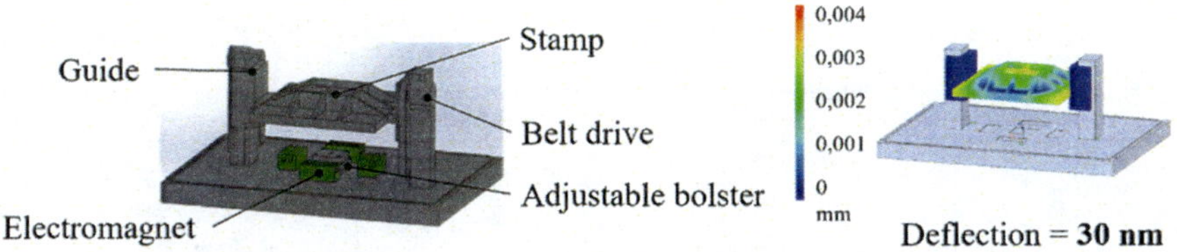

Fig. 6. Investigated concept with magnet actuators and deflection analysis

range of 100 μm and a static resolution <0.001% of the measuring range are used for position measurements according to the high accuracy requirements.

Both concepts have advantages and disadvantages. When embossing with piezo actuators, high positioning accuracies and fast response times are possible, but only small embossing paths. In addition, several components in the power train result in a large total deflection of the device during embossing. This deflection is very high compared to the total path and therefore rather unsuitable for the intended project. In contrast to the piezo actuators, the electromagnets are very cheap and the potential path is significantly greater. Due to the principle, only the stamp plate is in the power train, which results in a significantly reduced deflection. However, it is unknown which positioning accuracy can be achieved by using electromagnets.

4 Proof of Concept

In order to determine the positioning accuracy that can be achieved using electromagnets, a test bench consisting of four electromagnets, springs and distance sensors has been set up. The setup is shown in Fig. 7. A position control system is implemented, control parameters as well as target values can be modified via a GUI.

The idea: Achieving small positioning step sizes by very fine increase of the magnet current and thus of the magnet force. For this purpose, magnets with a maximum force of 3.5 kN and servoamplifiers are used, which allow minimum current steps of $\Delta I \approx$ 1 mA. The achievable accuracy and the maximum path range are strongly dependent not only on the current but also on the parameters of the initial air gap of the magnet, spring preload and spring stiffness. These values were optimized experimentally.

The preload of the springs ensures a constant counterforce at the beginning. As soon as the magnetic force exceeds the resulting spring force due to the current increase, a change in position results, which is detected by the capacitive sensors. Within the closed-loop control system the detected position is compared to the preset position and, if necessary, the current preset of the electromagnet suggestive is increased until the preset position is reached.

In first experiments, step sizes in the submicrometer range were realized. Figure 8 shows the position of the individual distance sensors in dependence of time. It can be seen that the stamp can be positioned without tilting with an accuracy of 100 nm.

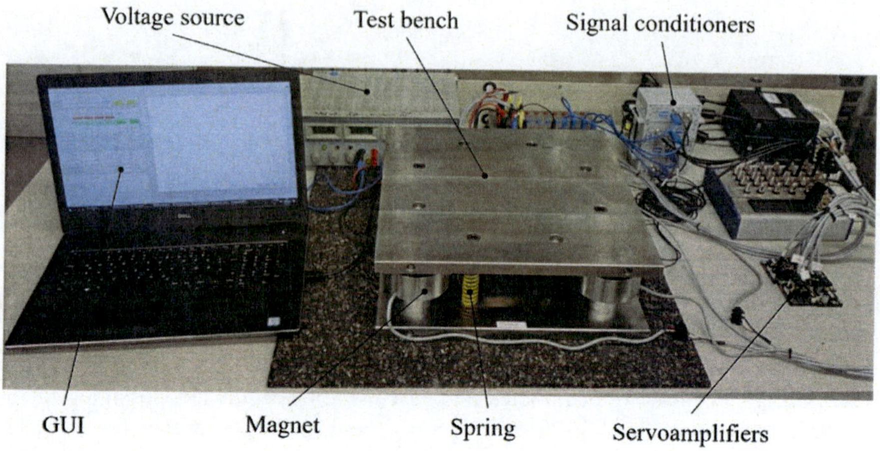

Fig. 7. Test bench set-up

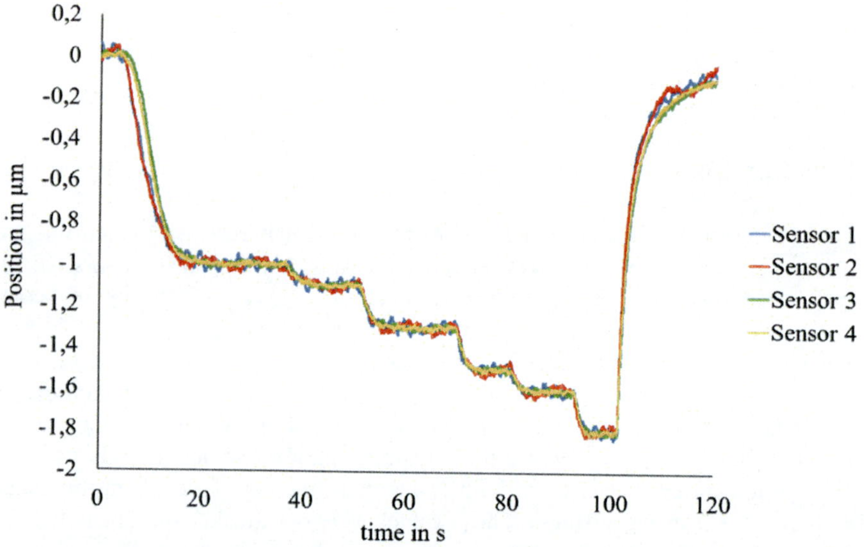

Fig. 8. Position control

5 Summary and Future Work

Based on the major challenges to the embossing process of smallest grating structures, concepts of possible embossing devices were developed, which show positioning accuracies in the submicrometer range and thus allow a reproducible production of microstructures with optical quality. Their potential suitabilities were firstly evaluated by simulating their individual deflection performance under load. Furthermore, it was shown experimentally by means of a first test setup that a positioning accuracy in the submicrometer range can be achieved with an electromagnet-spring system and a suitable control system.

Next, in addition to the reduction of environmental influences, the accuracy and the travel range are to be further increased. To this end, the test bench consisting of the magnet-spring system will be simulated and examined with regard to the optimum operating point depending on the parameters initial air gap and spring preload as well as spring stiffness. The first test structures are transferred and the achievable molding quality of the device is analyzed. The final step will be the construction of the embossing device, which can be used for a structure transfer at an exactly defined position on a component by embossing technology.

Acknowledgement. Funded by the Deutsche Forschungsgemeinschaft (DFG, German Research Foundation) under Germany's Excellence Strategy within the Cluster of Excellence PhoenixD (EXC 2122, Project ID 390833453).

References

1. Behrens, B.-A., Doege, E.: Handbuch der Umformtechnik, 3rd edn. Springer, Berlin (2016)
2. Behrens, B.-A., et al.: Method to emboss holograms into the surface of sheet metals. In: Shemet Conference 2013, vol. 549, pp. 125–132. Key Engineering Materials, Belfast (2013)
3. Saptaji, K.: Mechanical micro-maschining. In: Handbook of Manufacturing Engineering and Technology, pp. 1089–1107. Springer, London (2015)
4. Kawasegi, N.: Ion beam machining. In: Yan, J. (ed.) Micro and Nano Fabrication Technology. MT, pp. 529–554. Springer, Singapore (2018). https://doi.org/10.1007/978-981-13-0098-1_16
5. Zheng, R., et al.: Injection Molding. Springer, Heidelberg (2011)
6. Bhushan, B.: Springer Handbook of Nanotechnology. 2nd edn, Springer, Berlin (2007)
7. Deshmukh, S., Goswami, A.: Hot embossing of polymers—a review. In: 10th International Conference of Materials Processing and Characterization, pp. 405–414. Taylor & Francis, Mathura, India (2020)

Model-Based Diagnosis of Feed Axes with Contactless Current Sensing

M. Hansjosten(✉), A. Bott, A. Puchta, P. Gönnheimer, and J. Fleischer

Karlsruhe Institute of Technology, Kaiserstr. 12, 76131 Karlsruhe, Germany
Malte.Hansjosten@kit.edu

Abstract. State of the art drive controllers, based on numerical and programmable logic controllers (NC and PLC), have not yet established standardized and easily accessible endpoints to capture status and process variables, such as motor current, torque or position values. Direct access to those data sources is limited to proprietary tools or licenses and requires modern control hardware. Besides, available data sources are limited to sample periods down to the NC or PLC cycle time, that varies between 1 and 10 ms. In this paper, we introduce a low-cost, low-tech and low-effort solution for monitoring feed axes based on contactless current sensing. We deploy split-core current transformers onto motor power cables of a variable frequency drive achieving sample rates of 50 kHz. This provides a retrofit solution for feed axes monitoring. Also, we outline the required signal processing to show the solution's potential for further applications like anomaly detection. As a result, we enable a low-cost monitoring solution for machine tools using a physic-based model.

Keywords: Machine tool · Feed axis · Contactless sensing

1 Introduction

Knowledge on machine availability and process capabilities is of utmost importance for high productivity and product quality. One possible solution to bringing transparency to the equipment's lifecycle and process insights is to install in-situ sensors. Application of additional devices in-situ and close vicinity to the relevant sources provides high fidelity and sensitive information on the subject. Nevertheless, these sensors require critical care for their placement, usually accompanied with additional mounting fixtures. On top of that, they are expensive and require specialist staff during installation.

Concerned with this fact, a lot of research has been conducted on the topic of using internal sensors of already installed automation equipment, such as drive controllers, variable frequency drives (VFD) and programmable logic controllers (PLC). In literature this kind of monitoring is called sensorless condition monitoring [1], describing the fact that no additional device is installed for system or process diagnostics. A major drawback using built-in sensor sources in machines arises from the abundance of different hardware providers involved. Accessing machine sensor data involves the drive and VFD, numerical control (NC)/PLC and machine manufacturers alike. End users or operators

M. Liewald et al. (Eds.): WGP 2022, LNPE, pp. 314–323, 2023.
https://doi.org/10.1007/978-3-031-18318-8_33

that want to capture this data or deploy an application based on this data are confronted with different protocols, software tools and licenses required to access their data and limited sampling rates. In recent years, due to the propagation of Industry 4.0, control over and accessibility to valuable data sources is of great interest of all involved parties.

A solution is proposed to decrease the parties involved in the dataflow of drive data, focusing on motor current and shaft speed. Figure 1 shows the state of the art (a) with an alternative data flow based on contactless current sensing (b) within the control cabinet of the installed drive technology. This paper focuses on the acquisition and preprocessing of the current data to gain necessary information and outlines a concept for a modular modelling framework for feed axis assemblies.

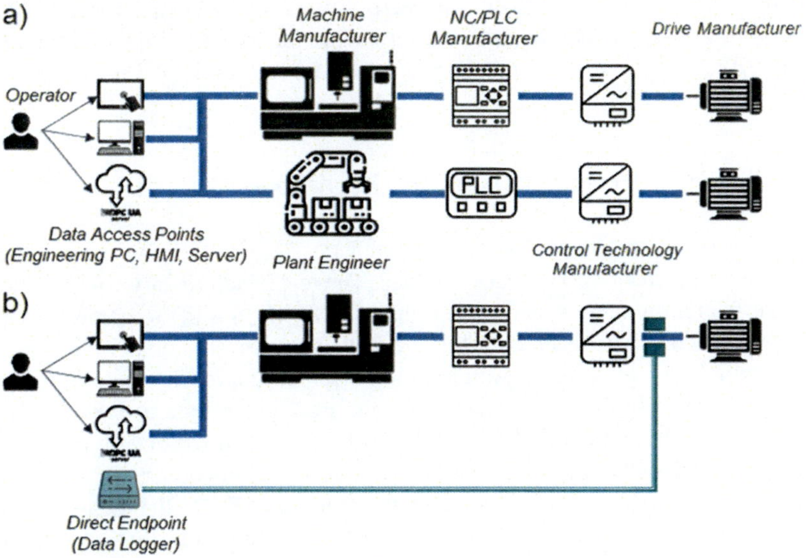

Fig. 1. State of the art dataflow within feed axis (a) and alternative data flow based contactless current sensing (b)

The approach presented uses conventional split-core current transformers to capture the current flow through drives' power cables to deliver process or component insights of the connected hardware. It is shown how these current signals can be processed to extract relevant information and how physics-based modelling of involved feed axis mechanics can be used to determine process forces. By implementing a modular architecture for such models with the possibility for data-driven fine tuning of model parameters a new monitoring approach in machine tools or production machines in general is created.

2 A Review of Current Based Condition Monitoring

There are many approaches to condition monitoring through investigations of the motor current [2]. These range from wear detection of motors and bearings [3, 4] to the characterization [5, 6] and change monitoring of system properties [7]. The different system properties can result from geometrical differences in the components or different forms of wear [5, 8]. This is the basis of current-based condition monitoring approaches widely used in industrial applications [9]. For these applications the current can be measured without additional sensors via the PLC [4, 5, 8–10] or through different kinds of external sensors [7, 11, 12]. Such additional external sensors are often costly and need expertise in their use and calibration [11, 12]. This is why in most applications current data is obtained from the control system. However, there are often considerable obstacles to user-oriented application due to licensing agreements and the wide variety of software interfaces. In addition, especially in brownfield applications, it is often very difficult to identify the relevant parameters on the control system, even if access to the data is available [13].

To address these challenges, this paper proposes a low-cost, low-tech and low-effort approach of contactless current measurement of a drive axes with automated data processing for the derivation of position and speed information and integration into a parameterizable model structure.

3 Contactless Current Sensing for Feed Axes

To test and validate the proposed system, contactless current sensors were integrated into an existing test bench for linear feed axes. The test bench mainly consists of a ball screw, a permanent-magnet synchronous motor (PMSM), a VFD to control the motor and a PLC for overall testbench control. The PLC also allows to obtain motor-data via OPC-UA, which is later used to validate the presented approach. As outlined before the sampling period in this case depends on the PLC cycle time as well as the used transmission protocol, which in the presented case amounts to around 40 ms.

The system presented in this paper uses conventional split core current transformers applied around two of the power cables running from the VFD to the motor to measure the motor current. The split core design of these sensors allows the application around the power cables without the need to remove or rewire them. The power cable represents the primary winding of the transformer around which the iron core with the secondary windings is applied. The alternating current in the power cable induces a magnetic field in the core which in turn induces an alternating current in the secondary winding. In a transformer the ratio of primary and secondary current is inverse proportional to the ratio of the number of primary and secondary windings. Since the primary winding in this case is a single cable, this simplifies the relation to the following Eq. (1), where N is the number of secondary windings.

$$I_{sec} = \frac{I_{prim}}{N} \tag{1}$$

Using a 680 Ω Resistor the resulting voltage drop over the resistor is measured with an analog-digital converter. Using Ohms Law and Eq. (1) this results in the following relation between the measured voltage and the current in the power cable:

$$I_{prim} = N \cdot \frac{U_M}{R} \tag{2}$$

To measure the voltage and log the resulting data, a Raspberry Pi with a specialized sensor shield is used. This allows the measurement of up to 8 Channels in parallel with a Sampling rate of 50kHz. The implementation of two sensors around different Phase-cables of the Motor allows to detect the direction of rotation (see Sect. 4.2).

4 Signal Processing

In order to input the measured data to a physics-based model of the feed axis, it first has to be preprocessed to extract the relevant information like the torque generating current, rotational speed, linear speed as well as an estimation of the feed axis position.

4.1 Extraction of Torque Generating Current from Sensors

To be able to determine the motor torque, the torque generating current has to be determined. The necessary correlations in electric motors are well documented in literature [14, 15]. Therefore, only a short overview will be given in this paper.

The most straight forward way to calculate an effective value of an alternating current is to the root mean square (RMS) value. However, this is an approximation, since it necessitates the application of a sliding window, thereby smoothing the data.

For a more dynamic current calculation the individual phase currents in the PMSM have to be considered. The PMSM in this work has three stator phase currents i_u, i_v and i_w inducing the rotating magnetic field which in turn moves the rotor. These three currents form the current vector $\vec{i}_{ges}$.

$$\vec{i}_{ges} = \frac{2}{3} * \left(i_u + i_v * e^{j120} + i_w e^{j240} \right) \tag{3}$$

This Vector can be either expressed in the stationary alpha-beta coordinate system:

$$\begin{pmatrix} i_\alpha \\ i_\beta \end{pmatrix} = \frac{2}{3} * \begin{pmatrix} 1 & -\frac{1}{2} & -\frac{1}{2} \\ 0 & \frac{\sqrt{3}}{2} & -\frac{\sqrt{3}}{2} \end{pmatrix} * \begin{pmatrix} i_u \\ i_v \\ i_w \end{pmatrix} \tag{4}$$

or in the rotor bound d-q-System (see Fig. 2).

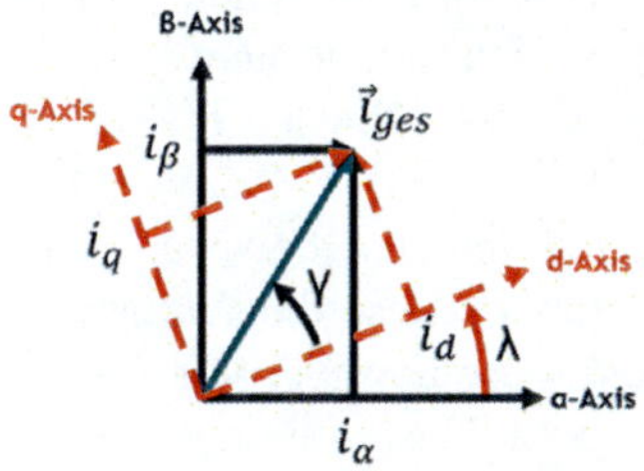

Fig. 2. Coordinate representation of a synchronous machine [16]

In case of the used star connection the sum of the three phase currents is always zero ($i_u + i_v + i_w = 0$), which means, that the current vector in the stationary coordinate system can be calculated by measuring just two-phase currents:

$$i_\alpha = i_u \ and \ i_\beta = \frac{1}{\sqrt{3}} * i_u + \frac{2}{\sqrt{3}} * i_v \tag{5}$$

Another important control variable to consider is the angle between the current vector $\vec{i}_{ges}$ and the d-Axis of the rotating coordinate system. The two main operating modes for PSMS are either $\gamma = 90°$ or $\gamma > 90°$. $\gamma = 90°$ represents the mode with maximum efficiency, since in this case the field weakening component of the current vector is zero ($i_d = 0$) and only the torque generating component is present [17]. Angle > 90° represents the field-weakening region. In this case the permanent magnetic field is weakened by the electromagnetic field generated by the i_d component of the current vector, which means the motor has to turn faster to reach the equilibrium of applied voltage and induced voltage, allowing for faster rotational speeds. In this work the motor is not run in the field weakening region, meaning $i_d = 0$ and therefore $|i_q| = \left|\vec{i}_{ges}\right|$. This allows us to calculate the absolute torque generating current without the knowledge of the angle between stationary and rotating coordinate system using the current-torque-constant K.

$$|M_{el}| = K * |i_q| \tag{6}$$

4.2 Extraction of Speed and Position Information

The rotational speed of a synchronous machine corresponds to frequency of the excitation field f_{el} divided by the number of pole pairs p:

$$n_{mech} = \frac{f_{el}}{p} \tag{7}$$

f_{el} is calculated by analyzing key-features of the current signal (like zero crossings and extreme values) and use those to calculate the rotational position and speed of the motor. To suppress existing noise the current signal is first filtered with a moving average. Subsequently, the positions of the zero crossings are determined by shifting the signal and determining the positions with change in the value range the quantity of zero crossings per time allows the determination of f_{el}. Two zero crossings also describe a rotation of the rotor by $\frac{180°}{p} = 36°$ allowing the direct estimation of the rotational position.

For determining position and velocity, however, only zero crossings are relevant which lie above a threshold value, because the stationary signal also oscillates around zero. The experienced amplitudes are however in a range < max(amplitude)/100. The application of the defined threshold thus eliminates the irrelevant zero crossings.

Within a section (between two zero crossings) lies an extremal point, which can be used as an auxiliary point in the interpolation of the discreet signal to determine a continuous corresponding rotary position. The use of the local absolute maxima allows a more robust determination of the position. The frequency of the excitation field can be determined from knowledge of the sampling frequency and the set of zero crossings per number of sample points. By considering two phases, the direction of rotation can be determined by observing the leading phase in each case. With this method an algorithm is implemented for automatically processing the voltage signal and extracting the position and velocity information.

5 Validation of Signal-Acquisition and -Processing

To validate signal-acquisition and processing, several test runs were carried out, with the setup described in Sect. 3. The calculated magnitude of the rotating current vector was compared to the mean RMS value of the individual measured phases as well as data acquired from the PLC and an additional data set. The additional data set is the mean current value of the three phases measured by a clamp meter and serves as additional validation. Since the magnitude of the rotating current vector is equivalent to the amplitude of the alternating current it was scaled by $\frac{1}{\sqrt{2}}$ to allow for easy comparison to the rms current values. Figure 3 shows clear similarities between the different current signals in value range and spread of the current for two exemplary velocities. The observable similarities in the shape of the current signal demonstrate the expected capability of the approach in measuring currents.

The validation of the obtained current signal allows in the next step a comparison of the determined positions and velocities with PLC data (also shown in Fig. 3). Additionally, the error between the signals is illustrated. For a rotational velocity of 200 $1/min$ the observable position error is between $\sim 5mm$ and $\sim 9mm$. The corresponding velocity error is here $\sim 4\ 1/min$. For 600 $1/min$ the minimum positional error is $\sim 1mm$ and the maximum $\sim 9.5mm$. The velocity error outside of the acceleration phase is around $\sim 3\ 1/min$.

6 Modeling a Feed Axis with Sensor Input

In order to use the extracted torque and speed information within in the context of feed axes, we pass the data stream into a physics-based model of feed axis components. The schematic structure of the applied model is shown in Fig. 4. The relationship between the different components is well documented in literature. Therefore, only the general structure of the model will be addressed.

The model uses the known physical relationship between the components and the specified characteristics of the individual data sheets. The necessary input values are the rotational motor speed (and acceleration as its derivation) and the torque generating current. Speed and acceleration are used to calculate the torque (or force) resulting from friction and dynamic effects for each component. In case of the ball rail a necessary intermediate step is to transform the rotational values to linear values by multiplication with the ball screw lead. The following calculation of axial dynamic loads also considers the mass of the feed axis sledge (and other passive masses). The resulting axial force is then an input for the calculation of the axial load torque of the ball screw. The required

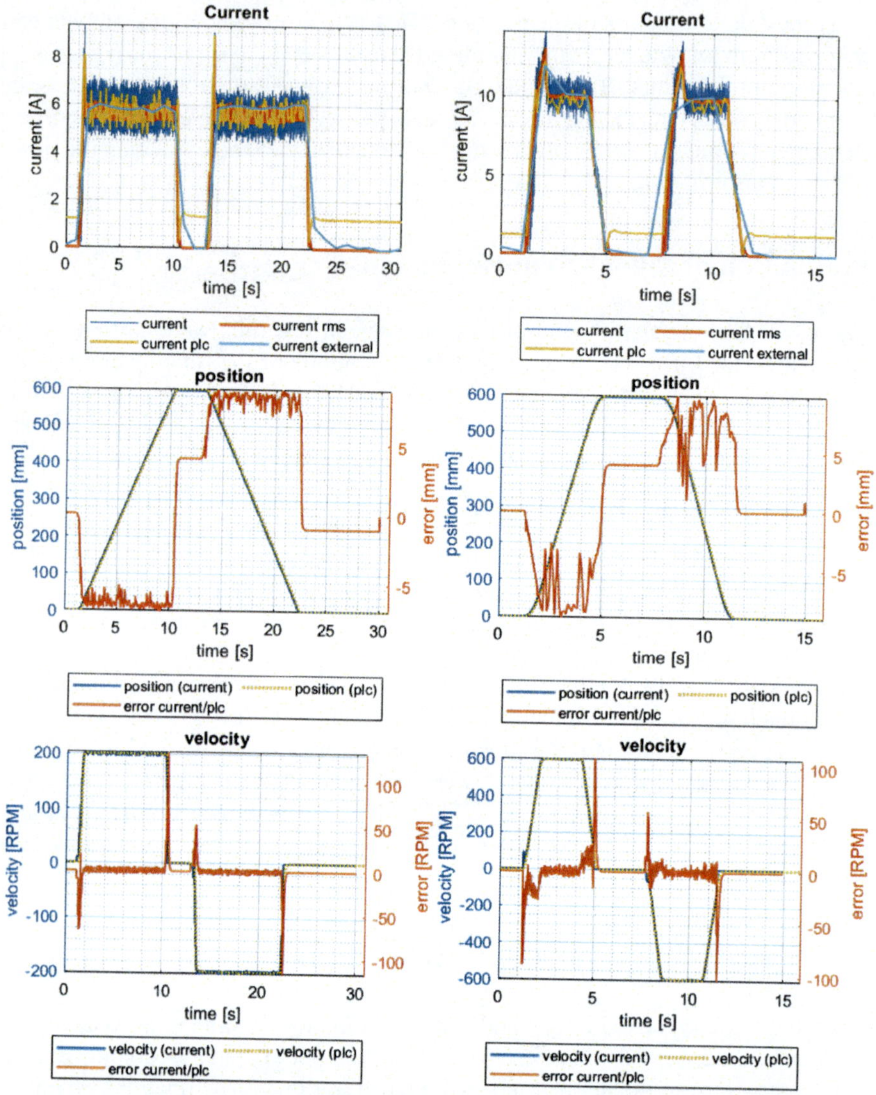

Fig. 3. Different signals derived from the system described in this paper, PLC and an additional validation measurement for a specified feed of 200 1/min (left) and 600 1/min (right)

torque M_{req} is then the sum of the required torque of ball screw and bearing. The calculation of the available motor torque M_{avl} allows to develop a relation between operating conditions and the resulting net torque:

$$\tilde{M} = M_{avl} - M_{req} \tag{8}$$

with $\tilde{M}$ as the net torque for the individual feed axis assembly.

In this context the net torque refers to the difference between torque delivered by the motor and estimated losses because of friction and acceleration loads. The net torque

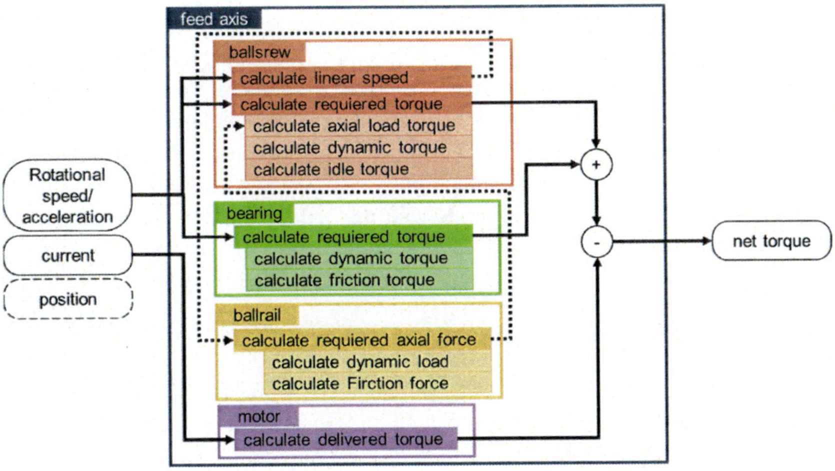

Fig. 4. Modular physics-based model of the feed axis and its components

enables the determination of the resulting force on the feed axis through the following equation.

$$F_{KGT} = \tilde{M} \cdot \frac{2\pi}{h} \tag{9}$$

The monitoring of this force allows the tracking of loads on the feed axis over its lifetime and thereby a more precise estimation of its condition and remaining lifetime. Since the model is physics based and purely analytical the computational power needed for its implementation is very low, thus allowing future implementation on the same Raspberry Pi single board computer used for the actual measurements allowing for a compact and easy integrable system.

7 Conclusion and Outlook

The system presented in this work provides a low-tech and low-cost possibility to gain access to motor current data. Because of the minimal effort needed for installation without the need for electrical rewiring, the approach is especially well suited for retrofit applications. This enables high resolution motor current monitoring in cases, where access to such data is restricted, either by proprietary or licensing limitations, or by technical limitations like old system architectures, that have no inherent capabilities to access such data or have no convenient data interfaces. The implementation of a physics-based model of the feed axis further enriches the informational value that can be provided.

However, while the effort for installation as well as the actual system cost is very low, the effort to build the physical model is rather high. This is a key issue for all applications utilizing current data to infer other parameters in feed axis and will be addressed in future works.

Another possibility for future use is the adaptation for anomaly detection schemes. Anomaly detection on motor-current data is a well-researched area [18]. Since the resolution of the measured current data is much higher than that of PLC-data typically used in such approaches it is reasonable to assume, that results can be improved.

In future works the first planned step in the development of this system is to run experiments with axial force sensing to validate the physical modelling approach outlined in Sect. 6. In addition, the physical model can be further refined by using the estimated positional values to allow for positional dependencies of axis stiffness.

References

1. Verl, A., Heisel, U., Walther, M., Maier, D.: Sensorless automated condition monitoring for the control of the predictive maintenance of machine tools. CIRP Ann. **58**(1), 375–378 (2009)
2. Han, Y., Song, Y.H.: Condition monitoring techniques for electrical equipment-a literature survey. IEEE Trans. Power Deliv. **18**(1), 4–13 (2003)
3. Zhou, W., Habetler, T.G., Harley, R.G.: Bearing condition monitoring methods for electric machines: a general review. In 2007 IEEE International Symposium on Diagnostics for Electric Machines, Power Electronics and Drives, pp. 3–6. IEEE (2007)
4. Schoen, R.R., Habetler, T.G.: A new method of current-based condition monitoring in induction machines operating under arbitrary load conditions. Electr. Mach. Power Syst. **25**(2), 141–152 (1997)
5. Zhang, Z., Wu, X., Liu, T., Liu, X.: Fault diagnosis of planetary gear backlash based on motor current and Fisher criterion optimized sparse autoencoder. Proc. Inst. Mech. Eng. Part C: J. Mech. Eng. Sci. 09544062211070160 (2022)
6. Corne, B., Knockaert, J., Desmet, J.: Misalignment and unbalance fault severity estimation using stator current measurements. In: 2017 IEEE 11th International Symposium on Diagnostics for Electrical Machines, Power Electronics and Drives (SDEMPED), pp. 247–253. IEEE (2017)
7. Nguyen, T.L., Ro, S.K., Park, J.K.: Study of ball screw system preload monitoring during operation based on the motor current and screw-nut vibration. Mech. Syst. Signal Process. **131**, 18–32 (2019)
8. Jamshidi, M., Chatelain, J.F., Rimpault, X., Balazinski, M.: Tool condition monitoring based on the fractal analysis of current and cutting force signals during CFRP trimming (2022)
9. Gangsar, P., Tiwari, R.: Signal based condition monitoring techniques for fault detection and diagnosis of induction motors: a state-of-the-art review. Mech. Syst. Signal Process. **144**, 106908 (2020)
10. Sato, R.: 3233 Wear estimation of ball screw and support bearing based on servo signals in feed drive system. In: Proceedings of International Conference on Leading Edge Manufacturing in 21st century: LEM21 2011.6, pp. _3233–1_. The Japan Society of Mechanical Engineers (2011)
11. Liu, X., Mao, X., He, Y., Liu, H., Fan, W., Li, B.: A new approach to identify the ball screw wear based on feed motor current. In: Proceedings of the International Conference on Artificial Intelligence and Robotics and the International Conference on Automation, Control and Robotics Engineering, pp. 1–5 (2016)
12. Yang, Q., Li, X., Wang, Y., Ainapure, A., Lee, J.: Fault diagnosis of ball screw in industrial robots using non-stationary motor current signals. Procedia Manufacturing **48**, 1102–1108 (2020)

13. Gönnheimer, P., Karle, A., Mohr, L., Fleischer, J.: Comprehensive machine data acquisition through intelligent parameter identification and assignment. Procedia CIRP, Elsevier, S. 720–725 (2021). https://doi.org/10.1016/j.procir.2021.11.121
14. Schröder, D., Böcker, J.: Elektrische Antriebe-Regelung von Antriebssystemen, vol. 2, pp. 978-3540896128. Springer, Berlin (2009)
15. Fuest, K., Döring, P.: Elektrische Maschinen und Antriebe. Wiesbaden. Vieweg+ Teubner Verlag, Germany (2000)
16. Imiela, J.: Verfügbarkeitssicherung von Werkzeugmaschinenachsen mit Kugelgewindetrieb durch modellbasierte Verschleissüberwachung. Berichte aus dem IFW, Hannover, Band 01/2006, Produktionstechnisches Zentrum GmbH, ISBN 3-939026-04-2, 164 S (2006)
17. Matevosyan, R.: Control vectorial del par motor de un motor brushless. Doctoral dissertation, Universitat Politècnica de València (2021)
18. Netzer, M., Palenga, Y., Gönnheimer, P., Fleischer, J.: Offline-online pattern recognition for enabling time series anomaly detection on older NC machine tools. J. Mach. Eng. Ed. Institution of the Wroclaw Board of Scientific Technical Societies Federation, S. 98–108. https://doi.org/10.36897/jme/132248

Measurement Setup and Modeling Approach for the Deformation of Robot Bodies During Machining

L. Gründel(✉), J. Schäfer, S. Storms, and C. Brecher

Laboratory for Machine Tools and Production Engineering (WZL) of RWTH Aachen University, Aachen, Germany

l.gruendel@wzl.rwth-aachen.de

Abstract. Conventional industrial robots (IR) represent a cost-effective machining alternative for large components. However, due to the serial kinematics and the resulting high tool deflections, they usually lack precision. Model-based simulation and control methods are used to increase the accuracy of IR regarding both planning and the process itself. The majority of the applied models include the compliances of the gears and bearings but neglect the deformations of the manipulator bodies. This paper introduces an approach to directly measure and evaluate the deformation of robot bodies in the presence of process forces. The measurement setup contains multiple Integral Deformation Sensors (IDS), which provide the change of length due to deformations of the respective body. Subsequently, the measurements are fed to a beam model (BM), which calculates the body's 3D Cartesian deflections. The presented approach is validated by static tensile tests on a conventional six-degree-of-freedom (DOF) robot manipulator.

Keywords: Robot machining · Deformation measurements · Beam theory

1 Introduction

Especially for small and medium-sized enterprises IR represent a lower investment risk compared to conventional machine tools. Furthermore, they offer flexibility in both positioning and applications and operate in wide workspaces. Therefore, robot machining has become more and more relevant for research and industry [1]. Nonetheless, IR exhibit a significantly higher compliance at the Tool Center Point (TCP) compared to machine tools [2] due to:

- non-preloaded drivetrains,
- space-optimized gearboxes with lower stiffness and larger backlash,
- more compliant components as bearings and robot bodies and
- an unfavorable mass distribution due to the serial structure.

The higher compliance at the TCP results in lower machining quality and an increasing tendency to chatter, which negatively affects the process stability [2].

M. Liewald et al. (Eds.): WGP 2022, LNPE, pp. 324–334, 2023.
https://doi.org/10.1007/978-3-031-18318-8_34

In addition to dedicated robot designs [3] or optimizations such as direct encoders at the joints [4], various model-based compensation approaches were developed over the recent years (cf. *Chapter 2*). The key element of any model-based compensation approach is the compliance model of the IR, which simulates the robot deflections due to process forces at the TCP. Since the main cause for deflections at the TCP is usually attributed to the stiffness of drivetrains and bearings, they are widely investigated metrologically and integrated in existing compensation approaches. The behavior of the bodies, on the other hand, is typically cumulated with the bearings or assumed to be ideally rigid [5, 6]. In addition, common measurement setups tend to identify the different stiffness parameters of the components together, which leads to errors.

The main contribution of this paper is the presentation of a measurement setup and modeling approach, which allows the decoupled measurement and simulation of the 3D deformations of beam-like robot bodies. Furthermore, the swing deformation is integrated into the compliance model via an extension of the Virtual Joint Method (cf. *Chapter 2*). The setup, containing IDS as described in [7] and [8], is validated and evaluated based on experiments on a conventional IR.

The paper is structured as follows: First, we discuss the state of the art regarding compensation, modeling and stiffness parameter identification approaches in Chapter 2. Then the BM is described in Chapter 3. Afterwards, the measurement setup is presented (cf. Chapter 3), followed by a validation and evaluation of the approach in Chapter 5. Finally, the results are summarized and an outlook on further research activities is given in Chapter 6.

2 State of the Art

In [9] a model-based process planning approach for robot machining is presented, which allows the process planner to avoid critical cutting parameters. The method is based on a static stiffness model. SCHNOES and ZAEH extend the stiffness model with a process force model. Hereby, the optimal workpiece placement and process parameters are determined and compensation offsets are calculated [10]. In [11] the authors present a model-based feed-forward control to compensate force-induced deviations at the TCP. Here, the model is based on the equations of motion derived from the Lagrangian equation coupled with a stiffness model. klimchik et al. Apply both, online and offline compensation, in order to increase the robots accuracy during machining [12]. The experiments with a KUKA KR270 showed a decrease of the maximum deviation by more than 90%.

In general, the stiffness respectively compliance models can be divided into three main groups [13].

- Finite Element Analysis (FEA)
- Structural Matrix Analysis (SMA)
- Virtual Joint Method (VJM).

FEA, as the most precise method, is mostly used in the final design phase of the IR, since the mesh fitting requires a lot of computation power. SMA follows the concept of FEA, but uses larger elements rather than finite elements. For example, the arm parts are represented as flexible 3D beams, which significantly reduces the computational effort. VJM is the most widely used approach for disturbance compensation in robot-based machining processes. It is based on the extension of the rigid multibody model by virtual joints describing the elastic deformations of the links, joints and actuators [13, 14].

Although the compensation of the static displacements is given a high priority in all the approaches mentioned above, the stiffness of the bodies is usually neglected. The models mostly include torsional and tilting stiffnesses of the joints, which can be identified in static tensile tests [5, 6]. Here, the force is applied via a pneumatic cylinder or a tension rod and the displacement is measured tactilely or via laser trackers. While these tests allow a proper excitation of the compliance parameters, the routine can only measure all parameters at once and therefore implies coupling errors between axes. Besides individual approaches such as [15], there are only a few methods that identify joint stiffnesses in a decoupled manner.

In [16] the beam-like robot bodies—the so-called swing (S) between axis 2 and 3 and the arm (A) between axis 3 and the wrist—are approximated with two ideal rigid bars and three torsion springs halfway along the body, respectively. In contrast to the widespread assumption that the robot bodies are negligible because they do not affect the total compliance of the IR, measurements in [16] show that the beam-like bodies (S and A, cf. Table 1) exhibit compliances in the same order of magnitude as the other components. RÖSCH identified the shown values with a static tensile test as described above using 3D-laser-scanning-vibrometry. The described measurement setup offers the identification of the bodies but also relies on coupled measurements of all components, which lead to coupling effects and therefore errors.

Table 1. Identified stiffness parameters of a KUKA KR240 in [Nm/rad], acc. to [16]

Joint	1	2	S	3	A	4	5	6
Bearing c_α	1,4e7	1,5e7	7,7e6	4,1e6	3,6e6	3,9e6	3,7e6	3,7e6
Bearing c_β	1,4e7	1,5e7	7,0e6	4,1e6	2,9e6	3,9e6	3,7e6	3,7e6
Gear c_γ	5,4e6	8,7e6	1,1e7	5,2e6	1,7e7	1,0e6	1,2e6	3,8e8

Summarizing the modeling and parameter identification approaches from the literature, there is no separate method for identification and validation of the link's stiffnesses. In addition, the impact of link deflections for any loading case was only evaluated by coupled measurements of the whole IR rather than for separate measurements at the links.

One method for separate measurements are IDS. In the field of machine tools, they are mainly used to measure thermal deformations as in [7] and [8]. The measuring principle of an IDS relies on a reference rod that is mounted on the surface of the machine structure (cf. Fig. 1). By supporting this rod with a fixed and a loose bearing, the rod can move axially in the loose bearing when the structure is deformed under an external load. A length gauge with a tactile measuring tip is attached to the structure on the side of the loose bearing to measure the displacement of the rod. In this way, the translational deformation of the structure between the attachment points of the IDS can be captured along the rod.

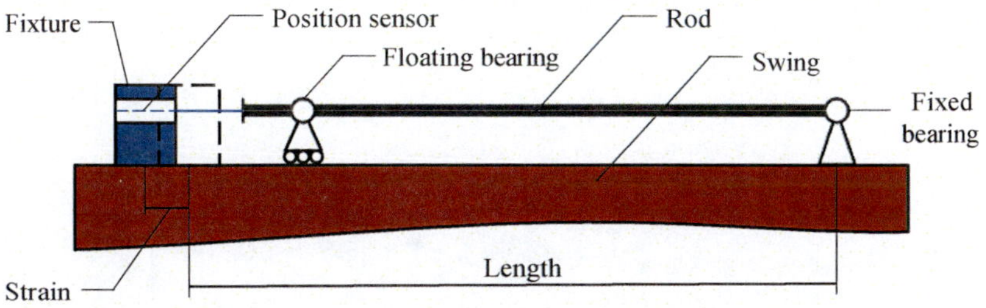

Fig. 1. Structure of an IDS

In the following chapter, the modeling approach is explained, followed by the experiments, which will present the usage of IDS to identify stiffness parameters.

3 Compliance Modeling

With the strain or contraction along its length each IDS provides a one-dimensional information about the deformation field of the machine structure. The detection of complex deformations due to tension, compression, bending, shear and torsion requires the installation of several IDS. In order to precisely calculate the deformation field of the structure based on measured IDS data, a suitable mechanical deformation model and a sufficient number of well-positioned sensors are necessary. The expected deformational behavior and the demands on accuracy determine the choice of the model. The required number of sensors depends on the model.

Since the swing can be considered as a beam-shaped component, a beam model is chosen to calculate the deformation field. Comparing bending stiffness and shear stiffness, the swing can be modeled based on the Euler-Bernoulli theory acc. to [17] (cf. Fig. 2). Therefore, the three-dimensional deformation field can be described using the deformation vector u acc. to (1).

$$\boldsymbol{u}(\boldsymbol{x}) = \begin{bmatrix} u_1 \\ u_2 \\ u_3 \\ \varphi_1 \\ \varphi_2 \\ \varphi_3 \end{bmatrix} = \begin{bmatrix} \overline{u}_1(x_1) - x_2 \cdot \varphi_3(x_1) + x_3 \cdot \varphi_2(x_1) \\ \overline{u}_2(x_1) \\ \overline{u}_3(x_1) \\ 0 \\ \varphi_2(x_1) \\ \varphi_3(x_1) \end{bmatrix}, \quad \text{with} \tag{1}$$

$$\varphi_3(x_1) = \overline{u}_2'(x_1) = \frac{\partial \overline{u}_2(x_1)}{\partial x_1} \tag{2}$$

$$\varphi_2(x_1) = -\overline{u}_3'(x_1) = -\frac{\partial \overline{u}_3(x_1)}{\partial x_1}. \tag{3}$$

where u_1, u_2 and u_3 describe the translational and φ_1, φ_2 and φ_3 describe the rotational displacements in the coordinate system corresponding to Fig. 3. $\overline{u}_1$ represents a longitudinal shift due to tension or compression, while $\overline{u}_2$ and $\overline{u}_3$ are transverse shifts because of bending. While φ_1 describes a rotation around the longitudinal axis due to torsion, φ_2 and φ_3 represent rotations around the respective transverse axis due to bending.

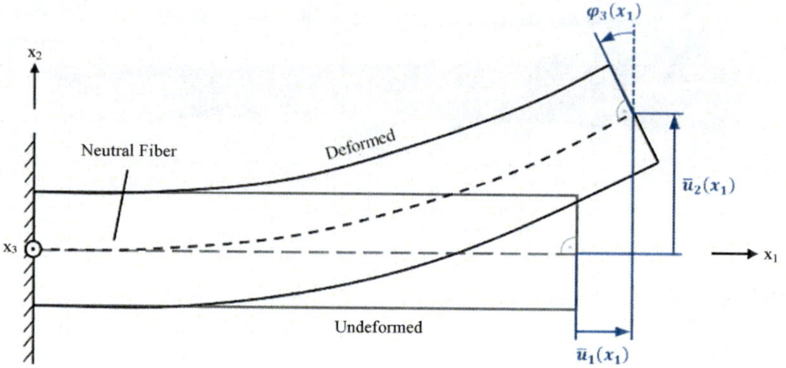

Fig. 2. Euler-Bernoulli beam model

The extension of the deformation field in (1) by deformations due to torsion allows a more precise prediction of the deformation behavior. The swing has a thin-walled closed cross-section. With this cross-sectional profile, a rotation around its longitudinal axis (torsion) and a translational displacement of the cross-section along the longitudinal axis (warping) occur as a result of torsional load. The swing can be modeled as a cantilever beam, with a fixed clamping at joint 2. Thus, torsion and warping are prevented at this point. At the free end, however, both deformations can occur unhindered. In order to consider torsion and warping in the deformation vector it is necessary to integrate additional terms. For this purpose, the kinematic assumptions for the deformation field within the Saint-Venant torsion theory can be used [18]. The extended deformation vector u_{est} reads:

$$\boldsymbol{u}_{\text{ext}}(\boldsymbol{x}) = \begin{bmatrix} \overline{u}_1(x_1) - x_2 \cdot \varphi_3(x_1) + x_3 \cdot \varphi_2(x_1) + \Psi(x_2, x_3) \cdot \kappa_1 \\ \overline{u}_2(x_1) - x_3 \cdot \varphi_1(x_1) \\ \overline{u}_3(x_1) + x_2 \cdot \varphi_1(x_1) \\ \varphi_1(x_1) \\ \varphi_2(x_1) \\ \varphi_3(x_1) \end{bmatrix}, \text{ with} \quad (4)$$

$$\kappa_1(x_1) = \varphi_1'(x_1) = \frac{\partial \varphi_1(x_1)}{\partial x_1}. \quad (5)$$

where Ψ is the warping function and κ_1 is the relative twist following (5), which is assumed to be constant. The coordinate system CF_{BM}, which the deformation vector of the swing refers to, is aligned with the averaged neutral fiber (ANF) in the middle sector of the swing as shown in Fig. 3. In order to integrate the swing deformation into the compliance model, the deformation is calculated at the target point, which marks the connection between swing and arm (cf. Fig. 3).

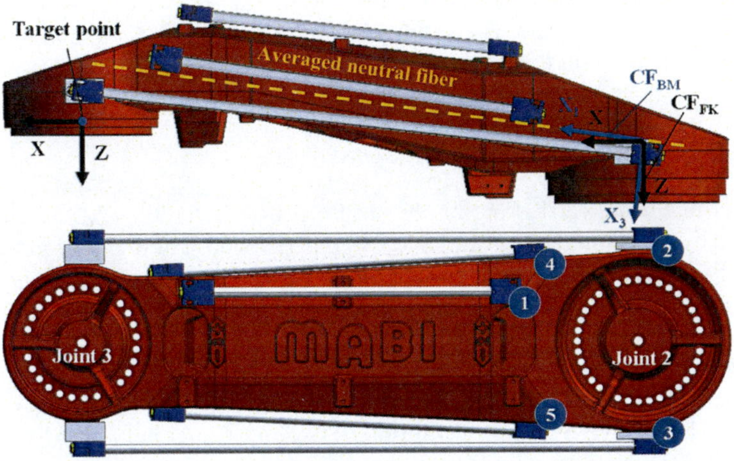

Fig. 3. Positioning of the IDS on the swing and coordinate system definition

The deformation behavior of the swing is determined using the extended Euler-Bernoulli BM in (4). Therefore, the four independent deformational DOF $\overline{u}_1$, $\overline{u}_2$, $\overline{u}_3$ and φ_1 are described as functions of the measured IDS data. For each of the mentioned deformational DOF a polynomial of third order is set up acc. to [8]. Subsequently, the derivations of these polynomials φ_2, φ_3 and κ_1 can be calculated acc. to (2), (3) and (5), as soon as the coefficients of the polynomials are identified. In order to calculate the coefficients of the polynomials mechanical boundary conditions are required. These are derived from the assumptions for cantilever beams and the IDS data. Since each IDS is aligned along the ANF, the relationship between the IDS_n data $_n$ and the translational deformation u_1 in (6) applies,

$$IDS_n = \int_{x_{1B,IDS_n}}^{x_{1E,IDS_n}} \frac{\partial u_1(\boldsymbol{x})}{\partial x_1} dx_1 \tag{6}$$

where $x_{1B,IDSn}$ and $x_{1E,IDSn}$ define the attachment points of an IDS. Hence, for the four independent deformational DOF four IDS are required, while a fifth sensor is installed to provide redundant information. Figure 3 shows the positioning of the deployed IDS. Having identified the coefficients, the polynomials $\overline{u}_1$, $\overline{u}_2$, $\overline{u}_3$ and φ_1 and their derivations φ_2, φ_3 and κ_1 can be determined. Subsequently, the deformation vector for the target point is calculated with the respective IDS data following (4).

The deformations of the swing at the defined target point can be interpreted as a translational and rotational displacement of the coordinate system at joint 3. First, the translational displacement is carried out. Therefore, the deformations calculated in

CF_{BM} are transformed into CF_{FK} (cf. Fig. 3). The body-fixed coordinate system CF_{FK} is aligned to the coordinate system at joint 3. Hence, their orientation coincides for any axis configuration and loading case. Extending the forward kinematics of the robot by the additional transformations, the deformation of the swing is integrated into the compliance model. Hence, the robot deflections at the TCP due to the compliance of the swing can be simulated. The validity of the measurement setup and the modeling approach is evaluated in the following chapters.

4 Experiments

The experiments are carried out on the six DOF IR MAX100 by MABI Robotic AG controlled by the Numerical Control (NC) Sinumerik 840d sl by Siemens AG (cf. Fig. 4). As already mentioned in Chapter 3, there are five IDS mounted on the swing of the IR. The force is applied with a Load Displacement Measuring Device (LDMD). The device applies compressive and tensile force curves at the TCP and allows the simultaneous measurement of the displacement and the force in the respective direction. In the following chapters, the applied forces are always defined in the base coordinate system shown in Fig. 4. The change of length within the IDS is measured with an Acanto position sensor by HEIDENHAIN GmbH.

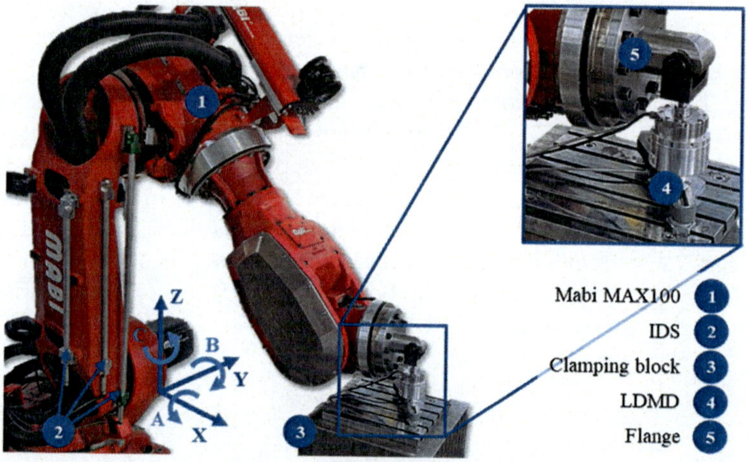

Fig. 4. Measurement setup

The IDS data is measured with a standard Industrial PC (IPC) by Beckhoff Automation GmbH & Co. KG, the IR configuration with the NC and the force and displacement with a National Instruments PC. During the measurements the data is pushed to a time series database (TimescaleDB) and is synchronized in time.

For the validation of the measurement setup and the BM and for the evaluation of the swing's influence on the total compliance of the IR eight measurement configurations were chosen (cf. Table 2). The poses cover the workspace in front of the workspace and enable different force directions. In the following section the approaches are validated and evaluated.

Table 2. Measurement poses

Pose No.	Axis No.						Force direction
	1	2	3	4	5	6	
1	−21	33	40	22	−73	−96	−Y
2	−2	11	69	1	−56	0	−X
3	−1	17	36	1	−55	0	Z
4	17	54	36	−17	−94	90	Y
5	−3	33	35	−2	20	93	Y
6	−26	5	53	2	31	24	−X
7	−20	16	15	0	59	20	Z
8	−34	32	11	0	47	124	−Y

5 Validation and Evaluation

First, the validity of the measurement setup is checked by superimposing the force data with the respective IDS data. In Fig. 5, the force and IDS data is shown for pose 1, where the force is applied in –Y direction of the base coordinate system. The force curve shows that the tensile area was not optimally reached. This behavior was also observed in other measurement poses and is attributed to the difficult-to-handle zeroing of the LDMD.

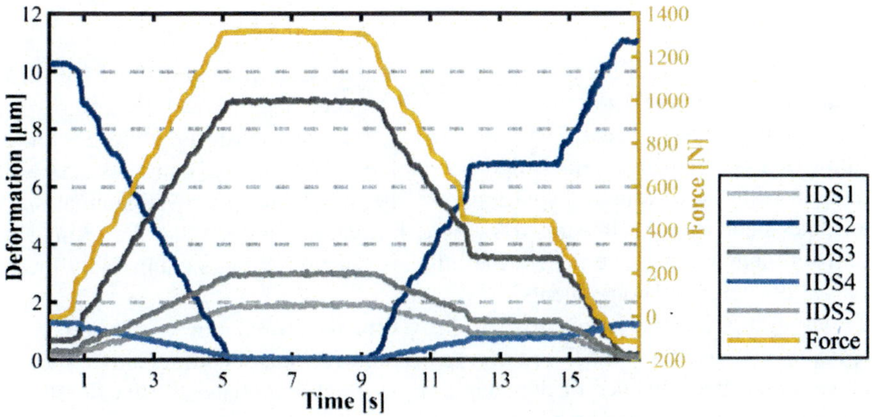

Fig. 5. Validation of the measurement setup with the respective deformations of the IDS according to the force applied in –Y direction in pose 1

Furthermore, the waviness of the force ramp as it rises and falls is attributed to friction in the LDMD housing. Nevertheless, the shown plots suggest a strong correlation between the force applied at the TCP and the measured deformations. Even the mentioned waviness is reproduced in the IDS data. As the IDS pairs 2, 3 and 4, 5 are positioned on opposite sides of the swing, the respective data shows antiproportional behavior. In addition, the shorter the IDS rod is, the less deformation is measured (cf. Fig. 3). Summing up, the measurement setup allows to excite the swing's deformations in the expected manner.

In Fig. 6, the resulting Cartesian displacements calculated with the presented BM are validated for pose 1. The data refer to coordinate system CF_{FK}, as shown in Fig. 3. As expected, the translational displacements in Z are the highest as the force is applied in the same direction at the TCP. The relatively small displacement in A suggests that the swing is stiffer in torsional direction. The bending of B is higher than in C direction, which seems intuitively correct after considering the swing model (cf. Fig. 3).

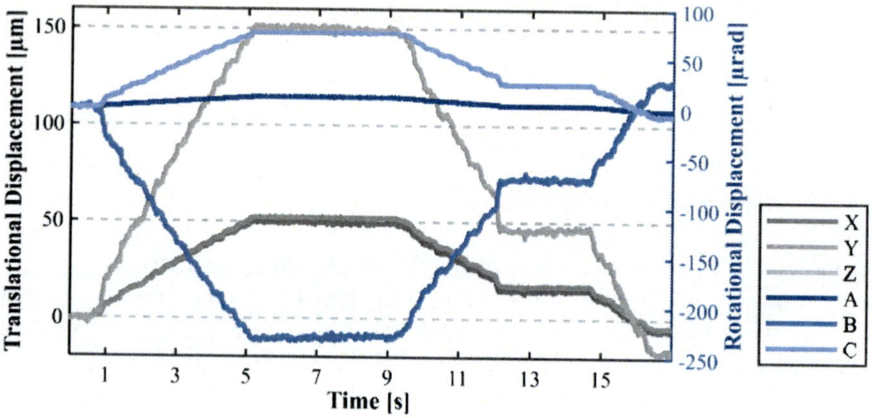

Fig. 6. Calculation of the Cartesian Displacements of the swing with the beam model in pose 1

Finally, the models for the total compliance at the TCP with and without the swing deformation are compared with the measured displacements. In Fig. 7 this comparison is shown exemplary for poses 1, 2, 6 and 8. As expected the overall compliance is higher in Y direction. Apart from a force-proportional offset, the simulated data shows a strong correlation with the measured displacements. The marginally improved simulation of the displacement with the integrated swing deformation suggests that the swing was successfully integrated. Nevertheless the influence of the swing is comparably low to the rest of the deformed components.

Summing up, the measurement setup proved to be a valid way to investigate the behavior of the swing separately. However, in order to make a sufficiently validated statement about the influence of the swing on the total compliance of the IR, further experiments should be carried out.

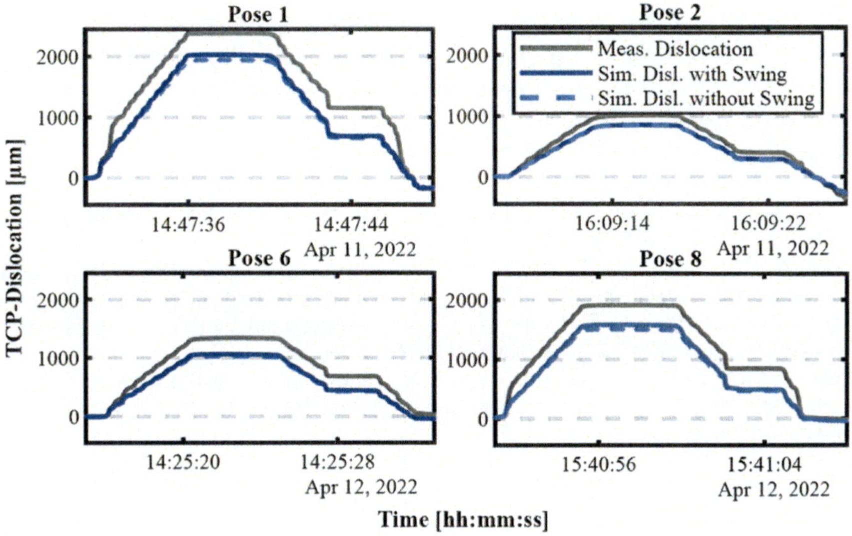

Fig. 7. Comparison of the measured and simulated dislocation of the TCP with and without the swing deformations included

6 Summary and Outlook

Due to an increasing relevance of conventional IR for machining tasks, model-based compensation approaches were developed during the past years. In contrast to the drivetrains and bearings the deformation of the robot bodies is often neglected and therefore not compensated. So far, there is no measurement setup to evaluate the body's influence on the total compliance separately.

Therefore, this paper presents both a measurement setup and a modeling approach for the beam-like bodies of an IR. The model is derived following the Euler-Bernoulli beam theory and the measurements are carried out with five IDS and a LDMD to apply forces and measure deflections at the TCP. The approach is validated on the swing of a conventional six DOF IR. The results show the intendent behavior of the measured deformation according to the applied forces. In addition, the BM shows valid results for the Cartesian deformation and the simulated total compliance at the TCP is marginally improved by integrating the swing. Nevertheless, further experiments will be carried out for a final evaluation.

Apart from smaller adjustments to the measurement setup and further stiffness experiments, the 6×6 compliance matrix of the swing is required in order to predict the deformation related to the applied force without IDS data. After a final evaluation of the developed prediction model, the compensation approach will be integrated as a feed-forward control.

Acknowledgements. The IGF-project 21926 N/2 (RoSiKo) of the research association FVP (Forschungsver-einigung Programmiersprachen für Fertigungseinrichtungen e.V.) was supported via the AiF within the funding program "Industrielle Gemeinschaftsforschung und—entwicklung

(IGF)" by the Federal Ministry of Economic Affairs and Climate Action (BMWK) due to a decision of the German Parliament. Furthermore, we gratefully acknowledge the support of the MABI Robotic AG and the support by D. Tipura and D. Vogel.

References

1. Verl, A., Valente, A., Melkote, S., et al.: Robots in machining. CIRP Ann. **68**, 799–822 (2019). https://doi.org/10.1016/j.cirp.2019.05.009
2. Pan, Z., Zhang, H., Zhu, Z., et al.: Chatter analysis of robotic machining process. J. Mater. Process. Technol. **173**, 301–309 (2006). https://doi.org/10.1016/j.jmatprotec.2005.11.033
3. Denkena, B., Bergmann, B., Lepper, T.: Design and optimization of a machining robot. Proc. Manuf. **14**, 89–96 (2017). https://doi.org/10.1016/j.promfg.2017.11.010
4. Möller, C., Schmidt, H.C., Koch, P., et al.: Machining of large scaled CFRP-Parts with mobile CNC-based robotic system in aerospace industry. Proc. Manuf. **14**, 17–29 (2017). https://doi.org/10.1016/j.promfg.2017.11.003
5. Cordes, M., Hintze, W.: Offline simulation of path deviation due to joint compliance and hysteresis for robot machining. Int. J. Adv. Manuf. Technol. **90**(1–4), 1075–1083 (2016). https://doi.org/10.1007/s00170-016-9461-z
6. Dumas, C., Caro, S., Cherif, M., et al.: Joint stiffness identification of industrial serial robots. Robotica **30**, 649–659 (2011). https://doi.org/10.1017/S0263574711000932
7. Baum, C., Brecher, C., Klatte, M., et al.: Thermally induced volumetric error compensation by means of integral deformation sensors. Proc. CIRP **72**, 1148–1153 (2018). https://doi.org/10.1016/j.procir.2018.03.045
8. Brecher, C., Klatte, M., Lee, T.H., et al.: Metrological analysis of a mechatronic system based on novel deformation sensors for thermal issues in machine tools. Proc. CIRP **77**, 517–520 (2018). https://doi.org/10.1016/j.procir.2018.08.245
9. Lienenlüke, L., Gründel, L., Storms, S. et al.: Model-based process planning for milling operations using industrial robots. In: 2018 3rd International Conference on Control and Robotics Engineering (ICCRE), pp 37–44. IEEE (2018)
10. Schnoes, F., Zaeh, M.F.: Model-based planning of machining operations for industrial robots. Procedia CIRP **82**, 497–502 (2019). https://doi.org/10.1016/j.procir.2019.04.331
11. Gründel, L., Lienenlüke, L., Storms, S., et al.: Robot-based milling operation. Machine learning algorithm for a model-based feed-forward torque control. WT Werkstattstechnik **109**, 352–357 (2019)
12. Klimchik, A., Bondarenko, D., Pashkevich, A. et al.: Compensation of tool deflection in robotic-based milling (2012)
13. Pashkevich, A., Klimchik, A., Chablat, D.: Enhanced stiffness modeling of manipulators with passive joints. Mech. Mach. Theory **46**, 662–679 (2011). https://doi.org/10.1016/j.mechmachtheory.2010.12.008
14. Klimchik, A., Wu, Y., Caro, S., Furet, B., Pashkevich, A.: Accuracy Improvement of robot-based milling using an enhanced manipulator model. In: Ceccarelli, M., Glazunov, V.A. (eds.) Advances on Theory and Practice of Robots and Manipulators. MMS, vol. 22, pp. 73–81. Springer, Cham (2014). https://doi.org/10.1007/978-3-319-07058-2_9
15. Pfeiffer, F., Hölzl, J.: Parameter identification for industrial robots. In: Proceedings of 1995 IEEE International Conference on Robotics and Automation, pp. 1468–1476. IEEE (1995)
16. Roesch, O.: Model-based on-line compensation of path deviations for milling robots. AMR **769**, 255–262 (2013). https://doi.org/10.4028/www.scien-tific.net/AMR.769.255
17. Timošenko, S.P., Goodier, J.N.: Theory of elasticity, 3rd edn. Engineering societies monographs. McGraw-Hill, New York (1951)
18. Mikeš, K., Jirásek, M.: Free warping analysis and numerical implementation. AMM **825**, 141–148 (2016). https://doi.org/10.4028/www.scien-tific.net/AMM.825.141

Determination of Tool and Machine Stiffness Based on Machine Internal and Quality Data

M. Loba(✉), C. Brecher, M. Fey, F. Roenneke, and D. -F. Yeh

Laboratory for Machine Tools and Production Engineering (WZL), RWTH Aachen University, 52074 Aachen, Germany
m.loba@wzl.rwth-aachen.de

Abstract. During machining, process forces cause form deviations on the workpiece depending on the interacting stiffnesses of all components involved. In order to avoid tolerance violations, it must be ensured that the resulting deflection of tool and workpiece are within the tolerance limit. At the same time, process parameters must be selected in such a way that a cost-efficient and productive production is possible. Stiffness can be determined experimentally by applying a specific force and measuring the resulting deformation. Due to the large variety of tools and tool holders as well as their combinations, it is generally too expensive to determine the stiffness in this way. During quality control, the workpiece geometry is measured, for example, with coordinate measuring Machines (CMM), so that resulting dimensional and form deviations can be determined. Furthermore, there exist approaches to predicting process forces based on machine internal data such as motor currents. In this paper, an approach is presented that enables a determination of the resulting system stiffness at the TCP based on machine-internal and quality data. During the milling process, machine internal data is recorded and a dexel-based material removal simulation (MRS) is performed. Therefore, the estimated force vector is calculated for each dexel. After the simulation, the virtual part is compared with the real part measured by a CMM to determine the deviation vector. By solving a linear equation system, the resulting stiffness is calculated. To reduce the influence of other effects, they are modeled in the MRS or usually reduced so that they are negligible.

Keywords: Stiffness modeling · Digital process chain · Quality data feedback

1 Introduction

With a share of over 40% of the gross output, material consumption is the highest cost fraction in Germany's manufacturing sector [1]. For this reason, resource efficiency is essential for the competitiveness of German companies, not only from an ecological but also from an economic point of view [2]. In order to prevent a waste of resources, a first part right production must be ensured. Quality predictive machining simulations enable a model-based prediction of the fulfilment of workpiece tolerances already in the process planning stage [3, 4]. At the same time, it is possible to determine the workpiece quality process parallel based on machine internal data and a material removal simulation (MRS)

M. Liewald et al. (Eds.): WGP 2022, LNPE, pp. 335–345, 2023.
https://doi.org/10.1007/978-3-031-18318-8_35

[5, 6]. Here, different influences such as clamping force deflections [7], thermal effects in the cutting zone [8], kinematic roughness [9] as well as the process force and its influence on the tool deflection [4] can be modeled [10]. To get trustworthy results, the models must be parametrized. Currently, measurements or expert knowledge are necessary for this. A solution that neither disturbs the running production nor requires complex modelling is therefore necessary. This paper presents an approach based machine-internal data and quality data enables the determination of tool stiffness in machine tools by using a MRS. For this purpose, process forces are predicted based on machine-internal data and contextualised with the help of MRS. Finally the tool stiffness is determined using the deviations between virtual and real workpiece.

2 State of the Art

2.1 Process Force Modelling

To model machining processes the prediction of process forces based on process parameters (e.g. cutting speed, depth or feed rate) is of essential importance. Process force models, such as those by Kienzle [11] or Altintas [12], are parameterized on empirically recorded data for particular workpiece-tool-material combinations. With the help of dynamometers, the process force can also be recorded. Dynamometers have a high frequency spectrum and measurement accuracy, but reduce the stiffness of the system by introducing additional compliance. To avoid these systematic disadvantages, there are already some approaches. Brecher et al. use spindle-integrated force sensors (SIFS), which determine process forces measuring displacement of the spindle bearings [5]. The SIFS does not negatively influence the stiffness of the overall system. At the same time, the proximity to the tool center point (TCP) allows accurate results to be achieved even in high-frequency range (up to 50 kHz). Postel et al. use acceleration sensors on the main spindle housing to predict process forces [13]. This approach is characterised above all by its fast implementability and low cost. Furthermore, Denkena et al. use strain gauges on the spindle slide to determine the process force as well as the tool stiffness [14]. All approaches have in common that the transfer functions between TCP and the sensor points must be known. The approaches mentioned allow for a tool-side estimation of the process force. By means of sensory zero-point clamping systems, the process force can also be predicted on the workpiece side [15]. These systems can be characterised by their rapid transferability to other machines. While the aforementioned solutions require an investment, approaches based on machine internal signals, such as axis currents or positions, offer a possibility for low-effort implementation. The sampling rate is usually limited by cycle time of the position control (~500 Hz), so that a mapping of dynamic effects is limited. Existing approaches for process force estimation by means of machine-internal data are often motor current-based [16] or are based on a prediction based on the difference of the linear and rotary measurement system [17]. These approaches often reach their limits, especially at low or reversing speeds (e.g. due to stiction); here, hybrid modelling using artificial neural networks can improve prediction [18, 19].

2.2 Stiffness Modelling

The resulting deviations due to process forces are always dependent on the interacting stiffnesses. Tool stiffness can be determined both experimentally and model-based. Experimentally, it is possible to determinate it using load-deformation-curves. A disadvantage of this method is that the measurement of all tool-holder-machine combinations is not economical. Approaches using substructure techniques circumvent this disadvantage. Here, the machine is initially measured with a dummy tool. The elements of the dummy model are decoupled up to the holder-machine interface. Afterwards the elements of the tool actual are coupled. The elements to be decoupled and coupled are often modelled as Bernoulli or Timoshenko beams [20, 21]. For an accurate stiffness prediction, in-depth knowledge of material and geometry of the tool, tool holder as well as the joints is necessary. Brecher et al. present in [22] different possibilities to simplify tool models. For example, tapered sections are modelled with a constant mean diameter and complex cutting edge geometries with an equivalent diameter. At the same time, a coupling of the entire tool by means of FE models [23] or pre-measured stiffness is also possible [24]. Furthermore, Denkena et al. determine the stiffness at the TCP by means of a soft collision and a process force measurement using strain gauges on the spindle slide [14].

3 Stiffness Prediction Based on Machine and Quality Data

3.1 General Concept

The general concept is shown in Fig. 1. Starting from a machining process, machine-internal data from the control (NC data stream) is sent to the MRS from [5]. In the MRS, the process force $\boldsymbol{F}_{cut}^{MCS}$ is determined in the machine coordinate system (MCS) by using the NC data and transformed to the tool coordinate system (TCS). Afterwards the dexel-based MRS generates a virtual workpiece. Besides the informations about the begin and end of the workpiece, each dexel also stores the actual axis position and the actual process force in the TCS $\boldsymbol{F}_{cut}^{TCS}$. The tool is modeled as a hull body represented by an Layered-Depth-Normal-Image (LDNI). By checking if a dexel is intersected by an LDNI a dexel gets cutted. The stiffness of the tool is initially assumed to be infinite. Besides the clamping force deflection is taking into account by a FE simulation [7]. Moreover, the geometric-kinematic machine error is measured previously and considered in the transformation calculation. To increase accuracy, of the tool stiffness prediction only features should be used where the deflection of the workpiece due to the process force is negligible. After production, the real workpiece is measured. By overlaying the real and virtual workpiece, the deviation vector $\boldsymbol{\varepsilon}_i^{WCS}$ is determined via the surface normals of the virtual workpiece. With the help of the stored axes positions the $\boldsymbol{\varepsilon}^{WCS}$ is transformed from the workpiece coordinate system (WCS) into the TCS. With the help of the force information of the virtual workpiece the tool stiffness is predicted by linear regression.

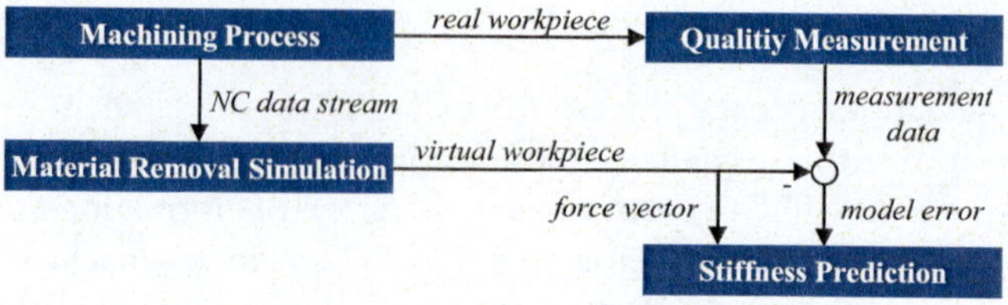

Fig. 1. General concept

3.2 Force Prediction

To determine the process force $\boldsymbol{F}_{cut}^{MCS}$ in the MCS, it is assumed that both the motor currents $\boldsymbol{I}_{Axis}$ and the difference between linear and rotary encoder $\boldsymbol{\Delta}_{Axis}$ are proportional to the process force $\boldsymbol{F}_{cut}^{MCS}$ [16, 17].

$$\boldsymbol{F}_{cut}^{MCS} \propto \boldsymbol{I}_{Axis} \propto \boldsymbol{\Delta}_{Axis} \tag{1}$$

In addition to the process force, axes jerk $\boldsymbol{j}_m$, acceleration $\boldsymbol{a}_m$, velocity $\boldsymbol{v}_m$, position-dependent $\boldsymbol{s}_m$ influences and static forces have an impact on $\boldsymbol{I}_{Axis}$ and $\boldsymbol{\Delta}_{Axis}$.

$$\boldsymbol{I}_{Axis} = \boldsymbol{K}_{I,F}\boldsymbol{F}_{cut}^{MCS} + \boldsymbol{I}_j(\boldsymbol{j}_1) + \boldsymbol{I}_a(\boldsymbol{a}_1) + \boldsymbol{I}_v(\boldsymbol{v}_1) + \boldsymbol{I}_{pos}(\boldsymbol{s}_1) + \boldsymbol{I}_{Stat} \tag{2}$$

$$\boldsymbol{\Delta}_{Axis} = \boldsymbol{K}_{\Delta,F}\boldsymbol{F}_{cut}^{MCS} + \boldsymbol{\Delta}_j(\boldsymbol{j}_1) + \boldsymbol{\Delta}_a(\boldsymbol{a}_1) + \boldsymbol{\Delta}_v(\boldsymbol{v}_1) + \boldsymbol{\Delta}_{pos}(\boldsymbol{s}_1) + \boldsymbol{\Delta}_{Stat} = \boldsymbol{s}_1 - \boldsymbol{s}_2 \tag{3}$$

With $\boldsymbol{s}_m$ as the axes position vector:

$$\boldsymbol{s}_e = (x_m, y_m, z_m)^T \begin{cases} m = 1 \rightarrow \text{Rotary Encoder} \\ m = 2 \rightarrow \text{Linear Encoder} \end{cases},$$

$$\boldsymbol{v}_m = \dot{\boldsymbol{s}}_m, \boldsymbol{a}_m = \ddot{\boldsymbol{s}}_m, \quad \text{and} \quad \boldsymbol{j}_m = \dddot{\boldsymbol{s}}_m$$

To eliminate the influences that are not due the process force an air cut is done. By loading the axes at standstill (see Fig. 3b) the proportionality constants $\boldsymbol{K}_{I,F}$ and $\boldsymbol{K}_{\Delta,F}$ are determined.

$$\boldsymbol{F}_{cut,I}^{MCS} = \frac{\boldsymbol{I}_{Axis} - \boldsymbol{I}_{Aircut}}{\boldsymbol{K}_{I,F}} = \boldsymbol{K}_{F,I} \cdot (\boldsymbol{I}_{Axis} - \boldsymbol{I}_{Aircut}) \tag{4}$$

$$\boldsymbol{F}_{cut,\Delta}^{MCS} = \frac{\boldsymbol{\Delta}_{Axis} - \boldsymbol{\Delta}_{Aircut}}{\boldsymbol{K}_{\Delta,F}} = \boldsymbol{K}_{F,\Delta} \cdot (\boldsymbol{\Delta}_{Axis} - \boldsymbol{\Delta}_{Aircut}) \tag{5}$$

With a kinematic transformation, the process force can be transformed into the TCS.

$$\boldsymbol{F}_{cut}^{TCS} = \boldsymbol{T}^{TCS,MCS}\boldsymbol{F}_{cut}^{MCS} \tag{6}$$

3.3 Stiffness Prediction

For the stiffness prediction $\boldsymbol{\varepsilon}_i^{WCS}$ is determined by the difference between the virtual (VP) and the real part (RP) for each point.

$$\boldsymbol{\varepsilon}^{WCS} = \boldsymbol{p}_{RP}^{WCS} - \boldsymbol{p}_{VP}^{WCS} \tag{7}$$

Afterwards $\boldsymbol{\varepsilon}^{WCS}$ is transformed into the TCS:

$$\boldsymbol{\varepsilon}^{TCS} = \boldsymbol{T}^{TCS,WCS}\boldsymbol{\varepsilon}^{WCS} \tag{8}$$

Now the linear equation system is formulated as:

$$\boldsymbol{F}_{cut}^{TCS} = \boldsymbol{K}\boldsymbol{\varepsilon}^{\mathrm{TCS}} \tag{9}$$

With the stiffness matrix

$$\boldsymbol{K} = \begin{pmatrix} k_{xx} & k_{xy} & k_{xz} \\ k_{yx} & k_{yy} & k_{yz} \\ k_{zx} & k_{zy} & k_{zz} \end{pmatrix} \tag{10}$$

Under consideration of Maxwell-Betti reciprocal work theorem:

$$k_{ij} = k_{ji} \tag{11}$$

(9) is determined with two error vectors. If more error vectors are available, (9) is solved by linear regression. Is just one error vector available is assumed that:

$$k_{ij} = 0 \,\forall\, i \neq j \tag{12}$$

4 Experimental Validation

4.1 Experimental Setup

For experimental validation, the specimen workpiece shown in Fig. 2 was machined from a C45 block on a Chiron FZ12S (see Fig. 3a). The workpiece has three reference surfaces A, B and C, for aligning the virtual with the real workpiece. The reference surfaces are machined in two steps to keep the process forces there as low as possible ($a_{e,ref,1}$ = 0.4 mm, $a_{e,ref,2}$ = 0.1 mm). Besides all features were down milled. To reduce thermal effects in the cutting zone coolant is used. The feed per tooth f_z, cutting depth a_p and width a_e used are listed for each feature in Table 1. In order to eliminate the influence of an incorrect measurement of the tool diameter or tool run out only surfaces are compared which have identically orientated surface normals. To produce the specimen workpiece end mills with diameters of 8, 12 and 16 mm in different holder types were used. The used combinations and the rotational speed n are listed in Table 2.

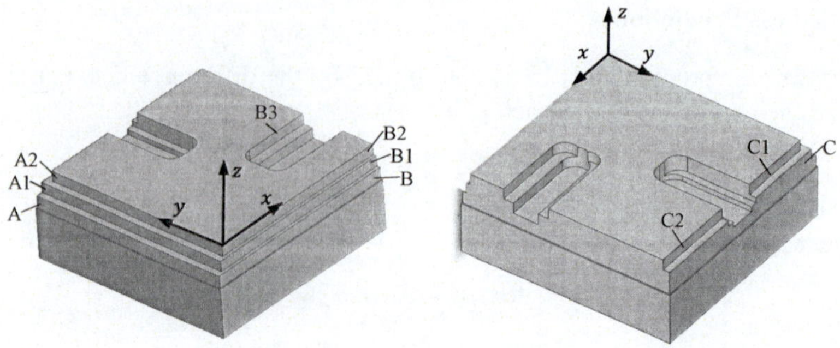

Fig. 2. Specimen workpiece

Table 1. Process parameter

	A1 (mm)	A2 (mm)	B1 (mm)	B2 (mm)	B3 (mm)	C1 (mm)	C2 (mm)
a_p	10	5	5	5	5	5	5
a_e	3.2	3.2	1.6	3.2	0.4·d	3.2	3.2
$f_{z,Ø8}$	0.034	0.034	0.034	0.034	0.034	0.026	0.017
$f_{z,Ø12}$	0.042	0.042	0.042	0.042	0.042	0.031	0.021
$f_{z,Ø16}$	0.075	0.075	0.075	0.075	0.075	0.056	0.038
Mark	+	×	Δ	∇	▷	□	◊

Table 2. Tool setup

Tool	Holder	$\frac{1}{min}$	l_{tool} (mm)	l_{free} (mm)
Fraisa Favora P8400391 d = 8 mm	Holex 304277 8 Weldon Chuck	6165	93.298	28.298
Fraisa Favora P8400501 d = 12 mm	Haimer A63.020.32 Collet Chuck	3960	138.742	38.742
Fraisa Favora P45317610 d = 16 mm	Schunk TENDO EC Ø20 Hydraulic Expansion Chuck	3380	129.157	49.157
		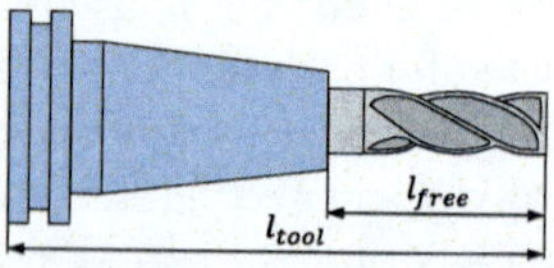		

4.2 Model Parametrization

To parametrize the force models load-deformation-curves are generated with steel dummys (see Fig. 3b). While inducing the force machine internal data is traced. With a correlation analysis, the signals from the force sensor and the machine internal control are aligned. Afterwards $K_{F,I}$ and $K_{F,\Delta}$ are determined by using linear regression. The results of one measurement is shown in Fig. 4. On the left side is the data parametrization for the motor current force model and on the right side for the difference of the position sensors. It is visible that I_{Axis} has a stronger hysteresis than Δ_{Axis}. This tends to make the force calculation less accurate. Forces are estimated too low during force increase and too high during force decrease. The results for $K_{F,I}$ and $K_{F,\Delta}$ are listed in Table 3. The other factors are compensated by an air cut.

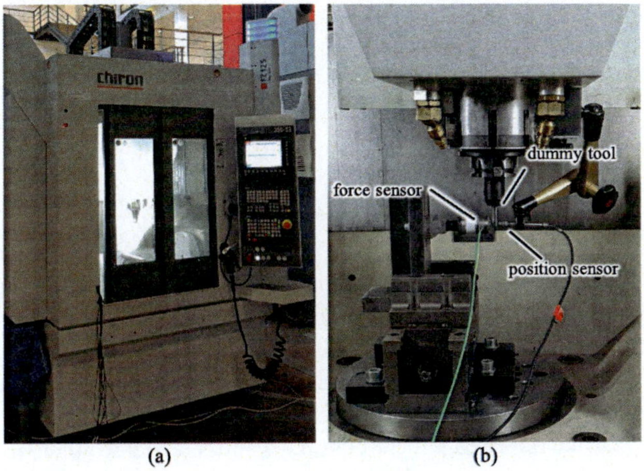

Fig. 3. Experimental machine (a) and measurement setup (b)

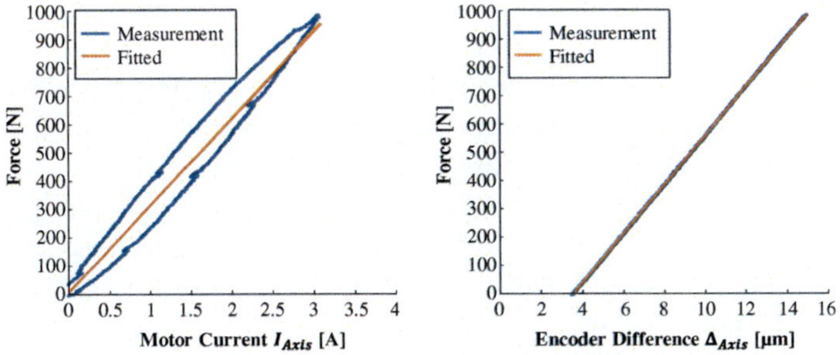

Fig. 4. Force model parametrization

4.3 Results

The specimen workpiece is produced three times for each tool. After machining the workpieces are measured on a Zeiss Contura CMM. The measurements were always

Table 3. Force model parameter

Factor	X-Axis	Y-Axis	Z-Axis
$K_{F,I}$	313 N/A	243 N/A	325 N/A
$K_{F,\Delta}$	86.1 N/μm	91.5 N/μm	106.0 N/μm

taken 2 mm below the highest engagement point of the tool. By using the 3D Normal Distribution Transform (3D-NDT) algorithm the transformation matrix for the alignment of the RP and VP is determined. Here only points from the reference surfaces are used. Afterwards all points of the VP are transformed and $\boldsymbol{\varepsilon}^{WCS}$ is determined, so that the stiffness can be predicted. The results for each force model, diameter, feature and part are shown in Fig. 5. On the left side are results from the motor current and on the right of the delta encoder model. To determine the stiffness, the force component in the surface normal direction of the feature was used. Each feature has its own mark (see Table 1). For each specimen workpiece one tool stiffness (blue, orange and yellow line) is predicted by linear regression. Besides there reference stiffness of the dummy model measured for each tool in x- (red line) and y-direction (green line).

It can be seen that good reproducible results are achieved especially for the Ø8 mm end mill. As a result of the higher stiffness for Ø12 mm and Ø16 mm, measured deviations are lower. This has a negative influence on the prediction, as many measured values are close to each other. At the same time, feature A1 here has a negative influence on the prediction. Here the tool is loaded on a length of 10 mm instead of 5 mm (see Table 1 a_p). Furthermore, another part of the tool is engaged at the measuring point of the feature. In future, it seems to make more sense to use a detailed stiffness model that considers further information like the engagement condition and then determines, for example, the Young's modulus or the effective surface moment of inertia.

5 Outlook

In this article is new approach to determinate tool-machine-stiffness based on machine internal and quality data in combination with a material removal simulation (MRS) is presented. Based on axis current signals and the difference of the linear and rotary encoders process forces are predicted. With the help of the MRS the forces get mapped to the workpiece. Afterwards the differences between the real measured and the virtual workpiece are determined. The concept is validated for three different milling tools. As the stiffness of the tool decreases, the accuracy of the prediction increases. Due to the use of correlating variables with process force and the deviation between real and virtual component, a transfer to other cutting processes with a defined cutting edge is possible in principle. For example, the diameter deviations can be used for turning or the deviation from the cylinder contour for drilling. In future, therefore it has to investigated if other force models or their parametrization could improve the prediction. Furthermore, it has to be examined which error sources have an influence on the prediction and how big their impact is. The influence of the orientation of the deviation vector $\boldsymbol{\varepsilon}^{WCS}$ has to be

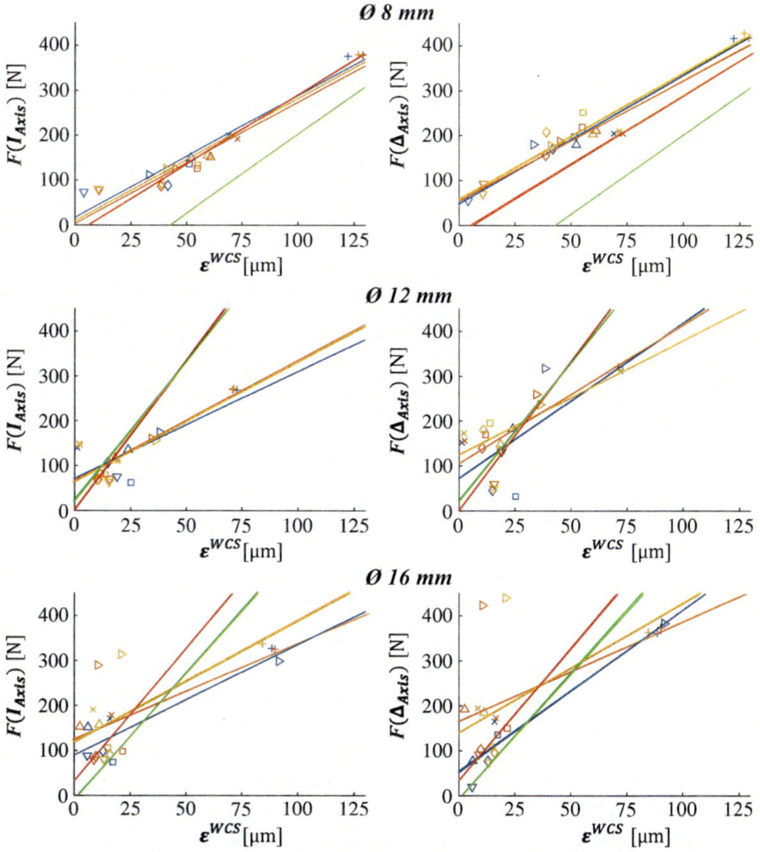

d	$k_{meas,x}$	$k_{meas,y}$	$k_{F(I),1}$	$k_{F(\Delta),1}$	$k_{F(I),2}$	$k_{F(\Delta),2}$	$k_{F(I),3}$	$k_{F(\Delta),3}$
8 mm	3.08 N/µm	3.52 N/µm	2.71 N/µm	2.87 N/µm	2.70 N/µm	2.68 N/µm	2. 73 N/µm	2.83 N/µm
12 mm	6.63 N/µm	6.20 N/µm	2.38 N/µm	3.46 N/µm	2.66 N/µm	3.09 N/µm	2.65 N/µm	2.56 N/µm
16 mm	5.87 N/µm	5.56 N/µm	2.46 N/µm	3.62 N/µm	2.14 N/µm	2.23 N/µm	2.70 N/µm	2.89 N/µm

Fig. 5. Validation results

investigated. Moreover, it should be focused if a breakdown of tool and machine stiffness in order to use the information gained for other machines.

References

1. Statistisches Bundesamt: Kostenstrukturerhebung im Verarbeitenden Gewerbe, im Bergbau sowie in der Gewinnung von Steinen und Erden. Fachserie 4 Reihe 4.3 (2019)
2. Schebek, L., Kannengießer, J., Campitelli, A., et al.: Ressourceneffizienz durch Industrie 4.0 - Potenziale für KMU des verarbeitenden Gewerbes, Berlin (2017)
3. Siebrecht, T., Kersting, P., Biermann, D., et al.: Modeling of surface location errors in a multi-scale milling simulation system using a tool model based on triangle meshes. Procedia CIRP 37, pp. 188–192 (2015). https://doi.org/10.1016/j.procir.2015.08.064

4. Brecher, C., Wellmann, F., Epple, A.: Quality-predictive CAM simulation for NC milling. Procedia Manuf. **11**, 1519–1527 (2017). https://doi.org/10.1016/j.promfg.2017.07.284
5. Brecher, C, Eckel, H.-M., Motschke, T., et al.: Estimation of the virtual workpiece quality by the use of a spindle-integrated process force measurement. CIRP Annals **68**, 381–384 (2019). https://doi.org/10.1016/j.cirp.2019.04.020
6. Königs, M., Brecher, C.: Process-parallel virtual quality evaluation for metal cutting in series production. Procedia Manuf. **26**, 1087–1093 (2018). https://doi.org/10.1016/j.promfg.2018.07.145
7. Knape, S., Königs, M., Epple, A., Brecher, C.: Increasing accuracy of material removal simulations by modeling workpiece deformation due to clamping forces. In: Schmitt, R., Schuh, G. (eds.) WGP 2018, pp. 72–80. Springer, Cham (2019). https://doi.org/10.1007/978-3-030-03451-1_8
8. Denkena, B., Schmidt, A., Henjes, J., et al.: Modeling a thermomechanical NC-simulation. Procedia CIRP **8**, 69–74 (2013). https://doi.org/10.1016/j.procir.2013.06.067
9. Denkena, B., Dittrich, M.-A., Huuk, J.: Simulation-based surface roughness modelling in end milling. Procedia CIRP **99**, 151–156 (2021). https://doi.org/10.1016/j.procir.2021.03.096
10. O'Toole, L., Kang, C.-W., Fang, F.-Z.: Precision micro-milling process: state of the art. Adv. Manuf. **9**(2), 173–205 (2020). https://doi.org/10.1007/s40436-020-00323-0
11. Kienzle, O.: Die Bestimmung von Kräften und Leistungen an spanenden Werkzeugen und Werkzeugmaschinen. VDI-Z: pp. 299–305 (1952)
12. Altintas, Y., Lee, P.: Mechanics and dynamics of ball end milling. J. Manuf. Sci. Eng. **120**, 684–692 (1998). https://doi.org/10.1115/1.2830207
13. Postel, M., Aslan, D., Wegener, K., et al.: Monitoring of vibrations and cutting forces with spindle mounted vibration sensors. CIRP Annals **68**, 413–416 (2019). https://doi.org/10.1016/j.cirp.2019.03.019
14. Denkena, B., Litwinski, K.M., Boujnah, H.: Detection of tool deflection in milling by a sensory axis slide for machine tools. Mechatronics **34**, 95–99 (2016). https://doi.org/10.1016/j.mechatronics.2015.09.008
15. Möhring, H.-C., Litwinski, K.M., Gümmer, O.: Process monitoring with sensory machine tool components. CIRP Annals **59**, 383–386 (2010). https://doi.org/10.1016/j.cirp.2010.03.087
16. Aslan, D., Altintas, Y.: Prediction of cutting forces in five-axis milling using feed drive current measurements. IEEE/ASME Trans. Mechatron. **23**, 833–844 (2018). https://doi.org/10.1109/TMECH.2018.2804859
17. Fey, M., Epple, A., Kehne, S., et al.: Verfahren zur Bestimmung der Achslast auf Linear- und Rundachsen G01L 1/04 (2016)
18. Königs, M., Wellmann, F., Wiesch, M., et al.: A scalable, hybrid learning approach to process-parallel estimation of cutting forces in milling applications. Schmitt R., Schuh G. (Publ.) **7**, 425–432 (2017)
19. Denkena B., Bergmann B., Stoppel D.: Reconstruction of process forces in a five-axis milling center with a LSTM neural network in comparison to a model-based approach. JMMP **4**, 62 (2020). https://doi.org/10.3390/jmmp4030062
20. Schmitz T.L., Duncan G.S.: Receptance coupling for dynamics prediction of assemblies with coincident neutral axes. J. Sound Vibr **289**, 1045–1065 (2006). https://doi.org/10.1016/j.jsv.2005.03.006
21. Schmitz T.L., Duncan G.S.: Three-Component Receptance Coupling Substructure Analysis for Tool Point Dynamics Prediction. J. Manuf. Sci. Eng. **127**, 781–790 (2005). https://doi.org/10.1115/1.2039102
22. Brecher, C., Chavan, P., Fey, M.: Efficient joint identification and fluted segment modelling of shrink-fit tool assemblies by updating extended tool models. Prod. Eng. Res. Devel. **15**(1), 21–33 (2020). https://doi.org/10.1007/s11740-020-00999-0

23. Albertelli P., Goletti M., Monno M.: An improved receptance coupling substructure analysis to predict chatter free high speed cutting conditions. Procedia CIRP **12**, 19–24 (2013). https://doi.org/10.1016/j.procir.2013.09.005
24. Matthias W., Özşahin O., Altintas Y., et al.: Receptance coupling based algorithm for the identification of contact parameters at holder–tool interface. CIRP J. Manuf. Sci. Technol. **13**, 37–45 (2016). https://doi.org/10.1016/j.cirpj.2016.02.005

Adaptable Press Foundation Using Magnetorheological Dampers

S. Fries(✉), D. Friesen, R. Krimm, and B.-A. Behrens

Institut für Umformtechnik und Umformmaschinen, Leibniz Universität Hannover, An der Universität 2, 30823 Garbsen, Germany
fries@ifum.uni-hannover.de

Abstract. Energy-bound forming machines such as forging hammers tend to vibrate due to abruptly applied process forces, which is particularly noticeable in form of intense vibrations of the machine environment. This paper presents a new concept of shock absorbers for forming machines, using dampers filled with magnetorheological fluids. Magnetorheological fluids are suspensions of magnetizable particles in a non-magnetizable carrier fluid. By applying a magnetic field, the internal structures and thus the rheological properties of the fluid can be varied. Using an evolutionary based control strategy, the damping can be adjusted depending on the excitation. The dependencies as well as challenges in the design process of magnetorheological dampers for forming machines are described. In addition, simulation results of foregoing studies concerning damper design and the evolutionary control strategy are presented.

Keywords: Forming machines · Magnetorheological dampers · Vibration damping

1 Introduction

Increasing stroke rates and the processing of high strength materials lead to increasing process forces, which cause intense vibrations on forming machines. This is particularly noticeable with forging presses, causing powerful shocks to the machine environment [1]. The resulting vibrations have negative effects on the surrounding area and the machine operators. For this reason press foundations are equipped with combinations of springs and absorbers, in most cases a combination of helical steel springs and viscous dampers [1]. Since conventional damper-fluids have an invariable viscosity and thus the damper provides constant damping coefficients, a compromise between a high and low damper viscosity must be found in the design process, limiting their usability only for a narrow operating range. This conflict can be avoided using dampers with variably controllable damping properties, which can be realised by using magnetorheological fluids as the damping fluid. Magnetorheological (MR) fluids are suspensions of solid, magnetisable particles in a non-magnetisable carrier fluid. The advantage of magnetorheological fluids over fluids with a constant viscosity, which are usually used in conventional shock absorbers, is the possibility of dynamically controlling their material properties. By applying a magnetic field, the internal structures of these fluids and thus the flow properties of the damper can be changed.

M. Liewald et al. (Eds.): WGP 2022, LNPE, pp. 346–354, 2023.
https://doi.org/10.1007/978-3-031-18318-8_36

2 Magnetorheological Dampers

The magnetorheological effect describes the reversible change in the flow and deformation behaviour of these materials in the magnetic field. In the initial state, the magnetic particles are almost isotropically distributed in the carrier medium. If the liquid is exposed to a magnetic field, the magnetic particles form dipoles, which lead to an anisotropic formation of the particles along the magnetic field lines. This results in a network of particles arranged in chains [2, 3].

If a force is now applied to the fluid from outside, the chains represent a resistance to the fluid flow [4]. The strength of the alignment along the magnetic field and thus the magnetic dipole moment depend, among other things, on the particle type and particle size. In this state, magnetorheological fluids obtain a high yield stress and thus a high shear strength as well as viscoelastic behaviour [4, 5].

Magnetorheological fluids are already being successfully utilized in shock absorbers in the automotive industry or for earthquake protection in buildings. Audi has been using a shock absorber system known as "Magnetic Ride" since 2006 in the Audi TT and later in the A3 compact car and the R8 sports car [6]. By means of the generated magnetic field, the damper characteristics can be adjusted to the respective driving situation. Dampers with magnetorheological fluids are also used in helicopter rotor blades. Damping-relevant instabilities only occur here in certain flight scenarios, which is why an adjustable damper system is used [7]. In the field of building protection, a magnetorheological damper was designed to protect bridges from earthquakes. The integration of this damper into already known insulation systems can significantly reduce the seismic response of these vibration-prone structures [8]. In [9], a magnetorheological damper for the protection of buildings during earthquakes is presented, which is designed for damping forces of up to 200 kN. Furthermore, a magnetorheological damper was developed for damping the cutting shock on a blanking press, which is installed between the table and the ram during shear cutting. The developed MR damper proved to be effective in damping the vibrations generated by the cutting shock and in limiting the excited frequencies [10].

3 Dependencies

When specifying the requirements for the magnetorheological press foundation, the hydraulic, magnetic and mechanical dependencies of the damper and vibration isolation must be matched to the intended loading scenario to ensure proper performance. The damper has a hydraulic and a magnetic force component (Fig. 1). The hydraulic force component primarily refers to the operation of the damper under currentless conditions and thus without a magnetic field, the damper ensures damping via the viscosity of the chosen MR fluid. The maximum damping force without a magnetic field thus represents the basic damping capacity.

The electromagnetic force component refers to the operation of the damper with an active magnetic field. The parameters that influence this state are the coil geometry (1–4) and the fluid gap (5/d) between the damper piston and the damper housing. A closed magnetic field is created by the magnetic flux (yellow lines) between the piston and the

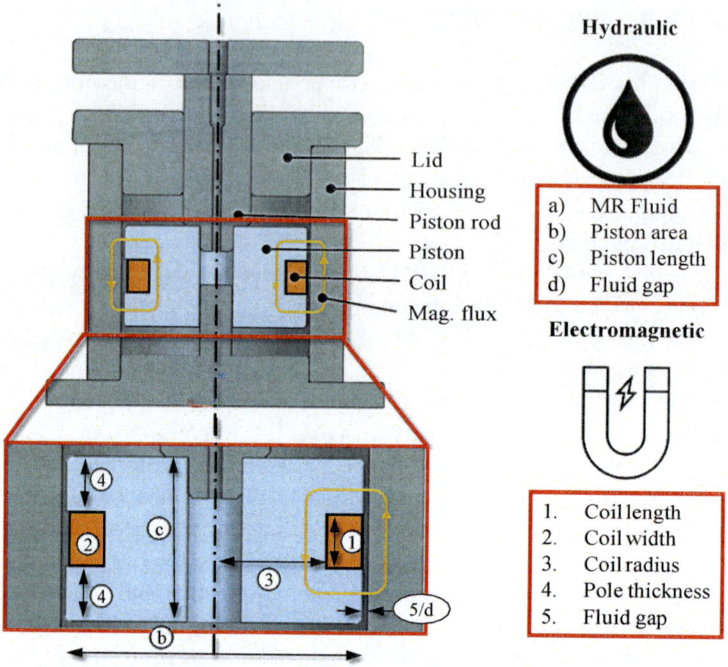

Fig. 1. Schematic representation of an MR damper

damper housing via the fluid gap (5/d). This area represents the active surface of the damper (4), as in this area the magnetic field locally magnetises the particles from the MR fluid and thus locally changes the viscosity of the fluid flow. On the one hand, the mechanical requirements of the magnetorheological press foundation refer to the weight of the machine it is going to be placed under, the carried out process on the machine and thus parameters like strokes per minute, process force and impact duration. On the other hand, they refer to the spring stiffness of the vibration isolation elements to be used parallel to the dampers and therefore the vertical natural frequency of the machine.

First, a quasi-static analysis was carried out to calculate the damper geometry, with the following boundary conditions: the piston of the MR damper moves with a constant velocity, the magnetic flux is fully developed and a simplified Bingham model is used to describe the fluid flow [11, 12]. With these simplifying assumptions, the complex non-linear problem, which has to be solved numerically by means of the Navier-Stokes equation, can be simplified to a flow between two plates. Furthermore, this simplification is only possible because the fluid gap between the damper piston and the damper pot is very small in relation to the overall diameter [11, 12]. Based on this simplification, the following formulas for the damper force of a magnetorheological damper result:

$$F_\eta = \left(1 + \frac{whv_0}{2Q}\right)\frac{12\eta QLA_p}{wh^3} \tag{1}$$

$$F_\tau = c\frac{\tau_0(H)LA_p}{h}\operatorname{sgn}(v_0) \tag{2}$$

$$F_{Damper} = F_{\eta} + F_{\tau} \tag{3}$$

w	Effective flow surface	L	Effective axial pole area/pole thickness
h	Width of the fluid gap	AP	Cross sectional area of the piston head
$v0$	Piston velocity	η	Dynamic viscosity
Q	Flow rate ($Q = A_P \cdot v_0$)	c	Empirical flow coefficient [2,07...3,07]

Formula (1) describes the viscous and thus non-variable force, which depends on the viscosity of the selected MR fluid. Formula (2) describes the MR force, which can be controlled on the basis of the magnetic field [12]. In order to take full advantage of the controllable range of such an MR damper, the challenge is to make the dynamic range of the damper and thus the adjustable damping force as high as possible. This can be achieved by making the electromagnetic force component of the damper higher than the fluid-dynamic or purely hydraulic force component.

Since the calculation of the geometric parameters only partially includes the electromagnetic properties such as coil geometry, wire diameter, etc., a Matlab tool was created which calculates and optimises the required geometric parameters of the damper based on the data of the MR fluid and the load case. The created tool simplifies the design, as optimal parameters for the damper design are calculated depending on the fluid properties of the selected MR fluid as well as defined boundary conditions. The tool is based on the simplification of a hydrodynamic fluid flow between two plates extended with an electromagnetic approach [13]. Since the geometric parameters of the damper are a multi-dimensional optimisation problem with many interdependent unknown variables, an optimisation approach has been implemented in the Matlab tool. The implemented optimisation approach is based on the optimisation function "fmincon" [14]. The objective of the optimisation is to maximise the dynamic range between the hydraulic and magnetorheological damping force.

4 Design

In Fig. 2, the damper design is visualized. The damper is designed with a piston utilizing two coils, which leads to an increase in the effective axial pole area (marked red in Fig. 2) and thus higher adjustable damping forces. A higher number of coils would increase the maximum possible damping force, but would also have a negative effect on the overall height. To avoid eddy currents, an opposite coil winding is used for both coils. To keep the air gap as constant as possible, a guiding sheath is attached around the piston, which is guided in the damper housing by a piston guide ring. The upper and lower piston areas of the damper are sealed off from each other by hydraulic sealing rings, so that the MR fluid can only flow through the fluid gap between the guiding sheath and the piston. The guiding sheath reduces the probability of the piston tilting under non-axial load and ensures a constant air gap throughout the relevant areas.

To compensate for the change in volume of the incompressible MR fluid due to piston movement and temperature fluctuations, a gas pressure spring needs to be installed in

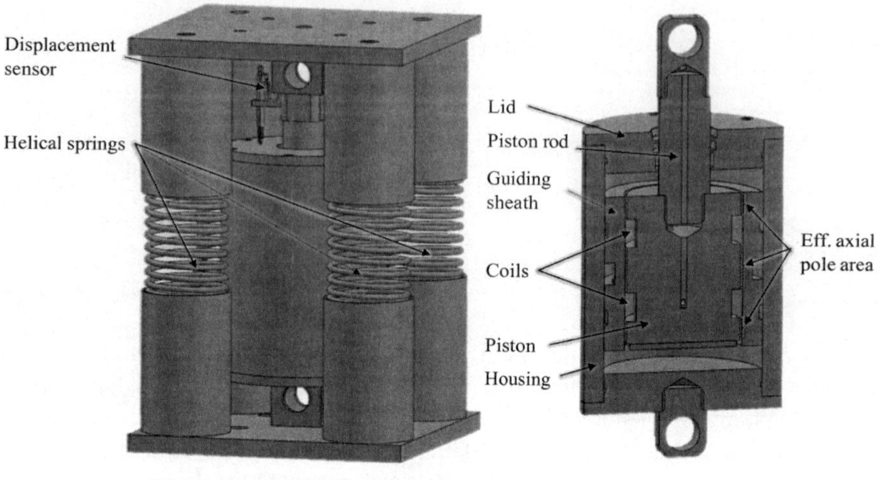

Fig. 2. Design of the magnetorheological press foundation

the lower piston chamber. The damper is placed centrally between four helical springs in order to minimise tilting under non-axial load. In addition, the damper is attached at the top and bottom via shock absorber eyes, so that it can be loaded in compression and in tension.

5 Testing

For the experimental investigation of the magnetorheological dampers, a test rig (Fig. 3 left) available at the IFUM, which generates shock loads by means of a hydraulic activation cylinder, is to be redesigned and adapted to the required conditions and purpose of use. The relevant parts of the test rig are visualized in Fig. 3 at the bottom right. The test rig consists of a hydraulic activation cylinder. This cylinder can maintain a hydraulically preloaded force up to a desired maximum release force, at which the force is abruptly released to zero. The piston of the activation cylinder is arranged under a load plate, which has several pre-tensioned disc spring assemblies. As soon as the sledge plate is moved onto the disc spring system, a continuous increase in force occurs. As soon as the maximum release force of the activation cylinder is reached, the activation cylinder releases the force and the loading plate abruptly hits the stops.

Acceleration and displacement measurements were carried out on this test rig in order to design the magnetorheological press foundation for test rig loads. The max. test frequency of the test rig is 60 strokes/min. The test rig is adapted to accommodate the designed MR press foundation. The dampers can then be tested by varying the maximum force of the activation cylinder. The maximum force can be varied by adjusting the preload of the disc springs and the hydraulic pressure of the activation cylinder. Acceleration and displacement measurements at maximum force (approx. 300 kN) were carried out on the test rig in its actual state. At the foot of the test rig, acceleration peaks of 300 g could be measured with a maximum displacement of about 1.5 mm and a vertical natural frequency of about 10 Hz.

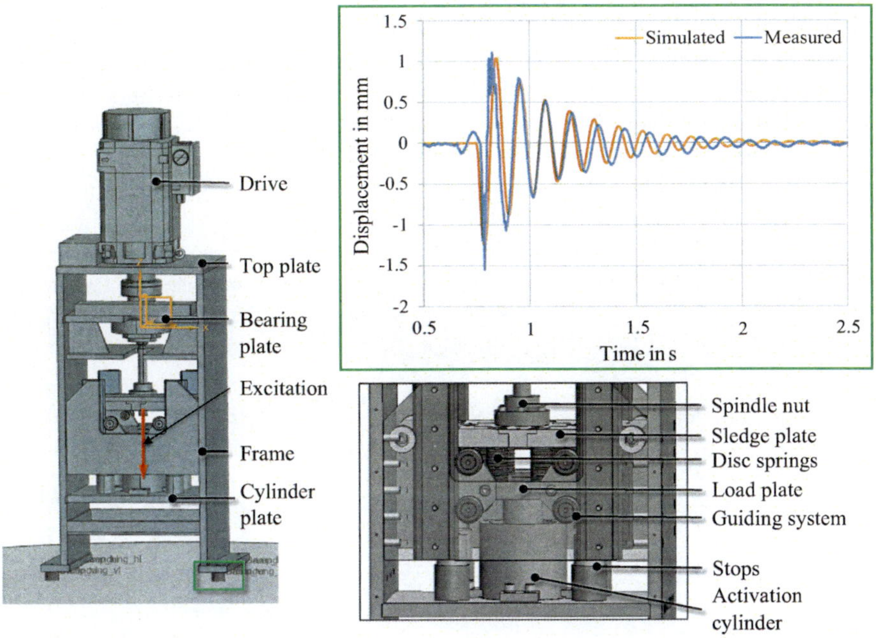

Fig. 3. Test rig with exemplary simulation results at maximum load

A multibody simulation (mbs) model of the current state of the test rig was made in Simcenter 3D. An impact force of about 300 kN on the load plate serves as the excitation. The stiffness values of the current viscoelastic shock absorbers, on which the test rig is currently placed, were determined via static calculations and the damping values on the basis of an evaluation of the vibration displacement measurement via the decay curve. As an example, Fig. 3, top right, shows a comparison of the simulated and the actually measured displacement at the foot of the test rig at maximum load. The simulation of an impact to the test rig at maximum load illustrates a good match of the amplitude and the dynamics. Deviations occur, among other things, due to the unknown non-linear stiffness characteristics of the viscoelastic shock absorbers the test rig currently stands on. This excitation has been taken over into the model of the nominal state for the design of the MR damper.

6 Control Strategy

To make the damper autoadaptive to different load scenarios, a control strategy based on the evolutionary optimisation algorithm CMA-ES is used. In previous research projects at the IFUM, this control strategy has proven to be effective [15–17].

A control loop has been designed for virtual design and testing of the control strategy. The optimisation algorithm is programmed in Simulink and is coupled with the MBS model of the test rig in Simcenter 3D. The coupling is done via an interface block in Simulink and a corresponding integration of the input and output variables in Simcenter

3D. After the test rig is subjected to shock loads in Simcenter, the resulting vibration displacement is measured at the damper and passed on to the optimisation algorithm as an output variable.

With maximum damping force there is also always maximum transmission of the impact to the environment, which is accompanied by increased vibration to the environment. In order to minimize this problem and to achieve an optimal ratio of a short decay time of the vibration and a minimal impact transmitted to the ground, the cost function of the optimisation algorithm requires the displacement of the test rig and the acceleration that is transmitted to the ground. The algorithm optimises the dampers fluid flow in a way, that the optimum of short decay times with the smallest possible forces on the environment is found. The control loop is carried out according to Fig. 4.

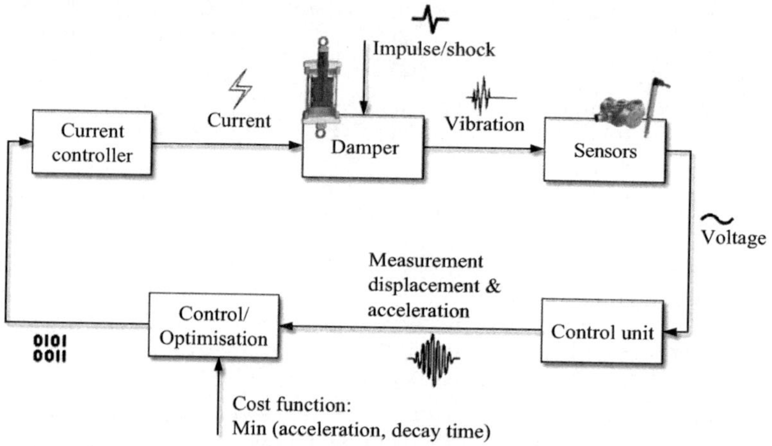

Fig. 4. Control loop

The impulse/shock is recorded with the displacement and acceleration sensors and processed according to the evaluation unit. The values are transferred to the control or the optimisation algorithm. The cost function provides a minimisation of the acceleration transferred to the ground and the achieved decay time with a pulse sequence adapted to the oscillation frequency. The optimisation algorithm provides a data set for the current controller. This iterative process is repeated until an optimal operating point is reached.

7 Summary and Outlook

This paper describes the development of an autoadaptive damping system based on magnetorheological fluids, which ensures damping optimised to specific loading conditions. It is investigated to what extent dampers based on magnetorheological fluids are suitable for damping the sudden vibrations resulting from impulse-like process forces, as they occur in particular in energy-bound forming machines. Special attention is paid to the evolutionary control strategy of the magnetic field, which provides an adaptation of the damping properties to the impact-like loads. A multi-body simulation model of the test rig, which represents the impulse-like loads of forming machines, was created to design the dampers and to test and optimise the evolutionary control strategy. Currently, the final procurement process of the damper's components takes place. Then one damper will be assembled in order to examine it on a hydraulic loading device. In parallel, the control system is being further optimised and the test rig is adapted for the practical application. Finally, a design of experiments follows, according to which the dampers will be examined under the test rig.

Acknowledgement. The IGF-project 20808N of the German Machine Tools' Association (VDW) is supported via the German Federation of Industrial Research Associations (AiF) within the framework of the Industrial Collective Research (IGF) program by the Federal Ministry of Economics and Technology based on a decision by the German Bundestag.

The Authors would like to thank the VDW and the members of the industrial consortium for supporting the research activities.

References

1. Doege, E., Behrens, B.-A.: Handbuch Umformtechnik, 3rd edn. Springer Vieweg, Berlin, Heidelberg (2016)
2. Zschunke, F.: Aktoren auf Basis des magnetorheologischen Effekts. Dissertation, University of Erlangen-Nuremberg (2005). Available online: https://opus4.kobv.de/opus4-fau/front-door/deliver/index/docId/190/file/DissertationFlorianZschunke.pdf. Last accessed 27 Apr 2022
3. Goldasz, J., Sapinski, B.: Insight into Magnetorheological Shock Absorbers, Springer, Cham (2015)
4. Bompos, D., Nikolakopoulos, P.: Experimental and analytical investigations of dynamic characteristics of magnetorheological and nanomagnetorheological fluid film journal bearing. ASME. J. Vib. Acoust. **138**(3), 031012 (2016)
5. Vicente, J., Klingenberg, D., Hidalgo-Alvarez, R.: Magnetorheological fluids: a review. Soft Matter **7**, 3701–3710 (2011)
6. Audi AG: Technology Portal. https://www.audi-technology-portal.de/de/fahrwerk/fahrwerksregelsysteme/audi-magnetic-ride. Last accessed 27 Apr 2022
7. Ngatu, G., Hu, W., Wereley, N., Kother, C., Wang, G.: Magnetorheology: Advances and Applications, pp. 307–341. RSC Publication, Cambridge (2014)
8. Kataria, N., Jangid, R.: Optimum semi-active hybrid system for seismic control of the horizontally curved bridge with magnetorheological damper. Bridge Struct. **10**(4), 145–160 (2014)
9. Jiang, Z., Christenson, R.: Hyperbolic tangent model for 200 kN large-scale magnetorheological fluid (MR) damper (2011). Available online: https://datacenterhub.org/re-sources/3879. Last accessed 27 Apr 2022
10. Ghiotti, A.; Regazzo, P., Bruschi, S., Bariani P.: Reduction of vibrations in blanking by MR dampers. CIRP Ann.—Manuf. Technol. **59**, 275–278 (2010)
11. Zhu, X.: Magnetorheological fluid dampers: a review on structure design and analysis. J. Intell. Mater. Syst. Struct. **23**(8), 839–873 (2012)
12. Yang, G., Spencer Jr, B.F., Carlson, J.D., Sain, M.K.: Large-scale MR fluid dampers: modeling and dynamic performance considerations. Eng. Struct. **24**(3), 309–323 (2002)
13. Hussein, S.: Systèmes de suspension semi-active à base de fluide magnétorhéologique pour l'automobile. Dissertation, Arts et Métiers ParisTech (2010)

14. Mathworks Matlab Documentation Homepage. https://de.mathworks.com/help/op-tim/ug/fmincon.html. Last accessed 27 Apr 2022
15. Behrens, B.-A., Krimm, R., Hilscher, S.: A new approach to compensate oscillations of pathlinked presses caused by inertial forces. Prod. Eng. **1140**, 361–368 (2016)
16. Behrens, B.-A., Krimm, R., Fries, S., Nguyen, T., Altan, L., Friesen, D.: Autoadaptive minimization of transfer system oscillations. In: Proceedings of the 9th Congress of the German Academic Association for Production Technology (WGP), pp. 121–129 (2019)
17. Marthiens, K.-O.: Autoadaptive Minimierung von Stößelschwingungen an Pressen. Dissertation, Gottfried Wilhelm Leibniz Universität Hannover (2012)

Implementation of MC-SPG Particle Method in the Simulation of Orthogonal Turning Process

P. Rana[1](✉), W. Hintze[2], T. Schall[1], and W. Polley[1]

[1] Mercedes-Benz Group AG, 70546 Stuttgart, Germany
pulkit.rana@mercedes-benz.com

[2] Institute of Production Management and Technology (IPMT), 21071 Hamburg, Germany

Abstract. In the automotive industry, due to fast changing markets and push for implementing novel materials, competitiveness relies increasingly on economical and short planning cycles of machining process. Digitalization has created opportunities for automotive industries to reduce time to market with the help of computer simulations of manufacturing process. In the past decade, particle methods like Smooth Particle Hydrodynamics (SPH), and Smooth Particle Galerkin (SPG), among many others, have been used to simulate machining process. The particle methods have an advantage over classical Finite Element Methods (FEM) as particle methods do not require remeshing or continuous mesh adaptation. Thus, particle methods eliminate the mesh entangling problems in simulation of large plastic deformations, such as machining. This paper describes the application of original SPG and the new Momentum-Consistent-SPG (MC-SPG) method in the orthogonal machining simulation of 1.4837D casted steel material. Furthermore, a study was conducted to understand how different SPG parameters affect simulation results, particularly force components, chip form and temperature. In the end, a Design of Experiment (DoE) was created to study the effects of cutting velocity, feed, and rake angle on force components. The simulation results were experimentally validated, and a good agreement (for cutting forces mean deviation 0–11% and feed forces 4–15%) was found between experimental and simulation results.

Keywords: Machining · Simulation · Finite element method

1 Introduction: Smooth Particle Galerkin Method

Smooth Particle Galerkin (SPG) is a mesh free method developed by Wu et al. [1] for deformation and failure analysis in solid-state mechanics. It is exclusively implemented in Software LS-DYNA. Wu et al. [2] claimed that SPG method can be used for impact and penetration simulations with material failure. This method is based on a mesh-free Galerkin approach to solve the partial differential equations of linear elastic problem. SPG uses a direct node integration method, which, based on the mathematical Eq. (1), can calculate the field variables and their derivatives at the same points [1, 2].

$$M^{\text{lump}} \times \ddot{U} = F_{\text{ext}} - F_{\text{int}} \tag{1}$$

M. Liewald et al. (Eds.): WGP 2022, LNPE, pp. 355–365, 2023.
https://doi.org/10.1007/978-3-031-18318-8_37

Here, M is the lumped mass matrix in diagonal form, $\ddot{U}$ is the vector that contains all particle accelerations. F_{ext} is the external force vector and F_{int} is the regulated internal force term [3].

Solving Eq. (1) can produce spurious-energy modes in the displacement field. To avoid these modes and to reach a stabilization effect, a non-residual stabilization term F_{stab} is added to the SPG formulation. Integration of this stabilization term separately would result in additional computing effort. The new equation system to be solved is shown in Eq. (2) [1, 3].

$$M^{\text{lump}} \times \ddot{U} = F_{\text{ext}} - F_{\text{int}} - F_{\text{stab}} \tag{2}$$

1.1 Momentum-Consistent Smoothed Particle Galerkin

The stabilization in SPG methods is accomplished without the use of a momentum equation residual and therefore they belong to the non-residual stabilization methods. In order to integrate these non-residual stabilization terms, multiple integration points that are matching at each particle are required, which are not computationally efficient. To combat these challenges, a new type of particle stabilization method, called MC-SPG, has been developed. In contrast to most other particle stabilization methods, in MC-SPG, smoothing is not based on residual or non-residual stabilization terms. This does not modify the system of equation and requires only one integration point per particle [4]. A second-order pulse-consistent velocity smoothing algorithm is used in MC-SPG, which is designed to provide accurate and stable results in thermal-structural coupling. The new smoothing algorithm eliminates the stabilization term of the discrete equation and again gives a system of equations the form shown in Eq. (1) [3].

1.2 Application of the SPG Method

Boldyrev et al. [5] simulated the orthogonal cutting of the aluminum alloy (AI6061-T6) using the SPG method. They received qualitative convergence of forces comparable to experimental results. Huang et al. [6] conducted simulations of self-piercing rivet using the SPG method and found a reasonable agreement between experimental and simulation results. Though a comparison between experimental and simulated forces was not performed, a detailed sensitivity analysis was carried out. They found that the choice of critical plastic strain, kernel update interval, dilation parameters, particle spacing, and mass-scaling factor has no significant effect on force results in simulations. Wu et al. [7] applied the SPG method to grinding process simulations and found a good match of reaction forces between simulation and experiment. The comparison of chip shape exhibits that discontinuous chips are formed in both, simulation and experiment. In addition, they found that refinement of the particle distance, kernel update interval and dilation parameters have either insignificant or no influence on force values [7, 8]. Liu et al. [9] implemented the SPG method for removal of blood clots using a high-speed rotating cutting tool. According to authors, a deviation between 3 and 20% was found in simulated cutting force and experimentally determined force. The feed forces were very low in both, simulation, and experiment. Pan et al. [10] implemented SPG modeling to the simulation of friction drilling process. They found a good match of feed force, torque, and temperature with experiments. In contrast, when using FEM, forecasts were very low. Rana et al. [11] implemented SPG method to orthogonal machining of a casted steel alloy and achieved a good agreement between simulated and experimental force components. For SPG, mean deviations of simulated forces with respect to measured forces amount to 12 and 7% for cutting and feed force respectively. They also compared the simulated chip form with experimentally obtained chip form and found a significant difference between both. However, thermal analysis of orthogonal machining process was not considered here.

This work presents important parameters that need to be used in SPG method for simulating an orthogonal machining process. A coupled thermo-mechanical 3D-orthogonal machining process is simulated, and simulated chip forms and force components are compared to experimental chips and forces.

2 Experimental Setup

The experiments were carried out on a Computerized-Numerical Control lathe Gildemeister CTX310. Hollow cylinders with an outer diameter d = 62 mm, a length l = 40 mm and a wall thickness of t = 1 mm were machined without lubrication under orthogonal conditions with inserts of tungsten carbide HW-K10. The cutting inserts have a clearance angle $\alpha_o = 7°$, rake angle $\Upsilon_o = 0°$ and 15° and cutting edge radii $r_ß$ = 24 µm and 33 µm respectively. The experimental width of cut corresponds to the wall thickness, thus set to a_p = 1 mm. A face-centered central composite design was implemented using the Minitab software to evaluate the effect of cutting velocity v_c, feed f and rake angle Υ_o, Table 1. The workpiece material was a casted steel alloy, 1.4837D, containing 11–12% of Nickel [12]. The experimental forces were measured with a 3-component dynamometer Kistler 9121 and were analyzed in DIAdem software. Figure 1 shows force components for $v_c = 180, f = 0.25$ and $\Upsilon_o = 0°$. In order to not distort the test results due to existing wear, a different part of cutting edge was used after each cut.

Table 1. Face-centered central composite design

Order	v_c	f	Υ_o	Order	v_c	f	Υ_o
4	180	0.35	0	7	100	0.35	15
8	180	0.35	15	6	180	0.25	15
1	100	0.25	0	5	100	0.25	15
2	180	0.25	0	9	140	0.30	0
10	140	0.30	15	3	100	0.35	0

3 Simulation Setup

The simulations were created in LS-PREPOST, a software from company Ansys. In the simulation model, only the part of the workpiece involved in the machining process is discretized with SPG particles. As SPG particles are computationally expensive, the finite elements were used for the bottom part of the workpiece to reduce the computation time [7]. The workpiece was discretized using hexahedral elements and divided into two parts, which use the same nodes at their point of contact. The division into an SPG and FEM part was achieved by defining different *SECTION keywords. Hence, for the upper part, *SECTION_SOLID_SPG was used which replaced the solid elements with SPG particles. The lower part of the workpiece was defined with keyword *SECTION_SOLID which created finite elements in that part. The cutting inserts were defined as rigid bodies with shell elements, since in this work, no tool wear was taken into account. The workpiece was fixed at bottom in all three directions, while the tool was provided with the cutting velocity. Figure 2 illustrates the discretized model of orthogonal machining in simulations.

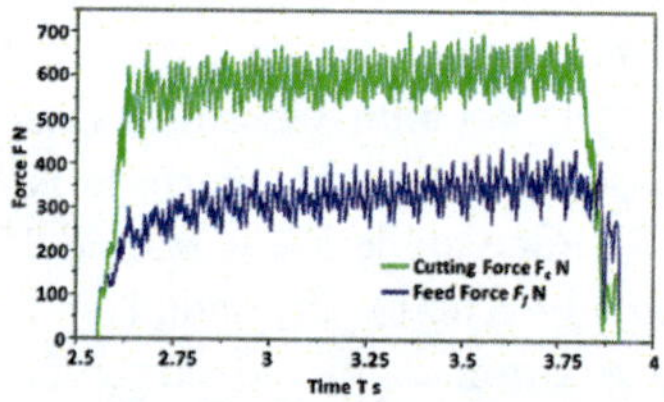

Fig. 1. Force components for $v_c = 180, f = 0.25$ and $\Upsilon_o = 0°$

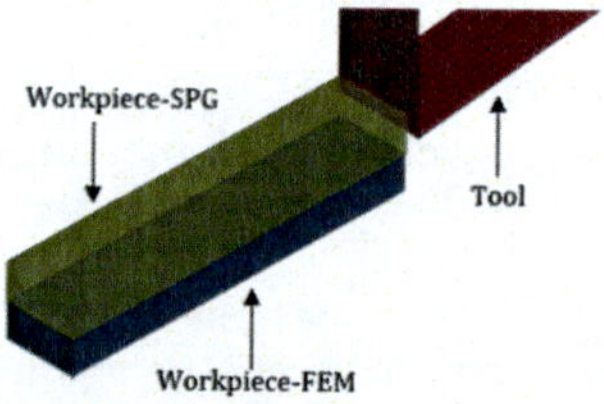

Fig. 2. Discretization of orthogonal cut

To do a coupled 3D-thermo-mechanical analysis, a thermal and mechanical material model for workpiece and tool was defined. Since material properties were assumed to be direction-independent in this work, the keyword *MAT_THERMAL_ISOTROPIC was used. The *MAT_RIGID keyword was used as mechanical material model of the tool because it was modeled as a rigid body. The workpiece uses the Johnson-Cook (JC) material model defined with keyword *MAT_JOHNSON_COOK. This material model is often used for the simulation of machining processes as well as for the simulation of processes with high strain, high strain rates and high temperatures [13, 14]. In LS-DYNA, an equation of state is additionally required to use the JC material model. Hence, a linear polynomial equation of state was used. The material properties of the workpiece used in simulations are same as mentioned in earlier research work [11].

To control the interaction between workpiece and tool, a contact type must be defined in simulations. For this reason, the keyword *CONTACT_AUTOMATIC_NODES_TO_SURFACE was used [10]. Shear friction model was used in simulations with a coefficient of friction value of 0.5. All simulations were conducted on Dell Precision T7810 tower with Intel(R) Xeon(R) CPU E5-2667 and 32 GB RAM computer.

4 Results and Discussions

In the following section, the effect of various SPG parameters such as ITB, IDAM and ISPLINE on simulation results are analyzed in Sects. 4.1 and 4.2. ITB is a para-meter used to choose the stabilization method, IDAM to choose the failure mechanism and ISPLINE to select the spline function [15]. In the end, the final simulation model results are experimentally validated.

4.1 Parameters Used: ITB = 0, IDAM = 1, ISPLINE = 0

In the first set of simulations, a failure strain (FS) value of 0.25 was used, which was equal to fracture strain. A stretching ratio (STRETCH) of 1.25 was defined. The resulting chip shape is shown in Fig. 3. It was observed that the combination of these parameters do not create chip bending in simulations. However, by assuming unrealistically high values for FS and STRETCH, a chip bend can be achieved. This is also illustrated in Fig. 3 for FS = 500 and STRETCH = 150. These parameters were selected randomly and were only intended to demonstrate the parameters that effect the results drastically. The creation of a chip bend only works reliably for $\Upsilon_{o} = 15°$. For $\Upsilon_{o} = 0°$, instead of a chip, a material accumulation is usually created in front of the tool.

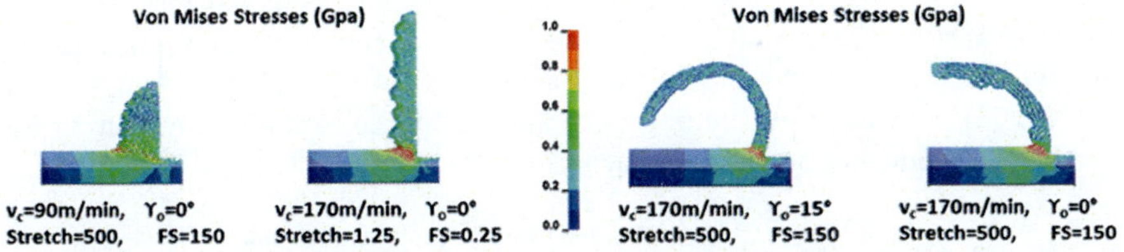

Fig. 3. Influence of the failure parameters on the chip form, $f = 0.2$ mm

4.2 Parameters Used: ITB = 3, IDAM = 13, ISPLINE = 14

For all the simulations mentioned in the previous sections, the parameter IDAM = 1 was used in keyword *SECTION_SOLID_SPG. In the following models, IDAM = 13, which is a modified version of failure mechanism, was defined. According to LS-DYNA, in new mechanism, the bond would break if shear strain is greater than FS, and either tension is higher than STRETCH or Compression is lower than 1/STRETCH.

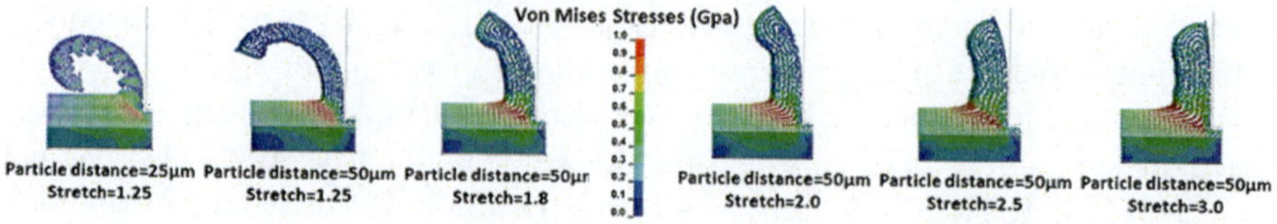

Fig. 4. Influence of STRETCH parameter on chip form

Table 2. Influence of STRETCH parameter on force components

Particle spacing (μm)	STRETCH	Cutting force F_c (N)	Feed force F_f (N)
25	1.25	360	160
50	1.25	410	165
50	1.80	610	250
50	2.00	650	285
50	2.50	720	350
50	3.00	750	355

In previous simulations, a cubic spline function (ISPLINE = 0) was used. It uses brick-shaped nodal support and in case of large body rotation (such as chip bending), the evaluation of the shape function changes with the rotation. To make the range equal to an ellipsoid, ISPLINE = 14 was defined in new simulations. Since the choice of identical dilation parameters creates a sphere in all spatial directions, the chip bending can be observed in simulations. Furthermore, this spline function is intended to better suppress zero-energy modes in pressure-dominant deformations. The comparison of chip form in Fig. 4 shows that the bending of chip decreases with the increase in STRETCH value or particle spacing. In the conducted simulations, a segmented chip was only achieved at the smaller particle spacing of 25 μm. In addition to affecting the chip shape strongly, STRETCH also influences the predicted forces, as shown in Table 2. A higher value of STRETCH increases both, cutting force and feed force. Additionally, cutting force decreases with reduction in particle spacing.

4.3 Final Simulation Models

The final simulation models were built with ITB = 3, IDAM = 13 and ISPLINE = 14. Furthermore, a value of 1.8 was used for all three dilation parameters. Due to the strong dependence of the simulation results on STRETCH value, the dilation parameters were determined empirically. The greatest attention was paid to cutting force, as that was the dominant force compared to other force components. For $\gamma_o = 0°$, STRETCH = 2.8 was determined, whereas STRETCH = 2.3 was used for $\gamma_o = 15°$. For the tool, a mesh size at the cutting edge rounding was determined by dividing the edge radius by five. A mesh size of 0.1 mm was chosen for the rake and clearance face. A particle spacing of 25 μm was used in the workpiece. It was 12 mm long and 0.05 mm wide, i.e. three particles were used over the width of the workpiece. The FEM part of the workpiece was 0.425 mm high. The height of SPG part varied with cutting parameters, hence the height of SPG part was 0.25 mm higher than the feed.

5 Comparison of Results

5.1 Force Comparison

To validate a simulation method in machining process, a good approximation of simulated forces with experimental forces is crucial. Both simulations and experiments were carried out according to the test plan mentioned in Table 1. In Fig. 5 (left), the cutting forces of individual experiments are compared to their simulative counterparts. The mean deviation of forces is between 0 and 11%. The deviation of 11% is an exception, which occurs at a rake angle of $\Upsilon_o = 0°$, a cutting velocity of $v_c = 100$ m/min and a feed of $f = 0.35$ mm. A comparison between the feed forces of simulations and experiments is also drawn in Fig. 5 (right). With a deviation of approximately 29%, an exception occurs again for a rake angle of $\Upsilon_o = 0°$, a cutting velocity of $v_c = 100$ m/min and a feed of $f = 0.35$ mm. For the remaining cutting parameters, the feed force deviation is in the range of 4–15%, which indicates a good approximation of the experimental forces in simulations. It is noticeable that the deviation in feed force is always higher for inserts with $\Upsilon_o = 0°$, which could be improved with an increase in STRETCH value.

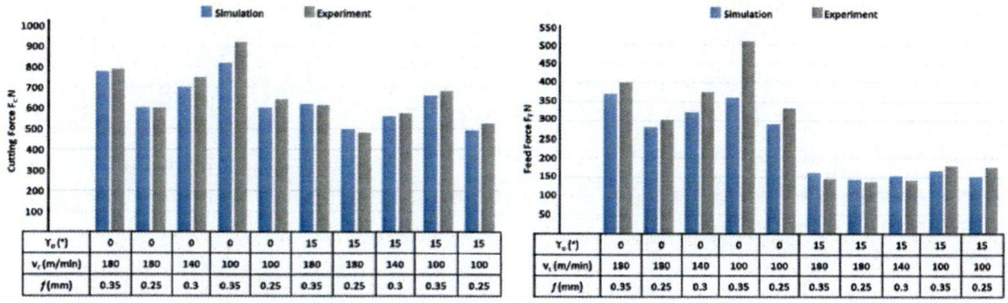

Fig. 5. Comparison of cutting (left) and feed (right) forces between simulations and experiments

Figures 6 and 7 show the influences of rake angle, cutting velocity and feed rate on cutting force and feed force, respectively. These influences are compared for both, experiment and simulation.

As depicted in figures, similar effects can be seen in experiments and simulations, but their characteristics differ. The forces decrease with increase in rake angle and cutting velocity. The significantly high effect of the rake angle is visible in feed force. With the increasing feed, an increase in forces can be observed, as also described by Klocke [16]. A high cutting velocity causes an increase in temperature, which in turn induces thermal softening of material. The softening of material can only be represented to a small extent in simulations and leads to a reduction in forces, which is always higher in experiments than in simulations [16].

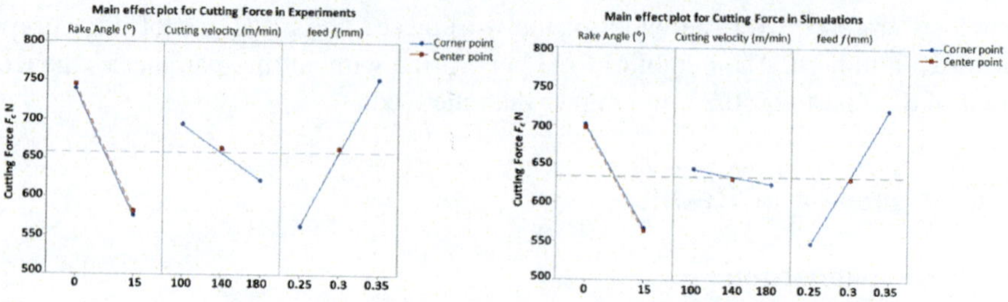

Fig. 6. Main effect diagram for cutting force for experiment and simulation

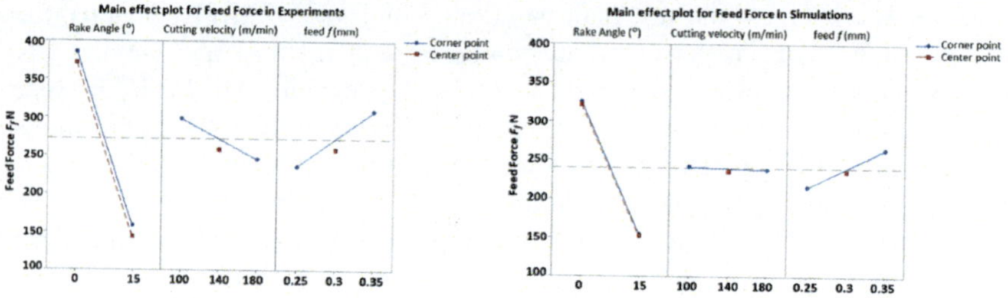

Fig. 7. Main effect diagram for feed force for experiment and simulation

5.2 Comparison of Chip Shape

For an optimal digital replication of machining processes, the simulation should also reliably predict chip formation. However, a comparison of chip shape is very difficult, as it greatly varies within the machining process. Due to this reason, the chips shown below should be seen as a guide, whereby indication of chip thickness is intended to show the scale. A comparison of a simulative and a real chip for rake angle $\Upsilon_o = 0°$ is shown in Fig. 8 (left). A similar curvature with segmented chip can be seen in both simulations and experiments.

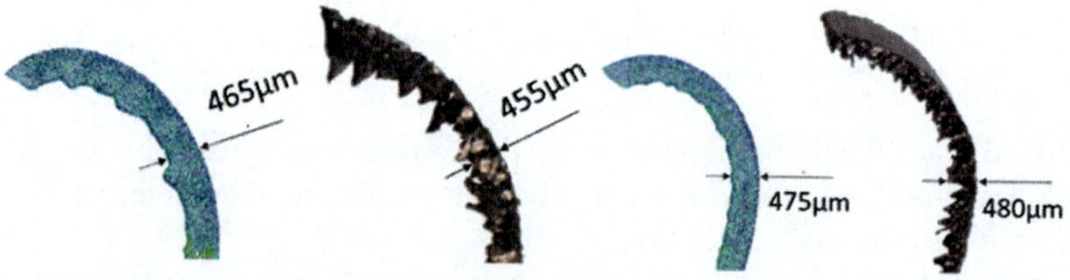

Fig. 8. Chip comparison: simulation versus experiment for $v_c = 100$ m/min, $f = 0.25$ mm and $\Upsilon_o = 0°$ (left) and for $v_c = 180$ m/min, $f = 0.25$ mm and $\Upsilon_o = 15°$ (right)

Figure 8 (right) compares the simulative and experimentally produced chips for cutting edge with a rake angle $\Upsilon_o = 15°$. Here, too, the chip form is comparable to the experimental chip form. However, the segmented chip is not formed in simulations. This can be improved by using a finer distance between the particles at the expense of simulation time.

5.3 Comparison of Chip Temperature

The simulated chip temperatures were not compared to the real temperatures, as they are very difficult to measure. Owing to this reason, it is only checked whether the temperature effects to be expected in reality also occur in simulation. Rana et al. [17] measured temperatures for same material and similar process parameters in a turning process, which were in the same range as simulated temperatures in this work. A comparison of the simulated temperatures for $\Upsilon_0 = 0°$ is shown in Fig. 9. To visually illustrate the differences, a maximum value of 650 °C is defined for the temperature scale. Like in real case scenario, the highest temperatures occur on the contact surface between chip and rake face. It can also be seen that the temperature rises with an increase in cutting velocity as well as feed.

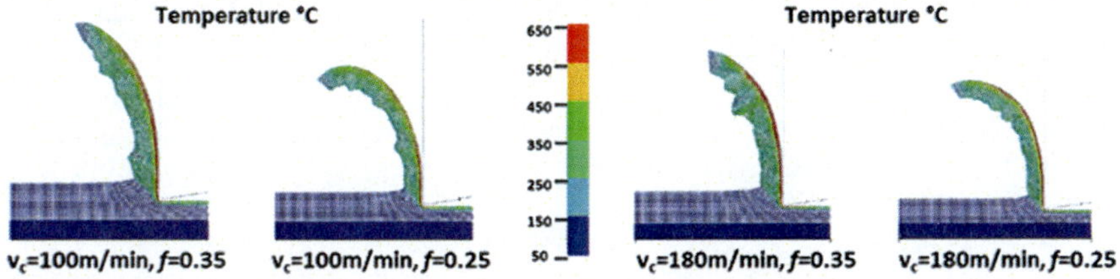

Fig. 9. Comparison of temperature for rake angle $\Upsilon_0 = 0°$

In Fig. 10, analogous to Fig. 9, the temperatures occurring in simulations with a rake angle $\Upsilon_0 = 15°$ are illustrated. The comparison between the two insert geometries shows that the chip temperature is higher for $\Upsilon_0 = 0°$ than for $\Upsilon_0 = 15°$. This result meets the expectation that the parameters feed, cutting speed and rake angle while increasing the cutting power lead to higher chip temperatures. Further validation of the simulated temperatures needs to be done by chip temperature measurements.

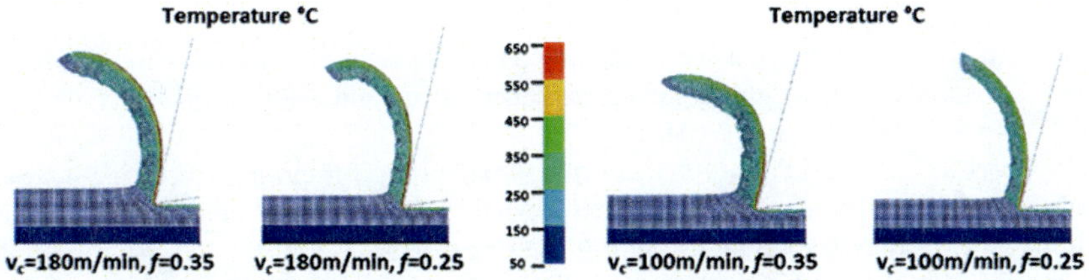

Fig. 10. Comparison of temperature for rake angle $\Upsilon_0 = 15°$

6 Conclusion

A thermo-mechanical model of orthogonal machining with the SPG and MC-SPG method for the material 1.4837D was built. Simulation results were validated against the experimental tests results. It was found that bending of the chip, with the original formulation of the SPG method, can only be achieved by assuming unrealistic values for the parameters controlling the bond breakage. Moreover, the simulated forces had a higher deviation from the experimentally measured forces. Adding to that, no realistic temperature distribution could be simulated. The problem of unrealistic temperature forecasting could be avoided by using the MC-SPG method. The comparison between final simulation models, with MC-SPG formulation, and experiments showed a good approximation of the force values. Furthermore, main effects of individual parameters on the forces of simulation and experiment were in good agreement. Additionally, the simulation predicted the effects of individual cutting parameters on expected temperature according to the relevant literature. The simulated chips represented a realistic form similar to experimental chip forms. However, the simulations could not predict segmented chips accurately for every cutting parameter, but this can be improved by using a finer particle spacing. The validation proves the applicability of the MC-SPG method for simulation of orthogonal machining.

Acknowledgements. The authors would like to thank Mr. Eckhard Zoch for helping in conducting the experiments, Mr. Pitt Held for helping in simulations and LS-DYNA Support Team for its dedicated technical support.

References

1. Wu, C.T., Guo, Y., Hu, W.: An introduction to the LS-DYNA smoothed particle Galerkin method for severe deformation and failure analyses in solids. In: 13th International LS-DYNA Users Conference (2014)
2. Wu, C.T.: Smoothed particle Galerkin formulation for simulating physical behaviors in solids mechanics, US 2015/0112653 A (2014)
3. Wu, C.T., Wu, Y., Lyu, D., Pan, X., Hu, W.: The momentum-consistent smoothed particle Galerkin (MC-SPG) method for simulating the extreme thread forming in the flow drill screw-driving process. Comput. Part. Mechan. **7**(2), 177–191 (2019). https://doi.org/10.1007/s40571-019-00235-2
4. Pan, X., Wu, C.T., Hu, W., Wu, Y.C.: Smoothed particle Galerkin method with a momentum-consistent smoothing algorithm for coupled thermal-structural analysis. In: 15th International LS-DYNA Users Conference (2018)
5. Boldyrev, I.S.: SPG simulation of free orthogonal cutting for cutting forces prediction. In: Radionov, A., Kravchenko, O., Guzeev, V., Rozhdestvenskiy, Y. (eds) Proceedings of the 4th International Conference on Industrial Engineering (2019)
6. Huang, L., Wu, Y., Huff, G., Huang, S., Ilinich, A., Freis, A., Luckey, G.: Simulation of self-piercing rivet insertion using smoothed particle Galerkin method. In: 15th International LS-DYNA Users Conference (2018)
7. Wu, C.T., et al.: Numerical and experimental validation of a particle Galerkin method for metal grinding simulation. Comput. Mech. **61**(3), 365–383 (2017). https://doi.org/10.1007/s00466-017-1456-6
8. Wu, Y., Wu, C.T., Hu, W.: Parametric and convergence studies of the smoothed particle Galerkin (SPG) Method in semi-brittle and ductile material failure analyses. In: 15th International LS-DYNA Users Conference (2018)

9. Liu, Y., Zheng, Y., Li, A.D., Liu, Y., Savastano, L.E., Shih, A.J.: Cutting of blood clots—experiment and smooth particle Galerkin modelling. CIRP Ann. **68**(1), 97–100, ISSN 0007-8506 (2019)
10. Pan, X., Wu, C.T., Hu, W.: A momentum-consistent stabilization algorithm for Lagrangian particle methods in the thermo-mechanical friction drilling analysis. **64**, 625–644 (2019)
11. Rana, P., Zielasko, W., Schuster, T., Hintze, W.: Orthogonal turning simulations for casted steel alloy using mesh free methods. In: Wulfsberg, J.P., Hintze, W., Behrens, B.A. (eds) Production at the Leading Edge of Technology. Springer Vieweg, Berlin, Heidelberg (2019)
12. Kaiser, T.: Entwicklung eines Nickel-reduzierten Austenits als Werkstoff im thermisch hoch beanspruchten Abgasturbolader. Dissertation Universität Clausthal, Universitätsbibliothek, Clausthal-Zellerfeld, Clausthal (2014)
13. Olleak, A.A, El-Hofy, H.A.: Prediction of cutting forces in high speed machining of Ti6Al4V using SPH method. In: Proceedings of the ASME 2015 International Manufacturing Science and Engineering Conference, vol. 1, Charlotte, North Carolina, USA. V001T02A018. ASME (2015)
14. Lampropoulos, A.D., Markopoulos, A.P., Manolakos, D.E.: Modeling of Ti6Al4V alloy orthogonal cutting with smooth particle hydrodynamics: a parametric analysis on formulation and particle density. Metals **9**, 388 (2019)
15. LS-DYNA. Keyword user's manual. vol. I. pp. 3240–3241 (2020)
16. Klocke, F.: Fertigungsverfahren 1—Zerspanung mit geometrisch bestimmter Schneide, vol. 9. Auflage, Springer Vieweg, Aachen (2018)
17. Rana, P., Hintze, W., Schall, T., Polley, W.: Study on the influence of the coating thickness in turning of a hard to machine material using FEM-simulation. In: Behrens, B.A., Brosius, A., Drossel, W.G., Hintze, W., Ihlenfeldt, S., Nyhuis, P. (eds) Production at the Leading Edge of Technology. WGP 2021. Lecture Notes in Production Engineering. Springer, Cham (2022)

Thermomechanical Multiscale PBF-LB-Process Simulation of Macroscopic Structures to Predict Part Distortion Recoater Collisions

K. Drechsel(✉), M. Frey, V. Schulze, and F. Zanger

Wbk Institute of Production Science, Karlsruhe Institute of Technology, Karlsruhe, Germany
kai.drechsel@kit.edu

Abstract. Process failure and part distortion are some of the main challenges in the additive manufacturing process laser powder bed fusion (PBF-LB). This is leading to increased part costs due to the need of post processing or a redesign and restart of a build job. Process simulation can enable engineers to predict possible build failures before the build job has been started. One of the primary problems with existing commercially available simulation approaches is the need for experimental data to calibrate the process model. To eliminate the need for calibration specimen, a new simulation technique was developed and is presented in this paper. Using a multiscale simulation, the calculation time can be significantly decreased compared to a single scale approach. On the micro scale, a high fidelity thermomechanical process model is developed to predict the associated inherent strains for the chosen process parameters and geometrical conditions. In addition, the material model is adapted to match the different phases present in the process. On the macro scale, a pure mechanical approach is used to predict part distortion and possible build failure due to recoater contact. In contrast to commercially available solutions, the scanning path is explicitly considered on both scales of the model to examine the influence of different scan strategies on the final parts properties. The simulation model were tested and validated against defined test specimen, which, as known from previous examinations, cause a recoater failure. All examinations were conducted with the commercially available aluminum alloy AlSi10Mg.

Keywords: Laser powder bed fusion · Process simulation · Thermomechanical simulation · Multiscale approach · Recoater contact

1 Introduction

Additive manufacturing (AM) enables engineers to realize geometrically complex structures without the need of special tools [1]. The ability to manufacture geometries directly from software tools [1], e.g. for topology optimization, increases the interest of the industry in AM processes like laser powder bed fusion (PBF-LB) [2]. Due to the large temperature gradients during the built job, undesired part distortion occurs [3]. In some cases, they reach a few millimeters requiring expensive trial and error [4]. Therefore,

M. Liewald et al. (Eds.): WGP 2022, LNPE, pp. 366–375, 2023.
https://doi.org/10.1007/978-3-031-18318-8_38

many research has been done simulating and compensating the part distortion [4–7]. The multi scale multi physics phenomena of the PBF-LB process require massive computation resources, when simulated directly [7–9]. Therefore, many researchers have investigated modelling approaches to reduce the computational effort while maintaining a sufficient accuracy [5]. A popular approach is the method of inherent strains, originally used for the simulation of welding processes [10]. This pure mechanical approach allows to calculate entire built jobs within a few hours [11]. To obtain the inherent strains, two methods, an experimental and a simulation based approach, are commonly used [6, 7]. With the experimental approach, the distortion of a cantilever is measured and the simulation is fit to match the experiment. The simulation approach utilizes a high fidelity thermomechanical model of the process to obtain the inherent strains directly from the simulation [7]. The calculation time for those models is often in the range of a few hundred CPU hours [12]. Therefore, a faster model to simulate a wide range of process parameters is desirable.

To reduce the part distortion and to stabilize the part during the built job, support structures are necessary in some cases [13]. However, those support structures increase the built volume and therefore increase the built time significantly due to the low built rates [14]. Hence, it is desirable to use a number of support structures as small as possible. In contrast, the risk of a costly built failure, e.g. due to a recoater collision with the distorted parts, increases with a reduction of support volume [15]. To the best knowledge of the authors, little research has been conducted in the area of recoater collisions. The aim of this work is to develop a pure simulation based, fast multi scale model to predict part distortion and recoater collisions for parts manufactured with AlSi10Mg.

2 Method

2.1 Simulation Model

To account for the multiscale phenomena occurring in the PBF-LB process, a twoscale finite element model (FE) was developed in this work. A sequentially coupled thermomechanical model was deployed on the micro scale to capture the laser material interaction and the melting, remelting and solidification behavior for AlSi10Mg. A pure mechanical model was deployed on the macro scale to capture the part distortion and recoater collisions. To couple the two scales, inherent strains were extracted from the micro model and fed into in the macro model. For both models the commercially available FE-software Abaqus 2021 was used. Abaqus provides a plug-in, especially designed for the simulation of additive manufacturing processes [16]. All simulations were done on a workstation with 128 GB ram and an AMD Ryzen 9 5950X processor.

Microscale. The aim of the model was to depict the melting, solidification and remelting. This is based on a realistic laser path and process parameters to simulate the residual stresses and corresponding inherent strains.

Thermal modelling: In the thermal model, the three dimensional heat conduction equation was solved. To account for the heat loss, radiation and convection were considered. The emissivity was set to $\varepsilon = 0.3$ [17] and in the top layer, the heat transfer coefficient was set to $\alpha = 7.9$ W/m^{2}K [18]. Three material states and one in between for the continuity of the release of the latent heat were considered. To distinguish between them, a solution dependent state variable (SDV) in time step i was set according to the current temperature and previous state. The algorithm to model the phase transfer is shown in Table 1 and was implemented through the subroutine USDFLD.

Table 1. Algorithm to distinguish between the modelled phases.

SDV_i	Value	Material state	Condition
	−1	Powder	If $T < T_{Solidus}$ and $SDV_{i-1} < 0$
	0	Liquid	If $T \geq T_{Liquidus}$
	1	Solid	If $T < T_{Solidus}$ and $SDV_{i-1} \geq 0$
		In-between	Linear interpolation between $T_{Solidus}$ and $T_{Liquidus}$ 0 for $T = T_{Liquidus}$ and 1 for $T = T_{Solidus}$

The material was initialized through the subroutine SDVINI with the state −1. Since individual powder particles cannot be resolved, the powder was modeled as a continuum as well. For continuity reasons, the specific heat capacity and density were identical to the solid material [7]. The conductivity of the powder was modeled using the relation

$$\lambda_{\mathrm{powder}} = \lambda_{\mathrm{solid}} \frac{(1-\varphi) n_{\mathrm{m}} a_{\mathrm{s}}}{\pi R_{\mathrm{n}}} \tag{1}$$

between powder and solid. The porosity was set to $\varphi = 0.4$, the radius of the sinter neck to $a_s = 4.6$ m, the average radius of the powder particle to $R_n = 42.5$ μm and the coordination number to $n_m = 12$ [19].

The laser material interaction was modeled with a concentrated flux. In comparison to the frequently used volumetric heat sources, like the Golak heat source [20], or surface heat sources, e.g. a Gaussian profile, the concentrated flux is directly assigned to the integration points. The longer the time step, the more integration points get a heat flux assigned at the same moment. The heat flux is calculated through the laser power, the scan velocity and simulation time step. The laser path was created using the open source slicer software SuperSlicer and turned into a readable input file by a python script.

Mechanical model. In the mechanical model, the three dimensional static impulse balance is solved. An isotropic stiffness matrix is used for the stress strain relation. The transient temperature field is propagated in the mechanical model. The thermal strains were calculated by

$$\varepsilon_{th} = \alpha(T) \Delta T 1 \tag{3}$$

with the temperature-dependent thermal expansion factor set to $\alpha = 23 \cdot 10^{-6}$ 1/K for solid and to $\alpha = 0$ for the powder and liquid, the unity matrix **1** and a reference temperature of 293.15 K. A von Mises plasticity with the extended power law

$$\sigma_f = R_{es}\left(1 + \frac{E\varepsilon_p}{R_{es}}\right)^N \quad (4)$$

to consider strain hardening [21] was implemented by the user subroutine UMAT. The same method as in the thermal model was used to distinguish between the three phases, while each phase got individual material properties assigned. In case of remelting, the plastic strains were set to zero.

Material model. The material behavior of AlSi10Mg was considered temperature dependent. For each of the three phases (powder, liquid and solid), the mechanical and thermal properties where set to the values shown in Table 2 [19, 22].

Table 2. Temperature dependent material properties.

Temperature	Conductivity	Specific heat	Yield strength	Young's modulus	Poisson's ratio
297.15 K	155 W/mK	0.74 J/gK	204 MPa	76 GPa	0.33
373.15 K	157 W/mK	0.75 J/gK	181 MPa		0.33
473.15 K	160 W/mK	0.84 J/gK	158 MPa		0.33
573.15 K	163 W/mK	0.92 J/gK	70 MPa	67 GPa	0.34
840.15 K	165 W/mK	1.10 J/gK	14 MPa		0.38
870.15 K	49 W/mK	1.29 J/gK	3000 MPa	1 GPa	0.33
2773.15 K	49 W/mK	1.29 J/gK	3000 MPa	1 GPa	0.33

The latent heat was set to 321 kJ/kg and is released continuously between $T_{Solidus} = 830.15$ K and $T_{Liquidus} = 870.15$ K. The Young's modulus for powder was set to 1 GPa as well.

Modell set up. To predict the inherent strains for a specific set of process parameters, the process was simulated on a square of 2 times 2 mm^2. Linear hexahedral elements with eight integration points and an edge length of 25 μm were used. To consider the surrounding powder and the baseplate, a 3.5 mm square of powder and a baseplate with a thickness 10 mm were modeled around the process field. The bottom side of the baseplate was hold at a constant temperature of 473.15 K. The elements were coarsened towards the outside, since a high resolution is not required in those regions. Six layers were simulated, to keep computational effort manageable. Each layer is deposited with progressive element activation [16]. To simulate the 12 s of recoating process and the 1.025 s of melting process, each recoating and melting process is considered in an individual step in the simulation with a fixed time increment of 1 s and 0.001 s respectively. After the last layer, the model is cooled down to 293 K. The inherent strains of the last three layers were extracted through a Python script, as these layers represent the build process away from the build plate much better than the first three.

Macroscale. The pure mechanical macro model uses the same material model and data as the micro model. Since no phase transformations were considered, the initial state of the material was set to 1. Similar to the micro model, the elements were activated with the progressive element activation. However, each layer in the simulation consists of three physical layers.

The inherent strains represent all inelastic strains occurring in a thermomechanical process [7] and can be defined as a second order tensor. Since the layer thickness in the PBF-LB process is relatively small, shear strains can be neglected [7]. The three remaining components are applied to the top layer in accordance to the chosen build strategy, e.g. stripes. The resulting deformations are calculated. Only a section of the build platform is simulated. To consider the rest of the base plate boundary conditions restricting the displacement of the base plate edges were implemented. The simulated geometries are shown in Fig. 1.

2.2 Experiment

Validation. To validate the simulation model, a cantilever, shown in Fig. 1 a) was built on a SLM 280HL machine. The cantilever had a height of 9 mm, a width of 10 mm and a length of 72 mm.

The process parameters, which were also used in the simulation model, are shown in Table 3.

Table 3. Process parameters used to build the cantilever and collision sample.

Power	Scan speed	Hatch distance	Layer thickness
350 W	1150 mm/s	170 μm	50 μm

The cantilever was cut on a height of 3 mm, marked in red in Fig. 1 a), to relax the residual stresses and as a result the cantilever bends upwards.

The same process was simulated in Abaqus and the maximum distortion was compared.

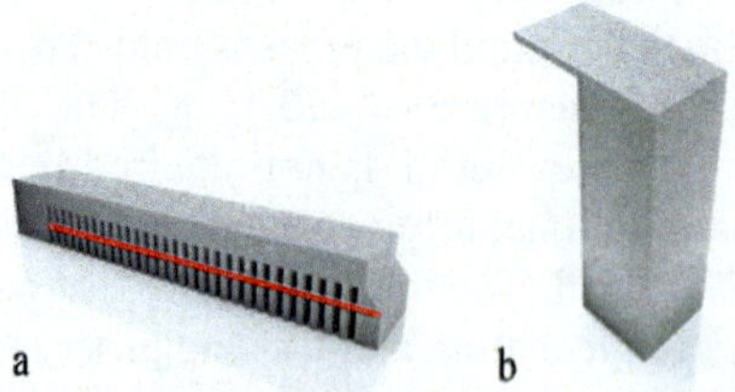

Fig. 1. Geometry of the cantilever in a) and the test specimen for the recoater collision in b).

Recoater Collison. The recoater collision was investigated by the test specimen shown in Fig. 1 b) [23]. It is prone to collide, since the overhang has no support structure. The test specimen, with a height of 20 mm and width of 10 mm and a length of 20 mm, was built with the same process parameters as the cantilever.

The build job was simulated using the same inherent strains as for the cantilever. To identify a recoater collision, the output file was processed with a Python script. The value of the displacement of the upper nodes in the top layer were taken for each simulated layer and compared to the thickness of the real powder layer. Due to the porosity of the powder bed, the real powder layer is higher than the set layer thickness [24]. If the displacement in build direction is larger than the real layer thickness, a collision is detected in that layer.

3 Results and Discussion

3.1 Microscale

The buildup of six layers was simulated in the thermal simulation. The total computation time was 97 min. Figure 2 a) shows the temperature in K near the end of the first layer exposure. Due to the different heat conductivities of powder and solidified material, an asymmetric temperature distribution can be seen.

Except for the rotated orientation of the scan vectors, the temperature history in the different powder layers are almost identical. In general, the overall temperatures in the upper layers are higher than near the base plate. This can be explained by the fact that the recently built layer, although already being solidified, has not completely cooled down to the build plate temperature yet. Therefore, a higher peak temperature can be achieved with the same energy input.For proper connection of the different layers, the depth of the melting zone should be greater than three times the layer thickness [25]. The remelting of already solidified material could be shown by examining the temperature over time on a fixed point, shown in Fig. 2 b). The temperature peak are a result of the heatsource moving through the measurement point.

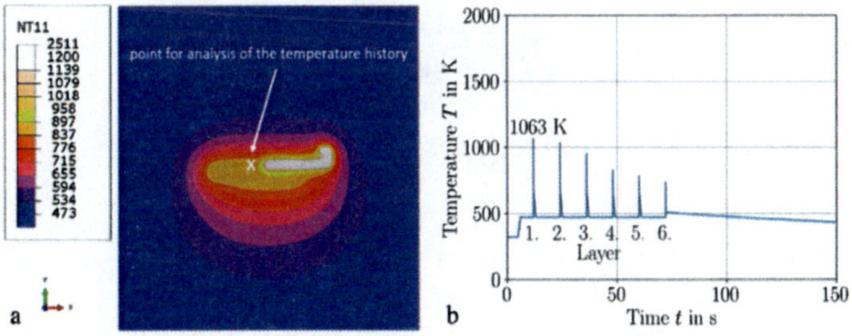

Fig. 2. Simulated temperatures within Abaqus: a) near the end of the first layer exposure in K, calculated with the micromodel; b) temperature history at one fixed node in the simulation over time.

Based on the temperature field of the thermal analysis, the thermally induced expansions are calculated. Compared to the thermal simulation, a complete static analysis requires significantly more computing time and more memory to complete. The total solution time for the static analysis was 15.3 h.

Figure 3 shows the prevailing von Mises stress at the same time as in Fig. 2 a). It can be seen that the stress response is delayed compared to the temperature amplitude. This can be explained by the presence of molten metal, which is considered stress-free. Likewise, plastic deformation is not possible in the melt. It should be noted that the melt pool has not been modeled as a fluid. So, circulations or instabilities are not considered in the simulation. However, a thermally induced expansion of the molten material can be detected, which finally causes the stresses in the solid during cooling. The stress across the scan direction reaches its maximum in the middle of the melt path.

Due to the chosen time increment, small gaps of unmolten powder, shown in Fig. 3, can occur. It could be examined that those can be avoided by decreasing the increment size. Since the influence of those parts can be neglected, the time increment can be raised to speed up the simulation process.

The stresses during cooling within a melt track exceed the yield strength of the material, as this drops significantly at elevated temperatures. The material deforms plastically and workhardens during cooling. This increases the stresses necessary for further plastic deformation. The powder can be considered free of stresses. The remelting of previous tracks in combination with the rotation of the laser trajectory leads to a uniform stress state in x and y direction. The equivalent plastic deformation was calculated to 0.45.

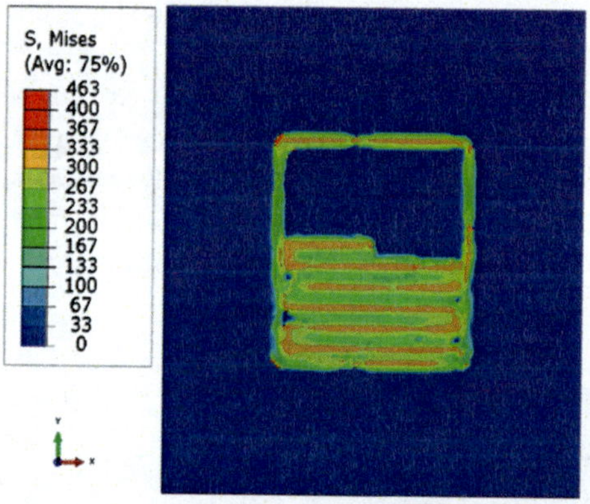

Fig. 3. Simulated von Mises stress in MPa at the same time as in Fig. 2 a)

The calculated inherent strains are shown in Table 4 by components. The shear strains can be neglected [6] and are therefore not listed. In x and z-direction large variations can be seen, especially in the first three layers. With rising layer number, the deviation minimizes for all six components. Because the whole part will be simulated, it is expedient to calculate the inherent strains based on the layers four to six to reduce the influence of the build plate.

Table 4. Calculated inherent strains by layer and component

Layer	ε_{xx}	ε_{yy}	ε_{zz}
1	-7.146×10^{-3}	-0.346×10^{-3}	13.697×10^{-3}
2	-5.684×10^{-3}	-0.327×10^{-3}	12.041×10^{-3}
3	-2.718×10^{-3}	-0.306×10^{-3}	8.801×10^{-3}
4	-0.705×10^{-3}	-0.271×10^{-3}	7.389×10^{-3}
5	-0.1056×10^{-3}	-0.254×10^{-3}	6.798×10^{-3}
6	-0.634×10^{-3}	-0.252×10^{-3}	6.579×10^{-3}

3.2 Macroscale

If the inherent strains $\varepsilon_{inh} = [-0.0498; -0.2584; 0.6913; 0; 0; 0]^T$ are used to simulate the part, the resulting stress and deformation shown in Table 5 are obtained.

Table 5. Simulated maximum stress and deformation at measurement point by components

σ_{xx}	σ_{yy}	σ_{zz}	u_x	u_y	u_z
276 MPa	286 MPa	416 MPa	80 μm	30 μm	1.74 mm

In good approximation, the von Mises stress is homogeneously distributed over the component. Due to the missing influence of further layers, the stress is reduced to a negligible value within the top 15 layers. The stress components in x- and y-direction are superimposed by a ladder-shaped structure due to the rotation of scan vectors within the xy-plane by 67 degrees.

The predicted deformation at the measurement point is 1.54 mm. It should be noted that the beam is subject to additional rotation along its longitudinal axis. Therefore, the z-displacement is not constant with respect to its cross section. Compared to the printed parts, the simulated deformations exceed the measured displacement of the manufactured cantilevers by 36.1%.This could be due to the necessary stiffness (for numerical stability) of the powder and liquid restricting the thermal strains more than in the real process. Another possible reason is the larger melt pool, which increases the local thermal strains and therefore increases the plastic deformation.

3.3 Recoater Collision

Recoater collisions cannot be detected in the build process of the cantilever. However, they can occur during printing of overhang structures. In the case of the simulated part, a collision was predicted in layer 162, which corresponds to a layer shortly after the overhang structure was started to be built.

The overhang structure of the test specimen collided indeed with the recoater during the experiment. Although the exact layer, in which the initial collision occurred could not be determined, the collision happened in the same geometrical feature of the test specimen as in the simulation.

As it can be seen in Fig. 4 b), the rubber lip was damaged, by the test specimen and had to be replaced after the built job. Nevertheless, the print job could be completed in this specific case. However, the quality of the part was impaired by damaged lip.

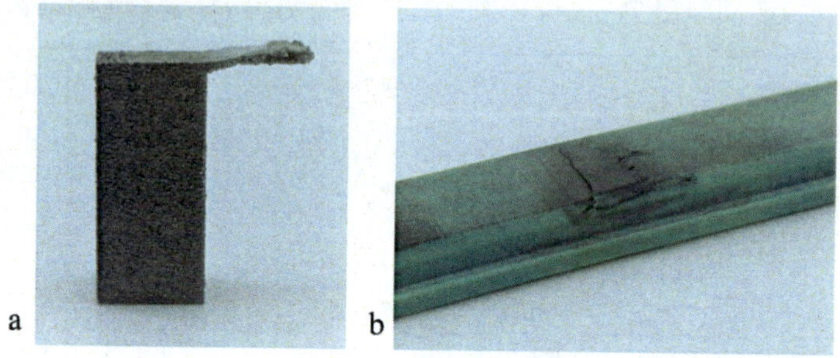

Fig. 4. Test specimen for the recoater collision (a) and the damaged recoater lip (b).

4 Conclusion

A sequential coupled two scale simulation model has been established to simulate part distortion and recoater collisions without the need of an additional experiment. The micro scale model is capable of simulating the inherent strains for given process parameters. The simplified heat source appears to be sufficient to get a good estimation of the resulting residual stresses and inherent strains. The computational effort can be reduced drastically in comparison to approaches with a Goldak heat source. On the macro scale, the prediction of a recoater collisions was demonstrated successfully.

References

1. Frazier, W.E.: Metal additive manufacturing: a review. J. Mater. Eng. Perform. **23**(6), 1917–1928 (2014)
2. Herzog, D., Seyda, V., Wycisk, E., Emmelmann, C.: Additive manufacturing of metals. Acta Mater. **117**, 371–392 (2016)
3. Mercelis, P., Kruth, J.P.: Residual stresses in selective laser sintering and selective laser melting. Rapid Prototyping J. (2006)
4. Buchbinder, D., Meiners, W., Pirch, N., Wissenbach, K., Schrage, J.: Investigation on reducing distortion by preheating during manufacture of aluminum components using selective laser melting. J. Laser Appl. **26**(1), 012004 (2014)
5. Alvarez, P., Ecenarro, J., Setien, I., Sebastian, M.S., Echeverria, A., Eciolaza, L.: Computationally efficient distortion prediction in powder bed fusion additive manufacturing. Int. J. Eng. Res. Sci. **2**(10), 39–46 (2016)
6. Setien, I., Chiumenti, M., van der Veen, S., San Sebastian, M., Garciandía, F., Echeverría, A.: Empirical methodology to determine inherent strains in additive manufacturing. Comput. Math. Appl. **78**(7), 2282–2295 (2019)

7. Keller, N. (2017). Verzugsminimierung bei selektiven Laserschmelzverfahren durch Multi-Skalen-Simulation. Doctoral dissertation, Universität Bremen (2017)
8. Bayat, M., et al.: Keyhole-induced porosities in Laser-based Powder Bed Fusion (L-PBF) of Ti6Al4V: high-fidelity modelling and experimental validation. Addit. Manuf. **30**, 100835 (2019)
9. Parry, L., Ashcroft, I.A., Wildman, R.D.: Understanding the effect of laser scan strategy on residual stress in selective laser melting through thermo-mechanical simulation. Addit. Manuf. **12**, 1–15 (2016)
10. Luo, Y., Murakawa, H., Ueda, Y.: Prediction of welding deformation and residual stress by elastic FEM based on inherent strain (report I): mechanism of inherent strain production (mechanics, strength & structure design). Trans. JWRI **26**(2), 49–57 (1997)
11. Zongo, F., Simoneau, C., Timercan, A., Tahan, A., Brailovski, V.: Geometric deviations of laser powder bed–fused AlSi10Mg components: numerical predictions versus experimental measurements. Int. J. Adv. Manuf. Technol. **107**(3–4), 1411–1436 (2020). https://doi.org/10.1007/s00170-020-04987-7
12. Li, C., Fu, C.H., Guo, Y.B., Fang, F.Z.: A multiscale modeling approach for fast prediction of part distortion in selective laser melting. J. Mater. Process. Technol. **229**, 703–712 (2016)
13. Klahn, C.: Laseradditiv gefertigte, luftdurchlässige Mesostrukturen: Herstellung und Eigenschaften für die Anwendung. Springer-Verlag (2015)
14. Bourell, D.L.: Perspectives on additive manufacturing. Ann. Rev. Mater. Res. **46** (2016)
15. Daňa, M., Zetková, I., Hanzl, P.: The influence of a ceramic recoater blade on 3D printing using direct metal laser sintering. Manuf. Technol. **19**(1), 23–28 (2019)
16. Dassault Systems. https://help.3ds.com/2021x/English/DSDoc/SIMA3DXANLRefMap/simaanl-c-amabout.htm?contextscope=cloud#simaanl-c-amabout. Last acsessed 04 May 2022
17. Li, Z., Li, B.Q., Bai, P., Liu, B., Wang, Y.: Research on the thermal behaviour of a selectively laser melted aluminium alloy: simulation and experiment. Materials **11**(7), 1172 (2018)
18. Ferrar, B., Mullen, L., Jones, E., Stamp, R., Sutcliffe, C.J.: Gas flow effects on selective laser melting (SLM) manufacturing performance. J. Mater. Process. Technol. **212**(2), 355–364 (2012)
19. Liu, C., et al.: Modeling of thermal behavior and microstructure evolution during laser cladding of AlSi10Mg alloys. Opt. Laser Technol. **123**, 105926 (2020)
20. Goldak, J., Chakravarti, A., Bibby, M.: A new finite element model for welding heat sources. Metall. Trans. B (Process Metall.) **15**(2), 299–305 (1984)
21. Martínez-Pañeda, E., Fuentes-Alonso, S., Betegón, C.: Gradient-enhanced statistical analysis of cleavage fracture. Eur. J. Mechan.-A/Solids **77**, 103785 (2019)
22. Uzan, N.E., Shneck, R., Yeheskel, O., Frage, N.: High-temperature mechanical properties of AlSi10Mg specimens fabricated by additive manufacturing using selective laser melting technologies (AM-SLM). Addit. Manuf. **24**, 257–263 (2018)
23. Cooper, K., Steele, P., Cheng, B., Chou, K.: Contact-free support structures for part overhangs in powder-bed metal additive manufacturing. Inventions **3**(1), 2 (2017)
24. Meiners, W.: Direktes selektives Laser Sintern einkomponentiger metallischer Werkstoffe. Dissertation, RWTH Aachen (1999)
25. Dilip, J.J.S., et al.: Influence of processing parameters on the evolution of melt pool, porosity, and microstructures in Ti-6Al-4V alloy parts fabricated by selective laser melting. Prog. Addit. Manuf. **2**(3), 157–167 (2017). https://doi.org/10.1007/s40964-017-0030-2

Digitization of the Manufacturing Process Chain of Forming and Joining by Means of Metamodeling

P. Brix[1](✉), M. Liewald[2], and M. Kuenzel[1]

[1] Mercedes Benz-AG, 71059 Sindelfingen, Germany
patrick.brix@mercedes-benz.com
[2] Institute for Metal Forming Technology, Holzgartenstraße 17, 70174 Stuttgart, Germany

Abstract. Manufacturing processes in sheet metal forming industry are subject to process-related variations, which can adversely influence the manufacturing costs and the quality of products. For example, during sheet metal forming of car body components, variations in material characteristics of the semi-finished product and in process parameters can lead to variations in the springback behavior of the sheet metal parts and therefore restrict the tolerances that can be realized with sufficient process reliability. In the assembly process, the springback variations of the individual sheet metal parts can also affect the dimensional accuracy of the joined sheet metal assembly and therefore the quality of the car body. In the course of digitizing the process chains in car body manufacturing, one of the objectives is to visualize such springback variations occurring after forming and joining processes of the individual sheet metal parts as well as of the sheet metal assembly at an early stage of development. On the one hand, this allows sheet metal parts and parameters with highest influence on the assembly to be identified and robustly designed, resulting in time and cost savings in hardware phase. On the other hand, tolerances for "less-important" parts of the assembly could be opened up, which may lead to additional cost and time reduction during die manufacturing. Against this background, the present paper provides an approach for modelling the manufacturing process chain of a forming and joining process considering variations in process parameters and material characteristics using finite-element-method and method of metamodeling. Here, metamodeling is used to predict the process behavior and thus reduce the required simulation effort. Based on the metamodels, Monte-Carlo simulation is carried out in order to perform variation and tolerance analysis.

Keywords: Sheet metal assembly · Manufacturing simulation · Variation

1 Introduction

Variations are an inevitable factor in mass-production manufacturing processes of sheet metal assemblies. Failure in handling such variations can adversely influence the manufacturing costs and the quality of products as wells as time-to-market. The analysis of

M. Liewald et al. (Eds.): WGP 2022, LNPE, pp. 376–385, 2023.
https://doi.org/10.1007/978-3-031-18318-8_39

occurring variations therefore gain a lot of attention in application research, especially in the automotive industry.

The variations in the sheet metal assembly process can be categorized into three main types, namely part variations, fixture variations and tooling variations [1]. This contribution focuses on the sheet metal part variations and their influence on the sheet metal assembly variation.

The main reason for sheet metal part variations in series production are varying material characteristics of the semi-finished product and changing process parameters. These varying material characteristics and process parameters influence the springback behavior and therefore lead to dimensional deviations of sheet metal parts. For this reason, stochastic sheet metal forming simulation is used to predict the part variations occurring during series production and their influence on the achievable part quality at an early stage of development [2].

In order to calculate such dimensional deviations of sheet metal assemblies in advance, Monte-Carlo (MC) based tolerance simulations are typically carried out. Such tolerance simulations are performed under the assumption of rigid bodies or by combining the linear finite-element-method (FEM) with the method of influence coefficient for non-rigid parts [3]. However, these simulation methods show some weaknesses in prediction accuracy, as they only assume the individual sheet metal part variations occurring during production, do not represent potential non-linear behavior of the manufacturing processes and have shortcoming in contact modeling [4]. This may lead to an overestimation of the sheet metal assembly variation and thus to tight tolerance specifications for the individual sheet metal parts.

Manufacturing process chain simulation using non-linear FEM provides an accurate prediction with respect to the dimensional state of a sheet metal assembly [5, 6]. However, the complexity and simulation effort of the manufacturing process chain simulation drastically increases when accounting for variations of parameters.

A solution approach for this is offered by the method of metamodeling, which allows to approximate the response of the manufacturing process chain simulation depending on the variation parameters and thus to reduce the simulation effort required. In addition, previous research activities show that the metamodeling method, including metamodel-based MC simulation provides high predictive accuracy with respect to springback variation for outer sheet metal parts [2] and to dimensional deviations of sheet metal assemblies [7].

Consequently, a combination of manufacturing process chain simulation and method of metamodeling could improve the numerical prediction accuracy of sheet metal part and assembly variations. The aim of the research work presented therefore, is to provide a metamodeling-based approach for continuous modeling the manufacturing process chain of sheet metal forming and assembly. This kind of modeling approach could improve the tolerance specification for sheet metal parts and assemblies and thus could lead to significant time and cost savings in the ramp-up of the body-in-white.

2 Simulation Methodology

In this contribution, a digital case study containing two sheet metal parts, which are joined to a sheet metal assembly, is presented (see Fig. 2). This digital case study was used to proof the ability of metamodeling the manufacturing process chain of sheet metal forming and sheet metal assembly. The non-linear FEM was used to perform the sheet metal forming and the sheet metal assembly simulation in order to predict the springback of the sheet metal parts after forming and of the sheet metal assembly after joining. The material characteristics Young's modulus and blank thickness of the sheet metal parts and the process parameters blankholder force and friction coefficient of the sheet metal forming processes were varied. The parameters were sampled and the manufacturing process chain simulation was performed to predict the springback of the sheet metal parts and the sheet metal assembly. Subsequently, the springback results were used to train metamodels for representing the process behavior depending on the varied parameters. Based on the metamodels, MC simulation was performed in order to predict the statistical springback variation for the sheet metal parts and the sheet metal assembly. Furthermore, multi-variate sensitivity analysis was carried out in order to identify the parameters with the highest influence on the springback of the sheet metal parts and the sheet metal assembly. The simulation approach is illustrated in Fig. 1.

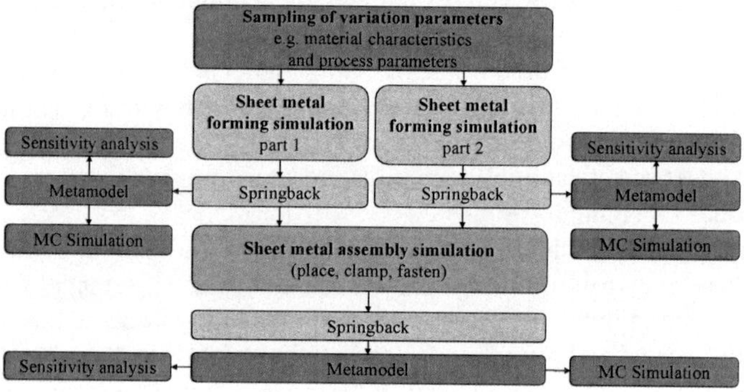

Fig. 1. Illustration of the simulation methodology

3 Simulation Setup

The simulation model is outlined in Fig. 2. An outer sheet metal part (OP, Al6014, t = 1.0 mm) and an inner sheet metal part (IP, Al5182, t = 1.15 mm) were formed, trimmed and joined by clinching to a sheet metal assembly. The FE solver LS-DYNA was used to perform the manufacturing process chain simulation. Here, after each simulation step a file was created containing information of the geometry, the blank thickness, the stress and strain state and the material history variables. These files served as input for each of the subsequent steps. As the complete simulation was performed with the same FE solver, mapping and morphing of the results was not required. All the above-mentioned information thus remained in the meshes and could therefore be transferred to all simulated process steps.

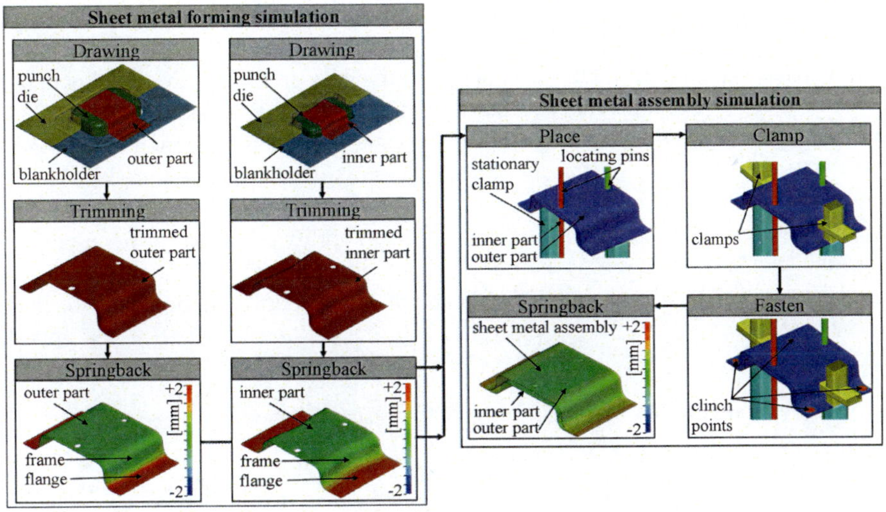

Fig. 2. Manufacturing process chain simulation model

3.1 Sheet Metal Forming Simulation

The sheet metal forming simulation of the outer and the inner sheet metal parts included positioning and gravity calculation of the blank, closing of the tools, forming, trimming and springback of the final sheet metal part with a 3-2-1 locating scheme. Fully integrated shell elements with seven through-thickness integration points were used. The size of a blank element was set to 1 mm resulting from a convergence study. In order to ensure the same meshes within the variation simulation, adaptive mesh refinement was deactivated. The contacts were modeled with a penalty based contact algorithm and Columb's law of friction. BarlatYLD2000 with the isotrop-kinematic hardening option was used as the yield criterion and the flow curve was described with the function of Hockett-Sherby.

3.2 Sheet Metal Assembly Simulation

The sheet metal assembly simulations were started with positioning of the sheet metal parts along two locating pins in the sheet metal assembly fixture. Here, gravity was taken into account to consider elastic deformations of the parts while positioning. After positioning, the sheet metal parts were clamped to their nominal position. One clamp was located between the flanges of the sheet metal parts and another clamp on the frame of the outer sheet metal part. The clamps were modeled as path-controlled rigid bodies. After clamping, the sheet metal assembly was created by clinching the formed sheet metal parts. Four clinching points, 8 mm from the outer edges of the flange, connect the sheet metal parts. A substitute model with beam elements representing the clinch points was used [8]. In the last step, the clamps were released and the springback of the sheet metal assembly was calculated with a 3-2-1 locating scheme. As in the forming simulation, a penalty based contact algorithm and Columb's law of friction were used for contact modeling. In addition, the same elasto-plastic material model as in the sheet metal forming simulation was used.

3.3 Variation Parameters

The main reason for sheet metal part variations in series production are varying material characteristics of the semi-finished product and changing process parameters. Therefore, the process parameters blankholder force and friction coefficient and the material characteristics Young's modulus and blank thickness were used as exemplary variation parameters for the sheet metal forming simulation of the outer and inner sheet metal part. The ranges were selected to ensure the manufacturability of the sheet metal parts. Six times the standard deviation of the normally distributed parameters corresponds to the defined range. Table 1 summarizes the mean, range, standard deviation and distribution for the parameters.

Table 1. Variation parameters for the sheet metal forming process

Process parameters	Mean	Lower	Upper	Standard dev	Distribution
Blankholder force OP [kN]	80	72	88	–	Uniform
Blankholder force IP [kN]	300	270	330	–	Uniform
Friction coefficient OP+IP	0.07	0.0675	0.0725	–	Uniform
Material characteristics	Mean	Lower	Upper	Standard dev	Distribution
Young's modulus OP+IP [GPa]	70	68	72	0.667	Normal
Blank thickness OP [mm]	1.00	0.985	1.015	0.005	Normal
Blank thickness IP [mm]	1.15	1.135	1.165	0.005	Normal

3.4 Metamodeling and Monte-Carlo Simulation

According to the last section, four parameters for every sheet metal forming process resulting in eight variation parameters for the manufacturing process chain simulation were defined. In this research, both linear and quadratic types of metamodels, radial-basis-function and feed-forward-neural-network were evaluated for representing the springback behavior after the sheet metal forming and the sheet metal assembly simulation. The minimum required amount of finite-element simulations considering eight variation parameters ($n = 8$) for building quadratic metamodels is 68 according to Eq. (1) [9].

$$\text{int}(0,75(n+1)(n+2)) + 1 \tag{1}$$

The D-optimal sampling method based on Space-filling was used and the uniform distribution was applied to the parameters in order to cover the design space as complete as possible. Based on the 68 simulations, the springback responses of the sheet metal parts after sheet metal forming and after sheet metal assembly were evaluated and used for training the metamodels. As a criterion for the metamodel accuracy, the root mean square error (RMSE) was calculated and compared for the different types of metamodels.

In the subsequent metamodel-based MC simulation, the variation parameters were sampled 10,000 times according to their defined probability distribution. Subsequently, the statistical results for the springback variation after sheet metal forming and after sheet metal assembly was evaluated. The complete simulation procedure was automatically performed using the software LS-Opt.

4 Results and Discussion

In the following, the results of the metamodel-based manufacturing process chain simulation will be presented. The metamodel accuracy for representing the springback after the sheet metal forming and after the sheet metal assembly process is presented, first. Subsequently, the results of the sensitivity analysis using sobol's variance-based sensitivity indices are outlined to show the influence of the investigated parameters on the springback after the sheet metal forming and after the sheet metal assembly process. Finally, the results of the metamodel-based MC simulation for prediction of the statistical springback variation conclude this section.

All the presented springback results are outlined for the example of one node in the middle of the flange and one node in the middle of the frame of the outer sheet metal part (see Fig. 2).

4.1 Metamodel Accuracy

The RMSE is displayed for the metamodel types linear and quadratic polynomial, radial-basis-function und feed-forward-neural-network in Table 2.

The highest accuracy for springback prediction after sheet metal forming and after sheet metal assembly was achieved with the quadratic polynomial and the neural-network metamodel. This was found in both the calculation of the springback of the flange and the frame of the outer sheet metal part. It was also observed, that the RMSE is always higher for the metamodels that predict the springback of the sheet metal assembly compared to the springback prediction for the sheet metal parts. However, with the maximum RMSE for springback prediction of 0.0377 mm for the linear polynomial metamodel, all the investigated metamodels showed a high accuracy.

Based on these results, the quadratic polynomial metamodel was chosen as the main metamodel for the following MC simulation due to the more general mathematical description of the model compared to the feed-forward-neural-network.

In addition, Fig. 3 visualizes the computed springback results from the finite-element simulation on the flange of the outer sheet metal part after the sheet metal assembly simulation versus the predicted springback by the main metamodel. The maximum deviation between the springback calculation of the finite-element simulation and the prediction of the quadratic polynomial metamodel was 0.08 mm for the flange and 0.003 mm for the frame of the outer sheet metal part.

4.2 Sensitivity Analysis

The results of the sensitivity analysis performed are displayed in Fig. 4. Here, each bar represents the influence of the variables on the springback of the flange and the frame of the outer sheet metal part after sheet metal forming and after sheet metal assembly.

Table 2. Comparison of RMSE (normalized by mean) for different metamodels

Metamodel sheet metal forming	Springback flange OP (mm)	Springback frame OP (mm)
Linear polynomial	0.0040 (0.23%)	0.00242 (0.42%)
Quadratic polynomial	0.0017 (0.10%)	0.00081 (0.14%)
Radial-basis-function	0.0025 (0.14%)	0.00091 (0.16%)
Feed-forward-neural network	0.0018 (0.10%)	0.00071 (0.12%)
Metamodel sheet metal assembly	*Springback flange OP* (mm)	*Springback frame OP* (mm)
Linear polynomial	0.0377 (4.43%)	0.0144 (6.13%)
Quadratic polynomial	0.0216 (2.53%)	0.0087 (3.68%)
Radial-basis-function	0.0280 (3.29%)	0.0110 (4.70%)
Feed-forward-neural network	0.0089 (1.04%)	0.0038 (1.62%)

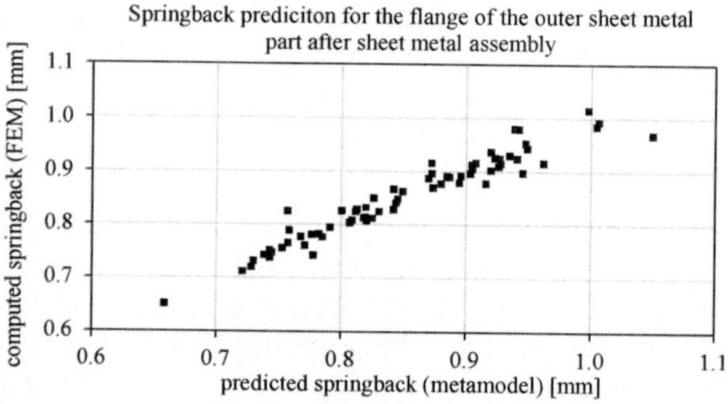

Fig. 3. Metamodel (quadratic polynomial) accuracy on the flange of the outer sheet metal part for springback prediction after sheet metal assembly

For the springback of the outer sheet metal part's frame and flange after sheet metal forming, the blankholder force had the highest influence (76.4 and 52.8%). This was followed by the initial blank thickness and the Young's modulus.

For the springback of the outer sheet metal part's frame and flange in the assembled state, the blankholder force of the inner sheet metal part had the highest influence (23.5 and 30.4%). In general, 74.5 and 67.7% of the influence on the springback of the outer sheet metal part's flange and frame in the assembled state resulted from the variables of the sheet metal forming process of the inner sheet metal part. This result indicates that the inner sheet metal part and its variation parameters contribute much more to the springback of the sheet metal assembly than the outer sheet metal part and its variation parameters. This is possibly due to the greater sheet thickness and the geometry of the inner sheet metal part, which result in a higher stiffness of the part.

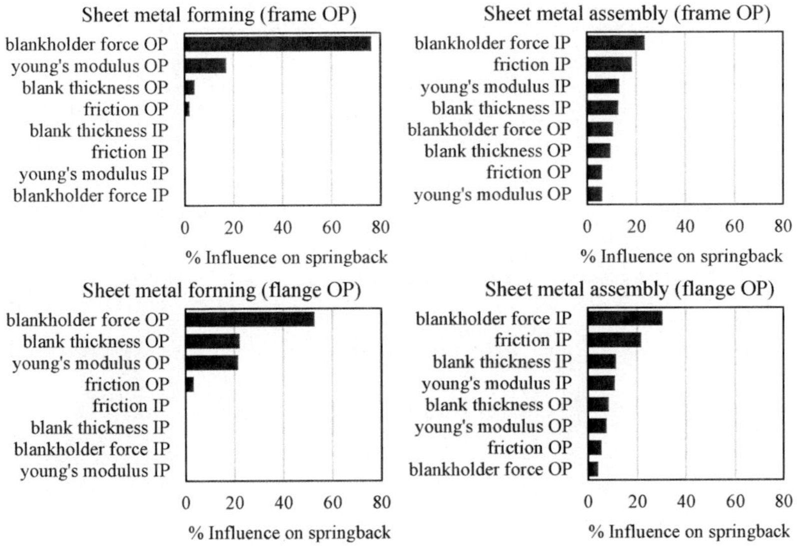

Fig. 4. Results of multi-variate sensitivity analysis (SOBOL) for springback of the outer sheet metal part's frame and flange after sheet metal forming and sheet metal assembly

4.3 Monte-Carlo Simulation

In accordance to the results presented previously, Fig. 5 displays the statistical springback results for the outer sheet metal part's flange and frame obtained from the metamodel-based MC simulation after sheet metal forming and after sheet metal assembly.

The mean springback of the outer sheet metals part's flange and frame after sheet metal forming was 1.74 and 0.57 mm. After the sheet metal assembly process, the mean springback was reduced to 0.85 mm for the flange and 0.27 mm for the frame. The compensation of 0.89 and 0.30 mm in the mean value might result from the clamping and joining of the sheet metal parts in their nominal position. The standard deviation, representing the springback variation was 0.035 mm for the flange and 0.0172 mm for the frame of the outer sheet metal part after sheet metal forming. In the assembled state, the standard deviation was higher for the flange (0.0526 mm) and remained the same for the frame (0.0179 mm). The higher standard deviation of the outer sheet metal part's flange in the assembled state might be due to the high standard deviation of the springback of the inner sheet metal part's flange (0.0448 mm, not displayed in Fig. 5).

Based on the obtained results from the metamodel-based MC simulation, it can be recommended to open tolerance for the frame of the outer sheet metal part (e.g. from ±0.3 to ±0.6 mm). The results also indicated that the springback variation of the sheet metal assembly can be reduced if the springback variations are optimized for the inner sheet metal part.

5 Conclusion and Summary

In this contribution, a metamodel-based approach for continuous modeling the manufacturing process chain of sheet metal forming and sheet metal assembly for prediction

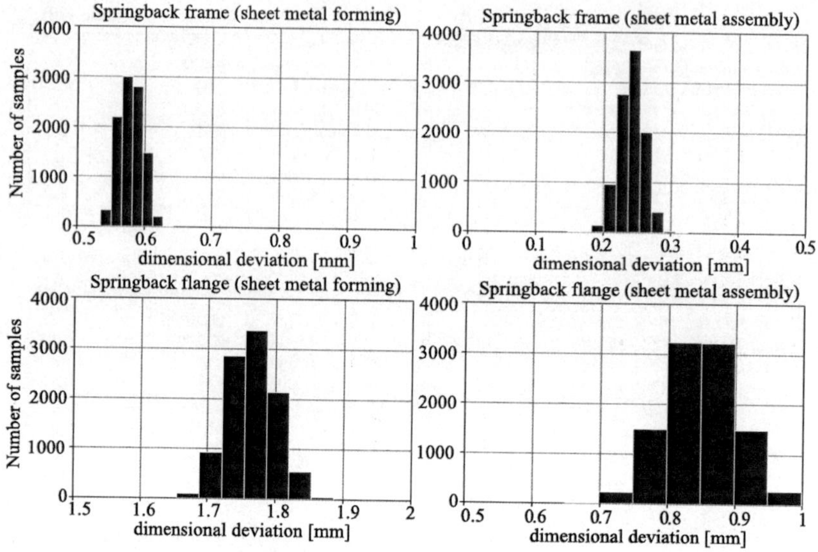

Fig. 5. Results of the metamodel-based MC simulation for the springback variation of the outer sheet metal part after sheet metal forming and sheet metal assembly

of sheet metal part and sheet metal assembly variation was presented. The modeling approach was based on manufacturing process chain simulation by the non-linear finite-element-method considering variations in process parameters and material characteristics. The method of metamodeling was used to perform a metamodel-based MC simulation for the prediction of sheet metal part and sheet metal assembly variations. The ability of the metamodeling approach was proofed on the example of a digital sheet metal assembly comprising an inner and outer sheet metal part and the sheet metal forming variation parameters blankholder force, friction coefficient, Young's modulus and initial blank thickness.

The investigated metamodels were trained with the springback results from 68 manufacturing process chain simulations. The results showed, that the polynomial quadratic metamodel had a high accuracy for springback calculation after sheet metal forming and after sheet metal assembly with RMSE of 0.00168 and 0.0216 mm. However, the analysis of the metamodel accuracy also showed, that for all investigated metamodels the RMSE was always higher for springback calculation of the sheet metal assembly compared to the springback calculation of the sheet metal parts.

MC simulation was performed on the metamodels for prediction of the statistical springback variation after the manufacturing processes. This metamodel-based variation analysis indicated compensation effects up to 0.89 mm for the mean value of the springback of the outer sheet metal part due to the sheet metal assembly process. However, the statistical springback variation of the outer sheet metal part was higher after the sheet metal assembly process compared to the state after sheet metal forming. This was due to the high springback variation of the inner sheet metal part after forming, which contributes more to the springback state of the sheet metal assembly.

Multi-variate sensitivity analysis was carried out in order to show the influence of the variation parameters on the springback after the sheet metal forming and the sheet metal assembly process and reinforced the obtained results from the variation analysis. The sensitivity analysis showed, that the blankholder force with 76.4% had the highest influence on the springback of the outer sheet metal part after sheet metal forming. However, after the sheet metal assembly process, up to 74.5% of the influence on the springback of the same point on the outer sheet metal part contributed from parameters of the forming process of the inner sheet metal part. This result indicated a much higher influence of the inner sheet metal part on the dimensional accuracy of the sheet metal assembly.

In future research, the presented simulation approach will be transferred to real car-body assemblies comprising manageable amount of sheet metal parts. By using the proposed simulation approach, metamodel-based optimization and variation analysis can indicate opening of tolerances for sheet metal parts and parameters with minor influence on the dimensional accuracy on the sheet metal assembly. This will lead to time and cost savings in die manufacturing and the ramp-up of car bodies.

References

1. Hu, M., Lin, Z., Lai, X., Ni, J.: Simulation and analysis of assembly processes considering compliant, non-ideal parts and tooling variations. Int. J. Mach. Tools Manuf **41**, 2233–2243 (2001)
2. Brix, P., Liewald, M., Eckstein, J.: Predicting springback variation and process-reliable tolerance limits of outer car-body panels by stochastic sheet metal forming simulation. IOP Conf. Ser. Mater. Sci. Eng. **1157** (2021)
3. Liu, S.C., Hu, S.J.: Variation simulation of deformable sheet metal assemblies using finite element methods. J. Manuf. Sci. Eng. **119**(3), 368–374 (1997)
4. Dahlström, S., Lindkvist, L.: Variation simulation of sheet metal assemblies using the method of influence coefficients with contact modeling. J. Manuf. Sci. Eng. **129**(3), 615–622 (2006)
5. Govik, A., Nilsson, L., Moshfegh, R.: Finite element simulation of the manufacturing process chain of sheet metal assembly. J. Mater. Process. Technol. **212**, 1453–1462 (2012)
6. Konrad, T.: Simulative Auslegung der Spann- und Fixierkonzepte im Rohbau, FAU Studien aus dem Maschinenbau 319. Universität Erlangen (2019)
7. Zheng, H., Upadhyay, K., Litwa, F., Paetzold, K.: A meta-model based approach to implement variation simulation for sheet metal parts using mesh morphing method. In: 13th European LS-DYNA Conference, Ulm (2021)
8. Kästle, C.: Simulationsmethode zur Beurteilung der Maßhaltigkeit von rollgefalzten Karosseriebaugruppen im Zusammenbau, Beiträge zur Umformtechnik 80. Universität Stuttgart (2016)
9. Stander, N., Basudhar, A., Roux, W., Liebold, K., Eggleston, T., Goel, T., Craig, K.: LS-OPT user's manual, version 7.0, Livermore Software Technology (2020)

Analysis of Cryogenic Minimum Quantity Lubrication (cMQL) in Micro Deep Hole Drilling of Difficult-to-Cut Materials

M. Sicking(✉), J. Jaeger, E. Jaeger, I. Iovkov, and D. Biermann

Institute of Machining Technology, TU Dortmund University, Baroper Straße 303, 44227 Dortmund, Germany

martin.sicking@tu-dortmund.de

Abstract. In modern manufacturing processes, the environmental impact becomes an increasingly important aspect. The aim of developing new coolant strategies is therefore an approach to increase the efficiency of the machining process while reducing coolant consumption. The priority is to optimize the supply of coolant to the tool-workpiece interface. In case of cryogenic machining, low-temperature liquefied gases are used to cool the tool's cutting edges and to decrease the overall process temperatures. The high cooling rates of this technology can reduce the thermomechanical loads for tools and workpieces especially in machining operations. Since cryogenic medium have no lubricating effect, additional lubrication strategies, e.g. Minimum Quantity Lubrication (MQL), are necessary to enhance the application limits of the cryogenic cooling technology. Nevertheless, in deep hole drilling, using small diameter twist drills, it is impossible to supply internal cryogenic coolant and MQL simultaneously. Therefore, this paper deals with a novel combination of cryogenic Minimum Quantity Lubrication (cMQL) by determining the lubricant's efficiency according to its solubility in liquid CO_2. The strategy leads to a significant increase in performance during deep hole drilling of difficult-to-cut materials and shifts the process limits in terms of tool life and feasible cutting parameters using environmentally friendly MQL techniques.

Keywords: cryogenic MQL · difficult-to-cut materials · small size diameter · twist drill · deep hole drilling · sustainable manufacturing

1 Introduction

Motivated by ecologic and economic demands, cutting processes are increasingly realized without emulsion-based coolant. In mass production, especially in automotive industry applications, it is required to design near dry cutting operations using minimum quantity lubrication. However, in drilling processes it is difficult to realize near dry machining, e.g., in machining of difficult to cut materials such as nickel-based alloy Inconel 718 and stainless steel X90CrMoV18. In such cases, the bottleneck is given by a combination of poor chip evacuation and low cooling rates [1].

M. Liewald et al. (Eds.): WGP 2022, LNPE, pp. 386–395, 2023.
https://doi.org/10.1007/978-3-031-18318-8_40

In addition, due to miniaturization trends in many industry sectors, an increasing demand for small-size parts and structures can be found. Hence, reliable manufacturing processes have to be designed to ensure economic value creation. Nevertheless, using small diameter twist drills, internal MQL supply becomes a major problem. Miniaturized cooling channels only enable the transfer of small amounts of coolant. Hence, using internally applied minimum quantity lubrication in demanding dimensions becomes a challenge. Consequently, insufficient lubricant supply could lead to increased tool wear and premature tool failure. Thus, the development of new strategies is necessary to extend process limits of near dry manufacturing processes. One attempt for reliable deep hole drilling processes is realized by high pressure coolant supply. Here a small amount of emulsion based coolant is applied at pressures of $p > 80$ bar. Unfortunately, energy consumption of the required equipment is relatively high, e.g. high pressure pumps [1, 2].

In order to further improve cutting processes, various studies on cryogenic coolants with a focus on process productivity, tool wear and workpiece quality, have been presented in recent decades. These technologies either use liquefied nitrogen (LN_2) or liquefied carbon dioxide (CO_2) to cool the cutting process [3]. The needed equipment defines differentiating factors between these two solutions. Since LN_2 is provided at temperature of $T = -196$ °C of coolant ducts through machine spindle and machine tool need to be insulated. In comparison, CO_2 technique can easily be retrofitted to existing machine tools. For that, the cryogenic coolant can be stored in gas cylinders at a pressure of $p = 57$ bar and at a room temperature ($T = 20$ °C) [2]. Consequently, liquefied CO_2 can be transferred to the point of action without the need for insulation. Expansion of liquefied CO_2 leads to cooling of the gas due to phase transformation and the Joule Thomson effect. To maximize this effect, pressure loss in gas pipes should be reduced to a minimum. Due to the expansion of CO_2, temperature of $T = -79$ °C are achieved [4, 5]. The relaxation of the pressure liquefied gas at the tool tip results in a snow jet containing dry ice crystals and cols CO_2 gas both of which are applied to the point of cutting. Since the lubrication effect of CO_2 is poor, cryogenic cooling can cause high friction at the tool-chip interface and increased tool wear. Thus, cooling techniques providing both cooling and lubrication have to be developed [6], e.g. by adding MQL liquids to the liquefied CO_2. Usually, both liquids are delivered to the chip formation or cutting zone by different pipes and applied individually [7–9]. However, internal supply of both liquids cannot be realized whenever dimensions of tools or tool interfaces with machine spindles do not allow appropriate piping.

A new approach to near dry machining and combining the advantages of CO_2 and MQL has been achieved by generating a single-phase solution from MQL fluids and liquefied CO_2. Thus, liquefied CO_2 is used as transport medium. Owing to the expansion of the CO_2 at the tip of the tools, MQL liquids get atomized and conveyed by the CO_2 flow into the cutting zone. This single-phase cooling and lubricating dissolution does not require special equipment, just one small pipe that easily can be integrated into machine tool. Most importantly, separation by centrifugal forces does not take place. Thus, cMQL can be used in combination with high spindle speeds as required in high speed cutting and machining with small diameter tools.

2 Experimental Setup

In the first place, solubility of MQL fluid in pressure liquefied CO_2 must be tested and optimized accordingly. In addition, cutting tests are required to evaluate the potential of the cMQL. Thus, in order to study effects and performance of cMQL supported metal cutting processes different research areas have to be taken in account. A critical deep hole drilling process using small size twist drills to machine difficult to cut materials has been defined to determine enhanced process limits. Preliminary test were realized with respect to probability of tool breakage due to insufficient lubrication.

2.1 Laboratory Setup to Evaluate Solubility of Different MQL-Fluids

Different MQL fluids were tested with respect to their miscibility with liquefied CO_2. The test rig and additional equipment are depicted in Fig. 1. CO_2 and MQL fluid are filled and mixed in a view cell.

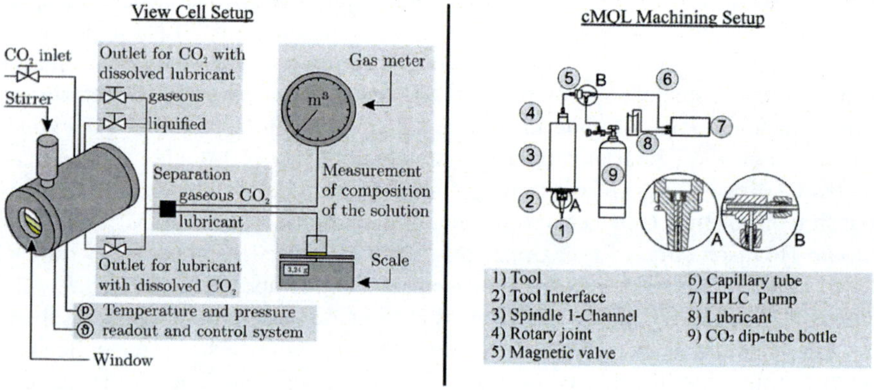

Fig. 1. Left: High pressure view cell; Right: Setup of machining system for cMQL

Pressure as well as temperature of components to be dissolved are controlled and can be adjusted individually. To measure phase equilibrium, samples can be drained off using various outlet valves placed in different positions of the cell. Thus, three different phases of fluids can be observed: CO_2 in lubricant, lubricant in liquefied CO_2 and lubricant in gaseous CO_2. Additionally, the mixing processes and the status of the system can be monitored and documented through an observation window at the front of the view cell. Measuring the phase-equilibrium of mixed fluids took place via a device in which CO_2 and lubricant are separated. In this process, the MQL sample is first deposited in a test tube. Due to the pressure drop, the CO_2 contained in the mixture becomes gaseous and escapes through a gas meter, which measures the contained CO_2 gas volume. Furthermore, the mass of separated lubricant is measured by a micro scale in order to analyze sample composition. At least, residual CO_2 dissolved in separated MQL fluid is eliminated e.g. by heating the sample. The contained CO_2 and lubricant mass can be determined from the volume of escaped CO_2-gas, in the mass of CO_2 dissolved in MQL fluid and the mass of the taken MQL fluid itself. Hence, the exact lubricant and CO_2 composition of the taken sample can be determined. During sampling, it is particularly important to keep pressure in the view cell constant, to avoid disturbance of phase equilibrium. Therefore, a hydraulic piston is used to compensate taken sample volume in the view cell.

As depicted in Fig. 2, a phase diagram can be created, showing the phase equilibrium of CO_2 and MQL fluid depending on temperature and pressure. Most importantly, these test do not only provide information on the maximum soluble amount of lubricant in CO_2 but also on the time needed to generate a saturated solution. These data must be considered, when designing a mechanical system to be integrated in production machine tools.

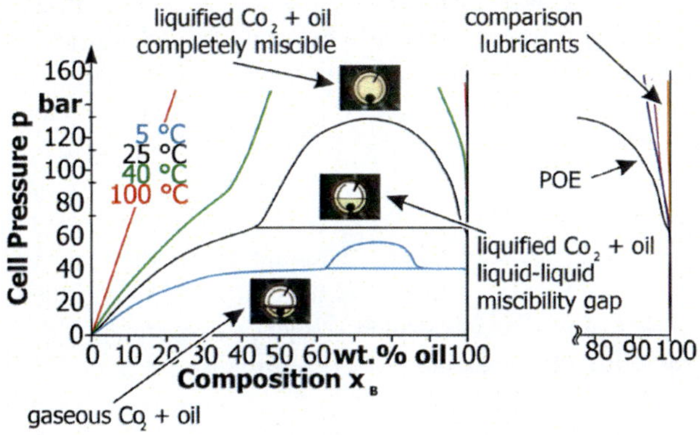

Fig. 2. Phase diagram of a system of polyolester (POE) and CO_2 compared to e.g. mineral oil [based on [10, 11]]

2.2 Machining Setup

As mentioned before, it is beneficial to generate single-phase solution in order to prevent separation of substance during discharge through pipes and machine spindle. On the other hand, MQL fluids not only need to enable the formation of single-phase systems, but also provide lubrication to improve performance of cutting processes. Therefore, a test setup was designed to evaluate the performance of various cMQL compositions in drilling small diameter holes using standard twist drills (Fig. 1). Injection of MQL fluid into CO_2 is controlled by a magnetic valve, which is mounted close to the rotary joint. The lubricant is injected through a second stainless capillary tube in a tee connector, just before the valve. Flow rate and pressure of MQL fluid are controlled by a HPLC-pump (High Performance Liquid Chromatography), which is used to feed the lubricant. The machine tool used enables spindle speeds of up to $n = 30000$ rpm and feed rated up to $v_f = 10$ m/min, thus, HSC and HPC machining processes can be realized.

2.3 Test Planning

This work focuses deep hole drilling in difficult to cut materials using small diameter twist drills. Hence, solid carbide drills with a diameter of $d = 1.4$ mm were used to drill through holes. Workpieces were made out of material 2.4668 (Inconel 718) and 1.4112 (X90CrMoV18) respectively. The tools enable boreholes with a length-to-diameter ratio of l/d = 8, thus, workpiece thickness of $s = 11.2$ mm was selected to fully exploit this ratio. The internal cooling channels of tools have a diameter of $d_{cc} = 0.2$ mm. Thus, low flow rates of CO_2 can be expected and no additional nozzles are required. In general, drilling test were realized under environmental conditions of $T_r = 20$ °C and $pCO_2 = 57$ bar which gives a mass flow of $\dot{m}$ $CO_2 = 2.2$ kg/h. Cutting parameters were set according to tool manufacture's recommendations. Therefore, feed was set to $f = 0.021$ mm. Nevertheless, in drilling 2.4668, cutting speed has to be reduced to $v_c = 25$ m/min whereas 1.4112 can be machined with a cutting speed of $v_c = 60$ m/min. Additionally, pilot holes were used with a depth of $t = 2$ mm. In order to enable a comparison of tool performance when applying cMQL with results achieved with conventional MQL lubrication, appropriate drilling tests were conducted as well. Drilling of nickel-based alloys tends to generate relatively long chips, which tend to jam in deep holes. Therefore, peck drilling was applied to improve chip removal and to minimize risk of premature tool failure. The number of strokes was set to $n = 1$ for cMQL and for eMQL to n = 3 to realize that the cutting edge is sufficiently lubricated. The increased number of machining strokes reduced the thermal load and consequently the tool wear. Evaluation of process and tool performance is based on feed force F_f and drilling torque T_D measured at different intervals and numbers of processes realized. To determine process characteristics the mechanical tool loads were measured at intervals of 25 holes in case of 2.4668 and 50 holes in case of 1.4112 since a higher tool life can be expected in drilling stainless steel compared to nickel based alloys. Furthermore, to evaluate tool wear, chip shape and burr formation a digital microscope was used.

3 Experimental Results

Based on results gained in preliminary tests the solubility of three polyolesters (POE) with different additive compounds have been examined with respect to their capability to develop single-phase systems. The most promising combinations were used to perform cMQL deep hole drilling operations into X90CrMoV18 and Inconel 718.

3.1 Solubility of Lubricants

Three different POE were tested and their solubility in CO_2 was evaluated using the pressure cell shown in Fig. 1. Objective of this study was to find out whether dissolution is depending on mixing respectively processing time. Therefore, tests were carried out at room temperature of $T_r = 20$ °C and a pressure of $pCO_2 = 57$ bar. The dissolution strategy was realized in order to observe short time characteristics, near to the conditions within the machine tool. Thus, stirring was done for $t = 5$ s only, stand still time under pressure was set to $t = 60$ s. These strategy was applied to determine in a first step, the thermodynamic equilibrium solubility. In the second step, it was investigated how this equilibrium can be reached at short contact times. These results are important to know since the mechanical set-up for real machining applications allows only short contact times and it is essential to know whether a single phase or two phases are present at the tip of the tool. The most important results of the solubility test (POE 1 to POE 3) are given in Table 1. For statistical reasons three samples (s.1 to s.3) were taken to determine the amount of MQL fluid dissolved.

Table 1. Measured solubility of lubricants in liquefied CO_2 after t = 65 s

T	POE 1	POE 2	POE 3
	65 s	65 s	65 s
s.1	5.2%	3.6%	0.1%
s.2	2.4%	6.0%	1.8%
s.3	2.8%	4.9%	0.0%
$\bar{x}$	3.5%	3.6%	0.6%

As it can be seen, the average value of solubility of POE 1 and POE 2 is almost constant and relatively high, whereas POE 3 is less soluble. However, it has to be considered, that the deviation of the data gained by three samples is quite high. Consequently, additional and intensive research work has to be done in order to improve reliability of these data. Nevertheless, it can be expected that performance of cutting processes applying either POE 1 or POE 2 will be better compared to processes using POE 3. In Fig. 3, effects and characteristics of dissolution processes are shown after a short stirring and relaxation time of the fluids.

Arrows indicate the different levels of fluids within the cell. The lower part of the cell is filled with MQL fluid or a single-phase solution of MQL fluid and CO_2 after stirring respectively. In the middle of the view cell, there is liquid CO_2 or a solution of liquid CO_2 and MQL fluid respectively, were samples are taken from. Obviously, since the volume of MQL fluid is higher after stirring, CO_2 is dissolved into POE 1 and POE 2 leading to a swelling of the liquid phases. This is especially important regarding machining processes, since there will be almost no specific relaxation time because of

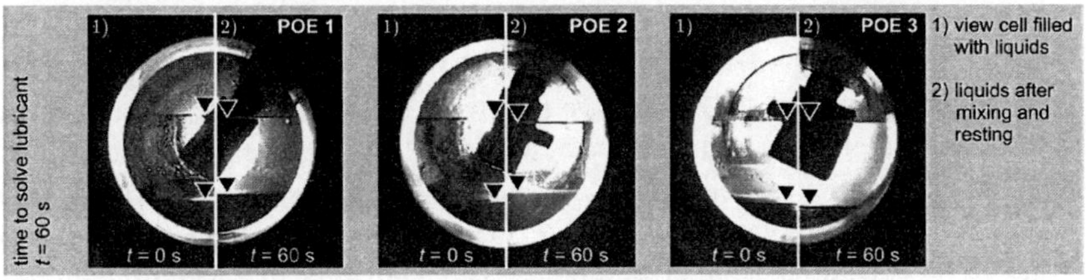

Fig. 3. Observed phase behavior of tested lubricants with liquefied CO_2

the limited length of the pipes providing the fluid through the machine spindle to the tip of tool. Thus, POE 3 probably will not be a suitable for cMQL.

3.2 Cutting Tests

In order to assess potentials of cMQL, deep hole drilling was realized in workpieces made of nickel-based alloy 2.4668 (Inconel 718) and in stainless steel 1.4112 (X90CrMoV18). For the tool life travel path tests, the standard cutting parameters were used (see Chap. 2.3). Furthermore, machining tests were done with different MQL systems. Beside internal cMQL, external (eMQL, $\dot{V}$ = 75 ml/h) as well as an internal double channel MQL system (iMQL, $\dot{V}$ = 75 ml/h) were applied. For internal cMQL the oil flow rate was set to $\dot{V}_{Oil}$ = 6 ml/h, which corresponds to a concentration of $\bar{x}_B$ = 0.5% of lubricant in the applied coolant mass flow. In order to further investigate potential of cMQL-technique, flow rate of MQL fluid was set to $\dot{V}_{Oil}$ = 30 ml/h ($\bar{x}_B$ = 1.5%) in a second test. The tool life travel path defined the performance of the different configurations. A summary of the results is given in Table 2. It can be seen that external eMQL is not suitable for deep hole drilling in the materials given.

Table 2. Tool life for different MQL systems ($\dot{V}_{Oil,\ cMQL}$ = 6 ml/h; $\dot{V}_{Oil,\ iMQL/eMQL}$ = 75 ml/h)

	cMQL			iMQL	eMQL
	POE 1	POE 2	POE 3	POE	POE
2.4668	1.72 m	1.29 m	1.11 m	<0.1 m	<0.1 m
1.4112	18.4 m	6.05 m	15.18 m	13.3 m	<0.1 m

In case of drilling X90CrMoV18 (1.4112), tool life significantly is enhanced by iMQL compared to eMQL technology. On the other hand, depending in the POE compound there are different effects, positive and negative, onto tool life when applying cMQL as compared to iMQL. Based on the findings on solubility as well as the tool life studies, the following investigations regarding cMQL were carried out using POE 1. As shown in Fig. 4 wear of cutting edges and formation of build-up edges significantly depend on the volume of MQL fluid applied in cMQL as well as in iMQL technology. As it can be seen, exceptional wear marks can be reached easily. It should be mentioned, that cutting tests were stopped once 2000 holes were drilled. Thus, when using cMQL with POE 1, the end of tool life was not reached yet despite of relatively high tool wear. The comparison of tool wear at length of feed l_f = 13.3 m shows, that influence of cooling with less amount of MQL fluid on tool wear is low compared to built-up edge. Nevertheless, by supplying a high volume of lubricant, not only wear rate can be lowered but also an improved process reliability can be achieved since short-shaped chips can be removed easily from high depth. This also relates to burr formation, which did not take significant effect even after 2000 drilled holes. That confirms, that a combination of an efficient cooling and lubrication has to be applied to improve drilling process, like it is in cMQL.

Changing the workpiece material to Inconel 718 (2.4468), thermomechanical load onto tools is much higher. Thus, in drilling nickel-based alloys, the effect on performance and tool life is massive when cMQL technology is realized (Table 2).

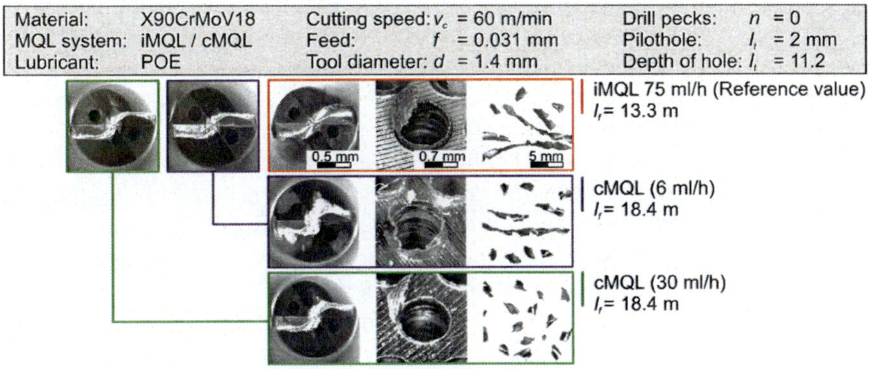

Fig. 4. Comparison of tool wear and burr formation in drilling X90CrMoV18

As compared to conventional MQL, cooling of cutting processes seems to be the key factor, especially when considering the challenging dimensions of the tool. When drilling with eMQL tool failure in most cases occurred, once 5 to 10 processes had been realized. Furthermore, chip formation was affected significantly by coolant technology chosen. Thus, drilling with eMQL generates long threaded chips, which tend to jam in chip flutes and cause tool failure (see Fig. 5 right side). Whereas the high cooling capacity of cMQL ensures formation of discontinuous chips. In addition, process capability and technological process parameters can be increased since discontinuous chips are formed. Hence, cutting speed was increased to 125% ($v_C = 31.25$ m/min) compared to manufacturer's recommendations (red graph). Accordingly, cutting Inconel 718 and applying cMQL productivity of processes can be increased by about 17% (cycle time). A detailed evaluation of process signals measured clearly indicates that in case of drilling Inconel 718, feed force (F_f) and drilling torque (T_D) are significantly smaller when cMQL is used. In addition, force and torque almost stay at the same level during depth of bore.

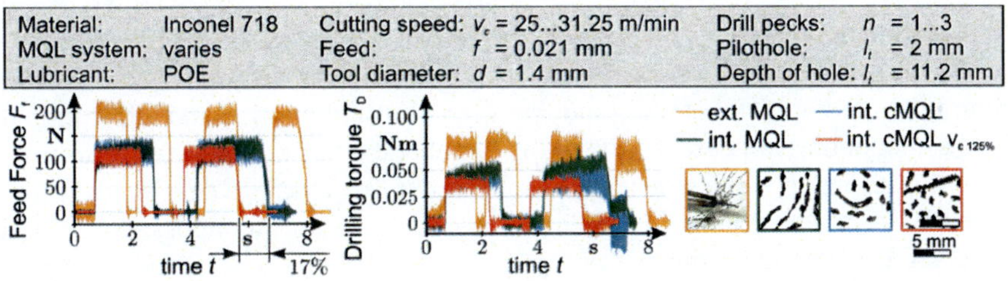

Fig. 5. Measured process force signals and collected chips during drilling Inconel 718

Figure 6 shows the results of drilling tests using a higher concentration of lubricant in the solution (blue) and a cutting speed of $v_c = 31.25$ m/min. As it was shown in drilling X90CrMoV18, a combination of high cooling and lubrication capacity increases tool life significantly. The most important criterion in this process seems to be stability of cutting edges, since breakouts occur in two out of three tools. Differences in the feed force at the beginning of the process can be attributed to the slight deviation of the tool shape. These deviations are within the manufacturer's tolerance. The cause of breakout in tool 1 can easily be found in the jammed cooling channel (Fig. 6 red drill). In case of tool 2, breakout appears after approximately 500 drilled holes. Despite this breakout, however, tool 2 was able to drill a total of 2000 holes. As expected, feed force (F_f) and drilling torque (T_D) were significantly higher after this tool failure occurred. In drilling test of tool 3, 2000 holes were drilled without any process disturbance. Thus, this test clearly shows the potential of cMQL technique. Comparing the chip shape at different feed travel path reflects the high process reliability, since no change can be recognized during the whole tool life. Furthermore, nearly no burr formation can be identified despite the high length of feed. Finally, the uniform wear of tool along the cutting edge, which was achieved with every tool during these tests, should be mentioned.

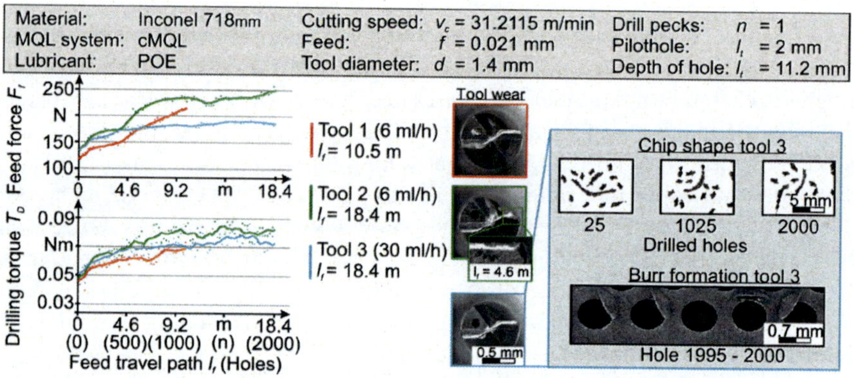

Fig. 6. Measured process force signals, tool wear, burr formation and collected chips during drilling Inconel 718 with cMQL application

4 Conclusion

Machining difficult to cut materials and applying MQL for cooling and lubrication purposes is a challenging process, owing to the fact that the technological window is quite narrow. Depending on workpiece material and machining task, e.g. deep hole drilling of Inconel 718, tool life is poor and reliable process design is almost impossible. To further improve MQL technique, liquefied cryogenic gas (CO_2) was used. Most importantly, MQL fluids, which form single-phase systems, were identified. This feature is prerequisite to enable internal supply of MQL to the tool tip. In doing so, process limits can be significantly increased. The performance of this technology was demonstrated in deep hole drilling of Inconel 718 and X90CrMoV18. In Addition, outstanding tool life can be achieved, if cMQL parameters are optimized with respect to amount MQL dissolved in CO_2 and conveyed to the chip formation zone. Thus, potential of the presented cMQL application is huge, since increased productivity, tool life and bore hole quality could be demonstrated. To further determine process limits and to use full potential of cMQL technique, additional investigations are required. In future investigations the focus will be set on interdependency between lubricant respectively MQL fluid and liquefied CO_2. Furthermore, relationship between process and machining technology must be investigated to gain fundamental process knowledge. Thus, e.g. solubility of different lubricants and their capability as lubricant in machining difficult to cut materials using cMQL will be considered as well. Therefore, drilling Inconel 718 with small diameter twist drills will be investigated by observing tool chip interface in varying cMQL conditions to analyze influence of CO_2 parameters, e.g. influence of pressure p_{CO2} and temperature T_{CO2}, on tool wear (e.g. width of wear mark), borehole quality (e.g. surface roughness, straightness deviations) and chip formation.

Acknowledgement. Gefördert durch die Deutsche Forschungsgemeinschaft (DFG, German Research Foundation) – Projektnummer 452408713. The authors would also like to thanks the project partner from the Chair of Particle Technology, Ruhr-University-Bochum.

References

1. Fratila, D.-F.: Environmentally friendly manufacturing processes in the context of transition to sustainable production. Comprehens. Mater. Proc. **8**, 163–175 (2014/05)
2. Busch, K., Hochmuth, C., Pause, B., Stoll, A., Wertheim, R.: Investigation of cooling and lubrication strategies for machining high-temperature alloys. Proced. CIRP, **41**, 835–840 (2016/12)
3. Jawahir, H.A., Biermann, D., Duflou, J., Klocke, F., Meyer, D., Newman, S., Pusavec, F., Putz, M., Rech, J., et al.: Cryogenic manufacturing processes. CIRP Annals **65**(2), 713–736 (2016)
4. Barber, C.R.: The sublimation temperature of carbon dioxide. Br. J. Appl. Phys. **17**(3), 391 (1966)
5. Pursell, M.: Experimental investigation of high pressure liquid CO_2 release behavior. Hazards Sympos. Ser. **158**, 164–171 (2012)
6. Clarens, F., Hayes, K.F., Skerlos, S.J.: Feasibility of metalworking fluids delivered in supercritical carbon dioxide. J. Manuf. Process. **8**(1), 47–53 (2006)
7. Astakhov, V.P.: Ecological Machining: near-dry Machining, pp. 195–223, Springer London, London (2008)
8. Pereira, O., Catalá, P., Rodriguez, A., Ostra, T., Vovancos, J., Rivero, A., Lépez-de Lacalle, L.: The use of hybrid co_2 + mql in machining operations. In: Procedia Engineering, vol. 32, pp. 492–499, Barcelona (2015)
9. Pereira, O., et al.: Internal cryolubrication approach for Inconel 718 milling. Proced. Manuf. **13**, 89–93 (2017)
10. Fahl, J.: Lubricants for use with carbon dioxide as refrigerant. Ki Luft- und Kaeltetechnik **34** (1998/08)
11. Hauk, Weidner, E.: Thermodynamic and fluiddynamic properties of carbon dioxide with different lubricants in cooling circuits for automobile application. Ind. Eng. Chem. Res. **39**(12), 4646–4651 (2000)

Check for updates

Friction Modeling for Structured Learning of Robot Dynamics

M. Trinh[1](✉), R. Schwiedernoch[2], L. Gründel[1], S. Storms[1], and C. Brecher[1]

[1] Laboratory for Machine Tools and Production Engineering, 52074 Aachen, Germany
m.trinh@wzl.rwth-aachen.de
[2] RWTH Aachen University, 52062 Aachen, Germany

Abstract. Due to their low rigidity compared to machine tools, industrial robots (IRs) are less suitable for dynamic processes such as machining. In order to benefit from the flexibility and large workspace of IRs for e.g. machining of large workpieces, a model-based feedforward control can be used to compensate the deviations of the tool center point. This control algorithm does not require additional components but a dynamics model of the IR, incorporating effects such as mass inertia and friction. This paper focuses on friction modeling of robotic drive trains comparing analytical, e.g. the LuGre model, and data-driven models, e.g. Long Short-Term Memory networks. The models are parametrized and trained using data from the first axis of a 6-degree-of-freedom IR and evaluated regarding their ability to model dynamic nonlinear friction effects like stick-slip. This serves as a basis for structured learning of robot dynamics by combining analytical and data-driven models.

Keywords: Friction modeling · Robot dynamics · Structured learning

1 Introduction

Friction is not a standalone process, but a complex interplay of different elementary mechanism at the micro level. Therefore, despite its importance in physical systems, it is often neglected or its complexity greatly reduced [1]. In some cases, this simplification is sufficient, in other ones, such as machining, a detailed modeling of dynamic and nonlinear friction effects is necessary. There are different analytical friction models (AFMs) such as the Coulomb Viscous (CV) or LuGre model [1, 2]. These models differ greatly in their level of detail, but are still based on simplifying assumptions. Data-driven models like artificial neural networks (ANNs) serve as alternatives, which possess the ability to model highly nonlinear systems while achieving high accuracies. Furthermore, the exponential increase of computing power in the last decades opens new possibilities in the application of ANNs. The goal of this paper is to develop a friction model (FM) that meets the high requirements stated above. For this purpose, a data-based friction model (DFM) is developed with the help of Long Short-Term Memory networks (LSTM), a type of recurrent ANN that is able to model highly nonlinear dynamic systems [3]. To validate this model, it is compared to different AFMs. The first vertical axis of the serial

M. Liewald et al. (Eds.): WGP 2022, LNPE, pp. 396–406, 2023.
https://doi.org/10.1007/978-3-031-18318-8_41

robot MABI MAX 100 with six DOFs from MABI Robotic AG is used for validation of the FMs. The resulting DFM can be integrated into the analytical dynamics model of the IR, forming a structured neural network, which combines the advantages of both approaches [4].

2 State of the Art

The dynamics model of an IR is shown in (1) and can be derived using the Newton-Euler or Lagrange method [5]. It can be used for control algorithms such as model-based feedforward control [6].

$$\tau = M(q) \cdot \ddot{q} + h(q, \dot{q}) + \tau_F \tag{1}$$

where τ is the joint torque, M is a positive-definite mass matrix, h are forces including centripetal, Coriolis and gravity terms which depend on the joint variables, the positions q, the velocity $\dot{q}$ and acceleration $\ddot{q}$. The last term τ_F includes all the friction influences experienced by a real system [5]. Modeling these influences is the focus of this work.

2.1 Friction in Industrial Robot Systems

The friction to be modeled is composed of several overlapping behaviors. The solution to the problem is to assume the robot axis as a homogeneous system which exhibits a specific friction behavior. Figure 1 shows the schematic structure of the first axis. Within the global system boundaries there are three subsystems. Due to the two gear stages, the subsystems have different transmission ratios or speed ranges. The highest speeds and lowest torques are applied to the electric motor during operation. Therefore, viscous friction will dominate here for low motor currents. At the output of the system, the lowest velocities and the highest torques or normal forces are observed. Thus, boundary and mixed friction will occur at high currents. The frictional resistance in the gear transmission behaves correspondingly. The system-specific friction behavior can deviate from classical ones due to the superposition of multiple effects.

2.2 Analytical Modeling of Friction

AFMs can be classified into two basic groups by distinguishing modeling in the macroscopic state of rest: static and dynamic models [11]. The static models are derived from Coulomb's considerations and assume that the system behaves statically neglecting microscopic processes as well as interactions at the contact surface. Because of these simplifying assumptions, static friction models possess a simpler structure and fewer parameters. This leads to a robust and simple implementation, which is why they are still commonly used in industrial applications. The dynamic models originated from the results of Dahl [12]. Here, it is assumed that small pre-sliding displacements, i.e. system state without macroscopic displacements, can occur even if a friction pair is macroscopically stationary. This is implemented in the AFM by the fact that the friction depends on the position in the pre-sliding regime (PSR) and on the velocity in the

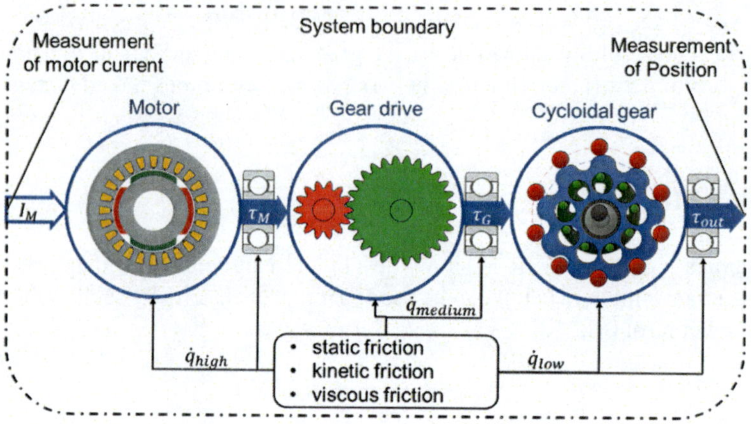

Fig. 1. System analysis of the first axis [7–10].

gross-sliding regime. The change between these regimes is devoid of discontinuities in the dynamic models, unlike the static ones. Dynamic models achieve this more realistic view with the help of a more complicated and thus error-prone parameter finding and calibration. In the following, two static (CV and Stribeck) and dynamic (LuGre and GMS) AFM are presented that are used in this paper. The Dahl model is omitted due to its inability to model viscous friction, which is supposed to be prevalent for the given system. For a detailed description of the models and a comparison, refer to [1, 2, 13].

Coulomb Viscous Model. The CV model (2) is composed of the Coulomb and the viscous model and is widely used in industrial applications [1]. It consists of the Coulomb μ and viscous friction parameter ν.

$$\tau_F = \mu * sgn(\dot{q}) + \nu * \dot{q} \tag{2}$$

Stribeck Model. The Stribeck model (3) combines the Stribeck function of the Stribeck effect with a linear model of viscous friction and is suitable for systems with few changes of velocity and direction. It possesses four additional parameters: f_c, f_s, δ and $\dot{q}_S$ (with $f_c, f_s, \dot{q}_S > 0$ and $f_s > f_c$).

$$\tau_R = S(\dot{q}) + \nu * \dot{q} \tag{3}$$

$$S = sign(\dot{q}) * \left[(f_c + (f_s - f_c) * \exp\left(- \left| \frac{\dot{q}}{\dot{q}_S} \right|^{\delta} \right) \right] \tag{4}$$

LuGre Model. The Lund-Grenoble (LuGre) model (5) combines the insights of the Dahl model for the PSR with the Stribeck model for the gross-sliding regime. It includes an internal state z, which describes the deflection of an introduced bristle model. This makes the parameter identification more difficult, since it is not measurable. Further parameters are σ_0 and σ_1 with $\sigma_0, \sigma_1 > 0$. With appropriate parameterization, this model is able to correctly describe friction in a variety of systems and kinematics.

$$\tau_R = \sigma_0 * z + \sigma_1 * \dot{z} + \nu * \dot{q} \tag{5}$$

$$\dot{z} = \dot{q} - \sigma_0 * \frac{\dot{q}}{S(\dot{q})} * z \tag{6}$$

GMS Model. Similar to the LuGre model, the GMS model (7) includes z, which, however, describes the position of newly introduced internal Maxwell slip elements. The major advantage is that by increasing the number of elements, even the friction in complex systems can be modeled realistically. Further parameters are (depending on the number of elements): k, ϑ, ρ and C (with $k_i, C, \rho_i > 0$, $\vartheta_i \geq 0$ and $\sum_{i=1}^{N} \rho_i = 1$).

$$\tau_R = \sum_{i=1}^{N} (k_i * z_i + \vartheta_i * \dot{z}_i) + \nu * \dot{q} \tag{7}$$

$$\dot{z}_i = \begin{cases} \dot{q}, & \text{stick for } z_i \leq \rho_i * S(\dot{q}) \\ sign(\dot{q}) C \left(\rho_i - \frac{z_i}{S(\dot{q})} \right), & \text{slip for } |\dot{q}| < 0 \end{cases} \tag{8}$$

2.3 Data-Driven Modeling of Friction

A few approaches for DFM can be found in the literature: [14] used feedforward neural networks (FFNNs) and a jump activation function, which led to a better performance than the Stribeck model. Hirose and Tajima [15] used a FFNN with an LSTM cell for detecting velocity changes. The hybrid network outperformed the LuGre model on a test rig for rolling friction. This paper uses LSTM networks, which, contrary to FFNNs possess recurrent network connections and a memory cell that remembers values over arbitrary time intervals [3]. Due to this characteristic, we assume that LSTM networks are able to model nonlinear and dynamic friction behavior. Compared to analytical models, data-driven ones can possess millions of parameters making it impossible for humans to understand. They are therefore black-box models. Furthermore, ANNs lack physical interpretability, because they are not based on physical equations that govern a system. Geist and Trimpe [4] define structured learning as a combination of analytical prior knowledge with data-driven techniques, therefore harnessing the advantages of both approaches. Similar approaches are called grey-box models or physics-informed neural networks [16]. Based on the results of this paper, further work will integrate DFMs into the robot's analytical dynamics model and analyze possibilities for structured learning of FMs.

3 Data Generation

All FMs, whether AFM or DFM, must be adapted to the existing system by parameterization or training using specially created and processed measurement data. The specific requirements of the respective models must be taken into account. The Stribeck model consists of the characteristic friction curve, which describes the friction behavior of the system in steady state, i.e. at constant velocities. Therefore, a trajectory with different constant velocities, called Steady-State-Trajectory (SST), is created in the range from [0.001, 55]°/s. These velocity values for joint 1 were determined in experimental results according to the amount of noise in the measured system response (motor current) and the rapid traverse in the given workspace ($|q_1| < 90°$). The current is filtered using a low-pass filter to remove high interference frequencies. The filtered current is multiplied with the motor constant and the gear ratio resulting in the motor torque. The theoretical torque is calculated using the recursive Newton-Euler algorithm [5] and subtracted from the measured torque leading to the frictional torque, which, in general, accounts for a large part of the total torque. The torques of the constant intervals are averaged and a new data set is generated from the averaged torques and velocities, which is used to identify the Stribeck model. The following trajectories are preprocessed accordingly.

For the dynamic AFMs two trajectories are necessary. The Stribeck parameters are identified using the SST. The effect of the dynamic parameters is dominant in the PSR, therefore the pre-sliding trajectory (PST) should represent this velocity range, where the influence of the position on friction prevails. Hence, low velocities and accelerations are selected, since the pre-sliding displacements are usually limited to a few degrees [1]. The universal trajectory (UT) is used to identify the CV model and to train the LSTM network. The former does not make special requirements, whereas the latter is a data-driven approach, whose accuracy depends on the quality of the data. Therefore, the UT should simulate the occurring friction behavior during application-related use of the robot for its possible velocity range. In addition, the UT on the robot must excite atypical and nonlinear friction behavior, as these friction phenomena are particularly difficult to model and thus provide a good metric for comparison. Since ANNs tend to develop a bias for situations they learn more often than other sections, there should be no repetitive intervals in the UT. The validation trajectory (VT) should also represent application-related and nonlinear friction effects. In order to prevent an unfair advantage of the CV and LSTM model in the final comparison of the friction models, the VT must differ from the UT by reasonable modifications. The final trajectories are shown in Fig. 2.

4 Friction Modeling

4.1 Parameter Identification of Analytical Models

The static and dynamic parameters can be determined in a coupled or decoupled manner. Since the parameters influence each other, higher accuracies were proven for the coupled method [17]. Nevertheless, the decoupled method is chosen in this paper to reduce the number of parameters to be identified simultaneously and to determine dynamic parameters in their relevant PSR. The basic procedure follows four steps: select a suitable trajectory, measure the real data, choose a formula to calculate the error between the measured data and the AFM and at last, minimize the error using an optimization algorithm. The MSE loss is usually chosen to calculate the error. Due to their simpler mathematical structure, the identification of static parameters is less demanding. Therefore, linear optimization algorithms, such as least squares, can be used [18]. Due to the nonlinearity of the dynamic models as well as possible complications in the integration of the differential equations, simple linear optimization algorithms and gradient methods cannot be applied. Instead, metaheuristic optimization methods such as genetic algorithms or glowworm-swarm optimization are used [1, 17]. In this work, a differential evolution algorithm is used. For reasons of comparability, the optimization algorithm is used for the identification of all AFMs.

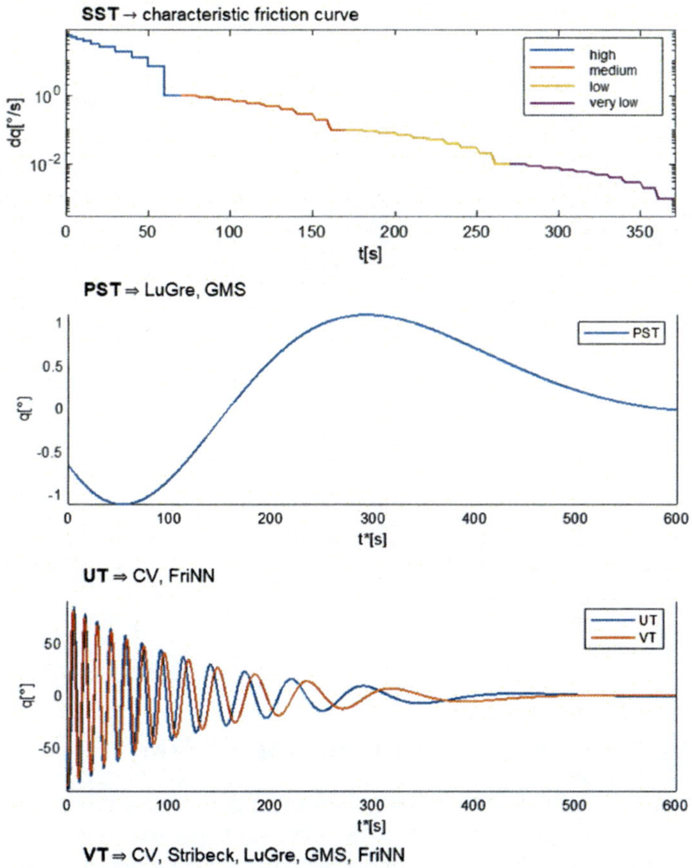

Fig. 2. Top: steady state trajectory (SST) with $\dot{q} \in [0.001, 55]°/s$ (logarithmic scale). Center: pre-sliding trajectory (PST). Bottom: universal (UT) and validation (VT) trajectories.

Using the measured data the characteristic friction curve is determined, which provides information about the search intervals of the static models. The parameters μ, f_c and f_s are estimated to a value smaller than one and $\dot{q}_S < 0.2°/s$. The slope of the entire curve can be estimated to $\nu = 0.05$. Therefore, the search interval is bounded above by 0.1. The limitation of the search intervals for the dynamic parameters is more difficult, because they are based on theoretical models (e.g. bristles). Therefore, no definite statements can be made in advance and the search intervals must be chosen accordingly large. For the GMS model, the number of elements N = 4 is chosen based on previous investigations, since this represents a reasonable compromise between modeling performance and number of parameters [13]. The constraints $f_s > f_c$ and $\sum_{i=1}^{N} \rho_i = 1$ are implemented by a linear constraint and taken into account during the optimization. Table 2 shows the results of the final identification. All friction models could be parameterized using the created data sets. Comparative values can be found in [19–22] (Table 1).

Table 1. Identified parameters for analytical models of first axis.

CV		Stribeck		LuGre		GMS	
μ	$7.165 * 10^{-1}$	f_c	$7.283 * 10^{-1}$	σ_0	294.140	k_i	93.360, 26.100, 40.480, $8.756 * 10^{-2}$
ν	$4,958 * 10^{-2}$	f_s	$7.771 * 10^{-1}$	σ_1	3.263	σ_i	6.732, 9.782, 2.527, $3.128 * 10^{-1}$
		$\dot{q}_S$	$5.042 * 10^{-3}$			ρ_i	$2.543 * 10^{-3}$, $3.141 * 10^{-3}$, $1.606 * 10^{-2}$, $8.815 * 10^{-1}$
		δ	2.000			C	73.120
		ν	$4.730 * 10^{-2}$				

4.2 Hyperparameter Optimization and Training of LSTM

An optimized form of grid search is applied for hyperparameter optimization (HPO) of the LSTM network using the UT. Table 3 lists the final hyperparameters of the resulting friction neural network (FriNN). The number of considered previous time points is called the sequence length. The sequence length of 10 shows that the FriNN has learned dynamic effects to increase accuracy. This is because certain friction processes or phenomena depend not only on the current input values, but also on the previous ones. The optimized FriNN is then trained using the UT.

Table 2. Optimized hyperparameters of FriNN.

Hidden layer	Hidden size	Batch size	Sequence length	Learning rate	Weight decay	Number of epochs
2	64	64	10	$2.69*10^{-3}$	$5.15*10^{-9}$	125

5 Evaluation

5.1 Comparison of Analytical Models and FriNN

The final identified AFM and trained FriNN are tested on the VT for a qualitative comparison, which can be seen in Fig. 3. In the top left, the predicted torques of the FriNN compared to the measured ones can be seen, showing a good fit. In this diagram, the VT is divided into four characteristic sections I to IV. In the bottom left, the qualitative MSE values for all FMs and their corresponding shares to the defined sections are shown. Here, the CV and Stribeck model show good results for sections I and II, which are characterized by high and medium velocities. Due to its ability to model the Stribeck effect, the Stribeck model performs better on section II. On the right of Fig. 3, sections III and IV are shown in detail for all FM, which mark the PSR and the lowest occurring velocities, respectively. The CV model shows multiple abrupt sign changes in III, which do not occur in the measured data. This is due to the different friction behavior in the PSR, which the CV model cannot account for. In IV the measured torque increases to a level higher than the Coulomb friction μ. Because of the low velocities, the viscous friction is negligible. Therefore, the CV model cannot model this friction behavior. In comparison, the Stribeck model shows similar difficulties with the sign changes and the atypical torque curve of section IV can only be roughly approximated.

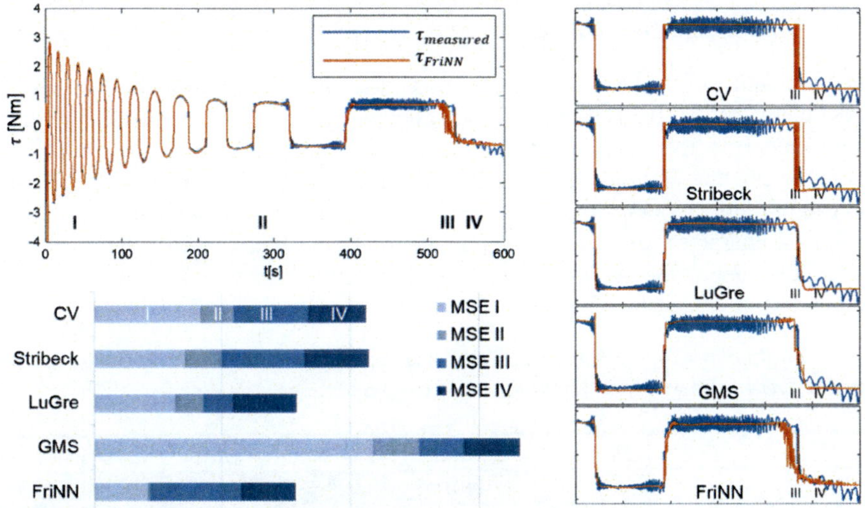

Fig. 3. Top left: results for FriNN using the validation trajectory. Right: detailed results for all models in section III and IV. Bottom left: MSE distribution over sections I–IV for all models.

For the first two sections, the LuGre model shows similar results to the Stribeck model. In contrast to the static friction models, it can almost correctly calculate friction in the PSR. However, for the atypical section IV it can only approximate the friction curve. The GMS model shows major differences compared to the previous FMs. High fluctuations can be observed in the transitions between direction changes of the torque.

Especially for section I, the calculated frictional torque is double the measured one, which can be seen in the increased MSE values. Since implicit numerical integration is stable, the problem must stem from the determined parameters. Presumably, the stiffnesses k_i of some elements could be responsible. In contrast to that, the PSR could be modeled. In direct comparison with the LuGre model, the curves in the low velocity range are less smooth.

FriNN can almost exactly model the occurring friction in sections I and II, showing the lowest MSE values of all FMs. In III the sign of the velocity oscillates strongly without the measured frictional torque reacting analogously. The deflections are smaller and less erratic than those of the static AFM, but FriNN models a premature change in the sign of the friction compared to the measured data. A possible explanation lies in the UT, where the transitions occur at higher velocities. This is also a possible reason for the deviations in the last section of the VT, where the measured frictional torque is much higher than expected despite the low velocity. Interestingly, in contrast to the AFMs, FriNN can model the asymmetric progression of friction during acceleration and deceleration. FriNN can model dynamic friction in the PSR and often calculates the transitions and gradients more accurately than the LuGre model. This confirms the selection of a recurrent network, since information about the previous system state is necessary for this. In FriNN, this information is not obtained by numerical integration, but is calculated and stored in each LSTM cell over all time points in the sequence.

5.2 Evaluation and Discussion

A final evaluation of the FMs is conducted under the following criteria: the characteristic values determined for the VT, the analysis of the torque curve in general and for specific friction scenarios, the effort of implementation and adaptation to the system and, to a lesser extent, the computing time. The R^2 loss is introduced, which describes the goodness of fit of a model compared to the measured values or their mean value. Table 3 contains the characteristic values and computation times. It can be seen that all FMs are capable of modeling the frictional torque of previously unknown trajectories.

Table 3. Final comparison of friction models using the validation trajectory. The colors indicate the performance from worst (dark red) to best (dark green).

	CV	Stribeck	LuGre	GMS	FriNN
MSE [Nm]	$2.28*10^{-2}$	$2.44*10^{-2}$	$1.44*10^{-2}$	$2.56*10^{-2}$	$1.86*10^{-2}$
R^2 [%]	96.68	97.52	98.54	94.40	98.11
Computation time [s]	$1.75*10^{-3}$	$1.41*10^{-2}$	$2.29*10^{-1}$	2.34	5.64

The CV model is the simplest friction model, with no special requirements for parameter identification and a medium fitting quality. In combination with its proven robustness, it is suitable for use cases where high accuracy at low velocities is not required. Despite its higher complexity, the Stribeck performs worse than the CV model. Special trajectories are required for this model and the measurement data must be processed into the system-specific friction characteristic. Therefore, an exclusive implementation of the Stribeck model does not seem recommendable. According to the characteristic values, the LuGre model achieves the highest goodness of fit. The implementation is complex due to the numerical integration and consideration of the model stability. Besides the multi-step process for identification of the Stribeck parameters, the dynamic parameters require an additional trajectory. The nonlinear differential equation limits the choice of optimization algorithms, although the actual process is done in a reasonable time due to the few parameters, provided that the correct search space can be identified. The LuGre model is the safe and proven choice when high accuracy is important, especially in the PSR. The GMS showed the highest implementation effort and the lowest robustness and can therefore not be recommended. The newly developed FriNN showed a high goodness of fit, with the exception of one particularly challenging friction scenario. Its accuracy was highest for medium velocities. A strength of this method is the modeling of system-specific friction behavior, which promises high flexibility. A drawback is the high computational capacity required for HPO.

6 Conclusion and Outlook

In this paper, different analytical friction models and the data-based FriNN were parametrized and trained for the first axis of a serial robot. In addition, methods for generation of optimized trajectories were introduced. Although all friction models were able to approximate a given frictional torque, large differences in the accuracies and the ability to reproduce specific friction phenomena were observed. Depending on the required conditions and accuracy, the CV model, the LuGre model and the FriNN model seem most promising. In further works, the training data for FriNN should be extended in order to not only cover use case-related friction behavior, but a range of challenging trajectories as well. Furthermore, a concept for friction modeling of the entire robot should be developed. Therefore, an extensive analysis should be carried out, on whether and how friction of the single axes are interdependent. Finally, an integration of FriNN into the analytical dynamics model of the IR is planned.

Acknowledgements. Funded by the Deutsche Forschungsgemeinschaft (DFG, German Research Foundation) under Germany's Excellence Strategy–EXC-2023 Internet of Production–390621612.

References

1. Ruderman, M.: „Zur Modellierung und Kompensation dynamischer Reibung in Aktuatorsystemen“. Universitätsbibliothek Dortmund (2012)
2. Freidovich, L., et al.: LuGre-model-based friction compensation. IEEE Trans. Control Syst. Technol. **18**(1), 194–200 (2010)
3. Hochreiter, S., Schmidhuber, J.: Long short-term memory. Neural Comput. **9**(8), 1735–1780 (1997)

4. Geist, A.R., Trimpe, S.: Structured learning of rigid-body dynamics: a survey and unified view from a robotics perspective. In: GAMM-Mitteilungen, vol. 44, no. 2 (2021)
5. Featherstone, R.: Rigid Body Dynamics Algorithms. Springer, New York (2008)
6. Gründel, L., et al.: Compensation of process forces with a model-based feed-forward control for robot machining. In: 4th International Conference on Control and Robotics Engineering (ICCRE), pp. 115–121 (2019)
7. Höfler Homepage. https://www.ahoefler.de. Last accessed 2022/05/08
8. Versandhandel Bodack Homepage. https://www.dasrad.info. Last accessed 2022/05/08
9. SycoTec. https://info.sycotec.eu. Last accessed 2022/05/08
10. Wikimedia Homepage. https://commons.wikimedia.org. Last accessed 2022/05/08
11. Piatkowski, T.: Dahl and LuGre dynamic friction models—the analysis of selected properties. In: Mechanism and Machine Theory, vol. 73, pp. 91–100 (2014)
12. Dahl, P.R.: A Solid Friction Model. Aerospace Corp El Segundo Ca (1968)
13. Al-Bender, F., Swevers, J.: Characterization of friction force dynamics. In: IEEE Control Systems Magazine, vol. 28, no. 6, pp. 64–81 (2008)
14. Selmic, R.R., Lewis, F.L.: Neural-network approximation of piecewise continuous functions: application to friction compensation. IEEE Trans. Neural Netw. **13**(3), 745–751 (2002)
15. Hirose, N., Tajima, R.: Modeling of Rolling Friction by Recurrent Neural Network Using LSTM, IEEE (2017)
16. Raissi, M., et al.: Physics-informed neural networks: a deep learning framework for solving forward and inverse problems involving nonlinear partial differential equations. J. Comput. Phys. **378**, 686–707 (2019)
17. Liu, D.-P.: Parameter Identification for LuGre Friction Model Using Genetic Algorithms, IEEE (2006)
18. Liu, L., Wu, Z.: Comprehensive parameter identification of feed servo systems with friction based on responses of the worktable. In: Mechanical Systems and Signal Processing, vol. 64, pp. 257–265 (2015)
19. Mata, V., et al.: Dynamic parameter identification in industrial robots considering physical feasibility. In: Advanced Robotics, pp. 101–119 (2005)
20. Zhang, S., et al.: Parameter estimation survey for multi-joint robot dynamic calibration case study. In: Science China Information Sciences, vol. 62, pp. 202–203 (2019)
21. Indri, M., et al.: Friction modeling and identification for industrial manipulators. In: IEEE 18th Conference on Emerging Technologies & Factory Automation (ETFA), pp. 1–8 (2013)
22. Grami, S., Bigras, P.: Identification of the GMS friction model based on a robust adaptive observer. Int. J. Model. Identification Control **5**(4), 297–304 (2008)

Potential of Ultra-High Performance Fiber Reinforced Concrete UHPFRC in Metal Forming Technology

K. Holzer[1](✉), F. Füchsle[1], F. Steinlehner[1], F. Ettemeyer[2], and W. Volk[1]

[1] Chair of Metal Forming and Casting, Technical University of Munich, Walther-Meissner-Strasse 4, 85748 Garching, Germany
katja.holzer@utg.de
[2] Fraunhofer Research Institute for Casting, Composite and Processing Technology IGCV, Lichtenbergstrasse 15, 85748 Garching, Germany

Abstract. In metal forming technology, faster development and production cycles are complicated by the production of deep drawing tools. Those are already mandatory even for small series. Compared to conventional concrete materials, ultra-high performance fiber reinforced concrete (UHPFRC) is characterized by its strength properties, which are close to the requirements for use in metal forming technology. The present work investigates the potential of UHPFRC in terms of strength properties and tool manufacturing to be used for prototyping as well as for small and very small series production in metal forming technology. For this purpose, the flexural strength and the compressive strength of UHPFRC materials is increased by adding carbon fibers of different lengths and different volume ratios. The material is evaluated by means of three-point bending tests and compression tests. Additionally, a novel indirect rapid tooling approach, a combination of fused deposition modeling (FDM) and room-temperature-vulcanizing (RTV) silicone molding, is introduced. This approach enables the casting of near-net-shape deep drawing tools with high dimensional accuracy.

Keywords: Ultra-high performance fiber reinforced concrete · UHPFRC · Indirect rapid tooling · Metal forming technology · Deep drawing

1 Introduction

Deep drawing enables fast and economical production of three-dimensional sheet metal components due to the high production rates. The high process forces require massive deep drawing tools, which are associated with high investments and long manufacturing times even for small series production [1]. An idea to face these challenges is the application of ultra-high performance concrete (UHPC) instead of steel for deep drawing tools. Compared to conventional concrete materials, UHPC is characterized by a small particle size of the cements and the addition of additives. This improves packing density, which accelerates the hydration reaction, and results in higher strength properties. UHPC reaches compressive strengths from about 150–200 N/mm^2, which are almost

M. Liewald et al. (Eds.): WGP 2022, LNPE, pp. 407–415, 2023.
https://doi.org/10.1007/978-3-031-18318-8_42

the same as those of typical tooling cast iron at 250 N/mm^2 [2]. This motivates the use of the material for application in metal forming technology. To further improve the flexural strength and prevent the formation of cracks, fibers are added to produce ultra-high performance fiber reinforced concrete (UHPFRC) [3, 4]. First use of high performance concrete in metal forming technology was made for production of sheet metal parts for prototype cars. The aim of Schwartzentruber et al. [5] was to reduce the manufacturing cost of deep drawing tools by switching from polymer resin concrete tools to high performance concrete tools with a gel coat. Kleiner et al. [6] successfully used UHPC dies for sheet metal hydroforming. The UHPC dies did not fail under internal pressure while hydroforming. Further, the friction between UHPC and sheet metal could be controlled such that the properties of the formed parts were of good quality. The development of UHPC tools also pursues the goal of lubricant-free forming. Guilleaume et al. [7] used deep drawing tools coated in a polymer concrete to make dry deep drawing conceivable.

The aim of this work is to open up the potential for the use of UHPFRC in metal forming technology. The first part deals with the optimization of the strength properties of UHPFRC by addition of carbon fibers. This work focusses on a suitable specimen preparation with optimized fiber volume ratio in order to increase the flexural strength. In the second part, an approach for near-net-shape casting of UHPC and UHPFRC is presented. Finally, an outlook on the use of UHPC as punch for sheet metal forming is given.

2 Optimization of UHPFRC Strength Properties

2.1 Specimen Preparation

The basic composition of the UHPFRC used in this work consists of binder, mixing water and a superplasticizer. Nanodur Compound 5941 from Dyckerhoff GmbH, Wiesbaden, Germany, was used as binder. This ready-to-use compound allows the production of UHPFRC without further additives. In addition to cement and synthetic silicic acid, Nanodur Compound 5941 contains quartz powder [8]. To the mixing water the superplasticizer ADVA Flow 375 (BV/FM) from GCP Germany GmbH, Lügde, Germany, was added. This PCR superplasticizer prevents the formation of agglomerates due to the steric effect and improves workability of the fresh concrete. Further, it ensures a homogeneous and pore-free microstructure of the final product. [2, 9]. In the course of this work, test specimen with different compositions were produced. The compositions given in added mass per volume ρ_i of each constituent as well as the resulting fiber volume ratio V_f are specified in Table 1. The compositions are easy to process and self-compacting. The use of a compaction process for de-aeration was not necessary [6, 10]. For the production of the test specimens, formwork elements were generated using the fused deposition modeling process (FDM). The elements were printed using PLA filament. To achieve a smooth specimen surface and avoid subsequent grinding of the hardened specimens, the inner surfaces of the wall elements were surface-ground. The elements of the formwork were fixed to each other with screws and sealed with plasticine at the contact surfaces. The formwork was designed in such a way that the filling side would not experience any force application during the testing. Each constituent was weighed before mixing.

The maximum tolerance was set at ±1 mass percent of the amount added. A commercial mortar bucket and trowel were used for the mixing process. Mixing was done by hand. The mixture was mixed for at least 12 min until a homogeneous consistency with dilatant behavior was obtained and poured into the formwork. In order to achieve the final strength of the UHPFRC after 48 h instead of 28 days, a heat treatment is carried out. The procedure is shown in Fig. 1. After the pouring, the UHPFRC specimens in the formwork are placed in a saturated atmosphere for 24 h at 20 °C. The specimens are then removed from the formwork and placed in a water bath. The water bath is heated from 20 °C to 90 °C within 1 h. Heat treatment is carried out according to Sagmeister [10] for 24 h at 90 °C. Cool down is done in the water bath to prevent dehydration and cracking of the surface. The specimens were stored in a water bath at 20 °C until use. In Fig. 2 light microscope and scanning electron microscope (SEM) images of the fracture surface of UHPFRC composition 3, fiber volume ratio of $V_f = 2\%$ of carbon fibers with 7 μm diameter and a length of 3 mm, are given.

Table 1. Composition of UHPFRC given in added mass per volume ρ_i of the constituents and the resulting fiber volume ratio V_f.

Composition	Binder	Water	Superplasticizer	Fiber type	Dimension (mm)	ρ_F (kg/m^3)	V_f (%)
	ρ_B (kg/m^3)	ρ_W (kg/m^3)	ρ_S (kg/m^3)				
1	2100.0	316.0	30.0	–	–	–	–
2	2100.0	316.0	30.0	Carbon	Powder	35.4	2
3	2100.0	316.0	30.0	Carbon	Ø 0.007 × 3	35.4	2
4	2100.0	316.0	30.0	Carbon	Powder	17.7	1
5	2100.0	316.0	30.0	Carbon	Ø 0.007 × 3	17.7	1
6	2100.0	316.0	30.0	Carbon	Powder	70.8	4
7	2100.0	316.0	30.0	Carbon	Ø 0.007 × 3	70.8	4

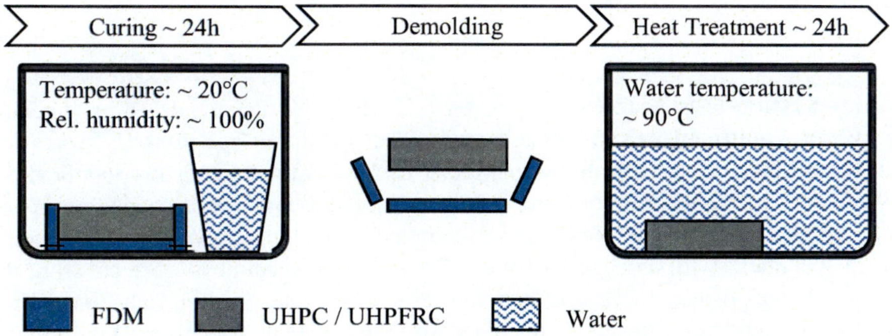

Fig. 1. Curing and heat treatment of UHPC/UHPFRC specimen.

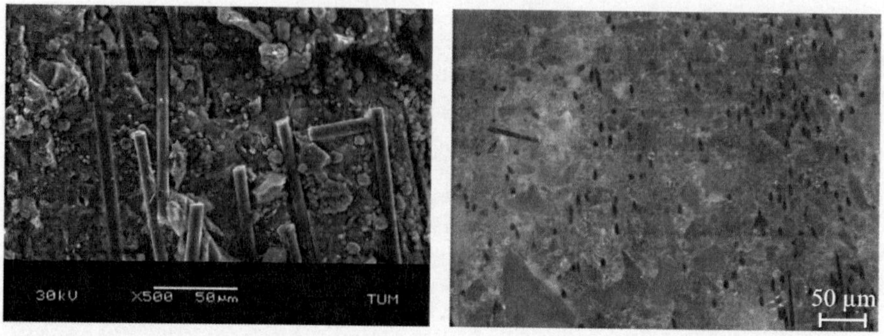

Fig. 2. Fracture surface of UHPFRC with carbon fibers 3 mm, $V_f = 2\%$; SEM 500x magnification (left) and light microscope with polarizing filter, 200× magnification (right).

2.2 Experimental Setup

Three-point bending tests were carried out on a Z020 table-top testing machine from ZwickRoell GmbH & Co.KG, Ulm, Germany. The control of the machine, as well as the recording of the force-displacement curves, was carried out with the software testExpert II from ZwickRoell GmbH & Co.KG, Ulm, Germany. The three-point bending test fixture used consists of an upper compression die and two lower supports. The bending fin of the upper part has a radius of 5 mm and is moved by the testing machine. The lower part has a fixed bearing and consists of a rail and two adjustable positioning slides. The radius of the lower supports is 5 mm. The tests were carried out in accordance with DIN EN 196-1 [11]. In contrast to the standard specifications, the specimen cross-section was set to 22.8 × 22.8 mm^2 to reduce the maximum test load [12]. The prisms have a length of 160 mm. The span of the supports was set to 100 mm. The tests were performed at a continuous test speed of 0.1 mm/s. The flexural strength f_f of the specimens was calculated according to formula (1) of DIN EN 196-1 [11] from the maximum compressive force F, the span of the supports l and the edge length b.

$$f_f = \frac{1.5Fl}{b^3} \tag{1}$$

Compression tests were carried out on a Zwick 1484 universal testing machine from Zwick GmbH & Co. KG, Ulm, Germany. The contact surfaces of the compression plates are surface-ground and aligned parallel to each other for uniform force application. System control and recording of the force-displacement curves was carried out using testExpert II software. The compressive strength of the concrete was tested on cube-shaped specimens with an edge length of 22.8 mm. The specimens were cut from the fractured halves of the three-point bending tests. A precision cutting machine with a diamond cutting disc was used for this purpose. The specimen was inserted such that it was centered in the fixture and the load was applied to the precision cut surfaces. The tests were performed at a constant speed of 0.01 mm/s. As termination criteria a load drop due to specimen fracture or maximum load of 180 kN were set. The compressive strength f_c of the specimens can be calculated according to formula (2) in accordance with DIN EN 12390-3 [13] using the cross-sectional area A_c or the edge length b and the maximum test load F.

$$f_c = \frac{F}{A_c} = \frac{F}{b^2} \tag{2}$$

2.3 Result and Discussion

In Fig. 3, the results of the three-point bending test of all compositions are shown. The median value of three tests per composition representing the force-displacement curve is displayed. Additionally, the mean value of flexural strength over three tests and the corresponding minimum and maximum values are given. For the compression test, the same data is given in Fig. 4. The graphs show a linear force increase up to a force drop. The UHPFRC shows hardly any plastic behavior and breaks brittle when reaching the maximum strength. The effect at the beginning of the compression test curves is caused by upsetting of the testing machine when aligning with the specimen.

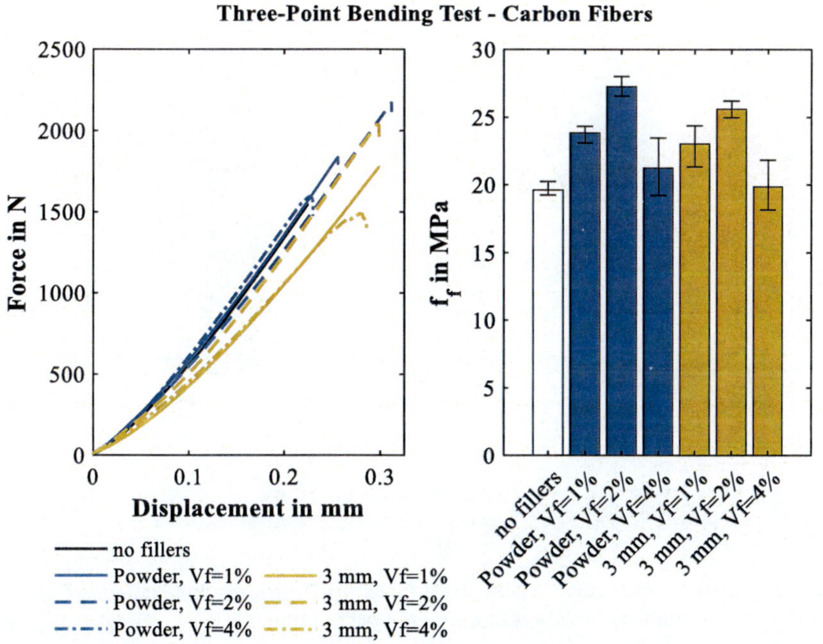

Fig. 3. Force-displacement curve of three-point bending tests (left) and resulting flexural strength of UHPFRC (right) with carbon fibers of different fiber volume ratio V_f.

The flexural strength increases for carbon fibers with a fiber volume content of $V_f = 2\%$, irrespective of the form of the carbon fibers present. This finding is in line with the state of the art [3]. The greatest increase of 38.9% is observed for carbon fiber powder with $V_f = 2\%$. If only a few fibers are added, a smaller effect can be observed. If, on the other hand, the fiber volume ratio is increased to $V_f = 4\%$, the distribution of fibers is poorer. Even small amounts of the fibers adhere to each other and form nest. This weakens the UHPC matrix and reduces overall material strength [14, 15]. The uneven distribution of the fibers is recognizable in Fig. 2 (right), as many fibers of the same orientation are close to each other. This is due to mixing by hand, which only introduces a low mixing energy, and due to the hydrophobic character of the fibers.

In compression, the shape of the carbon fiber makes a difference. In Fig. 4 an improvement of the behavior in compression can be see for carbon fiber powder. The effect is most pronounced, with an increase in compressive strength of 11.6%, for a fiber volume ratio of $V_f = 2\%$. On the opposite, the addition of 3 mm fibers reduces the overall compressive strength independent of fiber volume ratio. The fibers seem to weaken the UHPC matrix and reduce the overall compressive strength. For a fiber volume ratio of $V_f = 4\%$ the compressive strength is reduced by 24.1% compared to UHPC without fillers.

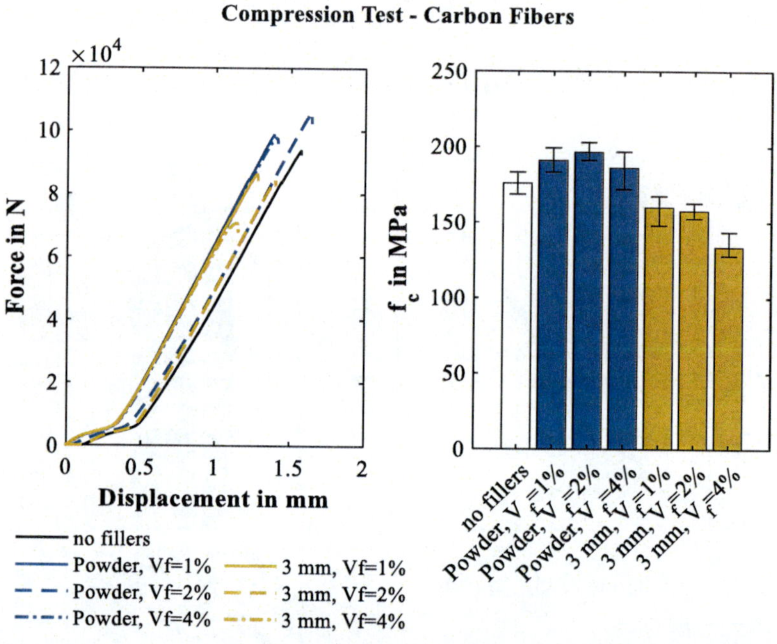

Fig. 4. Force-displacement curve of compression tests (left) and resulting compressive strength of UHPFRC (right) with carbon fibers of different fiber volume ratio V_f.

3 Indirect Rapid Tooling of UHPC Deep Drawing Tools

3.1 Approach

To take full advantage of UHPFRC as a material for deep drawing tools, a novel approach for shaping is needed. The aim is to enable rapid near-net-shape casting of drawing tools with simultaneous freedom in shaping. The approach presented here is classified as indirect rapid tooling, since the actual tool is not manufactured additively. In contrast to rapid manufacturing and rapid prototyping, the rapid tooling approach produces negatives of the end product rather than positives. Rapid tooling includes both the production of series tools (direct tooling) and prototype tools (prototype tooling). The formwork is produced in a first additive step, followed by a second non-additive step. The actual deep drawing tool production is not additive. By accelerating the formwork production and the near-net-shape approach, a fast production of cast tool elements is made possible [16, 17].

3.2 Formwork Manufacturing

Following the indirect rapid tooling approach, the formwork is made of a combination of an additively manufactured hard mold and a room-temperature-vulcanizing (RTV) silicone mold. At the beginning, a negative of the formwork is required. An existing deep drawing tool can be used for this purpose. In the context of this work, the formwork negative is produced using the fused deposition modeling (FDM) process. The procedure for manufacturing the formwork of a punch as present is shown in Fig. 5.

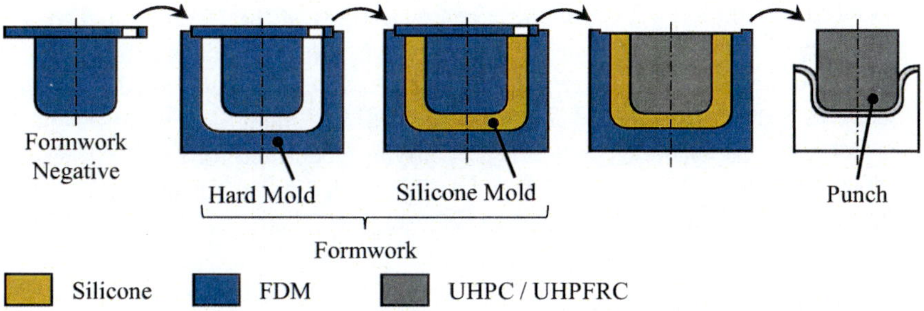

Fig. 5. Procedure for manufacturing formwork and UHPC/UHPFRC punch.

Since the surface shows process characteristics due to the filament deposition, the formwork negative was reworked by grinding, manually and with a disc sander, and coated with epoxy resin. After the hard mold is additively manufactured by FDM, the formwork negative is inserted and cast with silicone. UHPFRC generally has low shrinkage of 0.05 to 0.06 mm/m. For this reason, the formwork must be designed in such a way that demolding of the part is possible. In addition to the design of demoldable hard molds, the use of flexible molding materials also enables non-destructive demolding of the UHPFRC part. Silicones in particular are characterized by very good reproduction accuracy combined with high dimensional stability and high mechanical strength. The non-adhesive surface and high tear strength allow easy demolding. In addition, the good chemical resistance of silicone makes it suitable for processing concrete materials. In the present work, the two-component silicone ZA 50 LT from Zhermack GmbH, Öhlmühle, Germany, was used. The silicone is mixed in a ratio of 1:1 and cures at room temperature by vulcanization. After curing, the silicone has a hardness of 50 ShA and a tear strength of 12 N/mm.

3.3 Tool Manufacturing

Five punches made of UHPC with no fibers (composition 1) were produced with the formwork. The dimensional accuracy and form tolerances were examined with the LH87 coordinate measuring machine from the WENZEL Group, Wiesthal, Germany. There are significant height deviations in the range of 0.5–1.5 mm. This is due to different filling heights and different material removal by face grinding. There is a larger average deviation of −558.4 μm in the filling diameter over five punches, compared to an average deviation of −279.5 μm on the drawing radius side. This is due to gravimetric effects on the silicone mold caused by the weight of the UHPC. For the prototypical test geometry, in which a punch and an open die are used, only the drawing radius is important for shaping the sheet metal, since there is no form closure between punch, sheet and die.

Additionally, there is significant shrinkage compared to the dimensions of the formwork negative. The dimensional accuracy was then examined along the manufacturing steps. Three punches were measured before and after heat treatment and no significant shrinkage was observed. The heat treatment of the UHPC has no influence on the diameter deviations of the final part from the formwork negative. The coordinate measuring machine was used to measure the diameter of the formwork negative after rework, the silicone mold and three UHPC punches. The deviations of the UHPC punch from the formwork negative can be attributed to the shrinkage of the silicone mold. The silicone shrinks onto the formwork negative, whereas the UHPC reproduces the silicone mold well. Since the silicone used already exhibits a very small theoretical shrinkage of 0.05% after 24 h, the shrinkage was subsequently compensated for by an allowance on the formwork negative. This resulted in an UHPC punch diameter deviation of +7.2 μm from the nominal dimension at the drawing radius side.

4 Summary and Outlook

The strength properties of UHPFRC were varied by adding carbon fibers in different fiber volume ratios. The addition of carbon fiber powder with a fiber volume ratio of $V_f = 2\%$ resulted in a maximum increase of 38.9% in the flexural strength.

Further, an indirect rapid tooling approach for casting near-net-shape UHPFRC deep drawing tools was presented. Without compensation, punches with a mean diameter deviation of −279.5 μm could be produced. With compensation of the shrinkage of the silicone, the deviation could be reduced to +7.2 μm. In subsequent works, the gravimetric effects of the weight of the UHPFRC will be further investigated, as this will become a challenge with larger volumes or with form fit while deep drawing.

An UHPC punch has already been used to produce three cups each with an initial plate diameter of 135 mm and 150 mm from DX56D+Z with a thickness of 1 mm. During the deep drawing tests, the punch was subjected to a maximum drawing force of 97 kN without any damage or breakage. The tests showed high repeatability. In addition, the cups showed a good surface finish on the punch side with no run-on or run-off channels. This confirms the potential of UHPFRC in metal forming technology. In the further course, the die and blankholder, especially their active surfaces, will also be made of UHPFRC and integrated into the drawing system.

References

1. Hoffmann, H.: Handbuch Umformen, 2nd edn. Handbuch der Fertigungstechnik. Hanser, München (2011)
2. Fehling, E., Schmidt, M., Walraven, J.C., Leutbecher, T., Fröhlich, S.: Ultra-high performance concrete UHPC. Fundamentals - design - examples. BetonKalender. Ernst & Sohn, Berlin (2014)
3. He, J., Chen, W., Zhang, B., Yu, J., Liu, H.: The Mechanical Properties and Damage Evolution of UHPC Reinforced with Glass Fibers and High-Performance Polypropylene Fibers. Materials (Basel, Switzerland) (2021). https://doi.org/10.3390/ma14092455
4. Mohammed, B.H., Sherwani, A.F.H., Faraj, R.H., Qadir, H.H., Younis, K.H.: Mechanical properties and ductility behavior of ultra-high performance fiber reinforced concretes: effect of low water-to-binder ratios and micro glass fibers. Ain Shams Eng. J. (2021). https://doi.org/10.1016/j.asej.2020.11.008
5. Schwartzentruber, A., Bournazel, J.-P., Gacel, J.-N.: Hydraulic concrete as a deep-drawing tool of sheet steel. Cem. Concr. Res. (1999). https://doi.org/10.1016/S0008-8846(98)00208-7
6. Kleiner, M., Curbach, M., Tekkaya, A.E., Ritter, R., Speck, K., Trompeter, M.: Development of ultra high performance concrete dies for sheet metal hydroforming. Prod. Eng. Res. Devel. (2008). https://doi.org/10.1007/s11740-008-0099-z
7. Guilleaume, C., Mousavi, A., Brosius, A.: Hybrid deep drawing tool for lubricant free deep drawing. In: Proceedings of the International Conference of Global Network for Innovative Technology and AWAM International Conference in Civil Engineering (IGNITE-AICCE'17): Sustainable Technology And Practice For Infrastructure and Community Resilience, Penang, Malaysia, 8–9 Aug 2017, p. 140003. Author(s) (2017). https://doi.org/10.1063/1.5008159
8. DYCKERHOFF GMBH: Dyckerhoff NANODUR® Compound 5941. …zur einfachen Herstellung von UHPC. https://www.dyckerhoff.com/nanodur (2017). Accessed 19 Apr 2022
9. Kashani, A., Provis, J.L., Xu, J., Kilcullen, A.R., Qiao, G.G., van Deventer, J.S.J.: Effect of molecular architecture of polycarboxylate ethers on plasticizing performance in alkali-activated slag paste. J. Mater. Sci. **49**(7), 2761–2772 (2014). https://doi.org/10.1007/s10853-013-7979-0
10. Sagmeister, B. (ed.): Maschinenteile aus zementgebundenem Beton, 1st edn. Praxis. Beuth Verlag GmbH, Berlin, Wien, Zürich (2017)
11. Deutsches Institut für Normung e. V.: Prüfverfahren für Zement. Teil 1: Bestimmung der Festigkeit. Beuth Verlag GmbH, Berlin ICS 91.100.10(DIN EN 196-1) (2016)
12. Lechner, P., Stahl, J., Ettemeyer, F., Himmel, B., Tananau-Blumenschein, B., Volk, W.: Fracture Statistics for Inorganically-Bound Core Materials. Materials (Basel, Switzerland) (2018). https://doi.org/10.3390/ma11112306
13. Deutsches Institut für Normung e. V.: Prüfung von Festbeton. Teil 3: Druckfestigkeit von Probekörpern ICS 91.100.30(DIN EN 12390-3) (2019)
14. Swamy, R.N., Mangat, P.S.: Influence of fiber geometry on the properties of steel fiber reinforced concrete. Cem. Concr. Res. (1974). https://doi.org/10.1016/0008-8846(74)90110-0
15. Thomason, J.L.: The influence of fibre length, diameter and concentration on the strength and strain to failure of glass fibre-reinforced polyamide 6,6. Compos. A Appl. Sci. Manuf. (2008). https://doi.org/10.1016/j.compositesa.2008.07.002
16. Gebhardt, A.: Understanding Additive Manufacturing. Rapid Prototyping, Rapid Tooling, Rapid Manufacturing. Hanser, München (2012)
17. Mennig, G.: Mold-Making Handbook, 3rd edn. Hanser eLibrary. Hanser Verlag, München (2013)

Smart Containers—Enabler for More Sustainability in Food Industries?

P. Burggräf[1], F. Steinberg[1], T. Adlon[2], P. Nettesheim[1](✉), H. Kahmann[2], and L. Wu[1]

[1] Chair for International Production Engineering and Management, University of Siegen, 57223 Kreuztal, Germany
philipp.nettesheim@uni-siegen.de

[2] Laboratory for Machine Tools and Production Engineering (WZL), Campus Boulevard 30, 52074 Aachen, Germany

Abstract. In recent years, Machine Learning (ML) applications for manufacturing have reached a high degree of maturity and deal as a suitable tool for improving production performance. In addition, ML applications can be used in many other areas of production to enhance sustainability within the manufacturing process. One specific area is the storage and transportation of bulk materials with Intermediate Bulk Containers (IBC). These IBCs are currently used solely for their primary purpose of storage and transportation for raw and finished goods. But for a major part of their handling cycle time these IBCs are a black box, and therefore do not add additional value to manufacturers. By equipping those containers with sensor technology, new data can be generated along the entire supply chain, taking the sustainability of production to a new level. Within the research project smart.CONSERVE we use this additional data to prevent waste of resources through storage of production goods in defective IBCs through predictive maintenance. In this publication, we describe how the use of such smart IBCs in the food industry increases supply chain visibility and reduces food waste by presenting a number of use cases that are possible due to the new data availabilities. Additionally, we provide insights into the transferability of these use cases to other industries and the many opportunities for manufacturers to develop new smart services and ML applications based on the collected data to increase sustainability.

Keywords: Artificial intelligence · Supply-chain · Sustainability · Smart solutions · Smart services · Machine learning applications

1 Introduction

In 2008, the global financial and economic crisis had a strong impact on the industry and supply chain. Despite the recovery of the economy in the following years, manufacturing companies in the EU are still under highly competitive pressure [1]. In addition, new products are being launched on the market with increasing frequency and product life cycles are becoming shorter. This further increases the competitive pressure on companies [2]. In order to remain competitive, they have to reduce their production costs

M. Liewald et al. (Eds.): WGP 2022, LNPE, pp. 416–426, 2023.
https://doi.org/10.1007/978-3-031-18318-8_43

[3]. Because of the crises in the food industry during the last decade, for example the bovine spongiform encephalopathy, dioxin or rotten meat, classical swine fever, or avian influenza, customers are increasingly focusing on the quality, origin, and conservation of food [4, 5]. Through a data-transparent supply chain, customers can be informed about product-related information accurately, when they are purchasing food [6]. To gain the trust of customers, the food enterprises need to increase the transparency of the supply chain, as trust is one key factor in the food industry [7].

Nowadays, in the competitive environment, the manufacturing industries are facing different challenges. For example, the problem of scheduling and order release, inaccurate demand forecasting, and inefficient production. The economic activity of a manufacturing enterprise is to process, refine, and produce the raw materials and semi-finished goods [8]. Since the companies that produce food fulfill the mentioned factors, they can be considered to belong to the manufacturing industry [9]. Due to this similarity not only the challenges of the manufacturing industry can be transferred into the BMLE-funded joint project "smart.CONSERVE Smart Container Services for Food Industries" but also the developed smart Intermediate Bulk Containers (IBCs) as a solution for more data-transparency and enabler for Artificial Intelligence (AI) solutions for increasing sustainability within the supply chain can be retransferred to other manufacturing branches.

In recent years, Machine Learning (ML) applications have become a powerful tool for optimizing production performance in various areas of production. For example, with the help of AI, decisions or suggestions can be made automatically for humans. Moreover, AI can be used to optimize the scheduling and order release to provide the load and demand-oriented scheduling proposals for production [10]. Moreover, ML algorithms have been developed in recent years for numerous use cases to improve demand forecasting. An example of the relevance of a versatile data basis is the research of Nikolopoulos and Fildes, which proved that the integration of weather forecast data has a positive effect on the forecast of beer demand [11]. As numerous decisions within the supply chain depend on demand forecasting, more accurate demand forecasting with AI, based on a solid database, can not only decrease the logistic costs but also strengthen customer satisfaction [2, 12, 13]. Additionally, a well-fitting forecasting algorithm is a basic element for effective logistic management to maximize the utilization of trucks [1]. In summary, this means that a high level of data availability is the basis for raising the sustainability of many processes connected to the supply chain. Therefore this paper tries to answer the following questions:

- What data does a smart IBC need to track to enable smart services and ML applications that increase sustainability in the supply chain?
- How can a technical solution look like to measure product-specific data inside the container and send it in real time to a cloud application to enable users to analyze the data?
- How can digitalization in the food industry increase sustainability and reduce food wastage?

The remaining parts of this article are as follows. Section 2 gives a short introduction to the current use of IBCs for the storage and transport of goods, the lack of data and the challenges for a smart solution in food industries. The necessary data for enabling AI, increasing sustainability and reducing food wastage is defined in Sect. 3. In Sect. 4 different concepts for tracking solutions as well as a model for transferring the data from the inside of the IBC to a cloud application are presented. Afterward, Sect. 5 gives the reader an overview of ideas for smart services and ML applications to increase sustainability and reduce food wastage. Section 6 gives a final conclusion.

2 Initial Situation

In industry reusable IBCs play an important role. They are suitable for a wide range of industry sectors to transport and store the goods, such as beverages, foodstuff, chemicals, hazardous goods [14]. In order to meet the requirements of different industries and products, various types of IBCs are developed, such as composite IBC, plastic IBC, foldable IBC, heated IBC, and metal IBC. Because of hygiene, flavor stability, and other food-safety-related reasons, the most common IBC within the food industry is the stainless-steel IBC. For beverage and foodstuff, the IBC is designed to meet the standards of hygiene and flavor stability and to protect the stored goods from being affected by other external factors such as UV-radiation. Thus, the quality of beverages and foodstuffs during transport and storage can be guaranteed. Depending on the characteristics of the goods the IBC can be equipped with other necessary systems, such as heating or cooling systems [15].

Today, these IBCs are only used to transport and store product materials. A collection of product and container-specific data via sensors in the sense of a smart IBC is not carried out. This is due to the strict micro bacterial safety requirements of the food industry. It requires that the IBC is completely closed so that no contaminants can get to the product, which also means that no sensor data can be transmitted from inside via cables. On the other hand, the IBC shields the mobile signals to such an extent that direct transmission via the mobile network is not possible. As a result, manufacturers have no information about their IBCs and the products stored in them after they leave the Company's warehouse. This leads to various issues for supply chain partners which must be checked manually or inquired by phone or email by the partners, such as:

- Where are the IBCs and which ones are ready for collection?
- What are the current stock levels of the customers?
- Are all product-specific storage conditions consistently maintained?
- Does the container still have enough pressure to ensure the quality of the product?

However, especially in today's world with increasing automation of production and logistics planning as well as end-consumer demands for a transparent and documented supply chain, the collection of product and container-specific data is of great importance. The end-consumer's demand for transparency is also underlined by the smart container eZaar which was developed for at-home storage of food and monitors the food stock to prevent wastage [16]. Another call for transparency lies in the fact that IBCs are reused several times. Therefore, IBC-owners strive to achieve the highest possible utilization of containers in order to minimize fleet size and thus their capital commitment. However,

IBCs are often not immediately reported back to the supplier as empty and therefore remain unused in the warehouse instead of being picked up for reuse.

Different from other industries, one of the factors that must be considered in the food industry is the best-before-date of products, which may not be exceeded and thus, has an influence on production and warehouse planning [17]. Because of the perishable nature of food, the long lead times for food production, the seasonal variations in production and consumption, and the variability in product quality and yield, the food supply chain has an increased level of complexity [18]. To provide consumers with high quality fresh products, the planning process for the availability of production capacity and material requirements, the production process, and the distribution planning process must be integrated to ensure that the product can be available on time [19].

To solve these problems, the enterprise Pack wise developed a smart cap that can be mounted to plastic IBCs. Through this smart cap the information about location, fill level, and the temperature of the IBC will be collected, transferred via mobile-network, and provided to users via a cloud application. This information can help to track the IBCs, reorder the products, and guarantee the quality of the goods. For example, when transporting liquids that react sensitive to temperature, fluctuations, pressure, and vibrations, the monitoring of these kinds of fluids is required.

Despite these preliminary works concerning the tracking of plastic IBCs, there is no solution available to detect product-related data (i.e. fill level, pressure) when using metal IBCs. One reason for this is that it is not possible to measure the fill level with a radar through metal. Also, the inside temperature and inside pressure can not be measured through outside-mounted sensors. On the other hand, when mounting the smart cap on the inner side of the lid the mobile signal is weakened too much, which hinders the transfer of the data to the cloud application. Therefore in this paper, we present a tracking system solution for metal IBCs to measure the mentioned product-related data we develop in the joint project smart.CONSERVE. In addition, we give a first insight into a bunch of use-cases for AI solutions and smart services for further research to meet the challenges of the manufacturing industry by using smart containers.

3 Approach

In order to be able to meet the aforementioned challenges, enable ML algorithms, and strengthen sustainability in the food industry, the joint project smart.CONSERVE aims to equip stainless steel IBCs with intelligent information and communication technology. For this purpose, we use sensor technology that is already used in a wide range of applications and apply it for the first time in stainless steel IBCs. In order to investigate the precise requirements for a smart IBC, we first recorded the relevant information in the literature, workshops, and interviews with food manufacturers and producers of IBCs. These requirements can be divided into the categories "data recording" and "technical solution". The evaluation identified the requirements, the resulting sensor, and its function as shown in Table 1.

Table 1. Overview of requirements and necessary sensor technology.

Requirement	Sensor	Function
Pressure monitoring	Pressure sensor	With the help of a pressure sensor, it should be possible to measure the pressure inside the IBC. This measurement allows the creation of a pressure curve and gives an early warning in case of pressure loss, in order to repressurize the container to ensure the quality of the product and to maintain the IBC before the next cycle
Temperature monitoring	Thermometer	With the help of a thermometer, it is possible to monitor the temperature inside the IBC. This measurement enables the creation of a temperature curve. In addition, early warnings will be made possible to prevent the goods from being destroyed because critical temperatures have been exceeded
Fill-level monitoring	Radar sensor	A radar inside the IBC allows the container fill level to be tracked. This enables suppliers and customers to have an overview of the current stocks
Location monitoring	GPS	The GPS receiver enables companies to track the position of their IBCs. Deliveries can be automatically notified in advance. In combination with the fill level, automatic container collection and optimized route planning are also possible. In this way, optimal utilization of truck capacity and a reduction of the container fleet through faster turnaround times can be achieved
Misuse monitoring	Brightness sensor	The brightness sensor detects the brightness inside the IBC. The incidence of light indicates an improper opening of the container. Additionally, it also indicates a faulty closure of the lid
Identification of mishandling	Accelerometer	A built-in acceleration sensor records the acceleration forces acting on the container in the x, y, and z axes. Improper handling can thus be identified

(*continued*)

Table 1. *(continued)*

Requirement	Sensor	Function
Automatic receipts posting	RFID	Using RFID, it is possible to automate container bookings in the ERP system. This prevents errors due to manual activities and accelerates processes

4 Technical Concepts

Thus, the objective of the project is to develop a smart food container that uses modular sensor technology to record the defined and relevant data from the transported goods. The development of the food container focuses in particular on the modification, taking into account the microbacterial requirements of the food industry, of existing and standardized stainless-steel IBCs, which can be individually expanded according to customer requirements, by combining them with available satellite sensors. Based on our interviews and workshops we formulated the following data-related requirements for our hardware solution:

- The solution needs to be able to measure product-related data inside the metal IBC
- The solution needs to be able to send the sensor data from the inner side of the container via mobile network to a cloud application

To meet these requirements and to maintain high flexibility for future sensor changes, we decided to use a satellite sensor concept with a sensor pack mounted inside the container to measure all the required information inside. These information are send via Bluetooth to the so called smart cap mounted on the outside of the IBC. This smart cap measures all relevant information on the outside and sends it together with the inside-information via mobile network to a cloud application.

For the sensor pack we developed three different modular concepts for the adaptive combination of satellite sensors. The first one is a so-called component exchange concept (Fig. 1). This means that starting from a standard product including battery, controller, and communication hardware, it is possible to mount further sensors on free slots on the controller. This can be done initially at the time of purchase or subsequently due to changed requirements and offers customers high flexibility to react to technological developments.

The second one is a bus concept. This concept means that it is possible to add a module to an existing series when one or more modules are added to an existing base. During the implementation, the controller will be considered as the basic construction. The sensors, the communication unit, and the battery will be assembled as modules. Figure 2 shows the principle of the bus concept. The modules can be added or replaced as required by means of standardized interfaces. The advantage of this concept is that there is not only the possibility to change and add sensors for new functions in the future

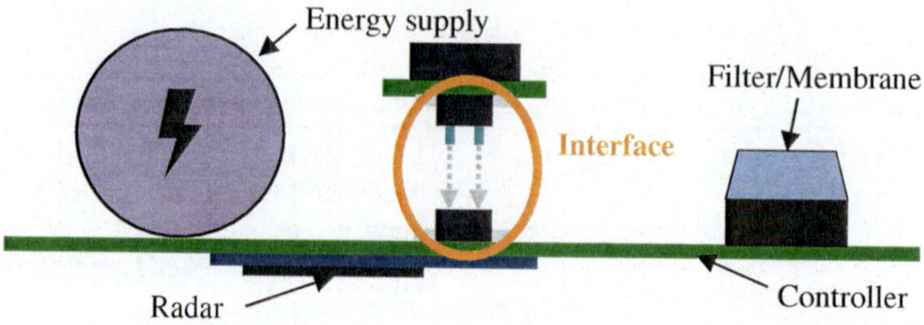

Fig. 1. Component exchange concept

as it is also possible in the first concept but also to renew the battery or the communication module whenever there is a new and more efficient technology available.

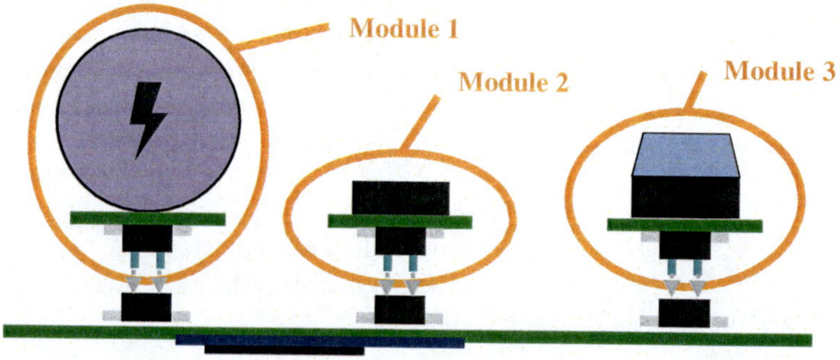

Fig. 2. Bus-concept

The third concept is software-based. This means that there is no physical modularity. The components of all variants will be used, and the use of Components is limited by software as it is done in Tesla cars [20]. The idea of the implementation is to integrate all components fix and install all sensors to enable a consistent production line. The functions can be added and activated by software. In this way, the assembly effort and the development effort for modularity will not arise, but there is no possibility for additional functions and future developments. Additionally, the costs for unused components must be covered. Figure 3 illustrates this principle.

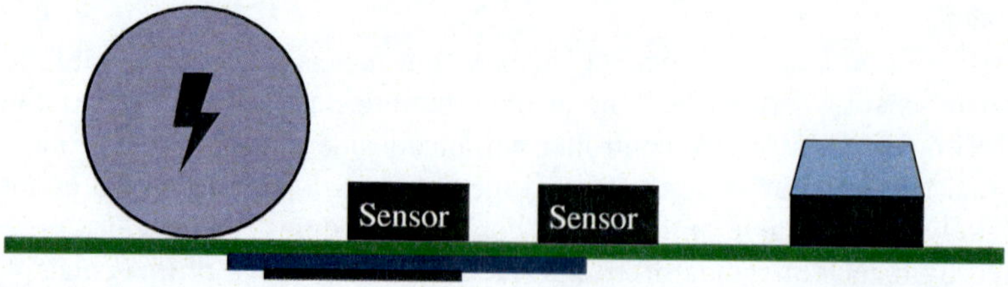

Fig. 3. Software-concept

In the further course of the smart.CONSERVE project, the three concept ideas will be tested for applicability in the food industry. In addition to general functional tests to determine the data quality of the different approaches, further criteria must be taken into account. Since the food industry is a highly price-sensitive industry, the expected lifetime of the different concepts must be determined in addition to the production costs in order to determine the annual costs for the use of the solution. Furthermore, it must be investigated whether the design according to the first two concepts poses a risk for microbacterial contamination. Based on the results, the appropriate concept will then be selected.

5 Enabling Sustainability with Smart Containers

By transferring the product- and container-specific data to a cloud application, it will be possible for customers and suppliers in the food industry, as well as companies in other industries that use metal-IBCs for the storage and transport of their products (i.e. fluids and granulated substances), to replace a previously existing data gap. The functions of the sensors described in Sect. 3 can already increase sustainability in the area of storage and transport in many respects. For example, the damage to goods due to violations of temperature- and pressure-limits can be reduced by continuous monitoring of the parameters, warnings, and subsequently early countermeasures. But in addition to monitoring data, companies can expand their existing business model with smart services in order to generate new revenues on the one hand and increase sustainability on the other. Also, there is the possibility to use the multitude of new data for machine learning applications. A selection of smart services and AI applications is presented below:

- Monitoring of the best before date
  - In industries with perishable products, such as the food industry, the digital twin can be used to track the product's best-before date. Customers can thus be warned in advance of the product's expiration and use it in good time. It is also possible to apply a minimum shelf life date that is linked to temperature or pressure. This enables companies to optimize inventory management by combining production planning, minimum resource shelf life and energy use during storage.
- Automation of container collection and route optimization
  - In all industries, location and level monitoring make it possible to automate the collection of empty IBCs. This data can be used in combination with historical data and an ML algorithm to forecast empty containers and plan optimal pickup routes. In this way, on the one hand, better utilization of truck capacity and thus a reduction in CO2 emissions can be realized. At the same time, the number of IBCs in circulation can be reduced, thus saving valuable raw materials. Due to the reduced space required for storing empty IBCs at the customer's site, the supplier can sell this service to its customers and generate additional revenue.
- Optimization of production planning through improved demand forecasting

  - The large amount of data from the smart containers enables suppliers to track their customers' inventories on a product-specific basis and to analyze usage behavior. Together with historical data, this enables suppliers to optimize demand forecasts using ML algorithms and to take this data into account in their own production planning. In this way, overproduction and thus a waste of resources and capital commitment due to planning inaccuracies can be reduced.

- Predictive maintenance

  - With conventional containers, it is not possible to detect defects such as pressure loss, which lead to damage and destruction of the goods, at an early stage. A gradual loss of pressure is often only detected after a long period of time when the goods are already at the customer's premises and the pressure has already fallen below the level required for product quality. With the help of smart IBCs, it is not

only possible to detect a creeping pressure loss in time to pressurize the container again, but predictive maintenance can also be realized, as the sensor technology can detect early signs of a pressure loss and trigger a maintenance request, thus avoiding food wastage due to defective containers.

6 Conclusion

In this work, a concept for smart metal-IBCs to enable ML applications and increase sustainability within the supply chain is presented. At first, several requirements for smart containers are defined and suitable sensors to meet the requirements are selected. Subsequently, it is shown how the data from the inside can be transferred to the cloud application via the smart cap. Besides three different concepts for the design of the modular sensor system are presented and the further project procedure for the selection of the concept suitable for the food industry is described. Finally, a selection of smart services and ML applications that will be enabled by smart IBCs and thus increase sustainability and quality and reduce food wastage within the supply chain are described. As the presented setup is not yet fully implemented this work focuses on the problem, the concept for the solution, and possible fields of further research to increase sustainability and reduce food wastage.

Acknowledgment. The project is supported (was supported) by funds of the Federal Ministry of Food and Agriculture (BMEL) based on a decision of the Parliament of the Federal Republic of Germany via the Federal Office for Agriculture and Food (BLE) under the innovation support programme.

References

1. Hart, M., Tomastik, M., Heinzova, R.: The methodology of demand forecasting system creation in an industrial company the foundation to logistics management. In: 2015 4th International Conference on Advanced Logistics and Transport (ICALT), pp. 12–17. IEEE (2015)

2. Afifi, A.A.: Demand forecasting of short life cycle products using data mining techniques. In: Maglogiannis, I., Iliadis, L., Pimenidis, E. (eds.) AIAI 2020. IAICT, vol. 583, pp. 151–162. Springer, Cham (2020). https://doi.org/10.1007/978-3-030-49161-1_14
3. Cheng, Y.-H., Hai-Wei, L., Chen, Y.-S.: Implementation of a back-propagation neural network for demand forecasting in a supply chain a practical case study. In: 2006 IEEE International Conference on Service Operations and Logistics, and Informatics, pp. 1036–1041. IEEE (2006)
4. Bastian, J., Zentes, J.: Supply chain transparency as a key prerequisite for sustainable agri-food supply chain management. Int. Rev. Retail. Distrib. Consumer Res. **23**, 553–570 (2013). https://doi.org/10.1080/09593969.2013.834836
5. Turi, A., Goncalves, G., Mocan, M.: Challenges and competitiveness indicators for the sustainable development of the supply chain in food industry. Procedia. Soc. Behav. Sci. **124**, 133–141 (2014). https://doi.org/10.1016/j.sbspro.2014.02.469
6. Wognum, P.M., Bremmers, H., Trienekens, J.H., et al.: Systems for sustainability and transparency of food supply chains—current status and challenges. Adv. Eng. Inform. **25**, 65–76 (2011). https://doi.org/10.1016/j.aei.2010.06.001
7. Astill, J., Dara, R.A., Campbell, M., et al.: Transparency in food supply chains: a review of enabling technology solutions. Trends Food Sci. Technol. **91**, 240–247 (2019). https://doi.org/10.1016/j.tifs.2019.07.024
8. Pollert, A., Kirchner, B., Polzin, J.M.: Duden Wirtschaft von A bis Z: Grundlagenwissen für Schule und Studium, Beruf und Alltag, 3rd edn. Dudenverlag, Mannheim, Leipzig, Wien, Zürich (2008)
9. Huang, Y., Wang, L., Liang, S.Y. (eds.): Handbook of Manufacturing. World Scientific, New Jersey, London, Singapore, Beijing, Shanghai, Hongkong, Taipei, Chennai, Tokyo (2019)
10. Jensen, T.: Whitepaper Use Cases for Industry (2020)
11. Nikolopoulos, K., Fildes, R.: Adjusting supply chain forecasts for short-term temperature estimates: a case study in a Brewing company. IMA J. Manag. Math. **24**, 79–88 (2013). https://doi.org/10.1093/imaman/dps006
12. Mircetic, D., Nikolicic, S., Maslaric, M., et al.: Development of S-ARIMA model for forecasting demand in a beverage supply chain. Open Eng. **610** (2016). https://doi.org/10.1515/eng-2016-0056
13. Zhu, X., Ninh, A., Zhao, H., et al.: Demand Forecasting with Supply—Chain Information and machine learning: evidence in the pharmaceutical industry. Prod. Oper Manag. 30:3231–3252. https://doi.org/10.1111/poms.13426
14. Biganzoli, L., Rigamonti, L., Grosso, M.: Intermediate bulk containers re-use in the circular economy: an LCA evaluation. Proced. CIRP **69**, 827–832 (2018). https://doi.org/10.1016/j.procir.2017.11.010
15. Saha, N.C., Ghosh, A.K., Garg, M., et al.: Food Packaging: Materials, Techniques and Environmental Issues. Lecture Notes in Management and Industrial Engineering. Springer, Singapore (2022)
16. Pila, R., Rawat, S., Singhal, I.P.: eZaar, the smart container. In: Shukla, B. (ed.) 2nd International Conference on Telecommunication and Networks—TEL-NET 2017: 10th11th August 2017, Amity University Uttar Pradesh, Noida, India. IEEE, Piscataway, NJ, pp. 1–5 (2017)
17. Prusa, P., Chocholac, J.: Demand forecasting in production logistics of food industry. AMM **803**, 63–68 (2015). https://doi.org/10.4028/www.scientific.net/AMM.803.63
18. Krishnan, R., Yen, P., Agarwal, R., et al.: Collaborative innovation and sustainability in the food supply chain evidence from farmer producer organisations. Resour. Conserv. Recycl. **168**, 105253 (2021). https://doi.org/10.1016/j.resconrec.2020.105253
19. Zhao, M.A., Setyawan, B.: Sales forecasting for fresh foods: a study in Indonesian FMCG. In: 2020 International Conference on Information Science and Communications Technologies (ICISCT). IEEE, Piscataway, NJ, pp. 1–9 (2020)

20. Wiegand, N., Imschloss, M.: Do you like what you (Can't) see? The differential effects of hardware and software upgrades on high-tech product evaluations. J. Interact. Mark. **56**, 18–40 (2021). https://doi.org/10.1016/j.intmar.2021.03.004

Investigation on the Influence of Geometric Parameters on the Dimensional Accuracy of High-Precision Embossed Metallic Bipolar Plates

M. Beck[1](✉), K. R. Riedmüller[1], M. Liewald[1], A. Bertz[2], M. J. Aslan[2], and D. Carl[2]

[1] Institute for Metal Forming Technology, University of Stuttgart, 70174 Stuttgart, Germany
maxim.beck@ifu.uni-stuttgart.de

[2] Fraunhofer Institute for Physical Measurement Techniques IPM, Georges-Köhler-Allee 301, 79110 Freiburg, DE, Germany

Abstract. The availability of effective and eco-friendly powertrain systems for electrification of passenger and commercial traffic is a crucial requirement for achieving current climate targets. With increasingly limited energy resources, fuel cell technology is gaining interest as an alternative to conventional electrical drives. Especially for heavy-duty and long-distance vehicles, where the required payload and range would require enormously heavy batteries, fuel-cell-technology offers a promising solution. Critical components of such modern fuel cells are metallic bipolar plates (MBPP) manufactured by high-precision embossing of thin metallic foils. The critical point is that even slightest fluctuations within the manufacturing process can lead to forming defects and result in unacceptable springback of metallic bipolar plates. Combined with the dimensional accuracy required for MBPP, extensive quality assurance and thus relatively low cycle times are inevitable in today´s production of these components. In this context, this paper deals with an approach to actively control the manufacturing process of MBPP based on numerical data sets. For this purpose material characterization of 0.1 mm stainless-steel foil (1.4404) was performed, allowing for comprehensive modelling of the embossing process and the springback behavior. In order to maintain a robust forming process aimed at increasing productivity, a numerical analysis was then conducted under variation of different geometric parameters using AutoForm R10. It was found that variation of selected geometric parameters such as channel width, channel height, draft angle and tool radii can remarkably reduce thinning and springback in MBPP production in compliance with tight tolerance specifications. Furthermore, the investigations show that active control of the lubrication conditions offers an additional possibility for subtle adjustments of the dimensional accuracy of produced components.

Keywords: Single-stage embossing · Numerical simulation · Metallic bipolar plate · Foil forming · Fuel cells

M. Liewald et al. (Eds.): WGP 2022, LNPE, pp. 427–438, 2023.
https://doi.org/10.1007/978-3-031-18318-8_44

1 Introduction

Fuel cells enable an efficient conversion of hydrogen and oxygen into electrical energy and heat while water is released as a reaction product. Thus, fuel cells are gaining increasing interest as an alternative to conventional electric drives. A single fuel cell consists of two electrodes which are separated by an electrolyte – usually by a membrane [1]. These membrane electrode assemblies (MEA) are then combined with bipolar plates to build fuel cell stacks. Here, the bipolar plates (BPP) are responsible for the electrical contacting of the individual MEAs, the supply of the reaction gases and the discharge of the resulting water. The BPPs therefore have to meet high requirements in terms of gas impermeability, electrical conductivity as well as corrosion resistance. In turn, this results in particularly high requirements for the dimensional accuracy of BPPs [2].

Graphite-based non-metallic BPPs have provided the foundation of fuel cell research for a long time. However, their suitability for the automotive industry is limited due to the complex machining process, high component thicknesses and material brittleness [3]. Therefore, metallic BPPs have recently gained more and more attention, since they offer higher thermal and electrical conductivities, better gas tightness, and more efficient material utilization at lower thicknesses [2, 3]. In general, two halves of the metallic bipolar plate, so-called monopolar plates (MPP), are manufactured separately by means of hollow embossing and are subsequently welded together into one bipolar plate in a subsequent step.

However, the hollow embossing of high-precision MPP made of metallic foils having thicknesses of less than 0.1 mm is a highly sensitive process and causes a variety of challenges for the manufacturing process. Defects such as cracks, wrinkles or insufficient dimensional accuracy due to springback occur even with the slightest fluctuations of material and process parameters and make extensive quality assurance of every MPP necessary. To this day, this issue results in long production cycle times and high costs during the manufacturing process. In particular, the geometric features of the filigree channel structures of the flow field such as channel width, channel height are the most important influencing parameters in this context and are in discrepancy with high productivity from a production engineering point of view.

Recent publications already do address this issue, but limit their investigations to very simplified laboratory-scale geometries with a limited number of channels or sections within the flow field of a MPP [4–8]. However, investigations on such a small scale allow only limited conclusions to be drawn on the overall quality of the formed components. Influences due to effects of interaction of numerous channels in the flow field of the plate in combination with directional and angular changes, as well as the resulting springback and thinning behavior are not sufficiently addressed in this way.

The main objective of this study is to identify a reasonably realistic experimental geometry to investigate the influence of various geometric features on the manufacturability of an MPP. This approach is intended to address the discrepancy between design specifications and feasibility from a manufacturing perspective and allows to derive recommendations for future measures. In particular, geometric parameters such as channel width, channel height, draft angle and tool radii were varied within a limited range of realistic values. Also input from industry is considered, ensuring carried out investigations are as close as possible to actual applications.

The basis of these investigations is an adapted as well as foil-specific material characterization, allowing comprehensive investigations to be carried out in forming simulations with AutoForm R10. In addition to investigated geometric influences, the possibility of process control through change in lubrication conditions in between individual strokes is examined. Such control is of particular interest due to the limited possibilities to influence part quality in the actual process. Based on these investigations, it should be possible to establish a basis for the future design process of metallic MPP in the long term.

2 Overall System for Active Process Control and Quality Assurance—"AKS-Bipolar"

Common quality assurance criteria in MPP production include the quality of the component surface finish and component edges as well as the dimensional accuracy (e.g. shape and position tolerances) of produced components. In terms of dimensional accuracy, manufacturers must meet very narrow tolerance limits within the range of micrometers, due to the very low material thicknesses of MPP of 0.1 mm and below. To this date, however, the detection of very small dimensional deviations in relation to the component size has been a major challenge, which, combined with the very extensive and obligatory quality control of individual MPP, leads to relatively low production cycle times in embossing of MPP. However, long-term objective of economic large-scale production of MPP is to achieve cycle rates of down to 1 Hz, which is currently still not achievable. In this context, the DFG-project "AKS-Bipolar" (*Active process control in the series production of high-precision embossed bipolar plates*) deals with exploring suitable approaches to actively control the manufacturing process of metallic MPP based on numerical data sets. The investigations presented in this paper were carried out as part of this project, which will terminate in September 2024.

By combining a comprehensive simulation toolchain with an inline-capable, full-surface 3D measurement of each manufactured component, "AKS-Bipolar" will realize a complete system for active process control and quality assurance for the series production of metallic MPP. A schematic representation of the intended demonstrator system within the framework of this joint project is depicted in Fig. 1.

For quality assurance objectives, a digital holographic sensor system specially developed for larger measuring fields is integrated into the production line. This system records the geometry of the produced components three-dimensionally during the production cycle in real-time having an extreme high measurement accuracy of 1 μm in height [9, 10]. The derived quality parameters such as channel geometry and overall flatness (springback) of the metallic MPP enable complete quality assurance and documentation. Due to the availability of highly accurate 3D-data, the simulation results can be compared and further adjusted thus creating a realistic digital twin of the process.

The simulation toolchain generates and optimizes the digital twin of the foil forming process by means of an artificial neural network, which is trained on the basis of extensive simulation data and 3D-measurements gained from the real manufacturing processes. In this way, recurring manufacturing issues and springback effects can be detected and suitable countermeasures such as the adjustment of the lubricant viscosity

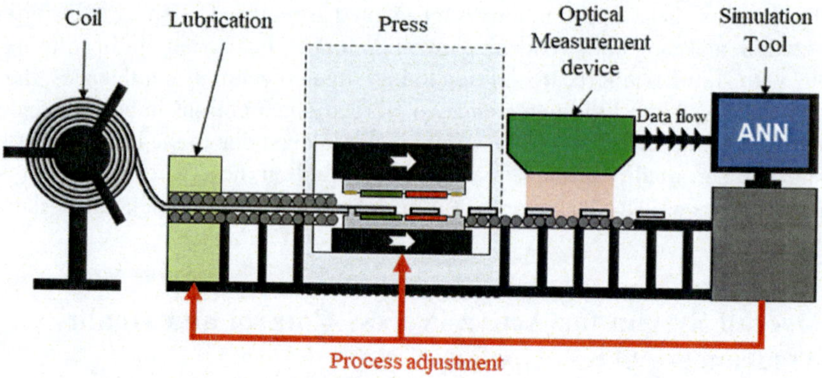

Fig. 1. Schematic representation of demonstrator of the DFG transfer project "AKS-Bipolar"

or press parameters can be specifically controlled. The investigations described below constitute the initial fundamentals for building the simulation toolchain of this system.

3 Material Characterization and Material Model

For the numerical investigations reported about in this contribution, stainless-steel foil (1.4404) with a thickness of 0.1 mm was used. To determine the material parameters required for the forming simulation uniaxial tensile tests according to DIN EN 6892 were first carried out using DIN 50125 Form H 20 × 80 specimens prepared in 0°, 45° and 90° rolling directions. The material properties obtained are listed in Table 1. In addition, bulge tests according to ISO 16808 were carried out allowing to gain data until a true strain value of 0.5 as shown in Fig. 2. For both tests, a strain measurement with GOM ARAMIS was carried out. Due to the high achievable value for true strain and expectedly low strains during the embossing of MPP no further extrapolation of the flow curve was needed. The flow curve thus was fitted directly with a polynomial approach to achieved experimental data.

Table 1. Material properties of stainless-steel 1.4404 / X2CrNiMo17–12-2.

Material	t (mm)	E (GPa)	Yield Strength (MPa)	UTS (MPa)	n	r_0	r_{45}	r_{90}
1.4404	0.10	200	238	556.2	0.43	0.72	1.06	1.22

In general, the value of the Young's modulus decreases when plastic strain in the sheet metal material is increased [11, 12]. Considering this effect is particularly important for the prediction of springback effects by means of forming simulations. Therefore, additional cyclic loading-unloading tests were performed as shown in Fig. 3 in order to determine this strain dependent reduction of the Young's Modulus for the steel foils considered. Here, the uniaxial tensile test specimens were loaded and unloaded at different strain levels, and the value of Young's Modulus was evaluated for each strain level accordingly. Finally, the curve for the strain dependent Young's modulus was obtained by fitting Eq. (1) to these values [12].

$$E = E_0 - (E_0 - E_a) * \left[1 - \exp(-\xi p)\right] \tag{1}$$

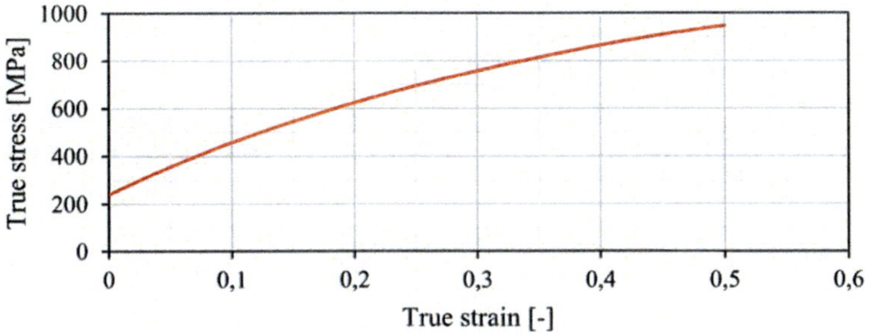

Fig. 2. Flow curve of stainless-steel 1.4404 (0.1 mm) based on uniaxial tensile and bulge tests

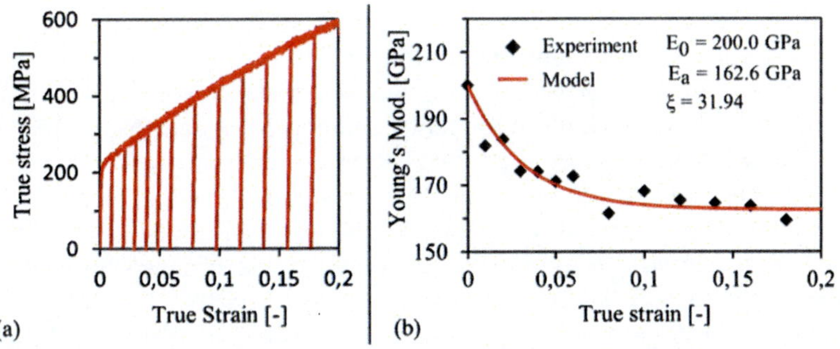

Fig. 3. (a) True stress diagram in cyclic loading-unloading test as a function of true strain, (b) strain dependent reduction of apparent Young's modulus.

For determining the forming limit behavior of the considered steel foil, scaled Nakajima tests were carried out using a punch with a reduced diameter of 20 mm compared to the Norm ISO 12004. The downscaling of the punch and the specimen's geometry were necessary to reduce the risk of wrinkling within the measuring area of the specimen. To reduce friction between punch and specimen, a layer of silicone with additional three layers of Teflon film and M-100 forming oil proved to be the most suitable solution. In this way, friction could be reduced tremendously, allowing for linear strain paths and thus reliable data for determining the Forming Limits Curve (FLC). For measurement of strains the optical measurement system ARAMIS was used. Corresponding specimens as well as acquired FLC are shown in Fig. 4.

The data sets obtained by such characterization tests were stored as a material model within AutoForm R10, whereby the strain rate-dependent Young's modulus as described in Eq. (1) was fed into utilized user-defined hardening model [13]. In this way, springback behavior of the metallic foil could be considered. For modelling of the yield surface, the BBC 2005 yield surface model was used.

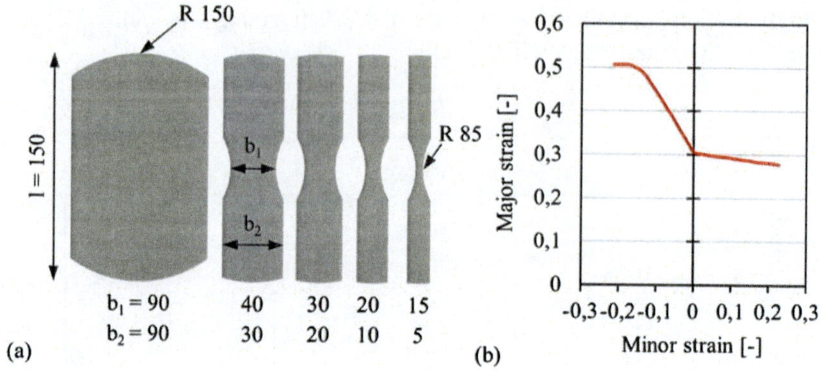

Fig. 4. (a) Scaled Mini-Nakajima specimens, (b) Forming Limit Curve (FLC) of 1.4404 stainless-steel (0.1 mm)

4 Numerical Investigation on the Influence of Geometric Parameters on the Dimensional Accuracy of MPP

In the subsequent sections, a numerical investigation of the influences of geometric parameters on the dimensional accuracy (springback) and material thinning of the MPP will be carried out based on a close-to-reality, complex MPP part shape. Hereby, characteristic geometric channel sizes are varied within a given framework in order to investigate the influences on chosen result variables by means of correlation coefficients. In used experimental geometry shown in Fig. 5, essential characteristics of a MPP were considered. Thus, the MPP design comprises inlet and outlet channels for both the reaction gases and the coolant. These channels are located outside the flow field of the MPP and contain additional grooves for appropriate seals. The flow field of the MPP design consists of five parallel channels with a constant distance of 0.6 mm in between and several changes of flow field direction. These changes of the flow field direction represent particularly critical areas during the hollow embossing process of metallic MPP. The designed MPP geometry as well as the geometric parameters of the embossing punch varied during the numerical investigations are shown in Fig. 5.

For each geometry variation, the die surface was designed with an offset of 0.1 mm based on the shown punch geometry. Thus, the punch, the stainless-steel foil and the active tool surfaces are in contact in every point at full closure of the embossing tool.

As a result, the process considered can also be referred to as hollow embossing with counter-pressure.

Parameters shown in Fig. 5 were varied within a fully parametric CAD-model in a range close to actual industrial application. The upper punch radius r_u and the lower punch radius rl were set dependent to each other with a difference of 0.1 mm. In this way, uniform curvature radii could be realized on the upper as well as the lower side of the MPP. Corresponding surface geometries were then extracted for forming simulation with AutoForm R10. The variation of geometric parameters is shown in Table 2.

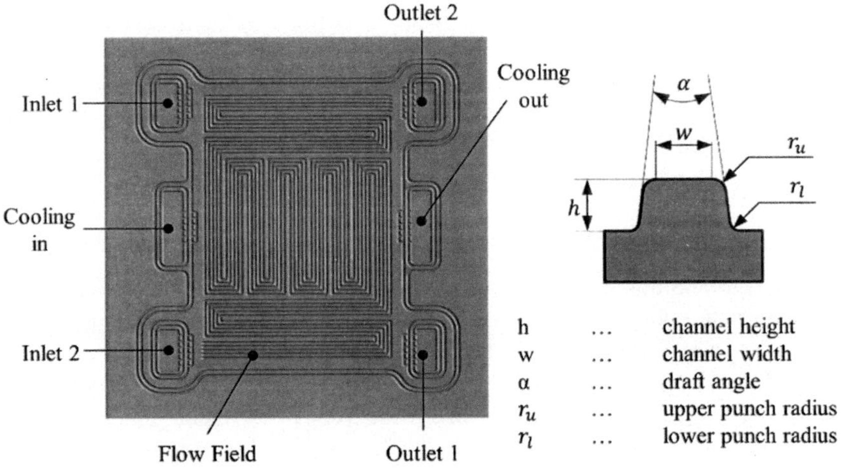

Fig. 5. Experimental geometry of a MPP and varied geometric parameters of the punch. Dimensions of shown metallic MPP are 100 mm × 100 mm

Table 2. Variation of geometric parameters of the punch in the design of the MPP.

Parameter	Symbol	Variation
Channel height	h	0.3 mm I 0.4 mm I 0.5 mm
Channel width	w	0.3 mm I 0.4 mm I 0.5 mm
Draft angle	α	0° I 10° I 20°
Upper punch radius	r_u	0.1 mm I 0.15 mm I 0.2 mm

Using a full factorial design, a total of 72 surface pairs of punch and die were created. The aim of the numerical investigations was to identify influences of relevant geometric parameters on the overall quality of corresponding part. Based on these findings, a geometry for a metallic MPP can be defined which can be manufactured in a single-step hollow embossing process. For the evaluation of the overall part quality, maximum thinning and springback within the flow field of the MPP are of particular interest.

The forming simulations in AutoForm R10 were performed based on the material model described in Sect. 3. For friction modelling, the Coulomb friction model was used with a constant friction coefficient of $\mu = 0.1$. Thick shell elements with eleven integration points over its thickness and an initial side length of 2.00 mm were used for meshing of the workpiece. Despite the relatively coarse initial mesh size of the metal foil, a very low threshold value for the contact penetration in combination with six refinement levels led to a fine meshing of the hollow embossed MPP during the simulation. A total number of approximately 1 million elements with a side length as low as 0.02 mm was used in the final simulation-step of each geometric parameter set. In this way, a sufficiently accurate representation of the channels within the flow field of the MPP was ensured by meeting convergence criteria in each simulation step.

Identified geometric parameter combination for the metallic MPP will then be used for further investigation of process control by modification of lubrication conditions between individual strokes. Thus, the possible control range due to change in the friction coefficient can be quantified.

5 Results and Discussion

5.1 Influence of Geometric Parameters of MPP on Its Overall Quality

For the evaluation of the simulation results with regard to the influence of the geometric parameters on the achievable part quality, particularly material thinning and springback behavior in the flow field area of the MPP were considered. Here, material thinning was evaluated based on the maximum thinning value, which usually occurred at sections containing a change in the direction of the flow field due to the interaction of many narrow radii. For the assessment of the springback behavior, the deviation between maximum and minimum displacement in Z-direction was used as an evaluation criterion. Deviation values at the outer edges of the MPP were not considered. Figure 6 shows simulation results for the parameter combination with the lowest value for thinning of 24% and a comparatively low dimensional deviation of 1.082 mm for springback. According to the FLC, no cracks occurred in this geometry variant, since the ratio between maximum computed major strain in an element and the corresponding acceptable major strain of the FLC is 0.62. In general, a failure can be expected at a ratio of maximum computed major strain to acceptable major strain of the FLC of approximately 1.0 or higher.

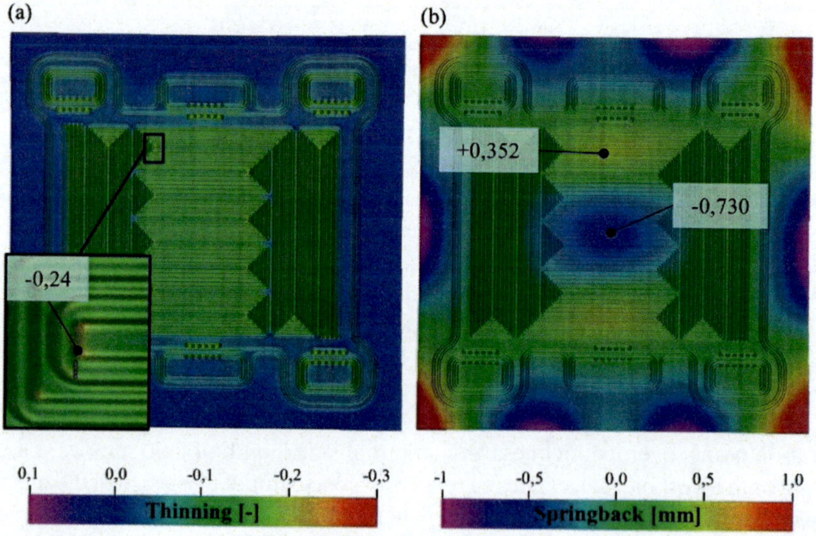

Fig. 6. (a) Thinning and (b) springback in metallic MPP with h = 0.3 mm | w = 0.5 mm | α = 10° | r_u = 0.2 mm

In addition, an analysis regarding the influences of the variation of individual geometric parameters on the quality of the MPP was carried out, using Pearson correlation coefficient. Respective correlation coefficients are shown in Table 3, where a value close to +1 or −1 indicates a strong correlation, whereas a value close to 0 indicates a low correlation and thus low dependence.

Significant correlations to the MPP´s quality was identified for the design parameters channel height *h*, channel width *w* and the upper punch radius r_u. In particular, the channel height shows a strong positive correlation to the unavoidable thinning within the flow field as well as a negative correlation with the springback deviation. Thus, an increased thinning of the material as well as a reduced springback behavior can be expected when increasing channel height *h* in the given geometry. A decrease in springback can be observed when increasing channel width *w*, as well as a decrease in thinning when increasing upper punch radius r_u. However, the Pearson correlation coefficient for the draft angle α is almost zero and indicates that there is no significant or explainable influence on the overall dimensional accuracy of the MPP.

Table 3. Correlations between geometric parameters and Springback and Thinning

Parameter	Springback	Thinning
Channel height *h*	**−0,41**	**0,90**
Channel width *w*	**−0,59**	−0,09
Draft angle α	0,04	−0,04
Upper punch radius r_u	−0,13	**−0,35**

Based on these results, the most suitable geometry in this study was composed of the smallest channel height *h*, the largest possible upper punch radius r_u as well as the largest channel width *w* within the given geometric parameter range. The influence of the individual geometric parameter combinations on thinning and springback are shown in Fig. 7.

Figure 7 shows that thinning and springback behavior of the MPP geometry considered in this investigation is significantly depending on the geometric parameters set. Here, single-step hollow embossing without occurrence of cracks is only possible for a channel height of h = 0.3 mm or a channel height of h = 0.4 mm combined with a relatively large radius of r_u = 0.2 mm. A channel height of h = 0.5 mm leads to cracks during the embossing process regardless of the parameter combination chosen. A comparison of achieved numerical results with actual measurements of MPP will be the subject of future investigations.

5.2 Process Control by Modification of Lubrication Conditions

Based on the identified geometry in Sect. 5.1 (see Fig. 6) an additional simulation with varying friction coefficient between $\mu_{min} = 0.05$ and $\mu_{max} = 0.30$ was conducted. In this way the approach for process control due to change in lubrication conditions, such as the adjustment of the lubricant viscosity can be investigated in more detail. Corresponding results of performed simulation are shown in Fig. 8.

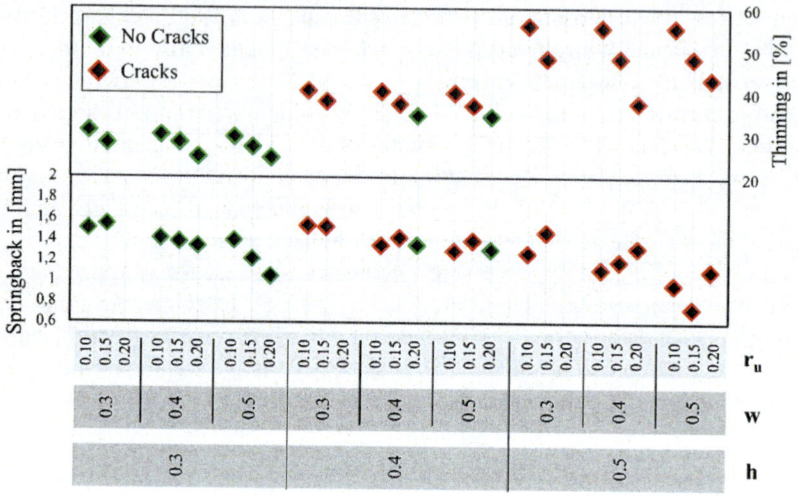

Fig. 7. Results for thinning and springback in case of different geometric sets of the MPP design

Depending on the position within the investigated flow field, the control range achieved for thinning is between 0% and 10,6%, whereas the control range for springback is between 0.131 mm and 0.400 mm respectively. As control range the difference between smallest and largest possible value within one numerical element is understood. However, when considering the actual distribution of control range for thinning within the flow field, then it is evident that significant adjustment is only possible in very limited areas. Thus, thinning is mainly dependent on the geometric parameters set.

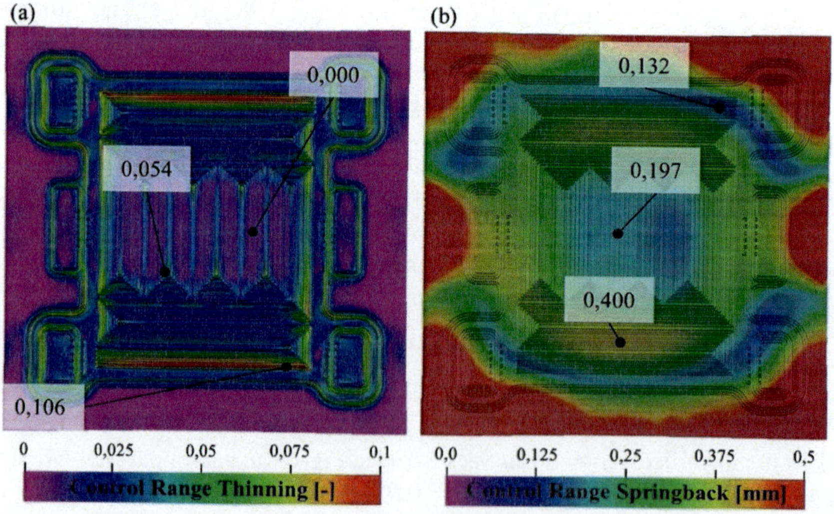

Fig. 8. Control Range for (a) thinning and for (b) springback within the embossed MPP for friction coefficients between $\mu_{min} = 0.05$ and $\mu_{max} = 0.30$

Such a high control range suggest that controlling the quality of the embossed metallic MPP by changing friction conditions is generally possible. However, the actual implementation and adjustment of the friction conditions between individual press strokes will be subject to further investigations within the scope of the project AKS-Bipolar.

6 Conclusion

This paper presents an initial research report of the project "AKS-Bipolar", which deals with exploring suitable approaches to actively control the manufacturing process of metallic MPP based on numerical data sets. By combining a comprehensive simulation toolchain with an inline-capable, full-surface 3D measurement of each manufactured component, "AKS-Bipolar" will realize a complete system for active process control and quality assurance for the series production of metallic MPP.

The investigations presented in this paper were carried out as part of this project and show how various geometric features influence the manufacturability of an experimental metallic MPP-geometry by means of single step embossing. For this purpose, process modelling and corresponding forming simulations were conducted with Auto-Form R10. Material data based on a foil-specific material characterization of a 0.1 mm thick 1.4404 stainless-steel foil were used as input for forming simulations, allowing comprehensive modelling of the embossing process. It was shown that an immense variation of the part quality, both in terms of thinning and springback is possible with slight changes to geometric parameters.

Additional simulations with variation of the friction coefficient enabled further adjustment of the overall part quality and showed that process control due to change in lubrication conditions as planned in project "AKS-Bipolar" is indeed possible to a certain extent. In this way, the results presented in this paper can serve as a basis for the future design of embossing processes of MPP as well as the design of a process control based on lubrication conditions.

Future research will focus on further improvement of the presented numerical model by adjustments based on acquired high-precision holographic 3D-measurement data. Thus, allowing an improvement of the design and embossing process of metallic MPP with regard to overall part quality and improving cycle times in mass production.

Acknowledgement. This work was supported within the Fraunhofer and DFG transfer programme. The authors like to acknowledge the German Research Foundation (DFG-Project 460294948) for financial support.

References

1. Kurzweil, P.: Brennstoffzellentechnik. Springer Vieweg, Wiesbaden (2012)
2. Porstmann, S., Wannemacher, T., Drossel, W.G.: A comprehensive comparison of state-of-the-art manufacturing methods for fuel cell bipolar plates including anticipated future industry trends. J. Manuf. Process. **60**, 366–383 (2020)

3. Bauer, A.: Experimentelle und numerische Untersuchungen zur Analyse der umformtechnischen Herstellung metallischer Bipolarplatten. Technische Universität Chemnitz, Chemnitz (2020)
4. Alo, O. A., Otunniyi, I.O., Pienaar, Hc.Z.: Manufacturing methods for metallic bipolar plates for polymer electrolyte membrane fuel cell. Mater. Manuf. Processes **34**(8), 927–955 (2019)
5. Bong, H.J., Lee, J., Kim, J.H., Barlat, F., Lee, M.G.: Two-stage forming approach for manufacturing ferritic stainless steel bipolar plates in PEM fuel cell: Experiments and numerical simulations. Int. J. Hydrogen Energy **42**(10), 6965–6977 (2017)
6. Zhang, R., et. al.: Investigation and optimization of the ultra-thin metallic bipolar plate multistage forming for proton exchange membrane fuel cell. J. Power Sources **464**, 229298 (2021)
7. Xu, Z., et. al.: Fabrication of micro channels for titanium PEMFC bipolar plates by multistage forming process. Int. J. Hydrogen Energy **46**, 11092–11103 (2021)
8. Zhang, P., et. al.: Investigation of material failure in micro-stamping of metallic bipolar plates. J. Manuf. Process. **73**, 54–66 (2022)
9. Fratz, M., Seyler, T., Bertz, A., Carl, D.: Digital holography in production: an overview. Light: Adv. Manuf. **2**(3), 283–295 (2021)
10. Fratz, M., et al: Inline application of digital holography. Appl. Optics **58**(34), 120–126 (2019)
11. Niechajowicz, A.: Apparent young modulus of sheet metal after plastic strain. Arch. Metall. Mater. **55**, 409–420 (2010)
12. Yoshida, F., Uemori, T.: A model of large-strain cyclic plasticity and its application to springback simulation. Int. J. Mater. Sci. **45**(10), 1687–1702 (2003)
13. Kubli, W., Krasovskyy, A., Sester, M.: Modeling of reverse loading effects including work hardening stagnatisson and early re-plastification. Int.J. Mater. Form. **1**, 145–148 (2008)

Investigation of Geometrical and Microstructural Influences on the Mechanical Properties of an Extruded AA7020 Tube

J. Reblitz[1(✉)], S. Wiesenmayer[1], R. Trân[2], and M. Merklein[1]

[1] Institute of Manufacturing Technology, Friedrich-Alexander-Universität Erlangen-Nürnberg, Egerlandstraße 13, 91058 Erlangen, Germany
jonas.reblitz@fau.de

[2] Fraunhofer Institute for Machine Tools and Forming Technology, Reichenhainer Straße 88, 09126 Chemnitz, Germany

Abstract. In the automotive sector, an important strategy for reducing CO_2-emissions is lightweight construction. In this regard, body-in-white parts offer a high potential for weight reductions by substituting conventional steel parts with tubular, aluminum-based components. Due to their high strength-to-weight-ratio and high crashworthiness, tube profiles are often used as safety-relevant car body components. Therefore, an exact determination of the material properties is necessary, in order to achieve a high prediction accuracy of FE-simulations. In contrast to the testing of flat semi-finished parts, there are only few standardized methods for the material characterization of tubular components. Furthermore, a profound knowledge regarding the influences of geometry and manufacturing process of the tube profiles on the material properties has to be gained. Thus, AA7020 tubes are analyzed in this research work. Tensile specimens are cut out of a tube by a laser-cutting machine. As a result, the cross-section of the samples is curved as well. To ensure a proper material testing, the clamping jaws for the tensile test are adjusted to the curvature of the samples. For the characterization of the tube properties, caused by the manufacturing process, optical measurements are performed. The mechanical properties of the different grain structures are determined by the thermo-mechanical simulator Gleeble 3500 in T6 condition. Therefore, tensile specimens with a reduced wall thickness are prepared. Afterwards the curved tensile samples are tested in T6 and W-temper condition. To validate the comparability of the determined material properties, a comparison is conducted with flat specimens prepared from the semi-finished tube profiles.

Keywords: Material characterization · W-temper condition · Tubes

M. Liewald et al. (Eds.): WGP 2022, LNPE, pp. 439–450, 2023.
https://doi.org/10.1007/978-3-031-18318-8_45

1 Introduction

The European Union aims to become climate-neutral by 2050. The focus is primarily on CO_2-emissions, which were reduced by 21% in 2019 compared with 1990 [1]. This was achieved mainly in the energy production, industrial and residential sectors. Only the emissions of the transporting sector have not been reduced. With a share of 26% of the whole CO_2-emissions of the EU, the greenhouse gas production has increased by 24% since 1990 [2]. The reasons for this are an increasing amount of traffic and the increasing engine power of registered vehicles [3]. Approaches for the reduction of CO_2-emissions include increasing connectivity of transport, an expansion of rail transport and the expansion of charging options for electrically powered vehicles [4]. Additionally, it is also necessary to reduce the energy consumption of the vehicles, whereby a reduction in vehicle weight is an important goal. The use of lightweight materials with a high specific strength as well as the use of tubular profiles are possible approaches to overcome these challenges. Due to their high stiffness and good crash properties, these are suitable for structural applications in car body construction [5]. Since these are safety-relevant components, a profound knowledge regarding the mechanical properties is required. However, there are only few standardized methods for the characterization of the material properties of tube profiles. Testing setups like the ring tensile test or the drift expanding test are only suitable for measuring the influence of weld seams and surface defects on the tube ductility. Furthermore, the tube characterization by tensile test requires a proper specimen preparation in order to not affect the stress or strain state. Therefore, in this research work, an extended understanding of the mechanical and geometrically conditioned tube properties is investigated with the aid of testing methods that are adapted to the semi-finished parts. In this context, the influence of the microstructure and the geometry of the tubular semi-finished parts made out of AA7020 on the mechanical properties in the tensile test is determined. Thus, the need of adapted testing methods for the characterization of profiles is to be evaluated. Since the tubular components are thick-walled, the local mechanical properties are analyzed for the different microstructure formations over the cross-section. As a result, a more precise characterization of the semi-finished parts for a tube hydroforming process could be derived. In order to extend the formability, tensile specimens are also characterized in W-temper condition. To validate the suitability of the applied testing methods, the mechanical properties of a curved and machined flat tensile specimen are compared.

2 Methodology

2.1 Specimen Preparation

For the characterization of the mechanical properties, tensile specimens are prepared out of tubular semi-finished parts. For this purpose, tubes with a diameter of 60 mm and a wall thickness of 5 mm are clamped at one side while the specimens are cut from the tube by a laser, as shown in Fig. 1a). To avoid melting at the inner diameter of the tube, a sheet is inserted which absorbs the scattered laser beams. Afterwards, the edges of the tensile specimens are milled to remove sections with a high surface roughness and a thermal influence. The generated specimen geometry is an adapted A_{50}

sample that can be tested by the thermo-mechanical simulator Gleeble 3500. According to this, the testing length measures 62.5 mm and the width 12.5 mm. In addition, two drillings are located at the clamping area to ensure a central positioning. Due to the tube geometry of the semi-finished parts, the tensile specimens have a curvature. To evaluate the mechanical properties of different grain structures over the wall thickness, the testing area of the samples is milled to a thickness of 1 mm.

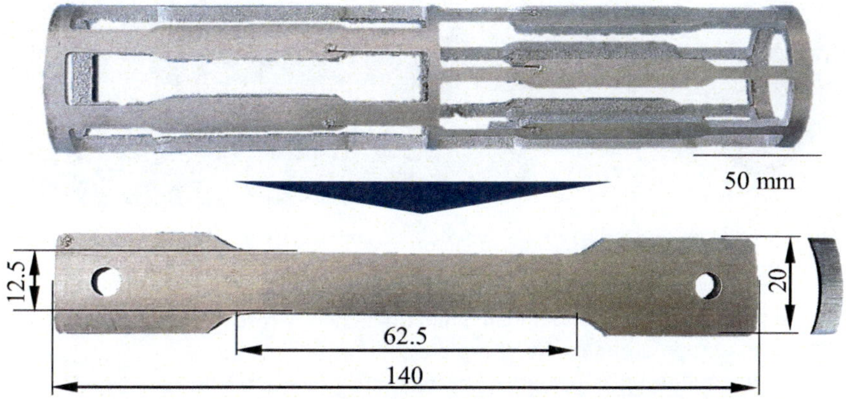

Fig. 1. Specimen preparation from an AA7020 tube by laser cutting

2.2 Methodology and Experimental Setup

For the tensile test, the clamping jaws were adjusted to the curved sample geometry, as shown in Fig. 2. This avoids a flattening of the specimens at the clamping as well as undesired influences on the stress state during the testing.

The material testing is performed by the thermo-mechanical simulator Gleeble 3500 at room temperature. In the middle of the experimental setup, the specimen with the curved clamping jaws is positioned and fixed by u-shaped separators. The optical measurement system Aramis from GOM GmbH detects the occurring strain distribution during the test. The illumination of the sample is achieved by two lamps. Therefore, a stochastic graphite pattern is applied to the sample surface to enable strain measurements. Since the test is not strain-controlled, marginal variations in strain rate may occur, which however are negligible.

2.3 Heat Treatment Conditions of the Tested Tensile Specimens

The evaluation of the mechanical properties of extruded AA7020 tubes over the wall thickness is performed in T6 condition. This state is induced by solution annealing, quenching and artificial ageing, which leads to the maximum strength [6] and a low formability. One approach to enhance the forming limit is W-temper forming. In this

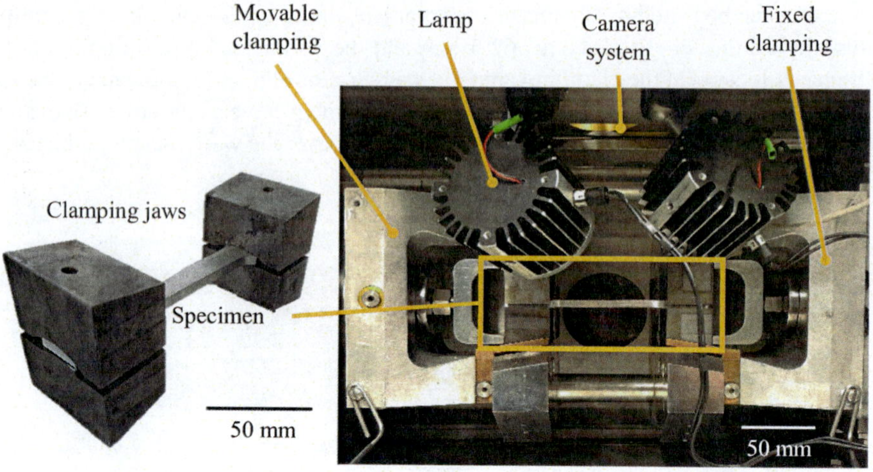

Fig. 2. Clamping jaws for a curved tensile specimen and experimental setup of the tensile test

context, the material is solution annealed and quenched before forming [7]. As a result, strength-increasing precipitates are completely dissolved, resulting in a supersaturated state. Since this condition is unstable, the forming process has to be carried out with only little delay. Hebbar et al. [8] observed a significant decrease in strength and an increase in fracture elongation for rolled semi-finished parts out of AA7020. They also determined a dependence between the mechanical properties and the temperature as well as the duration of the solution annealing.

Based on preliminary studies, in this paper the parameters for the solution annealing were set to 460 °C and ten minutes with a subsequent natural ageing time of 45 min.

3 Results and Discussion

3.1 Local Properties of the Tube Profile in T6-Temper Condition

Optical characterization in axial tube direction and over the cross-section. For the characterization of the tube properties in T6 condition, optical measurements of the grain structure are performed in axial direction and over the cross-section. The aim is to show that the material properties of thick-walled tubes vary across the cross section due to the manufacturing process. This could influence the forming behavior in a subsequent hydroforming process. According to Fig. 3b) there might be an orientation of the grain structure in axial direction which could be caused by the shear stress of the extrusion process [9]. Consequently, an anisotropic material behavior is expected. The black spots that can be seen in axial direction might be pores. In addition, the grain structure over the cross-section was analyzed as shown in Fig. 3c). In this regard, a various grain size and orientation was detected. At the inner and outer diameter, narrow sections with a fine grain structure are visible, whereas in the middle of the wall thickness a more coarse-grained section with orientations in circumferential direction was found. This can be attributed to the shear stress of the extrusion process caused by changes in the cross-section. For an extruded tube made out of AA7075, Nazari Tiji et al. [10] observed an elongation of the grain in circumferential direction to seven times the value of the wall thickness direction. One reason for the finer grain structure at the surface might be recrystallization induced by shear stresses and high temperatures [9]. Thus, various material properties over the cross-section are expected. Therefore, tensile tests are performed for different layers in wall thickness direction.

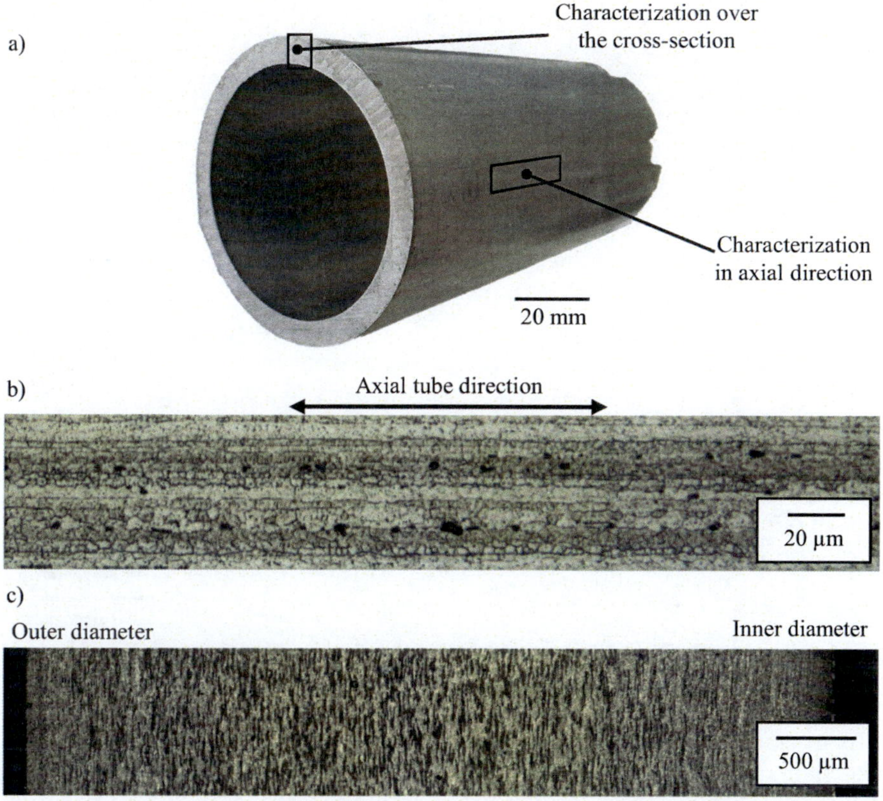

Fig. 3. Optical characterization of the tubular semi-finished parts located according to a) in b) axial direction and c) over the cross-section

Investigation on the mechanical properties over the cross-section. For the characterization of the mechanical properties over the wall thickness, three layers with different grain structures are defined. According to Figs. 3 and 4a) layer 1 seems to represent a section with grain orientations in circumferential direction next to the fine structure at the outer diameter. Layer 2 is in the middle of the wall thickness and shows weaker orientations in circumferential direction compared to layer 1. The third layer is located at the inner diameter with a finer grain structure. Khadyko et al. [11] also detected a fine grain structure at the surfaces of an extruded AA6063 profile, whereas the layer beside is more coarse-grained than the one in the middle of the wall thickness. The associated grain size of layer 3 measures 3.7 μm in wall thickness direction and 5.8 μm in circumferential direction. In the coarser layer 2, grain sizes of 8.2 μm over the sheet thickness and 13.7 μm in circumferential direction were detected.

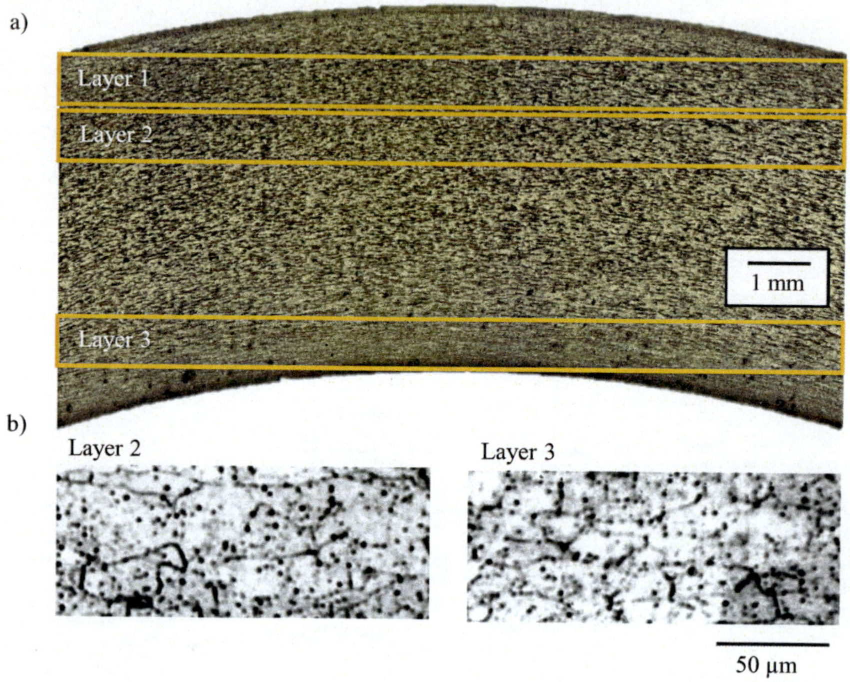

Fig. 4. a) Position and b) grain structure of the layers tested in a tensile test

To generate several layers, the testing area of the tensile specimens is milled to a thickness of 1 mm each. In conformity with the defined layers, the specimen cross-section is flat, see Fig. 5.

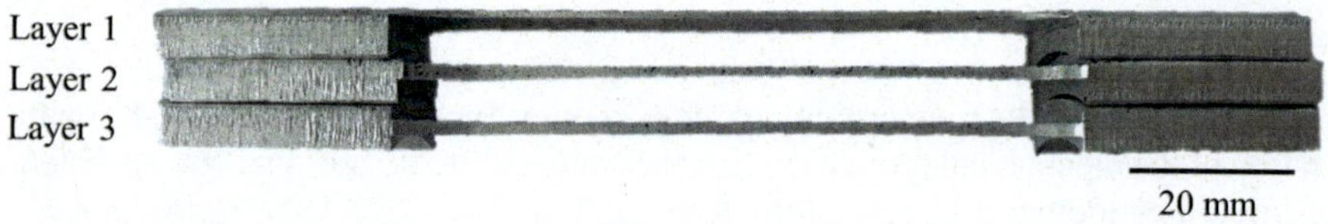

Fig. 5. Preparation of tensile specimens to investigate local mechanical properties over the cross-section

Due to the tube geometry of the semi-finished parts more than one grain section is covered by a flat specimen. However, qualitative statements on the local forming behavior of the tube components can be derived based on the mechanical properties shown in Fig. 6. Since the material properties are determined in the necking area with a length of 5 mm, no quantitative comparisons to other tensile test are possible. As a consequence, especially the elongations are distinguished by higher values than for the averaging over the standard range of 50 mm length. This method was chosen to reduce the influences of the milling process on the material testing. According to Fig. 6, the yield strength of the different layers is between 342 MPa for layer 1 and 364 MPa layer 3. A similar trend is visible for the ultimate tensile strength that increases from 385 MPa to 415 MPa. The higher strength at layer 3 might be caused due to the fine grain structure. As a result of the increased number of grain boundaries, dislocation movements are hindered and thus a reinforcement is induced [9]. The lowest values of the strength are observed for layer 1. The orientations transverse to the testing direction might be a reason for that. For the uniform elongation with values between 7.8% and 10.3% no clear trend is observed. The fracture elongation decreases form 19.6% for layer 1 to 17.9%. Thus, the sections with a low strength show a high ductility. The strain hardening exponent reaches values between 0.09 and 0.10. The knowledge of the local material properties across the wall thickness can be helpful for a tailored design of forming processes such as hydroforming. Furthermore, the locally different mechanical properties indicate the need of a tube specific material characterization. Consequently, flat blanks are not suitable for the determination of the anisotropic material properties of tubular components. However, with the determined values only qualitative statements are possible. Since the testing area Deviations can be induced by the milling of the tensile specimens and the flat specimen cross-section in conjunction with the low wall thickness.

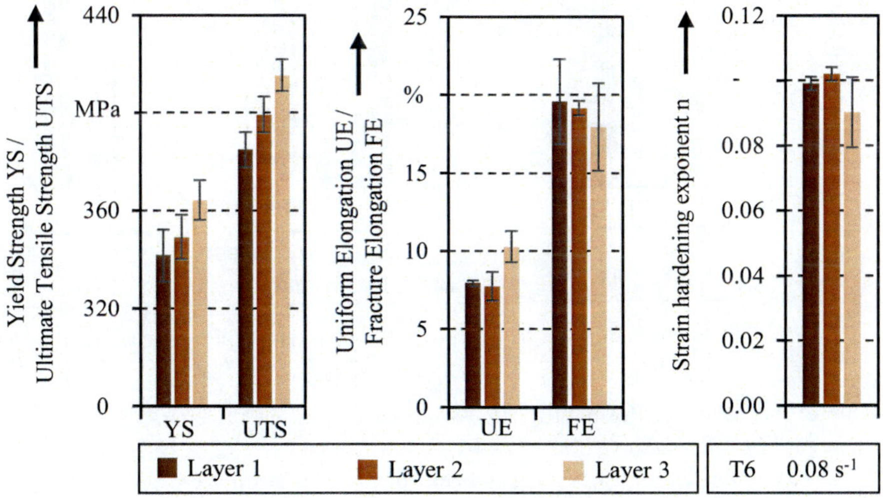

Fig. 6. Mechanical properties of different layers over the cross-section

3.2 Evaluation of the Mechanical Properties of T6 and W-temper Condition

Due to the low formability of 7xxx aluminum alloys in T6 condition, a comparison with the W-temper state is performed. For the determination of the material properties, the whole length of 50 mm is analyzed according to ISO 6892-1. The investigated mechanical properties are shown in Fig. 7 with three different strain rates for the more ductile W-temper condition. Initially, a comparison of the heat treatment conditions is carried out at strain rate 0.08 s^{-1}. Compared to the T6 state, the W-temper condition exhibits significantly lower strengths as well as higher elongations and strain hardening exponents. While the yield strength in T6 state shows a value of 357 MPa the W-temper condition has a value of 115 MPa. The tensile strength assumes values of 410 MPa and 224 MPa, respectively. For the uniform and fracture elongation, notable higher values can be achieved in W-temper condition compared to T6 state. Therefore, a significant increase of the forming limit is reached by solution annealing and quenching. In this context, the higher strain hardening exponent is also advantageous. In summary, the application of the W-temper state results in an increased formability due to lower strengths, higher elongations and an increased strain hardening exponent.

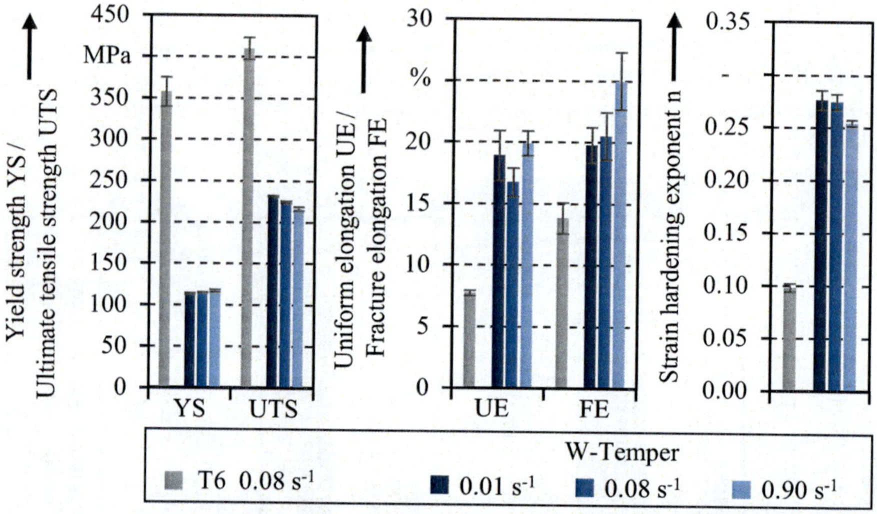

Fig. 7. Comparison of the mechanical properties of T6 and W-temper condition

Due to the enhanced formability in W-temper condition, this state is investigated closer regarding the strain rate sensitivity. For this purpose, values of 0.01, 0.08 and 0.90 s^{-1} were considered. It has already been shown in preliminary work that the PLC effect occurs in this state, whereas the influence becomes weaker with increasing strain rates [12]. Thus, the yield strength rises slightly with increasing strain rate from 114 MPa to 117 MPa, while the ultimate tensile strength decreases from 232 MPa to 216 MPa. A similar behavior has already been observed for AA5083 [13] and AA7075 [14]. For the uniform elongation, no trend is evident, but the fracture elongation increases from 19.8 to 25.0%. As a result of the decreasing ultimate tensile strength, the strain hardening exponent decreases slightly for higher strain rates. Consequently, the PLC-effect leads to an increase of the ultimate tensile strength while the fracture elongation is reduced. Due to the weaker PLC-effect at high strain rates, in this range a more ductile material behavior exists [15].

3.3 Comparison of the Mechanical Properties of a Curved and a Flat Tensile Specimen

For the evaluation of the comparability of the mechanical properties, generated by the adapted testing procedure for curved tensile specimens, a comparison with a flat sample is performed. In preliminary investigations [12], the stress and strain states for different sample geometry have been analyzed numerically. No significant deviations were found in the major and minor stress. Additionally, the testing force over the length elongation was identical for equivalent cross-section areas. Consequently, the investigated experimental setup was considered suitable for testing curved tensile specimens. Subsequently, this will be validated experimentally for the W-temper condition. For this purpose, the curved tensile specimens are milled flat to a thickness of 3.8 mm in the testing area, see Fig. 8.

Fig. 8. Geometry of the flat and the curved tensile specimen

As shown in Fig. 8, the flat tensile specimen is only milled in the testing area. The comparison is conducted on basis of the mechanical properties shown in Fig. 9.

Both the yield strength and the ultimate tensile strength show no significant deviations between the flat and the curved tensile specimen. The values are 116 and 114 MPa for the yield strength as well as 235 MPa and 232 MPa for the ultimate tensile strength, respectively. For the elongation, on the other hand, there are slight deviations. The uniform elongation shows values of 16.9% and 18.8%, whereas the elongation at break is 19.2% and 19.8%. One reason for those deviations might be the milling of the flat tensile samples, which can cause minor notch effects. Furthermore, narrow sections of the surface layers were removed. Especially the section at the outer diameter has a higher ductility. Additionally, there are high standard deviations for the elongations. Therefore, no significant deviations can be detected for different cross-sections. Consequently, there is a sufficient conformity between the experimental and numerical results. In this context, reliable material properties are generated by the adapted procedure for testing curved tensile specimens. Thus, the results of the tensile tests are comparable with those of flat specimens for the evaluated sample dimensions.

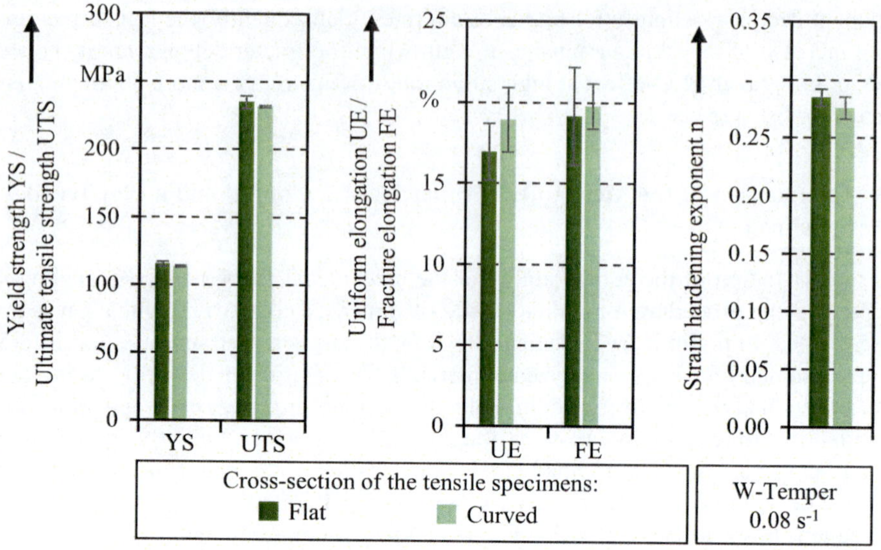

Fig. 9. Comparison of the mechanical properties of a flat and a curved tensile specimen in W-temper condition

4 Summary and Outlook

7xxx aluminum alloys are characterized by a high strength-to-weight ratio, which can be further elevated by the application of tubular semi-finished parts. Such components are predestined for a use as safety-relevant structural components. However, since there are hardly any standardized methods for testing tubular profiles, a method adapted to the tube geometry was described. Therefore, tensile specimens are cut out of an AA7020 tube by laser. To avoid undesired influences on the stress state during the material testing, clamping jaws adapted to the curvature of the samples were manufactured. For a better evaluation of the influences of the extrusion process on the semi-finished parts, optical measurements of the grain structure were prepared and analyzed. Three areas with different microstructures were identified. These include fine-grained areas at the surfaces, elongated grain structures next to the surface layers and coarser grained, less orientated microstructures in the middle of the wall thickness. To make qualitative statements regarding the material characteristics of the various layers, tensile specimens were milled to a thickness of one millimeter and tested for each section. In this context, a decreasing strength and increasing ductility were observed with increasing grain orientation in the circumferential direction. In order to increase the low formability of AA7020, the W-temper forming was investigated. Thus, a significant reduction of the flow resistance and an increase of the ductility were achieved. Due to the higher formability, the strain rate sensitivity of the W-temper condition was analyzed. As a result of the PLC-effect, the ultimate tensile strength decreases and the fracture elongation increases for higher strain rates. Finally, the suitability of the applied procedure for the preparation and testing of the tensile specimens from tubes was evaluated. This was realized by a comparison of the investigated mechanical properties of a flat and a curved tensile specimen. In this context, a good agreement of the strength parameters was obtained as well as a slight deviation in the strains, which could be attributed to the milling process of the flat specimens. Therefore, the described procedure for testing curved tensile specimens is a suitable method for accurate mechanical properties. These could be implemented in a material model for the numerical design of a hydroforming process. However, further research is necessary to investigate the local mechanical properties of the semi-finished products. In particular, the targeted preparation of specific microstructural sections needs to be examined in more detail. One approach is the preparation of the tube layers by a turning process with subsequent cutting of the specimen geometry. Furthermore, the transferability of the results on the comparability of the flat and curved tensile specimens to other specimen dimensions needs to be investigated. In order to generate lightweight tubular components, the tube bulge test as a proper tool for the characterization of the associated hydroforming process has to be analyzed. Finally, the material properties obtained in this way have to be compared with those from the adapted tensile test.

References

1. European Environment Agency: https://www.eea.europa.eu/ims/total-greenhouse-gas-emission-trends. Last accessed 20 Apr 2022
2. European Environment Agency: https://www.eea.europa.eu/themes/climate/eu-greenhouse-gas-inventory. Last accessed 20 Apr 2022
3. Statistisches Bundesamt: https://www.destatis.de/Europa/EN/Topic/Environment-energy/CarbonDioxideRoadTransport.html. Last accessed 20 Apr 2022
4. European Comission: https://transport.ec.europa.eu/news/efficient-and-green-mobility-2021-12-14_en. Last accessed 20 Apr 2022
5. Hashimoto, N.: Application of aluminum extrusions to automotive parts. Kobelco Tech. Rev. **35** (2017)
6. Totten, G.E., MacKenzie, D.S.: Handbook of Aluminum – Volume 1: Physical Metallurgy and Processes. Marcel Dekker, New York (2003)
7. Degner, J.: Grundlegende Untersuchungen zur Herstellung hochfester Aluminiumblechbauteile in einem kombinierten Umform- und Abschreckprozess. FAU University Press, Erlangen (2020)
8. Hebbar, S., Kertsch, L., Butz, A.: Optimizing Heat treatment parameters for the W-temper forming of 7xxx series aluminum alloys. Metals **10**(1361), 1–15 (2020)
9. Ostermann, F.: Anwendungstechnologie Aluminium, 3rd edn. Springer Vieweg, Berlin, Heidelberg (2014)
10. Nazari Tiji, S.A., et al.: Characterization of yield stress surface and strain-rate potential for tubular materials using multiaxial tube expansion test method. Int. J. Plast **133**(102838), 1–75 (2020)
11. Khadyko, M., Dumoulin, S., Hopperstad, O.S.: Texture gradients and strain localization in extruded aluminum profile. Int. J. Solids Struct. **97–98**, 239–255 (2016)
12. Reblitz, J., Reuther, F., Trân, R., Kräusel, V., Merklein, M.: Numerical and experimental investigations on the mechanical properties of milled specimens from an AA7020 tube. Key Eng. Mater. **926**, 1949–1958 (2022)
13. Clausen, A.H., Borvik, T., Hopperstad, O.S., Benallal, A.: Flow and fracture characteristics of aluminium alloy AA5083-H116 as function of strain rate, temperature and triaxiality. Mater. Sci. Eng. A **364**, 260–272 (2004)

14. Reuther, F., Lieber, T., Heidrich, J., Kräusel, V.: Numerical investigations on thermal forming limit testing with local inductive heating for hot forming of AA7075. Materials **14**(1882), 1–15 (2021)
15. Halim, H., Wilkinson, D.S., Niewczas, M.: The Portevin–Le Chatelier (PLC) effect and shear band formation in an AA5754 alloy. Acta Mater. **55**, 4151–4160 (2007)

Metallic Plate-Lattice-Structures for a Modular and Lightweight Designed Die Casting Tool

B. Winter[1](✉), J. Schwab[2], A. Hürkamp[1], S. Müller[2], and K. Dröder[1]

[1] Institute of Machine Tools and Production Technology, Technische Universität Braunschweig, Langer Kamp 19b, 38106 Braunschweig, Germany
benjamin.winter@tu-braunschweig.de

[2] Chair of Casting Technology, Friedrich-Alexander-Universität Erlangen-Nürnberg, Dr.-Mack-Straße 81, 90762 Fürth, Germany

Abstract. Conventional die casting tools are often oversized with regard to their mechanical stability, which results in increased energy requirements for kinematic processes and tool temperature control systems. In addition, these die casting tools often lack flexibility and modularity. One way to counter these disadvantages are lightweight structures, which can be built up in a modular manner depending on the application. Due to their larger cross-section compared to conventional lattice structures, plate lattice structures (PLS) offer higher mechanical load-bearing capacity. The aim of this work is to investigate such PLS with regard to their manufacturability in an additive manufacturing process. The specimens made from 1.4545 are first examined for dimensional accuracy and defects using computed tomography. Subsequently, an elastic model of these structures is numerically generated and calculated to obtain the structural stiffnesses. With this knowledge, a simplified numerical feasibility study was carried out.

Keywords: Die casting tools · Lightweight design · Additive manufacturing · Plate lattice structures · Feasibility study

1 Introduction

The majority of tool components for conventional die casting tools today are still specifically designed and manufactured by subtractive processes. These manufacturing principles require a high input of resources and the resulting die casting tools are inefficient due to their high weights and require a high input of energy for a proper temperature management [1]. Furthermore, die casting tools largely lack flexibility and are thus limited in their application [2, 3].

One approach to flexibility and modularisation is the standardisation of individual components or assemblies that form the basic structure of the die casting tool for different applications and do not have to be exchanged. The analysis for standardisation shows not only an increase in modularisation, but also a possible cost reduction of the die casting tool of up to 18% [4]. This standardisation creates modularity and fundamentally reduces the use of resources, but the weight of the die-casting tools is not reduced, which keeps the thermal inertia and the high thermal and kinetic energy input.

M. Liewald et al. (Eds.): WGP 2022, LNPE, pp. 451–460, 2023.
https://doi.org/10.1007/978-3-031-18318-8_46

One possibility for the lightweight design of die casting tools in combination with the modularity mentioned above is the use of exchangeable lightweight structures (e.g. lattice structures) in the mold frame [1]. However, since a high degree of stiffness with simultaneously high bearing loads is required in this area under cyclic stress, appropriate structures such as plate lattice structures (PLS) show great potential. These structures are based on atomic lattice structures and are connected to each other via their nodes by thin plates (see Fig. 1).

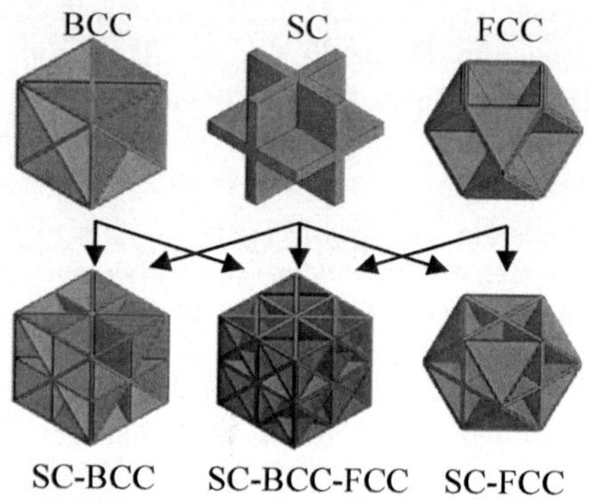

Fig. 1. Basic plate lattice structures and possible mixtures

Due to their larger cross-section compared to truss lattice structures, PLS offer high mechanical load-bearing capacity at the same relative densities [5]. The structures are divided into simple cubic (SC), body-centered cubic (BCC) and face-centered cubic (FCC) geometries (see Fig. 1), which alone show a strong anisotropy. By mixing the individual structures, direction-dependent properties can be adjusted up to isotropy [6]. It is also possible to combine the structures in different sizes to obtain adjusted properties [7].

Due to the manufacturing-related complexity of these geometries, subtractive manufacturing is not possible. Some published works have already dealt with the additive manufacturing of such structures and have successfully manufactured and investigated them using steel [8], plastic [9] (also reinforced [10]) and carbon [11], among others. Different processes such as selective laser melting (SLM), stereolithography or selected laser sintering were used.

The aim of the work is to demonstrate a way using such lightweight structures in a die casting tool, not only to reduce the weight of the tools, but also to obtain modularity and reduce thermal inertia and the required energy input during the process. The proposed concept consists of a subtractive manufacturing of the mold insert with integrated cooling channels and an additive manufacturing of the mold frame (see Fig. 2), which lies below and around the mold insert, using lightweight structures such as PLS, and to design it in a modular way so that different mold inserts can be used.

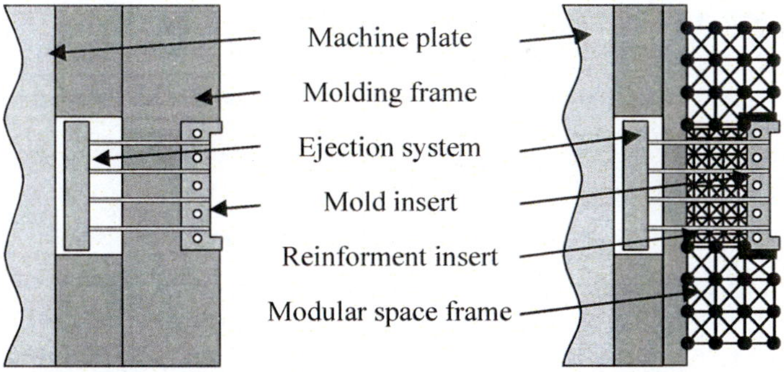

Fig. 2. Basic structure of a conventional (left) as well as a modular and lightweight designed (right) die casting tool

2 Method

For the development of a lightweight die casting tool, a basic understanding of the manufacturability and the resulting mechanical and optical properties of such PLS is obtained first. Due to the high compressive stresses during the die casting process, the focus of the work was put on the structures that can absorb high forces with low deformation in the normal direction. These are, among others, the SC-, the BCC- and the combination of both to the SC-BCC-structure shown in Fig. 1.

For the investigations, a high strength steel 1.4545 (15-5PH) is used since it provides the possibility to be used for investment casting as well, which represents an alternative manufacturing process. PLS were additively manufactured by selective laser melting (SLM) and optical investigations were carried out. After subsequent heat treatment, the structures were again optically investigated. Afterwards, an elastic model of these structures with different relative densities is established in order to obtain a numerical solution for the structural stiffness. A feasibility study was then carried out using a simplified die casting tool.

2.1 Manufacturing

For the following investigations, solid reference blocks with a size of $40 \times 40 \times 40\ \text{mm}^3$ and the plate lattice structures SC, BCC, and SC-BCC with a relative density of 0.5 and the same dimensions were additively manufactured by SLM (see Fig. 3). To achieve an elastic isotropy in the SC-BCC-structure, the BCC wall thickness t_{BCC} must be in the ratio $t_{BCC} = \sqrt{2}t_{SC}$ to the SC wall thickness t_{SC} [8].

The PLS were manufactured using a SLM125 machine from SLM Solutions Group AG. A substrate plate made of 1.4545 was used as a base onto which the structures were applied. During the process, the printing space was put under an inert gas atmosphere using argon to prevent oxidation of the steel. In addition, the gas flow can flush out dirt adhesions such as melting particles. The SLM process was carried out using the parameters shown in Table 1 and the 1.4545 metal powder from SLM Solutions Group AG.

Table 1. Parameters for the selective laser melting process

Parameter	Dimension
Scanning time per layer (s)	8
Substrate plate temperature (°C)	100
Laser power (W)	275
Layer height (μm)	50
Metal powder size (μm)	10–45

Subsequently, the produced PLSs were separated from the substrate plate and the outer surfaces were grinded plane-parallel to each other to avoid wedging or tilting of the geometries during further testing.

Finally, the manufactured structures were subjected to a heat treatment in order to achieve a homogeneous material condition in all structures and to ensure comparability (shown in Fig. 3). The structures were first subjected to a solution annealing at 1050 °C for one hour followed by air quenched. The structures were then annealed at 450 °C for one hour followed by air quenched again to allow for the removal of material stresses resulting from the solution annealing.

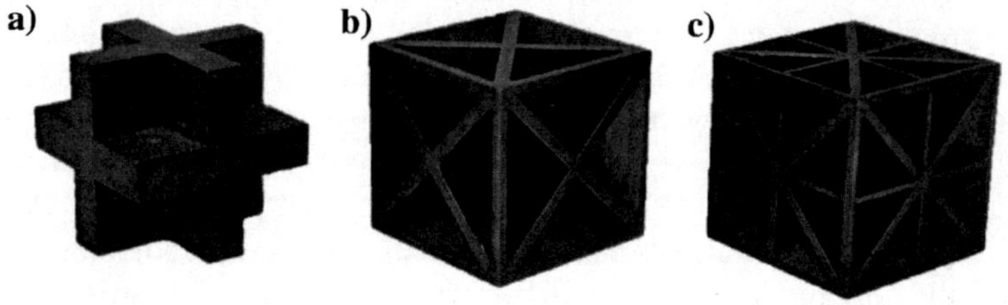

Fig. 3. Printed structures through selective laser melting: a) SC b) BCC c) SC-BCC

2.2 Characterization

After manufacturing, an optical characterization was carried out in order to analyse defects and strong distortions caused by the heat treatment. For the detection and measurement of such effects, the structures were scanned by the CT scanner type FF35 CT of the company Yxlon International GmbH. The scan creates a reconstruction of the component, which is used for the examination.

For the scan, the structures are placed on three carbon rods. In that way, the X-ray radiation has to penetrate as little solid material as possible during the rotating scan in order to obtain a high resolution of the component reconstruction. This procedure performs a three-dimensional scan of the structures and creates an image stack in each space direction. In addition, a 3D model can be created with which the surfaces and wall thicknesses can be measured.

2.3 Numerical Investigation of PLS

The numerical investigations of the structures were conducted with the software COMSOL MULTIPHYSICS 6.0. Initially, the individual structures are considered under uniaxial compressive loading. Subsequently, a simplified die casting tool with PLS is examined with regard to the deflection of the mold. The investigations are intended to provide information on the structural stiffness with the material used.

An elastic material model for the 1.4545 steel is created in both present investigations. A Young's modulus of 191 GPa (as built) respectively 209 GPa (heat-treated) is used from the manufacturer is used [12]. The increase in Young's modulus after solution annealing is a typical characteristic of maraging steels [13]. For both conditions, a Poisson's ratio of 0.3 and a density of 7800 kg/m^3 are assumed.

For the uniaxial compression investigations, only one-eighth of the entire geometry is considered due to the symmetrical design. This inserts a symmetry condition on the three cut sides. The load is applied displacement-controlled by an attached rigid body with a displacement of 0.025 mm. No friction applies between the bodies. The upper body is meshed with a structured mesh, while on the PLS a tetrahedral mesh with an element size of 1.65 mm is implemented.

For the feasibility study, a simplified tool setup is used to evaluate the deflection in a die casting mold through different setups. Conventional die casting dies consist of a frame and a mold insert which includes the cavity. In this finite element analysis only one tool half was considered. Such a design is compared to a tool setup where the area behind the mold insert is replaced by a 3 × 3 × 3 PLS with an edge length of 120 mm in each spatial direction. As external dimensions a size of 280 × 280 × 260 mm^3 was chosen. The resulting part in the cavity has a cuboid shape with 120 × 120 × 10 mm^3 size. Ejectors, cooling channels and ventilation were not included in the model.

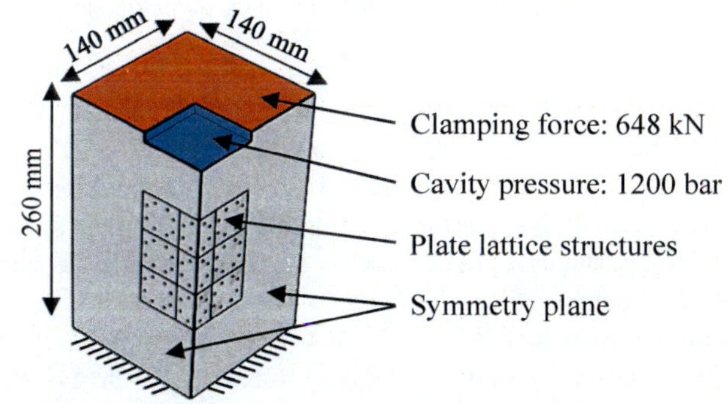

Fig. 4. Simplified die casting tool for the feasibility study

As shown in Fig. 4, just one fourth of the assembly is used and the cut planes are considered symmetric. At the bottom side, a fixed support is applied. Furthermore, a cavity pressure of 1200 bar and a clamping force of 648 kN was used. For the frame, the hot work steel H11 with a Young's modulus of 190 GPa [14] was assumed. Between the PLS and the frame a frictionless penalty contact was applied. To discretize the model, a tetrahedral mesh with a maximum mesh size of 1 mm was chosen.

3 Results and Discussion

First, the results of the CT-Scan examination are presented and discussed. Here, the focus is on possible defects in the material as well as the distortion of the geometries after heat treatment and the deviations from the target geometry. Based on the model setup, the relationship between the PLS and the geometric stiffness will be shown. With these results, a first feasibility study for the use of these structures in the lightweight die casting tool is carried out.

3.1 CT-Scan

The result of the CT scan of a SC-BCC structure with a relative density of 0.5 is shown in Fig. 5. Thereby the outer surfaces and some inner surfaces are well reproduced. However, due to the high density of the material and the size of the body, surfaces with low resolution are also visible.

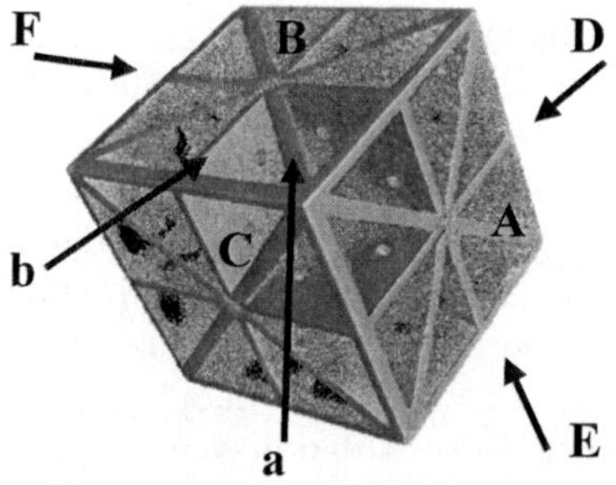

Fig. 5. CT-scanned SC-BCC structure containing investigated geometric elements

In order to obtain a statement about the distortion after heat treatment and the dimensional accuracy after the entire manufacturing process, a surface mesh was created on the scanned body. These measurements were carried out for both untreated and heat-treated structures. This mesh was used to measure the distance and angle of the opposing surfaces to each other. The surfaces *A* to *C* can be seen in Fig. 5, while the surfaces *D*, *E* and *F* are concealed opposite the surfaces *A*, *B* and *C*. The wall thicknesses of the structures were also measured. In the example of the SC-BCC structure in Fig. 5, *a* denotes the wall thickness of the BCC part and *b* denotes the wall thickness of the SC part. The results of the measurements are shown in Table 2 using the SC-BCC structure as an example.

The results show no significant distortion of the structure after the heat treatment. The opposite sides are still almost plane-parallel with a maximum deviation of 0.06° and the distances remain with a maximum deviation of 0.1 mm identical. The wall thicknesses also remain almost unchanged with a difference of 0.012 mm. However, the manufactured structures have generally smaller dimensions compared to the CAD model, especially between sides C and D by 0.543 mm. The wall thicknesses are also slightly thinner measuring 0.078 mm. These results show the necessity to consider the deviation for the dimensional accuracy of the subsequent lightweight die casting tool.

Table 2. Comparison of the dimensional accuracy of a SC-BCC structure to the CAD model

Geometry	CAD model	Untreated	Heat treated	Deviation
Side $A - F$ (mm)	40	39.893	39.974	−0.026
Angle $A - F$ (°)	180	179.912	179.871	−0.129
Side $B - E$ (mm)	40	39.830	39.952	−0.048
Angle $B - E$ (°)	180	179.747	179.804	−0.196
Side $C - D$ (mm)	40	39.401	39.457	−0.543
Angle $C - D$ (°)	180	179.898	179.896	−0.104
Wall thickness a (mm)	2.380	2.324	2.310	−0.070
Wall thickness b (mm)	1.683	1.612	1.605	−0.078

In the subsequent x-ray investigations, no defects could be detected. Nevertheless, as shown in Fig. 6a, in some cases strong surface roughness was identified. In addition, some edges showed burr formation (see Fig. 6b). Both effects can be weak points during subsequent mechanical loading. For subsequent investigations, the process parameters such as the reduction of the layer thickness and the laser power are changed for this reason, and structures are manufactured and investigated again.

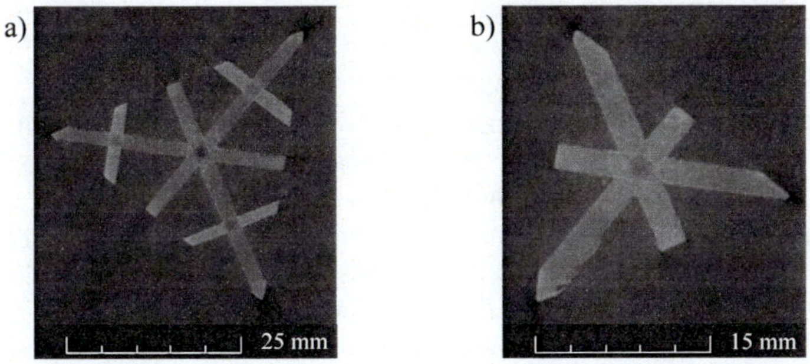

Fig. 6. X-rayed SC-BCC structure with characteristics like a) surface roughness and b) burr

3.2 Numerical Analysis

For the numerical investigations, the SC, BCC and SC-BCC structures with a relative density of 0.3, 0.5 and 0.7 were calculated using linear elastic material behavior and the structural stiffness before and after heat treatment is plotted (see Fig. 7). These are intended to represent the effective stiffness of a single structure with respect to its volume. For reference, the stiffness of an untreated and a heat-treated block with a relative density of 1 are shown.

Due to the largest cross section of the SC structure in the load direction, it consequently has the highest stiffness over all relative densities investigated. The BCC and

SC-BCC structures have identical stiffnesses in the uniaxial stress state. The advantage of these structures, however, is their isotropic behavior with regard to shear forces.

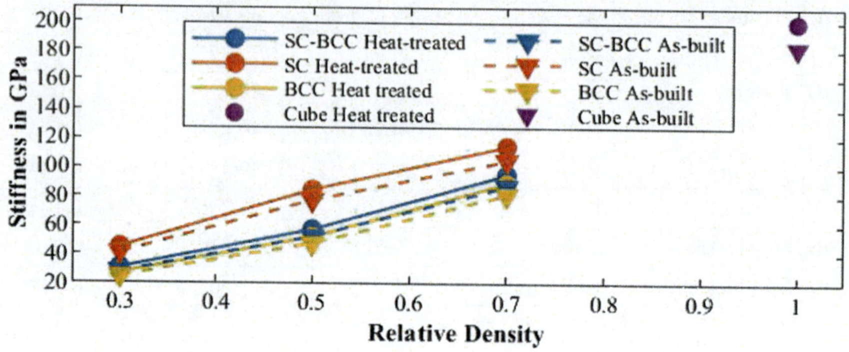

Fig. 7. Geometric stiffness of the investigated structures as a function of the relative density

As expected, the stiffness of all structures decreases with decreasing relative density. A non-linear progression between the examined measuring points is evident. This can also be found in the literature [6]. In addition, there is an increase in stiffness of the heat-treated samples, which approaches more and more the stiffness of the untreated structures with decreasing relative density. Thus, heat treatment appears to have a greater effect at higher relative densities. In the numerical investigations shown here, a geometrically conditioned Young's modulus of all considered structures was obtained and used for further designs of the targeted die casting tool.

From these investigations, it is evident that SC structures have the highest stiffness due to their constant cross section in the load direction, but they are susceptible to shear forces [6]. For this reason, the BCC and SC-BCC structures were included, which have a more isotropic behavior. Since not only pure compressive forces occur in the die casting process, this property must also be taken into consideration. For this reason, the SC-BCC structure with a relative density of 0.5 was used as an example for the feasibility study. The comparison of the massive tool setup with the PLS tool setup results in the higher deflection of the PLS tool. The resulting deformation of a cut plane through the tool is visible in Fig. 8.

Between the two setups, a difference of the maximum deflection of 36% occurs, which means that the PLS setup is less stiff than the massive setup. However, the reduction of the stiffness goes along with the reduction of volume and mass of 50%, respectively. This applies for the area behind the mold insert, for the whole simplified tool the mass reductions results in 9% compared to the massive setup. In general this correlation is obvious. By removing material, the stiffness decreases. Nevertheless a PLS can still be used in a die casting tool setup. The use is dependent on the desired properties regarding the geometric dimensioning and tolerances. Only the domain behind the mold insert was replaced, however it is also possible to substitute other parts of the tool with lightweight structures, for instance the frame. There also other setups like the usage of topology-optimized domain are conceivable. This is the subject of further investigations of the authorship.

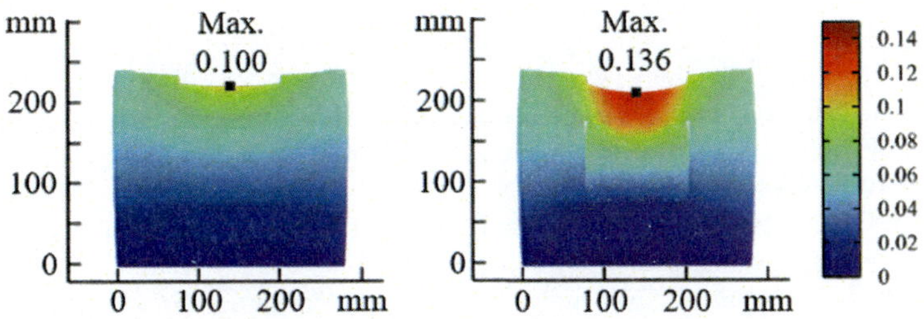

Fig. 8. Comparison of the deformation between a massive tool setup (left) and a lightweight tool setup with SC-BCC PLS (right)

4 Conclusion

The lightweight design and modularity possibilities of die casting tools have been studied poorly so far. Thus, although standardization of some components or assemblies can increase modularity, die casting tools still have high weight and consequently high thermal inertia. One possible lightweight design strategy can be the use of lightweight structures for components such as the mold frame, which is usually oversized. PLS have high stiffnesses with high bearing loads. Thus, they offer great potential as lightweight structures. For use in die casting tools, these must be tested for their suitability.

In order to obtain a more precise understanding of those structures, specimens were additively manufactured by the SLM process and examined by a CT scan. The results showed no defects, but there are strong burrs and large surface roughness on several surfaces. These can be weak points for the later use in die casting tools. The investigation also shows a high dimensional accuracy of the structures after heat treatment compared to the untreated structures. However, these are slightly smaller than the desired geometry. As a conclusion, the manufactured structures must first be tested under compression loads to obtain a basic understanding. Afterwards, the structures should be manufactured again using different process parameters and an offset in order to correct both the shrinkage and the possible weak points in the structures. Subsequently, these must be compared with the structures presented here in order to be able to evaluate the influence.

The numerical investigations have shown a correlation of the geometric stiffness to the relative density of different structures. The heat treatment results in an increase in stiffness, which is greater with increasing relative density. In addition, the SC structure has the highest stiffness due to its constant cross section in load direction. However, during a die casting process, shear forces and thermal stress also occur, which require an isotropic behavior of the structures. For this reason, a feasibility study was carried out based on a simplified die casting tool and using an SC-BCC structure with a relative density of 0.5. Although this results in a larger displacement of the tool by 36%, this is generally within the acceptable range and the use of PLS is possible. Subsequent work will also investigate these loading cases in more detail and include them in the evaluation for selecting a suitable structure design.

Acknowledgement. The research project "Design, analysis and principle demonstration of incrementally manufactured, modular lightweight die casting tools" is funded by the Deutsche Forschungsgemeinschaft (DFG, German Research Foundation) – 428178388.

References

1. Müller, S., Müller, A., Rothe, F., Dilger, K., Dröder, K.: An initial study of a lightweight die casting die using a modular design approach. Int. J. Metalcast. **12**(4), 870–883 (2018). https://doi.org/10.1007/s40962-018-0218-3
2. Brecher, C., Özdemir, D. (eds.): Integrative production technology. Springer, Cham (2017). https://doi.org/10.1007/978-3-319-47452-6
3. Queudeville, Y., Ivanov, T., Nußbaum, C., Vroomen, U., Bührig-Polaczek, A.: Decision and design methodologies for the layout of modular dies for high-pressure-die-cast-processes. Mat. Sci. Forum **618–619**, 345–348 (2009). https://doi.org/10.4028/www.scientific.net/MSF.618-619.345
4. Queudeville, Y., Vroomen, U., Bührig-Polaczek, A.: Modularization methodology for high pressure die casting dies. Int. J. Adv. Manuf. Technol. **71**(9–12), 1677–1686 (2014). https://doi.org/10.1007/s00170-013-5582-9
5. Berger, J.B., Wadley, H.N.G., McMeeking, R.M.: Mechanical metamaterials at the theoretical limit of isotropic elastic stiffness. Nature **543**, 533–537 (2017). https://doi.org/10.1038/nature21075
6. Tancogne-Dejean, T., Diamantopoulou, M., Gorji, M.B., Bonatti, C., Mohr, D.: 3D Plate-lattices: an emerging class of low-density metamaterial exhibiting optimal isotropic stiffness, Adv. Mater. **30**(45) (2018). https://doi.org/10.1002/adma.201803334
7. Xue, R., Cui, X., Zhang, P., Liu, K., Li, Y., Wu, W., et al.: Mechanical design and energy absorption performances of novel dual scale hybrid plate-lattice mechanical metamaterials. Extreme Mech. Lett. **40** (2020). https://doi.org/10.1016/j.eml.2020.100918
8. Tancogne-Dejean, T., Li, X., Diamantopoulou, M., Roth, C.C., Mohr, D.: High strain rate response of additively-manufactured plate-lattices: experiments and modeling. J. Dynam. Behavior Mater. **5**(3), 361–375 (2019). https://doi.org/10.1007/s40870-019-00219-6
9. Andrew, J.J., Schneider, J., Ubaid, J., Velmurugan, R., Gupta, N.K., Kumar, S.: Energy absorption characteristics of additively manufactured plate-lattices under low-velocity impact loading. Int. J. Impact Eng. **149** (2021). https://doi.org/10.1016/j.ijimpeng.2020.103768
10. Andrew, J.J., Verma, P., Kumar, S.: Impact behavior of nanoengineered, 3D printed plate-lattices. Mater. Des. **202** (2021). https://doi.org/10.1016/j.matdes.2021.109516
11. Crook, C., Bauer, J., Guell Izard, A. et al.: Plate-nanolattices at the theoretical limit of stiffness and strength. Nat. Commun. **11** (2020). https://doi.org/10.1038/s41467-020-15434-2
12. SLM Solutions Group AG, Material datasheet Stainless steel 15-5PH/1.4545/A564. https://www.slm-solutions.com/fileadmin/Content/Powder/MDS/MDS_Fe-Alloy_15-5PH_0919.pdf. Last accessed 07 May 2022
13. Monkova, K., Zetkova, I., Kučerová, L., Zetek, M., Monka, P., Daňa, M.: Study of 3D printing direction and effects of heat treatment on mechanical properties of MS1 maraging steel. Arch. Appl. Mech. **89**(5), 791–804 (2018). https://doi.org/10.1007/s00419-018-1389-3
14. Rothman, M.F.: High-temperature property data: ferrous alloys. ASM Int. (1988)

New Approaches in Machine Learning

Impact of Data Sampling on Performance and Robustness of Machine Learning Models in Production Engineering

F. Conrad[1](✉), E. Boos[1], M. Mälzer[1], H. Wiemer[1], and S. Ihlenfeldt[1,2]

[1] Technische Universität Dresden, 01062 Dresden, Germany
felix.conrad@tu-dresden.de
[2] Fraunhofer Institute for Machine Tools and Forming Technology, Reichenhainer Strasse 88, 09126 Chemnitz, Germany

Abstract. The application of machine learning models in production systems is continuously growing. Hence, ensuring a reliable estimation of the model performance is crucial, as all following decisions regarding the deployment of the machine learning models are based on this aspect. Especially when modelling with datasets of small sample sizes, commonly used train-test split variation techniques and model evaluation strategies encompass a high variance on the model's performance. This difficulty arises, as the available amount of meaningful data is severely limited in production engineering and can lead to the model's actual performance being greatly over- or underestimated. This work provides an experimental overview on different train-test splitting techniques and model evaluation strategies. Sophisticated statistical sampling methods are compared to simple random sampling, and their impact on performance evaluation in production datasets is analysed. The aim is to ensure a high robustness of the model performance evaluation, even when working with small datasets. Hence, the decision process for the deployment of machine learning models in production systems will be improved.

Keywords: Data sampling · Train-test-split · Performance evaluation · Usable artificial intelligence

1 Introduction

The successful training, validation and testing of machine learning (ML) and deep learning models requires significant amounts of meaningful data. Amounts that are sufficient enough to compensate data uncertainties due to missing values, measurement errors, anomalies, imbalanced data classes or skewed sample. Otherwise, a reliable performance evaluation of the resulting model is not ensured. Nevertheless, the handling of small data samples is a crucial part in some areas of applied data science due to limited data access. Machine tools are mostly developed as individual or custom constructions. Hence, their production behaviours and anomalies differ and data acquisition must happen at the individual machine [1]. Overall, the availability and accessibility of enough

M. Liewald et al. (Eds.): WGP 2022, LNPE, pp. 463–472, 2023.
https://doi.org/10.1007/978-3-031-18318-8_47

meaningful data of production processes and machine tools for ML-based applications is not entirely ensured. This problem does not necessarily apply to raw data but to comparable experiments and processes.

There are different approaches aiming to compensate this deficiency, for instance by either synthetically enlarging the dataset, e.g. with data from simulation models, or deliberately taking data uncertainty into account by modelling probabilistic ML-models [2]. These approaches come with different sets of advantages and disadvantages. However, the strategic handling of small data sets is already able to improve model reliability. In particular, data sampling techniques and model evaluation strategies are prominent methods to decrease model performance variance and increase model robustness. Due to the effects of sampling errors on statistical representations, skewed or even inadequate distributions of features within the data split between training, validation, and testing data are common consequences. This leads to inaccurate model evaluation scores, which follow by greatly over- or underestimating the actual performance. Best practices developed for sufficiently large datasets are often applied to data analysis tasks containing small datasets without a thorough scrutiny of the methods [3–5]. This can lead to crucial errors, especially when working with small datasets. The same principles apply to the evaluation and deployment of ML-based application in product engineering and machine tool monitoring.

2 Related Work

3 Evaluation Strategies

The following section presents a selection of studies investigating the influence of these evaluation strategies. Most of them are related to hyperparameter optimisation as well as feature engineering.

Vabalas et al. [6] investigate the behaviour of holdout (HO), simple cross-validation (SCV) and nested cross-validation (NCV) for HPO and feature selection on synthetic random data for a binary classification. On average HO and NCV accurately estimate the predictive power of 50% of the classifier for all dataset sizes (minimum 20), whereas k-fold CV strongly overestimates the performance for all sample sizes (maximum 1000). Tsamardinis et al. [7] compared the biases of CV, 'Tibshirani and Tibshirani' method and NCV induced by the HPO on classification datasets. The CV overestimates the performance on datasets of up to 100 samples. In contrast, the NCV had no bias even for a dataset size of 20.

Rao et al. [8], analysed the overfitting behaviour of SCV in the HPO and feature selection is shown for synthetic datasets up to 10.000 samples and benchmark datasets up to 728 samples. Leave-one-out CV (SCV with number of splits = number of samples) performs slightly better but cannot completely prevent the overestimating behaviour and is very computationally intensive. Varma et al. [9] show an overestimation of the true performance of SCV and leave-one-out CV performing HPO on synthetic data random data for a binary classification with a sample size of 40. The NCV is proposed as a solution for an unbiased performance evaluation.

Dobbin et al. [10] use only HO and investigate the optimal train test-ratio based on the training set size and classification accuracy for classification problems, resulting in a training proportion of 40% - 80% for a wide range of conditions.

Although HO and SCV are the most commonly used evaluation strategies, these studies show that the HO and SCV strategy can lead to wrong performance calculations, especially for small datasets. SCV has the problem that it tends to overfit due to data leakage [6]. With the HO strategy, no statistical analysis can be carried out. Therefore, it remains unclear what level of uncertainty the performance result has. Repeating the HO with different data splits can remedy this problem, leading to the NCV, which shows a superior performance calculation in all studies.

3.1 Data Sampling

ElRafey et al. [11] divided the data sampling methods into four categories, which are adopted in this paper. **Simple random sampling** is the most common approach in ML, as it is the standard in commonly used libraries. **Stratified sampling** divides the population into several non-overlapping subpopulations (strata) based on the labels. Then small samples from these strata are selected and combined into the final subsample. A major disadvantage is that a new hyperparameter (HP) is inserted in finding the optimal strata in regression problems, so this strategy is not universally applicable for all datasets. For classification, stratified sampling is commonly used with the classes as strata. **Cluster sampling** is similar to stratified sampling, but a clustering divides the dataset into subdatasets. From these clusters, samples are drawn to get the final subsample. Likewise, additional HPs are introduced with the clustering algorithms. Clustering has also been shown to distort the original distribution [12]. **Density-biased sampling** involves biased sampling intending to preserve the original distribution in the subsamples. Various methods attempt to fulfil this objective. Joseph et al. [13] has shown that the CADEX and DUPLEX algorithm cannot satisfy this objective, while the in [13] introduced method "SPlit" is able to do so. Additionally, SPlit is compared with random sampling in an HO evaluation scheme and reaches a better performance with a smaller variance.

4 Methods

4.1 Datasets

A total of three datasets were investigated in this study. As seen in Table 1, the three selected datasets from recent literature and public data repositories vary w.r.t. the sample size, and all have multiple target properties for the prediction. All target properties were handled as a single regression problem, resulting in eight predictions (tasks). The datasets Gou-2019 and Hu-2021 contain information about the chemical composition and material manufacturing process. The target properties are the mechanical properties of steel (Gou-2019) and aluminium (Hu-2019). The dataset Coraddu-2016 provides information about the actual operating point of a naval propulsion plant and the state of decay of the compressor and turbine in this system.

In this study a maximum of 2000 samples are used for modelling, to address the problems with small datasets. The rest of the dataset is taken for testing the evaluation strategies and data-sampling methods. There are many studies within this range in the domain of process engineering, a selection of them are [14–18]. Because of the smaller dataset size of Hu-2021, the maximum dataset size for modelling is set to 700 for the tasks tensile and yield strength and 600 for the task elongation.

Table 1. Overview of investigated datasets ordered by size and their properties in the published version

Alias + Source	Domain	Targets	Size	Features
Guo-2019 [3]	Predictive quality – steel manufacturing	Tensile strength	63,162	27
		Yield strength		
		Elongation		
Coraddu-2016 [4]	Predictive maintenance – naval propulsion	Compressor decay state	11,934	16
		Turbine decay state		
Hu-2021 [5]	Predictive quality – aluminium manufacturing	Tensile strength	896	27
		Yield strength	860	
		Elongation	783	

4.2 Models and Hyperparameter Optimization

For all experiments, the gradient boosting regressor (GBR) from sklearn[1] version 0.24 and the same HP space is used (c.f Table 2). The hyperparameter optimisation was performed on all datasets using a grid search with 200 random iterations. Additionally, a Bayesian HPO with 25 iterations was solely conducted for the real dataset. In this study, no feature engineering was conducted.

Table 2. Hyperparameter space used for the experiments

HP	N estimators	Max depth	Min samples split	Min samples leaf	Learning rate
Min value	10	2	0.001	0.001	0.01
Max value	400	11	0.01	0.01	0.9

4.3 Data Sampling Strategies

For each of the four categories of data sampling strategies (see Sect. 2.2), one method is selected for the comparison in this study. Simple random sampling is implemented

[1] Python library „scikit-learn“, https://scikit-learn.org/stable/.

with sklearn. For stratified sampling, the target value is divided into 25 bins. The data is then sampled by equally random sampling from these bins until the desired number of samples is reached. The cluster sampling is performed with K-Means (sklearn) clustering using 25 clusters and default HP. The data is then sampled by equally random sampling from these clusters until the desired number of samples is reached.

The method "SPlit" is chosen for the density-biased sampling. It has shown that it can preserve the original distribution very well in the subsamples. More detailed information on this method can be found in [13]. This method can be used via the R-package "SPlit" and is included in this study via the 'rpy2' package. The default HPs are used.

4.4 Evaluation Strategies

For this study, the two most commonly used evaluation strategies HO and SCV are investigated. Additionally the NCV is included, as an evaluation strategy which is more computational expensive but also very robust against data-leakage. The principle of these evaluation strategies is depicted in Fig. 1. For the HO strategy, the train-validation-test split is set to the proportion 60% - 20% - 20%. The SCV is conducted with 5 folds in all experiments. In the nested cross-validation, different train-test sets are created as an 'outer loop'. The number of outer splits is set to five for all experiments. For each of these five train-test sets, the train data is used for the HP optimization via a CV, hence the name, nested cross-validation. The test data from the outer loop is used for the performance evaluation. The number of inner splits is set to five for all experiments. In the final step, the best model is refitted on the whole model dataset in all evaluation strategies.

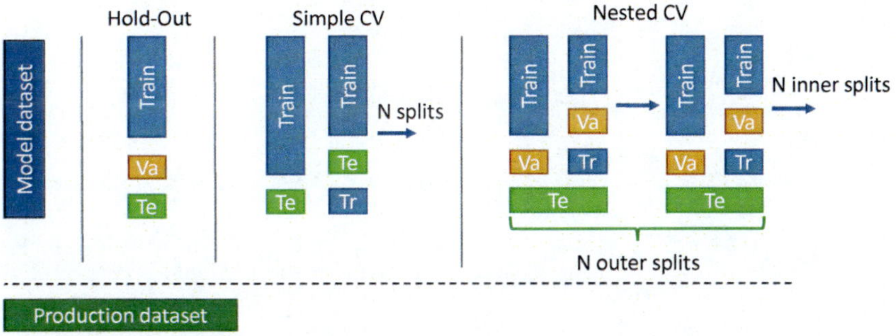

Fig. 1. Comparison of the different evaluation strategies used in this study. The dataset used for modelling (model dataset) is marked in dark blue, the train subset (Train) in blue, the validation subset (Va) in orange and the test subset (Te) in green. The data, which is not used in the modelling phase, is marked in dark green (production dataset)

5 Results

At first, the evaluation strategies for synthetic data are investigated, with all data splits performed using random sampling. The second step investigates how well the different data sampling methods can perform representative data splits. Finally, the evaluation strategies are implemented with the different data sampling methods and examined for their robustness and performance.

5.1 Comparison of Evaluation Strategies in Synthetic Data

The analysis of the evaluation strategies is first carried out with synthetic data. For this purpose, the feature values are randomly drawn from the uniform distribution U [0, 1]. The labels are alternately assigned 0 and 1, resulting in an even class distribution, and the features do not contain any meaningful information. The major advantage of this synthetic dataset is that the actual underlying accuracy of the models is 50%, and no model can perform better or worse. Therefore, the evaluation strategies can be assessed against the actual underlying accuracy of 50%.

The results presented in Fig. 2 show that the mean of the NCV and HO achieve the underlying accuracy of 50% for all dataset sizes. In contrast, the SCV gives an overly optimistic accuracy result. For small datasets with 50 and 100 samples, SCV estimates a precision of almost 60%. In addition, for a dataset with 5000 samples, all power estimates are above 50%, with a mean of 51.5%.

While NCV and HO both obtain a mean of around 50% for all dataset sizes, the NCV score is much more stable across the runs and shows far fewer outliers. This is achieved by the outer loop, which has the additional advantage that the stability of the performance can be evaluated over different train-test splits but at the cost of computational effort.

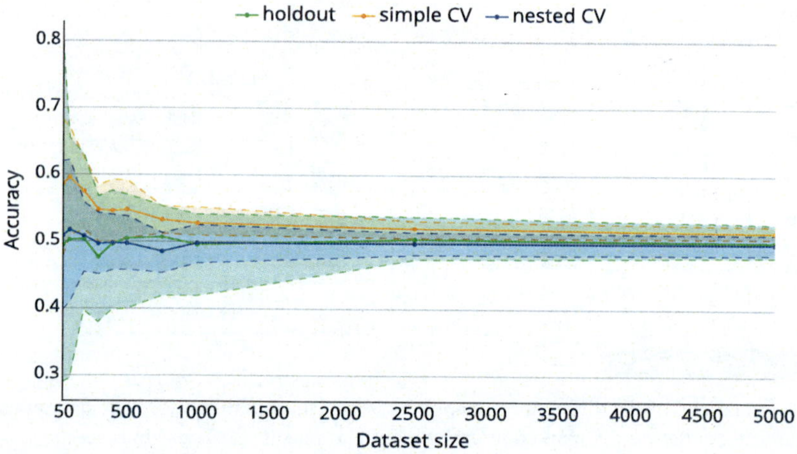

Fig. 2. Evaluation Strategies on synthetic dataset. Thick lines show mean test accuracy and dash lines show 5th and 95th percentile for 40 runs

5.2 Comparison of Data-Sampling Methods in Real Datasets

For the evaluation of the sampling strategies in real datasets, the dataset was split into two parts (train and test) using the sampling strategies. The training data was used to train a default GBR, so no HPO was performed. The aim is to obtain a small training dataset that is representative of the whole dataset. When a representative training dataset is found, the performance of the model is good, even with a small training dataset. Hence, the data-sampling strategies can be compared by their performance on the test dataset. For this comparison, the performance is normed to the mean performance of the random split for each task. The results from all tasks are summarized and shown in Fig. 3.

The results show that the data sampling SPlit is the only method which can consistently outperform the simple random split. The k-means and stratified data sampling performs consistently worse than the random split. Stratified random sampling performs particularly poorly with smaller datasets. Overall, the larger the train size, the smaller the impact of the sampling method. It is to mention that the stratified and cluster sampling are not tuned here and are only tested with one set of HPs. Therefore, better performance with HP tuning is possible. However, HP tuning is declined to achieve universal data selection in this process. Due to the worse performance than the random split and the large number of additional HP, cluster sampling is not used in the following investigations.

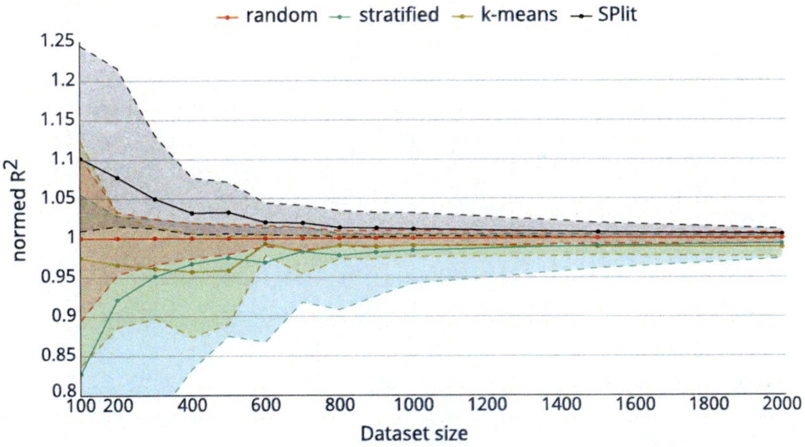

Fig. 3. Data sampling Strategies on real datasets. The test score is normed to the random split score. Thick lines show the mean normed R^2, the dashed lines shows the 10th and 90th percentile

5.3 Comparison of Evaluation Strategies with Different Data Sampling Methods on Real Datasets

For the assessment of the evaluation strategies on real datasets, the datasets were split into two portions. The first portion was investigated with the tree evaluation strategies (see Fig. 1: model dataset), while the second portion was left for testing (see Fig. 1: production dataset). With the large production portion, the true predictive power of the models can be set. This true predictive power is then compared with the performance estimation from the evaluation strategies, leading to a relative score.

$$R^2_{relative} = \frac{R^2_{evaluationstrategy}}{R^2_{productiondataset}}. \tag{1}$$

A relative score above 1 indicates an over-optimistic estimation of the performance from the evaluation strategy.

It is to mention, that in all evaluation strategies the performance is evaluated on 80% of the model dataset. A refit is necessary to get the score in the production dataset because for methods using cross-validation, there is one set of optimal HPs, but there are as many models with these HPs as folds. For the refit, the whole model dataset is used.

In addition, for the investigation on the real datasets, the NCV is modified with the data-sampling methods SPlit (NCV-SP) and stratified sampling (NCV-ST). For this setup only the splits from the outer loop modified to save computational effort, the inner splits are still randomly sampled.

The results presented in Fig. 4a show significant overfitting for the NCV-SP evaluation strategy when using the random search as HPO. This can be caused by the biased sampling used in the SPlit method. Outliers in the dataset can be underrepresented in the test set to achieve the same distribution in train and test set, which can cause an overoptimistic performance evaluation. As for the synthetic data, the HO strategy has the biggest spread. So in a single run, the performance can be greatly over- or underestimated, which explains the very low relative R^2 for the dataset size of 100. The NCV and HO have a relative value below 1, which is due to the refit after the evaluation. Since the refit leads to a 20% larger training set for the prediction of the production dataset. The SCV does not show such behaviour, which indicates slight overfitting. The NCV-ST shows no significant different behaviour than the NCV and is not shown in Fig. 4a and b for better readability.

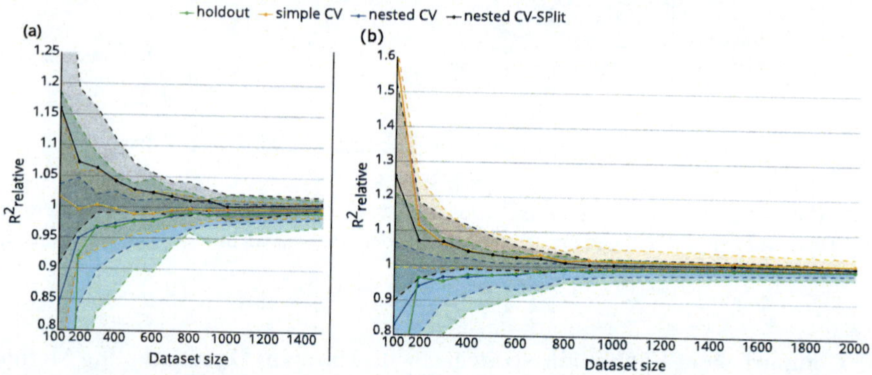

Fig. 4. Comparison of evaluation strategies on real datasets, for (**a**) random HPO and (**b**) Bayes HPO. Thick lines show the mean $R^2_{relative}$. The dashed lines show the 10th and 90th percentile

For the Bayesian HPO, the HO, NCV and NCV-SP show no different behaviour than in the random HPO, see Fig. 4b. In this case the SCV, shows a significant overfitting behaviour. For the dataset size of 100, the mean relative R^2 is 1.6, so an overestimation of the performance of 60%. For the dataset size of 1000 more than 10% of the runs with SCV, still have a relative score of over 1.05.

To compare the performance in the production dataset, the test score on the production dataset ($R^2_{\text{production dataset}}$) is normed to the mean performance of the NCV for each task. The performance in the production set shows no significant difference in the strategies that use cross-validation for the HPO, see Fig. 5. So for the NCV-SP, although overestimating the performance in the evaluation of the model dataset, the actual performance itself is the same. For the HO strategy, because no cross validation was used for the HPO, the performance especially for small datasets is worse.

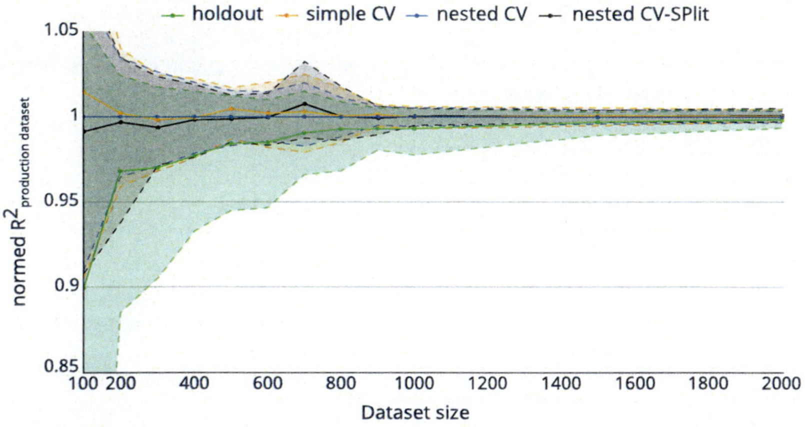

Fig. 5. Comparison of the production performance of the evaluation strategies. The score is normed to the NCV score. Thick lines show the mean normed R^2,the dashed lines shows the 10th and 90th percentile

6 Conclusion

Evaluating the true performance of machine learning models is a crucial part of the usage of ML in product engineering and machine tool monitoring. However, commonly used evaluation strategies like HO and SCV show different drawbacks, primarily with small datasets.

Although the HO strategy is safe against data leakage from train to test data, it does not provide a statistical view of the performance. Hence, the robustness of this strategy is small, and outliers can lead to a strong over- or underestimation of the performance. In comparison, SCV can tackle this problem with a statistical view of the performance. Nevertheless, it can suffer data leakage from train to test data, especially when used in small datasets and extensive HPO. NCV is the superior evaluation strategy, inducing no bias and giving a statistical view on the performance, hence an accurate and robust performance evaluation. However, this comes at the cost of computational effort and thus is only recommended for small datasets. Sophisticated data-sampling strategies can provide better representations in subsamples of the whole dataset, but this can also introduce overfitting evaluating the performance. Thus, no investigated data-sampling method could enhance the robustness of the evaluation strategies compared to simple random sampling.

Acknowledgement. This research was supported by the German Research Foundation (DFG) within the Research Training Group GRK2250/2.

References

1. Reuß, M., Verl, A.: Ermittlung der Auswirkung des statistischen Verhaltens baugleicher Werkzeugmaschinen. In: Internationales Forum Mechatronik, Cham (2011)
2. Ramos, F., Possas, R.C., Fox, D.: BayesSim: adaptive domain randomization via probabilistic inference for robotics simulators. arXiv preprint arXiv: 1906.01728 (2019)
3. Guo, S., Yu, J., Liu, X., et al.: A predicting model for properties of steel using the industrial big data based on machine learning. Comput. Mater. Sci. **160**, 95–104 (2019)
4. Hu, M., et al.: Prediction of mechanical properties of wrought aluminium alloys using feature engineering assisted machine learning approach. Metall. Mater. Trans. A. **52**(7), 2873–2884 (2021). https://doi.org/10.1007/s11661-021-06279-5
5. Coraddu, A., Oneto, L., Ghio, A., et al.: Machine learning approaches for improving condition-based maintenance of naval propulsion plants. J. Eng. Maritime Environ. **1**, 136–153 (2016)
6. Vabalas, A., Gowen, E., Poliakoff, E., et al.: Machine learning algorithm validation with a limited sample size. PLOS ONE **14**, e0224365 (2019)
7. Tsamardinos, I., Rakhshani, A., Lagani, V.: Performance-estimation properties of cross-validation-based protocols with simultaneous hyper-parameter optimization. Int. J. Artif. Intell. Tools **24** (2015)
8. Rao, R.B., Fung, G., Rosales, R.: On the dangers of cross-validation. an experimental evaluation. In: Proceedings of the 2008 SIAM International Conference on Data Mining, pp. 588–596. Society for Industrial and Applied Mathematics (2008)
9. Varma, S., Simon, R.: Bias in error estimation when using cross-validation for model selection. BMC Bioinformatics **7**, 1–8 (2006)
10. Dobbin, K., Simon R.: Optimally splitting cases for training and testing high dimensional classifiers. BMC Med. Genom. **4**(31) (2011)
11. ElRafey, A., Wojtusiak, J.: Recent advances in scaling-down sampling methods in machine learning. WIREs Comput. Statist. **9**, e1414 (2017)
12. Zador, P.: Asymptotic quantization error of continuous signals and the quantization dimension. IEEE Trans. Inf. **28**, 139–149 (1982)
13. Joseph, V.R., Vakayil, A.: SPlit: an optimal method for data splitting. Technometrics 1–11 (2021)
14. Xiong, J., Zhang, G., Hu, J., Wu, L.: Bead geometry prediction for robotic GMAW-based rapid manufacturing through a neural network and a second-order regression analysis. J. Intell. Manuf. **25**(1), 157–163 (2012). https://doi.org/10.1007/s10845-012-0682-1
15. Denkena, B., Bergmann, B., Becker, J., et al.: Time series search and similarity identification for single item monitoring. In: Congress of the German Academic Association for Production Technology, pp. 479–487. Springer, (2021)
16. Schwarzenberger, M., Drowatzky, L., Wiemer, H., et al.: Transferable condition monitoring for linear guidance systems using anomaly detection. In: Congress of the German Academic Association for Production Technology, pp. 497–505. Springer (2017)
17. Lawbootsa, S., et al.: Linear bearing fault detection in operational condition using artificial neural network. In: ITM Web of Conferences (vol. 24) p. 01004. (2019)
18. Li, C., et al.: Similarity-measured isolation forest: anomaly detection method for machine monitoring data. IEEE Trans. Instrum. Meas. **70**, 1–12 (2021)

Blockchain Based Approach on Gathering Manufacturing Information Focused on Data Integrity

T. Bux(✉), O. Riedel, and A. Lechler

Institute for Control Engineering of Machine Tools and Manufacturing Units, University of Stuttgart, Seidenstr. 36, 70174 Stuttgart, Germany
tobias.bux@isw.uni-stuttgart.de

Abstract. This paper presents a blockchain-based approach to machine data capturing and storage using Hyperledger Fabric with a low-cost, esp32 microcontroller. Our research focuses on the immutability of captured data to ensure that data on the blockchain is valid for use by customers and manufacturers. This requires an embedded implementation to sign collected data before it reaches modern IT environments like IoT gateways or programmable logic controllers (PLC). We discuss the challenges and advantages of using Hyperledger Fabric compared to an already validated implementation based on Ethereum technology. The proposed method demonstrates a solution to overcome the challenges that result from the Hyperledger Fabric communication protocol in an embedded environment. We thus illustrate both the technical design of the implemented logic on the microcontroller and the implementation of the distributed messaging protocol within Hyperledger Fabric. In our study, we validate the implemented method and perform a quantitative comparison with existing solutions from information technology. At last, we discuss the limitations of the proposed method and give an outlook on approaches that can potentially solve them.

Keywords: Immutability · Data exchange · Blockchain · Data integrity

1 Introduction

With the current transformation in industry to more decentralized and interconnected processes, commonly referred to as Industry 4.0, the amount of collected and needed data is steadily increasing. This collected data often is used to improve production flow, facilitate maintenance or product quality. A lot of it is gathered and used on-site of the production, in an isolated environment that can be assumed to be secure.

With Industry 4.0 on the rise, businesses start to share more data than ever. This includes production data being sent to a supplier for maintenance purposes or data that is sent to a customer to prove integrity and quality during the manufacturing process. In these cases, data is being shared outside of the isolated environment where it is gathered and used.

M. Liewald et al. (Eds.): WGP 2022, LNPE, pp. 473–483, 2023.
https://doi.org/10.1007/978-3-031-18318-8_48

By sending data from one business to another, both parties have an interest to prove the integrity of the data, which is often difficult in current facilities since data passes through multiple applications and machines before being stored immutable.

Simultaneous to the rise of Industry 4.0, blockchain technology is becoming more popular as well. Mostly known from the emergence of cryptocurrencies, the blockchain technology offers a lot of potential for secure and transparent data storage and interactions. Whilst comparatively new, blockchains also offer potential for the industrial sector, not only for business-customer interactions, but business-to-business interactions as well. Using blockchain for secure data transfer and storage can be very beneficial, but the advantages apply only if collected data is gathered valid and transparent as well. Since this aspect of blockchains is still in its beginnings, a lot of tasks are not completely solved. One of these tasks is the transfer of machine data to a blockchain, which necessitates securing data integrity on-site. In the following pages a solution for this issue is presented.

2 State of the Art

For the most part, studies involving blockchain in an industrial environment tend to be about feasibility for certain use cases and often do not involve data security on the lower hardware and software level [1]. Studies and implementations that have data traceability and integrity as their focus usually stop their argument at a blockchain level. For example, Leal et al. [2] look at end-to-end traceability in pharmaceutical manufacturing. They use the properties of a blockchain to prove traceability, but do not consider the path between the sensor and the blockchain as a potential source of corruption. Coming from an analytical perspective, Wu et al. [3] have developed a model that aims to make data integrity measurable and verifiable in an IoT environment. The approach only describes blockchain implementations that work with a proof of work consensus procedure. The approach of being able to draw a correlation to the authenticity and integrity of data from the mining time is therefore not applicable to consensus processes that can be used in industry. Even if this were the case, the path between sensor and blockchain is again not considered.

However, Singh [4] considers data integrity at the data capture level and defines potential challenges in maintaining data integrity. One of the challenges he addresses is the mutability of data and thus the destruction of integrity. His solution only focuses on the use of a blockchain and thus, again, does not include the path from data creation to the blockchain.

Korb et al. [5] recognized back in 2019 that the path of data from the generating sensor to the blockchain is a critical time period for data integrity. They ported an Ethereum client to a microcontroller to sign blockchain transactions as close to hardware as possible, giving them the integrity properties of a blockchain. Their proof of concept shows that it is possible to protect data from corruption from the moment it is generated. Limitations in the work of Korb et al. concern the blockchain implementation used. The features of the Ethereum Blockchain, such as the proof of work consensus or permissionless participation, prevent it from being useful in most production environments. The upcoming Proof of Stake consensus of Ethereum has the potential to change this but has not been fully integrated into the main blockchain. Nevertheless, the permissionless participation is a problem regardless of the consensus algorithm.

In summary most publications rely on the properties of a blockchain when arguing data integrity. The Korb et al. approach, on the other hand, shows that earlier signing and transaction creation can lead to higher data integrity. Therefore, in this paper, the idea of Korb et al. is applied to an industry-ready blockchain solution, implemented under the same conditions and subsequently evaluated.

3 Hyperledger Fabric and Ethereum

In the work by Korb et al. [5], Ethereum was used to transfer data from a microcontroller to a ledger. Ethereum is an open-source, public, permissionless blockchain, meaning everyone has access to the blockchain and can interact with it. It is known as one of the first to implement smart contracts, an idea that was first introduced by Nick Szabo [6].

Smart contracts enable developers to create distributed applications, which allows users to create and interact with publicly available, interactive contracts or other automated interactions. Ethereum uses Ether as its currency, which can also be sent or received by using smart contracts, making it a good platform to use for customer-business interactions. To use Blockchain technology to send manufacturing data in a business-to-business environment two main challenges need to be considered:

Confidentiality: When data is sent between businesses, such as transactions or production data, it is necessary to keep this data private between the two parties. This results in the need for permissioned networks, to make sure that public participants cannot interact with the network.

Performance: With the intention of sending a lot of data, like manufacturing data, comes the necessity of high transaction throughput and low confirmation latency to make sure that data can be captured in a frequency that is typical for a production environment.

To provide this, Hyperledger Fabric was created with a business-to-business approach in mind. Fabric is a permissioned blockchain with modular architecture. It is part of the Hyperledger project that was started by the Linux Foundation, with initial contributors such as IBM and Digital Asset. Like Ethereum it allows the execution of Smart Contracts, called Chaincode in Fabric. Notably, it allows components, like consensus and membership services, to be modular and for its smart contracts to be written in general-purpose programming languages like Java, Go and Node.js. Using custom consensus protocols, it does not require a native cryptocurrency and does not incentivize mining operations, thus reducing operational cost to the same level as other distributed systems. These factors make Hyperledger more applicable in an industrial setting than, for example, Ethereum.

Fabric also has a transaction architecture called execute-order-validate, which contrasts with most blockchain applications that use order-execute methods [7]. It separates the transaction flow into three steps:

1. Execute a transaction and check its correctness, thereby endorsing it
2. Order transactions via a (pluggable) consensus protocol
3. Validate transactions against an application-specific endorsement policy before committing them to the ledger.

With application-specific endorsement policies, transactions only need to be executed/endorsed by concerned nodes. This allows for parallel execution increasing overall performance and scale of the system.

In summary, Fabric identified the challenges Ethereum faced in a production environment and solved them by adding a permission-based approach, channels for data integrity and a consensus algorithm that requires less energy than a typical proof of work algorithm. Therefore, we made the decision to extend the work of Korb et al. [5] by a Hyperledger Fabric specific implementation.

4 Explanation of the Technical Design

The aim of this work is easily explained. Similar to the solution of Korb et al. [5] we want to save verifiably unchanged data from a sensor to a Blockchain, specifically a Hyperledger Fabric Blockchain. As already described by Korb et al. [5] there are some prerequisites and challenges that come with this rather simple statement:

In order to send data to a blockchain, it must be processed, and therefore a device is needed to do so. This device then becomes the next logical attack point in the data flow and therefore has to be as secure as possible. It needs to have unchangeable data processing logic to guarantee internal data integrity and a certificate mechanism to communicate with the downstream data flow. Therefore, no general-purpose operating system should be used, since the freedom that comes with it allows too many possible attacks on both the processing logic and the certification mechanism [8]. A hardware-oriented approach without a general-purpose operating system could be able to fulfill the requirements. An example would be a microcontroller that is programmed once and then sealed, with communication through either a WIFI module or wires connected to GPIO pins. The communication logic of the WIFI module has to be set to outgoing initiated traffic only to prohibit incoming requests. Since machine data must be captured at the source, a small microcontroller also helps with integration into the machine. It can be sealed inside of the machine and connected directly with the machine's sensors, avoiding tampering on the connection between sensors and microcontroller. This leads to the problem of memory and computational resources, which both are notoriously small on microcontrollers. Due to the potentially rather large requirements on a device to be a client in a blockchain network, this gave us two options:

We could (1) increase the capabilities of the microcontroller, which would lead to either a more complex operating system and/or a bigger size, which we want to avoid as mentioned before. We could also (2) introduce another device in the data flow that handles the blockchain communication.

By using another device that acts as a proxy for the blockchain, we can alleviate the requirements for our microcontroller, but we would also introduce another attack vector. If we can make sure that the data is intact when signed for the blockchain, the second option seems preferrable. In Fig. 1 a technical design for the described solution is presented.

For us to make sure that the data reaches the blockchain intact, the data must be formed into a transaction which the microcontroller signs. If the transaction is changed, it becomes invalid and therefore no false data can get added to the ledger. This still leaves

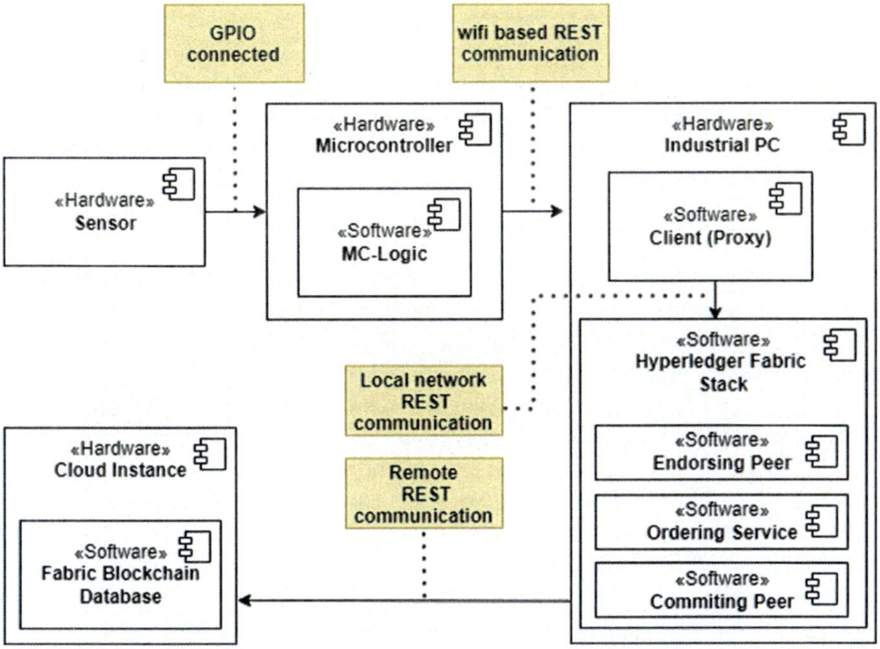

Fig. 1. UML Component Diagram describing the technical design of the presented solution

us with the possibility of data not being added or being lost due to tampering with the second device. But since we can monitor the data on the blockchain, this issue alerts us to the tampering. A benefit is that this second device can potentially act as a proxy for multiple microcontrollers since the microcontrollers will be the bottleneck due to the signing process without multithreading. During the signing process, no other action can be performed on the microcontroller. In theory, it would be possible to split the signing process in independent parts, interrupt between them and read new values that can stored for the next transaction. Nevertheless, without parallel execution, the signing process itself would take more time. This would lead to a decrease in transaction throughput and to delaying the processing of the additional gathered data.

Having a proxy device also means that any change in the blockchain's communication structure does not necessarily require a change in the microcontroller's logic if it can still verify the data it is signing. Since sealing the microcontroller in a production environment to protect it from tampering, any changes to its logic would mean an entire replacement of either the seal or the microcontroller. Therefore, avoiding a future change in its logic is also a point of consideration.

This leaves us with the conclusion of using a sealed microcontroller for data gathering and signing the blockchain's transactions, and a second device which is used as a proxy to communicate with the blockchain itself.

5 Implementation

In Sect. 5.1 the hardware and software structure of our setup will be listed. Subsequently, the sequence of the program and the data exchange are explained based on a sequence diagram.

5.1 Architectural Description of the Setup

As already described, minimum performance with almost no operating system functionality is prioritized to ensure a minimal manipulation potential. For that reason, the NodeMCU Kit with an ESP8266 module is used for this implementation, as it was the case in the work of Korb et al. [5].

Sensors can be connected to general purpose input and output (GPIO) pins to enable the collection of sensor data. On the microcontroller there is the potential to connect up to 16 digital GPIOs, but only one analog input. The used NodeMCU Kit has a processor which runs at 80 MHz, a flash memory of 4MB and WIFI-access due to the integrated ESP8266. The program for the microcontroller is written in the Arduino C/C++ language and deployed via USB. This makes it easy to hinder further change to the logic by simply removing access to the USB port, for example by physically sealing or removing it.

Out of these facts two challenges arose that we needed to solve:

The first challenge is that Hyperledger Fabrics SDK is not available for C/C++ [9]. This problem was resolved by manually porting the SDK to the Arduino C/C++ language.

Our second challenge was that 4MB of flash space is not enough space to store all the necessary files needed to handle Hyperledger identities, a ported SDK, the firmware and the program itself. Therefore, the Hyperledger Fabric transaction flow is handled by an external client, which communicates with the microcontroller using HTTP.

As previously mentioned, this has the added benefit that the microcontroller is not affected by changes in the connected Hyperledger Fabric Network and will not need to be adjusted. At the same time, this adaptation does not interfere with our plan of signing transactions on the microcontroller, as we are still able to do that.

5.2 Program Flow and Communication Structure

As shown in Fig. 2, the communication structure includes five different systems. The microcontroller as well as the client are one component in the standard Hyperledger structure and were separated for this work. The endorsing peers, the ordering service as well as the committing peers are standard implementations according to Hyperledger Fabric. The following section describes the program flow and data flow in more detail.

Similar to other Arduino projects, on the microcontroller a setup function is implemented that runs during startup followed by a loop function that contains our code which will be looped until exited. During setup the seed for randomness is set by an unconnected analog GPIO pin, resulting in an unpredictable seed for later cryptographic functions.

The input GPIOs for the sensors are configured and the connection to the local WIFI is established. Then the local time and date is requested from the client for accurate synchronized timestamps later on. The mentioned loop can be categorized into three

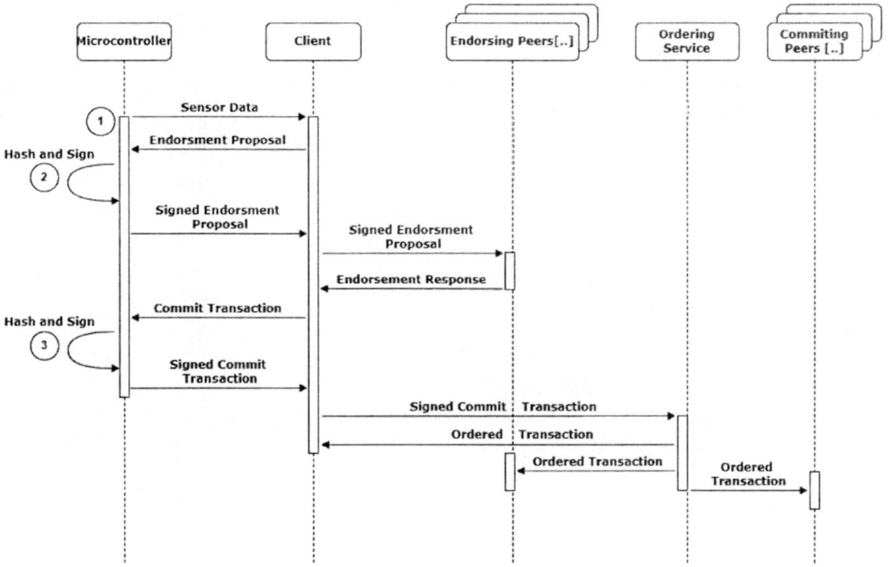

Fig. 2. UML sequence diagram describing the transaction creation procedure

steps which correlate to the three messages of the client needed for a successful addition to the blockchain.

The first step is performed in a set interval. The sensor data is read through the GPIO pins and put into a ring buffer with the timestamp of the read time. Then, if the WIFI is connected, a HTTP POST command is being send to the client which includes the captured data into an array.

The client then creates the Hyperledger Fabric endorsement proposal, which is a request to execute a certain Chaincode. In this case it is to update an array on the blockchain we use to store our data. This proposal is then sent back to the microcontroller for signing. The microcontroller checks if the originally sent data is correctly contained in the endorsement proposal. This step is important to secure data integrity due to the client running on an untrusted system. If the proposal is verified, step two is executed.

In the second step the received endorsement proposal is hashed using SHA-256 and signed using the configured private key and the secp256r1 curve.

The signed endorsement proposal is then sent back to the client using another HTTP POST request. The client then sends out the proposal to the endorsing peers and gets their responses. If successful it then generates the commit message, which is the request to update the ledger, and sends it to the microcontroller.

The third step is similar to the second, since the received commit proposal on the microcontroller is hashed, signed and being send back to the client, which then distributes the message to the ordering service to commit the message to the blockchain.

The microcontroller software does not control if the data was correctly added to the blockchain. It simply attempts to send its data continuously. Should one attempt fail, the data is still in the ring buffer with its timestamp for the next attempts, until it is overwritten. This makes it easier to tolerate inconsistencies in the WIFI connection or other outside effects.

The described software was modeled after the offline-signing tutorial of the Hyperledger Fabric GitHub [10]. The adjustments made were to remove the actual signing process and to separate the software into microcontroller and client parts. The signing part of the tutorial now takes place on the microcontroller. The constant comparison of the proposals and the data stored on the microcontroller is an addition, implemented to secure data integrity within the changed protocol.

6 Evaluation

For evaluation purposes the microcontroller was attached to a distance sensor by using the GPIO pins directly. Afterwards data was captured and stored to a blockchain as fast as the microcontroller was able to handle it. For evaluation, the results of Korb et al. are compared to timestamps taken during the running process of our implementation. The average times of actions displayed in Table 1 are taken from Korb et al. [5]. Due to the differences of the protocol used to insert data into the blockchain, the necessary steps differentiate. All numbers presented are time values in milliseconds:

Table 1. Average times in milliseconds for tasks during transaction creation for an ethereum blockchain

Acquire data and build unsigned transaction	Hash	Sign	Build signed transaction	Establish server connection	Positive server response	Total
8.6	9.7	1391.8	40.7	471.7	396.1	2316.6

Results were taken from the work of Korb et al. [5]

The average numbers for the implementation in this work, displayed in Table 2, are gathered under the same conditions as the data from Korb et al. [5]. Due to not having strong outliers, a simple average was built to compare values:

Even though the tasks necessary for both protocols are different due to the deviating protocols there are findings that need to be mentioned. First, the total time to create and send a transaction has improved with the proposed solution. One reason for this is the fact, that the signing process of a Hyperledger transaction is about ten times faster than it was for Ethereum. This might come from implementation differences because the hardware itself is identical. Another reason for the faster process can be found in the transaction sending process. While Ethereum uses a synchronous REST Call and therefor waits for a response, the Fabric approach just sends the finished transaction to a committing peer and starts to gather data again. Nevertheless, Fabrics protocol uses synchronous communication during the exchange of the proposal and the transaction, which is not necessary with Ethereum. The performance of the industrial PC leads to minimal downtime while waiting for proposal and transaction creation.

Table 2. Average times in milliseconds for tasks during transaction creation for a hyperledger fabric blockchain, implemented in this work

Acquire Data	Connect and create Payload	Send payload and receive proposal	Hash	Sign proposal	Send proposal and receive transaction	Hash	Sign transaction	Send transaction	Total
~0	21	151	3	127	113	7	126	72	620

In summary the implementation presented in this paper takes an average of 620 ms to acquire and send data to a Fabric Blockchain. This is comparable to lower end OPC UA Servers and therefor industrially applicable.

7 Summary and Future Work

In this paper an approach was given to capture machine data and store it in a Hyperledger Fabric Blockchain without losing data integrity on the way. First, the concept of our hardware and software and the requirements thereof were shown, afterwards the implementation was presented. Machine data was captured using a microcontroller and sent to the blockchain by using a client running on an industrial PC, but still the signing process of the data took place on the microcontroller to make sure the data is correct.

To improve the presented solution, a standalone microcontroller approach might be the next step. This would require multiple adjustments:

First, the Hyperledger SDK would need to be further changed to be executable on a small microcontroller. The Microcontroller nevertheless would most likely require more memory than available on the presented NodeMCU.

The second adjustment would be to make the Hyperledger Fabric parameters for the necessary connections editable over WIFI without compromising the data integrity of the controller. This would be necessary since in industrial use case scenarios members of a network might be added or removed and therefore endorsing and committing peers need to be added/removed from in order to get valid transactions.

The first issue can be solved by using different hardware and programmatic expanses, the second issue otherwise is a conceptional issue and therefore, no concrete solution can be suggested.

Regarding the solution proposed in this paper, more validation in terms of data integrity and security is necessary. What was shown in this paper is the design and the feasibility of the proposed approach. The next step of the authors is to use defined threat modelling techniques to compare this proposed solution to standard IIoT-solution for data exchange.

References

1. Mushtaq, Haq, I.U.: Implications of blockchain in industry 4.O. In: 2019 International Conference on Engineering and Emerging Technologies (ICEET), pp. 1–5 (2019). https://doi.org/10.1109/CEET1.2019.8711819
2. Leal, F., Chis, A.E., Caton, S. et al.: Smart pharmaceutical manufacturing: ensuring end-to-end traceability and data integrity in medicine production. Big Data Res. **24**, 100172. ISSN 2214–5796 (2021). https://doi.org/10.1016/j.bdr.2020.100172
3. Xia, W.U., Kong, F., Shi, J., Bao, L., Gao, F., Li, J.: A blockchain internet of things data integrity detection model. In: Proceedings of the International Conference on Advanced Information Science and System (AISS '19). Association for Computing Machinery, New York, NY, USA, Article 21, pp. 1–7 (2019). https://doi.org/10.1145/3373477.3373498
4. Singh, M.: Blockchain technology for data management in industry 4.0. In: Rosa Righi, R., Alberti, A.M., Singh, M. (eds.) Blockchain technology for industry 4.0. BT, pp. 59–72. Springer, Singapore (2020). https://doi.org/10.1007/978-981-15-1137-0_3

5. Korb, T., Michel, D., Riedel, O., Lechler, A.: Securing the data flow for blockchain technology in a production. IFAC-PapersOnLine **52**(10), 125–130 (2019), ISSN 2405-8963, https://doi.org/10.1016/j.ifacol.2019.10.012
6. Szabo, N.: Smart Contracts: Building Blocks for Digital Markets. p. 28 (1996)
7. Androulaki, E., Barger, A., Bortnikov, V., Cachin, C., Christidis, K., De Caro, A., Enyeart, D., Ferris, C., Laventman, G., Manevich, Y., Muralidharan, S., Murthy, C., Nguyen, B., Sethi, M., Singh, G., Smith, K., Sorniotti, A., Stathakopoulou, C., Vukolić, M., Weed, S., Cocco, Yellick, J.: Hyperledger fabric: a distributed operating system for permissioned blockchains (2018). https://doi.org/10.1145/3190508.3190538
8. Geer, D.E.: Playing for keeps: will security threats bring an end to general-purpose computing? Queue **4**, 9 (November 2006), 42–48 (2006). https://doi.org/10.1145/1180176.1180193
9. https://hyperledger-fabric.readthedocs.io/en/release-2.2/fabric-sdks.html. Last visited on 12 May 2022
10. https://hyperledger.github.io/fabric-sdk-node/release-1.4/tutorial-sign-transaction-offline.html. Last visited on 12 May 2022

Utilizing Artificial Intelligence for Virtual Quality Gates in Changeable Production Systems

A.-S. Wilde[1(✉)], M. Czarski[1], A. Schott[1,2], T. Abraham[1,2], and Christoph Herrmann[1,2]

[1] Institute of Machine Tools and Production Technology, Chair of Sustainable Manufacturing and Life Cycle Engineering, Technische Universität Braunschweig, Langer Kamp 19b, 38106 Braunschweig, Germany
a.wilde@tu-braunschweig.de

[2] Fraunhofer Institute of Surface Engineering and Thin Films, Bienroder Weg 54E, 38108 Braunschweig, Germany

Abstract. The demand for individualized products with high quality and low costs is a challenge for manufacturers. Classic production engineering approaches can no longer meet these requirements and often require long and complex start-up cycles, while producing scrap and limiting product changeovers. This challenge becomes more acute when factoring in recycled products and materials. One solution to this is the use of Changeable Production Systems, which can be planned, controlled and monitored on the basis of data and data-based algorithms. This paper presents a strategy for constructing and utilizing virtual quality gates (VQG) in the context of Changeable Production Systems. Their logic is based on artificial intelligence and they are placed at the edge of the network to facilitate fast decision making.

Keywords: Changeable production systems · Virtual quality gates · Edge computing

1 Introduction and Motivation

Production systems of the future have to cope with fast changing customer requirements and an increasing demand for individualized products [1]. Challenged by competitive market environments combined with technological developments, an ongoing change in production systems is necessary to remain economic and environmental. For companies aspects like rapid deliveries, new production processes and a market-driven flexible manufacturing system (volume and product configuration) are key elements to become an agile company organization [2]. Changeable Production Systems (CPS) are one concept to overcome these challenges and an enabler to establish an economic as well as sustainable production system under varying circumstances [3].

Improving the sustainability of production systems became a strategic goal for most companies in the past years. This is accompanied by and also resulting from a political

M. Liewald et al. (Eds.): WGP 2022, LNPE, pp. 484–493, 2023.
https://doi.org/10.1007/978-3-031-18318-8_49

shift towards more sustainability, which for example becomes visible in the European Green Deal [4]. Emerging concepts like transforming the linear product life cycle into a circular life cycle in the context of Circular Economy (CE) [5] result in further requirements for production systems. Emerging technological challenges are for example the implementation of new recycling and remanufacturing processes and the processing of recycled materials with fluctuating qualities [6, 7]. To overcome these challenges, production systems have to become more and more changeable. CPS are a promising approach to gain the changeability needed from future productions systems.

To monitor and control the process chains of a CPS to maintain a high level of process quality despite the high level of changeability, a high level of transparency is necessary, which can be achieved with a consistent data-based approach along the process chain. One approach to monitor and control the quality of different production processes is the implementation of virtual quality gates (VQGs) [8]. VQGs are virtual decision points in a process or a process chain, which can be used to monitor and control quality-relevant parameters [9]. Depending on the prediction of the VQG regarding the quality of one or multiple process steps, decisions are made on the further processing of the part.

Against this background, a concept is proposed that enriches CPS by implementing VQGs. Therefore, the paper is structured as follows. At first the concepts of CPS, VQGs and Cyber-physical production systems (CPPS) are introduced (Chap. 2). Afterwards, a strategy to implement VQGs in CPS is described and a framework derived (Chap. 3). In the next chapter an exemplary application is described to illustrate the benefits and the applicability (Chap. 4). Finally, the results are summarized and an outlook is given (Chap. 5).

2 Theoretical Background

2.1 Changeable Production Systems

Production Systems that can adapt quickly to modified and varying requirements are defined as Changeable Production Systems (CPS) [10]. Main characteristics of CPS are the physical and logical system design for changeability, the product family formation and modelling as well as an integrated product and manufacturing platform development [11, 12]. Understanding the implementation of CPS as an iterative process, key aspects are the setup of physical and logical systems considering core features of modularity, integratability, sustainability and diagnoseability [13]. An exemplary approach combining injection molding with additive manufacturing is presented in [14]. Further use cases are described in e.g. [15, 16].

The main benefits of the concept are the changeability of the systems to varying manufacturing processes, e.g. caused by varying qualities of recycled materials. Therefore, Changeable Production Systems are a beneficial approach to optimize production, remanufacturing and recycling processes. However, one drawback of these systems is the increasing complexity, especially when dealing with changing production environments.

2.2 Virtual Quality Gates

Quality is a determining factor that effects the economy of production processes as well as customer satisfaction. To ensure that the quality of semi-finished and finished products meets predefined requirements, quality control is put in place. Typically, quality controls are physical, offline control processes of a certain part, conducted after relevant manufacturing steps. In recent years different approaches for online quality monitoring were presented [17]. Digitalization and industry 4.0 force the development of sensor and measurement techniques as well as simulation methods that result in a large amount of data. To generate value from this, data mining methods combined with data driven models are employed and offer great opportunities for further progress, e.g. for quality prediction [18–20]. Thus, traditional physical quality inspection methods can successively be replaced by the implementation of virtual quality gates (VQG) [8].

The idea of VQGs is to systematically divide a defined manufacturing chain into different virtual, quality-relevant decision points. Along these VQGs, the product quality is checked, ensuring that required quality states are achieved before any further manufacturing steps are approved [9]. Different application examples of VQGs show promising approaches to improve quality and significantly reduce scrap parts towards a zero-error manufacturing system [21–23].

For the implementation of VGQs the broader concept of cyber-physical production systems (CPPS) can be used. CPPS typically contains four core elements: (I) a physical system, (II) the data acquisition, (III) a cyber system and (IV) the decision support [24]. Modifications and implementation of approaches for CPPS in different fields can be found in literature [25–27]. The concept of CPPS can be transferred to VQGs, where the physical process is monitored through a virtual quality control, as shown in Fig. 1. Here, the physical world contains all physical processes like injection molding, drilling and additive manufacturing. Acquiring data from different sources automatically and manually (II) the connection from the physical world (I) to the cyber word (III) is built up. Data is used to describe the physical world and feed data-based models or simulation tools. The models deduce critical process states or make quality predictions, which are then returned to the physical world. The closing of the loop is achieved by a decision support system (IV), which helps to optimize the physical world. The described CPPS can be used to form a VQG.

The growing demand for CPPS in various industries does not fit with the current cloud computing trend, where data is transferred to a central data center, computed and then transferred back. Edge Computing has emerged as a solution to support cloud infrastructures for time-critical applications, large data sources and privacy concerns [28]. With edge computing the data is not transferred, but is processed at the edge of the network as close to the data source as possible. This unlocks the following potentials: (1) reducing the latency of data-based applications, (2) increasing the sustainability by performing simple computations on edge devices, (3) reducing bandwidth issues by removing data transfers, (4) allowing intelligent orchestration of local computation resources [29, 30].

3 Framework for VQGs in CPS

In order to address the challenges when implementing VQGs into CPS, a flexible digitalization strategy is necessary. This strategy revolves around maintaining VQGs with Edge Computing in a constantly changing environment. Due to frequent changes of the

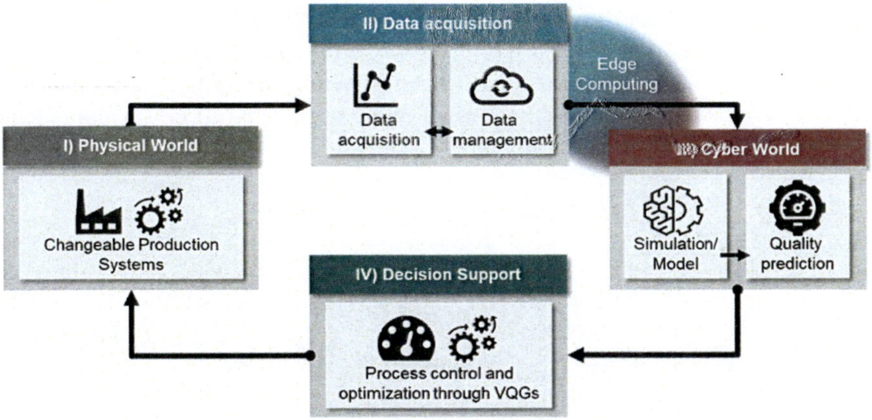

Fig. 1. Cyber-physical production system for VQGs

processes, VQGs are deployed onto edge devices. For an automatic installation, updating and maintenance of the software a framework that stretches all the way from the field to the cloud layer of a factory is proposed.

3.1 Field Layer

The field layer is at the shopfloor of a factory and therefore the closest to the various sources of data from a physical perspective. As depicted in Fig. 2, on this layer the data is acquired directly from the machines (e.g. process data, tool-integrated sensors), from product-integrated sensors or from external sensors (e.g. environment conditions). The acquired data is used to implement VQGs at the end of each process in the process chain and, if secondary materials are used or used products are remanufactured, a VQG at the beginning of the process chain is implemented to determine the material or product quality.

As the production system is changeable and the included processes have to be adaptable to different requirements, the VQGs have to have the same level of changeability. On the one hand, this is achieved by a high changeability of the underlying models to adapt easily to process changes. On the other hand, the output of the VQG of a previous process is used as in input into the current VQG. This connection ensures that the different VQGs are aligned with the entire process chain and meet the changing process conditions, which secures the functionality of the system even if e.g. the process order is changing.

To ensure fast, error-free data processing, edge computing is utilized to transform the data as close to the source as possible. Especially, for time critical processes the processing on the edge devices is beneficial since it is closest to the machine control and can quickly change production parameters. Additionally, if the number of production machines scales, the additional load of data is low since it is already aggregated at the edge. The first task for the edge device is to collect the data from the source using different communication protocols and transforming the data into a processable format.

Depending on the task a first preprocessing might occur e.g. taking the average of data in a certain time interval. If necessary, an AI model is used to analyze the data and derive insights. For the training of the AI model previously collected or historical data can be utilized for a higher accuracy of the model. The model is wrapped in a communication layer, e.g. an API Server, for the inference.

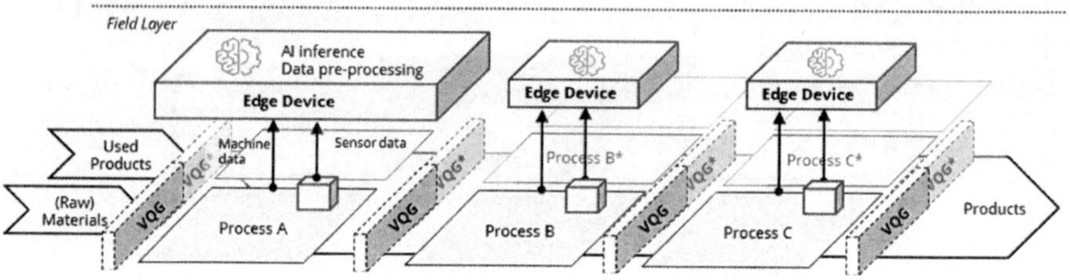

Fig. 2. Implementing VQGs on the field layer

3.2 Fog Layer

The fog layer is composed of local, high-performance hardware on the production site. In contrast to the devices on the edge layer, it has far superior performance and enables faster computations. The edge server, or a local cluster of multiple servers, is relatively close to the data source and thereby also limits the amount of traffic on the whole network. Since it is still on-site, privacy issues that would have come up when using cloud-based solutions can be omitted. To enable the changeability and flexibility to govern CPS we need the following elements (Fig. 3):

The AI Model Management contains a registry that hold different AI Models and different versions thereof. The models should be made accessible using interfaces like an API in order to allow stringing together multiple models. To further strengthen the flexibility, technologies like containers are used to develop platform independent applications. This facilitates a quick exchange of models (cyber world) in a CPS. From a developer's point of view there will only be a single codebase to work from for all environments. To deploy one of these models to an edge device the developer only needs to pull the selected container from the registry.

Supervision of process chain is especially important in the context of CE, where the supervision of the process chain as a whole becomes an important task. Variances in the material quality due to recycled materials might cause issues in a single process. These inaccuracies might be resolvable in the next process step with a slight adaption of the process chain.

As discussed before, CPS offer a lot of flexibility but also increase the complexity of a manufacturing system. An approach to benchmark the models is necessary in order to decide which model is best suited for a given task. Before the models form VQGs on the Edge they are tested with local data to see how well they perform. This does also come into play when variants of a new product are produced.

With Service orchestration various automated, computational tasks (services) can be managed. These include interfaces to the user like dashboards and supervision applications as well as resource intensive processes such as AI training and (long-term) data storage.

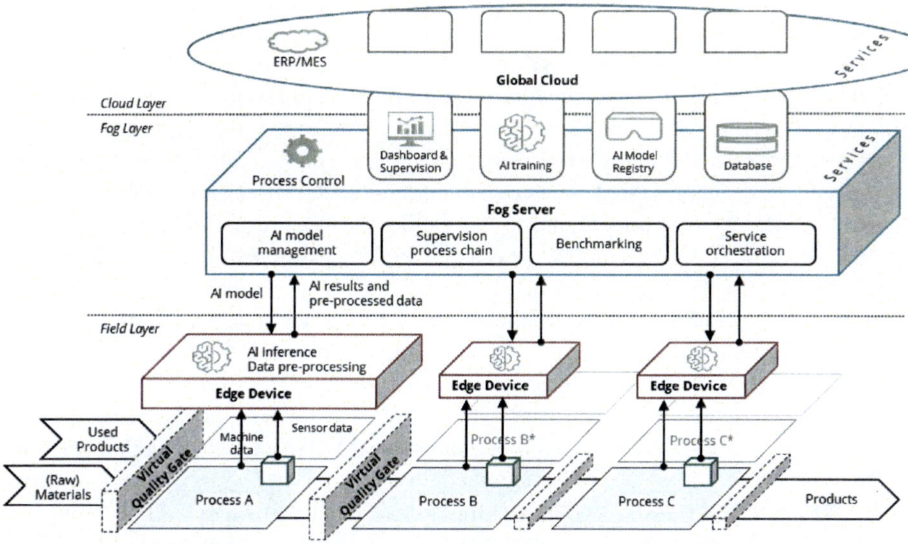

Fig. 3. Devices and services on different network layers

3.3 Cloud Layer

All the described services extend to the cloud, which can be seen as an extension of the local edge server but is not at the same physical location as the data sources. Using a cloud service allows to scale the CPS to a multi-site operation, where data and VGQs can be freely shared. Concepts like transfer learning or federated learning can improve the accuracy of the AI models. The cloud layer can be extended to the whole product life cycle to consolidate all product data. The consolidation allows the analysis and improvement of all life cycle stages in the context of CE, e.g. with information about a products use stage remanufacturing processes can be optimized.

With the implementation of these elements the CPS is able to maintain and update the algorithms on each edge device without the need for human intervention. Additionally, it is able to quickly adapt its VQGs if the process is modified.

4 Exemplary Application for CE

The exemplary application contains four different types of process chains within the exemplary CPS, as shown in Fig. 3. The order of the processes of the CPS is easily changeable and the process parameters are adaptable. For the exemplary process chains no reconfigurations of single processes are necessary. Thus, the setup time for a new process chain is assumed to be t = 0. All processes could therefore be executed simultaneously.

The starting point for the exemplary application is a defined process chain (I) to produce product P. To maintain a high process quality, VQGs are included after each process steps. The underlying models of the VQGs can be trained based on available historical data from the process or have to be developed based on current process data. The models are used to find an optimal set of process parameters.

A second product P*, which is a variant of product P, is introduced. For an optimal production of product P*, the process order has to be adapted and slight changes of processes are necessary, leading to process chain (II). As no historical data is available for this new process chain, the process parameters for each process have to be derived, which can be achieved by implementing the VQGs for this process chain. In order to determine the new VQGs, the models developed for process chain (I) and stored in the registry can be used as a starting point. With the beginning of operation, new sensor data as well as manually gathered quality data will be made available. The data can be used to benchmark the VQG models and qualify it to work for the new process chain. The best model is transferred onto the edge device to derive the quality of process steps without additional, manual measurements. Over time there will be more data from varying processes available leading to more generalized VQGs models, which results in more accurate predictions with fewer manual data collection necessary.

The presented framework also enables the implementation of process chains in the context of a circular economy. Here, the requirements for the changeability of the CPS are high because of e.g. changing material qualities, which will require ongoing process adjustments that can be made prescriptively by the VQG models. An exemplary application is shown in process chain (III), where primary and secondary materials are used. A VQG at the beginning of the process chain is added to monitor the material quality, which can differ depending on the material mix. This VQG can be used to adapt the following processes and derive the optimal process parameters more accurate.

Another exemplary use case is a remanufacturing process chain (IV). Product P_{Used} is remanufactured after the first use stage of the product. The resulting remanufactured product P_{Re} can be used in a second use phase. In addition to the VQGs of the processes, which are adapted as described above, additional VQGs are implemented. One VQG is installed before the first remanufacturing process to identify the product and material quality, which determines the optimal process chain for remanufacturing. To optimize the determination of the product quality and the process parameters, further VQGs can be implemented in the first use stage of the product to gather detailed information about the product state.

For all use cases the developed framework allows a quick adoption to new production processes and the reuse of gained insights in data-based models in the remanufacturing stage of the CPS. Therefore, a high transferability of the derived models is given, resulting in a reduction of the computing power and the implementation time of new models (Fig. 4).

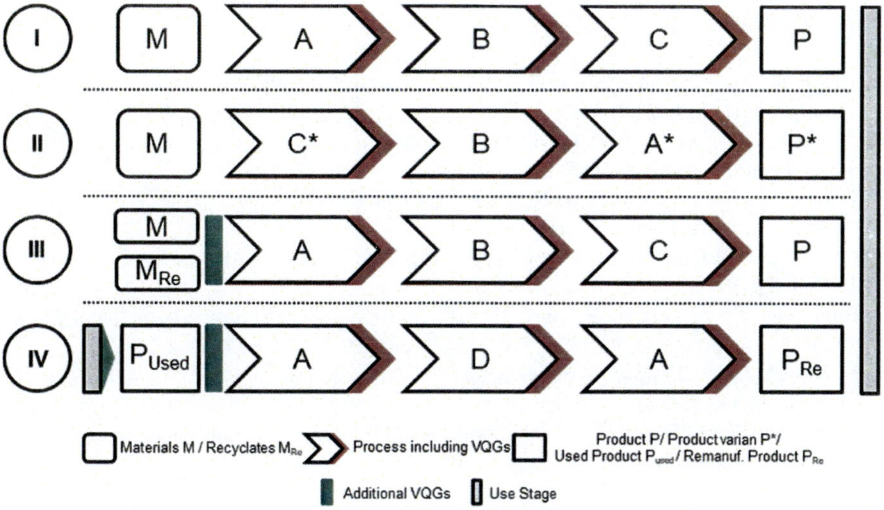

Fig. 4. Exemplary application for different process chains

5 Summary and Outlook

In this paper a framework for integrating VQGs in CPS is presented with the aim to supervise the quality of manufacturing processes as part of changeable process chains. A flexible digitalization strategy to acquire, process and store the necessary data to implement VQGs is derived from the field layer to the cloud layer. The underlying models of the VQGs are based on artificial intelligence and are placed at the edge of the network to facilitate fast decision making. The derived layer structure enables a fast transfer of the models to adapt quickly to changing production processes. The framework is applied to an exemplary CPS, where a product P and a product variant P* are produced, a product P is produced with primary and secondary materials and a used product P is remanufactured.

The framework may help to overcome the increasing complexity of CPS due to an increasing product individualization. Furthermore, the framework may help to react to increasing process fluctuations in the context of CE due to e.g. the use of recycled materials with varying material qualities.

In future work the framework and its components need to be applied to a real Changeable Production System to validate the results and to investigate the effectiveness of the strategy. The applicability for varying processes in accordance to the example applications will be examined as a next step. Furthermore, the influence of integrating life cycle data, e.g. from a products use phase, into the VQGs, especially for optimizing remanufacturing and recycling processes, will be examined.

Acknowledgement. This research and development project is funded by the German Federal Ministry of Education and Research (BMBF) within the funding initiative "HTS 2025—Hightech-Strategie 2025" (funding code: 16KIS1276). The authors are responsible for the content of this publication.

References

1. Reinhart, G.: Handbuch Industrie 4.0: Geschäftsmodelle, Prozesse, Technik. Carl Hanser Verlag GmbH Co KG (2017)
2. Pasek, Z.J., Koren, Y., Segall, S.: Manufacturing in a global context: a graduate course on agile, reconfigurable manufacturing. Int. J. Eng. Educ. 742–753 (2004)
3. ElMaraghy, H.A., El Maraghy, H.A. (eds.): Changeable and reconfigurable manufacturing systems. Springer, London (2009)
4. Claeys, G., Tagliapietra, S., Zachmann, G.: How to make the European Green Deal work. JSTOR (2019)
5. Ramadoss, T.S., Alam, H., Seeram, R.: Artificial Intelligence and Internet of Things enabled circular economy. Int. J. Eng. Sci. **7**, S. 55–63 (2018)
6. Korhonen, J., Honkasalo, A., Seppälä, J.: Circular economy: the concept and its limitations. Ecol. Econ. **143**, S. 37–46 (2018)
7. Araujo Galvão, G.D., de Nadae, J., Clemente, D.H., et al.: Circular economy: overview of barriers. Procedia CIRP **73**, S. 79–85 (2018)
8. Lee, J., Noh, S.D., Kim, H.-J., et al.: Implementation of cyber-physical production systems for quality prediction and operation control in metal casting. Sensors (Basel, Switzerland) **18**(5) (2018)
9. Filz, M.-A., Gellrich, S., Turetskyy, A., et al.: Virtual quality gates in manufacturing systems: framework, implementation and potential. J. Manuf. Mater. Process. **4**(4), S. 106 (2020)
10. Wiendahl, H.-P., ElMaraghy, H.A., Nyhuis, P., et al.: Changeable manufacturing - classification, design and operation. CIRP Annals **56**(2), S. 783–809 (2007)
11. Bortolini, M., Galizia, F.G., Mora, C.: Reconfigurable manufacturing systems: literature review and research trend. J. Manuf. Syst. **49**, S. 93–106 (2018)
12. Yelles-Chaouche, A.R., Gurevsky, E., Brahimi, N., et al.: Reconfigurable manufacturing systems from an optimisation perspective: a focused review of literature. Int. J. Prod. Res. **59**(21), S. 6400–6418 (2021)
13. Koren, Y., Gu, X., Guo, W.: Reconfigurable manufacturing systems: Principles, design, and future trends. Frontiers of Mechanical Engineering **13**(2), S. 121–136 (2018)
14. Gaub, H.: Customization of mass-produced parts by combining injection molding and additive manufacturing with Industry 4.0 technologies. Reinf. Plast. **60**(6), S. 401–404 (2016)
15. Lorenzer, T.: Wandelbarkeit in der Serienfertigung durch rekonfigurierbare Werkzeugmaschinen
16. Lotter, B., Wiendahl, H.-P. (Hrsg.): Montage in der Industriellen Produktion. Ein Handbuch für die Praxis; mit 18 Tabellen. Springer Vieweg, Berlin, Heidelberg (2012)
17. Gao, R.X., Tang, X., Gordon, G., et al.: Online product quality monitoring through in-process measurement. CIRP Annals **63**(1), S. 493–496 (2014)
18. Lieber, D., Stolpe, M., Konrad, B., et al.: Quality prediction in interlinked manufacturing processes based on supervised & unsupervised machine learning. Procedia CIRP **7**, S. 193–198 (2013)
19. Arif, F., Suryana, N., Hussin, B.: A data mining approach for developing quality prediction model in multi-stage manufacturing. Int. J. Comput. Appl. **69**(22), S. 35–40 (2013)
20. Hürkamp, A., Gellrich, S., Ossowski, T., et al.: Combining simulation and machine learning as digital twin for the manufacturing of overmolded thermoplastic composites. J. Manuf. Mater. Process. **4**(3), S. 92 (2020)
21. García, V., Sánchez, J.S., Rodríguez-Picón, L.A., et al.: Using regression models for predicting the product quality in a tubing extrusion process. J. Intell. Manuf. **30**(6), S. 2535–2544 (2019)
22. Tercan, H., Guajardo, A., Meisen, T.: Industrial transfer learning: boosting machine learning in production, S. 274–279 (2019)

23. Gellrich, S., Beganovic, T., Mattheus, A., et al.: Feature selection based on visual analytics for quality prediction in aluminium die casting, S. 66–72
24. Thiede, S.: Environmental sustainability of cyber physical production systems. Procedia CIRP **69**, 644–649 (2018)
25. Leiden, A., Herrmann, C., Thiede, S.: Cyber-physical production system approach for energy and resource efficient planning and operation of plating process chains. J. Cleaner Prod **280**, S. 125160 (2021)
26. Kannengiesser, U., Frysak, J., Stary, C., et al.: Developing an engineering tool for Cyber-Physical Production Systems. e & i Elektrotechnik und Infor-mationstechnik **138**(6), S. 330–340 (2021)
27. Wang, T., Wang, X., Ma, R., et al.: Random forest-bayesian optimization for prouduct quality predection with large-scale dimensions in process indus-trial cyber-physical systems. Internet of Things J. **7**(9), S. 8641–8653 (2020)
28. Shakarami, A., Ghobaei-Arani, M., Masdari, M., et al.: A survey on the computation offloading approaches in mobile edge/cloud computing environment: a stochastic-based perspective. J. Grid Comput. **18**(4), S. 639–671 (2020)
29. Varghese, B., Wang, N., Barbhuiya, S., et al.: Challenges and opportunities in edge computing, S. 20–26
30. Shi, W., Pallis, G., Xu, Z.: Edge computing [scanning the issue]. Proc. IEEE **107**(8), S. 1474–1481 (2019)

Analytical Approach for Parameter Identification in Machine Tools Based on Identifiable CNC Reference Runs

Philipp Gönnheimer(✉), Robin Ströbel, and Jürgen Fleischer

Karlsruhe Institute of Technology, wbk Institute of Production Science, 76131 Karlsruhe, Germany
philipp.goennheimer@kit.edu

Abstract. As a result of the steadily growing importance of data-driven methods such as digital twins, approaches for automated parameter identification in production equipment are becoming increasingly important. Previous work has shown that AI-based approaches for classification are increasingly reaching their limits. As a result of new developments, CNC reference runs with a high information content that can be specifically identified via an ID can be generated. In this context, it was possible to achieve an oscillation state on a test machine that is particularly well suited for identification. In this paper, an analytical approach is presented which, in addition to classification, can assign the signal to the respective source and therefore establish interdependencies between signals. Here, on the test machine tool, with successfully excited oscillations, all signals could be classified and assigned via the ID with high accuracy. If the oscillation state cannot be reached, classification accuracies of over 90% could be achieved, depending on the motion generation.

Keywords: Automation · Digital manufacturing system · Machine tool

1 Introduction

After ten years of Industrie 4.0, many companies are taking advantage of the potential created by the large number of data-driven approaches and profitable use cases. Use cases such as predictive maintenance and process optimizations, when based on data from the machine control system, not only bring advantages in the area of Overall Equipment Effectiveness (OEE), but can also be implemented inexpensively without additional hardware such as external sensors [1].

For the use cases mentioned, signals such as motor currents and axis positions are of particular importance in the area of machine control systems. Depending on the heterogeneity and age of the machines and systems, the provision type and form of the control signals vary greatly. It may be that the data is provided by an OPC UA server with an easily understandable node structure. Yet it is also possible that no common communication protocol is used and the data is unstructured or difficult for the operator

M. Liewald et al. (Eds.): WGP 2022, LNPE, pp. 494–503, 2023.
https://doi.org/10.1007/978-3-031-18318-8_50

to understand. Especially in brownfield production environments, data extraction and identification are in many cases a major challenge for the operator [1, 2].

The majority of companies state that a lack of technical knowledge, but primarily a lack of human resources and capacities, is a major obstacle to the use of data for Industrie 4.0 applications. In order to allow companies to make use of such use cases more easily, more quickly and more simply, and to be able to benefit from data, a tool would be needed that supports the operator in extracting and assigning machine control signals and, in the best case, automates this process [2].

2 State of the Art

There are already approaches that deal with the classification of time series data, as is the case with signals such as motor currents and axis positions, and are applied in different areas [3, 4]. In [5], for example, signals on a CAN bus system were automatically classified and assigned to four signal classes. In addition to further limitations, including manual pre-processing for an ML-based approach, these are, however, not transferable to machines and systems with regard to the requirements and framework conditions of production.

Therefore, in preliminary work, a system was developed that allows the extraction of machine control data from various data sources in the machine environment and uses an ML-based approach to identify and assign these signals to sought-after signal classes [6–8]. In the context of this preliminary work, it has been shown that while an ML-based approach can provide good results for some signals, it is highly dependent on the amount of training data as well as the target machine and its environment. Additionally, while ML-based approaches already achieve high accuracies in many cases, they do not represent a perfect solution. Also, an approach for automated machine reference runs was developed that generates information-rich data sets [9].

Since the developed ML-based approach and also other compared ones have potentials and can achieve good results, but do not represent an optimal solution and also have disadvantages, it is reasonable to investigate the potential of an analytical identification of machine control signals in more detail.

3 Own Approach

Therefore, the objective of the work underlying this publication is to develop an analytical approach to identify machine control signals such as motor currents and axis positions. To generate a scalable approach, an information model will be introduced, which is used to generate unique reference runs for the axis and spindles of a production system. Based on an approach to the minimization of time domain-based distortion of axis movements [9], position signals as well as speed signals of a spindle can be linked to the respective source via an ID. Using analytical relationships between those and signals, such as speed or load, corresponding signals can be found, and thus, be assigned to the ID. Hence, the aim of the presented six-step approach is to successively find related signals, classify those and link them to the source. In the first three stages, unrelated signals and signals required for later stages, are determined. Based on this, spindle signals are found in the fourth stage, and axis signals are found in the last two stages.

4 Boundary Conditions and Reference Run Generation

4.1 Experimental Setup

The runs and thus the signals were created on the I4.0 test machine tool of the wbk. This is a retrofitted DMC 60H – HDM machine tool, which was equipped with a new SINUMERIK control system and multiple sensors for I4.0 integration. The 4-axis machine consists of translatory axes (X, Y, Z), a rotary axis (B) and a main spindle. The basis of the present work are 100 common machine tool signals, which were recorded by the SPS with a sampling frequency of 500 Hz. Here, among other things, motor signals such as current, encoder position or control deviation as well as binary signals of the control system are recorded.

4.2 Overview and Information-Model

The presented approach should be applicable for production systems of any size, i.e. it should be scalable. To ensure this, unique reference runs must be created for all axes on the basis of information provided by the user. Therefore, the information model presented in Fig. 1 represents the main component for reference run generation. The model is initialized using the user data, which contains information such as the type of axes (translational, rotational or spindle) or the controller-internal axis identifier (X, Y, Z, B, …). During this process, a unique binary sequence is assigned to each axis representing an ID. Subsequently, the information model is used to automatically create reference runs according to [9]. For each axis, support points are derived from the IDs, through which the specified path of the reference run is fitted. The NC codes are then created using the individual movement specifications of the axes of a machine. To ensure minimum downtime, all axis movements are executed simultaneously.

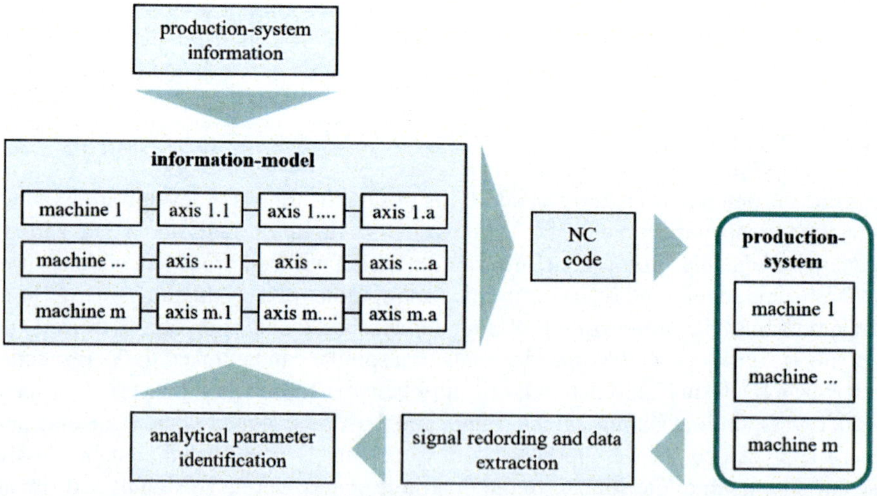

Fig. 1. Information flow during the parameter identification process

The machine-specific NC codes are executed on the machine specified by the user data in the real production system. During execution, the reference run and therefore the machine signals are recorded and extracted. The now available signals represent the input data of the analytical approach for parameter identification. Finally, information obtained about the identified signals can be fed back into the information-model via the IDs.

4.3 Reference Runs

The unique reference runs of all axes and main spindles are created using the IDs assigned by the information model according to [9]. The reference runs as the basis of the analytical approach are interpolated using cubic splines. Analogous to [9], an axis motion of 10 mm is specified for translatory and 2° for rotatory axes. The movement of the main spindle is specified by piece-wise constant speeds in a value range of 200 rpm. Due to the movement type developed in [9] using positioning axes which approach 250 points along the movement path, the oscillating movement shown in Fig. 2 occurs. The high information content as a result of the global course and the superimposed local oscillations lead to particularly well-suited runs for parameter identification.

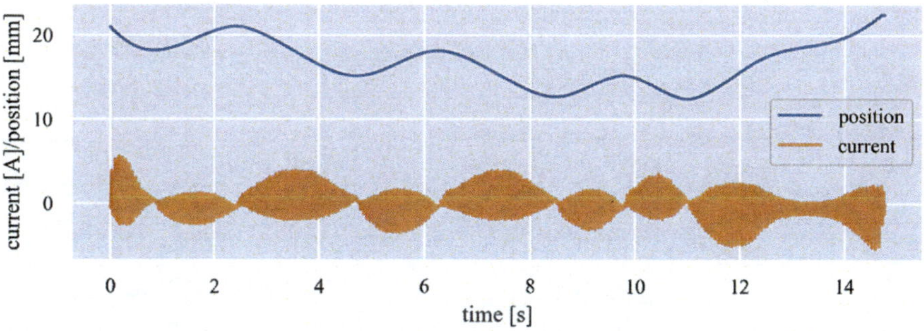

Fig. 2. Oscillating motion of a translational axis (blue: DES_POS; orange: CURRENT)

5 Analytical Approach to Parameter Identification

5.1 Approach to Signal Classification and Assignment

The processing of signals by the analytical approach involves two parts. On the one hand, the signals are assigned to a class, which is equivalent to a classification problem. On the other hand, they are assigned to an axis, if possible. Since the position course given by the reference run is based on the IDs, these can be derived from the position signals. Furthermore, if the ID of a position signal is known, all signals belonging to this axis, such as current or speed, can also be assigned. As can be seen in Fig. 3, the signal classes must be determined and a relationship to the associated axis must be established via the position signals. Signals without reference to an axis, such as noise, in contrast, only have to be classified.

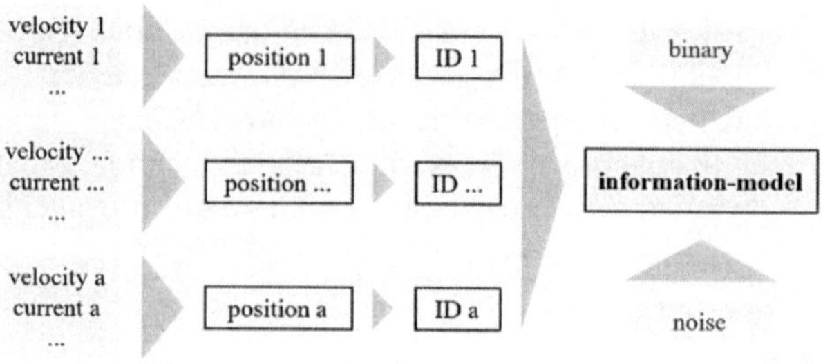

Fig. 3. Multilevel approach to classification and assignment

5.2 Target Classes of the Analytical Approach

For the determination of the classes, relevant target classes were defined on the basis of 100 common control signals. A distinction was made between stand-alone signals, which stand for themselves and cannot be assigned to an axis, spindle signals and axis signals. The derived target classes can be seen in Table 1. The naming of the target classes is based on the class names used in the data acquisition by a "Siemens Industrial Edge" [10].

Table 1. Target classes for signal classification

Stand-alone	Axis	Spindle
NULL	ENC1_POS	SPEED_SPINDLE
BIN	ENC2_POS	ENC_POS_SPINDLE
NAN	CTRL_POS	DES_POS_SPINDLE
CONST	DES_POS	CURRENT_SPINDLE
CYCLE	CMD_SPEED	TORQUE_SPINDLE
POWER	VEL_FFW	LOAD_SPINDLE
NOISE	CTRL_DIFF	
UNKNOWN	CURRENT	
	TORQUE	
	LOAD	

The stand-alone signals include signals which contain only the value null (NULL), binary values (BIN), non-numerical values (NAN), constant values (CONST), or a cycle signals (CYCLE). On the test machine, the power signals (POWER) have a low information content, so that they are only suitable for classification. The classes of noise signals (NOISE) and unknown signals (UNKNOWN) are also not related to an axis. The position signals of the axes are grouped into direct (ENC1_POS) and indirect position measurement (ENC2_POS). In addition, a distinction is made between the controller input position (CTRL_POS) and the commanded position (DES_POS). The commanded speed (CMD_SPEED) and the feedforward velocity (VEL_FFW) represent the speed of an axis. Furthermore, control deviation (CTRL_DIFF), current (CURRENT), torque (TORQUE) and load signals (LOAD) are assigned to the respective class. The defined target classes of the spindle signals correspond to those of an axis. Here the speed signals (SPEED_SPINDLE), the encoder signals (ENC_POS_SPINDLE) and the position signals (DES_POS_SPINDLE) are assigned to a collective class due to the specified target speed.

5.3 Analytical Correlations Used

For the presented approach, five relevant correlations between the individual signal classes were identified. By Cor. 1. And Cor. 2 the velocity signals can be linked to the position signals. Cor. 3 represents the correlation of the control deviation. A current or torque signal can be assigned to a position signal by Cor. 4. If an additional load signal is to be assigned, Cor. 5 can be used (Table 2).

Table 2. Analytical correlations

Number	Correlation
Cor. 1:	$VEL\_FFW \sim \frac{d}{dt}(DES\_POS)$
Cor. 2:	$CMD\_SPEED \sim \frac{d}{dt}(ENC1\_POS)$
Cor. 3:	$CTRL\_DIFF \sim CTRL\_POS - ENC2\_POS$
Cor. 4:	$-\frac{d^2}{dt^2}(ENC1\_POS) \sim CURRENT \sim TORQUE$
Cor. 5:	$\left\lvert\frac{d^2}{dt^2}(ENC1\_POS)\right\rvert \sim \lvert CURRENT\rvert \sim \lvert TORQUE\rvert \sim \lvert LOAD\rvert$

5.4 Analytical Approach to Parameter Identification

The developed approach consists of six stages, which can be divided into three groups. The first three steps deal with the determination of the stand-alone signals as well as the classes on which later steps are based. In Fig. 4, the gray block is used to illustrate the pre-processing, which thus forms the general structure. Subsequently, the signals of the spindles are processed in the spindle stage (green). The last two stages deal with the processing of the axis signals (blue).

5.5 Preprocessing

Within the pre-processing, the goal is to classify easy-to-determine classes as early as possible. Accordingly, in stage one the classes NULL, BIN, NAN, CONST, CYCLE and spindle position signals are determined successively using thresholds. Furthermore, the position signals of axes are isolated, SPEED_SPINDLE is determined and the ID

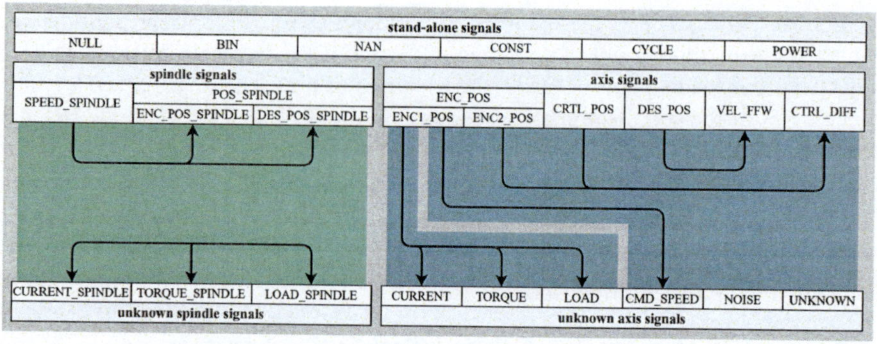

Fig. 4. Overview of stages, classes and their connections

is calculated for both. Stage two deals with the separation of the axis position signals into DES_POS, CTRL_POS, ENC1_POS and ENC2_POS. The ENC1_POS and ENC2_POS signals are split using the fact that oscillations at the end of the drivetrain are more pronounced than at the beginning due to compliance, which leads to higher values when a moving standard deviation is applied to the calculated jerk. In the third stage, signals are assigned to classes POWER, VEL_FFW and CTRL_DIFF. Furthermore, all signals that have not yet been assigned are separated into potential spindle or axis signals.

5.6 Spindle Stage

In spindle stage, spindle signals are to be classified and assigned using correlations. Using the derivation, the position signals belonging to SPEED_SPINDLE can be found. Based on the deviation DES_POS_SINDLE or ENC_POS_SPINDLE is assigned. According to Cor. 4, the rapid change of the spindle speed leads to impulse-like changes of CURRENT_SPINDLE and TORQUE_SPINDLE. These impulses are related to each other for the signals of one spindle and can therefore be used for grouping. Furthermore, the ID can be determined from the direction of these impulses. On the test machine a subdivision of CURRENT_SPINDLE and TORQUE_SPINDLE could be made after normalization due to a higher moving standard deviation of TORQUE_SPINDLE. Finally, the corresponding LOAD_SPINDLE signal can be found as a result of Cor. 5.

5.7 Axis Stage

The axis signals are processed in two stages. In axis stage one, kinematic relationships are used. Thus, the VEL_FFW, CMD_SPEED and CTRL_DIFF signals are determined and assigned using Cor. 1, Cor. 2 and Cor. 3. Axis stage two is used to classify the CURRENT, TORQUE and LOAD signals and assign them to the respective axis. For this purpose, the signals belonging together are grouped using Cor. 5 and then the LOAD signal is separated. On the test machine, the TORQUE signal precedes the CURRENT signal, which can be used for classification. Finally, the group ID can be determined either directly as a result of direction changes in CURRENT or TORQUE signals or using Cor. 4. Subsequently, all remaining signals are assigned to NOISE or UNKNOWN class.

6 Validation Criteria

The analytical approach is to be used for classification of signals as well as their assignment to the respective axes. Therefore, different accuracies are introduced for evaluation. The classification accuracy $C_1[\%]$ indicates the percentage of correctly classified signals. Signals of the class UNKNOWN are therefore not included. However, these signals are included in the classification accuracy $C_2[\%]$. . The assignment accuracy A [%] describes the percentage of signals that were correctly assigned to an axis or spindle, when an assignment is possible.

7 Results and Discussion

7.1 Training Dataset

The dataset, on which the approach is based, contains 30 unique recordings with 100 signals each. For the evaluation, as with the following data sets, all signals were evaluated simultaneously. In case of the learning data set it was possible to classify all signals correctly and to assign them to the correct axis.

7.2 Datasets Without Reliably Developed Oscillations

Within the scope of the project, 100 data sets were generated according to [1] under variation of the reference run generation, whereby the oscillations were not reliably developed. Accuracies C_1 of 91.333% to 97.5% were achieved. Since additional steps have to be performed for the axis assignment, Z is significantly lower than C_1. For the data set with the highest C_1 value, A is 85.611% and for the lowest C_1 value, Z is 74.611%. For the data set which was created analogous to the training data set, the accuracies are in between. Thus, it can be stated that with unreliably pronounced oscillations classification accuracies over 90% can be achieved. The more developed the oscillations are, the higher accuracies can be achieved. This effect is significantly stronger for Z (Table 3).

Table 3. Accuracies of the data sets without reliably developed oscillations

	$C_1[\%]$	$C_2[\%]$	$A[\%]$
Lowest C_1	91,133	94,200	74,611
Highest C_1	97,500	98,333	85,611
Training conditions	93,067	97,100	78,722

7.3 Validation Datasets

To validate the approach, three datasets were recorded analogous to the training dataset. By shifting the block change to the end of the braking ramp, the oscillation state could be reliably forced. This results in significantly higher accuracies, so that C_1 is above 98.8% and C_2 is above 99.2%. As before, the assignment accuracy A of 96.833% to 99.778% is below the other accuracies. Thus, it can be shown that high accuracies are reproducible on the experimental machine if the oscillatory axis movement exists (Table 4).

Table 4. Accuracies of the validation datasets

	C_1 [%]	C_2 [%]	A [%]
Validation dataset 1	99,300	99,567	96,833
Validation dataset 2	98,800	99,200	96,278
Validation dataset 3	99,800	99,800	99,778

8 Conclusion and Outlook

In the present work, an approach for classification and assignment of signals in machine tools based on reference runs has been developed. This approach makes use of the analytical correlations between signals. The aim is to classify as many of the given signals as possible, resulting in 24 target classes, among which 16 classes can be assigned to an axis or spindle. Classification accuracies of over 98.8% and assignment accuracies over 96.278% were achieved on the test machine tool during a validation. If the vibration state cannot be reached, accuracies above 90% can be achieved. Thus, the approach is a promising solution for parameter identification when reference runs can be performed.

In order to ensure a machine-independent approach, further tests must be carried out with regard to generalization. If necessary, individual rules should be adapted. Furthermore, the approach must become independent of the type of signal acquisition. Since the AI-based approach presented in [6] is independent of reference runs, the approaches will be merged in future works. Based on the presented results, a promising approach would be to adopt the general rules of the pre-processing and replace rules based on the reference runs by an adapted AI. Based on this, the knowledge gained about the analytical correlations could be used to obtain a general and independent set of rules. It should be noted, however, that only the reference runs provide a connection to the axes of the machine. If this is omitted, only a classification is possible.

Acknowledgements. We extend our sincere thanks to the German Federal Ministry of Economic Affairs and Climate Action (BMWi) for supporting this research project 13IK001ZF "Software-Defined Manufacturing for the automotive and supplying industry" (SDM4FZI).

References

1. Netzer, M., Begemann, E., Gönnheimer, P., Fleischer, J.: Digitalisierung im deutschen Maschinen und Anlagenbau – Aktuelle Studie zur Digitalisierung im deutschen Maschinen- und Anlagenbau und Bedarfsanalyse. In: wt Werkstatttechnik online, vol. 111; 07/08, pp. 525–530. VDI Verlag (2021)

2. Gönnheimer, P., Netzer, M., Lange, C., Dörflinger, R., Armbruster, J., Fleischer, J.: Datenaufnahme und -verarbeitung in der Brownfield Produktion - Studie zum Stand der Digitalisierung und bestehenden Herausforderung im Produktionsumfeld. In: ZWF Zeitschrift für wirtschaftlichen Fabrikbetrieb, vol. 117, no. 5. De Gruyter (2022). [in publication]
3. Wang, Z., Yan, W., Oates, T.: Time series classification from scratch with deep neural networks: a strong baseline. In: International Joint Conference on Neural Networks (IJCNN), pp. 1578–1585. arXiv (2016)
4. Ismail Fawaz, H., Forestier, G., Weber, G., Idoumghar, L., Muller, P.A.: Deep learning for time series classification: a review. In: Data Mining and Knowledge Discovery, vol. 33, no. 4, pp. 917–963. Springer Science and Business Media (2019)
5. Hozhabr Pour, H., Wegmeth, L., Kordes, A., Grzegorzek, M., Wismüller, R.: Feature extraction and classification of sensor signals in cars based on a modified codebook approach. In: Progress in Computer Recognition Systems, vol. 977, pp. 184–194. Springer, Berlin (2019)
6. Gönnheimer, P., Puchta, A., Fleischer, J.: Automated identification of parameters in control systems of machine tools. In: Production at the Leading Edge of Technology, pp. 568–577. Springer, Berlin (2020)
7. Gönnheimer, P., Karle, A., Mohr, L., Fleischer, J.: Comprehensive machine data acquisition through intelligent parameter identification and assignment. In: Procedia CIRP, vol. 104, pp. 720–725. Elsevier (2021)
8. Gönnheimer, P., Hillenbrand, J., Heider, I., Baucks, M., Fleischer, J.: Enabling data-driven applications in manufacturing: an approach for broadly applicable machine data acquisition and intelligent parameter identification. In: Production at the Leading Edge of Technology. Springer, Berlin (2022). [submitted]
9. Gönnheimer, P., Ströbel, R., Netzer, M., Fleischer, J.: Generation of identifiable CNC reference runs with high information content for machine learning and analytic approaches to parameter identification. In: Procedia CIRP. Elsevier (2022). [in publication]
10. Siemens AG: SINUMERIK. Analyze my workpiece/capture. In: Operating Manual, pp. 23–24 (2020)

Application Areas, Use Cases, and Data Sets for Machine Learning and Artificial Intelligence in Production

J. Krauß[1](✉), T. Hülsmann[1], L. Leyendecker[2], and R. H. Schmitt[2,3]

[1] Fraunhofer Research Institution for Battery Cell Production FFB, Bergiusstraße 8, 48165 Münster, Germany
jonathan.krauss@ipt.fraunhofer.de

[2] Fraunhofer Institute for Production Technology IPT, Steinbachstraße 17, 52074 Aachen, Germany

[3] Laboratory of Machine Tools and Production Engineering WZL, RWTH Aachen University, Campus-Boulevard 30, 52074 Aachen, Germany

Abstract. Over the last years, artificial intelligence (AI) and machine learning (ML) became key enablers to leverage data in production. Still, when it comes to the utilization and implementation of data-driven solutions for production, engineers are confronted with a variety of challenges: What are the most promising application areas, scenarios, use cases, and methods for their implementation? What openly available data sets for the training of ML and AI solutions do exist? In this paper, we motivate the challenges of applying AI and ML in production and introduce an extended taxonomy of application areas and use cases, resulting from a comprehensive literature review. In addition, we propose both a process model and a concept for an ML-Toolbox that are tailored to cope with the specific challenges of production. As a result, from an extensive study, we present and launch a comprehensive collection of currently more than 130 datasets that we make openly available online to serve as a continuously expandable reference for production data. We conclude by outlining three key research directions that are decisive for a widespread adoption of real-world ML. The contributions of this paper establish a foundational development framework that allows to identify suitable use cases, gain experience without having suitable in-house data at hand, improve existing data-driven solutions and promote applied research in this challenging field of ML in production.

Keywords: Machine learning · Artificial intelligence · Data sets · Big data · Production · Smart manufacturing

1 Challenges of Real-World ML Applications in Production

In the last decade, machine learning (ML) has gained tremendous importance in diverse fields, driven by fast paced developments in algorithms and model architectures, decreasing costs of both sensors and computing hardware. With the help of ML algorithms,

M. Liewald et al. (Eds.): WGP 2022, LNPE, pp. 504–513, 2023.
https://doi.org/10.1007/978-3-031-18318-8_51

information extracted from raw data can enhance both human and autonomous decision making [1]. The increasingly growing data volume is further fueling interest in production-related ML applications [2]. The utilization of ML in production processes aims to improve product quality and production rates, to monitor and forecast the condition of machines, and to optimize process parameters [3]. In contrast to entirely virtual systems, in which ML applications are already widespread today, production processes are characterized by the interaction between the virtual and the physical world. Data is recorded using sensors and processed on computational entities and, if desired, actions are translated back into the physical world via actuators [4]. This poses major challenges for the application of ML in production engineering systems.

By summarizing key process, data and model characteristics, Fig. 1 illustrates the field of tension that ML applications in production are exposed to. The combination of high reliability requirements, high risk and loss potential, the multitude of heterogeneous data sources and the non-transparency of ML model functionality lead to a gap between industrial ML applications in production systems and the progress in academic research and the virtual space. In particular, production data comprises a variety of different modalities, semantics and quality [1]. Furthermore, production systems are dynamic, uncertain and complex [1], and engineering and manufacturing problems are data-rich but information-sparse [5]. Besides that, due the variety of use cases and data characteristics, problem-specific data sets are required, which are difficult to acquire, hindering both practitioners and academic researchers in this domain [2].

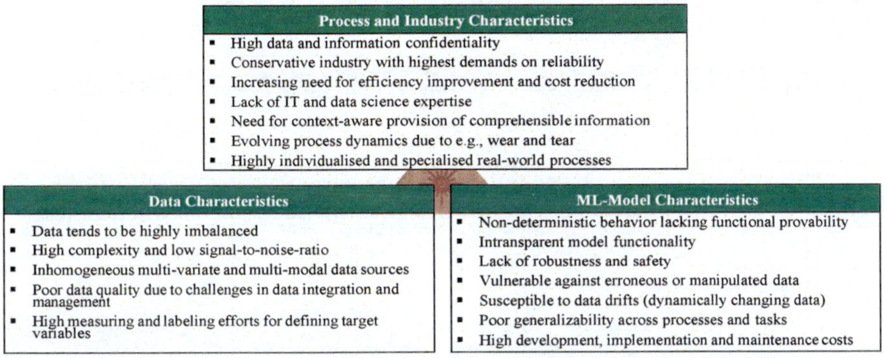

Fig. 1. The challenges for ML applications in production engineering result from the encounter of process, data and ML model characteristics

To provide structure to the field of production-related ML applications, and thus to promote development and research in this challenging domain, in this paper, we provide (1) an overview of **application areas and scenarios for ML in production**, (2) a comprehensive and expandable **collection of public production data sets**, and (3) a **production-specific development framework** comprising a process model and an ML-Toolbox. Thereby, we aim to establish a production-related framework that helps to answer the four key questions that engineers, and data scientists face in almost any data projects: *What is the problem to be solved? What data is needed? Which procedure is to be followed? Which tools and algorithms are most suitable in solving the given* *problem?* We aim to facilitate access to data by publicly hosting a collection of currently over 130 public production datasets to which we are encouraging to contribute.

2 Application Areas and Publicly Available Datasets

To structure the multifaceted research and development field of ML in production, we first introduce a new taxonomy of application areas in production and their corresponding application scenarios. We then present a new overview of open production data sets and provide insights into the availability of such datasets for different application areas and scenarios.

2.1 Application Areas

Identifying suitable use cases for the application of ML in production is often a difficult task for process engineers that do not have a computer science background [6]. To aid this process, we identified important production related application areas, which provide a starting point for use case selection. Each application area is further divided into specific application scenarios that describe concrete ML applications in production. An overview is found in Fig. 2. While some application areas have a direct connection to production processes, others cover production adjacent fields like logistics or the factory building. The areas and scenarios can also serve as a starting point for finding relevant open data sets for specific use cases.

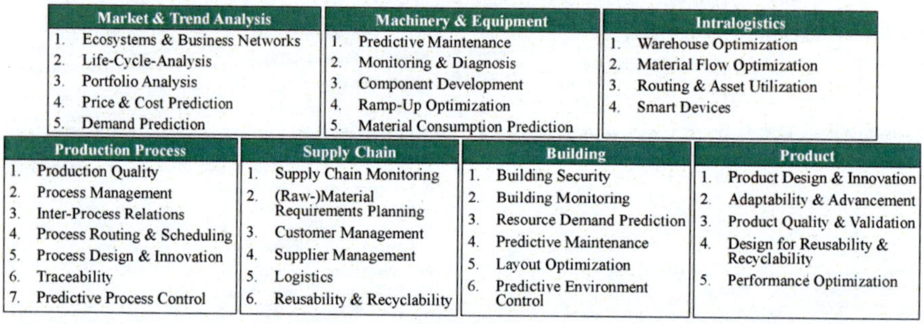

Fig. 2. Taxonomy of application areas and application scenarios for ML/AI in production

2.2 Collection of Public Production Data Sets

The performance of an ML algorithm or model is in many cases directly related to the quality and amount of available training data. Application specific data therefore is the starting point of any ML project. Because of this, a variety of data search engines and overviews of publicly available data sets exists. However, there is a distinct lack of up-to-date reference datasets for production applications. In contrast to this, there are areas such as object detection or speech recognition, where comprehensive, open, and widely used data sets exist [7–9]. Although some data set overviews offer the possibility to filter the data sets according to domains and use cases, they are often distributed across multiple different sources and difficult to find. The information provided also varies and, in most cases, lacks production-specificity. Some overviews of manufacturing data sets already exist. However, these originate from smaller projects that are not updated frequently and contain only a limited number of data sets [2, 10], or research institutions that only make their own data sets available [11–15].

For this reason, in a comprehensive search, we identified data sets that are linked to the production domain and conceptualized a standardized template for storing the collection of data sets in a tabular fashion. We enquired about which features about data sets are of particular importance in the context of production technology. Besides the *Name* and a short *Description* of each data set, the overview contains information about the *Year of Publication*, the *Industrial Sector*, the *Application Area* and *Application Scenarios*. Additionally, *File Type*, *Data Set Size*, *Data Modality*, *Learning Task*, *Number of Instances* and *Number of Features* are specified. Information on *Availability* and *License* are also provided. The underlying data is not stored in the table, but reference is made to the original source with a *Link*. The production data set overview can be accessed via *bigdata-ai.fraunhofer.de/production_datasets* (QR-code in Fig. 3).

Fig. 3. Extract and access link to the overview of publicly available production datasets

Figure 4 (left) visualizes the yearly count of datasets contained in our overview depending on their publication date. It can be seen that the number of published datasets has increased significantly since 2014. For the future, an upward trend is observable. While this is a welcome trend, this also emphasizes the need for a structured overview of datasets across different data set providers. Figure 4 (right) shows the distribution of the data sets across the individual industry sectors. The corresponding distributions across application areas and data modalities are visualized in Fig. 5. These statistics show which industry sectors and application areas are strongly or weakly represented, and which data modalities are common.

Data sets from metals and plastics domain, as well as electronics dominate the available datasets. A focus should thereby lie on increasing the share of open data sets from other fields in the future.

The most represented application areas in the overview, are the ones directly related to the production processes. This is to be expected, as these are the obvious first targets for the introduction of ML in the production environment. However, a holistic view

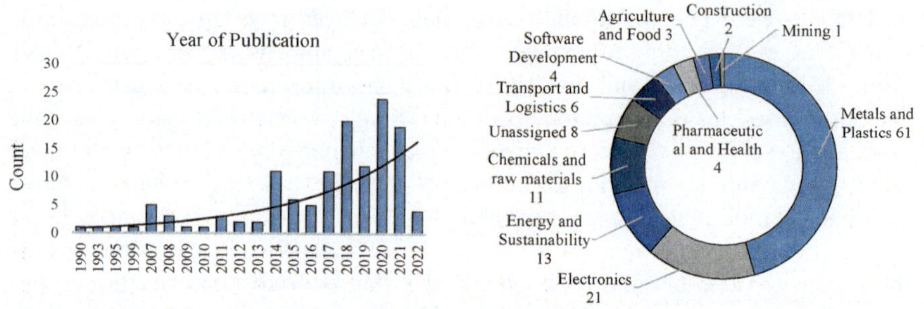

Fig. 4. Quantity of production datasets contained in the table depending on their year of publication (left) and distribution of data sets across industry sectors (right)

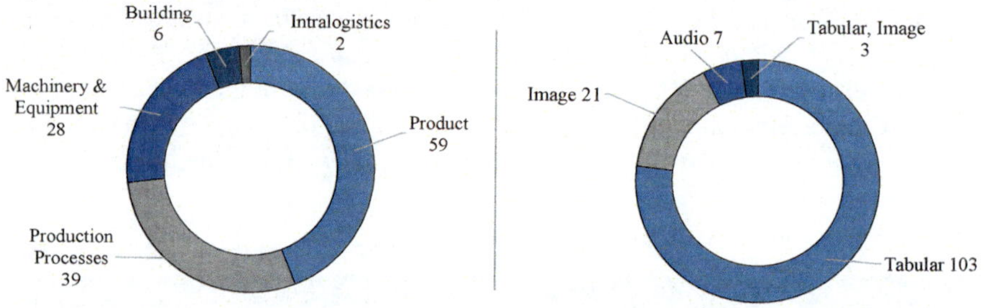

Fig. 5. Distribution of data sets across application areas (left) and data modalities (right)

of production, linking data from the production line as well as the building and the logistic will become increasingly more important in the future. A greater number of corresponding open data sets is therefore required.

3 Development Framework for ML Applications in Production

Once a use case and a suitable data set have been selected, following a development methodology and using dedicated tools is decisive. Because production engineering applications are special due to their criticality and inherent data characteristics, in this section, we present both a comprehensive process model and the concept of a toolbox, that have been specifically tailored for production-related ML applications.

3.1 ML Pipeline as a Process Model for ML Applications in Production

A further key component for developing ML applications is, besides the use case identification arising from both the chosen dataset and a corresponding objective, a standardized development methodology. Motivated by the previously outlined characteristics and resulting challenges of ML applications in production, a domain-specific process model called "Machine Learning Pipeline in Production" [16, 17] has been developed (see Fig. 6). This process model is derived from the CRISP-DM methodology [18] that takes particular account of the conditions and challenges in production technology. It is because of this precise, domain-specific focus that we have selected this process model. The basic procedure is briefly outlined in the following.

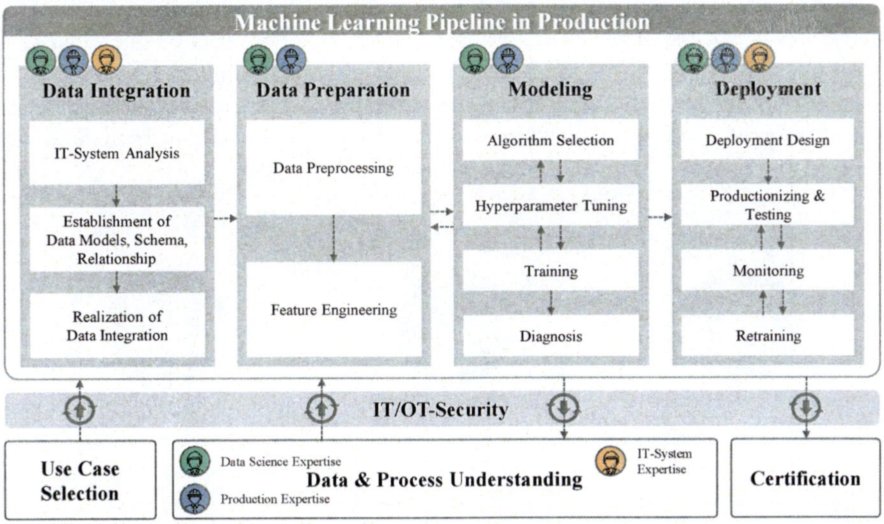

Fig. 6. Machine Learning Pipeline in Production: process model for developing ML applications in production (building on [16, 17])

Starting with the selection of the use case, data science, IT-system, and production experts need to build a common data and process understanding. On this basis, *Data Integration* deals with the digitization of the production process and corresponding management of the acquired data. In data preparation, first, the raw data is pre-processed (e.g., by imputing missing values, encoding nominal, or normalizing numerical features) to improve the quality of the data. The aim of *Feature Engineering* is then to further reduce the dimensionality of the data while maximizing its informational content by selecting certain features or constructing new ones. *Modeling* involves the iterative process of selecting, training, tuning and ensembling of machine learning algorithms with regard to an application-specific performance metric. Once a best-performing model or ensemble of multiple models has been found, *Deployment* deals with integrating the ML models into the production system including the design of the decision support system, productionizing, and testing of all software components, and defining a maintenance strategy. The certification aspect comprises solution approaches as to how ML applications in production can be qualified and finally certified regarding the criteria of transparency, robustness, safety and adaptivity. IT/OT-Security addresses the need for securing digital production systems against unauthorized intruders.

3.2 ML-Toolbox to Bring ML into Production in a Structured Way

In order to implement the presented ML pipeline in a strategic way and to accelerate common tasks in the different stages of the pipeline, the concept of an ML-Toolbox has been developed (Fig. 7). It sets a basis for the required tools, methodologies and techniques that can aid the adoption of ML in an organization in a standardized way. While the actual realization of the toolbox and its content will differ between organizations, it should always provide fundamental tools and methodologies for common tasks such as use case selection, data acquisition, model training, validation, and deployment. The concept of the ML toolbox consists of three main compartments covering all phases of the ML pipeline: *Data Interfaces, Data Science Tools,* and *Application Area Interfaces.* Each compartment contains an array of different tools such as standardized methodologies, software tools, libraries, checklists, best practices, and interfaces. In the following, the compartments and their content is introduced in more detail.

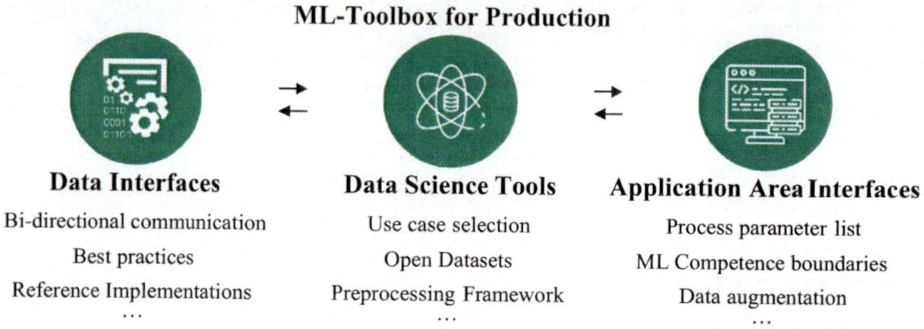

Fig. 7. The ML-toolbox consists of enabling ML tools from three compartments.

Data Interfaces: It ensures the availability of the required raw data and resulting output, as well as compatibility with existing systems. This includes *bi-directional interfaces* to digital twins, simulations or IIoT devices, realized by a single unified access point. In addition to the input data from the machines, the results obtained using data science tools are also accessible by other systems using this interface and may be used to directly influence the running production. To facilitate the integration of proof-of-concepts into real production environments, it also contains *methodologies, best practices,* and *reference implementations* for the deployment of trained models.

Data Science Tools: These include several tools that simplify and speed up common ML and AI activities in production, from selecting promising use cases to interpreting the results. The contained *use case selection tool* enables process experts with little data science knowledge to identify the most impactful use cases within their domain. Prototyping of ML applications without real production data is enabled by the list of *open production datasets* introduced in this paper. A data *preprocessing framework* with flexible implementations of frequently used preprocessing techniques for different types of data is also part of this compartment. Closely related to this, a *data quality assessment tool* provides guidance on possible applications and recommends suitable preprocessing operations. *Best practices* and *checklists* assist the training and *parameter tuning* of ML models, while included pre-trained models from different domains provide a starting point for new models.

Application Area Interfaces: These user-oriented tools are more specific to their respective application area and provide an interface between the process experts of that area and the data scientists. This includes *parameter lists* specific to the process under consideration. These consist of the most important process parameters, their detailed descriptions, and the relations between them, especially regarding the quality of the final product. *Guidelines* help to set the *competence boundaries* of an ML system: What decisions can safely be made by the system and what actions need to be approved by a human? This includes methodologies for *assessing the potential risk* of employing automated ML systems in a particular application area. Further parts of this compartment are data *visualization tools* for common graphical representations, tailored to the application area, as well as data *augmentation tools* that assist the labeling of raw datasets.

4 Need for Further Development

Based on the previously outlined challenges of utilizing ML in production, that are leading to the large gap between academic research and industrial application, in this section, we want to point out the most important research directions. These development directions also result from the authors' many years of experience in the field of digitization and data-driven optimization in industrial production processes.

1) **Sharing Manufacturing Data and Identification of Data-Sparse Areas**
One key aspect, which results from the multitude of specific and therefore difficult to interrelate production processes, is the need to make data available to the public or at least to the respective ecosystem (e.g., the company or factory). Currently, most production companies consider data as sensitive information that cannot be shared due to privacy and secrecy concerns [2]. Therefore, a rethink is required from a rather conservative perspective of treating operational data confidentially towards an open-source mentality. Possible incentives comprise both the exploitation of network effects and data-driven efficiency gains, the change in corporate culture and the improvement of public perception and attractiveness for potential new employees from data-engineering disciplines [19]. A widespread example of unused potential in the production domain is the machine manufacturer who, as soon as he delivers his machine to the customer, no longer has access to the data [19].
2) **Standardizing Data Management in Production Systems**
With the aim of making processes smarter, companies are digitizing their production lines and collecting large amounts of data. For a long time, the approach was to collect highly-distributed data on a large scale and to analyze it in the future with the help of ML algorithms [20]. Many companies have now found that the quality of the data collected in this way is too poor to extract reliable analysis results. This problem is particularly due to an inadequate data management. Data from different sources is not synchronized, and transformed in a standardized way, but dumped in unstructured data lakes [21]. In addition to the definition of industry-wide data management standards, we consider a change in strategy in industrial data projects as required. Data collection should be preceded by the conceptualization and implementation of data models and a data management plan in the future.

3) **Qualification and Certification of ML Applications in Production**
The combination of highly complex and non-transparent model functionalities, high reliability requirements and data quality problems often cause ML applications in production processes to fail [22]. Besides the technological distrust, ML introduces new potential risks, and is therefore more likely to be utilized in systems where its benefits are considered worth the increase of risk [23]. In order to operate in high-stakes environments, such as in production settings, ML systems need to be highly robust, but yet are oftentimes brittle in the face of real-world complexities [24]. To increase the trustworthiness of industrial ML applications among its stakeholders, it will therefore be decisive to be able to prove the fulfillment of a set of qualification criteria. We consider the following four criteria to be of primary importance: *transparency*, *robustness*, *safety* and *adaptivity*. For promoting a widespread adoption, it is therefore necessary to develop algorithms and methods regarding these criteria and to define independent development standards and certification procedures [25].

5 Conclusion

Despite fast paced developments in the area of ML algorithms and an increasing availability of data and computing power [1], industrial production technology is currently still struggling to fully exploit ML-enhanced optimization potentials. We attribute this finding to the specific challenges that are characteristic for ML applications in production: highest reliability requirements, high risk and loss potential, the multitude of heterogeneous real-world data sources and the non-transparency of ML model functionality. To expand and strengthen development and research efforts in this demanding domain, in this paper, we have presented an overview of application areas and application scenarios for ML in production, a comprehensive and expandable collection of openly available production data sets, and a production-specific development framework comprising a process model and an ML-toolbox.

References

1. Wuest, T., Weimer, D., Irgens, C., et al.: Machine learning in manufacturing: advantages, challenges, and applications. Prod. Manuf. Res. **4**, 23–45 (2016). https://doi.org/10.1080/21693277.2016.1192517
2. Jourdan, N., Longard, L., Biegel, T., et al.: Machine learning for intelligent maintenance and quality control: a review of existing datasets and corresponding use cases. https://doi.org/10.15488/11280
3. Kim, D.-H., Kim, T.J.Y., Wang, X., et al.: Smart machining process using machine learning: a review and perspective on machining industry. Int. J. Precis Eng. Manuf.-Green Tech. **5**, 555–568 (2018). https://doi.org/10.1007/s40684-018-0057-y
4. Monostori, L., Kádár, B., Bauernhansl, T., et al.: Cyber-physical systems in manufacturing. CIRP Ann. **65**, 621–641 (2016). https://doi.org/10.1016/j.cirp.2016.06.005
5. Lu, S.C.-Y.: Machine learning approaches to knowledge synthesis and integration tasks for advanced engineering automation. Comput. Ind. **15**, 105–120 (1990). https://doi.org/10.1016/0166-3615(90)90088-7

6. Krauß, J., Dorißen, J., Mende, H., et al.: Machine learning and artificial intelligence in production: application areas and publicly available data sets. https://doi.org/10.1007/978-3-662-60417-5_49
7. Russakovsky, O., Deng, J., Su, H., et al.: ImageNet large scale visual recognition challenge. https://doi.org/10.48550/arXiv.1409.0575
8. Panayotov, V., Chen, G., Povey, D., et al.: Librispeech: an ASR corpus based on public domain audio books. https://doi.org/10.1109/ICASSP.2015.7178964
9. Galvez, D., Diamos, G., Ciro, J., et al.: The people's speech: a large-scale diverse English speech recognition dataset for commercial usage (2021)
10. Lee, S., Jeon, M.: Awesome Public Industrial Datasets. https://github.com/makinarocks/awesome-industrial-machine-datasets
11. National Aeronautics and Space Administration: Data from NASA's Missions, Research, and Activities (2021). https://www.nasa.gov/open/data.html
12. Fraunhofer Institute for Digital Media Technology IDMT Datasets. https://www.idmt.fraunhofer.de/en/publications/datasets.html
13. Harvard Dataverse. https://dataverse.harvard.edu/dataverse/harvard
14. Institute of Electrical and Electronics Engineers IEEE. IEEE DataPort. https://ieee-dataport.org/datasets
15. UCI Machine Learning Repository—Center for Machine Learning and Intelligent Systems UC Irvine Machine Learning Repository. https://archive.ics.uci.edu/ml/index.php
16. Krauß, J.: Optimizing Hyperparameters for Machine Learning Algorithms in Production, 1st edn. Apprimus Wissenschaftsverlag, Aachen (2022)
17. Krauß, J., Pacheco, B.M., Zang, H.M., et al.: Automated machine learning for predictive quality in production. Procedia CIRP **93**, 443–448 (2020). https://doi.org/10.1016/j.procir.2020.04.039
18. Azevedo, A., Santos, M.F.: KDD, SEMMA and CRISP-DM: A Parallel Overview (2008)
19. Otto, B., Mohr, N., Roggendorf, M., et al.: Data sharing in industrial ecosystems: driving value across entire production lines (2020)
20. Wang, J., Zhang, W., Shi, Y., et al.: Industrial big data analytics: challenges, methodologies, and applications (2018)
21. Khan, M., Wu, X., Xu, X., et al.: Big data challenges and opportunities in the hype of Industry 4.0. In: 2017 IEEE International Conference on Communications (ICC), pp. 1–6. IEEE (2017)
22. Multaheb, S.A., Zimmering, B., Niggemann, O.: Expressing uncertainty in neural networks for production systems. Automatisierungstechnik **69**, 221–230 (2021). https://doi.org/10.1515/auto-2020-0122
23. Delseny, H., Gabreau, C., Gauffriau, A., et al.: White paper machine learning in certified systems. https://doi.org/10.48550/arXiv.2103.10529
24. Hendrycks, D., Carlini, N., Schulman, J., et al.: Unsolved Problems in ML Safety. arXiv (2021)
25. St. Clair, A.L., Smogeli, O., Odegardstuen, A., et al.: Trustworthy industrial AI systems: safer, smarter, greener. Group Technology & Research, Position Paper 2019 (2019)

Function-Orientated Adaptive Assembly of Micro Gears Based on Machine Learning

V. Schiller(✉) and G. Lanza

wbk Institute of Production Science, Karlsruhe Institute of Technology (KIT), Kaiserstr. 12, 76131 Karlsruhe, Germany
vivian.schiller@kit.edu

Abstract. The complexity of products is increasing and key functions can often only be realized by using micro components. The requirements of high-precision components often reach technological manufacturing limits. This is of particular importance for micro components with complex geometries, such as micro gears, where manufacturing deviations are relatively large compared to the component size and therefore have a large influence on the functional characteristics of the assembled product. In this paper, an approach is presented to predict and optimize the functional characteristics of assembled micro gear pairs in terms of Noise, Vibration and Harshness (NVH), based on optical in-line measurements of the entire topography of the gears. The overall quality is optimized by individually selecting the gears to be assembled with regard to minimising predefined NVH parameters. For implementation, a large number of possible combinations must be predicted. It is proposed to develop a meta-model with machine learning (ML) methods, which enables the near-real-time prediction of the NVH parameters of micro gear pairs, based on the optical in-line measurements.

Keywords: Micro gear · Quality control · Machine learning · Optimisation · Assembly

1 Motivation

The ongoing trend towards miniaturization has led to an increase in the complexity of products and key functions can often only be realized by using high-precision components. Micro components are crucial components in diverse, complex products that promise increasingly high growth in different industries [1]. The most common mechanical micro components are micro gears, which are used in a wide range of applications, e.g. in the automotive industry, mechanical engineering, measurement technology, robotics, aerospace technology, medical technology or consumer industries [2, 3]. Due to the small component size and the complex three-dimensional geometry, the production of micro gears while maintaining the required quality represents a major challenge for manufacturers. The quality requirements reach technological limits of production processes and the relatively large manufacturing deviations compared to the component size have a major impact on the functional properties of the assembled products [4]. A trade-off

M. Liewald et al. (Eds.): WGP 2022, LNPE, pp. 514–523, 2023.
https://doi.org/10.1007/978-3-031-18318-8_52

between reducing production costs by maintaining an increased throughput at higher deviations and avoiding scrap parts with tighter tolerances exists [5]. To meet this challenge, companies can consider two different approaches. On the one hand, an optimal allocation of tolerances and, on the other hand, an adaptive control of the production system [4, 6].

A potential solution towards adaptive process control might be to develop a metamodel with machine learning (ML) methods, which enables the near-real-time prediction of functional characteristics of assembled micro gear pairs in terms of Noise, Vibration and Harshness (NVH), based on optical in-line measurements of the entire topography of the gears. The development of a corresponding functional model enables the implementation of selective or individual assembly strategies. The overall quality can then be optimized by selecting the gears to be assembled with regard to minimizing predefined NVH parameters.

2 State of the Art

2.1 Micro Gears

There is no clear uniform definition of micro gears available. According to VDI 2731, micro gears are defined as gears that have two of the following three characteristics [3]:

1. Characteristic external dimensions (e.g. diameter or edge length) < 20 mm
2. Module < 200 µm
3. Structural details < 100 µm

The evaluation of gear measurements is carried out for all measuring methods by means of line- and point-based gear characteristics defined in DIN 21771 and VDI 2607. The most common geometric characteristics of gears are the following: profile total deviation F_α; profile form deviation $f_{f\alpha}$; profile slope deviation $f_{H\alpha}$; helix total deviation F_β; helix form deviation $f_{f\beta}$; helix slope deviation $f_{H\beta}$; cumulative pitch deviation F_p; single pitch deviation f_p [7, 8].

In order to extract all relevant characteristics of gears, function-oriented parameters can be determined in addition to geometrical parameters. Opposite to geometric parameters, all teeth on the circumference of a gear have an effect on the measurement result. A method for deriving function-oriented parameters is the single flank rolling test, in which the gear to be tested is rolled against a master gear. The master gear is to serve as an ideal partner and is to be manufactured accordingly with minimal deviations. The axes of the two gears are fixed. Both axes are equipped with rotary encoders and the deviations of the nominal positions from the driving (master gear) to the driven gear (test gear) are measured. The drive speed must be selected slow enough during the test so that no dynamic effects occur. The single flank rolling test provides four parameters as a result, which characterize the non-uniform kinematic transmission (see Fig. 1). The single flank rolling deviation F_i' is the deviation of the actual rotational positions compared to the nominal rotational positions. It is calculated as the difference of the largest leading and the largest lagging rotational position deviation. The single flank tooth-to-tooth deviation f_i' is the largest occurring rotational position deviation, which are determined individually within the rotational angle of a single tooth meshing. The long-wave component f_l' is determined by averaging out a polyline in which the short-wave components are suppressed and then calculating the difference of the largest leading and the largest lagging rotational position deviation. To ensure comparable evaluation results, the calculation should be based on an uniform cut-off wavelength. The short-wave component f_k' results from the differences between the recorded test data and the calculated averaging polyline [9].

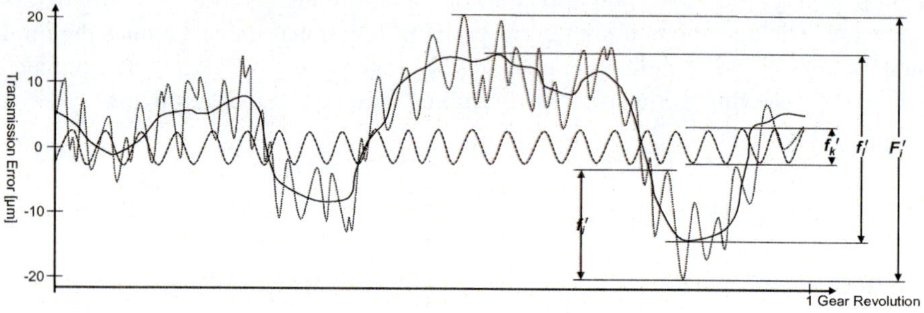

Fig. 1. Evaluation of the single flank rolling test parameters

2.2 Quality Control Loops

Quality control loops describe functional product model-based control strategies for optimizing production, aimed at reacting to process deviations and increasing product quality. These are based on a closed-loop approach at the organizational level of a production system and are suitable for production processes where the technological limits have been reached and no error reduction can be achieved by means of conventional methodologies [10, 11]. Developments in the context of Industry 4.0 in terms of near-real-time information processing, new sensors for individual component tracking and sensor technology for production-integrated measurements of quality data enable the recording of component-specific quality data [12]. Various control loop concepts for increasing product quality can be implemented with the quality controller taking process parameters, measurements, control strategies and a product model into account (see Fig. 2) [13]. For existing control loops, different assembly and adaptation strategies are possible as production-related reaction measures to increase quality despite production deviations. The main idea lies in the individual over-fulfillment of a component feature, wherein the possibility exists to compensate for a quality-critical feature deviation of a second component. By compensating the associated components, tighter tolerance ranges can be realized compared to conventionally assembled components. [12, 13].

In the adaptive manufacturing control loop, target values in upstream production processes of individual components are statistically adjusted to meet quality requirements. Processes can be controlled based on models that link the results of function-orientated in-line or in-process measurements with process parameters [10].

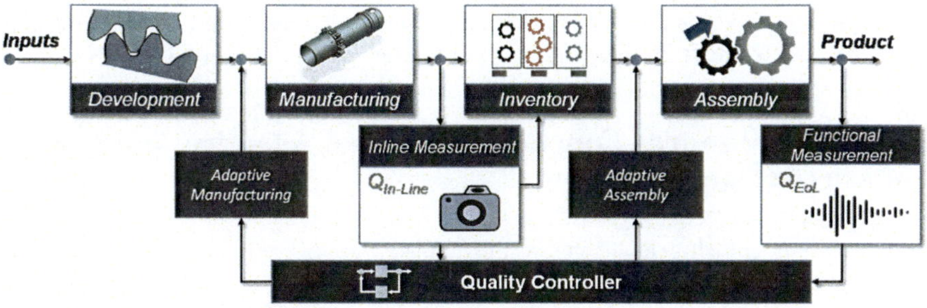

Fig. 2. Quality-based control cycles based on [10]

In the quality control loop of adaptive assembly, suitable partners are selected to compensate for deviations [12, 14]. Adaptive assembly is used as a generic term that includes individual and selective assembly strategies. Tan & Wu discuss two methods for pairing optimization: Direct Selective Assembly (DSA) and Fixed Bin Selective Assembly (FBSA) [15]. FBSA belongs to selective assembly strategies, which is a method of dividing individual components into several tolerance classes based on their individual deviation from a specific target value and then pairing them with a corresponding component of another tolerance class. When components are divided into classes, information about the exact values of the measurements is lost with both the number of classes and the width of the classes determining the degree of information loss [16]. DSA on the contrary is an algorithm that finds an optimum for the available parts and each part is assigned exactly one other part and thus belongs to individual assembly strategies. For individual assembly, the components are not divided into tolerance classes. The exact measurements are linked to a unique component ID and accordingly no information is lost. Individual assembly strategies thus lead to an increase in product quality compared to selective assembly strategies, but are also associated with higher costs, as individual traceability and storage of the components must be ensured and implemented into the production system [17].

2.3 Functional Prediction

To enable adaptive assembly strategies, models are needed that enable functional predictions based on production data. With the help of these models, the assembly strategies can be implemented by identifying how the quality of different component combinations differ from one another. For instance, Berthold et al. investigate gear rattling at the drive train of an internal combustion engine, using simulations to study the cause of the noise emission and the relevant operating points [18]. Furthermore, the simulation is validated against physical measurements and the results show that the simulation is suitable to perform parameter optimization. Wu & Yan et al. use a functional product model to sort components and group them into different quality levels in order to predict the gear machining precision [19]. This group assignment then serves as a label and is tested for significance with further machine data. Subsequently, different regression algorithms are applied and compared on a previously defined quality metric. Denecke uses regression with polynomials and KNN to predict the flow coefficients for labyrinth seals of thermal flow machines [20]. The aim is to derive measured and literature data of the flow coefficient from the geometry parameters of the labyrinth seal.

3 Approach Towards Function-Orientated Adaptive Assembly of Micro Gears

Geometric manufacturing deviations, as the main sources of excitation of gear vibrations, play a central role for the final product quality with respect to NVH [21]. This is of particular interest for micro gears where the manufacturing deviations are relatively large in relation to the component size [9]. One possible solution is to develop an approach for intelligent optimization of micro gear assembly in terms of quality control loops based on geometric measurement data. For developing adaptive assembly strategies, many approaches can be found in the literature for - compared to the assembly of gears - simple geometric relations. The quality delta of gears is non-linear and depends on a large number of parameters. Therefore, algorithms designed to optimize the quality must meet these requirements. In the case of individual assembly, the output of the control algorithm is an instruction as to which gears are to be assembled together. No approaches exist yet in the literature that take into account the trade-off between maximum part usage and maximum quality. For selective pairing, there are groups to which parts are assigned. These groups are each assigned to at least one pairing group and within these groups, the components are randomly selected for subsequent assembly. The literature on selective assembly deals with linear problems, a small number of dependent parameters, or with known, simple tolerances. There is no approach yet that addresses the complex and highly non-linear constraints of micro gears. Further, no approaches have yet been identified in the literature for the development of functional models for micro gears. In order to address the gaps identified in the literature, this paper proposes an approach towards functional models of assembled micro gear pairs and thus enable adaptive assembly strategies of micro gears to be studied and devolved.

Gauder et al. developed a general approach for function-oriented quality assurance of micro gears with respect to noise behavior [22]. However, the focus of the publication is on the uncertainty evaluation of optical 3D gear measurements and the evaluation of the information content of areal parameters in comparison to nominal parameters. Building on the work of Gauder et al., a detailed approach towards adaptive assembly strategies for function-oriented pairing of micro gears is presented below (Fig. 3). In order to keep the results general, a reference data set consisting of micro gears with different numbers of teeth and modules in the range of 0.2–0.5 mm is used as a basis. When creating the reference data set, the micro gears must be taken randomly from the production processes so that the gears contained in the reference data set reflect the real production deviations that occur over time. The surfaces of the micro gears are then measured using optical focus variation technology with a sensor of type μCMM from Bruker Alicona. When selecting a suitable sensor, both the suitability for 100% in-line measurements and the resolution necessary for measuring the surfaces of micro gears are relevant. To enable in-line integration, the trade-off between the measurement uncertainty and measurement time must further be taken into account [23].

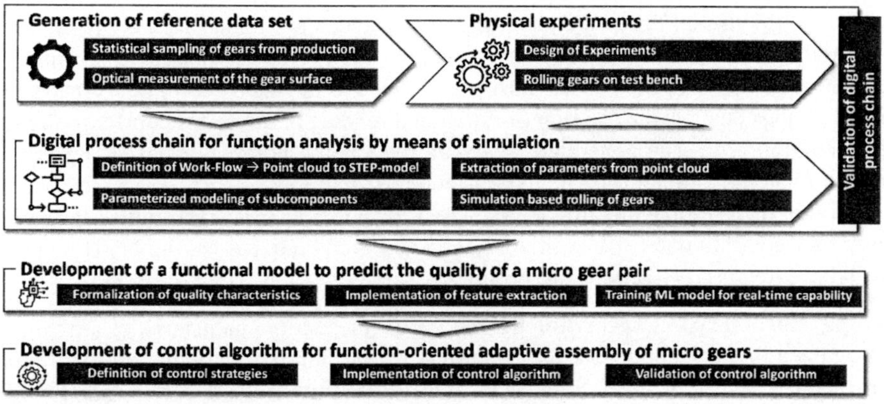

Fig. 3. Approach towards the development and validation of function-orientated adaptive assembly strategies for micro gears

The measured point clouds then serve as an input into a digital process chain. This process chain is divided into two main parts. On the one hand, it is used to convert the point clouds into STEP models in order to simulate the rolling of two gears using commercial simulation software. The output of the simulation is the angular position of the driven gear in relation to the angular position of the driving gear. The complexity of the simulation can be extended as desired and, in addition to the tooth geometry, other functionally relevant component information can be taken into account (see Fig. 4).

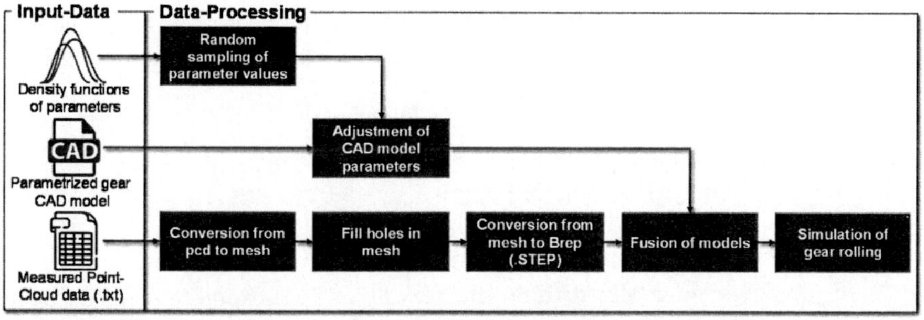

Fig. 4. Simulation approach to incorporate tolerance chains across multiple components using parameterized models

On the other hand, the digital process chain is used to extract geometric parameters according to DIN 21771, VDI 2607 or VDI 2612, and function-oriented parameters according to VDI 2608 from the point clouds.

On the basis of the parameters extracted from the point clouds, rolling tests can then be planned, carried out and evaluated by means of Design of Experiments. The experiments are carried out on a purpose-built micro gear test bench from Frenco GmbH, which is suitable for gears with modules < 0.5 mm. The measurements performed on the test bench serve to compare different simulation programs as well as to validate the developed digital process chain. Once validated, the digital process chain can be used to efficiently develop control strategies in terms of costs, material and time. The generation of learning data on the basis of virtual and deviated gear models using Skin Model Shapes enables the training of functional models of micro gears prior to the start of production [24]. The generated gear data reflects the nominal shape as well as the process-related manufacturing deviations and the random scatter.

In order to develop a functional model to predict the quality of assembled micro gear pairs, relevant parameters must first be defined which characterize the NVH behavior. A total of 2 feature extractions are proposed.

Rolling characteristics according to VDI 2608 - these four characteristics are selected analogously to the function-oriented values from the standard, with the only difference that in the given application two gears with deviations roll off each other [9]. The calculation for the single flank rolling deviation F_i' is done according to Eq. 1, with v_{mag} representing the angular velocity.

$$F_i' = max(v_{mag}) - min(v_{mag}) \tag{1}$$

Using the known nominal rotation speed v_{nom} and number of theeth n_{theeth}, the tooth meshing time t_{tooth} must first be calculated (Eq. 2) in order to subsequently determine the single flank deviation (Eq. 3).

$$t_{tooth} = \frac{n_{theeth}}{v_{nom}} \tag{2}$$

$$f_i' = max((max(v_{mag}) - min(v_{mag}))(t_{tooth})) \tag{3}$$

To determine the long-wave component f_l' (Eq. 4), a polyline v_{poly} must be calculated by averaging over a window width of three tooth meshes as suggested in VDI 2608.

$$f_l' = max(v_{poly}) - min(v_{poly}) \tag{4}$$

The short-wave component f_k' can subsequently be calculated according to Eq. 5.

$$f_k' = max(v_{mag} - v_{poly}) - min(v_{mag} - v_{poly}) \tag{5}$$

Integrated jerk over normalized time - the jerk j_{mag} of a movement is the derivative of the acceleration with respect to time (Eq. 6) and the integrated jerk j_{int} is obtained using Eq. 7 and accordingly serves as an approximate for the total energy of impacts. Since the time durations of various number of teeth ratios differ, it is necessary to divide by the duration $t_{duration}$.

$$j_{mag} = \frac{d^2 v_{mag}}{dt^2} \tag{6}$$

$$j_{\text{int}} = \frac{\int_0^{t_{max}} j_{mag} dt}{t_{duration}} \tag{7}$$

The use of a functional model based on ML methods is suitable for predicting the defined characteristics in a running production environment, as ML models are be able to predict the function and make decisions with regard to the product and the assembly process without the use of costly and time-consuming physical test benches or simulation models. Both function prediction by measurement and function prediction by simulation are not applicable due to time and cost constraints. To illustrate, the number of possible combinations of two times 100 gears is 20,000 for the operation of one gear stage in both directions of rotation. The geometric and functional parameters derived from the point clouds can be used as input into the functional-orientated model, and various ML models such as Artificial Neural Network, Support Vector Regression, K-Nearest Neighbor Regression or Decision Tree Regression can be compared with each other in order to select the most suitable one. Once a suitable functional model has been identified, it can be used to investigate the challenges of selective and individual assembly strategies of micro gears mentioned above and suitable solutions can be developed. The overall data processing work-flow is visualized in Fig. 5.

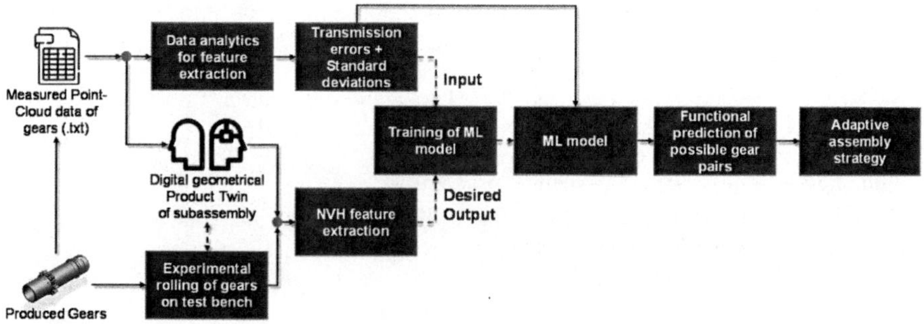

Fig. 5. Data work-flow for the development and use of adaptive assembly strategies

4 Summary

Within the scope of this work, an approach for predicting the functional characteristics of micro gear pairs is developed, which enables adaptive assembly strategies. To increase flexibility as well as reduce both costs and time for future developments, a digital process chain for generating reference datasets is proposed. Furthermore, parameters are defined that can be used to characterize and quantify the quality of gear pairs with respect to NVH. It is proposed to develop function-oriented models based on ML methods that allow the near-real-time prediction of the quality parameters and, accordingly, the use and development of both individual and selective assembly strategies.

The focus of future investigations will be on the proof-of-concept regarding the capability of function-orientated models, which use parameters derived from the measured point clouds as input to predict the defined NVH parameters. These will be determined both by means of physical experiments and simulations, thereby enabling the digital process chain to be validated at the same time. Further, the nominal geometric parameters according to DIN 21771 and VDI 2607 are studied in terms of information content with respect to their usability in combination with function-orientated models. To this end, an aerial evaluation of the entire flank surfaces with suitable parameters according to Pfeifer et al. [25] or Goch [26] might be required to adequately describe the defects occurring on the flank surfaces.

Acknowledgments. This research and development project is funded by the Deutsche Forschungsgemeinschaft (DFG, German Research Foundation)—Project-ID 431571877. The authors thank the DFG for this funding and intensive technical support.

References

1. Arndt, O., Hennchen, S.: Wertschöpfungs-und Wettbewerberanalyse für den Spitzen-cluster. MicroTEC Südwest **8** (2011)
2. Slatter, R.: Mikroantriebe für präzise Positionieranwendungen. Antriebstechnik **42**(6) (2003)
3. VDI 2731: Microgears—Basic Principles. Beuth Verlag, Berlin (2014)
4. Lanza, G., Haefner, B., Kraemer, A.: Optimization of selective assembly and adaptive manufacturing by means of cyber-physical system based matching. CIRP Ann. **64**(1), 399–402 (2015)
5. Lanza, G.: Resilient production systems by intelligent loop control. In: Excellezenzcluster Integrative Produktionstechnik für Hochlohnländer. Aachen (2016)
6. Dantan, J.-Y., Eifler, T.: Tolerance allocation under behavioural simulation uncertainty of a multiphysical system. CIRP Ann. **70**(1), 127–130 (2021)
7. DIN 21771: Zahnräder, Zylinderräder und Zylinderradpaare mit Evolventenverzahnung – Begriffe und Geometrie. Beuth Verlag, Berlin (2014)
8. VDI 2607: Rechnerunterstützte Auswertung von Profil- und Flankenlinienmessungen an Zylinderrädern mit Evolventenprofil. Beuth Verlag, Berlin (2000)
9. VDI 2608: Einflanken- und Zweiflanken-Wälzprüfung an Zylinderrädern, Kegelrädern, Schnecken und Schneckenrädern Handbuch Messtechnik II. Beuth Verlag (2001)
10. Wagner, R., Haefner, B., Lanza, G.: Function-oriented quality control strategies for high precision products. Procedia CIRP **75**, 57–62 (2018)
11. Mease, D., Nair, V., Sudjianto, A.: Selective assembly in manufacturing: statistical issues and optimal binning strategies. Technometrics **46** (2004)
12. Colledani, M., et al.: Design and management of manufacturing systems for production quality. CIRP Ann. **63**(2), 773–796 (2014)
13. Schmitt, R., Niggemann, C., Isermann, M., Laass, K., Matuschek, N.: Cognition-based self-optimisation of an automotive rear-axle-drive production process. J. Mach. Eng. **10**(3), 68–77 (2010)
14. Tsutsumi, D., et al.: Towards joint optimization of product design, process planning and production planning in multi-product assembly. CIRP Ann. **67**(1), 441–446 (2018)
15. Tan, H.Y., Wu, C.F.: Generalized selective assembly. IIE Trans. **44**(1), 27–42 (2011)
16. Ebrahimi, D.: Integrated quality and production logistic performance modeling for selective and adaptive assembly systems, Ph.D. Thesis, Politecnico di Milano (2014)
17. Meyer, A., et al.: Concept for magnet intra logistics and assembly supporting the improvement of running characteristics of permanent magnet synchronous motors. Procedia CIRP **43**, 356–361 (2016)
18. Berthold, B., Hierlwimmer, P.: Vorhersage und Analyse von Getrieberasseln. ATZ Automobiltechnische Zeitschrift **118**(9), 60–65 (2016)
19. Wu, D., Yan, P., Guo, Y., Zhou, H., Chen, J.: A gear machining error prediction method based on adaptive Gaussian mixture regression considering stochastic disturbance. J. Intell. Manuf. 1–19 (2021)

20. Denecke, J.: Rotierende Labyrinthdichtungen mit Honigwabenanstreifbelägen: Untersuchung der Wechselwirkung von Durchflussverhalten, Drallverlauf und Totaltemperaturänderung. Ph.D. Thesis, Universität Karlsruhe (2008)
21. Klocke, F., Brecher, C.: Zahnrad- und Getriebetechnik. Auslegung - Herstellung - Untersuchung - Simulation. Hanser, München (2017)
22. Gauder, D., Wagner, R., Gölz, J., Häfner, B., Lanza, G.: Funktionsorientierte Qualitätssicherung von Mikrozahnrädern hinsichtlich des Geräuschverhaltens.tm. Technisches Messen **86**(9), 469–477 (2019)
23. Gauder, D., Gölz, J., Biehler, M., Diener, M., Lanza, G.: Balancing the trade-off between measurement uncertainty and measurement time in optical metrology using design of experiments, meta-modelling and convex programming. CIRP J. Manuf. Sci. Technol. **35**, 209–216 (2021)
24. Schiller, V., Gauder, D., Gölz, J., Bott, A., Wannenwetsch, M., Lanze, G.: Generation of artificial learning data to train functional meta-models of micro gears. In: 17th CIRP CAT Conference, Metz (2022)
25. Schmitt, R., Pfeifer, T., Naplerala, A.: 3D-Abweichungsanalyse komplexer Zahnflankengeometrien: Die Spreu vom Weizen trennen. QZ Qualität und Zuverlässigkeit **50**(10), 89–91 (2005)
26. Goch, G., Ni, K., Peng, Y., Guenther, A.: Future gear metrology based on areal measurements and improved holistic evaluations. CIRP Ann. **66**(1), 469–474 (2017)

Data Mining Suitable Digitization of Production Systems – A Methodological Extension to the DMME

L. Drowatzky(✉), H. Wiemer, and S. Ihlenfeldt

Institute of Mechatronic Engineering, Technische Universität Dresden, 01062 Dresden, Germany
lucas.drowatzky@tu-dresden.de

Abstract. In many conventional areas in mechanical engineering, such as mechanical design, there are process models for engineers like VDI 2221 that guide through the process with methodological support, provide criteria for evaluating the results and thus ensure quality. Generalized process models such as CRISP-DM, KDD and SEMMA already exist for Data Mining, as well as DMME, DAPLOM or ISO 17359 specifically for production engineering. However, these only focus on the sequence of the necessary tasks in several phases without naming adapted methods or without considering aspects of data analysis. Furthermore, the transferability to new use cases or the reuse of the developed solutions has not yet been addressed. In this paper, based on the stages of the DMME, adapted methodical guidelines for enabling machines to acquire data that is suitable for Data Mining are provided. The methods focus on the identification and prioritization of analysis goals and the design of measurement chains and experiments for the acquisition of training data based on the process and the machine structure. In terms of reusability, approaches to transfer the results into templates will be discussed. The methods are applied in a condition monitoring project for a concrete mixing machine.

Keywords: Usable artificial intelligence · Data mining workflow · Condition monitoring · DMME · Digitization · Data mining orientated engineering · Data mining in production technology · AI-ready engineering · Data mining process model

1 Introduction

In the context of "Industry 4.0" and "Internet of Things" (IoT), the importance of using machine and process data has increased significantly. Many companies have realized the technical and economic potential for the application of data mining (DM) in the production environment, but have little or no experience in developing their own applications. In order to enable these companies to autonomously upgrade existing machines or digitize new machines, it is essential to provide methodological support in addition to process models. A major challenge in the engineering sector, in contrast to other domains, is that often no data basis for data mining is available and must first be generated. Specific

M. Liewald et al. (Eds.): WGP 2022, LNPE, pp. 524–534, 2023.
https://doi.org/10.1007/978-3-031-18318-8_53

requirements for data acquisition must be met, such as visibility of effects or freedom from disturbances, otherwise the predictive accuracy of data analysis algorithms cannot be satisfactory. Since the validation of the data set based on the predictive accuracy of a data model is done only at the end of a DM development process, adapted methods must be provided to assure the quality of the results of each step. This paper proposes a methodological support for the part of digitization and dataset generation as an extension of the DMME process model.

2 Existing Process Models for Data Mining

For data mining without reference to production engineering, the three process models KDD (Knowledge Discovery in Databases) [1], SEMMA (Sample-Explore-Modify-Model-Assess) [2] and CRISP-DM (Cross-Industry Standard Process for Data Mining) [3] are often mentioned. In the meantime, modifications and supplements of these models exist in order to provide corresponding support for engineering projects. All process models divide projects for data-driven analysis into different phases and tasks, which must be processed sequentially or iteratively. The most relevant process models are presented below.

2.1 KDD, SEMMA and CRISP-DM

KDD and SEMMA each begin with a phase for reviewing the entire existing data set, followed by the selection of data, the phase of data pre-processing through to model building and thus the analysis. Finally, the results are evaluated and interpreted. The support of the user is limited to the subdivision of a project into the individual phases as well as the definition of at least one goal, which must be achieved in each phase. Due to the abstraction of the process model with respect to general applicability for many domains, no specific instructions or methods for the fulfillment of the goals are provided.

CRISP-DM was developed for the industrial context and extends the previous process models KDD and SEMMA. Here, too, an existing data set is assumed. However, the process model explicitly adds a phase for the definition of an economic analysis objective (*Business Understanding*) at the beginning. Furthermore, the phases are subdivided into tasks, thus providing the user with somewhat better support. Nevertheless, specific methods for the fulfillment of the individual objectives are missing. While KDD and SEMMA are sequential, CRISP-DM is characterized by an iterative structure. This enables an intermediate control of results and, if necessary, the repetition of individual tasks to achieve the intermediate goals. In the engineering context, a phase for data generation/digitization is missing on the one hand, and a precise listing of suitable methods for processing all tasks on the other.

2.2 DMME, DAPLOM and ISO 17359

As an extension to the CRISP-DM, the process model DMME (Data Mining Methodology for Engineering Applications) [4] exists as well as DAPLOM [5] as an extension of the DMME to processes. In the area of general condition monitoring of machines, the standardization ISO 17359 [6] exists as a process model. Figure 1 shows the three development guidelines with the sequence of the individual project phases. Essentially, all models include phases for defining the economic objective with cost-benefit analysis. This is followed by phases for data generation (highlighted in gray in Fig. 1) with investigation of the cause-and-effect relationships and fault symptoms, as well as the development and implementation of data acquisition with final execution of tests. Thus, the first fundamental deficit of the missing phase for the generation of data sets of CRISP-DM, KDD and SEMMA is already solved. Finally, phases for data pre-processing, the development of analysis algorithms and validation follow.

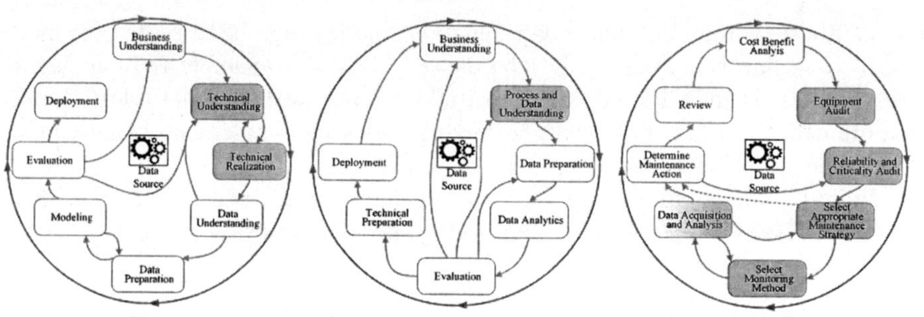

a) DMME b) DAPLOM c) ISO 17359

Fig. 1. Comparison of engineering-DM-process models DMME, DAPLOM and ISO 17359 (highlighted in gray are the phases related to digitization and data generation)

This article will focus on the digitization of plants and machines for the creation of an AI-ready data basis. This is addressed in the DMME by the *Technical Understanding* and *Technical Realization* phases, in the DAPLOM model by the *Process and Data Understanding* phase and in ISO 17359 starting with the *Equipment Audit* phase up to *Data Acquisition and Analysis.* Since DAPLOM is an extension of the DMME with a focus on manufacturing processes, the tasks hardly differ and are therefore not considered separately.

Figure 2 compares the phases and tasks of the DMME with those of ISO 17359 and assigns the methods of the ISO 17359 standard to the tasks. The ISO standard addresses the system analysis and measurement system development without going into more detail on data analysis. However, specific methods for fulfilling the tasks are named. The DMME, in turn, covers the holistic workflow including data mining, but so far without specific methodological support. By extending the DMME with the methods of ISO 17359, among others, the second deficit of the lack of methodological support for the digitization of plants in a way that is suitable for data mining can be remedied for users.

3 Concept for Method-Supported Process Model DMME

The tasks of the DMME for the design of a digitization solution are now to be supported by suitable methods in the following. The ISO standard 17359 forms the basis for this,

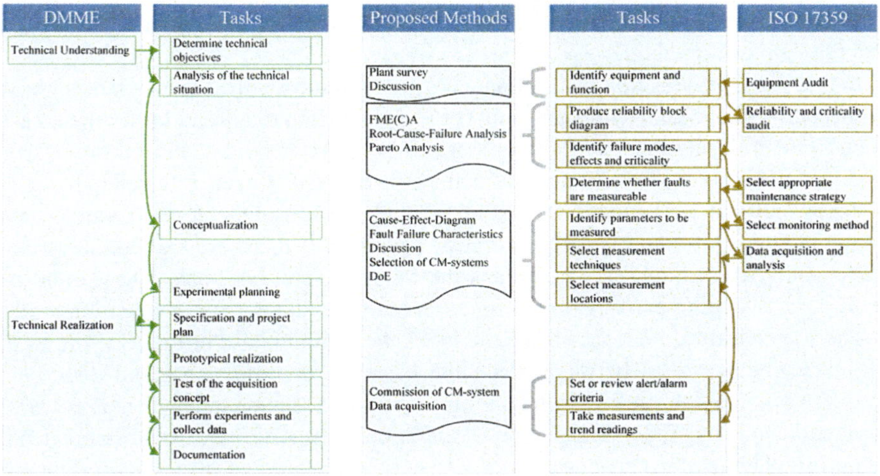

Fig. 2. Comparison on phases, tasks and proposed methods targeting digitization between DMME and ISO standard 17359

as it already specifies many methods. Alternatively, established methods from practical examples or other standards are researched, which could be successfully applied. Figure 3 shows the methods to be used for the design phase *Technical Understanding* of the DMME and its tasks. The first task *Determine technical objectives* does not exist explicitly in ISO 17359, but for the determination of technical objectives with an adapted method of FMEA or FMECA, reference is made to the standard DIN EN 60812 [7] or ISO 13379-1 [8]. These standards include the listing of problems and defects from the process (Process FMEA) or from the machine (System FMEA) as well as criteria for evaluation in order to prioritize them. The result of this step is then at least one technical objective or a technical use case as an analysis task.

In the task *Analysis of the technical situation*, ISO 17359 proposes a component-by-component detailing of the preceding FMEA/FMECA. This method specifies occurring errors, but criteria for the description of error patterns or error symptoms are missing. A method that can fulfill this requirement is found in the ISO standard 13379-1 with the FMSA (Failure Mode and Symptom Analysis). The result of this step is a list of possible faults with effects and symptoms for the analysis application for all relevant process and machine components, as well as an evaluation via a monitoring priority number (MPN). The error symptoms represent a first possibility for the identification of measured variables for the data acquisition. However, the method cannot identify further influencing and disturbing variables on the measurand. Frequently, the fault symptoms such as vibration frequencies or force/torque changes are also subject to other influences such as the process load or environmental conditions. For the selection of suitable sensors and the design of tests, in order to be able to make a reliable prediction of the fault conditions later, further investigation is required. This can be done, for example, by means of the Ishikawa method or root-cause-diagram. In different projects, e.g. in the development of a condition monitoring application for a thermal power station [9], exactly these methods are used for this purpose. The result of the task is an assignment of physical measurands from the FMSA, as well as influence and disturbance variables from the Ishikawa method.

In the *Conceptualization* task, a final concept for data acquisition is to be developed. For this purpose, the fault symptoms must be translated into requirements for the measuring equipment. Using the example of monitoring wear-related faults in rolling bearings [10, 11], a method for quantifying fault patterns is frequently used, referred to in the following as fault pattern analysis. As a rule, the fault patterns depend on parameters such as the shape/size of components or process variables. If these are known, an estimate can be made, for example, for the measuring range or sampling rates. These then form the basis for checking existing data sources or designing a new data acquisition system.

The *Experimental Planning* task concludes the design phase of digitization. In this task, experiments are to be planned on the basis of all influencing and disturbance variables as well as the various fault conditions in order to have a labeled database for the subsequent analysis model. The statistical method DoE has already been established for this purpose [12]. An experiment plan is derived by varying and combining influencing and disturbance variables as well as the various fault conditions, which should later enable a reliable prediction of permissible and impermissible conditions.

In addition to methodical user support, however, the consideration of reusability aspects is also becoming increasingly relevant. Currently, individual solutions are often developed that are difficult or impossible to transfer to the same or similar machine types. The majority of the results of the investigation are independent of the specific machine and are suitable for reuse. The assumption here is that the respective partial results can be transferred into a template specific to the application. If another project with a similar use case is planned later, only an adaptation is required instead of a complete redevelopment.

After the initial task *Determine technical objectives*, various objectives are described in the project. Each objective represents its own use case, for each of which a template can be created. The subsequent description of the error states with their symptoms is generic and depends only on the rough machine structure. If the structure is similar, this result can be adopted directly later.

The requirements analysis for data acquisition in the *Conceptualization* task is system-dependent, but the qualitative error patterns remain the same and mathematic calculations remain valid. Since the requirement analysis often depends directly on quantitative system or process parameters, the estimation can be transferred by substitution. Thus, the result can be reused without major effort and a designed measurement system can be tested and scaled with little effort.

The same applies to the experimental plan from the Task *Experimental Planning*. The defined test variables are mainly generic, so that the structure of the experiment plan can be retained. Only the quantitative experimental parameters have to be adapted.

4 Application on a Predictive Maintenance Usecase of a Concrete Mixing Machine

As part of an industrial research project, a predictive maintenance application is to be developed to increase the availability of mixers. The machine used is a smaller KKM-L 30 laboratory mixer from Kniele. Since the company manufactures mixers for different

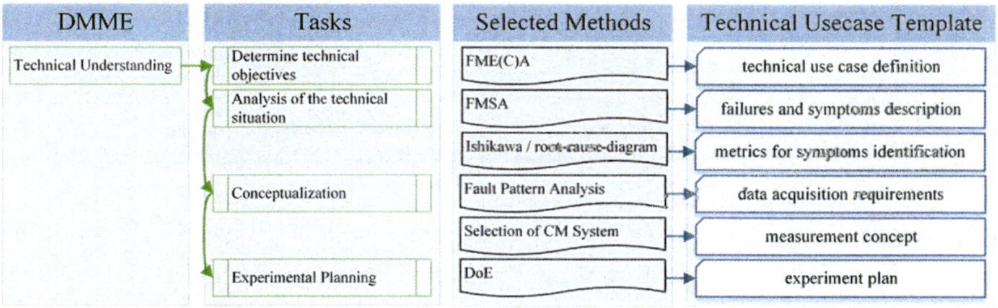

Fig. 3. Methodical approach to design a use case-orientated, reusable DM-suitable digitization

applications and in different sizes, transferability aspects play an important role. The availability-critical part of the mixer is essentially limited to the drive system of the stirrers. A schematic representation can be found in Fig. 4.

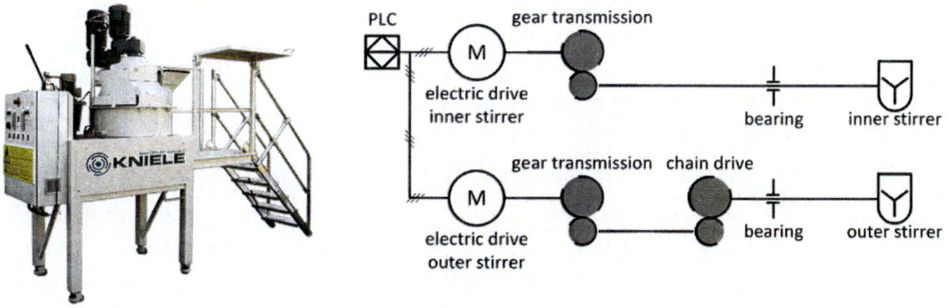

Fig. 4. Similar mixer Kniele KKM-L 100 [13] and schematic representation of the drive system of the mixer

Determine Technical Objectives – FME(C)A:

Using the example of the mixer, availability-critical mechanical faults mainly occur on the guides and transmissions of the stirrer drives as well as the scrapers and blades during operation. Prioritization is carried out according to standard 13379-1 using the FMEA/FMECA method according to categories such as frequency of occurrence, operational safety, costs, influence on product quality and downtime. The FME(C)A aims to list and prioritize machine and process problems. The evaluation in these categories is based on experience reports. A tabular form of this analysis is illustrated in Table 1.

Analysis of the technical situation – FMSA:

Next, an FMSA is performed for the task Analysis of the technical situation according to the ISO 13379-1 standard. The selection of the technical objectives from the previous task is now further detailed by breaking it down into relevant assemblies and components. Subsequently, fault condition types, causes and symptoms are assigned to the corresponding components. The evaluation of the monitorability by means of the monitoring priority number (MPN) is carried out qualitatively by estimation and discussion

Table 1. Determine technical objectives derived from ISO 13379-1

Issue with - Availability - Performance - Quality	frequency of occurrence	downtime	maintenance costs	Impact on safety	Impact on product quality	priority
failure of outer stirrer	+	++	++	+	++	720
failure of inner stirrer	+	++	++	+	++	720
marginal lubrication chain drive	+	+	-	o	o	288
failure of bearings	o	++	++	+	+	432

and is similar to the RPN of the FMEA. A high MPN is the measure for good monitorability and relevance. In Table 2, the FMSA is applied as an example for a selection of defects.

Analysis of the technical situation – Ishikawa Diagram:

The FMSA has assigned symptoms to each fault condition. Measured variables can already be derived from this, but they are also influenced by other machine parameters. In the example of the mixer system, the vibration behavior of the agitators and bearings as well as the respective drive forces are of particular interest, since these allow conclusions to be drawn about the fault conditions. However, the vibrations of a neighboring machine falsify the measurement of the bearing vibrations. If there is mixing material in the mixer cone, this also has an influence on the drive forces. In order for a data analysis model to be able to reliably distinguish an anomaly from a normal condition, such influences are identified by means of an Ishikawa diagram. Figure 5 shows such an Ishikawa diagram for the detection of the bearing wear condition by means of vibration measurement.

Conceptualization – Fault Pattern Analysis:

The fault condition symptoms of the FMSA must now be converted into requirements for the design of the measuring system. For this purpose, worst-case estimates are made for the various operating states. Using the example of the bearings, the maximum speeds of the stirrers in nominal operation are used and the characteristic fault frequencies of the bearings are calculated as in [10]. From this, requirements such as measuring range and sampling rate can be derived. Figure 6 illustrates that for a bearing.

Conceptualization – Selection of CM-System:

The measured variables torque and vibration were identified to detect the faults in the bearings and stirrers. Since a PLC and intelligent frequency inverters are already installed, the speed and motor current as a proxy for the torque can be recorded at high frequency from these data sources. For measuring the vibration at the bearing, an acceleration sensor is retrofitted for each bearing, which is also read out at high frequency by means of an IO unit via the PLC. Both measuring systems meet the defined requirements of fault pattern analysis.

Experimental Planning – DoE:

Finally, the corresponding experimental variables are extracted from the Ishikawa diagram. A guideline for the development of such experimental designs is discussed in

Table 2. FMSA applied on a concrete mixing machine

part	failure mode	failure effect	failure cause	failure symp-toms	likelihood of detection	likelihood of severity	confidence levels of diagnoses	confidence levels of prognoses	monitoring priority number
outer stirrer	rupture	inhomogeneous mixture, contamination	wear, jamming	vibration; torque	4	3	3	3	108
inner stirrer	rupture	inhomogeneous mixture, contamination	wear, jamming	vibration; torque	4	3	3	3	108
bearings	marginal lubrica-tion	lack of guidance, breakdown of the bearing	lack of maintenance	torque	2	2	1	1	4
	wear	lack of guidance, breakdown of the bearing	pitting	vibration	3	2	2	2	24
…	…	…	…	…	…	…	…	…	…

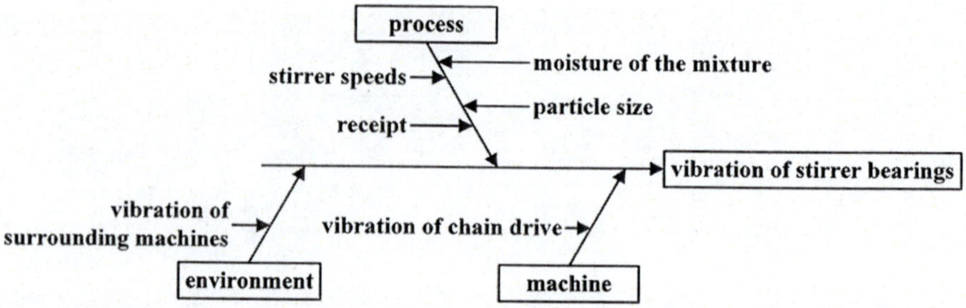

Fig. 5. Ishikawa diagram to detect dependencies of failure symptoms to process, machine and environment parameters

rolling element diameter	DW	23,50 mm
pitch diameter	DT	4,76 mm
nominal contact angle	$\propto$	0 °
number of rolling elements	z	9
shaft frequency	f_N	10 Hz

outer race defect frequency	f_{od}	35,89 Hz
inner race defect frequency	f_{id}	54,11 Hz
rolling element defect frequency	f_{re}	47,34 Hz

$$f_{od} = \frac{z * f_N}{2 * DW} * \left(1 - \frac{DW}{DT} * \cos \propto\right)$$

$$f_{id} = \frac{z * f_N}{2 * DW} * \left(1 + \frac{DW}{DT} * \cos \propto\right)$$

$$f_{re} = \frac{f_N * DT}{DW} * \left(1 - \frac{DW^2}{DT^2} * \cos^2 \propto\right)$$

Fig. 6. Determining requirements for a vibration sensor to detect pitting at a SKF bearing type 6002 [14]

[12]. In the case of the mixer, these are the controls, directions of rotation and speeds of the individual stirrers, the filling level and the fault conditions, which can be varied. The quantified range of experimental parameters and their gradation are taken from the operating ranges of the system and from discussions with the technologists. The expected database should contain a comprehensive combination of good and fault states after the experiments have been conducted. A possible experimental design is excerpted in Table 3.

Table 3. Illustrative DoE for faulty bearings and stirrer while varying process parameters

Labels	Process Parameters	Units	Levels			
			0	1	2	3
A	inner stirrer motor	rpm	200	1000	1500	2200
B	inner stirrer direction		0		1	-1
C	outer stirrer motor	rpm	150	500	1000	1600
D	outer stirrer direction		0		1	-1
E	mixer filling level	%	0	40	80	110
F	faulty inner stirrer		0			1
G	faulty outer stirrer		0			1
H	faulty inner bearing		0			1
I	faulty outer bearing		0			1

a) List of process parameters and their levels

Run	A	B	C	D	E	F	G	H	I
0		0	0	2	0	0	0	0	0
1		0	1	2	0	0	0	0	0
2		0	2	2	0	0	0	0	0
3		0	3	2	0	0	0	0	0
4		0	0	2	0	3	0	0	0
5		0	1	2	0	3	0	0	0
6		0	2	2	0	3	0	0	0
7		0	3	2	0	3	0	0	0
8	0	2		0	0	0	0	0	0
..	...	...		...	...	...	...	...	...

b) DoE e.g. as OVAT-approach

5 Conclusion

In this paper, established process models for data mining in production engineering were presented. Subsequently, methods adapted to the DMME process model for the design phase of the digitization solution were assigned and their application was illustrated on a project for monitoring a mixer. The methodical support thus offers an improved methodology, especially for inexperienced users. In addition, a simplified reusability and transferability can be achieved through the discussed template approach. In future work, the transfer approach of the obtained failure mode behavior models will be investigated and validated as well as methods for testing the data acquisition concept and the data quality are to be developed for the *Technical Realization* and *Data Understanding* phases, for example by applying the V-Model for Data Quality [15] and Exploratory Data Analysis.

Acknowledgements. The authors would like to thank the BMWi for supporting this work under grant number KK5023201LT0 as well as the project partners Kniele GmbH and Bikotronic-Industrie-Elektronik GmbH for their technical support.

References

1. Fayyad, U., Piatetsky-Shapiro, G., Smyth, P.: From data mining to knowledge discovery in databases. AI Mag. **17**(3) (1996)
2. Azevedo, A., Santos, M.: KDD, SEMMA and CRISP-DM: a parallel Overview. In: IADIS European Conference Data Mining (2008)
3. Wirth, R., Hipp, J.: CRISP-DM: towards a standard process model for data mining. In: Proceedings of the 4th International Conference on the Practical Applications of Knowledge Discovery and Data Mining. Springer, London (2000)
4. Wiemer, H., Drowatzky, L., Ihlenfeldt, S.: Data mining methodology for engineering applications (DMME)—a holistic extension to the CRISP-DM Model. MDPI Appl. Sci. **9** (2019)
5. Harman, D., Buschmann, D., Scheer, R., Hellwig, M., Knapp, M., Schmitt, R.-H., Eigenbrod, H.: Data analytics production line optimization model (DAPLOM)—a systematic framework for process optimizations. In: Proceedings of the 11th Congress of the German Academic Association for Production Technology (WGP), Dresden, Sept 2021
6. International Organization for Standardization: ISO 17359:2018 Condition monitoring and diagnostics of machines. General Guidelines
7. Deutsches Institut für Normung e.V.: DIN EN 60812—failure mode and effects analysis (FMEA), IEC 56/1579/CD:2014
8. International Organization for Standardization: ISO 13379-1:2012 Condition monitoring and diagnostics of machines—data interpretation and diagnostics techniques. Part 1: General Guidelines
9. Mushiri, T., Mhazo, T. K., Mbohwa, C.: Condition based monitoring of boiler parameters in a thermal power station. In: Procedia Manufacturing 21—15th Global Conference on Sustainable Manufacturing, pp. 369–375 (2018)
10. Tandon, N., Choudhury, A.: A review of vibration and acoustic measurement methods for the detection of defects in rolling element bearings. Tribol. Int. **32**(8), 469–480 (1999)

11. Mendel, E., Mariano, L.Z., Drago, I., Loureiro, S., Rauber, T.W., Varejao, F.M., Batista, R.J.: Automatic bearing fault pattern recognition using vibration signal analysis. In: IEEE International Symposium on Industrial Electronics (2008)
12. Antony, J.: Design of Experiments for Engineers and Scientists, 2nd edn. Elsevier
13. Kniele GmbH: Labormischer KKM-L. Available: https://www.kniele.de/de/mischersysteme/labormischer-kkm-l. [Online], 25 Apr 2022
14. SKF: 6006-2RS1—Rillenkugellager. Available: https://www.skf.com/de/products/rolling-bearings/ball-bearings/deep-groove-ball-bearings/productid-6006-2RS1. [Online], 25 Apr 2022
15. Wiemer, H., Dementyev, A., Ihlenfeldt, S.: A holistic quality assurance approach for machine learning applications in cyber-physical production systems. MDPI Appl. Sci. **11** (2021)

An Implementational Concept of the Autonomous Machine Tool for Small-Batch Production

E. Sarikaya(✉), A. Fertig, T. Öztürk, and M. Weigold

Institute for Production Management, Technology and Machine Tools (PTW), Otto-Berndt-Straße 2, 64287 Darmstadt, Germany
e.sarikaya@ptw.tu-darmstadt.de

Abstract. The increasing demand for customized and complex products with small batch sizes confronts the manufacturing industry with new challenges, which can only be handled by flexible and dynamic manufacturing processes. As a major part of the process chain, autonomous machine tools can contribute to satisfying these requirements. Although there are many contributions on autonomous machine tools in research, the development of a self-learning, autonomous AI-integrated machine tool has been implemented neither in the industry nor in research. This paper proposes an industry-oriented concept of a self-learning machine tool. The system architecture consists of the existing CAD, CAM and CNC process chain and extends it with a knowledge base and an intelligent CAPP system for domain knowledge representation and decision making. Process knowledge is represented by using a continual learning process simulation approach for small-batch production.

Keywords: Autonomous machining · Process planning · Machine tool

1 Introduction

Modern CNC machine tools for metal cutting are highly automated production systems, which require a high degree of process understanding and process planning with correspondingly high user qualifications for their economical and cost-efficient use [1]. Particularly, in the case of small batch sizes of complex and expensive products, production is carried out with conservatively selected process parameters that are far below the productivity optimum. This conservative strategy is based on the fact that manufacturing relies on the technical understanding and experience of the machine operators. Furthermore, no sustainable process understanding can be achieved by the user due to highly varying customized products. The increasing demand for customized and complex products with smaller batch sizes require more flexible, reliable and dynamic production processes [2].

The integration of Industry 4.0 technologies in machine tools already offers the user initial approaches to support the optimization of selected process parameters through the visualization of measurement data from machine-internal or external sensors in

M. Liewald et al. (Eds.): WGP 2022, LNPE, pp. 535–544, 2023.
https://doi.org/10.1007/978-3-031-18318-8_54

combination with warning thresholds on the control panel [3, 4]. However, to enable optimization of the entire process planning, the information from the measurement data must be linked with other preceding and subsequent sub-processes, such as computer-aided manufacturing (CAM) and quality inspection.

The technical functionality of an autonomous machine tool is often represented in two superordinate control loops, an inner loop and an outer loop [2, 5, 6]. The inner control loop comprises real-time capable systems for process monitoring and -control and aims at short-term autonomy of the machine tool. The outer control loop covers medium-term autonomy through intelligent process planning systems [7].

The maturity level of process monitoring and real-time control that is applied in series production is very high and already state of the art [4, 8, 9]. Much effort has also been done in research for small-batch production for the inner loop with model-based approaches [10–12]. However, due to the large number of influencing factors in machining, the models only have a limited range of validity, which does not allow them to be transferred to other workpieces, machines or processes [7]. The outer control loop, which satisfies medium-term autonomy, has the lowest degree of maturity due to its high complexity and technological requirements. Previous researches concentrate either on applied subtasks, such as automated feature recognition [13], tool selection [14] and technology parameter optimization [15], or conceptual approaches that describe the overall system but provide little insight into the respective modules [2]. Hence, there is a lack of applied concepts in industrial manufacturing. From this, a need for industrial-oriented research activities can be identified. Hence, this paper focuses on the outer control loop and proposes an approach for autonomous process planning based on a continual learning process model.

Since model-based approaches already provide promising results for process monitoring and control, they can also be considered for process planning in small-batch manufacturing. However, applied models are usually machine-specific which limits the transferability to other systems. Therefore, this paper follows a machine-independent machine learning (ML) approach for process modeling. In addition, due to the multifactorial nature of process planning, artificial intelligence (AI)-based methods that incorporate evolutionary techniques are often used for decision making. [16]. This paper proposes a novel concept that brings both advantages together. A continual learning model-based approach is introduced to represent process knowledge while an evolutionary algorithm is used for decision making.

2 Knowledge Representation

Humans are particularly characterized by their cognitive abilities with regard to mental perception and thought processes. They can perceive and process signals from the environment and thus react to them. The ability to learn also enables them to improve their knowledge and skills in order to adapt to environmental changes. The development of an autonomous machine tool will only be possible when this intelligence can be mapped to a computer system. In the manufacturing industry, this can be achieved through a combination of cognitive science, automation technology and computer science [6]. For this purpose, a comprehensive knowledge base is applied with interacting modules which continuously learn to enhance the knowledge but also to be flexible against environmental changes. The knowledge base must cover product information and requirements, manufacturing resources and a comprehensive process knowledge. Most of these data are already well-structured in the feature-oriented standard STEP-NC, formally also known as ISO 14649 [17], which has been proposed for intelligent machining systems. STEP-NC offers a comprehensive data model for CAD/CAM systems and CNC machines and is therefore considered for data modeling.

2.1 Process Planning System

The intelligent computer-aided process planning (CAPP) is one of the most challenging tasks for implementation of an autonomous machine tool. CAPP bridges the gap between design and manufacturing and has the main task of transforming the product design into detailed instructions regarding the cutting tools applied, the manufacturing strategy and the process conditions. Due to its complexity, AI models are often used to seek the optimal solution during process planning [16]. According to Al-wswasi et al. [18], Li et al. [19] and Dittrich [7], recent CAPP systems show the following shortcomings:

1. The input and output functions of CAPP are considered independently, although they can affect each other.
2. Many systems include huge databases to store knowledge that are necessary for decision-making. Here, a lack of information could cause a failure. Furthermore, in the age of big data, databases could become increasingly larger.
3. Similarity-of-feature relationships are used to store previous machining knowledge. Thus, unfamiliar features could lead to suboptimal process planning.

In addition, further shortcomings have been identified by the authors of this paper:

4. The proposed CAPP systems do not allow process planning according to different target variables such as productivity, efficiency, and quality.
5. Simulation-based approaches concentrating on small-batch production are usually machine-specific and do, therefore, not allow the transfer to other machines.
6. Many knowledge representation methods are based on predefined logics or rules. These methods show a lack of adaptability and self-optimization.

To overcome these shortcomings, this paper presents an approach using an integrable, automatable, and continuously learning simulation model which interacts with the process planning module. At the same time, the model serves as a sparse representation of process domain knowledge. To assist the process planning module, the simulation predicts the target variables including workpiece quality, machining time and efficiency. With this new approach, the CAPP system can iteratively verify various configurations during process planning by getting feedback about the target variables.

2.2 Feature Representation

Since machining operations strongly depend on the geometrical elements of the part, a feature-based approach, which describes the topology and geometry of the part, is utilized for part representation [19]. This approach is widely used to cover the information gap between the product design, the manufacturing operations and the quality inspection [13]. For this purpose, many researchers have developed various methods for automated feature recognition in recent years, which can be divided into five categories: Syntactic pattern recognition, graph-, hint- and logic rule-based as well as artificial neural network [18].

However, automatic feature recognition is a recent research topic with unsolved challenges, such as feature intersection [18]. Therefore, a human-assisted feature detection method based on syntactic pattern recognition is applied in this paper [20]. The recognized features are represented in the STEP-NC based data model [17]. The feature-based data model provides all necessary design information, such as geometrical dimensions and raw material information. Quality requirements such as geometrical tolerances and surface finish are complemented according to Zhu et al. [21].

2.3 Cutting Tool and Technology Parameters

Cutting tool selection is a crucial part towards process planning of a machining operation which requires considerable experience. However, selecting the optimal cutting tools and technology parameters cannot be accomplished simply based on an individual's familiarity or experience. Thus, systems are required that determine the usability of all available tools multifactorially [22]. Therefore, a relational tool database (ToolDB) has been established which contains the master data of all available cutting tools including tool type and their geometrical properties as well as their related range of appropriate cutting parameters. Table 1 lists the minimum requirement of tool information and technology parameters that is needed by the CAM module for creating milling, drilling and tapping operations.

Table 1. Minimum required tool attributes and technology parameters by CAM for milling, drilling and tapping operation

Tool properties	Technology parameters
Tool type	Cutting speed/Spindle speed
Tool diameter	Feed per tooth/Feedrate
Shaft diameter	Ramp angle
Tool length	Feedrate for ramps
Cutting edge length	Depth of cut
Corner radius	Width of cut
Drill-point angle	Peck depth (only for drilling)
Drill-chamfer angle	
Thread pitch	

2.4 Machine Tool

In the presented concept, internal machine tool data which is directly acquired from the control unit is used to generate process knowledge. The control unit consists of an industrial PC from Beckhoff which comes with the TwinCat software environment equipped with I/O modules, CNC kernel, programmable logic controller (PLC) and human machine interface (HMI). A function block has been implemented in the PLC, which merges the data from these modules and writes the fused datastreams into a database. The time-series based machine tool data is related to data from other sources such as quality inspection. Therefore, utilizing the TimescaleDB a relational time-series database has been applied. TimescaleDB is an open-source extension of PostgreSQL and provides a powerful database for data-intensive applications [23]. Since not every signal undergoes cyclic changes, the data points have been divided into high-frequency data with a sampling rate of 1000 Hz and event data. The high-frequency data consists of sensor and drive data from the I/O module and of CNC data. These data are buffered in the memory with a FIFO (First In – First Out) ring buffer and written as data packets to the database with a time interval of 0.5 s. PLC and HMI data are considered as event data and are only written to the database when changes occur. Data about the active tool, for example, is only gathered after a tool change. The data are then matched to each other using a unique unix timestamp.

2.5 Quality Inspection

In order to be able to assess the autonomous process planning concerning the quality target variable, part-specific quality data must be recorded. In order to automatically match the data from quality assurance digitally to the respective workpiece and process, it is necessary to enrich the obtained quality information with metadata about the workpiece according to STEP-NC [17, 21]. Therefore, each feature-based measurement is enriched with a unique identifier and subordinated hierarchally to the unique identifier of the workpiece and the respective feature element. Depending on the applied manufacturing measuring equipment, the data is either merged via a user interface in the case of manual testing or automatically transferred when automated testing is involved (e.g. with a coordinate measuring machine). The data is finally merged in the database with the corresponding data from the process chain.

3 Intelligent Machining System Architecture

The intelligent system architecture of the autonomous machine tool leans on the existing CAD, CAM and CNC process chain which is expanded by an inner and outer control loop. The inner loop is a real-time capable process control loop for process adaption which allows the compensation of uncertainties and disturbances during machining. The outer loop, or experience control loop, consists of a self-learning CAPP module which is continuously retrained in the cycle of each produced workpiece. Since this paper focuses on the medium-term autonomy, only the aspects of the outer control loop will be discussed in the following. It includes the knowledge base which enables the structured extraction and build-up of process knowledge, which is conventionally only acquired by individual employees through extensive professional experience. Hence, the overall system of the learning machine tool continuously improves itself and can immediately apply the enhancements during the planning of a new product.

3.1 Experience Control Loop

The experience control loop is responsible for autonomous process planning based on customer-specific CAD information and product requirements. The geometric information is firstly processed by a feature recognition algorithm that examines the workpiece topology and breaks it down into its geometric features such as planar faces, pockets, slots and round holes [17]. According to STEP-NC, each geometry has its self-describing attributes which fully characterize the geometric feature. Based on these attributes, a hierarchical information model has been developed which is needed from CAM software for the tool selection. Figure 1 exemplary shows the information model of the closed pocket feature which allows the process planning module to reduce the number of applicable tools for the desired operation [20].

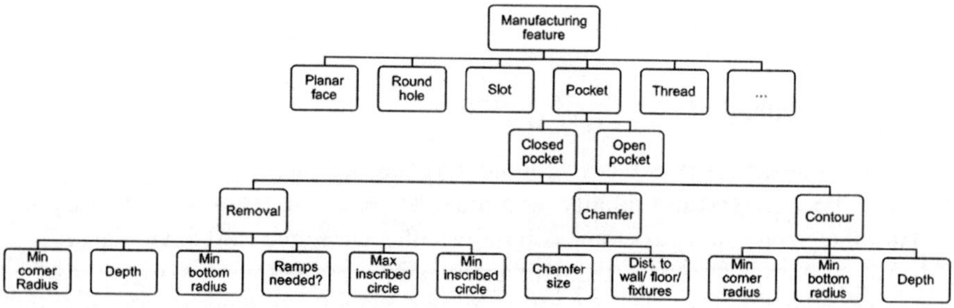

Fig. 1. Geometric feature attributes needed for proper tool selection from CAM on the example of a closed pocket

As depicted in Fig. 2, final tool selection together with appropriate cutting parameters is performed by the process planning module using an evolutionary algorithm. Typically, each cutting tool comes with manufacturer specifications about the range of recommended process parameters. This already provides a predefined range in which the evolutionary algorithm can optimize the technological parameters. The output of the algorithm describes the strategy, cutting tools used and technological parameters for the CAM software which creates the operation. Within the virtual process simulation loop, indicated with green in Fig. 2, the created operation is processed in the ML-based simulation model to predict the target variables consisting of workpiece quality, machining time and efficiency. The model-based prediction is used as a feedback to guide the evolutionary algorithm towards an optimal design solution.

This procedure is performed for all recognized geometric features of the workpiece. The post-processor subsequently generates the NC program for the machine tool. When the machine changes the state into the operation mode, the real-time capable inner control loop takes over. Here, machine data is processed in real-time to compensate

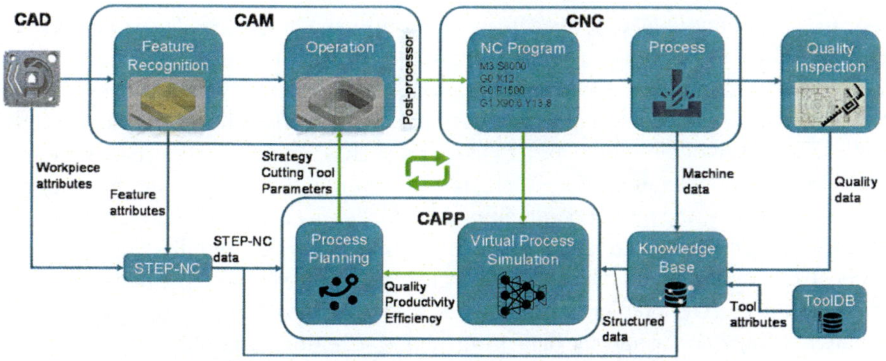

Fig. 2. Intelligent system architecture of the autonomous machine tool

process uncertainties but also transferred to the knowledge base, consisting of a relational time-series database, for more time-intensive tasks. After the machining process, quality inspection is conducted either manually or with the use of production metrology such as a coordinate measuring machine. In either case, the quality data must be incorporated to the database to be compared against the predicted target variables by the virtual simulation model. Finally, in the cycle of each produced workpiece, the ML-based process model of the intelligent CAPP system is retrained.

3.2 Virtual Process Simulation Model

Process planning is conventionally conducted by CAM systems which don't consider process or machine specific characteristics [7]. For this reason, it is essential to integrate a virtual process simulation model into the intelligent system architecture. An offline process simulation, as depicted in Fig. 3, provides insight into predicted process and target variables. Thus, the acquired information can be used by the evolutionary algorithm to evaluate multiple scenarios to iteratively find the optimal solution for process planning. However, virtual process simulation usually comes with two main challenges. On the one hand, the parametrization of process models is becoming more and more sophisticated with the increasing complexity of today's machine tools, on the other hand, these models must be constantly maintenanced manually according to changes in process and machine behavior. To meet these challenges, this contribution presents a ML-based approach through the integrated interaction between CAM and CNC, for virtual process simulation.

The virtual process simulation model, shown in Fig. 3, is based on the NC program generated by the post-processor after process planning. The virtual NC kernel (VNCK) of the machine tool builds on this. It interprets the G-code and generates the target trajectory in the interpolation cycle of the machine. The resulting path trajectory and spindle speed, which are later also supplied to the machine's servo controllers as target values, are used for material removal simulation. The cutter engagement analysis provides valuable process information, which are commonly used to develop process force models, for example [10]. This includes, for instance, the removed volume of material, the wrap angle, which gives the surface of the tool that actually touches the material, and the depth of cut, which is also derived from the tool contact surface. These data provide a sufficient basis to develop a process model which predicts process signals. Due to the previously mentioned challenges regarding model complexity and ever-changing boundary conditions, a continual learning approach has been applied to predict the sensor signals. The ML model predicts process signals such as spindle load, drive axis current as well as control deviations. The same data is acquired during machining. After each machining process, the model parameters are adapted according to the prediction error of the model. The continual learning model is fully integrated into the control system, so that it can automatically be retrained after each process, without the intervention of manual effort. The predicted process variables are then further used together with the cutter engagement analysis data from material removal simulation to predict the workpiece quality with a second ML model. This is also a continual learning model, since the model parameters are adjusted as soon as quality data enter the database.

Hence, the virtual process simulation provides the target variables quality, productivity and efficiency. The target values are considered during process planning to achieve an optimal compromise.

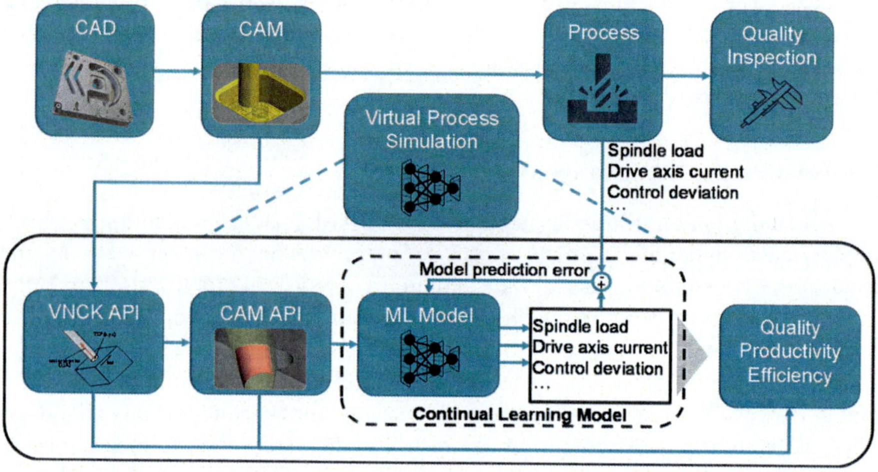

Fig. 3. Functional overview of the virtual process simulation model

4 Conclusion and Outlook

The intelligent system architecture of an autonomous machine tool extends the existing CAD, CAM and CNC process chain by an inner and an outer control loop. The inner control loop compensates process uncertainties in real-time, while the outer control loop is responsible for offline process planning. The maturity level of real-time monitoring and control is very high and represents the state of the art in CNC machining. However, this does not apply to intelligent process planning. Especially in small-batch production, intensive research must be conducted to meet the requirements for the manufacturing of individualized and complex products. This paper shows an industrial-oriented implementational concept of the intelligent process planning system. The individual modules have already been successfully implemented and must be holistically validated in the further process. Furthermore, the concept involves a new approach for process domain knowledge representation based on a continual learning simulation model, which is used by an evolutionary algorithm for decision making. Sufficient data must be collected to be able to train the ML model. In future works, the proposed approach will be validated by applying the intelligent CAPP system in a real manufacturing environment.

Acknowledgement. This research and development project "AICoM" is funded by the German Federal Ministry of Education and Research (BMBF) within the "The Future of Value Creation - Research on Production, Services and Work" program (02P20A064) and managed by the Project Management Agency Karlsruhe (PTKA). The authors are responsible for the content of this publication. The authors would like to thank the "AICoM" project partners (https://lernendewerkzeugmaschine.de/) for their content support.

References

1. Weck, M.: Werkzeugmaschinen 4. Springer, Berlin (2006)
2. Dittrich, M.-A., Denkena, B., Boujnah, H., Uhlich, F.: Autonomous machining—recent advances in process planning and control. J. Mach. Eng. **19**, 28–37 (2019). https://doi.org/10.5604/01.3001.0013.0444
3. Ceratizit: Full process control with ToolScope. Available at: https://cuttingtools.ceratizit.com/int/en/services/toolscope.html?referrer=. Accessed 03.05.2022.321Z
4. DMG: MPC 2.0—Machine Protection Control. Available at: https://de.dmgmori.com/produkte/digitalisierung/integrated-digitization/production/technologiezyklen/mpc-2-0-machine-protection-control. Accessed 03.05.2022
5. Schmucker, B., Trautwein, F., Semm, T., Lechler, A., Zaeh, M.F., Verl, A.: Implementation of an intelligent system architecture for process monitoring of machine tools. Procedia CIRP **96**, 342–346 (2021). https://doi.org/10.1016/j.procir.2021.01.097
6. Park, H.-S., Tran, N.-H.: Development of a smart machining system using self-optimizing control. Int. J. Adv. Manuf. Technol. **74**(9–12), 1365–1380 (2014). https://doi.org/10.1007/s00170-014-6076-0
7. Dittrich, M.-A.: Autonomous Machine Tools—Definition, Elements and Technical Implementation, 1st edn. TEWISS, Garbsen (2021)
8. Teti, R., Jemielniak, K., O'Donnell, G., Dornfeld, D.: Advanced monitoring of machining operations. CIRP Ann. **59**, 717–739 (2010). https://doi.org/10.1016/j.cirp.2010.05.010
9. Abellan-Nebot, J.V., Romero, S.F.: A review of machining monitoring systems based on artificial intelligence process models. Int. J. Adv. Manuf. Technol. **47**, 237–257 (2010). https://doi.org/10.1007/s00170-009-2191-8
10. Witt, M., Schumann, M., Klimant, P.: Real-time machine simulation using cutting force calculation based on a voxel material removal model. Int. J. Adv. Manuf. Technol. **105**(5–6), 2321–2328 (2019). https://doi.org/10.1007/s00170-019-04418-2
11. Brecher, C., Wetzel, A., Berners, T., Epple, A.: Increasing productivity of cutting processes by real-time compensation of tool deflection due to process forces. J. Mach. Eng. **19**, 16–27 (2019). https://doi.org/10.5604/01.3001.0013.0443
12. Kushwaha, S., Gorissen, B., Qian, J., Reynaerts, D.: In-Process Virtual Quality Monitoring. SSRN J. (2020). https://doi.org/10.2139/ssrn.3724122

13. Zhang, X., Nassehi, A., Newman, S.T.: Feature recognition from CNC part programs for milling operations. Int. J. Adv. Manuf. Technol. **70**(1–4), 397–412 (2013). https://doi.org/10.1007/s00170-013-5275-4
14. Carpenter, I.D., Maropoulos, P.G.: A flexible tool selection decision support system for milling operations. J. Mater. Process. Technol. **107**, 143–152 (2000). https://doi.org/10.1016/S0924-0136(00)00707-X
15. Leo Kumar, S.P., Jerald, J., Kumanan, S., Aniket, N.: Process parameters optimization for micro end-milling operation for CAPP applications. Neural Comput. Appl. **25**(7–8), 1941–1950 (2014). https://doi.org/10.1007/s00521-014-1683-0
16. Leo Kumar, S.P.: State of the art-intense review on artificial intelligence systems application in process planning and manufacturing. Eng. Appl. Artif. Intell. **65**, 294–329 (2017). https://doi.org/10.1016/j.engappai.2017.08.005
17. ISO International Organization for Standardization. ISO 14649-1: Industrial automation systems and integration—physical device control; Data model for computerized numerical controllers. Part 1: Overview and Fundamental Principles 25.040.20, 35.240.50 (2003)
18. Al-wswasi, M., Ivanov, A., Makatsoris, H.: A survey on smart automated computer-aided process planning (ACAPP) techniques. Int. J. Adv. Manuf. Technol. **97**(1–4), 809–832 (2018). https://doi.org/10.1007/s00170-018-1966-1
19. Li, X., Zhang, S., Huang, R., Huang, B., Xu, C., Zhang, Y.: A survey of knowledge representation methods and applications in machining process planning. Int. J. Adv. Manuf. Technol. **98**(9–12), 3041–3059 (2018). https://doi.org/10.1007/s00170-018-2433-8
20. ModuleWorks: Digital Manufacturing. Available at: https://www.moduleworks.com/digital-manufacturing/?area=. Accessed 03.05.2022
21. Zhu, W., Hu, T., Luo, W., Yang, Y., Zhang, C.: A STEP-based machining data model for autonomous process generation of intelligent CNC controller. Int. J. Adv. Manuf. Technol. **96**(1–4), 271–285 (2018). https://doi.org/10.1007/s00170-017-1554-9
22. Arezoo, B., Ridgway, K., Al-Ahmari, A.M.A.: Selection of cutting tools and conditions of machining operations using an expert system. Comput. Ind. **42**, 43–58 (2000). https://doi.org/10.1016/S0166-3615(99)00051-2
23. TimescaleDB: Timescale Docs. Available at: https://docs.timescale.com/timescaledb/latest/#welcome-to-the-timescaledb-documentation. Accessed 03.05.2022

Benchmarking Control Charts and Machine Learning Methods for Fault Prediction in Manufacturing

S. Beckschulte[1(✉)], J. Mohren[1], L. Huebser[1], D. Buschmann[1], and R. H. Schmitt[1,2]

[1] Laboratory for Machine Tools and Production Engineering, WZL of RWTH Aachen University, Campus-Boulevard 30, 52074 Aachen, Germany
s.beckschulte@wzl.rwth-aachen.de

[2] Fraunhofer Institute for Production Technology IPT, Steinbachstraße 17, 52074 Aachen, Germany

Abstract. This paper examines and benchmarks different approaches in their ability to detect and predict faults in manufacturing processes, based on real-world use cases and with respect to their differing dataset properties. Knowing about the occurrence of faults becomes more and more important in manufacturing due to increasing quality demands and legal guidelines. In addition, the complexity of manufacturing processes is constantly increasing. This stems from a higher product variance resulting from individual and customized products as well as additional external influences such as human errors, environmental factors and tool wear. As a result, today's process data is often no longer normal distributed. Furthermore, data volume steadily increases, thereby opening new opportunities for data-driven analytics approaches. Frequently applied control charts for statistical process control (SPC) often lack the ability to deal with multiple variables and non-normal distributed data at the same time, since multivariate and nonparametric control charts are underrepresented in past research. Consequently, there is a need for new process control methods in manufacturing that are suitable for large amounts of data and cover diverse and dynamic distribution models. Therefore, machine learning models have been recognized as feasible approaches to meet these requirements. For comparison a Hotelling's T^2 control chart, a K-Chart, an Isolation Forest, an ARIMAX model and a Neural Network have been implemented. We evaluate each method by missed detection rate (MDR), false alarm rate (FAR) and whether signals occurred before or after the faults. Real-world data sets of a commercial vehicle manufacturer serve as benchmarking basis.

Keywords: ARIMAX · Control chart · Fault prediction · Hotelling's T^2 · Isolation forest · K-chart · Machine learning · Neural network · Production · SPC

M. Liewald et al. (Eds.): WGP 2022, LNPE, pp. 545–554, 2023.
https://doi.org/10.1007/978-3-031-18318-8_55

1 Introduction

Today, the most popular methods for fault detection and process monitoring are control charts [1]. Therefore, they are a widely applied tool in SPC to detect abnormal behavior in order to improve and optimize processes. This success of traditional control charts is based on stable and constant processes in the past [2, 3]. The underlying idea of most traditional control charts assumes that sampled data follows an a-priori known and normally, independently, and identically distribution. However, these characteristics no longer reflect the majority of today's manufacturing processes. While more and more data being available within manufacturing, the amount of data pushes currently established methods for handling and processing data to their limits. Additional challenges such as customers demand for higher individuality of products result in more variants and more complex manufacturing processes. Product lifespans are decreasing and new products are coming onto the market more frequently and quickly. This reduces the time window for products to reach a profitable zone and requires faster and more successful ramp-up processes [4]. With further increasing expectations for product quality, it is becoming more difficult for manufacturers to be profitable [5]. This enforces more efficient processes and fewer faults in production.

Further, this increase in process complexity is often addressed by monitoring more variables simultaneously, which results in more features being recorded. While control charts exist that are capable of handling multiple variables, there are few novel control charts that do not require the assumption of a known distribution [6]. The application of an unsuitable control chart, whether in its univariate or parametric specification, fails to take into account the interrelation and correlation structure of monitored variables. Hence, it is found to be inefficient, time consuming and misleading while resulting in a high FAR [3, 7].

Consequently, new approaches for process control and identification of faults are required to process large amounts of variables and data while learning the complex interrelationships of manufacturing processes. Hence, machine learning models present a feasible choice, as they have shown promising results in similar applications so far [8, 9]. Currently, a wide range of approaches is available for anomaly detection that are suitable for applications in manufacturing. To identify the most promising approaches, a benchmark has to include comprehensive metrics, but also investigate the effects of use case properties on the performance. This requires benchmarking a variety of approaches from control charts, machine learning and their combination on diverse real manufacturing data sets.

2 State of the Art

Until now, traditional control charts have often been used to monitor product quality in the field of manufacturing. Their success is mainly based on stable processes with little variance and low complexity, which is no longer the case in modern manufacturing processes [2, 10]. Today's manufacturing industry incurs a constant changing environment of large scale, high process complexity and many uncertainties [11].

Nevertheless, multivariate parametric control charts represent the major research progress of the last decades, which require the assumption of knowing the distribution in advance. These known and common distribution models are often unsuitable for modern manufacturing processes and lead to poor performance [6]. Contrary, multivariate nonparametric control charts are a new topic and thus cover more novel approaches. In order to investigate differences in performance of a parametric and nonparametric approaches, two control chart approaches are included in this benchmark.

The *Hotelling's* T^2 method, introduced by Hotelling in 1931, is selected as a multivariate parametric control chart (Fig. 1). This approach is considered to be a traditional control chart [13]. It examines the Mahalanobis distance of each point to the center of successful operations. The Mahalanobis distance is a unitless value, where a value of 0 represents the center of the distribution and a large value indicates the distance to each principal component axis. Thereby, it measures the distance between a point and the mean of the distribution shown as summarized standard deviations. The calculation of the standard deviation assumes a normal distribution, hence the method's parametric property. Consequently, this method considers the correlation of the data, unlike the Euclidean distance [14].

A *K-Chart* is chosen as a more novel and multivariate nonparametric control chart. This approach uses a Support Vector Data Description (SVDD) to create a decision boundary and calculates a distance for each point, which is then plotted on a control chart [12]. SVDD is a machine learning technique for single-class classification and outlier detection. Thus, historical training data is approximated by a decision boundary based on support vectors, which classifies values into two categories of in-control or out-of-control. For each new data point, the kernel distance is calculated as a measure of the distance between the kernel center and the monitored data point. This approach is very similar to the Hotelling's T^2 control chart and is regarded as its nonparametric counterpart since no distribution is assumed in advance [12, 15].

Anomaly detection can either be formulated as an unsupervised learning task or a regression task incorporating a synthetically generated anomaly score. Following selection will cover an unsupervised learning approach as well as regression algorithms to compare both (Fig. 1). An *Isolation Forest* examines how often data must be randomly subdivided to completely isolate points. Since it is not able to model temporal correlations, an *ARIMAX* model and a *Neural Network* are chosen due to their ability to analyze complex and extensive correlations.

3 Methodical Approach

First, four suitable use cases and corresponding data sets are investigated and selected. The selection is based on choosing heterogeneous combinations with respect to the total number of faults and the variance of the included variables per each data set. Thereby, data sets with diverse properties are identified for benchmarking. A detailed data description, as well as the selection process is described within the next section of this paper.

The selection and implementation of control charts and machine learning approaches is based on the State of the Art. It includes a univariate parametric and a multivariate nonparametric approach. Both approaches are then applied to each of the four chosen data sets. Besides the exceeding of control limits, other common control chart rules for identification of out-of-control points are adopted, such as the recognition of trends

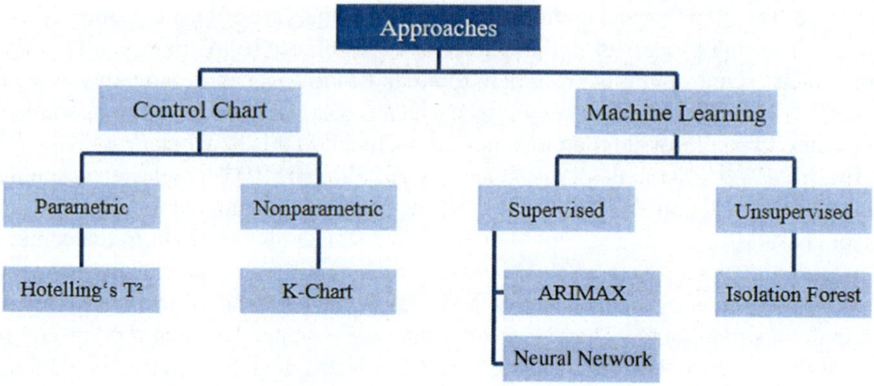

Fig. 1. Selecting approaches for benchmarking

or shifts. Similarly, machine learning approaches for anomaly detection are being pursued alongside. Again, opposing approaches are included in the form of supervised and unsupervised models.

Different data preprocessing methods and hyperparameter settings are applied iteratively to improve the results. After hyperparameter optimization and preprocessing optimization, multiple metrics are applied to evaluate every combination of the approaches and four data sets. Metrics include the number of faults not detected, the number of false alarms, and the time signals occurrence relative to faults. Finally, results are combined in a comprehensive benchmark that includes two metrics, one for the detection of faults and one for their prediction. Variations in performance are investigated based on different data set properties. Based on the benchmark, an assumption can be derived regarding which approach is best suited for the selected use cases.

4 Use Case Selection

The examined use cases originate from the final assembly of a commercial vehicle manufacturer, which performs a series of manual assembly steps along the production line. During assembly, pneumatic torque screwdrivers with integrated measurement sensors are used, enabling logging of torque and angle of rotation for quality control and verification purposes. These screwdrivers log the data in separate data tables. An identification number is used to link the individual data to the specific screw connection. Due to a final end-of-line check with data tracking, error data is available for every screw connection performed. For the conducted experiments, only data sets of torque screwdrivers were selected, which have an extensive and approximately similar data history. An initial data review indicated a limit of at least 40.000 generated data points per screwdriver within one year to obtain an constant data history for the use case. After filtering, a selection of 76 out of 382 data sets (19.9%) remained for the experiments.

Further, a variety of different properties is included in the final selection, which consists of the number of faults and the variance of torque and angle of rotation. A data set can consist of either a low or a high number of faults and a small or large variance of variables. In each case, the mean values from the 76 data sets were formed. The value for the number of faults is 5.22%, and the variances in torque and angle of rotation are 114.13 Nm^2 and 317.83°. This combination of properties can be represented by four groups, which are shown in Table 1.

Table 1. Data set groups

		Variance in both variables	
		Small	Large
Number of faults	Low	Group 1	Group 2
	High	Group 3	Group 4

One use case is selected for each of the four groups. The final selection of use cases is shown in Table 2 and their corresponding data sets will function as a basis for following benchmarking.

Table 2. Final selection of use cases

Group	Use case	Data points last 365 days	Variance torque [Nm^2]	Variance angle of rotation [°]	N.i.O. rate [%]
Group 1	UC 1	98.336	0.02	7.70	0.22
Group 2	UC 2	217.751	398.86	1087.84	4.27
Group 3	UC 3	74.939	0.04	52.15	5.93
Group 4	UC 4	295.617	474.61	1430.95	6.00

5 Evaluation

First, this chapter addresses the definition of three metrics for evaluation. Second, an example of the histogram detecting shifts for the K-Chart is highlighted.

All implemented approaches send a signal indicating abnormal behavior, which is interpreted as an instance close to a fault. A signal can be sent either before a fault, at exactly the same step or after. Therefore, a time window is defined for the following evaluation which matches signals with faults. Since all selected data sets contain a fault rate smaller than 10%, this window includes five steps before and after each point of interest. This avoids too much overlap of faults within each window when the faults are assumed to be evenly distributed. As a result, signals can be assigned more precisely to faults and vice versa. Based on this, three metrics will be evaluated for each combination of approach and data set. The metrics include the MDR, FAR and a histogram which shows when signals have been sent in relation to faults.

The MDR represents how many faults have not been detected within the defined window. Here, for every fault, it is checked whether there is a signal between five steps before or after the fault. If there is no signal present, the count of false negative values (FN = false negative) is increased by one. In the end, the number of false negatives is divided by the total number of faults examined (P = positive), which results in the MDR, as seen in Eq. (1). A similar proceeding is conducted for the FAR, where for every signal it is checked, whether a fault occurred within the defined window. If this is not the case, the count of false positive (FP = false positive) values is increased by one and divided by the number of instances that are not faults (N = negative), as seen in Eq. (2).

$$\mathrm{MDR} = \mathrm{FN}/\mathrm{P} \tag{1}$$

$$\mathrm{FAR} = \mathrm{FP}/\mathrm{N} \tag{2}$$

However, MDR and FAR do not provide any information on the ability to predict faults before they occur. Hence, a histogram is added as an additional metric to provide insight into when signals appear in relation to the presence of faults. The histogram bins are the defined window ranging from −5 to + 5. These numbers represent the time difference between the actual fault and the recognized signal. For each signal within the window of every fault, this relation is analyzed and transferred. If a signal within the window can be directly related to another fault that occurs before or after the point under investigation, then it will not be considered. A direct and unambiguous relation between a signal and fault is present if a signal is sent at the exact time as the fault.

As an example the following will investigate signals based on the detection of shifts for the K-Chart. Here, it is assessed if seven consecutive points fall above or below the mean distance of in-control points. These findings are illustrated in Fig. 2, for all four selected use cases. From the results it can be observed that use case 1 and 2 contain signals that occurred before actual faults. While there are almost no signals after faults in use case 1, additional signals are present after faults in use case 2. Since, the number of signals before faults is significant larger, a tendency to predict faults could still be identified for the second use case. Nevertheless, this does not apply to use case 3 and 4, where no capability to predict faults could be recognized, since signals are almost evenly distributed before and after the presence of faults.

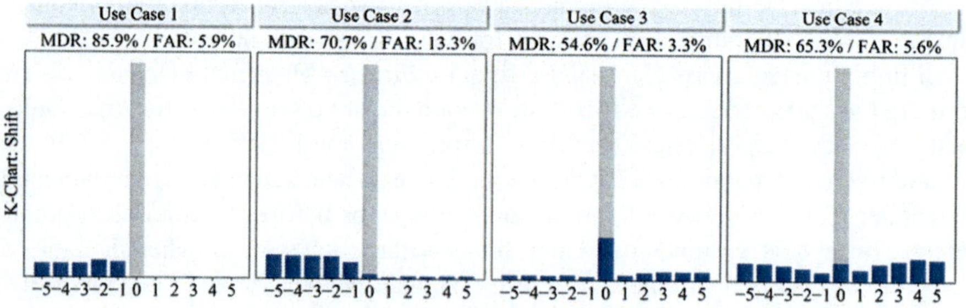

Fig. 2. Histogram shift for K-chart

6 Benchmarking Approaches

The following gives an overview of all implemented and evaluated approaches for the detection and prediction of faults. For this two new metrics are considered, which reflect the capability to detect and predict faults. The *F1-score* includes the ability to detect faults without sending false signals, where a perfect score is represented by 1 and the worst score by 0. As seen in Eq. (3), if either a lot of faults are not recognized or a lot of false alarms are sent, the F1-score punishes these extremes in both cases.

$$\mathrm{F1} = 2 * ((Precision * Recall)/Precision + Recall) \tag{3}$$

$$\mathrm{Precision} = \mathrm{TP}/(\mathrm{TP} + \mathrm{FP}) \tag{4}$$

$$\mathrm{Recall} = \mathrm{TP}/(\mathrm{TP} + \mathrm{FN}) \tag{5}$$

The ability to predict faults is numerically evaluated by the percentage of faults that have been predicted in the test data. The 20% test data was not used for training, contains the latest data from the collected data sets and was used for evaluation. A predicted fault consists of at least one signal within 5 steps before the fault occurs, while no signal is sent in the following 5 steps. This is examined for each combination of approach and use case, where the number of predicted faults is divided by the total number of faults included in the test data. For each of the metrics F1-score and percentage of predicted faults, a heat map is developed that includes all approaches and all four selected use cases. The results for the F1-scores represents the ability of approaches to detect faults within the specified window (Fig. 3). Regarding the control charts, the best F1-scores are identified for signals based on control limits. Further, the multivariate nonparametric control chart (K-Chart) generally shows better fault detection results than its multivariate parametric counterpart (Hotelling's T^2). For direct machine learning approaches, the best results are shown by approaches based on the current observation. Here, the Isolation Forest has good results for use cases 1 and 3, which share the property of a low variance in both variables torque and angle of rotation. The ARIMAX model only resulted in a sufficient F1-score for the first use case. Overall, three methods can be identified that perform outstandingly well across all use cases, including the Neural Network (architecture: fully connected layers, input layer, 100, 100, 20, output layer) based on the current observation, the control limit signals of the K-Chart and the control limit signals of the Hotelling's T^2 control chart. In summary, several approaches proved suitable for detecting faults within the given window, and near-perfect F1-scores were obtained across all data sets.

The heat map below shows with the percentage of faults that could be predicted within the test data for the four selected use cases (Fig. 4). While this illustrates the ability to successfully send signals before faults occur, it does not account for faults that were detected with a delay. Without providing a comprehensive picture, this metric is still sufficient to indicate which approaches are more promising for this task with respect to the given use case. First, it can be stated that no approach in combination with a use case achieves a perfect score by predicting all faults. However, it is uncertain how many faults per use case can be predicted in theory, since faults can also occur randomly. Considering the control chart approaches, the best results in terms of fault prediction can be

	Hotellings T2: Control Limit	Hotellings T2: Shift	Hotellings T2: Trend	Hotellings T2: Combined	K-Chart: Control Limit	K-Chart: Shift	K-Chart: Trend	K-Chart: Combined	Isolation Forest	Multidimensional Isolation Forest	ARIMAX	Multidimensional ARIMAX	Neural Network	Multidimensional Neural Network
UC1	88.5%	4.7%	16.3%	27.4%	100.0%	8.3%	16.0%	22.2%	99.8%	0.0%	85.5%	5.2%	99.8%	46.1%
UC2	97.7%	19.6%	37.2%	53.0%	97.9%	18.4%	30.5%	33.5%	55.6%	1.3%	29.0%	15.0%	99.8%	44.4%
UC3	95.9%	47.5%	57.5%	68.0%	99.9%	68.6%	57.2%	73.7%	96.3%	0.0%	7.7%	16.0%	99.9%	68.3%
UC4	95.7%	17.8%	31.3%	45.8%	98.9%	39.4%	23.8%	45.7%	18.2%	0.0%	0.1%	17.6%	99.8%	52.4%

Fig. 3. Heat map F1-scores

obtained from combined signals of control limits and pattern recognition. While these signals of the Hotelling's T^2 control chart achieve overall the best results for use cases 3 and 4, combined signals of the K-Chart outperform all other approaches for use cases 1 and 2. Therefore, the Hotelling's T^2 control chart provides the best performing approach for use cases with a low number of total faults, while the K-Chart is most promising for use cases with a relatively high number of total faults. Among the direct machine learning approaches, only the ARIMAX model, which is based on a multidimensional input over the last observations, showed adequate results for fault prediction. Here, multidimensional input refers to the addition of another temporal dimension. This contains the last 10 data sets in time, which are reshaped in a one-dimensional vector in order to be processed by the algorithms. All direct machine learning approaches based only on the current observation failed to predict faults. While this total failure also applies to the Isolation Forest based on a multidimensional input of the last observations, an insignificant number of faults were predicted by the Neural Network for the same input. However, the Neural Network's ability can be further questioned if the associated histogram is taken into account, which shows that more signals occurred after faults than before.

	Hotellings T2: Control Limit	Hotellings T2: Shift	Hotellings T2: Trend	Hotellings T2: Combined	K-Chart: Control Limit	K-Chart: Shift	K-Chart: Trend	K-Chart: Combined	Isolation Forest	Multidimensional Isolation Forest	ARIMAX	Multidimensional ARIMAX	Neural Network	Multidimensional Neural Network
UC1	1.1%	6.7%	10.7%	15.6%	0.0%	16.3%	13.3%	24.4%	0.0%	0.0%	0.0%	18.1%	0.0%	0.7%
UC2	0.9%	7.8%	12.6%	18.9%	0.8%	19.9%	8.8%	24.7%	0.0%	0.3%	2.4%	19.5%	0.4%	3.4%
UC3	4.8%	12.9%	7.3%	20.2%	0.4%	8.7%	7.4%	12.6%	0.0%	0.0%	1.7%	13.6%	0.3%	6.1%
UC4	3.9%	18.4%	8.9%	26.1%	0.1%	11.9%	7.7%	14.8%	0.0%	0.0%	0.0%	20.6%	0.4%	5.9%

Fig. 4. Heat map prediction of faults

7 Summary and Discussion

This work has successfully demonstrated the possibility to predict faults based on given real application data and with currently available approaches. Furthermore, differences in the performance of the individual approaches, but in conjunction to the use cases data properties, have been identified. Thus, the results can serve as a decision support for selecting the most promising approach under the given circumstances. However, this work experiences limitations.

A total of 14 different signals were evaluated based on control limits and pattern recognition for control charts, as well as different inputs for machine learning algorithms. While this considers a high number of different approaches, there are many more methods for predicting faults. Accordingly, it is very likely that approaches capable of achieving better results have not been investigated in this paper.

It is impossible to determine exactly which signal is intended for which fault. Therefore, a time window of 5 points before and after the observation was defined, which establishes this relationship vaguely. Since the total fault rate for all data sets is well below 10%, this order of magnitude was chosen to avoid as much overlap of faults as possible, assuming a uniform distribution. This allows a more unambiguous assignment of signals to faults. Nevertheless, this approach does not investigate whether faults can be predicted even more than 5 steps in advance or what effects signals with more than 5 steps delay have on the metrics used. Still, the applied metrics were able to measure performance in terms of fault detection and prediction and were also able to highlight variations in performance for different use cases. When assessing the ability to detect faults, unrecognized faults and signals sent in the absence of a fault are of particular interest. These are individually represented by the MDR and FAR, where almost perfect scores for both metrics could be identified by multiple approaches. During benchmarking, these metrics are combined in the F1-score for a comprehensive and transparent overview. The second part of benchmarking analyzes the percentage of predicted faults within the test data, which reflects an initial assumption for the general ability to predict faults. However, a high percentage of predicted faults is not desirable if too many false alarms occur. As a result, the metric must be considered in the overall context and does not include a sole presentation of the performance. Although none of the approaches predicts all faults consistently in every use case, the results are still promising. Besides, it is uncertain how many faults can be predicted in theory, since they can also occur randomly and without underpinning reason.

Besides the abovementioned limitations regarding exploration of feasible algorithms, as well as the performance assessment of benchmarked approaches, limitations are also imposed by the number of investigated use cases. Even though, the use case selection aimed at representing an as high as possible data variance, the overall amount of total data points is considerably small in comparison to modern research within machine learning. The multitude of existing circumstances on the shopfloor might therefore not fully be represented. This stems from the fact, that most commercial vehicle manufacturer still lack capabilities in data gathering and management due to insufficient digitization. Accompanied by this fact, data quality in the use cases is degraded by misclassification, which are not explainable even by experts at the company. Nonetheless, the implemented and benchmarked approaches show applicability in general and can be considered for use even due to data quality shortcomings.

Acknowledgement. The authors gratefully acknowledge the financial support of the research project "value chAIn" (project number 19S21001B) funded by the German Federal Ministry for Economic Affairs and Climate Action (BMWK) based on a resolution of the German Bundestag and the project supervision by TÜV Rheinland Consulting GmbH.

References

1. Farokhnia, M., Niaki, S.T.A.: Principal component analysis-based control charts using support vector machines for multivariate non-normal distributions. Commun. Stat. Simul. Comput. **49**(7), 1815–1838 (2020)
2. Bai, Y., et al.: A comparison of dimension reduction techniques for support vector machine modeling of multi-parameter manufacturing quality prediction. J. Intell. Manuf. **30**(5), 2245–2256 (2018). https://doi.org/10.1007/s10845-017-1388-1
3. Jin, J., Loosveldt, G.: Assessing response quality by using multivariate control charts for numerical and categorical response quality indicators. J. Surv. Stat. Methodol. **9**(4), 674–700 (2021)
4. Schiller, M., Heftrich, C., Engel, B.: Remote production. Procedia CIRP **99**, 242–243 (2021)
5. Mizgan, H., Ambruş, O.: Challenges in the automotive JIS/JIT production of steering wheels involving traceability. Acta Marisiensis. SeriaTechnologica **18**(2), 42–46 (2021)
6. Koutras, M.V., Triantafyllou, I.S. (eds.): Distribution-Free Methods for Statistical Process Monitoring and Control. Springer, Cham (2020). https://doi.org/10.1007/978-3-030-25081-2
7. Tran, K.P. (ed.): Control Charts and Machine Learning for Anomaly Detection in Manufacturing. SSRE, Springer, Cham (2022). https://doi.org/10.1007/978-3-030-83819-5
8. Jakubowski, J., Stanisz, P., Bobek, S., Nalepa, G.J.: Explainable anomaly detection for Hot-rolling industrial process. In: 2021 IEEE 8th International Conference on Data Science and Advanced Analytics, pp. 2–3. IEEE, Porto (2021)
9. Maboudou-Tchao, E.M., Silva, I.R., Diawara, N.: Monitoring the mean vector with Mahalanobis kernels. Qual. Technol. Quant. Manage. **15**(4), 459–474 (2018)
10. Beckschulte, S., Kiesel, R., Schmitt, R.H.: Manuelle Fehleraufnahme bei Mass Customization. ZWF **116**(4), 188–192 (2021)
11. Cheng, Y., Chen, K., Sun, H., Zhang, Y., Tao, F.: Data and knowledge mining with big data towards smart production. J. Ind. Inf. Integr. **9**, 1–13 (2018)
12. Sun, R., Tsung, F.: A kernel-distance-based multivariate control chart using support vector methods. Int. J. Prod. Res. **41**(13), 2975–2989 (2003)
13. Hotelling, H.: The economics of exhaustible resources. J. Polit. Econ. **39**(2), 146–147 (1931)
14. Mahalanobis, P.C.: On the generalized distance in statistics. In: Proceedings of the National Institute of Sciences, pp. 49–55. Natural Institute of Science of India, Calcutta (1936)
15. Kakde, D., Peredriy, S., Chaudhuri, A.A.: Non-parametric control chart for high frequency multivariate data. In: 2017 Annual Reliability and Maintainability Symposium (RAMS), pp. 1–6. IEEE, Orlando (2017)

Enabling Data-Driven Applications in Manufacturing: An Approach for Broadly Applicable Machine Data Acquisition and Intelligent Parameter Identification

Philipp Gönnheimer(✉), Jonas Hillenbrand, Imanuel Heider, Marina Baucks, and Jürgen Fleischer

wbk Institute of Production Science, Karlsruhe Institute of Technology, Kaiserstr. 12, 76131 Karlsruhe, Germany

philipp.goennheimer@kit.edu

Abstract. Due to the ongoing trend of digitization and the strong increase in the number of Industrie 4.0 use cases, the use of machine tool process data in data-driven applications, namely process or condition monitoring, is on the rise. However, the provision of data such as motor currents or the positions of a machine's axes—which are essential to these applications—can in many cases only be achieved under high individual effort. This is largely due to two aspects. Firstly, due to the heterogeneous nature of communication infrastructures and information models in the machine environment, there is no one-size-fits-all solution for the acquisition of data. Secondly, in cases where the denomination of the sought-after parameters is not known in advance, an implementation of data-driven applications becomes challenging. Thus, the objective of the research work presented is the development of a system capable of extracting time series data from a variety of machines and related data sources and determining their identity (e.g. current, position, etc.) based on their characteristics using an intelligent approach. A prototypical software application with a modular structure is presented, which envelops functionalities from the discovery and acquisition of data to its intelligent identification.

Keywords: Machine tool connectivity · Intelligent parameter identification

1 Introduction

The increasing digitization and interconnectedness of automation systems adds to their complexity, which constitutes a challenge for machine and plant operators. Applications in the fields of process optimization and predictive maintenance gain further traction. However, their successful deployment depends on the provisioning of real-time data (e.g. control parameters such as axis positions and motor currents) that can be located on different data sources. Typically, a data scientist seeking to set up a data-based application must task a technician with the data acquisition. The diversity of today's manufacturing

M. Liewald et al. (Eds.): WGP 2022, LNPE, pp. 555–563, 2023.
https://doi.org/10.1007/978-3-031-18318-8_56

systems, which often employ machines and controllers from different manufacturers and of differing age, makes the provisioning of machine data for intelligent applications a challenging endeavour. An array of proprietary communication protocols must be dealt with to enable connection to a satisfactory number of different data sources. Moreover, even when efforts towards standardising communication are being made, for example in the context of the widespread implementation of the OPC UA standard, adopters of data-based applications face the difficulty that lies within the use of non-standardised, often vendor-specific nomenclature of control parameters. This may be remedied by companion specifications such as umati [1], which adapt the OPC UA information model depending on the use case. However, it cannot be assumed that a companion specification covers all parameters needed in a data-based application. These factors have led to the status quo, in which intelligent applications cannot yet be deployed on a broad scale and only exist as insular solutions tailored towards specific machines and use cases.

The paper aims at introducing an approach that facilitates the deployment of data-based applications on a broad scale. For this purpose, a prototypical solution, including a GUI, has been designed. Its emphases are:

- Acquisition of time series data from a variety of data sources
- Identification of time series data based on their characteristics

The paper is structured as follows: After this introductory first section, Sect. 2 shall discuss the state of the art, i.e., "Which approaches for machine data acquisition, and which approaches for identification of time series data are currently in use?". Based on the state of the art, an outline of the proposed concept is given in Sect. 3, illustrating its novel approach. Section 4 will then break down the approach into detail and showcase the functionality of the developed application in its entirety—from the extraction of time series data from a data source to its intelligent identification. The publication's results as well as an outlook on further research are discussed in Sect. 5.

2 State of the Art

This section gives an overview of the current state of machine data acquisition and intelligent identification of time series data. Moreover, an outlook onto approaches combining these disciplines, i.e. approaches that generally enable the deployment of intelligent, data-based applications is given.

2.1 Machine Data Acquisition

The acquisition of machine tool controller data is the prerequisite for the deployment of intelligent applications. However, this basic link within the construct that is Industrie 4.0 is easily overlooked. It appears research and industry interests are predominantly focussed on developing solutions whose functionalities assume existing data as a starting point [2]. Moreover, diverse manufacturing environments, which employ a variety of different machines, programmable logic controllers (PLCs) and communication protocols, constitute a challenge for any data acquisition approach [3]. These factors have led to the current situation of industrial data acquisition, which can be broken into two categories. The first category is represented by applications (monitoring, planning etc.) provided by machine manufacturers, into which data acquisition is integrated. Such apps, albeit convenient and reliable solutions, are only applicable to specific products and use cases. The second category are software packages solely dedicated to data acquisition. These are widely applicable solutions which distinguish themselves through the range of supported communication protocols/standards. The flexibility of such solutions however requires greater know-how on the part of the user.

A remedy to this status quo lies within the standardisation of information exchange. Communication standards such as OPC UA, MTConnect and MQTT make it easier for developers to gain access to machine data. However, the limiting factor to the deployment of such standards to the shopfloor is the ability or inability of the equipment at hand to support them. Considering the age of machinery that is currently in use in Europe and Northern America, which ranges between 10 and 20 years [4], data acquisition cannot solely rely on these newer standards.

2.2 Automated Parameter Identification

As part of a feasibility study for automated parameter identification, an approach for a rule-based identification process was developed. A set of five rules is used to assess the probability that a data sample belongs to a predefined class. In the case of high probability, the sample is assigned to this class. The rules are based on plausibility checks in five areas: comparison of an expected signal curve with the actual signal curve in G-code (only applies to NC axes), comparison of the value range with a previously defined target range, comparison of the gradient with a target value, correlation with other parameters and the name of the variable recorded in the sample. The plausibility check requires extensive domain knowledge, which makes this approach challenging to integrate.

Based on this feasibility study for parameter identification, a three-stage identification process combining both intelligent and rule-based approaches was developed. On the first stage an initial classification of the signals is conducted. To this end, three different artificial neural networks and a random forest model were tested, and parameter studies were also carried out to identify the parameter configurations with which the AI models achieve the best results. On the second and third stages the signals are further classified and assigned to the respective axes of the machine tool. The results of this method are discussed in [5] and [6].

2.3 Holistic Solutions for the Provisioning of Industrial Data

Holistic applications that deal with the provisioning of industrial data in general and take on tasks ranging from the initial discovery and extraction of data to its translation into a uniform information model have not yet established themselves on the market, though some approaches do exist. Rockwell Automation introduces such concepts in two US patents [7] and [8]. The proposed solutions intend the discovery of a manufacturing plant's data sources (e.g. industrial controllers) and the data items stored therein as well as their transformation into a uniform information model. The data items are then made available via a search engine. A critical feature of these approaches is the desired functionality to access data items distributed throughout a manufacturing plant from the outside through a client application without the need of physical proximity to the data sources. The contextualisation of data into an information model results from its denomination (data tags). Detection of dependencies between data items is also envisaged. This is done by examining data tags and their mutual referencing in control programs.

While these approaches are promising, presently no state-of-the-art application exists, that would provide a sufficient solution to the combined challenge of acquiring data from a variety of data sources and automatically identify them based on their characteristics. Therefore, the following sections propose a conceptual design and a prototypical implementation for such a system.

3 General Concept

This section illustrates the general concepts that the authors propose for an integrated solution, providing functionalities for data acquisition and auto-identification. As will be shown, the integrated solution joins two separate software projects that are currently in development at wbk Institute of Production Science. These projects deal with data acquisition and classification of machine tool data respectively and can be deployed as standalone solutions. However, a productive use case results from the combination of their functionalities.

3.1 Proposed Concept for Machine Data Acquisition

The machine data acquisition landscape is dominated by two categories of applications. On the one hand are machine-specific apps, whose operation does not require extensive know-how. On the other hand are broadly applicable software suites, whose deployment and operation do require expertise. The herein proposed approach seeks to bridge the gap between these two types of applications. The aim is to create an application which enables the user to extract data from a variety of data sources without the need of any prior knowledge or training. This is achieved through the bundling of solutions for industrial communication under one consistent framework. Due to the availability of open-source libraries, Java was chosen as a programming language for this data acquisition approach. As shown in Fig. 1, an array of libraries which serve as "drivers" for a specific data source are incorporated into one framework. This framework includes an http server, serving as the connecting point for client applications such as the auto-identification app introduced in Sect. 3.2. The use of a REST API as connecting point ensures an easy integration of data acquisition capabilities into the workflows of client applications. Via this connecting point, a client can request a listing of a data source's parameters (e.g. all nodes of an OPC UA server or every topic published by an MQTT broker). To establish a connection, the user must provide the necessary access data depending on the data source. In the case of a UA server, its endpoint URL is required, an MQTT broker requires hostname and port—login credentials may also be required.

From the listing of parameters of the data source, the user can then choose a subset of parameters which they would like to subscribe to. A subscription records the parameters' values with a sample rate and a number of samples as specified by the user. The resulting time series are then transferred to the client application's http client as JSON objects.

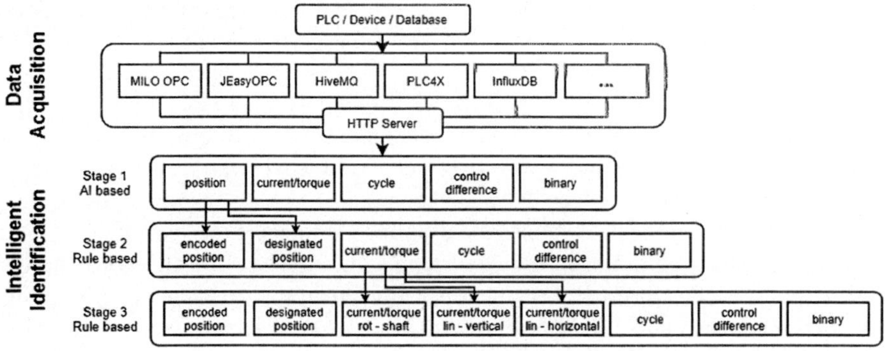

Fig. 1. Proposed concept for machine data acquisition and parameter identification.

3.2 Proposed Concept for Automated Parameter Identification

The concepts for intelligent identification of control parameters proposed in the following are incorporated into one software application that operates on time series data—for instance as provided by the above-mentioned data acquisition app. The automatic identification of the parameters is divided into three levels (see Fig. 1). The first stage is an AI-based classification into the classes "position", "current/torque", "control difference", "cycle" and "binary". The second and third stages follow a rule-based approach that determines correlations among the signals and assigns them to individual axes.

Stage 1

For the AI-based identification on stage 1, three different artificial neural networks (ANN) and one random forest classifier were tested. The ANN architectures that were used are the recurrent Long-Short Term Memory (LSTM) network, a residual neural network (ResNet) and a fully convolutional network (FCN). The ANNs can process time series data, whereas the random forest classifier uses specific characteristics of data, such as the minimal difference between two data points, the occupied bandwidth, the sum of all absolute differences between two data points, the range, the minimum of the second derivative, the number of maxima of the derivative etc. The four AI models assign each signal to one of the five classes "binary", "position", "cycle", "control difference" and "current/torque".

Stage 2

One goal of the second stage of the classification approach is to assign individual signals within the "position" and "current/torque" classes to one another to identify pairs of target position and actual position signals as well as pairs of current and torque signals. The second goal is to classify the position signals as target position and actual position signals.

To identify the actual and target position signals belonging to one axis or the shaft as well as current and torque signals of one axis or the shaft, the approach uses the Spearman correlation to pair signals from the position class as well as from the "current/torque" class. From all signal pairs with a correlation coefficient higher than 0.99, the signal pair with the highest correlation coefficient is used for further classification.

The prerequisite for the identification of the target position and actual position signals is their synchronised sampling, as the identification approach relies on the time difference between the target and the actual signal (the target signal always being slightly ahead). The identification process uses the dynamic time warping (DTW) algorithm to distinguish the target position from the actual position. To avoid long computation times, only characteristic segments of signals are examined, which are the starting and braking moments. To extract these locations in the signals, the standard deviation of samples of five data points each is formed. A segment is extracted around the sample with the highest standard deviation and the DTW algorithm is applied to the two segments of a pair of position signals. The DTW algorithm matches the two segments to each other and, as a result, outputs a path that connects the respective matching points of the segments. The alignment of the path can be used to determine which signal precedes the other. Based on this, the position signals can be classified into the new classes "designated position" (leading signals) and "encoded position" (following signals).

Stage 3

On stage 3 of the classification process, the "current/torque" signal pairs and the target position of the same axes as well as the corresponding control difference and the target position get assigned to one another. Based on that, the current and torque signals of the shaft and those of the linear axes can be distinguished. The current and torque signals belonging to the linear axes can be subdivided even further into horizontal or vertical axis signals.

For the pairwise assignment of the target position and the current torque pairs as well as the target position and the control difference, the approach is again to identify the signals with the highest Spearman correlation and assign them to one another. When assigning the target position signals and the "current/torque" signal pairs to each other, only one signal from the signal pair is used for correlation. The second signal is then added respectively. The signals in the one current and torque signal pair that is uncorrelated with a position signal can then be classified as the shaft current and torque, since the shaft as a rotating axis does not have position signals.

This step is followed by the classification of horizontal and vertical axes. As the vertical axis must work against gravity, its mean absolute current exceeds those of horizontal axes. Since the current signals cannot be distinguished from the torque signals and the range of the two signals is different, it is necessary to take both signals in the "current/torque" signals pairs into account to make the pairs comparable. For this purpose, firstly, the absolute mean values of the two individual signals within one pair and, secondly, the mean of the two absolute mean values are calculated. The one signal pair with the highest common mean value can be assigned to the vertical axis. Using the absolute value of the means is necessary as the sign depends on the direction of movement.

4 Showcase of the Prototypical Solution

Following the introduction of the general approach, this section illustrates the complete process chain consisting of data acquisition and auto-identification. As an exemplary data source, an OPC UA simulation server is used, which, under a set of simulated nodes, publishes signals from a machine tool data set recorded during the machining of a part. Among others, the recording consists of torque, current, encoded position, designated position, and control difference signals of 6 axes of a machine tool.

In this use case, all user interaction takes place inside the GUI of the auto-identification app that serves as the overarching user interface for data acquisition as well as the identification process, including the possibility of the input of domain knowledge. The data acquisition is initiated via the identification app's http client.

Along with OPC UA, data acquisition functionalities are available for OPC DA, MQTT, select PLCs and InfluxDB. Support for additional data sources is in development.

As shown in Fig. 2, the connection to the data source (here: the OPC server) is established through the auto-identification app, which connects to the data acquisition app's http server and ultimately to the data source itself. Upon the setting up of the connection, a query of every node is executed. The resulting list of nodes is returned to the user, who can then choose a subset of nodes for subscription. A subscription is executed for as long as the user specifies and with the user's desired sampling rate. Upon completion of the data acquisition app's subscription, the resulting time series are returned as JSON object, and the auto-identification app comes into action.

After choosing an AI model from the selection presented in Sect. 3.2, the user may provide information about the machine from which the data originates. Such domain knowledge includes the type and number of axes (linear, rotary) as well as the expected signals, such as position, torque and current signals etc. The entry of this additional information is optional and serves to increase the accuracy of the identification. The results of the identification are compared with the added information and deviations are displayed after the identification. Finally, settings for the identification model can be made. On the data side, this includes the number of data points per signal sample to be classified and the limit value for the standard deviation of a signal below which a signal is classified as inactive. On the model side, it can be specified how many samples of a signal are to be evaluated and how many predictions of a class must match in order to assign the signal to this class. In addition, the user can specify the minimum prediction probability that a sample needs to be assigned to a class and, for the cycle signal, how much it may deviate from an ideal cycle signal, which increases incrementally. For levels 2 and 3, the minimum correlation coefficient for the individual correlation pairs to be identified can be specified. The higher these parameters are, the higher is the probability that the signals and correlation pairs have been identified correctly. However, the number of signals and correlation pairs that remain unidentified increases, since more prediction probabilities and correlation coefficients may not exceed their specified minimums. After the data and parameters have been selected, the classification is carried out automatically and the results including the prediction probabilities are displayed on the GUI. For the data set used in this work, the model was able to identify all active signals on stages 1 and 2 correctly with prediction probabilities of at least 96%, except for the control difference signals which were classified as current or torque signals. Furthermore, the rule for the assignment of the signals to individual axes could not be executed since not all correlated pairs could be identified on stage 2.

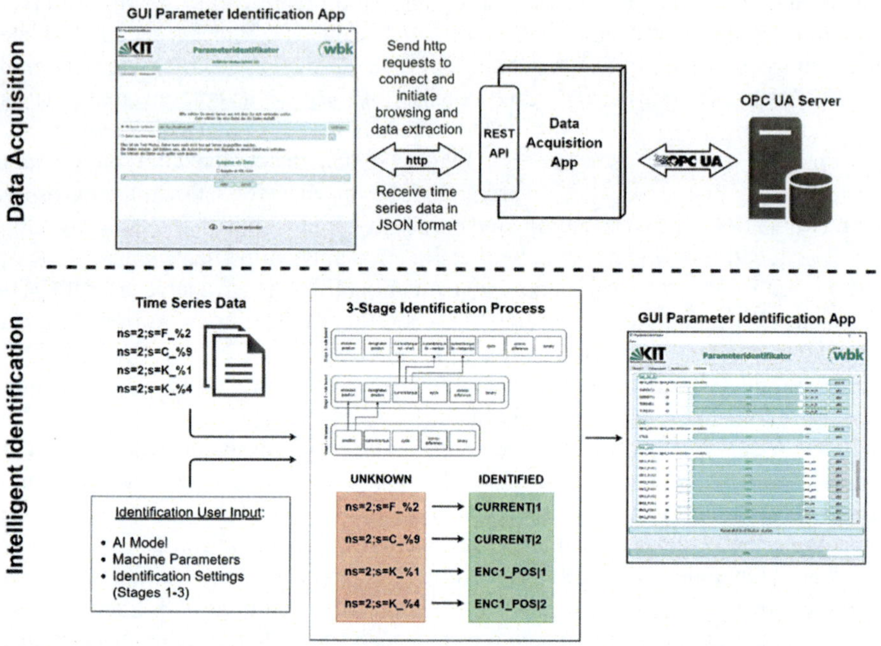

Fig. 2. Example use case: data acquisition combined with intelligent parameter identification.

5 Conclusion

As illustrated in Sect. 4, the capabilities of acquiring controller data and conditioning it to be used in further applications could be integrated into one application. This integrated range of functions is a novel approach, since currently existing applications mostly cover the respective functions only. Moreover, the application is designed in a manner which allows users with no prior knowledge of a specific machine or data source and its control parameters to deploy software solutions which depend on these parameters. In summary, the paper's aims were to:

- Showcase the challenges that constrain the widespread adoption and use of intelligent, data-driven applications
- Propose and implement a prototypical solution to these challenges.

The next steps in the further development of the auto-identification app are improvements in the identification of data on stages 2 and 3 whose stage 1 identification was incomplete. Additionally, further improvement of identification through user input of domain knowledge is sought. For machine data acquisition, further steps would be the integration of additional types of data sources as well as functions for searching networks for connected data sources.

References

1. VDW e.V.: OPC UA for Machine Tools Part 1: Machine Monitoring and Job Overview OPC Unified Architecture Information Models (2020)
2. Teng, S.Y., Touš, M., Leong, W.D., How, B.S., Lam, H.L., Máša, V.: Recent advances on industrial data-driven energy savings: digital twins and infrastructures. Renew. Sustain. Energ. Rev. **135** (2021)
3. Lenz, J., Wuest, T., Westkämper, E.: Holistic approach to machine tool data analytics. J. Manuf. Syst. **48**, pp. 180–191 (2018)
4. Guerreiro, B.V., Lins, R.G., Sun, J., Schmitt, R.: Definition of smart retrofitting: first steps for a company to deploy aspects of industry 4.0. In: Advances in Manufacturing, vol. 1, pp. 161–170. Springer International Publishing (2018)
5. Gönnheimer, P., Puchta, A., Fleischer, J.: Automated identification of parameters in control systems of machine tools. In: Production at the Leading Edge of Technology, vol. 1, pp. 568–577. Springer, Berlin Heidelberg (2020)
6. Gönnheimer, P., Karle, A., Mohr, L., Fleischer, J.: Comprehensive machine data acquisition through intelligent parameter identification and assignment. Procedia CIRP **104**, 720–725 (2021)
7. Bliss, R.E., Reichard, D.J., et al.: United States patent application publication: crawler for discovering control system data in an industrial automation environment (2016)
8. Dorgelo, E.G., Gordon, K.G., et al.: United States patent: indexing and searching manufacturing process related information. Patent No.: US 8,285,744 B2 (2012)

Data-Based Measure Derivation for Business Process Design

M. Schopen(✉), S. Schmitz, A. Gützlaff, and G. Schuh

Laboratory for Machine Tools and Production Engineering (WZL), RWTH Aachen University, Aachen, Germany
M.Schopen@wzl.rwth-aachen.de

Abstract. Highly competitive markets urge manufacturing companies to enhance the performance of their business and production processes. As part of business process improvements, companies examine as-is-process models for weaknesses (process analysis) and then derive measures for an improved to-be-process (process design). The process design phase is typically conducted manually in workshops, which makes it liable to a set of challenges. First, it is highly dependent on the participants' knowledge about measure types and their suitability for addressing process weakness types. This knowledge is often heterogeneous and deficient. Second, participants with proprietary positions expose the workshop-based approach to subjective influences. Third, open brainstorming on potential measures is time-consuming. Fourth, the impact of measures on the business process is barely quantifiable at the time of process design without using process data. Existing approaches for event log-based weakness detection use this data to semi-automate the process analysis phase. Their outputs are process weakness types detected in event logs on business processes. For the phase of process design, such semi-automated approaches are not available. This paper presents a concept for data-based measure derivation for business process design. Therefore measure types for process design, that impact the process flow and are context-independent, are identified in literature. In a second step, they are allocated to process weakness types that they can address. In the third step, the measure types' impact on business processes is quantified using event log information. Thereby, the concept enables semi-automated measure identification and process performance related decision support for business process design.

Keywords: Process design · Decision support · Process mining

1 Introduction

Manufacturing companies face a great challenge given the increasing demands of globalisation and the complex, dynamic markets [1]. This consequently increases the pressure on production processes, which can be subsumed into the category of business processes following the definition of [2]. The design of effective and efficient business processes hence is an essential factor for the operational success of manufacturing companies [2] [3]. It also serves as the foundation for business process improvement, alongside process

M. Liewald et al. (Eds.): WGP 2022, LNPE, pp. 564–573, 2023.
https://doi.org/10.1007/978-3-031-18318-8_57

analysis [2]. Business process improvement, as defined by [4], is the practice of continuously evaluating, analysing, and improving business processes that are critical to an organization's success. Within business process improvement there are several phases, such as process identification, process discovery, process analysis, process redesign, process implementation, and process monitoring [2]. In the design phase, the existing business process structures are redesigned. This is done with the aim of improving business process performance parameters, such as process time [5].

For process design, conventionally methods such as workshops and interviews are deployed [6], which inhere a series of challenges. By using workshop-based methods, the collection and analysis of information within process design is highly time-consuming [7, 8]. Furthermore, the results of these methods in the process design phase are influenced by the subjective descriptions and impressions of the participants [3]. Thus, the evaluation of the improvement measures depends to a large extent on the methodological experience and process knowledge of the participants. In addition, the effects of a measure cannot be systematically quantified in workshops without the use of process data [9].

For process analysis, there already exist approaches that counteract these challenges. For example, the identification of process weaknesses can partly be automated by applying process weakness model algorithms to event logs [10]. This helps to enhance current process mining solutions. Process mining uses data stored in event logs, which are readily available in today's information systems, to monitor and improve business processes [11]. Event logs are sets of events that are captured during process execution. The constitutive components of an event log are the case-ID of the process execution, the activity name of the recorded event as well as its start and/or end time in the form of timestamps [12].

For the phase of process design, a data-based solution for derivation of improvement measures does not yet exist. Analogue to process analysis, existing process mining solutions do not provide assistance for process design as they do not incorporate domain-specific expert knowledge [8]. This paper develops improvement measure models that incorporate domain-specific knowledge to enable data-based methodical support of the process design phase and address the aforementioned challenges of current process design approaches. This paper is structured as follows. In Sect. 2 related work is reviewed, before explaining the solution approach in Sect. 3. Based on this approach the measure models are developed in Sect. 4 prior to summarizing the results and giving an outlook to future research in Sect. 5.

2 Related Work

In literature, several approaches for measure derivation in business process design can be found. A series of both creative and heuristic-based approaches were examined regarding their suitability to solve the challenges elaborated in Sect. 1.

Creative process design approaches such as workshops and interviews focus on the interpersonal interaction to initiate and amplify the generation of ideas for designing the business process [2]. The challenges of these methods include a lack of precise and systematic approaches to problem solving due the limited knowledge of the participants [3]. This leads to subjective influences and inconsistent results [14], making them unsuitable for the automated development of measures, as explained in Sect. 1.

The creative approaches are contrasted with heuristic-based approaches, which are classified as analytical process design methods [2] and have already been utilized to address the challenges of creative approaches. Instead of creative solution finding, these approaches apply pre-developed measure types to improve business processes [2]. Focusing on a heuristic-based approach, [15] developed 29 best practices (BP) as clearly defined business process design measures, which were subsequently categorised in terms of their impact on process costs, flexibility, quality and time. Niedermann [16] enhances this approach by adding methodological knowledge regarding weakness and measure types using data-based methods. Despite [13] additionally addressing the identification of weaknesses within a business process using event logs, there is no quantification of the impact business process weaknesses have on business process improvement or prioritization of actions to increase process performance. Cherni et al. [17] uses event logs to for weakness identification as well, but, like [13], lacks prioritization of measure types. Kumar and Rong [18], on the other hand, quantifies the impact of business process weaknesses on business process performance, in contrast to [13]. Kumar and Rong [18] develops a mathematical calculation model for six of [15] best practice measures which also quantifies the potentials and further prioritizes measure types to increase performance. Nevertheless, [18] overlooks the use of event logs and hence does not address all challenges.

None of the analysed approaches for standardized and data-based process design can fully address the challenges of conventional process design. Although there are standardized "best practice" process design measures, they lack a distinct allocation to business process weakness types. Additionally, an event log based quantification of potential benefits of process design measures with regard to business process performance is missing. Since none of the approaches holistically addresses these aspects, the potentials of standardization and databased support in the phase of business process design can be assessed as not fully exploited. Therefore, the central challenge is the formalization of domain-expert knowledge on business process design measures. This formalization needs to result in measure type models with a standardized allocation to generic business process weakness types from data-based methods for process analysis [10]. Additionally the models need to include the quantification of the measure type's impact onto business process performance.

3 Research Methodology

The formalization of domain expert knowledge into measure type models for process design is necessary for a systematic, automated and objective event log based procedure. This formalization is carried out in three steps. First, business process design measure types are identified in the literature, redundancies are eliminated and the relevance of the resulting measures is ensured. Subsequently, the measure types found are assigned to the weakness types introduced in [10]. Closing, the impact of a measure type onto the business process is assessed and quantified, which transforms the measure type into an improvement measure model applicable for process design.

To identify improvement measure types for process design, a systematic literature review is conducted in three steps (cf. Sect. 4.1), the results of which are consolidated with the help of an interdependency analysis according to [19]. The aim of this analysis is to identify mutual dependencies and duplicates within the list of identified measure types in order to aggregate those with a high degree of content similarity. The weighting of the interdependencies is done pairwise and in four levels, whereby the rating 3 stands for a high, 2 for a medium, 1 for a low and 0 for no interdependency between the measures considered. An aggregation is made for those measures that showed a high mutual dependency. The results are then evaluated in terms of their relevance for a standardised and objective derivation of business process design measure models based on event logs in the context of process mining. This involves checking whether the measure type is clearly defined, whether it has an influence on the process structure and whether it can be applied context-independently.

The remaining aggregated measure types are then assigned to the weakness types defined in [10]. For this purpose, first the weakness as well as the improvement measure types' influence on the process structure are analysed. If a match of the effects is found, the corresponding measure type is considered suitable for improving the respective process weakness type. This step was conducted in workshops with experts from the field of business process management who are regularly involved in business process improvement.

Following this, the impact of the measure types on business process performance is determined to enable an objective and context-independent indication of a measure type's benefit based on event logs. To do this, mathematical equations were formulated that describe the effects of improvement measure types on process time.

The quantification is based on a consideration of process time as a proxy for business process performance. The representative use of process time for business process performance is thereby justified by the composition of the business process performance term. Business process performance consists of the three aspects process time, process cost and process quality [20]: Process time is considered as a key performance characteristic and can be measured directly from event logs. Process costs however are not directly derivable from event logs and can therefore not be determined context-independently and objectively. As a result, taking costs into account would contradict the requirements of this paper. Additionally, the process costs are often calculated by multiplying a cost rate with the corresponding process time. Thus, process time is the constitutive factor also for process costs and double counting is to be avoided. The execution quality of a process indicates whether and to what extent a recorded process deviates from the originally planned chain of activities. Deviations are considered as performance losses that are expressed in the form of additional process time and are therefore included in the analysis of the latter. Consequently, process time represents all aspects of business process performance that are to be taken into account in this work and is therefore used as a proxy for business process performance hereafter.

For the calculation of the process time, the process cycle time is used. The results presented in this paper are based on weakness and measure types in sequential occurring events. The implicit exception to this is parallelization, which is characterised by the parallel occurrence of events. This approach is sufficient for a cycle time calculation, since in this context it is not the position of weaknesses and measures that is decisive, but only their timely effects [6].

For the quantification of the measure types, a universal model of process time is developed, which considers business processes as a chain of successive events with execution and transition times (see Fig. 1). This universal process time model generally follows the logic of [21]. From this model, the additional times are excluded, as these are calculative times meant only for planning purposes. Hence, additional times are not to be included in the evaluation of actual process times. The developed process time model is based on the structure of event logs in order to use them as a data source for later calculation. The required variables and the resulting time model are presented in Fig. 1.

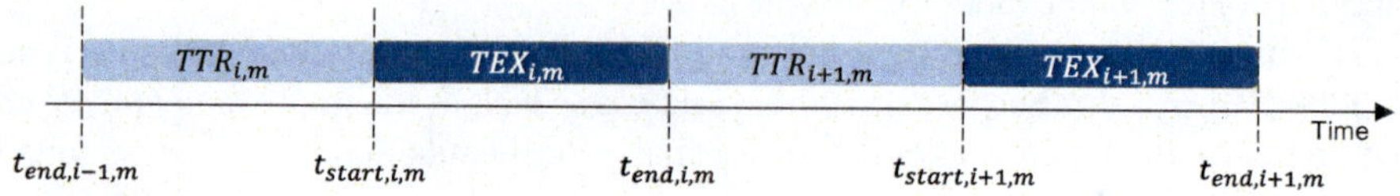

with:
i = *event*
m = *case-ID*
$t_{start,i,m}$ = *start timestamp of an event i with case-ID m*
$t_{end,i,m}$ = *end timestamp of an event i with case-ID m*
$TEX_{i,m}$ = *execution time of an event i with case-ID m*
$TTR_{i,m}$ = *transition time of an event i with case-ID m*

Fig. 1. Process time model

The process time model developed in this way possesses all the properties to be utilised subsequently in the quantification of improvement measure effects.

4 Findings

4.1 Collection of Potential Improvement Measure Types

To identify relevant measure types for process design, a systematic literature review was conducted following the logic of [22] on the portals Science Direct, IEEE Xplore, Scopus, Web of Science, Taylor & Francis and Google Scholar. The following boolean term was used as the search string: (business process) AND (design OR reengineering OR improvement) AND (measure OR pattern OR heuristics OR best practice OR suggestion), which returned 4166 contextual publications for further review. The result was a list of 292 individual measure types, which in a first step were then checked for duplicates and consolidated using an interdependency analysis, leaving 81 types. As a further restriction, the results had to fulfil three requirements to be considered as relevant. First, the identified measure types need to be defined unambiguously in literature to allow clear delineation. Second, relevant measure types need to have an influence on the structure of business processes in terms of execution or transition times. Third, measure types need to be generically applicable in business process improvement, independently from process and company context. This serves the concept's central idea to formalise methodological knowledge generically. Context-dependent knowledge is explicitly not integrated into the model, but considered in the specific application of the model. The remaining number of relevant improvement measure types could thus be reduced to six. The results are briefly presented below as definitions of the measure types relevant for business process improvement:

1. **Avoidance of Iterations** [23, 24]
   Rearrangement of the process structure so that backloops and activity repetitions are eliminated
2. **Elimination** [24]
   Elimination of activities that do not contribute to value creation
3. **Parallelization** [25]
   Parallel execution of suitable activities that were previously executed sequentially
4. **Acceleration** [26]
   Reduction of the execution time of an activity to a shorter recorded execution time of the activity
5. **Integration** [26]
   Combination of separate activities into a superordinate one in order to reduce interfaces and transition times
6. **Reduction of Transition Time** [24, 27]
   Reduction of waiting and transport times between activities.

4.2 Assignment of Weakness and Measure Types

The identified six generic improvement measure types were then assigned to the weakness types in [10] (see Table 1). The assignment was made on the basis of a process-structural analysis of the weaknesses' and measures' impact:

The weakness type of *redundant activities* can be counteracted by **avoidance of iterations**. In this context, an iteration is considered to be any repetition of an activity. By paying attention to an error-free and complete initial execution of an activity, further executions become superfluous and redundant activity executions can be avoided. The same applies to *backloops*, where, in contrast to redundant activities, a whole chain of activities is repeated as a block. Iteration avoidance can thus prevent unnecessary activity execution. To avoid *unintentional activities*, **elimination** of these activities is performed in business process improvement, thus completely removing the activities from the business process. *Parallelizable activities* that are executed sequentially in the as-is state of a business process can be counteracted through **parallelization** in order to shorten the total throughput time. Activities that have an *unsuitable execution time* can be improved by **acceleration** of the execution. This ensures that an activity is always performed with maximum efficiency in the shortest possible time. In order to reduce *transition times*, the improvement measures **integration** and **reduction of transition time** are suitable. Integration combines two previously sequential activities into one superordinate activity, so that the transition time between the two activities is effectively reduced to zero. Another approach to improve transition time weaknesses is a reduction of transition times, whereby the layover and transport times between two activities can be reduced, but not eliminated.

Table 1. Allocation of improvement measure types to weakness types

Business Process Weakness Types [10]	Business Process Improvement Measure Types: Avoidance of Iterations	Elimination	Parallelization	Acceleration	Integration	Reduction of Transition Time
Redundant Activities	⊗					
Backloops	⊗					
Unwanted Activities		⊗				
Parallelizable Activities			⊗			
Unsuitable Scopes				⊗		
Transition Times					⊗	⊗

The table shows that the measure type **avoidance of iterations** is suitable for improving the two weakness types *redundant activities* and *backloop*. This is due to the high degree of similarity of the weaknesses' process-structural impact. Simultaneously, two measure types exist for the improvement of *transition times*.

4.3 Quantification of the Impact of Measure Types on Process Time

The previous chapters generated relevant, suitable and generally applicable measure types for process design. To obtain the necessary information on potential benefits of these measure types, their effects on the process time had to be determined next.

The first step in this process was a qualitative analysis of the possible impacts of each measure type with regard to their effects on process time. The obtained knowledge about the possible manifestations of the measure types was then combined into a generally applicable pattern. If such a generalisation was not immediately possible, assumptions were formulated under which the statements would be valid. The qualitative findings on the time-related effects of the measure types were then translated into simple mathematical formulations. This first required the definition of the necessary variables. In the subsequent quantification, these variables were used to mathematically describe the temporal effects of the measure types independent of business process and company context.

The results are presented below using the example of the improvement measure type avoidance of iterations for backloops. First, the effects of the improvement measure type were recorded qualitatively using the time model developed above. For example, Fig. 2 illustrates that by eliminating the repeated execution of a process section, the cycle time of the entire business process is reduced by the sum of the execution and transition times within that process section.

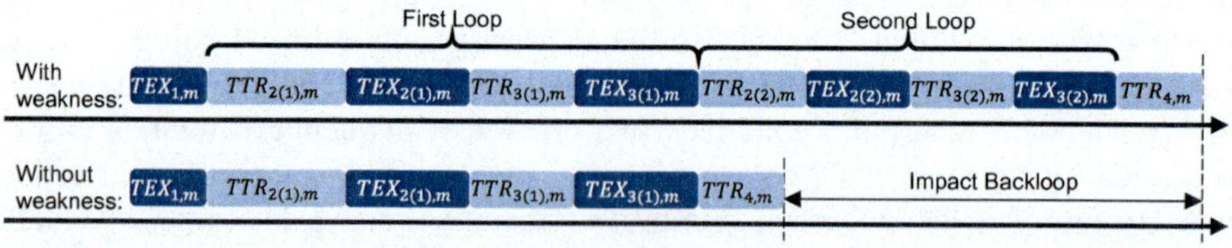

Fig. 2. Effect of the measure type avoidance of iteration to counteract a backloop

As shown in Formular 1, these qualitative results were then quantified by transferring the findings into mathematical formulas for the time-related effects of the measure types:

$$T_{avoidance\ of\ iteration,backloop,I^*,m^*} = \sum_{i \in I^*} \left(t_{end,i,m^*} - t_{end,i-1,m^*}\right) \tag{1}$$

with:

I^* = Set of events within the backloop
m^* = case-ID in which the backloop occurs
$T_{avoidance\ of\ iteration,backloop,I^*,m^*}$ = Time potential through avoidance of the backloop.

An overview over all derived measure quantifications is shown in Table 2.

Table 2. Quantification of business process improvement measure types

Business process improvement measure	Quantification formula
$T_{integration,i^*,m^*}$	$t_{start,i^*,m^*} - t_{end,i^*-1,m^*}$
$T_{reduction\ of\ transition\ time,i^*,m^*}$	$\left(t_{start,i^*,m^*} - t_{end,i^*-1,m^*}\right) - p\% - \underset{m \in M}{Quantile}\left(t_{start,i^*,m} - t_{end,i^*-1,m}\right)$
$T_{avoidance\ of\ iteration,backloop,I^*,m^*}$	$\sum_{i \in I^*} (t_{end,i,m^*} - t_{end,i-1,m^*})$
$T_{avoidance\ of\ iterations,redundant\ activity,i^*,m^*}$	$t_{end,i^*,m^*} - t_{end,i^*-1,m^*}$
$T_{elimination,i^*,m^*}$	$t_{end,i^*,m^*} - t_{end,i^*-1,m^*}$
$T_{acceleration,i^*,m^*}$	$\left(t_{end,i^*,m^*} - t_{start,i^*,m^*}\right) - p\% - \underset{m \in M}{Quantile}\left(t_{end,i^*,m} - t_{start,i^*,m}\right)$
$T_{parallelization,I^*,m^*}$	$\sum_{i \in I^*} (t_{end,i,m^*} - t_{end,i-1,m^*})$

The summarising result of the presented research thus consists in seven business process design measure models, which contain the information needed to determine which measure type is applicable in which context as well as what effects on process time can be expected from their application in a specific business process improvement scenario. It is noticeable that the quantification formulas for the measures "backloop" and "parallelization" as well as "avoidance of iterations, redundant activity" and "elimination" are identical. This is because their effects on the process structure, in particular on the execution and transition times, are identical. However, these measures need to be listed separately, as their definition is substantially different (s. Sect. 4.1) and they are founded on different process weaknesses.

5 Discussion and Outlook

Increasing competitive pressure makes it necessary to continuously improve the effectivity and efficiency of business processes. Designing enhanced to-be processes is the decisively value-adding step in business process improvement. However, the results of conventional workshop-based methods are subjective and non-repeatable as they are highly dependent on the experience and skills of the participants. Besides, workshops and interviews make process design projects costly and time-intensive, which hampers their regular execution.

The data-based and automated identification and evaluation of potential improvement measures can help solve these challenges. For this purpose, application independent improvement measure models were derived, assigned to the generic weakness type models of business process improvement and quantitatively evaluated with regard to their impact on process time. The resulting set of six measure models enable event log-based and standardized derivation of measure proposals including impact quantification as a decision support to reduce efforts while increasing the objectivity and speed of process design.

This approach formalizes domain-specific expert knowledge about process design analogue to the formalization in [10] for the process analysis phase to enable further automation. Future research should address the incorporation of the weakness and measure models into a data-based decision support in form of a superordinate calculation model. The calculation of such a model also requires a process time based central metric to determine overall business process performance as a central decision variable.

Acknowledgement. Funded by the Deutsche Forschungsgemeinschaft (DFG, German Research Foundation) under Germany's Excellence Strategy—EXC-2023 Internet of Production—390621612.

References

1. Quick, D., Eissen, J., Genge, N.: How Intelligent Processes Differentiate Best-Run Businesses in the Digital Economy, pp. 4–5 (2019)
2. Dumas, M., La Rosa, M., Mendling, J., Reijers, H.A.: Grundlagen des Geschäftsprozessmanagements, 1st ed. Springer, Berlin (2021)
3. Schuh, G., Gützlaff, A., Schmitz, S., Van der Aalst, W.M.P.: Data-based description of process performance in end-to-end order processing. CIRP Ann. Manuf. Technol. **69**(1), 381–384 (2020)
4. Povey, B.: The development of a best practice business process improvement methodology. Benchmarking Qual Manage Technol **5**(1), 27–44 (1998)
5. Hammer, M., Champy, J.: Reengineering the Corporation, 1st ed., pp. 34–38. Harper Business, New York (1994)
6. Schmelzer, H.J., Sesselmann, W.: Geschäftsprozessmanagement in der Praxis, 9th edn. Carl Hanser Verlag, München (2020)
7. Kerremans, M.: Market Guide for Process Mining. Gartner Inc., Stamford (2019)
8. Bergener, P., Delfmann, P., Weiss, B., Winkelmann, A.: Detecting potential weaknesses in business processes. Bus. Process. Manag. J. **21**(1), 25–54 (2015)
9. Netjes, M.: Process improvement: the creation and evaluation of process alternatives, pp. 1–4. Technische Universiteit Eindhoven, Eindhoven (2010)
10. Schuh, G., Gützlaff, A., Schmitz, S., Schopen, M., Bröhl, F.: Event log-based weakness detection in business processes. In: IEEE International Conference, pp. 734–738. IEEE, Singapore (2021)

11. Van der Aalst, W.M.P., et al.: Removing Operational Friction Using Process Mining: Challenges Provided by the Internet of Production (IoP), pp. 1–4 (2021)
12. Reinkemeyer, L.: Process Mining in Action, 1st ed., pp. 3–10. Springer Nature Switzerland, Cham (2020)
13. Gross, S., Yeshchenko, A., Djurica, D., Mendeling, J.: Process Mining Supported Process Redesign: Matching Problems with Solutions, pp. 24–31 (2020)
14. Hierzer, R.: Prozessoptimierung 4.0, 2nd ed. Haufe Lexware, Freiburg (2020)
15. Reijers, H.A., Liman Mansar, S.: Best practices in business process redesign: an overview and qualitative evaluation of successful redesign heuristics. Omerga **33**(4), 283–306 (2005)
16. Niedermann, F.: Deep Business Optimization (2015)
17. Cherni, J., Martinho, R., Ghannouchi, S.A.: Towards improving business processes. In: Procedia Computer Science, pp. 279–284. Elsevier B.V., Amsterdam (2019)
18. Kumar, A., Rong, L.: Business workflow optimization through process model redesign. In: IEEE International Conference, pp. 1–17. IEEE, Portland (2020)
19. May, G., Chrobok, R.: Priorisierung des unternehmerischen Projektportfolios. Zeitschrift für Führung und Organisation **70**(2), 108–114 (2001)
20. Schuh, G., Boos, W., Kuhlmann, K., Rittstieg, M.: Bewertung von Geschäftsprozessen in indirekten Bereichen. Zeitschrift für wirtschaftlichen Fabrikbetrieb **105**(4), 265–270 (2010)
21. Gummersbach, A., Bülles, P., Nicolai, H., Schieferecke, A., Kleinmann, A., Hinschläger, M., Mockenhaupt, A.: Produktionsmanagement, 4. Auflage, Verlag Handwerk und Technik, Hamburg (2008)
22. Kitchenham, B., Charters, S., Budgen, D., Brereton, P., Turner, M., Jørgensen, M., Mendes, E., Visaggio, G.: Guidelines for performing systematic literature reviews in software engineering, Version 2.3 (2007)
23. Gadatsch, A.: Grundkurs Geschäftsprozess-Management: Analyse, Modellierung, Optimierung und Controlling von Prozessen, 9th edn. Springer Vieweg, Wiesbaden (2020)
24. Schuh, G.: Change Management—Prozesse strategiekonform gestalten, 1st edn. Springer, Berlin (2006)
25. Wagner, K.W., Patzak, G.: Performance Excellence, 3rd edn. Hanser Verlag, München (2020)
26. Schuh, G., Kampker, A., Stich, V., Kuhlmann, K.: Prozessmanagement. In: Schuh, G., Kampker, A. (eds.) Strategie und Management produzierender Unternehmen, 2nd ed., pp. 327–382. Springer, Berlin (2011)
27. Zur Mühlen, M., Hansmann, H.: Workflowmanagement. In: Becker, J., Kugeler, M., Rosemann, M. (eds.) Prozessmanagement, 7th ed. Springer, Berlin (2012)

Improving a Deep Learning Temperature-Forecasting Model of a 3-Axis Precision Machine with Domain Randomized Thermal Simulation Data

E. Boos[1(✉)], X. Thiem[1], H. Wiemer[1], and S. Ihlenfeldt[1,2]

[1] Technische Universität Dresden, 01062 Dresden, Germany
eugen.boos@tu-dresden.de

[2] Fraunhofer Institute for Machine Tools and Forming Technology, Nöthnitzer Strasse 44, 01187 Dresden, Germany

Abstract. With the continuous rise of industry 4.0 applications, artificial intelligence and data driven monitoring systems for machine tools proved themselves as highly capable alternatives to classical analytical approaches. However, their precision is limited to a number of crucial aspects. One of the main aspects revolves around the lack of meaningful data, which leads to imprecise and false model predictions. This issue is closely linked to production processes and machine tools in production engineering, as the available amount of meaningful real data is strongly limited. The usage of simulation models to acquire additional synthetic data is able to fill this lack. This work looks into improving the prediction accuracy of a deep learning model for temperature forecasting of a 3-axis precision machine by combing and comparing real process data with domain randomized simulation data. The used thermal simulation model is based on finite element models of the machine assemblies. Model order reduction techniques were applied to the FE models to reduce the computational effort, increasing the simulation-to-reality gap. The approach is evaluated on unseen real data, demonstrating the underlying potential of the inclusion of synthetic data from simulation models of machine behavior.

Keywords: Condition monitoring · Synthetic data · Domain randomization · Deep learning forecasting · Digital twins

1 Introduction

In machine tool engineering as well as robotics, the correct positioning of the tool center point (TCP) is at most importance for ensuring a functioning workflow, as well as meeting component quality standards. This is especially the case for high precision tasks such as milling and grinding. However, despite regular calibrations of the TCP, temperature induced deformations during different operation tasks lead to offsets of the TCP [1]. In dependence of the magnitude, these offsets lead to rejections of high precision components. One possible way to counteract is the usage of digital twins such

M. Liewald et al. (Eds.): WGP 2022, LNPE, pp. 574–584, 2023.
https://doi.org/10.1007/978-3-031-18318-8_58

as simulation models, which run in parallel to the machine tool and predict temperature induced displacements based on real load cases [2]. With this information, the TCP can be adjusted in real time. As advantageous as this approach is, it still has shortcomings. In particular, the computational time necessary, for a high precision finite element (FE) thermo-mechanical simulation of a machine tool, limiting the complexity of the model to meet the real time computation requirements. As well as the limitation to solely one prediction step, restricting additional features, which profit from a multi-horizon forecasting.

These shortcomings can be eliminated with the usage of Deep Learning (DL) forecasting models [2]. Two of the currently best performing DL forecasting models are DeepAR [3] and the Temporal Fusion Transformer (TFT) [4]. Both models are based on Recurrent Neural Networks (RNN) and specially designed for probabilistic time series multi-horizon forecasting. However, as powerful as these models can be, they are limited to two crucial aspects: the historic data itself, and the available amount of data [5]. As the prediction is limited to past events, previously unknown process behaviors and errors will not be detected. Whereas, not enough data of the overall process and possible error cases leads to imprecise and false predictions. These issues are closely linked to production processes and machine tools in production engineering. Machine tools are mostly individual or custom constructions—even if structurally identical, they often differ in features, equipment level and purpose. Therefore, their production behaviours are not fully comparable and data acquisition must happen at the individual machine [6]. In total, the availability and accessibility of enough high-quality data of production processes and machine tools for Artificial Intelligence (AI) based applications is not given. The collaboration of a physical simulation based digital twin to synthetically generate data for the AI model can fill the lack of accessible real data [7]. Unfortunately, discrepancies between simulations and the real world make the transfer challenging. This reality gap [8] postulates that even the most complex simulation is not able to represent reality in all its details, due to inaccurate parametrization, model simplifications and unmodeled physical effects, for instance such as wear-and-tear and local environmental influences. The synthetic data source does not represent reality well enough to be used in a direct equivalence with real world data [9]. Bridging this gap is crucial in the expansion of available machine tool data sets and filling the existing lack of data. In robotics and robotic learning, Domain Randomization (DR) techniques are used to enable a comparability of graphical simulated image data with real data to further expand the learning data sets [8]. The goal of these approaches is to adjust the data distribution of simulated data sets to either match the distribution of real data [10] or let the real data distribution be a subset of the synthetic data distribution [11].

This study looks into the transfer of DR techniques to improve the temperature-forecasting accuracy of a TFT model with the inclusion of synthetic data from a thermal simulation model of a 3-axis precision machine. It serves as groundwork and a proof of concept for transferring this study's findings to the TCP offset correction.

2 Related Work

Mapping real world interactions and behaviors within simulated environments is a well-known problem solving strategy. Especially the FE method offers a tool for engineers to recreate complex mechanical and thermal relationships to further improve insight. However, when it comes to transferring behaviors from simulated environments into the real world, results are taken with caution due to discrepancies between the two. Model simplifications, inaccurate parametrization and unmodeled physical effects make transferring behaviors from simulations challenging [8]. These differences are known as the reality gap. One possible approach to make the simulation closely match the physical reality are further system identifications and high-detailed simulations. However, AI-based approaches are in need of high amounts of data. High quality simulations might be helpful, but do not fully solve the issue at hand. Nevertheless, including data from high-quality graphical simulation with other approaches can reduce the number of real labeled samples [12].

When it comes to low level simulations, adding noise to the system parameters [13] can improve the realism of the simulated results. However, it also adds another layer of difficult to evaluate uncertainty, as well as might lead to unfeasible results, which both are unacceptable in production engineering. Approximate Bayesian Computation (ABC) [14] has been one of the main methods to leverage simulation uncertainties. Hereby, parameter settings are accepted or rejected if they are within a certain specified range determined by real observations. The set of accepted parameters approximates the posterior for the real parameters. However, ABC methods have a high computational effort, making them almost unusable in combination with complex FE simulations.

Simulation randomization and transfer learning techniques such as Domain Adaption (DA) have been a powerful tool in robotics and Reinforcement Learning (RL) [15]. With the inclusion of probabilistic simulation parameters, the simulation data distribution is adjusted in a way to fit the real data distribution. Classical optimization routines [16], Bayesian optimization methods [15], as well as the usage of Generative Adversarial Networks [17] have shown promising results. Another form of simulation randomization techniques is the Domain Randomization (DR) (see Sect. 3.2). DR describes the concept of highly randomizing simulation parameters in order to cover the real distribution of the real-world data in spite of the bias between the simulation model and the real world [8]. The goal is, by providing enough simulated variability, to adjust the data distribution in simulated data sets so that the distribution of real data is a subset of it. This follows the assumption, that if an AI-model at training time is capable of working with the randomized simulation data, it is able to generalize the learned data features onto the real data set at testing time. Bayesian Domain Randomization via probabilistic inference showed promising results in robotics simulation environments [18].

Most works in this field focus on computer vision tasks and robotics. For time-series data, transfer learning techniques such as DA and DR have been used for learning temporal latent relationships in health data [19], to perform speech recognition [20] and anomaly detection [21]. However, their effective implementation into AI-applied production engineering problems has be reserved yet.

3 Machine Set Up, Methods and Data Basis

3.1 Temperature Prediction of Demonstrator Machine

The machine in Fig. 1 (left) is selected as the demonstrator for this work. It is a three axes Cartesian serial kinematic in lightweight construction. It consists primarily of aluminum alloy plates which are clamped with the help of tie rods [22]. This machine has the special ability to compensate errors in six degrees of freedom. Three ball screw axes drive the z-slide. Flexure bearings connect the ball screw nuts and the guide carriages to the z-slide. Therefore, the z-slide can be tilted around X- and Y-direction in small angles up to 10 mrad. Two parallel linear direct drives actuate the y-slide. Flexure bearings connect the y-slide to the guide carriages of the linear guideways at the z-slide. In this way, the y-slide can be tilted around the z-direction up to 1.6 mrad. The rotatory axes are named as virtual since they are no rotatory axes in the usual sense [23]. The corresponding simulation model is physically based (see Fig. 1(b)). The temperature field of the machine is calculated using a reduced FE model. The finite element model of the machine is divided in submodels. These submodels do not include relative movement between machine components (e.g. between x-slide and y-slide). The systems of the submodels are reduced by model order reduction methods (Krylov subspace [24]) to obtain a fast computational model. Technological load data, e.g. axis positions, velocities, motor current from the machine control are used as input data for the model. These data are used to calculate the heat sources, due to friction and power loss in various machine components. In addition, the thermal conduction at the interfaces between the submodels and to the environment is calculated. The environmental temperature is measured with a Pt100 resistance temperature sensor. Empirical models are used to calculate the heat sources and thermal conductions.

Fig. 1. Machine tool set up (left), and corresponding simulation model (right).

3.2 Applied Domain Randomization

In robotic vision tasks such as object localization, object detection, pose estimation, and semantic segmentation, DR is mainly applied to *domain features*. These are features which describe the domain in which the robot has to perform in, rather than key features of its main task. Lighting exposure, number of light sources, textures, color, noise and field of view of the camera are examples for possible domain features from vision tasks [8]. Therefore, visual DR provides enough simulated domain variability of visual parameters to better adjust to real world key parameters. Transferring this concept onto time series data and physically simulation models, such as FE models, is however, challenging as domain-based features do not exist in the same sense. Each parameter of a physical model is linked to a static, dynamic, linear or nonlinear physical relation to the real world and hence randomizing those means randomizing key parameters, which are mapped onto a time series. The simulation model used in this work has a total of roughly 240 adjustable key parameters, such as material, engine and profile linear guide parameters, as well as emission coefficients, scaling factors for convection and radiation to environment and many more. For this proof of concept, out of the total adjustable parameters, 12 different scaling factors were chosen and randomized uniformly by $\pm$ 20% of their initial value. The scaling factors for the power loss approaches of the machine components bearings, profile rail guides, ball screw spindles to nuts, synchronous motors and linear direct drives were selected. They are often recommendations by manufacturers, which offer simplified model adjustments. Hence, their level of uncertainty is equivalently high and therefore suitable for randomizing.

3.3 Real and Synthetic Data Basis

The data categories used in this work can be split into two spate sources—real and synthetic data sources. The real data was collected from experimental setups, where the machine was analyzed during test runs with different load cases. All in all experimental data of 5 days was available, each day with testing times of roughly 10 h. The machine data—axis positions, velocities and motor current of all 5 engines, was collected with a sampling rate of 100 Hz. The target value is the temperature development within the machine. To completely map the temperature development of the machine, it is equipped with overall 48 Pt100 temperature sensors with a sampling rate of 0.1 Hz. Due to redundancies in the temperature curve development, of the 48 build in sensors the 2 with the most unique temperature development where chosen as the target prediction value of the TFT model—one (X1) in the center of the x-slide, and the other one (Z1) in the center of the z-slide. Sensor X1 is placed rather closely to the engine of the x-slide and highly correlates with its usage. Whereas, the substantial size of the z-slide, in comparison to the x and y-slide, buffers the heat development of the Z1 sensor. The synthetic data was collected from the simulation model (see Sect. 3.1). Hereby, real historic machine data was used as input for the simulation model. Due to the limited amount of actual real historic machine data, its usage as input values was applied multiple times, however, with different randomized scaling factors to generate deviating temperature predictions (see Sect. 3.2). As a consequence of the different sampling rates between machine and temperature data, a direct mapping between input and target data was not feasible. The used AI-based forecasting model (see Sect. 3.4) requires input and target to be the same sampling rate. Therefore, for the usage with the AI-model the machine data was cut to a sampling rate of 1 Hz, and the temperature data was stretched to 1 Hz. This turned the temperature curve into a floor and ceiling function-like curve. Additionally, the simulation based temperature curve is more continuous due to the higher resolution in comparison to the Pt100 sensors. Further obstructing the uniformity between machine and temperature data, as well as real and synthetic data. To tackle this uniformity the real and simulated temperature curves were transformed into a B-spline, improving the continuous curvature of the temperature, and hence the comparability of the sampling rates between input and target data.

3.4 Model Architecture and Learning Strategy

The core DL model used in this work is specialized on forecasting. The inputs are time series sequences (machine data) and the output of the model is the next possible time step of the target time series (temperature). The deviation between the prediction and the actual event allows evaluating current and possible future states of the machine. DeepAR [3] and the Temporal Fusion Transformer (TFT) [4] are currently the best performing probabilistic forecasting models in regards to tabular or time series data. Within this work the TFT model was used, as it outperformance DeepAR in regards to multi-horizon time series forecasting [4]. The TFT uses recurrent layers for local processing to learn temporal relationships, and interpretable self-attention layers for long-term dependencies.

Table 1. Teaching strategies and data source distribution

	TFT (real data)	Amount [samples]	TFT (simulation data)	Amount [samples]
Training	Real data	67.802	Simulation data	13.536.968*
Validation	Real data	34.501	Real data	34.501
Testing	Real data	34.501	Real data	34.501

The TFT models used in this work were designed to evaluate the last 720 time steps (in seconds) of machine data as input to predict the next 180 time steps of the temperature development of two selected temperature sensors (see Sect. 3.3). The loss function used in this work is the quantile loss, with quantile values for $q_2, q_{10}, q_{25}, q_{50}, q_{75}, q_{90}$ and q_{98}. Based on these input and prediction time windows a hyperparameter optimization was done with the usage of solely real data. Additionally, an early stoppage was implemented to finish the learning process automatically, if the validation accuracy does not change over a set number of epochs. This set up was used to train and compare two separate TFT models—one trained, validated and tested only on real data, and the one trained on simulation data, but also validated and tested on the same real data samples (see Table 1). Hence, a comparison between the two TFT models is possible, where one was trained on limited amounts of real data, in contrast to one which was trained on significantly more simulation data. For teaching the real data TFT overall 67.802 samples of real data were used. In contrast, for the simulation data TFT roughly 13.5 million samples of simulated data existed, where for reach training epoch 128.000 samples were randomly selected. The real data samples for validating and testing both models were identical to further support a fair comparison.

4 Experimental Results

Additionally to the forecasting of the absolute temperature values, separately trained TFT models for forecasting the temperature derivative were introduced. The reasoning behind this addition is due to the simplicity of the temperature curve, which can be approximated with the 1st order delay elements. To further demonstrate the potential of the proposed approach, as well adjust the complexity of the forecast to the application on the TCP offset correction, the temperature derivate curve was included. The curve itself is the derivative of the temperature B-Spline curve (see Sect. 3.3).

4.1 Absolute Temperature Forecasting Performance

Tables 2 and 3 show the performance results of the two different TFT models respectively for each of the two different sensors. Both models, trained with either real or simulation data, reach high performance results for the MAE, MSE as well as the R_2 values of the corresponding next time step forecast, as well as for the 180th time step prediction (see Sect. 3.4). Although the simulation data TFT does score better, the performance difference to the real data TFT is small. The same observation applies to the evaluation of the quantile values shown in Table 2.

Table 2. Evaluation metrics of the temperature forecast

	MAE	MSE	R2–1	R2–180
TFT RD—Z1	2.54e−3	1.70e−5	1.000	1.000
TFT SD—Z1	8.01e−4	1.49e−6	1.000	1.000
TFT RD—X1	1.97e−2	7.34e−4	0.998	0.996
TFT SD—X1	3.06e−3	2.94e−5	1.000	0.998

4.2 Temperature Derivative Forecasting Performance

The forecast of the temperature derivative demonstrates the difference in performance between the two TFT models. Figure 2 shows the prediction curve of the respectively next time step for sensor Z1 and Fig. 3 the results for sensor X1. For both sensors the real data TFT (left) performs in significant deviations from the actual temperature derivative

Table 3. Pinball loss results of the included quantiles of the temperature forecast

	q2	q10	q25	q75	q90	q98
TFT RD—Z1	1.64e−4	5.60e−4	1.13e−3	1.68e−3	8.46e−4	2.48e−4
TFT SD—Z1	5.47e−5	3.36e−4	4.56e−4	4.27e−4	2.28e−4	6.30e−5
TFT RD—X1	2.53e−3	5.87e−3	8.02e−3	4.63e−3	2.28e−3	5.60e−4
TFT SD—X1	6.13e−5	1.47e−4	1.97e−4	2.09e−4	1.20e−4	4.29e−5

curve. Whereas, the prediction curve of the simulation data TFT (right) clearly overlaps with the actual curve. Additionally the level of uncertainty, displayed by the quantile values, is strictly smaller with the simulated data TFT. This is especially visible between the results of sensor X1. Figure 4 shows the prediction of the next 180 time steps (180 s) from the time point t_0 onwards. This comparison shows another level of performance difference between the two models. Whereas, the simulation data TFT (right) is able to follow the actual curve, as well as predict the next time steps with little uncertainty, the real data TFT (left) is not able to provide a reliable prediction. The evaluation metrics shown in Tables 4 and 5 provide a similar comparison outlook. Especially the R_2 values show a significant performance difference between the real data TFT and the synthetic data TFT.

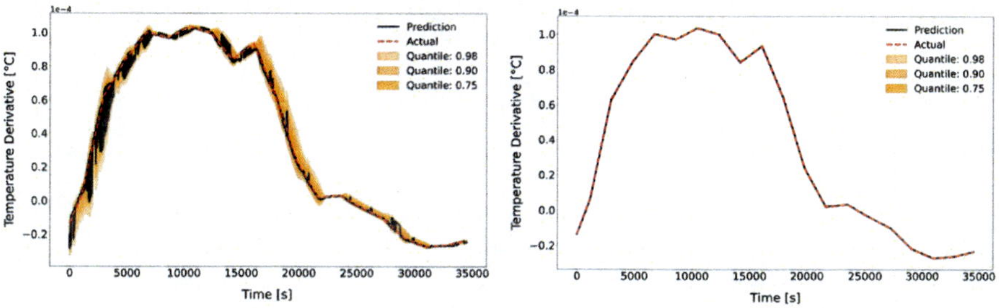

Fig. 2. Complete derivative prediction results for sensor Y1 of the real data TFT (left), and simulation data TFT (right).

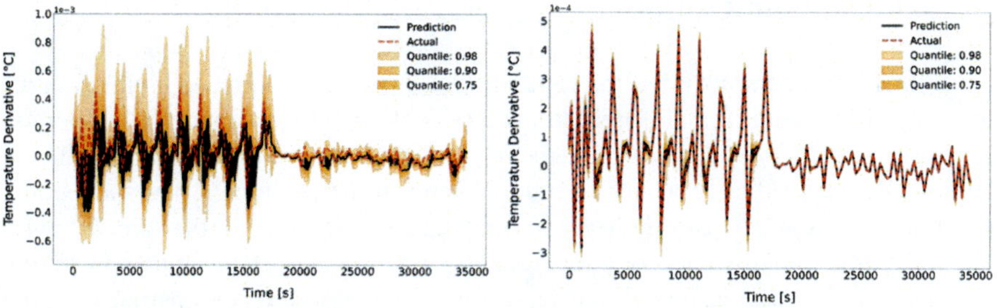

Fig. 3. Complete temperature derivative prediction results for sensor X1 of the real data TFT (left), and simulation data TFT (right).

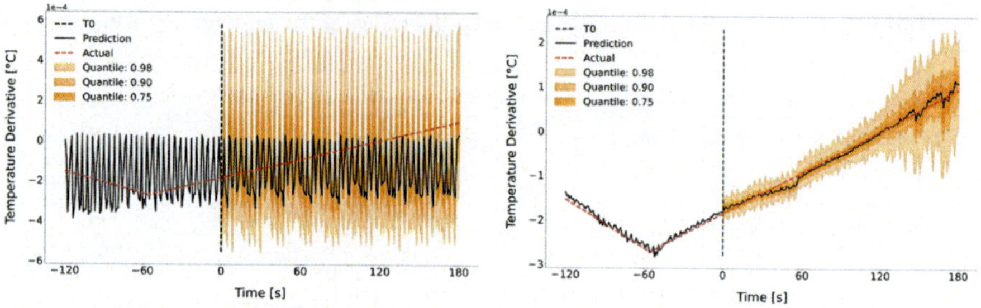

Fig. 4. Temperature derivative prediction of the next 180 time steps for sensor X1 of the real data TFT (left), and simulation data TFT (right).

Table 4. Evaluation metrics of the temperature derivative forecast

	MAE	MSE	R2–1	R2–180
TFT RD—Z1	1.49e−2	7.78e−4	0.997	0.993
TFT SD—Z1	8.32e−4	6.57e−6	1.000	1.000
TFT RD—X1	8.67e−1	1.56e−0	−0.08	−0.581
TFT SD—X1	5.12e−2	1.61e−2	0.999	0.124

Table 5. Pinball Loss results of the included quantiles of the temperature derivative forecast

	q2	q10	q25	q75	q90	q98
TFT RD—Z1	7.99e−4	3.13e−3	4.17e−3	6.15e−3	3.02e−3	7.85e−4
TFT SD—Z1	4.14e−5	1.01e−4	9.99e−5	1.34e−4	9.46e−5	3.46e−5
TFT RD—X1	3.48e−2	1.45e−1	2.08e−1	3.80e−1	3.22e−1	1.74e−1
TFT SD—X1	6.58e−4	1.56e−3	1.57e−3	2.46e−3	1.78e−3	6.94e−4

5 Conclusion and Outlook

All in all this work demonstrated the capability of domain randomized simulation data from thermal FE-models for DL forecasting tasks. Although a basic machine temperature forecast is possible with solely real data, the forecast of more complex machine behaviors is limited. Besides offering a more accurate deterministic forecast, the inclusion of simulation data also decreases the level of uncertainty. Even though the forecast within this work is limited to the machine temperature development, it shows the opportunity it offers for the usage as a feasible TCP offset correction method. It demonstrates a certain superiority in comparison to the simulation model itself. The DL-forecasting model works faster and is able to predict multiple time steps ahead. These two advantages allow a more reliable forecast and therefore a potentially improved adjustment of the TCP offset. Additionally, it eliminates the necessity of a real time simulation model, as simulation models are further simplified to meet this requirement. As the computational effort of accumulating machine behavior from simulations is out-sourced from the operational time to the development phase, more complex simulations can be used and hence further improve DL prediction accuracy.

The next step will be the adoption to the TCP correction, as well as a direct performance comparison between simulation based and DL-based models. Future direction to further improve the accuracy of the DL-models include the addition of optimization routines, instead of set uniformly distributed simulation parameters, to further adjust the distribution of the simulated data to the real data distribution.

Domain randomization is a promising research direction for production engineering toward bridging the reality gap between machine behavior and simulation results.

References

1. Großmann, K., et al.: Thermo-Energetische Gestaltung von Werkzeugmaschinen: Eine Übersicht zu Zielen und Vorgehen im SFB/Transregio 96, Zeitschrift für wirtschaftlichen Fabrikbetrieb (2012)
2. Mayr, J., et al.: Thermal issues in machine tools. In: CIRP (2012)
3. Salinas, D., et al.: DeepAR: Probabilistic Forecasting with Autoregressive Recurrent Networks (2019)
4. Lim, B., et al.: Temporal Fusion Transformers for Interpretable Multi-horizon Time Series Forecasting (2020)
5. Kiangala, K.S., Wang, Z.: Initiating predictive maintenance for a conveyor motor in a bottling plant using industry 4.0 concepts. Int. J. Adv. Manuf. Technol. **97**(9–12), 3251–3271 (2018). https://doi.org/10.1007/s00170-018-2093-8
6. Reuß, M., et al.: Ermittlung der Auswirkung des statistischen Verhaltens baugleicher Werkzeugmaschi- nen, Internationales Forum Mechatronik (2011)
7. von Rueden, L., et al.: Combining machine learning and simulation to a hybrid modelling approach: current and future directions. In: Advances in Intelligent Data Analysis XVIII, Lecture Notes in Computer Science (2020)
8. Tobin J., et al.: Domain randomization for transferring deep neural networks from simulation to the real world. In: IEEE/RSJ International Conference on Intelligent Robots and Systems (2017)
9. Koos, S., et al.: Crossing the reality gap in evolutionary robotics by promoting transferable controllers. In: GECCO'10 (2010)
10. Lee, S., et al.: StRDAN: synthetic-to-real domain adaptation network for vehicle re-identification. In: IEEE Conference on Computer Vision and Pattern Recognition (2020)
11. Zhao, W., et al.: Sim-to-real transfer in deep reinforcement learning for robotics: a survey. In: IEEE Symposium Series on Computational Intelligence (2020)
12. Richter, S.R., et al.: Playing for data: ground truth from computer games. In: European Conference on Computer Vision. Springer (2016)
13. Tan, J., et al.: Sim-to-real: learning agile locomotion for quadruped robots (2018)
14. Beaumont, M., et al.: Approximate Bayesian computation in population genetics. Genetics (2002)
15. Wilson, G., et al.: A survey of unsupervised deep domain adaptation. ACM Trans. Intell. Syst. Technol. (2020)

16. Chebotar, Y., et al: Closing the Sim-to-Real Loop: Adapting Simulation Randomization with Real World Experience (2019)
17. Rao, K., et al.: RL-CycleGAN: reinforcement learning aware simulation-to-real. In: IEEE/CVF Conference on Computer Vision and Pattern Recognition (CVPR) (2020)
18. Ramos, F., et al.: BayesSim: adaptive domain randomization via probabilistic inference for robotics simulators (2019)
19. Purushotham, S., et al.: Variational adversarial deep domain adaptation for health care time series analysis. In: International Conference on Learning Representations (2017)
20. Hosseini-Asl, E., et al.: Augmented cyclic adversarial learning for low resource domain adaptation. In: International Conference on Learning Representations (2019)
21. Vercruyssen, V., et al.: Transfer learning for time series anomaly detection. In: Proceedings of the CEUR Workshop (2017)
22. Ihlenfeldt, S., et al.: Innovatives mechatronisches Systemkonzept für eine hochdynamische Werkzeug- maschine. In: Bertram, T., Corves, B., Janschek, K. (eds.) Mechatronik 2017
23. Ihlenfeldt, S., et al.: Simplified Manufacturing of Machine Tools Utilising Mechatronic Solutions on the Example of the Experimental Machine MAX. Springer International Publishing (2020)
24. Freund, R.W.: Model reduction methods based on Krylov subspaces. Acta Numerica (2003)

Game-Theoretic Concept for Determining the Price of Time Series Data

J. Mayer[1(✉)], T. Kaufmann[1], P. Niemietz[1], and T. Bergs[1,2]

[1] Laboratory for Machine Tools and Production Engineering (WZL), RWTH Aachen University, Campus-Boulevard 30, 52074 Aachen, Germany
j.mayer@wzl.rwth-aachen.de

[2] Fraunhofer Institute for Production Technology IPT, Steinbachstraße 17, 52074 Aachen, Germany

Abstract. The digitization of manufacturing and the economic potential from the utilization of data sets result in the monetarization and trading of process data. In data marketplaces, a suitable pricing mechanism ensures the definition of a fair value of data for the data consumer and the data owner. Game theoretic concepts ensure this fairness by determining the value functions of the participants in a transaction, which is based on the quality and quantity of data. In this paper, cooperative games are used to model data transactions. First, empirical methods are used to realistically determine the value of data sets. Using the example of a data set of a grinding machine, a function is set up depending on the quality dimensions, with the help of which declarations about the usability of the data set for regression models are possible. This is followed by the pricing of a data set based on the Kalai-Smorodinsky solution. It offers the advantage that the quality of a data set does not have to be assumed as a variable and that pricing is possible individually for already existing data sets whose quality is not to be adjusted further. Finally, the chosen approach is compared with the common method for price discovery, the Stackelberg game, with respect to the concession of the actors, based on the negotiation result to the maximum possible benefit. In the outlook, potential further developments of the approach based on Kalai-Smorodinsky are discussed.

Keywords: Data monetization · Data pricing · Game theory

1 Governance

The governance of a data marketplace deals with the applicable decision rights as well as the formal and informal control mechanisms on the sharing platform [1]. In addition to defining the technical characteristics of the platform, it regulates the cooperation and competition among marketplace participants through established principles [2]. The effective mechanisms of platform governance to maintain data security and integrity incentivize the sharing of authentic data. Monitoring of platform input and output and pricing of provided assets complete platform governance. The pricing of data is subject to various challenges, making it difficult for data sharing to reach market maturity. In this

M. Liewald et al. (Eds.): WGP 2022, LNPE, pp. 585–599, 2023.
https://doi.org/10.1007/978-3-031-18318-8_59

paper, following the introduction to the challenges of data pricing and the fundamentals of game theory, a concept for pricing manufacturing data based on data quality for use in machine learning (ML) models is presented. This concept, which is presented on the example of data of an industrial grinding machine, forces the transferability of a data pricing method for any equipment of the production engineering for the use case of a ML model of the linear regression.

2 Basics of Pricing

Fair pricing in the context of industrial data requires provision for the benefits of both seller and buyer, fair consideration of the increments of benefits, respect for data quality and equal bargaining power of both parties, and maximum willingness to pay of the buyer and cost of the seller. Game-theoretic approaches allow observing the requirements. One method widely used is the Stackelberg game, which is briefly described in Sect. 2.1.

2.1 Stackelberg-Game

In the simplest case, the Stackelberg game is a strategic game with two players, in which the so-called Stackelberg leader (data owner) makes his decision first and the Stackelberg follower (data consumer) decides with reference to it [3]. Since multiple data owners participate in a data marketplace and they offer their data through the data marketplace operator, the data marketplace operator can be used as a Stackelberg leader to reduce complexity [4].

The Stackelberg game proceeds in two stages. In the first, the marketplace operator determines the pricing strategy, after which the data consumer defines its buying strategy in the second stage. The solution to the Stackelberg game represents a Nash equilibrium. Value functions are needed to determine the payoff of participants at different quantities and quality levels of data at different prices. A data consumer's value depends first on the price of the data set and second on the quantity and quality of the data. The value of the data owner is increased by the selling price and decreased by the cost of data aggregation and preparation.

The Stackelberg game has certain drawbacks when it comes to pricing. Designed to model a competitive situation in a market, the Stackelberg game is less well suited for cooperative interactions between data owner and data consumer [5], in which both parties each seek a contract conclusion, as opposed to strict competition. Without this, neither can generate value. Additionally, players are disadvantaged depending on the turn order. As soon as several players compete for the quantity of a good, the Stackelberg leader gets an advantage over the follower [6]. Another disadvantage is the assumption of rationality among the players. This rationality leads to an attempt to anticipate the strategy of the follower, so that suboptimal payoffs may result [7]. The cooperative game promises to remedy the drawbacks of the Stackelberg game and is described below.

2.2 Cooperative Games

In cooperative games, players can make binding agreements among themselves [5]. An outside entity or contract ensures that these agreements are enforced and a strategy is chosen. A cooperative game is defined by the set of potential players N, the payoff space P with payoff vectors $u = (u_1, \ldots, u_n)$, and the conflict point c. For the price negotiation for an economic good, this conflict point represents a lack of agreement and is defined as $c = (0/0)$. There is a bargaining problem if there exists a $u \in P$ that promises a higher payoff than the conflict point c for all players. For a bargaining problem between two players with value functions u_1, u_2 and the conflict point $c = (c_1, c_2)$ with player-specific payoffs c_1, c_2, the problem can be represented two-dimensionally using the value limit $H(P)$ of the bargaining space. The value limit $H(P)$ is the set of all pareto-optimal payoff pairs in P on which the optimal payoff pair lies. This can be determined, for example, using the Kalai-Smorodinsky (KS) solution. This solution, unlike the Nash approach, allows the comparison of individual value gains [8]. Both the Stackelberg game and the cooperative game are based on the value-based pricing of data. The value of data in the present context refers to its use in ML models, the outcome of which depends heavily on the quality of the data with which the algorithm is trained, in addition to the quantity [9]. The quality of the prediction of an ML model increases when the data quality increases. The latter will be discussed in more detail in the following.

2.3 Data Quality

Data quality is characterized by multidimensionality. Quality dimensions (QD) can be divided into objective and subjective. Objective QD are considered measurable, while subjective dimensions depend heavily on the preferences of the data consumer and cannot be measured directly [10]. Objectively measurable criteria according to Wang and Strong include accuracy and completeness [11], which are the focus of this paper. There are different ways to model the overall quality from the different influencing factors. Yu and Zhang recommend the principle of integrated quality for the use case of pricing, in which the influence of the QD on each other is represented [12]. Assuming that the QDs negatively influence each other to a certain degree, the result is:

$$q_{im}^{I} = q_{i(m-1)}^{I} + q_{im}(1 - q_{i(m-1)}^{I}) \tag{1}$$

It is difficult to validate a relationship between QD and the resulting overall quality of a dataset because there is no metric for capturing or quantifying the overall quality of a dataset. Therefore, similar to Niyato et al. [3], this work makes the assumption that a data set is only as high quality as the results it can produce. To this end, the effects of data quality on the goodness of a regression model are examined. The underlying data set represents an external cylindrical grinding process, in which shafts were machined by plunge grinding. By varying selected grinding parameters (workpiece hardness, cooling lubricant exit velocity, bond hardness of the grinding tool), the effects on surface roughness, circularity deviation and shaft end diameter of the workpiece as well as radial wear of the grinding wheel were investigated.

To investigate the functional relationship between the QD and the quality of the ML model, specially developed for this application, the expression of the quality dimensions is degraded incrementally and randomly. After each iteration, the model goodness is recalculated using the root mean square error (*RMSE*). The determined data points are visualized by the light blue color and the course of the point cloud is approximated by functions. The exact shape of the curves depends on parameters such as the original quality of the data set used and the correlation between the independent variables and the target variable (see Fig. 1). Therefore, no general correlation can be formed here.

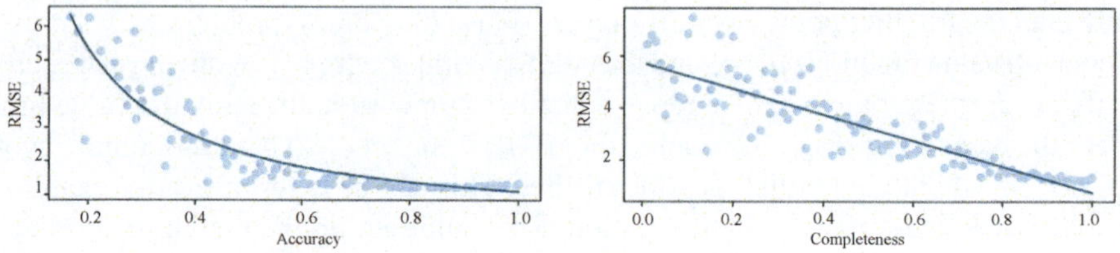

Fig. 1. Correlation between *RMSE* and QD accuracy and completeness

A hyperbolic course of the accuracy-*RMSE* or a linear course of the completeness-*RMSE* is assumed for the definition range of the QD on [0, 1]. The following functions with the constants α, β, γ, δ are obtained:

$$RMSE_A = \frac{1}{\alpha A + \beta} \tag{2}$$

$$RMSE_C = \gamma - \delta C \tag{3}$$

Before a functional dependence between the quality of the ML model and the QD can be found, the influence of the QD among each other shall be considered similar to Yu and Zhang [12]. For this purpose, the completeness is reduced gradually, and the resulting accuracy of the data set is calculated after each step. A linear relationship between accuracy and completeness follows (see Fig. 2).

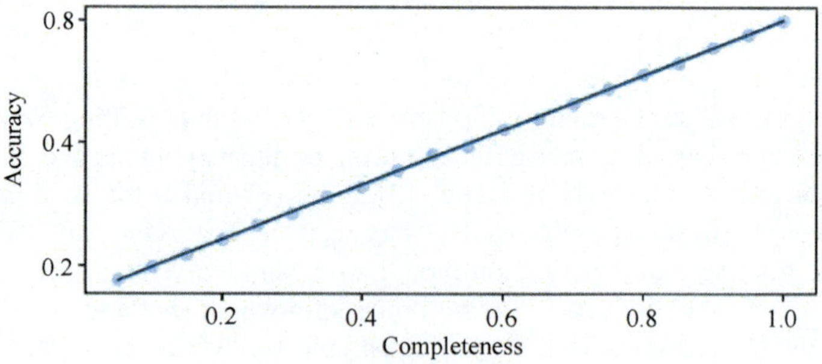

Fig. 2. Correlation between accuracy and completeness

To map the functional relationship between the quality of the ML model and the QD of the data set, further conditions have to be defined. The quality of the regression model shall be defined on the interval [0, 1], where 1 represents a model of highest quality. This means that the functions on the interval [0, 1] should be monotonically increasing. Thus, a change of sign is required for the functions $RMSE_A$ and $RMSE_C$, since the functions represent the course of the model quality when the quality dimension decreases, which is why they decrease monotonically. Consequently, the quality of the model should grow with increasing QD. Furthermore, for all partial functions g of which the quality function Q is composed, $g(1) = 1$ shall hold. This is necessary, since the quality of the ML model should take a value of 1 at maximum QD. To ensure this, the function $RMSE_A$ must be extended by a term φ, which shifts the function, but without influencing the slope of the function.

1. The following applies to the accuracy as a function of the completeness:

$$A(C) = C \tag{4}$$

2. For the quality of the regression model applies as a function of accuracy:

$$Q(A) = -\frac{1}{\alpha A + \beta} + \varphi \tag{5}$$

3. For the impact of completeness on the quality of the ML model holds:

$$Q_C(C) = C \tag{6}$$

Consequently, the accuracy is negatively affected if the completeness is not equal to 1. For this reason, the expression $A(1 - C)$ is subtracted from the accuracy of the data set. The result is a quality function of the ML model depending on the accuracy and completeness with the normalization factor of 0.5:

$$q(A, C) = \left[\left(-\frac{1}{\alpha A - A(1 - C) + \beta} + \varphi\right) + C\right]\frac{1}{2} \tag{7}$$

This three-dimensional function is visualized in Fig. 3.

2.4 Research Gap

Pricing models can be distinguished as static and dynamic. Static methods are characterized by a lack of consideration of the individuality of data sets [13]. Dynamic methods, on the other hand, take into account not only the individual metrics of the various data sets, but also the demands on the data and the price of the data consumer and data owner. Game theoretic approaches, for example, can be classified as dynamic pricing models. A screening of existing scientific approaches illustrates the increasingly relevant topic of a data marketplace and the lack of a holistic evaluation of existing methods. Some authors push the correlation between data quality and ML model accuracy as the driving element of pricing [4], others the data quantity and model accuracy [14]. The Stackelberg game, for example, models the price of a data set after determining a value function [4]. Existing approaches currently lack reference to the pricing of manufacturing data, consideration

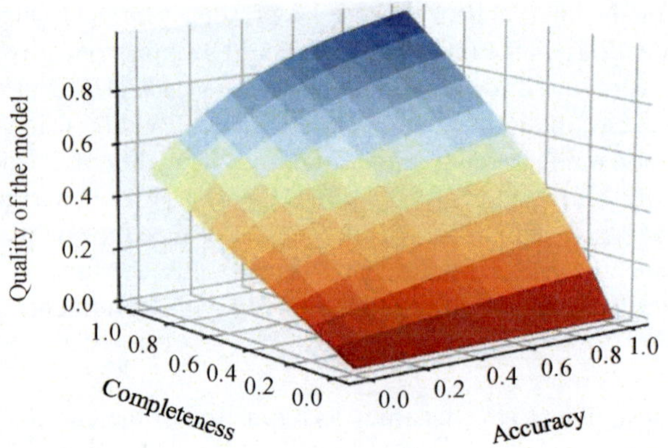

Fig. 3. Graphical illustration of the quality function

of multidimensional data quality for the use case in ML models, and multidimensionality of the value function with respect to data quality and quantity. Pricing is subject to challenges typical of mechanical engineering, such as the process-related boundary conditions, the accuracy of the measurement results of the sensors, sensor calibration and position, and the diversity of the technical data itself, which increase the comparability between individual data sets and thus the difficulty of pricing. In this paper, the first two aspects of the research gap (pricing of manufacturing data, multidimensionality of data quality) are addressed.

3 Value Function

Regardless of the game-theoretic pricing approach, value functions are needed for both data consumers and data owners. To define these functions, manufacturing companies were surveyed in several stages. In the first part of this survey, the value of the data consumer was determined using a method according to Halter and Mason [15]. The second part of the survey used selected questions to determine the respondents' risk aversion to monetary gains from data sales. Due to the subjectivity of the respondents, a generally valid value function cannot be determined. Respondents were presented with different courses of action at different payoffs a, b, c, d, where $a > b > c > d$. Within the equally likely situations, an indifference value is to be chosen for which a decision maker considers two courses of action to be equivalent (Table 1). In the present case, the payoffs represent data of different quality levels from 0 to 1, where the data are identical in content, size, and all other metrics. The chosen indifference value is taken as a given value for the next game.

To generate a value function graphically requires two anchor points, $[0, u(0)]$ and $[c', u(c')]$, which can be chosen arbitrarily, and a scaling point [15]. To determine the scaling point, the value of any indifference value can be freely chosen. It follows:

$$u(c) = \frac{1}{2}u(c') \tag{8}$$

Table 1. First part of the survey

Game 1				Game 2			
		Decision-making options				Decision-making options	
		A1	A2			A1	A2
Equiprobableevents	E1	d	c	Equiprobableevents	E1	Indifference value 1	c'
	E2	a	b		E2	a	b
Game 3				Game 4			
		Decision-making options				Decision-making options	
		A1	A2			A1	A2
Equiprobableevents	E1	0	Indifference value 1	Equiprobableevents	E1	Indifference value 1	Indifference value 2
	E2	a'	Indifference value 2		E2	a''	Indifference value 3

$$u(a') = u(c) + u(c') - u(d) \quad (9)$$

$$u(a'') = u(a') + u(c') - u(c) \quad (10)$$

Based on the determined four indifference values, the value functions can be created. It holds that $u(d) = 0$ since the value d in the survey is zero. For the further procedure, the value functions are approximated by a function (dark blue graph), which is subsequently the basis for the value function of the data consumer (Fig. 4).

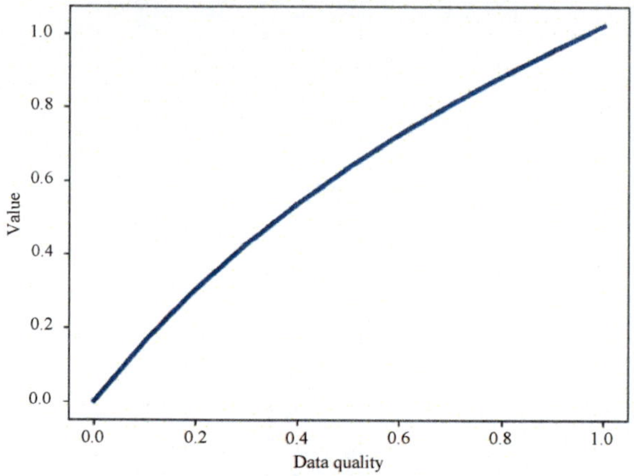

Fig. 4. Graphical illustration of the value functions of the first part of the survey

For the value function of the data owner, the value generated by monetary gains and consequently the Arrow-Pratt measure of risk aversion with regard to negotiations with data sets is relevant. With the help of a survey, individual negotiation situations are simulated at defined probabilities of occurrence (Table 2), in which the participants can either accept a fixed offer for the data set or negotiate with another company for the price. In the negotiation case, there is also the potential situation of a breakdown without monetary payout.

Calculating the expected value of a negotiation and comparing it to the fixed payoff serves to define risk behavior. Choosing a low expected value symbolizes risk affinity. In the underlying survey, all participants exhibited risk-neutral to risk-averse behavior (concave course of the value function). For the further procedure, the commonly used isoelastic value functions (CRRA) represent this concavity well [16]. It holds:

$$u(x) = x^{1-r}, r \geq 0 \quad (11)$$

Table 2 can now be used to calculate the measure of risk aversion of the respondents [17]. Due to the small size of the survey and the small distance between the individual values of risk aversion, the mean value is used ($\overline{r} = 0.191$). In Fig. 5 the course of the function $u(x) = x^{1-r}$ is drawn, which results for $\overline{r}$.

Table 2. Second part of the survey

Situation	Height fixed offer	Negotiation		Number of respondents		
		Successful	Not successful	Fixed offer	Negotiation	Indifferent
1	400 €	1000 € (50%)	0 € (50%)	–	5	1
2	300 €	1000 € (50%)	−250 € (50%)	–	6	–
3	100 €	1000 € (25%)	−250 € (75%)	6	–	–
4	6 €	1000 € (35%)	−400 € (65%)	1	5	–

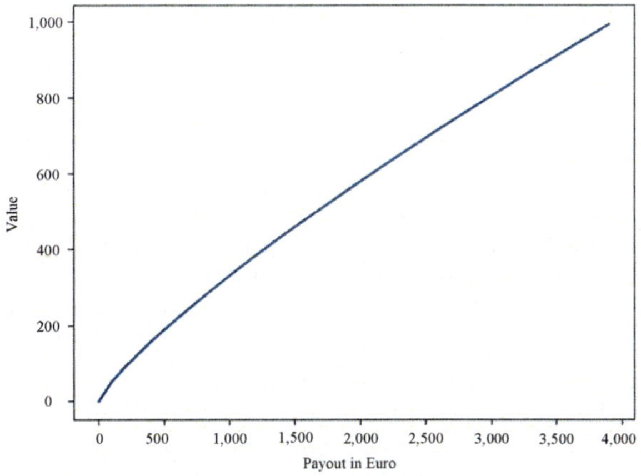

Fig. 5. Graphical illustration of the value functions of the second part of the survey

To determine the progression and to maintain the comparability of the value functions of data owner and data consumer, the value is normalized to the interval [0, 1]. Otherwise, the price would influence the value functions excessively. In the normalization, the ideal value of the datum is ascribed the greatest possible value. In the function shown in the figure, this price is symbolically 4000 €. However, it can be captured realistically by a survey. The defined functions serve as a basis for finding the value functions for data owner and data consumer. When setting up the value functions, other aspects such as the costs of data aggregation are taken into account in addition to the empirically determined function curves. The following functions form the cornerstone for pricing using game-theoretic concepts of any kind.

3.1 The Value Function of the Data Owner

The value of the data owner increases with the level of the sales price. Since normalization also influences the price of the data set, an average value is formed from the ideal value from the data owner's point of view and the data consumer's maximum willingness to pay:

$$p_v = \frac{p_v^o + p_{max}}{2} \tag{12}$$

Furthermore, following Niyato et al. [3] and Liu et al. [4], the costs for aggregation and preparation of the data have to be considered. In these sources, these costs are always considered as constants. In this paper, however, we force quality-based pricing with a cost function to be normalized of the form $k_i = kq_i$ with k_i for the cost and q_i for the quality of a dataset i [12]. Substituting the normalized price into the function $u(x) = x^{1-r}$ and taking into account the costs, we obtain the value function of data owner i as a function of the price p_i:

$$u_i^o(p_i, q_i) = \left(\frac{p_i}{p_v}\right)^{1-r} - \frac{kq_i}{p_v} \tag{13}$$

3.2 The Value Function of the Data Consumer

The value of the data consumer is positively influenced by the quality of the data set and negatively influenced by the price. Using the value function of the quality determined from the survey values, the value function of the data consumer i is obtained:

$$u_i^c(p_i, q_i) = \gamma + \ln(\delta + \varepsilon q_i) - \frac{p_i}{p_v} \tag{14}$$

The parameters $\gamma, \delta, \varepsilon$ serve to scale the empirically determined value function. The survey conducted for this paper resulted in the parameters $= 0.8$, $\delta = 0.42$, $\varepsilon = 0.8$. These values will be used in the further course.

4 Implementation of the Kalai-Smorodinsky Solution

According to KS, the negotiation problem between data owner and data consumer is solved with the point $u^* = (u_i^{c*}; u_i^{o*})$ if applies to this point:

$$\frac{u_i^{o*} - c_2}{u_i^{c*} - c_1} = \frac{m_2 - c_2}{m_1 - c_1} \tag{15}$$

To obtain the coordinates of point m, a function is set up that describes the course of the boundary of the payoff space. The boundary of the bargaining space can be determined by the value functions of the data owner and data consumer. First, the value function of the data consumer is solved for p:

$$\frac{p_i}{p_v} = \gamma + \ln(\delta + \varepsilon q_i) - u_i^c \tag{16}$$

It follows by substituting into the value function of the data owner:

$$u_i^o(u_i^c) = (\gamma + \ln(\delta + \varepsilon q_i) - u_i^c)^{1-r} - \frac{kq_i}{p_v} \tag{17}$$

The straight line equation $(c, m) = c_2 + \frac{m_2}{m_1} u^o$, which runs through $c = (c_1, c_2)$ and $m = (m_1, m_2)$ can be determined by finding the coordinates of each point. The coordinates m_1 and m_2 of the point $m = (m_1, m_2)$ can be determined by setting one of the two value functions to zero:

$$0 = (\gamma + \ln(\delta + \varepsilon q_i) - u_i^c)^{1-r} - \frac{kq_i}{p_v} \tag{18}$$

By rearranging the equation:

$$m_1 = u_i^c(u_i^o = 0) = \gamma + \ln(\delta + \varepsilon q_i) - \left(\frac{kq_i}{p_v}\right)^{\frac{1}{1-r}} \tag{19}$$

And analog for m_2:

$$m_2 = u_i^o(u_i^c = 0) = (\gamma + \ln(\delta + \varepsilon q_i))^{1-r} - \frac{kq_i}{p_v} \tag{20}$$

For an exemplary calculation, a data quality of $q_i = 0.8$ required by the Data Consumer, costs for data preparation and provision of $k = 500$ € and a value assumption of $p_v = 3000$ € are applied. Furthermore, it is applied $p_{min} \leq p_i \leq p_{max}$ and $p_{min} > k_i$. The KS solution straight line equation is also known. The payoff space modeled in this way with solution u^* is shown in Fig. 6. A maximum willingness to pay of $p_{max} = 2500$ € was chosen. The colored area represents the subset of the payoff space determined by the defined constraints. Here, irrational points (e.g., prices above p_{max}) were considered for finding the solution of the bargaining game. Then, the solution was checked to see if it satisfied the defined constraints. This procedure reinforces the KS specific axiom of individual monotonicity as well as the fairness of the solution [18].

After the solution of the bargaining game is found, the value function of the data owner can be rearranged to p_i to determine the price. The result is:

$$p_i = \left(u_i^o + \frac{kq_i}{p_v}\right)^{\frac{1}{1-r}} p_v \tag{21}$$

From this follows with $u_i^{o*} = 0.395$ and $r_o \neq 1$ the price $p_i^* = 1351.314$ €. After calculating the price for a fictitious dataset, the results of the negotiation when varying the parameters such as risk aversion of the data owner and quality of the dataset are considered in the following. The left side of Fig. 7 presents the price of a data set and the profit of the data owner depending on the quality of the data set. Both increase as a function of increasing quality, but level off due to the costs of data preparation, which also increase. The difference in value between data owner and data consumer at different levels of risk aversion of the data owner is plotted on the right side of Fig. 7. The lower the risk aversion of the data owner, the smaller the difference in value and the fairer the result.

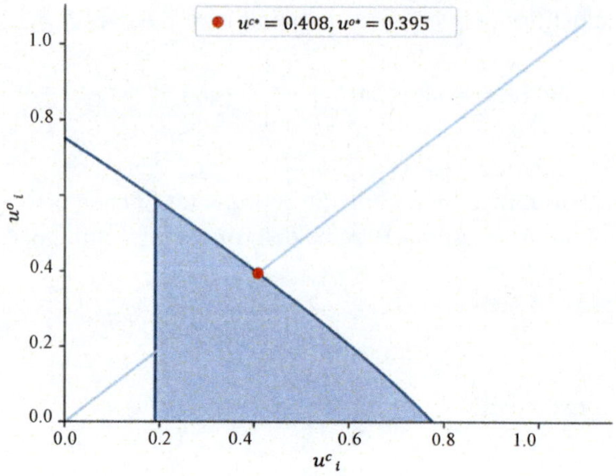

Fig. 6. Graphical illustration of the KS solution

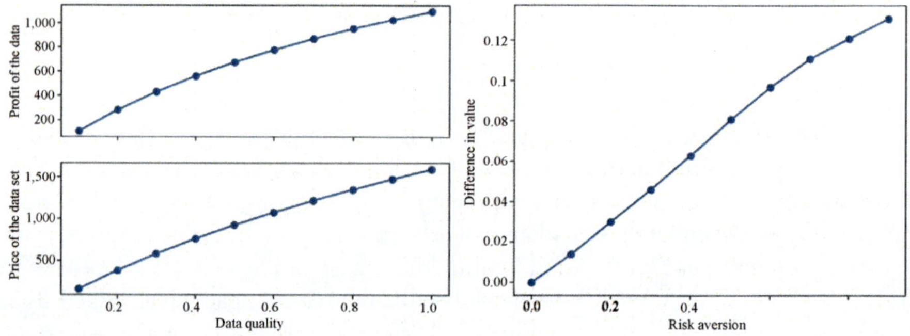

Fig. 7. Price and profit depending on data quality as well as the difference in value depending on different levels of risk aversion of the data owner

The dark blue line in Fig. 8 represents all combinations of QD accuracy and completeness with the overall data quality value $q_i = 0.8$. It can be seen that higher quality data are costly to produce and therefore have higher prices. For higher quality data, the dark blue line would become significantly shorter as the number of possible combinations of QD for a defined overall data quality value decreases.

5 Comparison to the Stackelberg Game

For a comparison of the two game-theoretic methods, data sharing must be modeled as a Stackelberg game. Except for the value functions, all parameters can be adopted. The Stackelberg game is solved by backward induction, so that the Stackelberg follower is considered first. The reason for the lack of suitability of the value function set up so far is explained on the basis of the data consumer:

$$\frac{\partial u_i^c}{\partial q_i} = \frac{\varepsilon}{\delta + \varepsilon q_i} = 0 \tag{22}$$

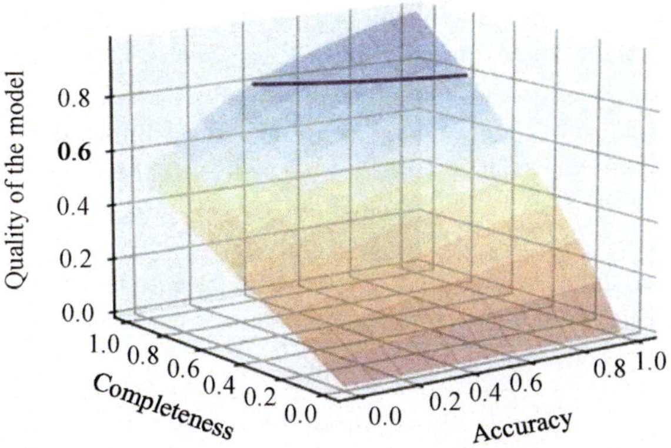

Fig. 8. All combinations of QD accuracy and completeness with the overall data quality value $q_i = 0.8$

The problem of this function is the independence of the optimal response q_i^* of the Data Consumer from the pricing strategy of the Data Owner. Therefore, the equation must be modified so that the response function depends on the Data Owner's price p_i:

$$u_i^o(p_i, q_i) = q_i\left(\left(\frac{p_i}{p_v}\right)^{1-r} - \frac{k}{p_v}\right) \tag{23}$$

$$u_i^c(p_i, q_i) = \gamma + \ln(\delta + \varepsilon q_i) - \frac{p_i}{p_v}q_i \tag{24}$$

With $\frac{\partial u_i^c}{\partial q_i} = 0$, the optimal response function of the data consumer now results with:

$$q_i^* = \frac{p_v}{p_i} - \frac{\delta}{\varepsilon} \tag{25}$$

By inserting the response function into the value function of the data owner and then deriving and searching for the zeros, the price p_i can be determined. For $p_{min} = 500$ € and $p_{max} = 2500$ € a price p^* of 1142.3 € follows. For the resulting quality, the reaction function of the data consumer is used. With the calculated pricing strategy, $q_i^* = 0.72$. For comparability with the KS solution, this must be formed at a data quality of $q_i^* = 0.72$, since this represents the optimal response of the Data Consumer to the Data Owner's pricing strategy when using the Stackelberg game. The result is a price $p^* = 1244.64$ €. Here the first advantage of the KS solution becomes apparent. The data consumer is able to purchase data with the desired quality. In the Stackelberg game, on the other hand, only a minimum quality requirement can be given. Thus, it is only conditionally suitable for quality-based pricing of data sets, since data sets have to be adjusted to the calculated quality after the price has been calculated. To experience the fairness of a negotiation, the concession of the players can be formed [19]:

$$\kappa_i = 1 - \frac{u_i}{max\, u_i} \tag{26}$$

The concession using the KS solution is $\kappa_i{}^c = 0.51$ for the data consumer and $\kappa_i{}^o = 0.5$ for the data owner. Here, the maximum value of the KS solution is represented by the point m. In the case of the Stackelberg game, this results in a value for the Data Consumer $u_i{}^c = 0.832$ with $q_i = 1$ and $p_i = p_{min}$. The maximum value of the data owner is $u_i{}^o = 0.345$ and is reached at $q_i = q_{min}$ and $p_i = p_{max}$. Thus, the concessions are $\kappa_i{}^c = 0.375$ and $\kappa_i{}^o = 0.382$. The difference in concessions is $\Delta\kappa_i = 0.01$ when using the KS solution and $\Delta\kappa_i = 0.07$ when using the Stackelberg game. Both methods are considered fair, since data consumer and data owner have to make similar concessions with respect to the outcome of the methods. In terms of price, the KS solution delivers a higher profit for the data owner.

6 Prospect

In this paper, a novel method and exemplary analysed in the field of industrial grinding data for pricing a dataset was presented which, compared to the Stackelberg game, offers the advantage that the quality of a dataset does not have to be assumed as a variable in quality-based pricing. This allows individual pricing for already existing data sets whose quality is not to be further adjusted. This allows SMEs with less available capital to acquire shares of datasets at a lower price and share in the economic benefits. Moreover, the data consumer can purchase data according to his needs. As a result of the KS solution, the price of a dataset can be determined for different quality levels. This concept enables the transferability of the procedure of a pricing of data of any equipment of the production engineering for the use case of the ML model of the linear regression. An extension of the methodology developed in this paper is to consider the data quantity when determining the price. The data quantity can be recorded either as a variable or as a constant. However, this increases the complexity of solving the negotiation game. On the other hand, data consumers with a low willingness to pay can also acquire data of high quality, since it is thus possible to buy shares of data sets. Furthermore, it has to be examined whether and in which dimensions the presented correlations change if other ML models are considered instead of regression models.

References

1. Tiwana, A., Konsynski, B., Bush, A.: Research commentary—platform evolution: coevolution of platform architecture, governance, and environmental dynamics. Inform. Syst. Res. (2010). https://doi.org/10.1287/isre.1100.0323
2. Al-Ruithe, M., Benkhelifa, E., Hameed, K.: Data governance taxonomy: cloud vs. non-cloud. Sustainability (2018). https://doi.org/10.3390/su10010095
3. Niyato, D., Abu, M.A., Wang, P., Kim, D.I., Han, Z.: In: IEEE International Conference on Communications (ICC), Kuala Lumpur (2016)
4. Liu, K., Qiu, X., Chen, W., Chen, X., Zheng, Z.: Optimal pricing mechanism for data market in blockchain-enhanced internet of things. IEEE Internet Things J. **6**(6), 9748–9761 (2019)
5. Holler, M.J.: Einführung in die Spieltheorie, 8. Auflage, Springer (2019), pp. 122 f., 199ff
6. Thon, M.P.: First-Mover and Second-Mover Advantage under Uncertainty. TU München (2018)

7. Rosenthal, R.: Games of perfect information, predatory pricing, and the chain store. J. Econ. Theor. **25**(1), 92–100 (1981)
8. Kalai, E., Smorodinsky, M.: Other solutions to Nash's bargaining problem. Econometrica **43**(3), 513–518 (1975)
9. Kessler, R., Gómez, J.M.: Implikationen von Machine Learning auf das Datenmanagement in Unternehmen. In: HMD Praxis der Wirtschaftsinformatik: Ausgabe 5, pp. 89–105. Springer (2020)
10. Zaveri, A., Rula, A., Maurino, A., Lehmann, J.: Quality assessment methodologies for linked open data. Semantic Web J. (2013)
11. Wang, R., Strong, M.: Beyond accuracy: what data quality means to data consumers. J. Manage. Syst. **12**(4), 5–33 (1996)
12. Yu, H., Zhang, M.: Data pricing strategy based on data quality. Comput. Ind. Eng. **112**, 1–10 (2017)
13. Sen, S., Joe-Wong, C., Ha, S., Chiang, M.: A survey of smart data pricing: past proposals, current plans, and future trends. ACM Comput. Surv. **46**(2), 1–37 (2013)
14. Lange, J., Stahl, F., Vossen, G.: Data marketplaces in different research disciplines: a review (2017). Retrieved fromhttps://doi.org/10.1007/s00287-017-1044-3
15. Halter, A.N., Mason, R.: Value measurement for those who need to know. Western J. Agric. Econ. **3**(2), 99–109 (1978)
16. Wakker, P.P.: Explaining the characteristics of the power (CRRA) value family. Health Econ. **17**, 1329–1344 (2008)
17. Kimball Miles, S., Sahm Claudia, R., Shapiro, M.: Imputing risk tolerance from survey responses. J. Am. Stat. Assoc. **103**, 1028–1038 (2008)
18. Chen, M.A: Individual monotonicity and the Leximin solution. Econ. Theor. 353–365 (2000)
19. Garcia, R.C., Contreras, J, De Lima Barbosa, M., De Silva Toledo, F.: Raiffa-Kalai-Smorodinsky Bargaining Solution for Bilateral Contracts in Electricity Markets (2020)

Method for a Complexity Analysis of a Copper Ring Forming Process for the Use of Machine Learning

F. Thelen[1], B. Theren[1(✉)], S. Husmann[2], J. Meining[3], and B. Kuhlenkötter[1]

[1] Ruhr-Universität Bochum, Universitätsstr. 150, 44801 Bochum, Germany
theren@lps.rub.de
[2] Bleistahl Services GmbH & Co. KG, 58300 Wetter, Germany
[3] K2 Digital Transformation GmbH, 45891 Gelsenkirchen, Germany

Abstract. The aim of the Industry 4.0 initiative is to secure Germany's future as an industrial location and to strengthen its competitiveness compared to other countries. In contrast to large companies, it is more difficult for medium-sized ones to implement the migration from Industry 3.0 to 4.0, as they do not have the financial and human resources to fully replace all systems currently in operation. Therefore, the migration needs to be executed evolutionarily by retrofitting existing production facilities so that new acquisitions can be avoided. In this paper, such a retrofit will be analyzed based on a machine for forming copper rings, which is part of a process for manufacturing valve seat inserts for combustion engines. Since production is carried out under a high cycle time and the rings meet tight tolerances, condition monitoring is to be implemented to detect failures at an early stage. For this purpose, the approaches Design of Experiments (DoE) as well as Machine Learning (ML) are considered. Both options are evaluated based on a complexity analysis using the environment concept of an intelligent agent by RUSSEL & NORVIG. Finally, suitable supervised ML algorithms are selected.

Keywords: Machine learning · Data analysis · Complexity analysis · Predictive maintenance · Industry 4.0 · Retrofit

1 Introduction

As a result of the rising popularity of Machine Learning (ML) research, developed methods are widely applied to various industrial fields [1]. At the same time, with rapid development of Industry 4.0 and IoT technologies, for example through new sensors, processing software or low-cost storage solutions, hurdles are set low as data collection costs are decreasing. Therefore, ML has quickly found its way into production technologies as well [1, 2]. However, particularly for small and medium-sized companies, implementation of ML use cases is associated with a high effort, mainly because guidelines as well as industry specific best practices are still missing [3].

This paper is dedicated to assess, whether a production process is suitable for implementing ML and which algorithms should be selected. Because a suitable generalized

M. Liewald et al. (Eds.): WGP 2022, LNPE, pp. 600–610, 2023.
https://doi.org/10.1007/978-3-031-18318-8_60

approach for this challenge is still missing, the analysis is performed on a machine forming copper rings for the infiltration in downstream sinter process to produce valve seat inserts. As a defective ring (deviations in diameter and wound volume) results in a different materials structure and a malfunctioning valve, high demands are placed on the product accuracy. The goal is the implementation of condition monitoring in the production process.

For supporting the application of ML, many guidelines such as the CRISP-DM [4] or the TDSP-Lifecycle [5] were established, which organize the implementation into phases from problem understanding to product deployment. Using ML methods for manufacturing process gets more and more attention [6]. As an example, ML use cases for the ring-rolling, the incremental sheet forming process and an hydraulic test bench were developed and tested [7–9]. However, concrete assistance on algorithm selection is given minimally. Therefore, this work proposes using the widely accepted concept of an intelligent agent by RUSSEL and NORVIG [10] as a base for performing a complexity analysis. The agent perceives the environment through sensors. The perception is then processed internally, which is most frequently invisible to the user, and suitable responses are developed, which are executed through the actuators. The environment can be further characterized by specific properties, which will be used in Sect. 3 to analyze the problem's complexity.

Conventionally, the determination of the physical relationships of a technical system is performed using Design of Experiments (DoE), that is why the ML approach will be compared to applying DoE in Sects. 4 and 5. Lastly, an appropriate supervised learning algorithm will be selected from linear methods, Ensemble Trees, Support Vector Machines and Neural Networks.

2 Machine Description

The machine subject of this paper produces copper rings for valve seat inserts in automotive combustion. Such a ring is shown in Fig. 1. After fabrication, a ring is fed into a secondary production process, in which sintering together with a powder green compact occurs forming the valve seat ring. In case of a defective copper ring, the correct functioning of the sintering process can no longer be guaranteed. However, if already sintered, defective rings can only be detected during final engine inspection, which is why the quality control of the rings needs to follow high demands. The rings quality is defined by a total of four **output parameters**: the inner diameter, roundness, weight, and gap size.

The ring production process is illustrated in Fig. 2. First, the copper wire is unwound from a coil (1) and fed through a lubricant wetting unit (2). Too high as well as too low lubricant dosage needs to be avoided, as overdosing leads to increased slippage and underdosing to increased wear of tools.

To remove the wires pre-embossing, the wire is then straightened by a row of rolls (3). The feed is generated by four additional rolls (4), whose pressure on the wire can be adjusted by changing their vertical distance. The rolls press the wire against the deflection tools (5), bringing the wire into a circular shape. In general, excessive pressure leads to wear of the deflection tools, however, due to the copper's flexibility, wear effects are

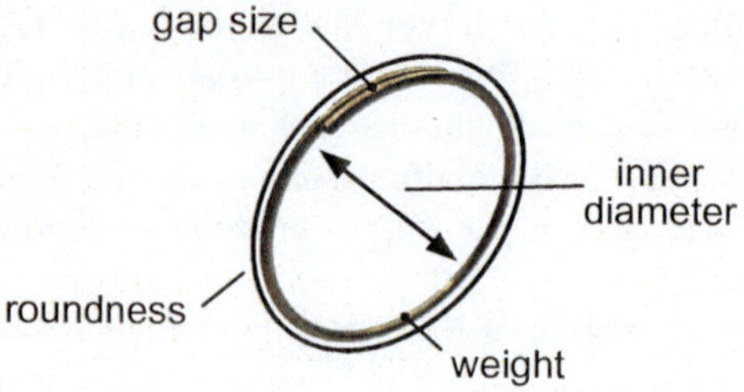

Fig. 1. A copper ring with illustrated output parameters

rather low. The two deflection tools exert the greatest influence on the geometry of the rings as their position relative to the wire can be adjusted by a total of nine degrees of freedom. Finally, a ring is separated from the remaining wire by two knives (6) and is then guided to the follow-up process in which the actual valve ring seat ring is produced. In conclusion, the described process can be adjusted by a total of three **input parameters**: the degree of wetting, the contact pressure, and the orientation of the deflection tools. A sensor concept exists for both the input and output parameters, but this is only considered in abstract form in the further course of the work.

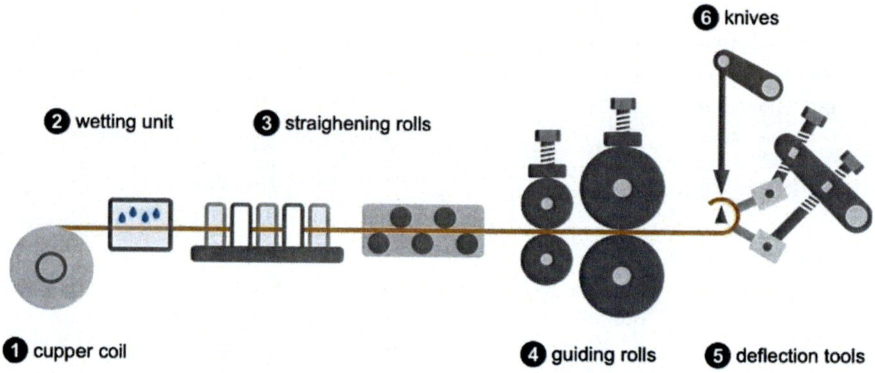

Fig. 2. Manufacturing process of the copper rings.

In addition to the parameters presented, environmental influences exist as well, which will not be investigated further in this work due to the accompanying increase in complexity. These include influences as the ambient temperature, humidity, or the responsible machine personnel. Eliminating the influence of such parameters can thus be relatively easily by keeping them as constant as possible during machine operation.

3 Complexity Analysis

In order to analyze the complexity of applying ML to the production process, the already presented concept of an intelligent agent by RUSSEL and NORVIG is used. Therein, the agent's environment is characterized by certain so-called dimensions, which are used here to assess the usability of the previously defined both input and output production parameters. In [10], each dimension is split into two categories, which are here expanded with a point system to evaluate ambiguous cases more precisely. The dimensions which characterize the environment of an intelligent agent are shown in the header of Table 1.

The environment is referred to as **observable**, if an agent's sensors give it access to the complete state of the environment at each point in time. Therefore, a production parameter is called observable if it can be measured entirely by sensory means. Observability is further differentiated here to distinguish whether situations can occur when measuring is only possible under certain conditions. If the subsequent state of the environment depends solely on the current state and the actions selected by the agent, the environment is called **deterministic**. Accordingly, a parameter is deterministic if it is not dependent on any other parameter. Further differentiation is here performed as well depending on the number of parameters influencing each other. If the actions of an agent can be divided into episodes, in which the agent first perceives the environment and then reacts to it, the environment is called **episodic**. An environment is called **static** if the state of the environment does not change while the agent is deliberating, otherwise it is called dynamic. Lastly, a distinction in terms of the number of perceptions an agent experiences is made. If there is a fixed number of possible perceptions, the environment is called **discrete**, otherwise continuous.

The result of applying these environment dimensions to the ring forming machine are shown in Table 1 for the input, and in Table 2 for the output parameters. Regarding observability, every parameter is generally observable, the input parameters though only partially, because machine conditions can occur in which measurements lose their significance. In terms of the contact pressure and deflection tools this can happen through an excessive use of lubricant or wearing of the tools. The degree of wetting was classified as even less observable because the uniformity of distribution of the applied liquid remains unknown.

Whether the input parameters are deterministic depends on the position of a parameter in the manufacturing sequence. As wetting occurs before any other processing step, the degree of wetting is fully deterministic. The contact pressure on the other hand is measured after wetting and therefore depends on it. As the orientation of the deflection tools is the last parameter in the sequence, it depends on both other parameters. Since the output parameters are generally influenced by every input parameter, the inner diameter, the roundness, and the weight are non-deterministic. Only the gap size is rated with 66% as it depends only on one out of three parameters. All parameters behave identically when episodic and static are considered.

Since all parameters can always be measured at fixed points in time and do not change while the agent is perceiving the environment, all input and output parameters are episodic and static. The degree of wetting and the orientation of the deflection tools are discrete since they do not change during the production of a single ring. The contact pressure is a continuous variable. The same criterion applies to the output parameters. Here, the inner diameter as well as the gap size are continuous, roundness and weight are discrete.

Table 1. Evaluation matrix of the input parameters

input parameters	observable	deterministic	episodic	static	discrete
degree of wetting	◑	●	●	●	●
contact pressure	◕	◑	●	●	○
orientation of deflection tools	◕	○	●	●	●

Table 2. Evaluation matrix of the output parameters

output parameters	observable	deterministic	episodic	static	discrete
inner diameter	●	○	●	●	○
roundness	●	○	●	●	●
weight	●	○	●	●	●
gap size	●	◕	●	●	○

By aggregating the ratings of the individual input and output parameters, the complexity of integrating ML into the ring forming process can be assessed. This is shown in Table 3. The fewer ratings are assigned to the parameters with respect to the previously discussed properties, the more complex a problem is to be classified. Accordingly, the dimensions episodic, static, and observable are indicating a rather low complexity. Measurements can therefore be performed once per cycle at a fixed time and evaluated by the algorithm until the next cycle. Thereby, observability is especially important, as an algorithm would otherwise miss relevant information needed to perform any appropriate actions. More than half of the parameters need to be measured continuously, which complicates the evaluation effort by the agent, since continuous measured variables must first be discretized at a suitable point prior to evaluation. A lower complexity is expected as well in terms of observability because every parameter apart from the degree of wetting was classified as observable. Because wearing of the tools can lead to invalid measurements, wearing needs to be monitored and sufficient lubricating maintained while generating the training data.

The highest complexity is generated by the interdependencies of the majority of parameters which means the environment is not deterministic. Therefore, no linearity between input and output parameters can be assumed without accepting large losses in accuracy.

Table 3. Evaluation matrix of the hole agent

	Observable	Deterministic	Episodic	Static	Discrete
viewed agent	◕	◔	●	●	◑

4 Effort Analysis of Conventional Solutions

Most commonly, DoE is performed to determine the physical relationships of technical systems. Therefore, the effort of performing a proper DoE on the presented production problem is discussed in this chapter and later compared to the ML approach.

Instead of varying one parameter while keeping the others constant to explore the laws of a technical system, DoE aims to reduce the experimentation effort significantly by strategically exploring the parameter space and determining the number of experiments to achieve statistical significance [11]. After the experiments, a regression function can be used to map the desired parameters relationships. First, a model of the problem at hand is created, in which all input, output and disturbance values are defined. The controllable factors are the variables that will be specifically changed during the experiments, while the uncontrollable factors are kept constant. The response variables are the output variables of the system [12, 13].

The experiments of the DoE are summarized in experimentation plans. Depending on the number of input combinations, a distinction is made between **full factorial** and **fractional factorial plans**. A full factorial plan includes all combinations n of the controllable factors, which are measured in several levels m and which must be defined prior to the experiments. The number of required trials N can then be calculated by Equation 1.

$$N = n^{\mathrm{N}} \tag{1}$$

The levels in which the variables are changed are defined by the assumed physical laws. If the correlations are considered linear, $m = 2$ may be suitable, for quadratic relations for example, $m = 3$ is chosen. Since the parameters of the ring forming machine were analyzed to be mostly non-deterministic, at least $m = 3$ would have to be selected. With eleven input parameters and an assumed quadratic correlation, $N = 177147$ experiments would have to be performed.

$$N = 2^{\mathrm{k-p}} = 2k + n_{\mathrm{O}} \tag{2}$$

By using a fractional factorial plan (Equation 2), composed for example in an orthogonal array, the number of experiments can be reduced to $N = 2074$ (with $p = 0$, $n_{\mathrm{O}} = 4$ [14], but without even considering the necessary preliminary screening experiments, the effort of realizing a solution with DoE is found to be far too high.

5 Effort Analysis of ML and Comparison to DoE

In Sect. 3, the complexity of the ring forming process was generally characterized as observable, non-deterministic, episodic, static, and discrete. As applying DoE was demonstrated being too complex, ML is considered as a realistic alternative. A high observability is a key criterion for both DoE and ML, because a model will otherwise operate on incomplete and unreliable data, which results in weak predictions. Whereas DoE is not applicable in the case of a non-deterministic process because of a too high number of experiments even for a quadratic model, ML defines its strengths in two ways. On the one hand, an algorithm finds the relation between parameters autonomously. On the other hand, a ML model is less constrained than a DoE model, as non-parametric algorithms such as Support Vector Machines (SVMs) or Neural Networks (NNs) do not require assumptions about parameters prior to training and are therefore more flexible. This aspect is further discussed in Sect. 6.

When considering the properties episodic, static, and discrete, either DoE or ML can be deployed. However, if a production process differs from this situation, realization with DoE becomes impractical. If e.g. discretization of the measured values is impossible in a continuous environment due to information loss, ML has to be applied.

In contrast to frequent perception, ML also has some notable disadvantages as well. The training of a ML algorithm for example produces a black box model, which means although the input parameters are mapped to the output parameters, the exact operations remain hidden from the user. That means, if a deeper analytical understanding of the system is desired, ML is not an alternative. But in the context of production systems, finding the exact system relationships is often less relevant compared to achieving a high prediction accuracy of the system model. Currently, prediction accuracy often also correlates with measurement accuracy, but also with the domain knowledge of the devices. In order to increase the accessibility of ML for small and medium sized companies, further knowledge needs to be gained at both points. Nevertheless, the effort required by ML should not be underestimated as the final accuracy depends on various parameters such as the data quality and the adjustment of hyperparameters. Optimizing these parameters is time-consuming, which must be considered before ML is deployed.

6 Selection of the ML-Algorithm

In this section, suitable ML-algorithms for controlling the ring forming process will be selected. For condition monitoring, an algorithm must be able to predict the quality of the rings based on the measured input parameters to perform classification into defective and faultless products, which indicates a supervised learning problem. In general, supervised ML can be divided into classification and regression. While classification deals with category targets, so-called classes, regression, on the other hand, deals with numerical targets that can take on continuous values [15]. The individual steps in the development of a classification algorithm are shown in Fig. 3.

When choosing classification, between data acquisition of input and output parameters through sensory and training, conversion of the output parameters into labels must be performed. Such an algorithm could then predict on the fly, whether the quality requirements are met, or a product is to be sorted out. However, this requires that the ring dimensions are already known during data collection and the dimensions are not subject to change. In addition, machine controlling would be impossible, since the information as to whether a ring meets the quality requirements is not sufficient for a control system.

By using regression instead, these disadvantages could be eliminated as continuous variables are predicted. Because the output is not categorical, no conversion prior to the training, but rather a posterior one after the prediction is necessary. Changing the ring dimensions later on would not require re-training, condition monitoring could be performed more accurately, and the continuous output data could be used by a control system. Therefore, a regression algorithm could perform more flexible and precise monitoring than classification, but on the downside, more training data is necessary.

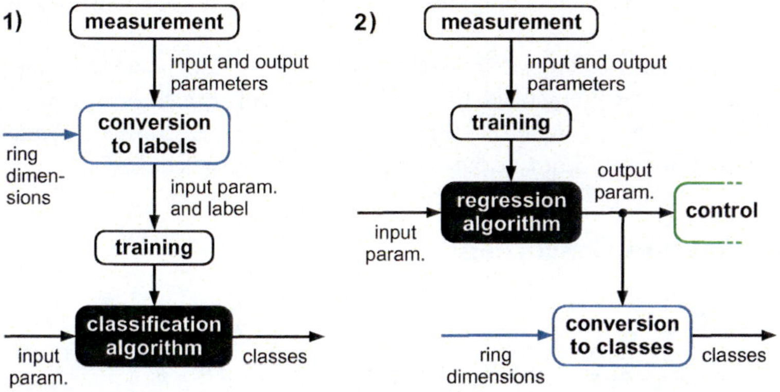

Fig. 3. Comparison of training and application of a classification versus a regression algorithm

A further division of supervised learning algorithms can be made into parametric and non-parametric algorithms. Parametric algorithms make certain model assumptions in advance of the training, with the advantage of training simplification and acceleration, but with the disadvantage of a lower maximum accuracy to be achieved [16]. Often the relationship between targets and features is simplified by assuming linearity. However, as a result of analyzing the machines complexity earlier, the relationship between input and output parameters cannot be assumed to be linear because the parameters were found to be mostly non-deterministic. As no assumptions about the parameter relationships are being made and the demand for dimensional accuracy of the rings is high, non-parametric algorithms are more suitable for the ring forming process, but are more computationally intensive, that is why longer training times must be expected.

Another constraint for a potential algorithm concerns the short cycle time of 80 parts per minute, which requires a high computational speed of the algorithm, since it has to predict the output within 750 ms. For this reason, lazy learning such as k-nearest neighbor algorithms, although they are non-parametric, are not suitable for the process at hand either. Those algorithms work directly with the data set provided and therefore need large datasets for achieving high accuracy, which also increases processing times significantly [17].

The last property considered here is the algorithms ability to handle multiple outputs. According to [18], solutions for multi-output regression can either be problem transformation or algorithm adaptation methods. In problem transformation methods, a multi-target regression problem is split into several single-target problems, which means training can be parallelized easily. In contrast, algorithm adaptation methods are working with multiple targets simultaneously and are therefore considering possible relationships among the output variables unlike transformation methods, resulting in smaller and faster models [18]. With respect to the present process, algorithm adaptation methods should be favored as they offer a higher potential in terms of speed and accuracy of calculations and the training of eleven independent models would be too time-consuming. Both Ensemble Trees and SVMs exist as multi-output models and can therefore be used to perform condition monitoring. NNs possess a special role, as they can deal with several outputs without fundamental modifications, simply by introducing multiple output neurons.

In principle, Ensemble Trees, SVMs and NNs can be used for conditional monitoring. However, it is difficult to predict which of these options will achieve the best performance, as this is highly dependent on the dataset and the learning process is highly statistic [18]. That is why all three algorithms should be tested for a final selection.

7 Summary and Conclusion

In this paper, the applicability of ML for a real-world production process was analyzed based on the environment concept by RUSSEL and NORVIG. The performed analysis steps are summarized in Fig. 4. The properties observable and deterministic were found to be the most important ones in case of the ring forming process. With a lack of observability, an algorithm is not able to access the complete state of the environment at each point in time, which results in weak predictions. A high observability is therefore important for applying either DoE or ML for a given production process. If the parameters of the process are mostly deterministic, the implementation of DoE will require less effort, as the parameter's relationships are not interdependent and a relatively small number of experiments can be conducted to analyze those relationships. If on the other hand, the parameters are found being non-deterministic, ML is more suitable in order to avoid a high number of experiments. Therefore, in case of the process at hand, the ML approach is preferred over DoE.

In terms of using classification or regression algorithms for a production process, the effort and benefit should be evaluated as performing classification involves fewer requirements for the training data, but an extension of a condition monitoring program in terms of machine control is not possible. With existence of non-deterministic parameters and multiple outputs, non-parametric and adaptation algorithms such as Ensemble Trees, SVMs or NNs were chosen, because possible relationships among the output variables are taken into account.

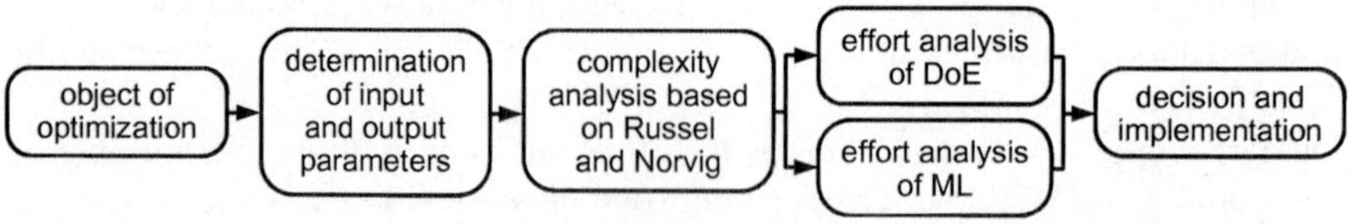

Fig. 4. Sequence of the proposed complexity analysis of an optimization problem

As ML in production offers many opportunities a broad spectrum of applications exists. In order to help companies decide whether a solution using ML is suitable and to help getting started with ML, further development of guidelines and best practices have to be made. This paper made a first step in this direction by proposing a complexity analysis based on the concept of an intelligent agent was proposed and applied to a real-world production process. However, further studies need to be conducted applying the concept to other processes in order to validate its applicability. In addition, the complexity analysis could be extended by further factors, e.g. the inclusion of the setting of the hyperparameters.

References

1. Weichert, D., Link, P., Stoll, A., Rüping, S., Ihlenfeldt, S., Wrobel, S.: A review of machine learning for optimization of production processes. Int. J. Adv. Manuf. Technol. **104**, 1889–1902 (2019)
2. Kang, Z., Catal, C., Tekinerdogan, B.: Machine learning applications in production lines: a systematic literature review. Computer. Ind. Eng. **149** (2020)
3. Mayr, A., Kißkalt, D., Meiners, M., Lutz, B., Schäfer, F., Seidel, R., Selmaier, A., Fuchs, J., Metzner, M., Blank, A., Franke, J.: Machine learning in production—potentials, challanges and exemplary applications. Procedia CIRP **86** (2019)
4. Chapman, P., Clinton, J., Kerber, R., Khabaza, T., Reinhartz, T., Shearer, C., Wirth, R.: CRISP-DM 1.0: Step-by-step data mining guide. CRISP-DM Consortium (2000)
5. Microsoft: The Team Data Science Process Lifecycle. https://docs.microsoft.com/de-de/azure/architecture/data-science-process/overview. Last accessed 2022/04/11
6. Fahle, S., Prinz, C., Kuhlenkötter, B.: Systematic review on machine learning (ML) methods for manufacturing processes—identifying artificial intelligence (AI) methods for field application. Procedia CIRP **93**, 413–418 (2010)
7. Fahle, S., Glaser, T., Kuhlenkötter, B.: Investigation of suitable methods for an early classification on time series in radial-axial ring rolling. ESSN: 2701-6277 (2021)
8. Fahle, S., Glaser, T., Kneißler, A., Kuhlenkötter, B.: Improving quality prediction in radial-axial ring rolling using a semi-supervised approach and generative adversarial networks for synthetic data generation. Production Eng. **16**(1), 175–185 (2022)
9. Neunzig, C., Möllensiep, D., Fahle, S., Kuhlenkötter, B., Möller, M., Schulz, J.: Approach to data pre-processing for predictive quality of hydraulic test results in a dynamic manufacturing environment. Automation 2022, VDI-Berichte 2399, pp. 425–438 (2022)
10. Russel, S., Norvig, P.: Artificial Intelligence: A Modern Approach, 3rd edn. Prentice Hall, New Jersey (2010)
11. Selvamuthu, D., Das, D.: Introduction to Statistical Methods, Design of Experiments and Statistical Quality Control. Springer, New Delhi (2018)
12. Siebertz, K., Van Bebber, D., Hochkirchen, T.: Statistische Versuchsplanung—Design of Experiments (DoE), 2nd ed. Springer Vieweg, Aldenhoven, Aachen (2017)
13. Montgomery, D.C.: Design and Analysis of Experiments, 10th ed. Wiley, Arizona (2020)
14. Kleppmann, W.: Versuchsplanung: Produkte und Prozesse optimieren, 10th edn. Carl Hanser Verlag, München (2020)
15. James, G., Witten, D., Hastie, T., Tibshirani, R.: An Introduction to Statistical Learning with Applications in R. Springer, New York (2013)
16. Murphy, K.P.: Machine Learning: A Probalistic Perspective. The MIT Press, Cambridge, London (2012)

17. Ertel, W.: Introduction to Artificial Intelligence. Springer, London (2011)
18. Borchani, H., Varando, G., Bielza, C., Larrañaga, P.: A survey on multi-output regression. WIREs Data Mining Knowl. Discov. **5**, 216–233 (2015)

Advancements in Production Planning

Prediction of Disassembly Parameters for Process Planning Based on Machine Learning

Richard Blümel(✉), Niklas Zander, Sebastian Blankemeyer, and Annika Raatz

Institute of Assembly Technology, Leibniz University Hannover, An der Universität 2, 30823 Garbsen, Germany
bluemel@match.uni-hannover.de

Abstract. The disassembly of complex capital goods is characterized by strong uncertainty regarding the product condition and possible damage patterns to be expected during a regeneration job. Due to the high value of complex capital goods, the disassembly process must be as gentle as possible and being adaptable to the varying und uncertain product's state. While methods based on data mining have already been successfully used to forecast capacity and material requirements, the determination of the product's or component's condition has become apparent in the recent past. Despite the rapid increase in sensor technology on capital goods such as aircraft engines and their use for condition monitoring due to countless interfering effects, it is only possible to react spontaneously to the product's condition. So far, we have concentrated on product condition-based prioritization of disassembly operations in a logistics-oriented sequencing strategy. In this article, we present an approach to predict disassembly process-planning parameters based on operational usage data using machine learning. With the prediction of disassembly forces and times, processes, tools and capacities can be efficiently planned. Thus, we can establish a component-friendly disassembly process adaptable to varying product conditions. In this article, we show the successful validation on a replacement model of an aircraft engine.

Keywords: Disassembly planning · Regeneration · Machine learning

1 Introduction

In order to expand the lifetime of complex capital goods, they must be maintained and overhauled regularly. That allows the monetary value to be continued into further service phases by maintenance, repair and overhaul (MRO) [1]. However, high stress on components and the resulting effects of wear on them lead to the partial or complete loss of their initially designed product properties. From an economic and ecological point of view, an efficient regeneration process can be useful, especially with regard to the regeneration of complex capital goods and their components [2]. The research of the Collaborative Research Center (CRC) 871: "Regeneration of Complex Capital Goods" addresses this issue and the process steps for the regeneration of lost component properties of such

M. Liewald et al. (Eds.): WGP 2022, LNPE, pp. 613–622, 2023.
https://doi.org/10.1007/978-3-031-18318-8_61

capital goods, using aircraft engine's high-pressure turbines (HPT) as an example [3]. Particularly with regard to the capacity planning of the regeneration process, a critical process step from an economic point of view is the component-protecting disassembly of undefined solidified connections between HPT blades and disks. The joints solidify during the engine's service life depending on operating parameters, e.g. operating hours, landing-takeoff (LTO) cycles, and environmental influences, e.g. flight routes over sea or deserts. After disassembling individual components, a detailed inspection is performed, and the regeneration effort is identified.

Due to the high number of turbine blades, the technologically complex component properties, and thus economic aspects, a component-protecting disassembly process of these components is of high relevance for the actual regeneration of a complex capital good. Damage to the blades must be avoided to prevent the need for spare parts and successful regeneration. Also, the disassembly of the individual blades, as the last component when disassembling the engine, must be performed fastest possible to prevent unnecessarily long slack times for components disassembled prior [1, 4]. Due to the uncertain condition of the blade-disk joint, the disassembly is carried out by manually or hydraulic hammering out. Performing the disassembly by highly qualified personnel ensures the adaptability to the unknown und varying condition of the operational solidified joint, in order not to irreparably damage or destroy the complex and sensitive components.

In this article, we present an approach to predict the effort for disassembly tasks based on the engine's usage. Using the example of HPT blades, we show the implementation and usage of a learning model that predicts tool dimension and disassembly time, increasing the workload information in order to plan and prioritize disassembly tasks [1]. However, since we cannot use components with a real product history for the experiment due to availability, the approach is validated on a replacement model.

2 Related Work

Due to environmental conditions, the turbine's hot exhaust gas mixture and high tensile loads during typical operation, its blading is subjected to three major influences: fatigue, corrosion and creep [5]. These influences can negatively affect the properties of the components, which requires regular maintenance and overhaul, as well as regeneration at the end of their life cycle. As a result of the service life and the previously mentioned influences, solidification mechanisms occur in the joint between the blades and the turbine disk. Usually, it requires a complex and costly disassembly process of these components during the regeneration of the capital goods. Depending on the degree of solidification, a defined breakaway force has to be applied to the disassembly object for the detaching movement of the components. Currently, different disassembly strategies are used for this process and the loosening of these solidified joints. The individual blades can be released from the turbine disk by hydraulic extraction using a special pulling device [6]. With this tool, the blade is dismantled hydraulically by pulling out in the axial direction of the turbine disk using a hook slide located between the gap of the turbine disk and the blade root. Due to the necessary support of the device on the disk and the positioning cycles after each individual disassembled blade, this disassembly strategy is considered to be very time-consuming. There is also an increased risk of component damage, such as pressure notches on the components, due to contact points between the disassembly objects and the device. Another strategy is to hammer the blades out of the turbine disk with hydraulic or manual hammer blows. This leads to an undefined high introduction of non-reproducible disassembly forces into the disassembly object, which can lead to irreparable damage to the components. Therefore, highly trained and qualified employees generally perform the disassembly tasks [7].

The time required for this disassembly step during regeneration depends on the degree of solidification between blades and disk. Due to the high number of blades (e.g. 64 in a IAE V2500 engine, according to the manufacturer), this is a critical process step that significantly influences the actual capacity planning of the regeneration. Therefore, it is highly relevant for capacity planning to obtain exact knowledge of the expected damage pattern on the components already in an early planning horizon. Also, due to the loss of knowledge of the blade disk joints condition, the planning of workforce, disassembly times and tools can only be adaptively performed when the engine has already been partially disassembled.

That challenge is addressed by predictive maintenance. Ran et al., for example, summarized in their work the primary purpose of predictive maintenance as the reduction of costs, elimination of unexpected downtime and the improvement of availability and reliability of systems [8]. As an approach, Eickemeyer developed a damage library to predict the effort for the regeneration of capital goods, such as aircraft engines [9]. He defined temporal model categories to optimally perform workforce, resources or time planning. The categories are divided into long-term, from a timeframe of several months up to a year, medium-term, of several weeks and short-term, of hours during regeneration. The database has information on 650 regenerations so far performed, containing data such as operating hours, LTO cycles or engine type. A Bayesian network processes that data to predict the regeneration effort for particular assemblies or components.

Based on Eickemeyer's research, we present our work on setting up a learning model to predict the regeneration effort of disassembling HTP blades with the engine's operational data as input. As aforementioned, the challenge of the disassembly is the loss of knowledge of the blade disk joints condition. Using the learning model, we achieve the prediction of disassembly tools and time before the initial disassembly sequence. In order to set up the learning model, we identified potential factors influencing the disassembly effort, using the response surface method as explained in the following.

3 Setup

The regeneration of turbine blades requires a component-protecting disassembly of the solidified joints between the individual blades and the turbine disk. In [10], we introduced and presented a replacement model of a solidified blade disk joint. Since no operational solidified turbine blade disk joints were available in our investigation, we used an external force by clamping the turbine disk segment with a defined clamping force F_{Cl}. That induces a contact pressure on the joint's contact surfaces, replicating the solidification. By varying the external force, we achieve a variation of different operation scenarios, like differing flight hours or LTO-cycles. That results in a solidification force $F_S(z)$

opposing the disassembly of the blade. Therefore, the disassembly force $F_D(z)$ must be greater than $F_S(z)$ to initiate a disassembly movement. However, $F_D(z)$ must not exceed a material-specific maximum to prevent damaging the blade root [11].

In order to ensure an automated, reproducible and component-protecting disassembly process of these components with an optimum cycle time, the further procedure is based on a vibration-aided disassembly. The vibration, induced by a piezo stack actuator, is superimposed on the disassembly movement, and it is positioned in an electrically operated linear drive on the disassembly [8]. As known from the literature, vibration superimposed on a movement reduce the coefficient of friction [12]. Using as a tool in the disassembly, it reduces the disassembly force required to detach the solidified connections between the turbine blades and disk [13]. In order to determine the degree of solidification and thus also the necessary capacity utilization prior to the actual regeneration, information on previous regenerations can be used as input data.

As shown by Eickemeyer in [9], using machine learning (Bayesian network) can predict the regeneration effort. However, in this work, we use the replicated model of the solidified joints [10] to train the learning model to predict disassembly tools and times. Based on the joint's condition, suitable feed motion and piezo parameters in order to ensure an optimum cycle time while ensuring a component-protective disassembly of the turbine blades are selected. In the first step, we use the response surface method (RSM) to identify and characterize the input variables, like the clamping force as the solidification replacement or the vibration's parameters [14, 15]. As a result, we obtain information on how they influence the disassembly force needed to dismantle the blades. In the second step, we use the RSM's result, to set up a learning model, able to predict the disassembly force based on the joint's condition. Integrated into the disassembly tool, it also sets the parameters to execute a component-friendly disassembly while achieving low disassembly times.

4 Results

As aforementioned, we use the RSM to identify and analyze the disassembly process of solidified HPT turbine blade disk joints. Depending on the solidification condition, a prediction of planning parameters becomes possible with the calculated optimal parameters.

4.1 Characterization of Disassembly Parameters

The initial step of the RSM analysis is the identification of the inputs and outputs. Since we aim to identify the influences on the disassembly force, we will set F_D as the system's output. The input factors are accordingly the clamping force F_{Cl}, representing the operational data, the piezo stack actuator vibration's characteristics, such as amplitude, frequency and waveform and the disassembly speed v_D, defined as disassembly length $l_D(\mathrm{z})$ per disassembly time. The piezo actuator allows the use of three waveforms, sinusoidal, triangle and sawtooth, representing a categorical factor, with the experimental design repeated for each category. Using a face centred composite design of experiments

Table 1. Levels of influential parameters in CCF

Factor	Low (−1)	Medium (0)	High (+1)
Clamping force F_{Cl} (N)	2000	3500	5000
Disassembly speed v_D (mm/s)	1	5.5	10
Frequency f_{Pi} (Hz)	10	35	60
Amplitude A_{Pi} (μm)	10	55	100
Waveform WF_{Pi}	Sinusoidal	Triangle	Sawtooth

(CCF), we performed 72 randomized experiments in order to investigate the relationship and interaction of each input (Table 1).

The following analysis using multiple linear regression (MLR), we can set up an equation which predicts the disassembly force depending on the inputs and their interactions, as in Eq. 1:

$$\begin{aligned}\widehat{F_D} = {} & \beta_0 + \beta_1 \cdot F_{Kl} + \beta_2 \cdot f_{Pi} + \beta_3 \cdot v_D + \beta_4 \cdot A_{Pi} + \beta_5 \cdot WF_{\mathrm{Sin}} + \beta_6 \cdot WF_{Tri} \\ & + \beta_7 \cdot F_{Kl} \cdot f_{Pi} + \beta_8 \cdot F_{Kl} \cdot v_D + \beta_9 \cdot F_{Kl} \cdot A_{Pi} + \beta_{10} \cdot F_{Kl} \cdot WF_{\mathrm{Sin}} \\ & + \beta_{11} \cdot F_{Kl} \cdot WF_{Tri} + \beta_{12} \cdot f_{Pi} \cdot v_D + \beta_{13} \cdot f_{Pi} \cdot A_{Pi} + \beta_{14} \cdot f_{Pi} \cdot WF_{Sin} \\ & + \beta_{15} \cdot f_{Pi} \cdot WF_{Tri} + \beta_{16} \cdot v_D \cdot A_{Pi} + \beta_{17} \cdot v_D \cdot WF_{\mathrm{Sin}} + \beta_{18} \cdot v_D \cdot WF_{Tri} \\ & + \beta_{19} \cdot A_{Pi} \cdot WF_{\mathrm{Sin}} + \beta_{20} \cdot A_{Pi} \cdot WF_{Tri} \end{aligned} \quad (1)$$

The individual factors are each influenced by a coefficient β_i in the regression equation, describing the factor's influence on the disassembly force. After the experimental procedure, we evaluated the results using the analysis of variance (ANOVA) [15]. Among other information, we also obtain a statement on whether the model is statistically significant. That examines whether the model can be applied as calculated, i.e. whether the input variables significantly influence the output variable as calculated. The determination of the validity in predicting the disassembly force according to the regressions equation (Eq. 1) is determined by the p-value. If the p-value is less than 0.05, the model can be considered significant, i.e. it is robust in predicting the disassembly force. With the calculated p-value lower than 0.001, we can assume that the model is valid.

In addition, we evaluate the goodness of fit by calculating the coefficient of determination R^2 and the adjusted-R^2. Adding more input variables to the equation always increases R^2, even if the variable has no influence. The adjusted-R^2 indicates the percentage of variation explained only by the inputs that actually affect the output. An R^2 value of 0.9714 and an adjusted-R^2 value of 0.9602 indicate a good model fit using MLR. Therefore, we can assume a sufficient precision of the prediction accuracy.

According to the RSM procedure, we calculated the optimal setting parameters to reduce the disassembly force in the next step. We perform a comparison with varying disassembly speed and vibration waveform to demonstrate the results. We particularly highlight these two factors, since the speed regards the time and capacity aspect, and varying the waveform showed a dependence on the reduction of the disassembly force in preliminary experiments. Table 2 shows the values of each input factor. The tests are executed with a disassembly with a clamping force of 4000 N and the shown parameters for the piezo stack actuator.

Table 2. Optimal parameters to minimize the disassembly force

Fixed parameters	
Clamping force (F_{Cl})	4000 N
Amplitude (A_{Pi})	100 μm
Frequency (f_{Pi})	60 Hz
Varying parameters	
Disassembly speed (v_D)	1 mm/s, 5.5 mm/s, 10 mm/s
Waveform (WF_{Pi})	Sinusoidal and Triangle

We performed the disassembly tests in a randomized order. Table 3 shows the results of 45 runs, each the mean value of the maximum disassembly force. In addition, we present the percentage reduction compared to without vibration.

Table 3. Result of the reduction of the maximum disassembly force

	$v_D = 1$ mm/s	$v_D = 5.5$ mm/s	$v_D = 10$ mm/s
F_D without vibration	2157 N	2139 N	2144 N
F_D w. sinusoidal vibration	1717 N (−20.4%)	1950 N (−8.8%)	1931 N (−9.9%)
F_D w. triangle vibration	1802 N (−16.5%)	1910 N (−10.7%)	2022 N (−5.7%)

We achieved the maximum reduction of the disassembly force at a sinusoidal waveform and a disassembly speed of 1 mm/s. The maximum reduction decreases with increasing disassembly speed when superimposing the triangle vibration. However, when using sinusoidal vibration, the influence of the disassembly speed is more complex. In addition, the waveform also influences the maximum reduction of the disassembly force, depending on the disassembly speed. Based on the results, we can develop a learning model in the following step that can predict process parameters for a component-protecting disassembly.

4.2 Learning Model to Predict Disassembly Parameters

In order to predict disassembly process parameters, we developed a learning model. Based on operational usage data of an aircraft engine, tool dimension for a component-protective and disassembly times for capacity planning are the primary determinants. However, as mentioned initially, we approximate these data through the replacement model. A variation of the clamping force represents different operational usage of the aircraft engine [10]. Using the experiment's data described in Sect. 4.1, we train another regression model to predict the disassembly force based on the pre-set clamping force, representing the aircraft engine's usage. In addition, we executed further disassembly runs with random levels of the influential factors to obtain a test subset to test the trained model, with the split between training and test data being 80–20%.

To evaluate the learning model, we calculate the coefficient of determination R^2 and the symmetric mean absolute percentage error (sMAPE). As discussed in the literature, they are used to evaluate machine learning studies [16]. An R^2 of 0.9248, close to 1, and sMAPE of 8.594%, close to 0%, indicate a good predictive performance of the learning model.

Integrated into the control of the disassembly device, tool parameters such as the values of the piezo stack actuator are automatically adjusted. The input parameters for the learning model are the maximum disassembly force according to material-specific limits and the clamping force set at the disassembly test rig, replicating the joint's solidification. Additionally, an operator has to specify the maximum disassembly speed. The disassembly speed, which mainly determines the time per disassembly operation, serves as the key influencing variable on the disassembly force. The learning model attempts to keep the disassembly speed as fast as possible while at the same time adhering to the force limit. It then calculates the difference between predicted and given maximum disassembly force by varying the piezo stack actuator's parameters, amplitude, frequency and waveform. The difference is necessary to determine by how much the predicted force is less than the given force. That results in the setting parameters for disassembly at maximum disassembly speed while the force limit is not exceeded and the difference is within a predefined safety interval.

Figure 1 shows an exemplary disassembly process of ten blades being successively disassembled. The disassembly speed is increased step by step up to the maximum speed (in the shown example 6 mm/s) while never exceeding the force limit by varying the adjustment parameters of the piezo stack actuator. If the force limit might be exceeded, either the speed is reduced, or the disassembly is stopped. That will prevent any possible damage to the blade root.

That results in the following disassembly scenario for capacity planning: Based on the known operational data of the engine, the learning model predicts reasonably accurate the disassembly force. If an engine is to be disassembled with an identical operational history to an already known and disassembled engine, the setting parameters already learned can be reused. However, assumed an engine is to be disassembled of an unknown type or a flight scenario that has not yet been disassembled. In that case, the disassembly force is calculated based on previously executed disassembly runs by the learning model. The example in Fig. 1 shows that only a few disassembly runs are needed until a target speed of 6 mm/s, set by the operator as the maximum disassembly speed, is reached. Thus, if the engine type or flight data is unknown, within a short approximation interval of, in our case, ten out of 64 blades, approx. 15% of the disassembly runs, the learning model adapts to the target disassembly speed. The data collected can then be integrated into the learning model's database to enhance and improve its performance.

For capacity planning, it follows that disassembly time and tool dimension can be predicted depending on the knowledge of the engine's condition. With a minimum number

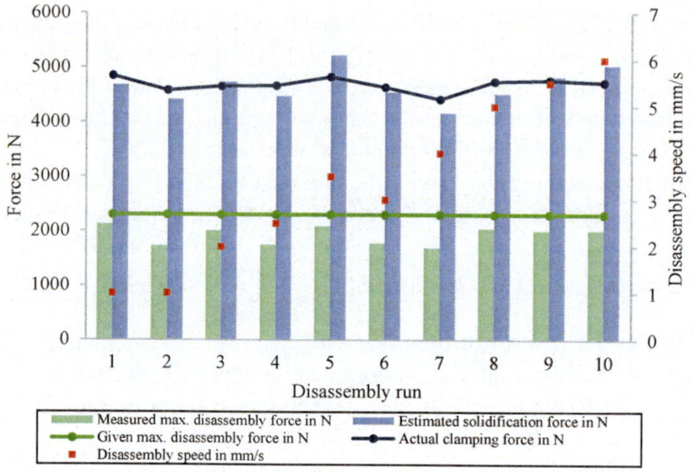

Fig. 1. Diagram of an exemplary disassembly process including ten runs

of disassembly runs to achieve the optimal setting parameters, an efficient disassembly process planning can be realized. That allows the disassembly process, characterized by a high degree of uncertainty, to become plannable and adaptable to the unknown product's condition.

5 Conclusion and Outlook

This paper presents the development of a learning model to predict disassembly parameters for optimal process and capacity planning. An aircraft engine's operation leads to a loss of knowledge of its assembly joint's condition. The exemplary investigated connection of the HPT blades and disks solidifies to an unknown extent. Therefore, it is challenging to predict the data, such as tool dimensioning and disassembly time, required for the disassembly process. Resources, machines or workforce can thus only be determined in short-term during disassembly.

In order to tackle that challenge, we developed a learning model which predicts the disassembly force to perform a component-protective disassembly. By adding the disassembly time as the crucial factor for time capacity planning, we were able to show its dependence on the disassembly force. That allows the planning of tools and temporal capacity based on the engine's operational data, such as flight hours, routes, or LTO-cycles.

We identified influential factors during the disassembly on needed dismantling forces using the response surface method. With the aid of superimposed vibration, we reduced the maximum needed disassembly force to overcome the solidification force induced by the joint's solidification. The subsequent multiple linear regression allowed the disassembly force to be described as a function of the influencing variables. These include the clamping force as a replacement model of operational solidification, the vibration's adjustment variables, and the disassembly speed. The reduction of the maximum disassembly force due to vibration, thus allowing an increase of the disassembly speed.

From the results of the RSM, we then developed the learning model to predict the disassembly force based on the clamping force as the replacement model for the joint's solidification. Based on that prediction, the control chooses optimal setting parameters for the piezo stack actuator adhering the material-specific maximum force limit (Fig. 1). That enables the execution of a component-protecting disassembly. The learning model can considers the target disassembly speed, set by an operator. Depending on the knowledge of the joint's condition in comparison with already disassembled blade disk joints, the learning model increases the speed as fast as possible. In our example, only a few disassembly runs were required. With the predicted knowledge of required disassembly duration and tool dimensioning and thus operating resources and workforce, the disassembly as the initial step in the regeneration chain can be planned so that resources are used optimally and slack times are prevented. Thus, disassembly process planning can already be carried out on a medium-term planning horizon, as the engine's operation data provides the database.

In future work, the application of the learning model should be applied and confirmed on components with a real usage history. Also, a comparison of multiple linear regression to artificial neural networks (ANN) showed an advantage in the predictive accuracy of the MLR. That was possibly due to the small database of training data. Other machine learning approaches such as ANN or Bayesian networks should be compared by expanding the amount of input and training data.

Acknowledgements. Funded by the Deutsche Forschungsgemeinschaft (DFG, German Research Foundation)—SFB 871/3-119193472.

References

1. Lucht, T., Heuer, T., Nyhuis, P.: Disassembly sequencing in the regeneration of complex capital goods. In: Nyhuis, P., Herberger, D., Hübner, M. (eds.) Proceedings of the Conference on Production Systems and Logistics : CPSL 2020. Hannover: Institutionelles Repositorium der Leibniz Universität Hannover, pp. 12–20. (2020). https://doi.org/10.15488/9642
2. Seliger, G.: Montage und demontage. In: Grote, K.H., Feldhusen, J. (eds.) Dubbel. Springer, Berlin, Heidelberg (2005). https://doi.org/10.1007/3-540-26620-8_127
3. Aschenbruck, J., Adamczuk, R., Seume, J.R.: Recent progress in turbine blade and compressor blisk regeneration. Proc. CIRP **22**, 256–262 (2014). https://doi.org/10.1016/j.procir.2014.07.016
4. Heuer, T., Lucht, T., Nyhuis, P.: Material disposition and scheduling in regeneration processes using prognostic data mining. Proc. Manuf. **43**, 208–214 (2020). https://doi.org/10.1016/j.promfg.2020.02.138
5. Bräunling, W.: In: Flugzeugtriebwerke, Springer, Berlin, Heidelberg (2001). https://doi.org/10.1007/978-3-642-34539-5
6. Hohmann, S.: Method and device for dismantling a turbine blade, ALSTOM Power N.V 1101 CS Amsterdam (NL), (European Patent No. EP1149662, 2001). https://data.epo.org/publication-server/document?iDocId=1925273. Last Accessed: 04 May 2022
7. Tao, W., Huapeng, D., Jie, T., Hao, Wa.: Recent repair technology for aero-engine blades, recent patents on engineering. **9**(2), 132–141 (2015). https://doi.org/10.2174/1872212109666150710184126

8. Ran, Y., Zhou, X., Lin, P., Wen, Y., Deng, R.: A survey of predictive maintenance: systems, purposes and approaches (2019)
9. Eickemeyer, S.C.: Kapazitätsplanung und-abstimmung für die regeneration komplexer Investitionsgüter, Dissertation Leibniz Universität Hannover, Garbsen: Berichte aus dem IFA (2014). ISBN: 978-3-944586-77-9
10. Bluemel, R., Raatz, A.: Experimental validation of a solidification model for automated disassembly. In: Herberger, D., Hübner, M. (eds.) Proceedings of the Conference on Production Systems and Logistics: CPSL 2021. Hannover: Institutionelles Repositorium der Leibniz Universität Hannover, pp. 339–348. (2021). https://doi.org/10.15488/11250
11. Middendorf, P., Blümel, R., Hinz, L., Raatz, A., Kästner, M., Reithmeier, E.: Pose estimation and damage characterization of turbine blades during inspection cycles and component-protective disassembly processes. Sensors **22**(14), 5191 (2022). https://doi.org/10.3390/s22145191
12. Littmann, W., Storck, H., Wallaschek, J.: Sliding friction in the presence of ultrasonic oscillations: superposition of longitudinal oscillations. Arch. Appl. Mech. **71**, 549–554 (2001). https://doi.org/10.1007/s004190100160
13. Mullo, S.D., Pruna, E., Wolff, J., Raatz, A.: A vibration control for disassembly of turbine blades. Proc. CIRP **79**, 180–185 (2019). https://doi.org/10.1016/j.procir.2019.02.041
14. Stöber, M., Müller, D., Thümmler, A.: Einsatz der response-surface-methode zur Optimierung komplexer Simulationsmodelle. Technical Report 05003 (2005). ISSN 1612-1376. https://doi.org/10.17877/DE290R-7704
15. Witek-Krowiak, A., Chojnacka, K., Podstawczyk, D., Dawiec, A., Pokomeda, K.: Application of response surface methodology and artificial neural network methods in modelling and optimization of biosorption process. Biores. Technol. **160**, 150–160 (2014). https://doi.org/10.1016/j.biortech.2014.01.021
16. Chicco, D., Warrens, M.J., Jurman, G.: The coefficient of determination R-squared is more informative than SMAPE, MAE, MAPE, MSE and RMSE in regression analysis evaluation. PeerJ. Comput. Sci. **7**, e623 (2021). https://doi.org/10.7717/peerj-cs.623

A New Approach to Consider Influencing Factors in the Design of Global Production Networks

M. Martin(✉), S. Peukert, and G. Lanza

wbk Institute of Production Science, Karlsruhe Institute of Technology, Kaiserstr. 12, 76131 Karlsruhe, Germany
Michael.martin@kit.edu

Abstract. Uncoordinated decisions that have a long-term impact on the production network lead to inefficient structures and limit the ability to change. However, the ability to change is a basic prerequisite for future decisions. At the same time, the world is becoming more volatile, uncertain, complex, and ambivalent. To counteract this, external and internal influencing factors must be considered in the early stages of planning global production networks (GPN). The design of GPN is on the one hand associated with a large number of degrees of freedom and on the other hand with a large number of influencing factors. Influencing factors can thereby be known and predictable, but also unknown and unpredictable. To make production networks capable to change in the long term, influencing factors and their effect on the network design must be considered. The combination of influencing factors with consideration of uncertainty still needs further research in the context of network design. Thus, this article aims to develop a method for network design that does not only take external and internal influences into account at an early stage but also leads to a network configuration that considers these influences and increases resilience. To achieve this, the influencing factors should first be represented in scenarios using the receptor theory. Subsequently, the scenarios can be incorporated into the optimization of the network configuration by choosing a solution from a predefined solution space. The process of solution selection and testing can be supported by a digital twin. The result is an initial concept that merges these different steps into a continuous process that can be used to design adaptable GPN in the future.

Keywords: Influencing factors · Network configuration · Uncertainty · Global production networks · Changeability · Resilience

1 Introduction

Today, manufacturing companies operate in an increasingly complex and constantly changing environment. This environment is characterized by volatility, uncertainty, complexity, and ambiguity (VUCA world) [1]. The influencing factors shaping the environment can be internal or external in origin [2] and quantitative or qualitative [3]. Many of the planning parameters used are subject to uncertainty [4]. Only a comprehensive

M. Liewald et al. (Eds.): WGP 2022, LNPE, pp. 623–632, 2023.
https://doi.org/10.1007/978-3-031-18318-8_62

consideration of the influencing factors can reduce the uncertainty associated with them [3, 5]. In addition, there are always new influencing factors that are not or only very difficult to predict (e.g. blockade in the Suez Canal, the Corona crisis, the semiconductor shortage, etc.).

The influencing factors are contrasted by network structures that have been built up over the last decades as well as resulting from short-term and uncoordinated decisions [2, 6]. Global Production Networks (GPN) are cumbersome to adapt and the complex decisions have a long-term impact [7]. Only continuous planning and proactive updating of decisions can reduce the complexity and uncertainty [8, 9]. To counteract the increasing complexity and uncertainty in the global environment, the flexibility and changeability of the considered GPN can be used [10]. This allows companies to foresee adaptation needs and changes at an early stage, which can then be quickly translated into reality. This could increase the resilience of the GPN.

Digital twins (DT) offer a way to continuously collect and provide data. A DT is the virtual and computerized counterpart of the real system [11]. With a DT, new measures can be tested easily and quickly with the help of simulations. However, the DT must be integrated into an overall concept to achieve maximum benefit. This allows shorter, proactive planning cycles to be realized and uncertainties to be minimized.

Combined with an approach from the software industry, the DevOps approach, continuous integration, updating, and deployment can also be realized for GPN. This approach is already the subject of research at the production and machine levels [24].

The presented research work pursues the goal of presenting a first overall concept, which takes up influencing factors and their uncertainty, enables a continuous and iterative procedure with the help of a DT, and can determine a suitable solution as a result of a need for change. The following research questions will be answered in this paper:

1. How can the drivers of change be quantified and their uncertainty be represented by scenarios?
2. How can the appropriate measures be selected from the solution space?
3. How the selected measures can be quickly tested and transferred into reality with the help of approaches from the software industry
4. How can the digital twin act as an enabler of the overall concept?

2 State of the Art

The investigation of the state of the art will be divided into two topics. On the one hand, influencing factors (Drivers of Change) and uncertainties related to their future development shall be considered. On the other hand, approaches for the design and function of GPN are to be considered with which the need for change resulting from the changing influencing factors can be met. From the deficits of the considered approaches, a research gap results, which shall be addressed with the methodology developed in Sect. 3.

Westkämper for example states that drivers of change refer to the turbulent environment in which a company operates. The drivers of change can occur both externally and internally [12]. For example, the drivers of change are applied in site selection [13]. Once the drivers of change are known, they can be analyzed and evaluated. The approaches of Gille and Zwißler [14] and Lanza et al. [15] attempt to categorize the change drivers and thereby enable quantitative analyses. A sound analysis of the drivers of change can reduce the uncertainties associated with them [15].

In addition to the described approaches that focus on the change drivers, further approaches consider change drivers and their uncertainties. Hawer models and classifies the fuzziness of planning parameters using a guideline to achieve the most accurate mapping of fuzziness, for example, through probability distributions. Using appropriate uncertainty propagation mechanisms, uncertainties of the planning parameters that depend on other influencing factors are identified. The identified fuzziness forms the basis for planning a factory's ability to change [16]. In his work, Möller develops a scalable, cross-stage, and valuation-oriented model of production that specifically focuses on uncertainties and defines a hierarchical, life-cycle cost model [17].

The approaches of Hawer and Möller focus on the production and factory level [16, 17]. Further approaches contemplate the network level. In their approach, Reinhardt and Krebs consider several factors that influence site selection. Their method allows the consideration of multidimensional, qualitative, and quantitative uncertainties. These are modeled with probability distributions and the fuzzy method and integrated into a model for structured monetary costing [3]. The paper by Schuh et al. presents a quantitative approach for uncertainty assessment. By evaluating consistent scenarios possible developments of external factors are transformed into a measure of uncertainty [6]. The scenario model is based on the work of Gausemeier [18]. Dobler et al. also describe an approach based on scenarios. It is used for the early detection of production technology deficits under uncertainty derived from megatrends [19].

In the following, approaches that try to counteract the uncertainties in the design of GPN by proposing different configurations or which propose measures to reduce the influence of the change drivers are, presented. Neuner distinguishes between certain and uncertain, uninfluenceable factors. On a more detailed level, a representation of the uncertainty behavior can be done either with or without exact probabilities. Neuner thereby uses the uncertainties to better configure international production networks. The result always represents a network variant or alternative [20]. Lanza and Moser present a model for dynamic multi-objective optimization of GPN. It evaluates the impact of influencing factors and optimizes the design of the GPN [21]. Moser presents an approach for migration planning in GPN in a volatile environment. The approach can be used to identify robust migration paths of the network configuration considering multidimensional drivers of change from the business environment [8]. Ude presents a decision support model for the configuration and evaluation of a globally distributed value network. Dynamics and uncertainties are taken into account in the input data by integrating a Monte Carlo simulation in the simulation model [22]. Schuh et al. present a systematic approach to determine the necessary level of agility in GPN. The necessary agility levels are based on the consideration of the volatility of the influencing factors and a cost-benefit decision [23]. In his work, Sager combines existing approaches to create a new approach for configuring GPN. To reduce time and cost, he applies a cyclic approach to develop alternatives to incrementally improve GPN [9].

Based on the analysis of the state of the art, it can be concluded that influencing factors are taken into account, but no approach continuously considers the uncertainty of influencing factors and transfers them into a DT. Uncertainty is often only used as an input into an optimization model which tries to reduce the uncertainty. Different approaches focus on one part of an overall solution. A combination into an overall concept is not done so far. Rather, Sager's integration into a cyclic approach is only done for a one-time optimization and not a continuous approach over the entire life cycle.

For this reason, a new concept is needed, which takes influencing factors and their uncertainties into account. To use the model over the entire life cycle the concept needs to be continuously useable. To not only treat a part of an overall solution, but to develop an overall concept, a DT must be provided, in which the status quo is represented and the step of selecting alternatives is taken into account. Only by an overall view and a harmonization of the steps, a continuous iterative concept can be created.

3 Conception of a Methodology for GPN Design

This concept of the methodology is presented in Fig. 1 and explained in more detail in the following sections. This is done by the different phases of the loop. The loop is based on the DevOps approach, which originates from the software industry [24]. It is intended to improve quality, increase speed and improve collaboration.

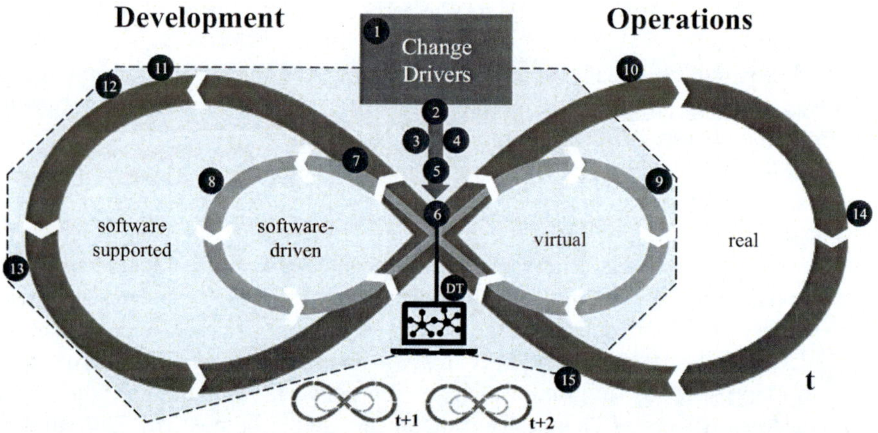

Fig. 1. The concept for continuous integration of change drivers into an iterative planning cycle for the design of GPN (adapted from [24])

The starting point of the network design methodology are change drivers that emerge from the external environment of a GPN or internally from the network itself. These change drivers are mapped to the GPN via the receptor theory [25] and the uncertainty of the change drivers is modeled via scenarios (Sect. 3.1). To derive the need for change from the scenarios a DT is needed in which the change drivers are considered and continuously updated. This DT is updated in case of future changes. On the one hand, this DT must be created and, on the other hand, it must be structured in such a way that changes can be implemented without increased effort (Sect. 3.2). With the help of a DT, a simulation model can be created. If the change drivers and the verification of these in the simulation model result in a need for change, this is examined in more detail in the next step and a solution is sought from a solution space (Sect. 3.3). With a further model, the best possible solution for the specific need for change is figured out (Sect. 3.4). This process is divided into a development phase and an operation phase both in reality and by the DT.

3.1 Change Drivers and Uncertainties

The starting point of the methodology for designing GPN are change drivers that have an external or internal origin (Fig. 2). External change drivers are, for example, new laws or market changes, while internal change drivers reflect the number of employees or the sequence of processes (1). All of these change drivers have an impact on the GPN targets. An established approach to create the link between change drivers and production is the receptor theory according to Cisek et al. [25]. With the receptor theory, the influencing factors can be transferred to GPN. The goal is to be able to derive quantitative statements about the receptors, which can then be used for further steps. By defining receptor key figures the change drivers can be consolidated and quantified (2) [26]. The fuzzy logic [27] supports the transfer of qualitative drivers into quantitative values. This logic has been applied several times in the context of production and has proven to be reasonable. For this reason, fuzzy logic shall also be used in this work to quantify influence factors. Fuzzy logic is relevant for the influencing factors that cannot be modeled using empirical data or probability distributions. In fuzzy logic, the first step is to transform linguistic variables into membership functions by fuzzification. Subsequently, with IF-THEN rules (so-called inference rules) the input variables can be transferred to an output variable. In the last step, the defuzzification enables the derivation of a quantitative value for the target variable which can be used in the scenarios of the receptor key figures.

If all influencing factors are available in a quantitative value, the influencing factors can be transferred to the receptor key figures. In this step, the uncertainties associated with the change drivers must be taken into account. This is to be made possible by scenarios. In addition to the quantitative and qualitative uncertainties, which can be taken into account e.g. via different distributions (3) or as described above via fuzzy logic (4) [3] a distinction should also be made between epistemic and aleatory uncertainty. While aleatoric uncertainty is already taken into account by stochastic distributions and cannot be reduced further, epistemic uncertainty can be reduced by information acquisition [5]. Through the Continuous Improvement and the Continuous Integration approach from Fig. 1, the epistemic uncertainty is reduced step by step. This can simplify the selection of solutions because fewer developments of the future have to be considered.

Once the uncertainties have been taken into account, scenarios are available for the various receptors (5). In the next step, these scenarios must be transferred to the DT (6). For this purpose, existing standards are to be used to enable transferability to other use cases. The transformation scenarios are to be integrated into a data model of the DT. One possibility for a standardized transfer of data to the DT is, for example, the Core Manufacturing Simulation Data model (CMSD) [28] or the idea of the Asset

Administration Shell (AAS) [29]. Such a standardized data format can support the fast creation and update with regard to the transformation scenarios.

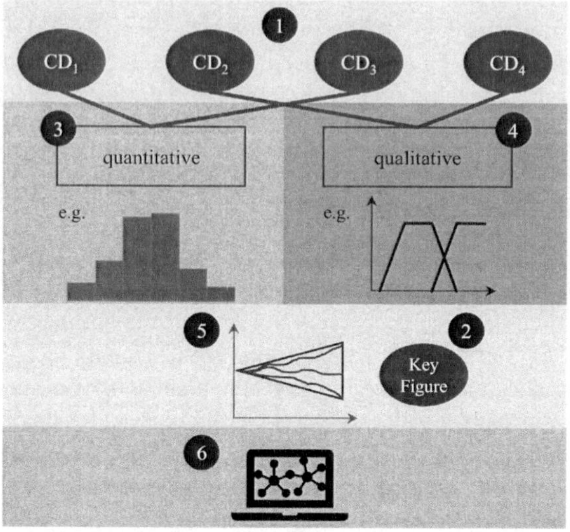

Fig. 2. Consideration of the uncertainties of the change drivers in scenarios of the receptor key figures and consideration in the DT

3.2 Mapping of the Current Status—Simulation Model

To be able to map the scenarios of the receptor key figures in the DT, they must first be available in their current form. This can be done by building on the data model from Sect. 3.1, which does not only contain the scenarios, but also other modules, the logic, and the parameters of a DT. The structure of the DT is thus already determined by the design of the data model. The concrete information and thus the parameterization of the DT takes place via the continuous improvement and updating of the data model. The later form of the DT can then be derived from this data model and instantiation with real values. The result is a DT that takes into account the previous external and internal change drivers. Future developments of the change drivers can then be compared with this model by scenarios.

The separation of the DT as a software solution from the hardware (the physical production network) enables a realistic check of changes within the DT without increased adaptation effort on the DT itself as well as in reality. This enables rapid trials of an alternative solution for coping with the scenarios. If the investigation reveals a more efficient solution in terms of the receptor key figures than the actual state, it can be tested digitally and then, if successful, transferred to reality (Continuous Deployment). However, as described above, this does not only apply to external change drivers, but also to internal ones. This means that the internal adaptations of the network, such as other types of transport between locations, are always mapped in their most current form. Finally, it should be mentioned that the availability of an up-to-date DT means that the need for change resulting from the drivers of change can be identified more quickly and solutions for this can be found. If a need for change arises that cannot be made possible by simple adjustments in the GPN, the task is to select the appropriate solution from a set of solutions, evaluate them and, if successful, transfer it back into reality. The need for change can be defined as the difference between the capabilities of the current state (CS) and the requirements of the target state (TS_n) resulting from the different scenarios.

$$NfC = CS - \sum_{1}^{n} TS_n \tag{1}$$

3.3 Measures for Network Adaption

If a need for change arises from the consideration of the scenarios within the DT (7), this must be addressed to maintain the functionality of the real system. For this purpose, a distinction should be made between the terms flexibility and changeability [30]. If only minimal adjustments or adaptations result from the consideration of the scenarios, which were already considered and taken into account in previous cycles, then these can be taken over directly both in the DT (software-driven) (8) and after a check (9) in the real system as well (10). An example of such an adjustment is a change in the process sequence that was already recognized in the previous cycle and is now only finally checked and transferred again due to its occurrence in the scenario. In such a case, the adaptation is within the flexibility corridor of the system. Another example of flexibility is the increase of the produced quantity of a location by adding further working shifts of the employees in the production.

The situation is different with a change that results in a change of the system. Here, the flexibility corridor is no longer sufficient and the system must be fundamentally adapted. Such an adjustment can be, for example, the opening of a new site, the extension of a site by further process steps, or the change of a transport relationship between the sites. If these adaptation options are feasible and can be implemented at the current time, these options define the system's changeability corridor. This can also be described as a solution space (11), in which different solutions can be found, with which the need for change can be compensated.

If there are no solutions in the solution space that can solve the need for change, the next step is to try to expand the solution space (12). This step is the most complex. And cannot be supported by the DT, because new possible solutions which are not considered yet need to be found. For a local company which not thought about a subsidy abroad, a new site in a new country would be an expansion of the solution space.

3.4 Selection of the Most Suitable Solution—Optimization Model

After the solution space is available and if necessary extended, the step of the solution selection follows (software-supported) (13). For this, again the receptor characteristics are to be consulted, which can serve as objective functions. For example, the goal can be to achieve a certain number of pieces under given boundary conditions, such as the costs. In addition, the costs themselves can also represent an objective function, which should be kept as low as possible. For the selection of a solution, the state of the art already shows approaches that focus on optimization models. What is new in the presented concept is the consideration of possibilities of extending the solution space without setting up a new DT, simulation model and optimization model (a), to consider the scenarios more strongly and thus to also map progressions of receptor key figures in the future (b), to include uncertainty more strongly (c) and to reduce it by an iterative procedure similar to Sager [9] (d). The optimization should therefore not only take place for one point in time but for different points in time, to represent a progression and find a solution s which fits today and in the future. Such an approach can be based on the migration planning of Moser, who specifies the conversion capability at different points in time to achieve the optimum [8]. In this way, the DT is not only created at a current point in time, but step by step the DT of the future are also already anticipated and updated. The DT can then be accessed at every point in time. Analogous to software development, a backlog is created and continuously adapted. Regarding the existing literature, a new optimization model needs to be created because the existing approaches lack the ability to continuously update the solution. This new model focuses on the scenarios and uncertainties of the change drivers in more detail. Furthermore, it should be in close interaction with the DT. Thus, optimal updates can be implemented at short notice.

In the last step of the method, the found solutions are brought into reality by implementing the changes at the existing GPN (10). The new configuration of the GPN is then working until the next NfC (14). Out of this configuration new data can continuously be collected and transferred back to the DT (15). In addition to running the loop in the current time plane (t), the loop can also be run in future time axes (t+1, t+2, etc.) to fully account for the scenarios. Thus, similar to Moser [8], future adjustments can already be taken into account and a pipeline for further changes can be built.

4 Conclusion

The environment in which globally operating companies find themselves is increasingly determined by growing uncertainty and complexity. Furthermore, companies must be able to react quickly to changes such as delivery problems, new laws, or market changes. Companies need to align their GPN to remain competitive. However, these have grown historically and decisions were made intuitively.

The presented concept should enable a continuous and early consideration of the influencing factors and their uncertainties. It uses approaches from software development that have become established there. In addition, the focus is on a DT, which enables a fast, proactive planning process that is initially carried out independently of the existing GPN. Changes in the environment or to the existing system are continuously integrated and the DT is updated. Possible solutions to meet a NfC can be tested and, if successful, transferred to reality. Further research is recommended to realize the individual sub-aspects of the overall concept. Especially the continuous consideration of the change drivers and their uncertainties requires further research. Also, the automated spanning of the solution space and the selection of a solution needs to be further investigated in the context of GPN.

Acknowledgement. We extend our sincere thanks to the German Federal Ministry for Economic Affairs and Climate Action (BMWK) for supporting this research project 13IK001ZF "Software-Defined Manufacturing for the automotive and supplying industry https://www.sdm4fzi.de/".

References

1. Mack, O., Khare, A., Krämer, A., Burgartz, T. (eds.): Managing in a VUCA World. Springer, Cham (2016). https://doi.org/10.1007/978-3-319-16889-0
2. Lanza, G., et al.: Global production networks: design and operation. CIRP Ann. **68**(2), 823–841 (2019)
3. Krebs, P., Reinhart, G.: Evaluation of interconnected production sites taking into account multidimensional uncertainties. Prod. Eng. Res. Devel. **6**(6), 587–601 (2012)
4. Burggräf, P., Schuh, G. (eds.): Fabrikplanung. Springer, Heidelberg (2021). https://doi.org/10.1007/978-3-662-61969-8
5. Morse, E., et al.: Tolerancing: Managing uncertainty from conceptual design to final product. CIRP Ann. **67**(2), 695–717 (2018)
6. Schuh, G., Prote, J.-P., Schmitz, T., Nett, S.: Quantitative Erfassung externer Unsicherheiten bei der Gestaltung von Produktionsnetzwerken. Zeitschrift für wirtschaftlichen Fabrikbetrieb **111**(12), 784–788 (2016)
7. Schuh, G., Prote, J.P., Dany, S.: Reference process for the continuous design of production networks. In: 2017 IEEE International Conference on Industrial Engineering and Engineering Management (IEEM), pp. 446–449. IEEE (2017)
8. Moser, E.: Migrationsplanung globaler Produktionsnetzwerke. Dissertation, Shaker
9. Sager, B.M.: Konfiguration globaler Produktionsnetzwerke. Dissertation, Technische Universität München (2019)
10. Hingst, L., Wecken, L., Brunotte, E., Nyhuis, P.: Einordnung der Robustheit und Resilienz in die Veränderungsfähigkeit (2022)
11. Kritzinger, W., Karner, M., Traar, G., Henjes, J., Sihn, W.: Digital Twin in manufacturing: a categorical literature review and classification. IFAC-PapersOnLine **51**(11), 1016–1022 (2018)
12. Westkämper, E.: Wandlungsfähige Produktionsunternehmen. Das Stuttgarter Unternehmensmodell. Springer Berlin Heidelberg (2009)
13. Meyer, T.: Selection criteria: assessing relevant trends and indicators. In: Abele, E., Meyer, T., Näher, U., Strube, G., Sykes, R. (eds.): In: Global Production. A Handbook for Strategy and Implementation. Springer, Berlin, Heidelberg (2008)
14. Gille, C., Zwißler, F.: Bewertung von Wandlungstreibern. Zeitschrift für wirtschaftlichen Fabrikbetrieb **106**(5), 310–313 (2011)
15. Lanza, G., Moser, R., Ruhrmann, S.: Wandlungstreiber global agierender produktionsunternehmen—Sammlung. Klassifikation und Quantifizierung. Wt Werkstattstechnik Online **4**, 200–205 (2012)
16. Hawer, S.: Planung veränderungsfähiger Fabrikstrukturen auf Basis unscharfer Daten. Dissertation, Technische Universität (2020)
17. Möller, N.: Bestimmung der Wirtschaftlichkeit wandlungsfähiger Produktionssysteme. Utz, München (2008)
18. Gausemeier, J., Fink, A., Schlake, O.: Szenario-management. Planen und Führen mit Szenarien. Hanser, München (1995)
19. Dobler, R., Hofer, A., Martin, M., Reinhart, G.: Prognose produktionstechnischer Defizite. Zeitschrift für wirtschaftlichen Fabrikbetrieb **116**(3), 100–105 (2021)

20. Neuner, C.: Konfiguration internationaler Produktionsnetzwerke unter Berücksichtigung von Unsicherheit, 1st edn. Gabler, Wiesbaden (2009)
21. Lanza, G., Moser, R.: Multi-objective optimization of global manufacturing networks taking into account multi-dimensional uncertainty. CIRP Ann. **63**(1), 397–400 (2014)
22. Ude, J.: Entscheidungsunterstützung für die Konfiguration globaler Wertschöpfungsnetzwerke. Ein Bewertungsansatz unter Berücksichtigung multikriterieller Zielsysteme, Dynamik und Unsicherheit. Shaker; Wbk Inst. für Produktionstechnik, Aachen, Karlsruhe (2010)
23. Schuh, G., Prote, J.-P., Franken, B., Ays, J., Cremer, S.: Dedicated agility: a new approach for designing production networks. In: 2018 IEEE International Conference on Industrial Engineering and Engineering Management (IEEM), pp. 1–5. (2018)
24. Neubauer, M., Ellwein, C., Frick, F., Fisel, J., Kampert, D., Leberle, U.: Kontinuität als neues Paradigma. Computer Automation **2022**, 28–31 (2022)
25. Cisek, R., Habicht, C., Neise, P.: Gestaltung wandlungsfähiger Produktionssysteme. Zeitschrift für wirtschaftlichen Fabrikbetrieb **97**(9), 441–445 (2002)
26. Stähr, T.J.: Methodik zur Planung und Konfigurationsauswahl skalierbarer Montagesysteme - Ein Beitrag zur skalierbaren Automatisierung. Dissertation, KIT (2020)
27. Zadeh, L.A.: Fuzzy sets. Inf. Control **8**(3), 338–353 (1965)
28. Lee, Y.-T.T., Riddick, F.H., Johansson, B.J.I.: Core manufacturing simulation data—a manufacturing simulation integration standard: overview and case studies. Int. J. Comput. Integr. Manuf. **24**(8), 689–709 (2011)
29. Grothoff, J.A., Wagner, C.A., Epple, U.: BaSys 4.0: Metamodell der Komponenten und Ihres Aufbaus, 1st edn (2018)
30. Wiendahl, H.-P., Reichardt, J., Nyhuis, P.: Handbuch Fabrikplanung. Konzept, Gestaltung und Umsetzung wandlungsfähiger Produktionsstätten, 2nd edn. Hanser, München, Wien (2014)

Pushing the Frontiers of Personal Manufacturing with Open Source Machine Tools

M. Omer(✉), T. Redlich, and J.-P. Wulfsberg

Helmut Schmidt University, 22043 Hamburg, Germany
mohammed.omer@hsu-hh.de

Abstract. The democratization of desktop 3D printing has opened the domain of manufacturing to the masses. Today individuals can design and manufacture a variety of products in their living rooms. However, scaling a product from prototype to production and setting up a small-scale manufacturing business is often hindered by the expensive machinery and high upfront capital investment required. This paper presents the findings of a unique experiment that was carried out to understand the process of prototyping a relatively complex product (in this case, a 3D printer) in a home setting and then scaling it up to a small-scale production (10 units). In order to partially automate the manufacturing processes, two open source machine tools (OSMT), whose blueprints are freely available in the internet, were built, namely a CNC laser cutter and a CNC milling machine. The experiment reveals the particularities of starting a small-scale production in a home setting and the potential of OSMT to affordably scale up production, while also highlighting the challenges of OSMT adoption.

Keywords: Open source machine tools · Sustainable manufacturing · Personal manufacturing · Desktop manufacturing · Open source hardware

1 Introduction

In recent years, personal manufacturing has seen an upsurge. The internet and digitalization processes have made desktop machines available to the masses. This phenomenon, however, is limited in scope to educational and recreational purposes—they have not yet brought about the next industrial revolution that many promised it to be [1]. Personal manufacturing is furthermore subject to spatial and resource limitations as well: The majority of users who can access machine tools are based in industrialized countries with people in developing and low resource countries lacking access.

Developments in the field of open source economics have the potential to solve these challenges. By making build instructions of desktop machine tools freely accessible on the internet, they can theoretically be replicated everywhere in the world and for production purposes. To expand the knowledge of challenges and potential of open source desktop machine tools, this paper presents the findings of an experiment that included the prototyping of a relatively complex open source 3D printer in a home setting and the subsequent up-scaling of the same to a small-scale production of ten

M. Liewald et al. (Eds.): WGP 2022, LNPE, pp. 633–641, 2023.
https://doi.org/10.1007/978-3-031-18318-8_63

units. In the process, two open source machine tools (OSMT), namely a CNC laser cutter and a CNC milling machine, were also built to partially automate the production process. Through a description of the prototyping and small-scale production processes, an understanding is generated about the potential of personal manufacturing based on open source principles. Challenges and suggested solutions to OSMT adoption are also highlighted.

2 Background

2.1 Personal Manufacturing

Manufacturing has traditionally been the provenance of large corporations and qualified professionals due to the costs incurred and the machinery and know-how required to produce products on a large scale [1]. However, this has changed with the advent of the internet and the digitalization of manufacturing. Starting in factories and production floors in the last few decades, the digitalization of manufacturing has since reached the desktops, basements, and living rooms of individuals. There is no universal definition for personal manufacturing, and it is referred to under various names in the literature, including desktop manufacturing, personal digital fabrication, and personal production. Neil Gershenfeld defines personal manufacturing as "the ability to design and produce your own products, in your own home, with a machine that combines consumer electronics with industrial tools" [5].

Today more and more people with creative inclinations are setting up small businesses from their homes by selling hand-made artifacts that are often customized with the help of laser engraving or CNC milling on online e-commerce platforms such as Etsy and Shopify. To date, Etsy has millions of sellers who have generated billions of dollars in revenue [4]. However, since many of these home based makers do not have access to their own machine tools due to the high upfront costs to invest in production machinery, they are forced to outsource the complex manufacturing steps to larger firms. The higher costs of outsourcing manufacturing and relative inflexibility in production planning, means they are only able to sell on a relatively small scale. To produce a product on a relatively large scale with limited resources, it is either an option to outsource the manufacturing to large manufacturing firms (usually abroad to countries such as China), also known as contract manufacturing, or to invest in an entire factory setup and set up an in house mass production facility [10]. Neither of these options are feasible for individuals with limited funds and capacities.

2.2 Machine Tools for Everyone?

In the last years, desktop machines have become more accessible through drastically sinking prices. In 2001, the cheapest 3D printer on the market cost $45,000, whereas today, entry level personal 3D printers cost between $400 and $750, making them affordable to individuals and small businesses [3]. Smaller models of laser cutters and CNC mills are nowadays offered by manufacturers to cater to non-industrial users. However, these commercial consumer machines are primarily intended for hobby purposes and have limited functionality and capabilities which make them unsuitable for production. Therefore, they are largely marketed towards hobbyists and technology enthusiasts, who want to use these machines for primarily recreational purposes [8].

These machines have become increasingly popular in the many makerspaces and fab labs, which are communal manufacturing spaces where citizens can access digital fabrication machines that enable users to transform digital designs into physical prototypes. Some have even called these fabrication spaces the start of the next industrial revolution that would create new modes of production [1]. However, recent large scale studies of makerspaces around the world have revealed that most of them are predominantly utilized for educational or prototyping purposes and seldom for production [6]. Even though they offer the opportunity to get first experiences with digital fabrication technologies and machines to manufacture prototypes, they do not represent a truly scalable production facility for individuals looking to scale their manufacturing business. Moreover, these machine spaces must be rented, are not available everywhere and therefore are not a truly sustainable model for long term production plans.

2.3 The Open Source Revolution

Open source hardware (OSHW) presents a potential solution to this problem by applying the principles of open source software (OSS) to tangible objects. Similar to OSS, a quickly increasing number of physical artifacts is made freely available for anyone to access, use, modify, distribute, and sell [2]. This includes machine tools whose manufacturing documentation are made available online, including build instructions, bills of materials (BOMs), electronic schematics, and CAD drawings. This subset of OSHW is known as open source machine tools (OSMT) which are a key enabling technology to democratize manufacturing technology [9].

OSMT designs can be found strewn across the internet, under keywords such as 'DIY' or 'homemade' machine tools. Out of the plethora of OSMT designs available online, there are only a few that can actually be replicated, since many designs are published with incomplete build information. There is, however, increasing awareness among OSHW practitioners for designing and documenting machine tools in such a way that users without technical backgrounds can also replicate them. Designs that are well documented are often replicated by users around the world, who form a rich online community that actively takes part in the development and testing of these designs, further improving them while helping each other in the building and developmental processes. Two representative examples of such projects are the MPCNC, an open source CNC milling machine that is mainly made from 3D printed parts, and the Fabulaser mini, an open source CNC laser cutter (see Fig. 1).

In order to increase the replicability of the machines, the designers have meticulously documented the steps required for building them. This includes the provision of a complete bill of materials, a build assembly manual, electrical wiring schematics, software configuration guide, a user guide, and manufacturing tutorials. The total material costs for the CNC mill amount to about $800 while the laser cutter costs about $1600, making both machines significantly cheaper than commercial offerings with the same capabilities. The increased affordability and easy to understand build guides have helped make these complex machine tools more accessible to users.

Fig. 1. **Left:** MPCNC, an open source milling machine [11]; **right:** Fabulaser mini, an open source laser cutting machine [7]

3 Methodology

The research design for this paper was divided into two parts. First, an OSMT product design: the 'Hypercube Evolution' created by the thingiverse user Scott3D was replicated in a home setting and the CAD design was further developed and modified [12]. Next, ten units of the product were produced to simulate a small-scale production in a home setting. To increase productivity and reduce labor intensive steps, two other open source CNC machines were also built, namely the abovementioned MPCNC and Fabulaser mini (see Fig. 1). The aim of this work was to understand how OSMT can help in overcoming the challenges of affordably setting up a small-scale production at home while also highlighting the associated hurdles with replicating these machine tools while maintaining the machine performance and precision.

3.1 Prototyping at Home

For the initial replication, an open source design for a 3D printer was chosen with a license that allowed modifications and commercialization of the product [12]. The replicability of the design has been confirmed by the number of people that had successfully rebuilt the machine with the freely available design files. With the native CAD files published on the internet, the printer design could be further developed and modified with ease (see Fig. 2). In particular, the safety and user-friendliness of the product were to be increased. The aim of the further developments of the design was to completely enclose the machine so that no moving parts were exposed as well as implement upgrades such as a direct drive extruder, an integrated cable chain, a temperature controlled exhaust fan among others.

Comprised of more than 170 individual, unique components, the 3D printer is a relatively complex product and therefore this experiment is unique compared to producing a single printed object and is more representative of a complex consumer product. The precincts of the replication were a limited budget and predefined user functionality and safety requirements that called for critical analysis of each component. The printer design can be subdivided into several modules and sub-modules as seen in Table 1.

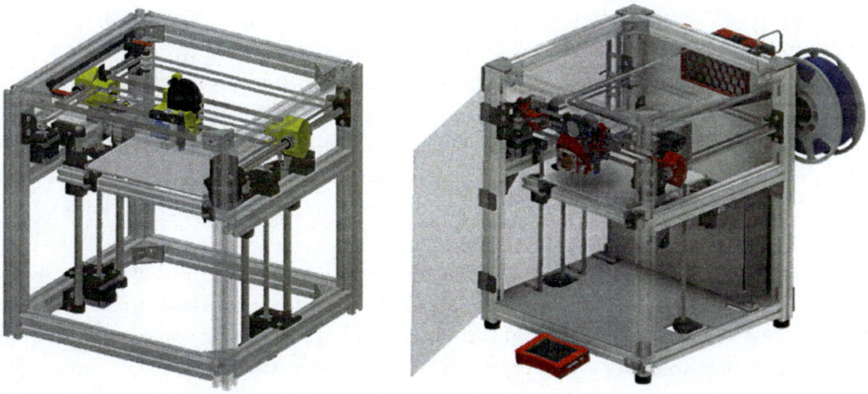

Fig. 2. Left: Original 3D printer design—hypercube evolution [12]; **right:** Modified and further developed design.

Table 1. Overview of manufacturing processes involved in manufacturing the 3D printer

Module	Raw material	Manufacturing process	Tools
Frame	Aluminium profiles	Cutting, drilling	Mitre saw and drill
Connecting parts	Plastic filament	FDM process	3D Printer
Base panel	MDF (wood) 3 mm	CNC laser cutting	Oscillating blade and drill
Back panel	DIBOND 3 mm	CNC milling	Oscillating blade and drill
Wiring	Wires, connectors	Stripping and crimping	Wire cutter, wire stripper, crimper
Electronics	PCB, resistors, SMDs etc	Soldering	Soldering iron

3.2 Scaling Production

In the next step, production was upscaled to ten units based on the first 3D printer prototype. Several bottlenecks were identified within the manufacturing workflow. Among these the most time consuming was the 3D printing of the connecting parts which could only be sped up by printing on multiple 3D printers simultaneously. The 3D printer itself was then used for printing parts for the next printer; in this way, each built printer was put into operation to print the parts of the next printer, setting up a so-called print farm.

The base and back panels of the 3D printer had complex cutouts and holes at specific locations. Manufacturing them was therefore highly time consuming, labor intensive, and prone to measurement errors. The process involved measuring the required cutouts accurately and marking them, followed by drilling holes and then cutting out the shapes with an oscillating blade. Unsteady hand movements made the cuts uneven and inconsistent, requiring further post processing and posed a safety hazard for the user. The manufacturing of the panels was therefore identified to benefit the most from CNC automation. To that end, the Fabulaser mini laser cutter and the MPCNC were replicated. The MPCNC parts were individually sourced from various shops and the 3D printed parts were printed on the already built 3D printer. It was built following a pictorial build manual provided by the designer. The laser cutter was bought as a DIY kit from the designer and assembled together by following the comprehensive IKEA-style build manual [7]. The panels of the ten 3D printers were then cut out using the two self-built open source machine tools (see Figs. 3 and 4).

A further bottleneck in building the printers was preparing the wiring harnesses. Preparing the wiring by cutting, stripping, crimping, soldering, and testing the individual components was highly time consuming and relatively prone to human errors. However, no open source machine tool could be identified to automate the process. Due to a lack of time, no new machines were developed for this purpose, however there is great potential to increase productivity and reduce human errors by using machines for automating the cables preparation process.

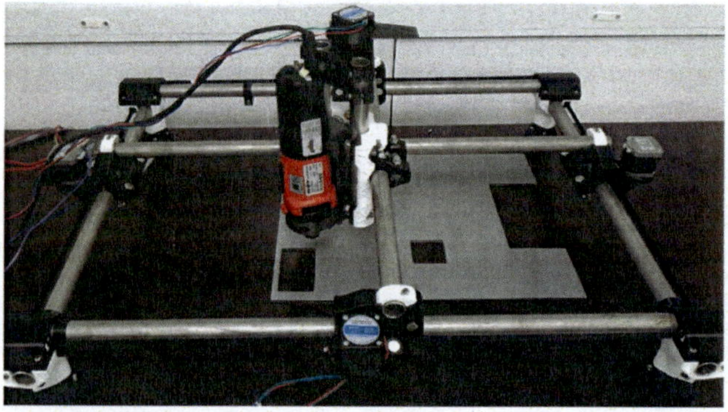

Fig. 3. The self-built MPCNC milling machine with the machined Dibond back panel.

Fig. 4. The self-built Fabulaser mini laser cutter with the cut out MDF base panel.

4 Results and Discussion

Building the 3D printer was more complex compared to building the CNC milling machine and the laser cutter, since no build manual was available for it. The printer had to be built by relying on build videos available online, seeking advice on community forums, and carefully studying the 3D printer CAD model. This shows that an open source machine tool with an accompanying build manual greatly facilitates replication. Finding out where to purchase the individual components and how to group them together to save money and reading other user's comments to determine if the vendor's parts were of good quality required time-consuming online searches and analysis. The further design and modification process to implement new features, however, consumed the most time, due to the constant trial and error process followed by testing and validation. When new components had to be integrated, when an alternative component was chosen, or when sudden issues arose, various community forums had to be searched to see if other users had encountered a similar issue.

The freely published native CAD design files allowed the modification and further development of the printer and reduced the time needed to get to a final product. This demonstrates one of the biggest advantages of OSHW, whereby the design sharing function allows users to skip the time consuming and expensive R&D required to start developing a product from scratch. Instead of constant reinvention, an existing design can be taken and adapted to one's specific requirements while further improving the design. When the design is then shared with the community, others can also benefit from the improved design. This drives innovation while allowing those who do not have the funds or resources for R&D to start producing a working product locally and directly contribute to income generation.

4.1 Key Challenges and Barriers with Replication of OSMT in Home Settings

There are several challenges that need to be considered in an OSMT replication process in a home setting. Firstly, users need engineering domain knowledge to safely make modifications to the design. Missing or false information within a build also requires extensive troubleshooting and efforts to figure out solutions through trial and error, additional research, or asking in community forums. Missing or insufficient documentation can hinder replication. Building machines furthermore requires some understanding and experience in assembling machines with precision and accuracy to guarantee the user's safety. As the use of power tools like angle grinders and Mitre saws can be unsafe and loud inside homes, a semi-detached workshop setup would be preferrable.

A further barrier is the amount of time that is necessary to complete the building process. An OSMT project can require the sourcing of hundreds of individual components, making the endeavor complex and time consuming. When some parts are unavailable locally, long shipping times need to be considered for their import. Missing a part can therefore drastically delay the build process. Moreover, low quality control with cheap components increases the chances of receiving faulty products. When a listed component is not available and an alternative component has to be sourced, modifications to the build are necessary which is associated with further expenses and delays in the process. Understanding these difficulties, some designers sell their machines as complete kits so that the end user does not have to spend time scouring the internet for components [7]. This is also a common business model for OSHW practitioners who publish their designs on the internet for free, whereby the sales of kits become a means of generating income for the designer.

The material cost of building the CNC mill and laser cutter amounted to about $2600, which is significantly lower than commercial machines on the market. to the option to choose the machine components and modifying the design allowed for building the machine tools to the specific spatial and budget constraints is a major advantage but requires some technical knowledge. The biggest benefits in using the CNC machine tools as seen in Table 2 were the drastic increase in productivity, the elimination of labor intensive and dangerous tasks, and a consistent and high-quality finish on components.

Table 2. Comparison of implementing CNC automation to speed up manufacturing

Part	Power tools	Time (min)	Automation	Time (min)
Base panel (3 mm MDF)—1 part	Oscillating blade and drill	60	CNC milling	2
Back panel (4 mm Dibond)—1 part	Oscillating blade and drill	45	CNC laser cutting	333

5 Conclusion

Self-built, open source machine tools can be much more affordable than commercial versions, facilitating the access to advanced manufacturing and automation technology that would otherwise be unobtainable. By seizing the new opportunities offered by open source and personal manufacturing, small businesses can flourish by increasing productivity, lowering intensive labor, and supporting revenue-generating activities. Sophisticated CNC machine tools enable microentrepreneurs to develop beyond low-value-added manufacturing processes and compete with larger firms with more money, resources, and machines, thus resulting in fairer and more inclusive industrialization. Moreover, individuals at home who would like to earn on the side or setup a small business can also affordably and sustainably progress from one off prototypes to a small-scale production.

3D printers are not the only technology that can democratize manufacturing. Consumer products are often a mixture of mechanical, electrical, and structural components manufactured from a plethora of parts and materials. Various kinds of OSMT would be required for individuals to set up a so-called open source microfactory that would allow them to produce complex products from their living rooms. Today ingenious makers, designers, and engineers are developing small scale OSMT which are easy to replicate for anyone and publish them on the internet for free. Projects like the MPCNC and the Fabulaser mini focus on creating truly open source machine tool designs. However, there are still many challenges for the large-scale adoption of open source machine tools for use in production. Some of these are incomplete documentation, lack of quality control and no standardization.

Acknowledgements. The authors would like to thank the Center for Digitalization and Technology Research of the Bundeswehr (dtec.bw) and the Bundeswehr IT (BWI) for their support.

References

1. Anderson, C.: Makers. RH Business Books, London, The New Industrial Revolution/Chris Anderson (2012)
2. Balka, K.: Open source product development. The meaning and relevance of openness. Zugl.: Hamburg-Harburg, Techn. Univ., Institut für Technologie- und Innovationsmanagement, Diss., 2011. Forschungs-, Entwicklungs-, Innovations-Management. Gabler, Wiesbaden (2011)
3. Chan, K., et al.: Low-cost 3D printers enable high-quality and automated sample preparation and molecular detection. PLoS ONE **11**(6), e0158502 (2016)
4. Church, E.M., Oakley, R.L.: Etsy and the long-tail: how microenterprises use hyper-differentiation in online handicraft marketplaces. Electron. Commer. Res. **18**(4), 883–898 (2018). https://doi.org/10.1007/s10660-018-9300-4
5. Gershenfeld, N.A:. In: Fab. The Coming Revolution on Your Desktop—From Personal Computers To Personal Fabrication. Basic Books, New York (2005)
6. Hennelly, P.A., Srai, J.S., Graham, G., Meriton, R., Kumar, M.: Do makerspaces represent scalable production models of community-based redistributed manufacturing? Prod. Planning and Control **30**(7), 540–554 (2019)
7. Ingrassia, D.: In: Fabulaser Mini. An Open Source Laser Cutter. http://fabulaser.net/. Accessed 28 Jan 2022
8. Mota, C.: The rise of personal fabrication. In: Proceedings of the 8th ACM Conference on Creativity and Cognition—C&C '11. ACM Press, New York, USA, pp. 279. (2011). https://doi.org/10.1145/2069618.2069665
9. Omer, M., Kaiser, M., Moritz, M., Buxbaum-Conradi, S., Redlich, T., Wulfsberg, J.P.: Democratizing manufacturing—conceptualizing the potential of open source machine tools as drivers of sustainable industrial development in resource constrained contexts. In: Herberger, D., Hübner, M. (eds.) Conference of Production Systems and Logistics (2022)
10. Reynolds, E.B., Samel, H.: Manufacturing startups. Mech. Eng. **135**(11), 36–41 (2013)
11. Zellers, R.: Introduction to The MPCNC—V1 Engineering Documentation 2022. https://docs.v1engineering.com/mpcnc/intro/. Accessed 12 May 2022
12. Scott_3D.: HyperCube Evolution (2017). https://www.thingiverse.com/thing:2254103. Accessed 12 May 2022

Aggregated Production Planning for Engineer-To-Order Products Using Reference Curves

F. Girkes[1](✉), M. Reimche[1], J. P. Bergmann[1], C. B. Töpfer-Kerst[2], and S. Berghof[3]

[1] Production Technology Group, Technische Universität Ilmenau, 98693 Ilmenau, Germany
florian.girkes@tu-ilmenau.de
[2] IWB Industrietechnik GmbH, Langenscheidtstraße 7, 99867 Gotha, Germany
[3] Berghof Group GmbH, Lindenstraße 2, 07426 Königsee, Germany

Abstract. The production of highly individualized engineer-to-order products has special characteristics that lead to a significant increase in the complexity of production planning and control. Therefore, aggregate resource planning is a dynamic and complex process that must always deliver reliable results. But without appropriate tools, these predictions can only be achieved with significant manual effort. Therefore, this paper presents a holistic method that predicts and schedules the required manufacturing resources for new customer orders based on a type representative by means of product modularization and data preparation of approximately identical historical manufacturing orders. This allows the actual processing status of the current customer project to be derived from the preplanning by means of a concurrent calculation in order to be able to initiate countermeasures at an early stage in the event of project delays and also to reduce the lead time of the customer order by preallocating the required production resources.

Keywords: Engineer-to-order · Aggregate production planning · Type representative

1 Introduction

Production companies are in a constant state of change. They are challenged to compete in global markets. Growing demands for individualized products with increasing quality and decreasing prices bring logistical performance, such as high delivery reliability or fast delivery and throughput times, into focus as a competitive factor. Delivery dates and manufacturing costs can be determined at an early stage and deviations from deadlines can be detected by a valid forecast of the required resources and their rough planning [1, 2]. In contrast, inaccurate forecasting can lead to missed delivery dates and manufacturing costs, resulting in a loss of customer confidence and subsequent costs for late deliveries [3]. This prediction is particularly relevant for mechanical and plant engineering, a typical example of the engineer-to-order process with predominantly a large number of individual parts and complex production processes. Here, resource

M. Liewald et al. (Eds.): WGP 2022, LNPE, pp. 642–651, 2023.
https://doi.org/10.1007/978-3-031-18318-8_64

rough planning includes not only manufacturing and assembly, but also upstream processes such as design, order planning, or the purchasing process of raw materials and purchased parts [4]. Furthermore, the products of a machine and plant manufacturer often consist of a large number of components that are individually designed in order to achieve a customized solution for the individual customer [5]. Thus, the product characteristics defined in the design process represent a unique selling point for the companies. In engineer-to-order processes, several strategies and concepts have been introduced that aim to predict the resource requirements of manufacturing processes. However, they are usually estimated without sufficient information on available capacity, mainly concern manufacturing processes, or include very general principles. Thus, a variety of approaches to manufacturing resource prediction have been proposed in the current literature, which consider different data and methods or algorithms [6, 7]. However, despite their importance, the needs of manufacturing companies operating under the engineer-to-order principle have rarely been considered [4]. In addition, it is clear that current production planning approaches have a strong preference for the short term planning level. Only a few approaches include multiple planning levels in their solution or consider the internal company supply chain holistically [2, 8]. Therefore, this paper presents and develops a holistic method that enables medium-term production planning for manufacturing companies according to the engineer-to-order approach along the entire customer order. The aim is to create a new form of dynamic process control and process monitoring for the current processing status of the current customer project already in the quotation phase and to significantly shorten the lead time of the customer order by scheduling the required production resources.

The presented method is based on product modularization and the allocation of the actually required production resources of the modules of comparable historical production orders. The modules are integrated into a predefined type representative according to the new customer order during the quotation process, so that the production resources and lead times can be forecast and scheduled in the planning environment of the manufacturing company. After scheduling, the type representative is designed as an aggregated, real reference curve corresponding to the desired customer order and presented as monetary value added over the order lead time. This easy-to-read form of visualization represents the core of the presented methodology and provides the basic prerequisite of the analysis for the holistic internal logistical supply chain.

Therefore, this paper is organized as follows: First, a brief overview of relevant work is given in Sect. 2. Next, the developed concept is presented and described in Sect. 3. Finally, Sect. 4 presents the conclusion and outlook of this research work.

2 State of the Art

Engineer-to-order (ETO) is a production strategy in which all development, engineering and production activities only start after a customer order has been confirmed [9]. The ETO environment is characterized by the following elements: Customized products and manufactured in small quantities. To achieve such product customization, ETO companies apply non repetitive processes that are labor intensive and require a highly skilled workforce [10]. The technical design therefore hardly includes the search for the optimal production process (i.e., the technical design effort cannot be amortized by many sold items), yet it must be possible to evaluate the profitability of the project as accurately as possible from the beginning [11].

In addition, several phases and parts of an ETO project are outsourced to specialized suppliers, increasing the number of companies involved in each project [11]. These ETO characteristics have a strong influence on the entire planning process, which also includes the scheduling of purchasing activities. Hicks et al. [4] identified that most ETO companies take a reactive approach to procurement, where functions are divided into departments and are predominantly bureaucratic in nature. This is also confirmed by Lalic et al. [12]. Since most ETO products are delivered by a project based approach, the management methods used by these ETO companies are inspired by the traditional order management literature. As an example, order lead times are determined by the production schedule, considering available production capacity, technical restrictions, due dates, and system status. The order sequence is determined according to the company's own rules in order to calculate the start and finish dates of the orders at the workstations [13].

Because the development, procurement, and production phases are often simultaneously executed, most ETO companies rely on effective collaboration and a dynamic planning process. According to the current state of the art, such traditional approach to ETO project planning does not take into account the iterative nature of most technical activities and the holistic view of project phases. As a result, project activities are often disorganized and schedule delays occur, resulting in late project deliveries and cost overruns [12].

3 Methodology

Cost curves, which show a monetary increase in the value of a new project over the time of its realisation, are a way for project manufacturers, e.g. the special machine-construction industry, to better monitor the cost progress of their projects. Cost curves can also be used as a tool for controlling new projects if a reference or ideal curve exists for a new project. Delays in the value growth of a project or an excessive increase in the value growth are signals that can detect potential deviations in project execution in the special machinery sector at an early stage. The effectiveness of this control instrument depends strongly on the accuracy of the fit between the reference curve and the cost curve of the new project. The methodology in the following describes a procedure for generating a suitable reference curve which over time, based on production data from projects already completed, identifies a suitable value development for the purchased material, for the working time, etc. The methodological procedure for the formation of an ideal reference curve, which serves as a cumulative representation of the value development of each individual assembly, contains eight steps as described below and additionally as overview in Fig. 1.

1. At the beginning all projects already completed by the company are assigned to a predefined number of product families. A product family is described by production characteristics, such as the number of processing steps, by the product segment and by the customer order [14].

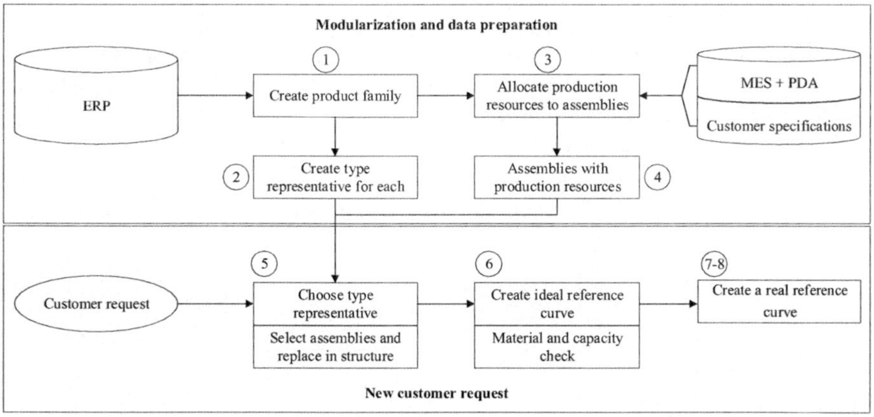

Fig. 1. Process flow chart for the creation of a real reference curve.

2. In the second step a fictive type representative is created for each product family defined in point 1 using the type representative method. According to Helbing, a type representative is a representative of a group and it represents all relevant elements of a project group with their characteristics and characteristic relationships [15]. A fictive type representative is a theoretical product that has not been manufactured, which was formed from the characteristics of the group. The newly generated bill of materials for the fictitious type representative contains all component groups including all associated in-house production and procurement parts and it forms the basis for the subsequent work steps.
3. In the next step, the development costs incurred in the past and the production costs (e.g. assembly costs) are assigned to each product family. At this point it is important to distinguish between expenses that can be assigned to an individual assembly and expenses that relate to the entire product. When each assembly group is assigned its actual reported costs, the average costs of the projects containing the relevant assembly groups identified below are included for the project as a whole.
4. Each assembly of the identified product family is then clearly described using product-specific characteristics. The characteristics have been selected in such a way that they reflect the customer requirements from the specifications with their influence on the costs and the required production time. For this purpose, the customer requirements with the cost-influencing factors are previously identified in already completed orders by means of a Pareto analysis and narrowed down to the most relevant factors.
5. In the fifth step, the customer project request is assigned to a product family. After the assignment, those assemblies are identified in the fictive type representative of the product family that have the highest match between the customer requirements of the requested project and the assembly characteristics of the fictive type representative. If one or more customer requirements are met by different assemblies, the most similar assembly is selected.
6. The identified assemblies with the associated routings and bills of material for the in-house production and procurement articles form the basis for generating an ideal

reference curve. In the first step, all the routings used for the identified assemblies are analysed with the aim of fully recording the resources (booking units) used for the production of the respective assembly and arranging them according to the value stream of the respective product family. In the second step, all required production and assembly hours (execution time-ex) for the production of the assemblies identified in step 5 are assigned to each resource and summed up per resource. Equivalently, the effort for product development and work preparation of the respective assembly is determined (see Fig. 2).

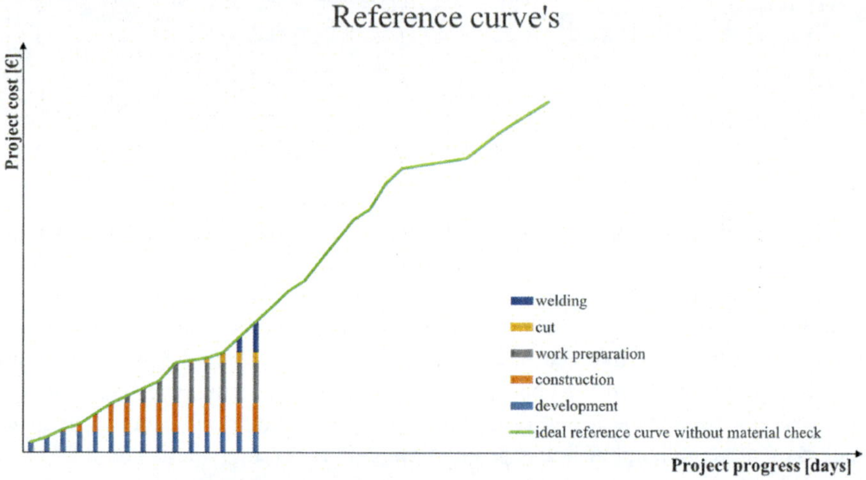

Fig. 2. Building an ideal reference curve.

7. The procedure for creating the ideal reference curve is continued with the analysis of the externally procured parts of the assemblies identified in step 5. Thus, the re-procurement times (rt) for the raw material and for the purchased parts of the relevant assemblies of the fictitious type representative are identified in order to determine the earliest times for the value added, which are defined as rt (e.g. the material is made available for the processing of the resource) and rt + ex (the effort for the material processing) (See Fig. 3). The creation of the ideal reference curve takes place under the assumption that the theoretically possible capacity of the resource required for the execution of the work step is available at all times (assumption of unlimited capacity).
8. Finally, the ideal reference curve for the new product is generated. An ideal reference curve does not consider any resource availability (no capacity constraint) and is initially plotted cumulatively along the value stream by means of forward scheduling of the value added points (duration of resource utilization by all relevant work groups), with the amount of value added points being calculated by multiplying the duration and the cost rate of the resource. Subsequently, the points of the generated curve are checked for plausibility. For this the material availability for the execution of the work on the respective resource is examined, while the procurement times of the raw material or purchase components from the step 5 with the identified demand times

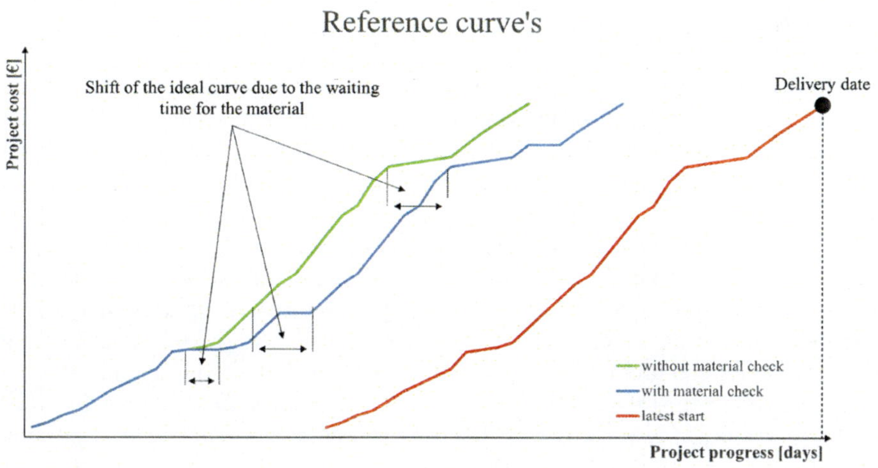

Fig. 3. Ideal reference curve with material check.

starting from the planned project start are compared. For rt > the time of need, then the ideal reference curve is shifted by the resulted difference starting from the time of need into the future. After checking the temporal plausibility of the ideal reference curve, the total value of all required raw material or purchased components, which are needed for the execution of the work of a resource, is plotted at the beginning of the processing. The step is completed with an addition of the ideal reference curve with the material values.

If the delivery date for the new project is known or specified, the company can use backward scheduling to determine the latest times for an increase in the value of the project and the latest project start (see Fig. 4, latest start). If, after the customer has placed the order, the company loads the required effort and takes into account the current resource availability, it receives the real reference curve for the value based control of the new project (see Fig. 4, with material and capacity check).

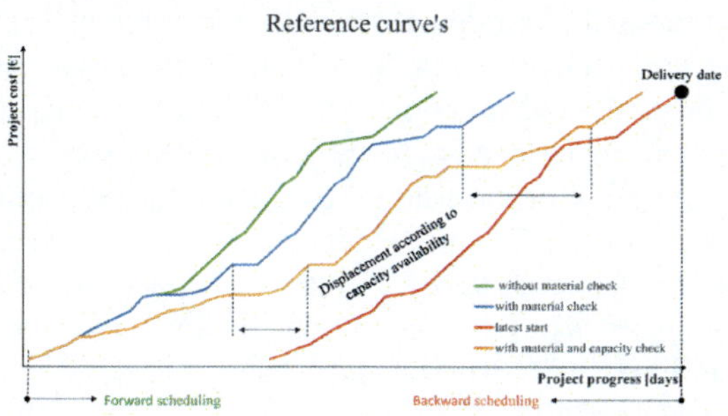

Fig. 4. Real reference curve with material and capacity check.

4 Case Study

The method described in Sect. 3 has been validated at a project manufacturer in the field of special machine construction whose core competencies include the customer specific production of conveyor belts. These conveyor belts (customer projects, in short: projects) have an average project duration of currently 60 days and are manually precalculated by an employee of the project manufacturer according to the customer's requirements.

The structure of the projects consists of four assemblies (ASM). Due to their required production resources and fixed order of allocation planning, these have been combined into a product family, which includes the following sequence of value creation: Quotation preparation, development and design, work preparation, cutting, turning, milling, and assembly. Afterwards, the fictive type representative of this product family has been created, which contains the bill of material structure and routings of all used articles from the assemblies, as well as its past resource requirements. For the scenario presented in the following, the project specific characteristics for the individual assembly groups were identified and served as the basis for the rough planning. Thus, two characteristics for each ASM (e.g., the total length and the total width) were used for selection.

Following the selection of the most similar ASMs, the ideal reference curve was created from their aggregated production resources and the project schedule by means of forward scheduling (see Fig. 5, without material check.) and backward scheduling (see Fig. 5, latest start.). The following assumptions have been made:

- The capacity utilization of all required booking units is 85%.
- Disruptions due to delivery date deviations in the external supply chain (e.g. due to the Corona pandemic) have been corrected by equating the planned procurement time (rt) with the goods receipt posting.
- The sequence in the routing serves as a template for the production process.
- The material requirements have been distributed accordingly to the required booking units by setting the longest rt of the articles of the respective assembly before the required processing step.

Subsequently, the ideal reference curve (forward scheduling, see Fig. 5, without material check) with the capacity availability and procurement times of the items was adjusted with the determined requirement times to obtain the real reference curve (see Fig. 5, with material and capacity check). The rt of the articles has thereby started after the project step development and construction. After material receipt, the incoming goods department then books the material costs (m€) to the project (see Fig. 5, m€ cost jump). Subsequently, after expiration of the waiting time (wt), the respective processing step (see Fig. 5, Process time, pt, continuous cost increase) can take place.

Due to the daily feedback via the ERP system of the conveyor belt manufacturer, the real resource requirements of the customer project could be recorded and compared with the real reference curve with material and capacity check (see Fig. 5, with material and capacity check, real project).

When predicting the manufacturing costs incurred for a new sales order, a difference of 9.3% (< 10%) of the manufacturing costs was achieved in the comparison of pre- and post-cost calculation. This shows that the described concept is suitable for the prediction

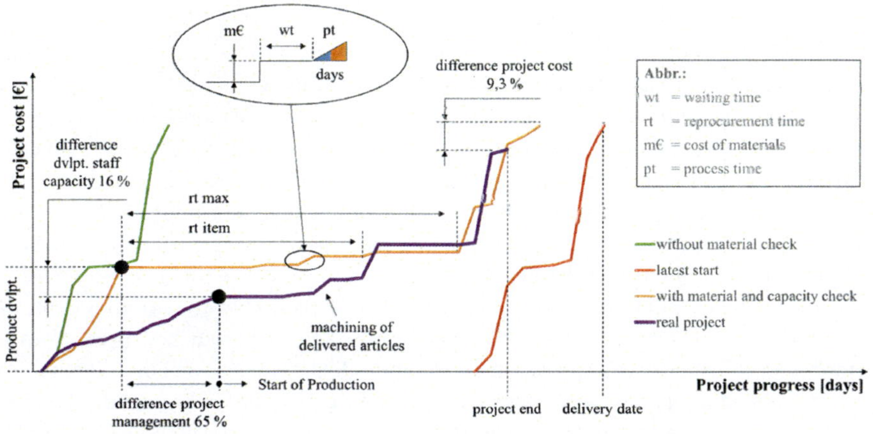

Fig. 5. Evaluation of a real project from the conveyor belt manufacturer.

of manufacturing costs based on a selected example. The agreement in the prediction of the selected AMS's in comparison to the real product manufacturing process is structured as shown below in Table 1.

The different cost increases after the product development of the real project to the ideal reference curve result from newly developed articles and work steps that have been brought forward with articles from stock. However, an intersection can be identified that reflects the planned rt with the real rt and will be further investigated in ongoing research project.

The required overall resources in product development, as an initial reference point for project progress control, have been estimated with a manufacturing cost variance of 16% (see Fig. 5, difference dvlpt. Staff capacity). The current time offset within project management of 65% shows a manipulated deviation for optimization in personnel planning. In personnel planning, it was not possible to clearly differentiate downtimes, e.g. due to illness, overtime reduction, etc. from past customer orders, as these were previously documented by the conveyor belt manufacturer in a traceable manner.

Table 1. Percentage match in aggregate project steps.

Quotation preparation	Development	Work preparation	Cutting	Turning	Milling	Assembly	Material costs
57%	89%	82%	114%	134%	42%	145%	109%

5 Conclusion and Outlook

This paper uses the example of a conveyor belt manufacturer in the field of special machine construction to show that it is possible to predict aggregate production planning on the basis of past customer projects. It could be shown on the basis of an example that the required production resources for new customer orders could be determined even before product development. The clear and easy to understand presentation of the target variables of production costs and order lead time as a real reference curve means that project progress can be monitored and countermeasures can be initiated at an early stage in the event of deviations from the planned schedule. However, this has so far required close comparability of the assemblies within the product family to the new customer project. Therefore, it is necessary to make this variance easier to estimate in the further course of the research project. The customer requirements from the product specifications must be classified more precisely and their influence and effects on the entire value creation process of the product family and the requested product must be determined. Furthermore, the occurring process disturbances (schedule deviations, personnel management) are to be analyzed and classified in order to incorporate them into the aggregated production planning. In this way, the order throughput time can be further specified and a more robust planning basis can be ensured.

Acknowledgement. The authors sincerely thank Thuringer Aufbaubank and Thurin-gian Ministry for Economic Affairs, Science and Digital Society for supporting the research project, AgiLief-Agile engineering and supply chain planning for new products through early pattern recognition, (2019VF 0030).

References

1. Jericho, D., Jagusch, K., Sender, J., Flügge, W.: Herausforderungen in der durchgängigen Produktionsplanung bei ETO-Produkten. ZWF **115**(12) (2020)
2. Cannas, V.G., Pero, M., Pozzi, R., Rossi, T.: An empirical application of lean management techniques to support ETO design and production planning. IFAC-PapersOnLine **51**(11), 134–139 (2018)
3. Schmidt, M., Nyhuis, P.: Das Hannoveraner Lieferkettenmodell Produktionsplanung und -steuerung im Hannoveraner Lieferkettenmodell: Innerbetrieblicher Abgleich logistischer Zielgrößen. 1st edn. Springer, Berlin, Heidelberg, pp. 7–26. (2021)
4. Hicks, C., Mcgovern, T.: Product life cycle management in engineer-to-order industries. Int. J. Technol. Manage. **48**, 153–167 (2009)
5. Christina, R., Felix, B.: Improving data consistency in production control. Proc. CIRP **41**, 51–56 (2016)
6. András, P., Dávid, G., Botond, K., László, M.: Manufacturing lead time estimation with the combination of simulation and statistical learning methods. Proc. CIRP **41**, 75–80 (2016)
7. Burggräf, P., Wagner, J., Koke, B., Steinberg, F.: Approaches for the prediction of lead times in an engineer to order environment—a systematic review. IEEE Access **8** (2020)
8. Valencia, T., Lamouri, S., Pellerin, R., Dubois, P., Moeuf, A.: Production planning in the fourth industrial revolution: a literature review. IFAC-PapersOnLine **52**(13), 2158–2163 (2019)
9. Powell, D., Strandhagen, J., Tommelein, I., Ballard, G., Rossi, M.: A new set of principles for pursuing the lean ideal in engineer-to-order manufacturers. Proc. CIRP **17**, 571–576 (2014)
10. Nääs I.: In: Advances in Production Management Systems. Initiatives for a Sustainable World. 1st edn. Springer International Publishing (2016)
11. Jünge, G., Alfnes, E., Kjersem, K., Andersen, B.: Lean project planning and control: empirical investigation of ETO projects. IJMPB **12**(4), 1120–1145 (2019)

12. Kjersem, K., Giskeødegård, M.: Planning procurement activities in ETO projects. In: Advances in Production Management Systems—Towards Smart and Digital Manufacturing, vol. 592. Springer International Publishing, Cham (2020). https://doi.org/10.1007/978-3-030-57997-5_65
13. Baker, K.., Dan T.: In: Principles of sequencing and scheduling. Wiley (2013)
14. Kletti, J., Schumacher, J.: Die perfekte Produktion. In: Manufacturing excellence durch short interval technology (SIT). Springer, Berlin, Heidelberg (2010)
15. Helbing, K.: Handbuch Fabrikprojektierung. Mit 331 Tabellen. 3rd edn. VDI-/Buch. Springer, Berlin, Heidelberg, Dordrecht, London, New York (2010)

Template-Based Production Modules in Plant Engineering

J. Prior[1](✉), S. Karch[2], A. Strahilov[2], B. Kuhlenkötter[1], and A. Lüder[3]

[1] Ruhr-University Bochum, Chair of Production Systems, Universitätsstr. 150, 44801 Bochum, Germany
prior@lps.rub.de
[2] let's dev GmbH & Co. KG, Alter Schlachthof 33, 76131 Karlsruhe, Germany
[3] Chair of Production Systems and Automation, Otto-V.-Guericke University, Universitätspl. 2, 39106 Magdeburg, Germany

Abstract. The example of the automatization ramp-up of the hydrogen electrolyzer production illustrates the complexity of plant engineering from scratch to production. Collaboration takes place in an interdisciplinary environment of several business and engineering units simultaneously. A complex chain of a multitude of process steps is created, as well as the exchange of data from various programs and simulations. With virtual commissioning (VC) and digital models of the plant, new possibilities arise to continuously validate the results of the planning process. As a result, there is a time advantage compared to traditional physical commissioning. There are always sudden market ramp-ups of products, for example, solar cells, car batteries, and fuel cells. Particularly due to the strong increase in demand for hydrogen electrolyzers, it is important to design the product and the production plant simultaneously. A plant manufacturer can respond to the increasing demand by pre-designing a configurable and scalable pool of production modules based on black-box approaches to the product in its various sizes, weights, etc. as is being investigated in the research project FertiRob. The aim of this paper is to present a concept of designing configurable production modules for a rough solution space. Templates created for a production module are shown, which can be accessed by a future plant configurator to customize the configurable modules and retrieve certain master data. Future research activities are aimed at securing the templates via a neutral data exchange platform (e.g., AutomationML, PLCopen XML, COLLADA) so that access is guaranteed via a configurator and all other engineering software.

Keywords: Production module templates · Plant engineering · Configurator

1 Introduction

The plant engineering process is becoming more complex because of the increasing product complexity, mass customization of products, and especially the decreasing product lifecycles of novel products [1]. By simulating the plant and products, the VC makes a great time advantage [2]. Due to a steady intensification of data connection among

M. Liewald et al. (Eds.): WGP 2022, LNPE, pp. 652–663, 2023.
https://doi.org/10.1007/978-3-031-18318-8_65

numerous tools in engineering, the concept of simultaneous engineering was created. Therefore, processes like the construction of plant modules and for example, the gripper or conveyors are done simultaneously. With many iteration loops and qualitative data exchange, the whole construction was increasingly harmonized with each other. This creates another time advantage [3].

Experience of the FertiRob consortium shows that today's internal engineering processes show that the plant manufacturer no longer starts from scratch with a new plant but selects similar projects which are adapted to the new product. As a result, all designs and programs of the plant modules must be revised, especially the semantics für the new use case. This process is prone to many mistakes thereby leads to data inconsistency. The research question that can be derived from this motivation is: How does the construction process in plant engineering can be optimized and process time reduced? To what extent can a production module be standardized, mapped, and parameterized by a user-centered lean template, especially without the final product.

The hydrogen electrolyzer product group is currently experiencing a massive market increase. Today, all electrolyzer systems of any manufacturer are still assembled under manual labor conditions [4]. As a result, the demand for automated production solutions is expected to increase shortly [5]. Due to the fact, that electrolyzer are currently developing continuously it is important to design the product and the plant simultaneously. Efficient preparation of plant design, for example using predefined production modules, can give plant engineers a competitive edge in the future. The approach presented in this paper, gives a collection of generalized production modules to the plant engineer, and a standardized basis to develop a new plant. These production modules are adaptable to a range of product variants through configurable (and lean-oriented in data and configurable parameters) templates with defined parameters in sizes, positions, paths, etc. This approach is intended further to provide time advantage, support in plant engineering and collaboration due to the standardized template.

One target is to develop a prototype of a configurable and template-based production module. The target is limited to the product group of hydrogen electrolyzers in the evaluation.

2 Basics of Plant Engineering

In addition to virtual commissioning as an important step in this approach, the lean date idea is also important and should be mentioned in the basics.

In case of the complexity and individuality, there are many different approaches to process diagrams of plant engineering. Nevertheless, an example from previous works [6, 7] is given (Fig. 1). This diagram can be seen as the basis for this work and depicts the process steps from the mechanical design over electric/fluid design to the PLC/Software. The VDI/VDE 3695 "Engineering of industrial plants Evaluation and optimisation of the engineering Fundamentals and procedure" [8] gives an overview of the main processes in plant engineering, and [6] describes the "production system development" process in more detail. The virtual engineering and commissioning are also mentioned as simultaneous steps. Especially the virtual engineering, the mechanical design, and the PLC are steps, that can be designed forwarded with this approach.

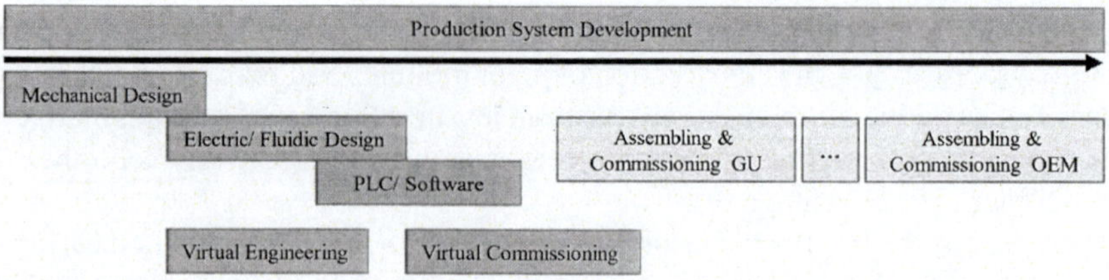

Fig. 1. Selection of processes in plant engineering, following [6, 9]

2.1 Virtual Commissioning

The approach of this paper is among others based on the principles of the VC. The simulation model required for VC is referred to as the mechatronic plant model in [10] and can be divided into the two sub-models extended 3D geometry model and the behavior model. The extended 3D geometry model includes the 3D geometry of the plants with mechanics, kinematics, parts transport, sensors, and interfaces. This sub-model represents the plant mechanics for VC. The behavior model, in turn, is required so that the mechatronic plant model can react to control outputs during VC in the same way as the real plant, pass on information to other components and set corresponding control outputs [10]. Furthermore, signals are continuously exchanged between the two sub-models via certain interfaces. The sub-models not only have to communicate with each other but also with the real controller.

In the context of this paper, VC plays a very important role in the modular template-based configuration of production plants. The major reason for this role is that VC has been continuously oriented towards the use of templates or standard modules/models since its beginning. The development and use of prepared behavioral models of individual components, such as valves, robots, servo drives, welding controls, etc., is already in daily use. Their use in practice has already been a state of the art for years and has proven its worth [11]. The most common tools for VCs offer prepared libraries with templates for behavior models. In general these tools also provide the possibility to create own behavior models [9], e. g. RF::Suite [12], WinMod [13], iPhysics [14], Tecnomatix/Process Simulate [15], ABB RobotStudio [16].

In recent years, companies have focused intensively on obtaining such behavior models directly from the responsible component manufacturer, in close cooperation with research institutes, thus eliminating the necessary of modelling and managing the models themselves. The main challenge was to store and distribute the models in a standardized exchange format without creating dependencies on specific tools, which each company uses in the internal process. This assumption is based on the numerous research projects that have addressed this issue, such as AVANTI [17], ENTOC [18], SPEAR [19].

Another initiative focusing on a standard based on is a comprehensive XML based object-oriented data modelling language, is the AutomationML e.V. [20]. The Component classification working group [21] concentrates not only on the technical description of components via the exchange open data format AutomationML but also, on the description and exchange of behavior models [22–24]. Similar to the research projects mentioned, the analysis of the application of standardized data formats such as Functional

Mock-up Interfaces (FMI [25], OPC UA [26, 27], etc.) are also part of the component classification.

2.2 Lean Engineering and Lean Information Management

Lean focuses on resource efficiency and pursues perfection through the elimination of the eight types of waste (Muda) in production or literally in all business processes: overproduction, inventory, overprocessing, transport, waiting, movement / motion, defects, and unused intellect/skills [28]. Planning principles and lean admin support lean engineering artifacts and processes. The focus is on ergonomic workplaces, a value stream-oriented machine design, one-piece-flow, or efficient material provision [29]. Especially the oversizing of plants includes significant potential for waste [30].

In the planning process of the engineering phase, it is important to create an efficient collaboration and coordination as early as possible. Standardization is the baseline for interdisciplinary work. In addition, in the context of lean information management a unified data format and data flow is a key factor for successful projects. With an efficient approach, shortening of the order throughput time can be realized [29].

3 Comparison Between Actual and a Template-Based Engineering Process of Production Plants

This section provides a summary of the comparison between the current design process and the template-based approach. The aim is to illustrate the intended use of templates in defined phases in the plant design process. Additionally, the presented process steps are the strongly reduced result of several sources (e.g., [31, 32]) and discussions with large plant manufacturing companies in Germany that have been active as service providers in this market for at least 20 years. The summary does not consider exceptions and describes selected contents. Not mentioned are for example, electric design, PLC and software design.

3.1 Mechanic Design

Current Engineering: Already completed projects are usually taken over by the same customer one-to-one. For this purpose, complete assemblies are copied as 3D CAD models with all associated components. On this basis, the 3D geometry model of the component is introduced and positioned. Based on the geometry differences between the previous and the new product component, an initial check of the mechanical design to be adapted is done, for example, a clamping device, is identified and performed. This verification is also performed considering the tools required to carry out the process, such as welding guns, grippers, etc.

Template-Based Engineering: Based on predefined rules and templates for the production modules, the CAD models are generated in the specific CAD tool, properly named and structured into (well-defined) parts and assemblies. Any CAD delivery specifications from the respective customer, if available, are integrated trough the rules.

Experiences from each project are fed back and can result in a revision of the currently valid standard. Thus, roles can be extended or adapted. A 100% ready-made solution is not advisable to ensure the flexibility of the modules.

3.2 Robot Programming

Current Engineering: As a rule, the robot programs are based on predefined standards. These standards include specifications such as naming the robot programs, their configuration files, function names, etc. In addition, robot programs have functions to be used to ensure customized logical routines, such as point-to-point motion, set spot welding/technologies, special functions, etc. Here, the robot program is generated manually in two steps. In the first step, the OLP robot program is programmed. With this, the basic framework is created and the points/paths to be traversed are defined and initial checks are carried out for process sequence, accessibility (without collisions), and cycle times. In the second step, the robot program is detailed. Logical dependencies between the entire process, which are controlled by the PLC program, and the robot, which takes over certain sub-process steps, are integrated into the robot program, such as interlock dependencies, releases, specification of subroutines, feedback to the PLC, etc.

Template-Based Engineering: The robot programs are also stored using the pre-bound production modules as templates and the rules. They already contain the robot program structure, parameterized functions, and the specified points or paths. The points are generated based on the rules (between the product and the production module) and predefined templates for the production modules. Thus, the OLP robot program as well as the detailed robot program can be generated. In principle, two approaches are conceivable, each one has their own advantages and disadvantages.

First, the robot program can be deposited as a finished robot program, which has given place holders. A customer-specific subsequent adaptation will still be possible and will take place in the respective robot program.

The other option is characterized by the fact that there is no robot program (no files), but the entire files with their content/code for the robot program are generated from the configurator based on the rules.

4 Concept of Template-Based Production Modules

In this chapter, the main part of this paper will be presented. The concept of template-based production modules is the part of a selection options of configurable modules that display nearly all processes in an automated production line for a specific product group, for example solar panels, TV devices, or electrolyzers, with its specific variants and sizes from different manufacturers.

This concept is to be part of a superordinate method in which nearly the whole production line can be built up of modules like a building-block approach. These blocks are standardized in construction, electricity, logistics, and other interfaces. To ensure maximum flexibility, the automated processes are robot-cantered designed. One module is designed for one production process.

To design a module, it is required to have references about the product structure, such as height, weight, etc. Based on a rough market analysis, the modules can be adapted for a specific product group. In a next step the modules need to be elaborated further and detailed. As mentioned above, this template-based concept takes part of the whole plant engineering process, from the construction phase of the robot cells through the kinematic simulation of the complete line. However, the final layout must be manually adjusted and checked.

The DIN 8580:2020-01 and the VDI 2860 (withdrawn and under revision) summarize the general handling and joining processes in manufacturing. These will be the fundament for the wording and organization of the modules. Selected handling and joining processes are depicted in Fig. 2.

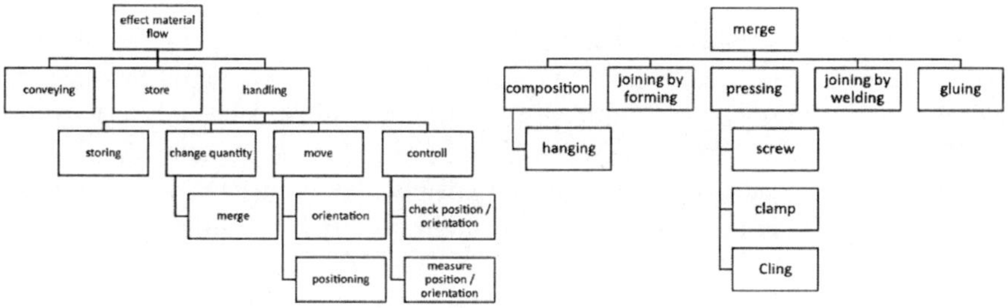

Fig. 2. Selection of merging [33] and handling [34] processes

For module selection, it is important to know the information about the range and weight of the objects handled. Based on these standardized processes the following naming convention could be adapted: "Module name—process robot 1—process robot 2—weight/range". The following concept will be elaborated for the handling and joining processes. The handling process is one of the most used for automation with robotics [35]. Due to the scope, the gripper will not be considered.

4.1 Process Steps for Template-Based Production Modules

The process steps for designing the template-based production modules are depicted in Fig. 3. After designing the modules and templates and working out the selection options of production modules, the engineer can choose the right modules for the specific product. When the product is in design phase the first step is to indicate the possible process steps to assemble the product. To find the right module the engineer indicates the individual components, that must be assembled automatically. Thus, the right modules can be selected. The preformed modules must be configurated through a template by passing individual parameters. Then the engineer has the first approaches for n modules (*n x modules*). Subsequently, the finished modules will be connected to a production line, and after the sequence control is created, the first approach of a complete kinematized production line is finished, as a part of the VC.

4.2 Designing the Template-Based Production Modules

As part of FertiRob consortium and expert interviews, the question of what a typical assembly module for an electrolyzer might consists of is addressed. There are some generally valid and logical facts. An assembly cell, especially when used with industrial robots, must necessarily have a fence. Only the mechanical interfaces of the fence have

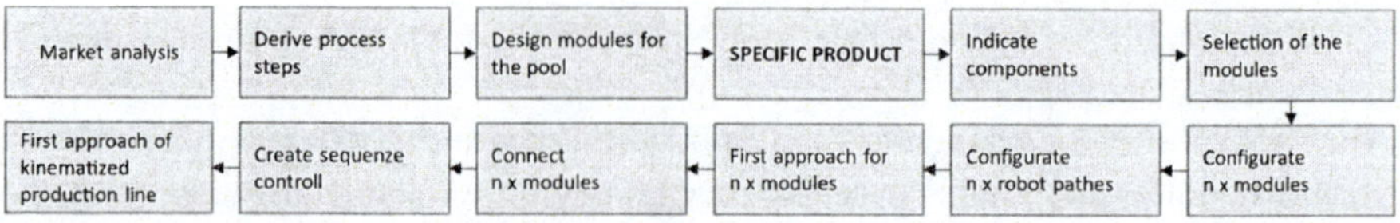

Fig. 3. Process flow for designing and configuring the modules

openings to avoid the risks of accidental collision between the plant operator and robots. It is necessary that the product enters the cell on one side and exits on the other side; this way it is possible to realize a series production typical of automated production. In addition, an opening for a component provision had to be designed. In FertiRob a electrolyzer is chosen that has an approximate size of a control cabinet with dimensions about 2 m in height and a surface of approximately 1 m^2 (e.g., [36]). Previous work has shown, that this variant of dimensions is often use [5]. The components should be provided for the robot paths at a fixed position. This can be achieved with component carrier systems with also fixed positions. During assembly operations, it is proven in practice to position the component with one robot and attach it with another robot. In the example (Fig. 4), a screwdriver robot with an integrated nut feeder is used. Like the product components, the product must also be located at a fixed point. The robots are positioned next to the part conveyor to achieve optimum use of their reach. Figure 4 shows a sketch of this assembly module and the template in which the variables for the configuration of the construction can be entered.

This module is created configurable with the aid of a few parameters, as is also known from parameterizable CAD design [31]. The most important properties, such as position and dimensions, can be configured. There are some constrains, from which further parameter emerge, like positions of the components and the product. Some positions can be derived from the configured design, while few others still must be entered manually. As an example, the position on x- & y-axis from the home position (black dot) of the part conveyor are given, as well as the length and width of the conveyor. When these variables are changed, the construction cell and the robot paths are changing, too.

In addition, it is planned that the template consists of a further not changeable restriction part, in which data for the later selection of the correct modules with the configurator are available, viz., manageable size, or the weight, of the construction units. It can also contain more detailed information, for example information about the gripping system and sensors.

Due to the fixed defined positions, robot paths can also be pre-designed. In this case, the production module is configured with CAD and then kinematized in for example ABB RobotStudio. However, other robot simulations are also suitable for this process. In this way, the robot path can already be planned with start and end positions as variables without content. The task of the engineer is therefore to determine a selection of parameters during configuration and check whether collisions occur.

The specific example is simplified so that the principle is illustrated slightly. The focus is on pre-designing as many steps as possible. To have a complete pool of production modules for the plant planning of a whole product group consisting also of different orders of variants of products, as PC-size, up to container-size products, as it is also known from the electrolyzer production [5], also for the individual production steps different orders of magnitude for the module must be developed. The module shown is suitable for a predefined range of sizes. Other cell arrangements outside of this range must be drawn up with other specific robots.

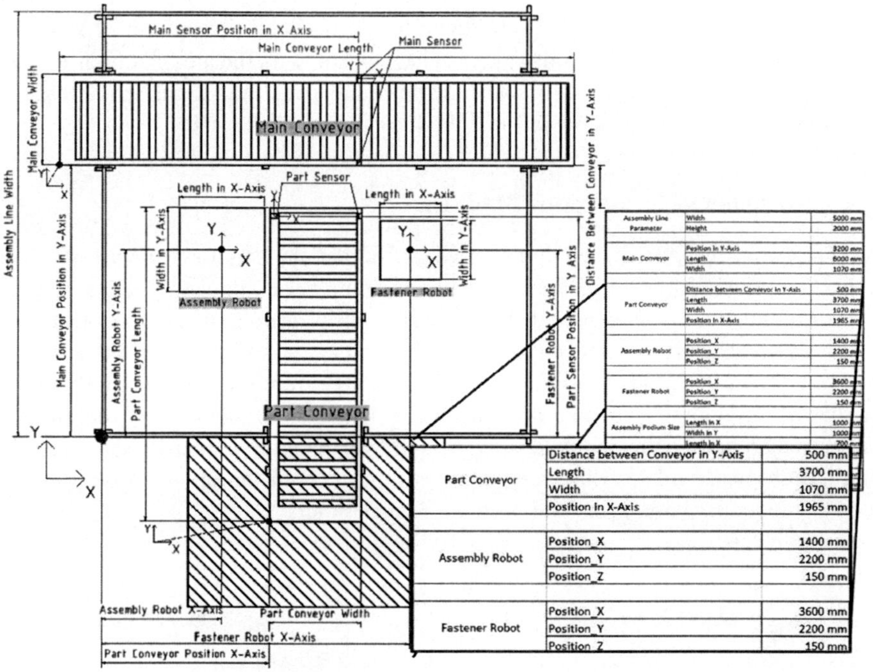

Fig. 4. Example of production module with a configurable template

5 Evaluation

Based on the production of the hydrogen electrolyzers product group, a configurable and template-based production module is developed for a specific production process step. The special feature of this module is that the size of the products varies to a large extent, and the production cell should cover a spectrum as wide as possible.

Electrolyzers are the devices to produce hydrogen by splitting water into oxygen and hydrogen using electric current. There are a few different variants on the market today. Nevertheless, the electrolyzers do not differ in the essential topology. They all have the same basic components, such as an electrolysis stack, components for gas drying, a compression device, a storage device, a water treatment, etc. Only the alkaline cycle in the alkaline electrolyzers and the better water purification in the proton exchange membrane electrolyzers differ. In addition, the main difference between the specific types of electrolyzers is still the size and weight, with the same power [5].

As the preceding work shows [5], especially the stacking process of electrolysis stacks is potential-rich for automation. Electrolysis stacks range in size from 5 cm × 5 cm to 50 × 50 cm and beyond. The stack consists roughly of a solid bottom and top plate, with the membrane electrode assembly (MEA) and the bipolar plates (BPP) alternating on a rail system in between. The process steps are stacking, component separation, leakage test, position check, pressing of the stack, and screwing. In the following, the stacking process is considered [37]. Following, the fuel cell production of the Fit-4-AMandA research project [37], this stack is also stacked with the aid of a mobile carrier system (ref. Fig. 5). The size of this stack can be changed, as can the carrier system. The aim is now to develop a stacking cell matched to the size of the stack.

5.1 Construction of the Stacking Modules

Essentially, as already described in the previous chapter, the cell must consist of a fence with several physical interfaces for the supply and delivery components. Thus, a supply for the top and bottom panels, MEA, and BPP supply are required. In Addition, an electrolysis stack carrier is fed in and out. This moves in a circuit to a press-station, that is not shown here. The individual MEAs and BPPs are placed alternately on this stack carrier after the lower plate has been deposited. Two robots are suitable. One robot must also handle the endplates. Figure 5 shows the module in different sizes. The configurations were made using templates as described in the previous chapter. The movement sequences and processes were also carried out.

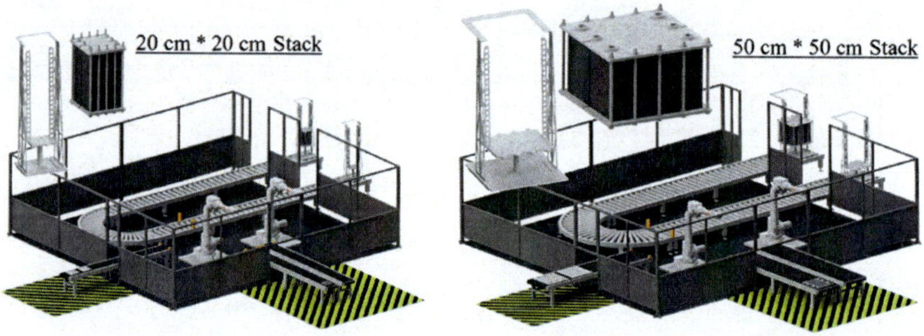

Fig. 5. Stacking modules in different sizes after filling their template

5.2 Evaluation from the Lean Perspective

The use of production modules with a configurable template is an option for strictly implementing and standardizing resource efficiency in plant engineering as well as in the production process to be planned. It is also a good foundation for the later VC.

A configurable interface realizes a lean, standardized database and information flow. The collaboration and coordination of engineering tasks in multidisciplinary engineering before a customer order reduce feedback loops and the order throughput time.

Related to the series production, the benefit is especially in the scalable templates to ensure a customer-specific sizing of the plants and to minimize space requirements and walking distances. The flow orientation and the arrangement of the modules without intermediate buffers can be ensured by layout restrictions.

The next step is to make the restrictions from the design and production process to be concrete. Examples such as the part lists, assembly sequences, degree of automation will sharpen the production modules.

6 Conclusion and Outlook

In this paper, a methodology was presented that could be used to support and automate some process steps in plant planning. This methodology has the potential to enable a time advantage through pre-designed and template-based configurable production modules.

In preliminary work, the focus of automated plant design was shifted from the production view to the product view. For this purpose, the topology of a product group was generalized based on market analyses and placed in a product template in the neutral data exchange format AutomationML [38]. This exchange format is particularly suitable for plant engineering due to the large number of engineers working on the plant at the same time as well as for the simultaneous work with different tools [39]. In future, a general valid template for production modules will be created so that the parameters and the pool of configurable modules can be centrally referenced and stored. This has also the advantage to support interdisciplinary work due to standardized templates and process steps. In the future, product and production module templates are linked with each other, creating a logic that automatically filters out the appropriate production modules after a specific product has been entered.

Acknowledgment. Parts of this work were supported by the Federal Ministry of Education and Research (BMBF) under grant number 03HY113A within the research project H2Giga—FertiRob.

References

1. Wang, Y., Ma, H.-S., Yang, J.-H., Wang, K.-S.: Industry 4.0: a way from mass customization to mass personalization production. Adv. Manuf. **5**(4), 311–320 (2017). https://doi.org/10.1007/s40436-017-0204-7
2. Wünsch, G.: Methoden für die virtuelle Inbetriebnahme automatisierter Produktionssysteme. H. Utz, München (2008)
3. Bullinger, H.-J. (ed.).: In: Wege aus der Krise. Springer, Berlin, Heidelberg (1993)
4. Sinnemann, J., Prior, J., Kuhlenkötter, B.: Skalierbare Elektrolyseurmontage. Zeitschrift für wirtschaftlichen Fabrikbetrieb **116**(12), 913–916 (2021). https://doi.org/10.1515/zwf-2021-0214
5. Prior, J., Bartelt, M., Sinnemann, J., Kuhlenkötter, B.: Investigation of the automation capability of electrolyzers production. In: Procedia CIRP, to be published
6. Biffl, S., Lüder, A., Gerhard, D. (eds.): Multi-Disciplinary Engineering for Cyber-Physical Production Systems: Data Models and Software Solutions for Handling Complex Engineering Projects. Springer, Cham (2017)

7. Strahilov, A. (ed.).: In: Digitaler Schatten von Produktionsanlagen als Big Data Quelle—Herausforderungen and Potential. Big Data Technologien in der Produktion von der Datenerfassung bis zur Prozessoptimierung (2017)
8. VDI/VDE 3695—Engineering of Industrial Plants Evaluation and Optimisation of the Engineering Fundamentals and Procedure (2020)
9. Strahilov, A., Hämmerle, H.: Engineering workflow and software tool chains of automated production systems. In: Biffl, S., Lüder, A., Gerhard, D. (eds.) Multi-Disciplinary Engineering for Cyber-Physical Production Systems, pp. 207–234. Springer, Cham (2017). https://doi.org/10.1007/978-3-319-56345-9_9
10. Kiefer, J., Ollinger, L., Bergert, M.: Virtuelle Inbetriebnahme—Standardisierte Verhaltensmodellierung mechatronischer Betriebsmittel im automobilen Karosserierohbau. vol. 07, pp. 40–46. (2009)
11. Strahilov, A.: Von virtueller Anlage zum Digitalen Schatten—Erfahrung aus der Praxis. München, Oct. 22 (2019)
12. RF::Suite|EKS InTec. [Online]. Available https://www.rf-suite.de/. Accessed 20 Apr 2022
13. WinMOD—virtual commissioning and more! [Online]. Available https://www.winmod.de/english/. Accessed 20 Apr 2022
14. Virtuelle Inbetriebnahme VIBN mit iPhysics|machineering.com. [Online]. Available https://www.machineering.com/. Accessed 20 Apr 2022
15. Siemens Digital Industries Software, *Technomatrix/Process Simulate.* [Online]. Available https://www.plm.automation.siemens.com/global/en/. Accessed 20 Apr 2022
16. ABB RobotStudio. [Online]. Available https://new.abb.com/products/robotics/de/robotstudio. Accessed 20 Apr 2022
17. AVANTI. [Online]. Available https://itea4.org/project/avanti.html. Accessed 20 Apr 2022
18. ENTOC. [Online]. Available https://itea4.org/project/entoc.html. Accessed 20 Apr 2022
19. SPEAR. [Online]. Available https://itea4.org/project/spear.html. Accessed 20 Apr 2022
20. AutomationML. [Online]. Available https://www.automationml.org/. Accessed 20 Apr 2022
21. Organisation—AutomationML. [Online]. Available https://www.automationml.org/organisation/organisation/. Accessed 20 Apr 2022
22. Strahilov, A., et al.: Improving the transition and modularity of the virtual commissioning workflow with AutomationML. vol. 4 (2016)
23. Süß, S., Hauf, D., Strahilov, A., Diedrich, C.: Standardized classification and interfaces of complex behaviour models in virtual commissioning. Proc. CIRP **52**, 24–29 (2016). https://doi.org/10.1016/j.procir.2016.07.049
24. Hundt, L., Wiegand, M., Lüder, A., Meyer, T.: Das AutomationML-Komponentenmodell Engineering-Informationen konsistent zusammenführen (2022)
25. Functional Mock-up Interface. [Online]. Available https://fmi-standard.org/. Accessed 20 Apr 2022
26. Home Page—OPC Foundation. [Online]. Available https://opcfoundation.org/ Accessed 20 Apr 2022
27. Henßen, R., Schleipen, M.: Interoperability between OPC UA and AutomationML. Procedia CIRP **25**, 297–304 (2014). https://doi.org/10.1016/j.procir.2014.10.042
28. Ohno, T.: In: Das Toyota Produktionssystem. Campus (1993)
29. Bertagnolli, F.: Lean Management. Wiesbaden, Springer Fachmedien Wiesbaden (2022)
30. Takeda, H.: The Synchronized production system—going beyond just-in-time through Kaizen: Kogan Page (2006)
31. Unverdorben, S.: Architecture framework concept for definition of system architectures based on reference architectures within the domain manufacturing (2021)
32. Sinnemann, J.: Methodik zur effizienten Energiesimulation von automatisierten Produktionsanlagen in der virtuellen Inbetriebnahme. Ruhr-Universität Bochum (2021)

33. DIN 8580:2020–01: Fertigungsverfahren—Begriffe, Einteilung (2019)
34. VDI, VDI 2860—Handhabungsfunktionen, Handhabungseinrichtungen; Begriffe, Definitionen, Symbol: Projekt 2023. [Online]. Available https://www.vdi.de/richtlinien/details/vdi-2860-handhabungsfunktionen-handhabungseinrichtungen-begriffe-definitionen-symbol. Accessed 15 Apr 2022
35. Reinhart, G., Flores, A.M., Zwicker, C.: Industrieroboter: Planung—Integration—Trends Ein Leitfaden für KMU, 1st ed. Würzburg: Vogel Buchverlag (2018). [Online]. Available https://www.content-select.com/index.php? id=bib_view&ean=9783834362360
36. Rittal GmbH & and KG, Co., 8808000 Anreih-Schranksystem VX25 *Basisschrank.* [Online]. Available: https://www.rittal.com/de-de/products/PG0002SCHRANK1/PG0026SCHRANK1/PGRP21063SCHRANK1/PRO70035?variantId=8808000. Accessed 23 Apr 2022
37. Fit-4-AMandA. [Online]. Available: https://fit-4-amanda.eu/events/fch-ju-programme-review-days-2019/. Accessed 20 Apr 2022
38. Prior, J., Penczek, L., Brisse, M., Hundt, L., Kuhlenkötter, B.: A method for mapping novel product groups in AutomationML as the first step for creating their virtual twin. In: IEEE International Conference on Emerging Technologies and Factory Automation (ETFA), vol. 27, to be published
39. Drath, R.: AutomationML—A Practical Guide. De Gruyter, Berlin, Boston (2021)

Lean Engineering and Lean Information Management Make Data Flow in Plant Engineering Processes

Sabrina Karch[1(✉)], Johannes Prior[2], Anton Strahilov[1], Arndt Lüder[3,4], and Bernd Kuhlenkötter[2]

[1] let's dev GmbH & Co. KG, Alter Schlachthof 33, 76131 Karlsruhe, Germany
sabrina.karch@letsdev.de
[2] Chair of Production Systems, Ruhr-Universität Bochum, Universitätsstr. 150, 44801 Bochum, Germany
[3] Chair of Production Systems and Automation, Otto-v.-Guericke Universität, Universitätspl. 2, 39106 Magdeburg, Germany
[4] Center of Digital Production, Seestadtstr. 27/TOP 19, 1220 Wien, Austria

Abstract. The plant engineering process is characterized by high complexity, diverse interfaces and multidisciplinary processes. Today, there is still no standardized reference process architecture. A proven procedure is to use and adapt existing planning documents with certain similarities to a new project. This involves lots of effort not only in the design phase but much more in the semantic adaptation, resulting from redesign and reprogramming. Due to these challenges and the increasing importance of data as gold of digital age, a continuous flow of data and information is becoming even more important in plant engineering. Creating a solid base of engineering and operation data free of waste supports effectiveness and efficiency during the entire product life cycle. A new approach is to design general and configurable production modules. Just as virtual commissioning brings considerable time and quality benefits, the design of defined process steps in advance of a customer order is intended to bring forward some of the tasks, standardize multidisciplinary work and unify the data base. The aim of this paper is to present the lean-based concept of configurable production modules. Thereby, a focus is especially on lean information management to achieve an effective as well as an efficient plant engineering process and to create the requirements for a lean production process. The concept of configurable production modules is applied to the example of the plant design process of automated production plants for hydrogen electrolyzers.

Keywords: Lean information management · Data flow · Lean engineering · Configurable production modules

M. Liewald et al. (Eds.): WGP 2022, LNPE, pp. 664–676, 2023.
https://doi.org/10.1007/978-3-031-18318-8_66

1 Introduction

Optimizing the factors of time, costs, quality and individuality (flexibility) is a significant competitive advantage of manufacturing companies in the digital age [1], beginning already in the production engineering phase of the production life cycle. The effectivity and efficiency in the engineering phase of automated production systems and the information engineering affects the entire life cycle of a production system [2].

Automatization and digitalization of components, as well as processes, enable a significant increase in productivity, simplify data collection [3] as the "gold of the modern age" [4] and are the basis of cyber-physical production systems, with the aim of self-organized, real-time production of customized products [2, 5]. Digitalization seeks to manage complexity, redesigns process flow as well as generates new products, services, or business models [5], for example digital twins.

Lean is a renowned system idea to focus on effectivity, efficiency and reduce complexity through a customer-centered approach and the reduction of waste in value streams [6, 7]. "Design instead of re-design!" and teamwork of product development and production engineering are the nursery of lean processes [6].

The triad of automation, lean and digitization has an inseparable correlation and offers varied fields of action in context of engineering of production systems. Optimization in the engineering can be addressed to the planning process itself (e.g., lead time of information) and also to the processes to be planned (e.g., cycle time of plants).

The purpose of the paper is to use respective strength of each discipline as a basis to identify levers for reducing complexity in the planning process, establish the conditions for an efficient production system and to confirm the importance of a lean information flow from the beginning of a product life cycle. All activities follow the vision of generating plant engineering at the push of a button. The purpose is further specified in two research questions.

- How can the application of lean-aligned and predefined, configurable production modules and templates optimize the plant engineering process?
- How can the focus on lean information flow support effectivity and efficiency from the beginning of a product life cycle?

First, basics and state-of-the-art of the engineering process are discussed, using the example of plant engineering, followed by digitization and the significance of lean engineering and lean information management. Based on this literature-centered study of today's processes, fields of action are derived and presented in the main part. First levers are considered using the example of plant engineering for hydrogen electrolyzers within the sub-project FertiRob of the BMBF lead project H2Giga. Finally, a summary and an outlook for further work is given.

2 Data Flow in a Plant Engineering Process

The production engineering and especially the plant engineering is a significant phase of the product life cycle of the production system [8]. This phase is usually divided into several sub-phases, in which layout, process flow, mechanics, electronics and software of the system are designed (Fig. 1) [9].

The arrangement of the plant in a production building as well as the value stream are detailed with 2D layouts and process descriptions in plant layout and process design.

In mechanical design, the detailed 3D geometric model of the production system is created respectively fixed with all corresponding mechanical design data and process restrictions for the following sub-phases. Some of these mechanical design data are 3D geometry models, 2D drawings, bill of materials (BOMs), process animations as native simulations or videos, layout plans, simulation models for virtual engineering, documentation, production process descriptions, offline robot programs, etc.

Starting from the BOMs, all parts and components, integrated into the production systems, serve as input to start the next sub-phase of electronical/fluidic design. In this, electrical data is created and important electrical decisions are made that have a major impact on the production system. In this process, an initial restriction of the available production data from the system is made because of the selected electrical components, such as electrical drivers, pneumatic drivers, various sensors and safety components.

In addition, the electrical components restrict the accuracy of the available data on the one hand and the frequency in which the data can be taken from the components via the communication technology on the other hand, e.g., ProfNet, Profibus, OPC UA, etc. on the other hand [2].

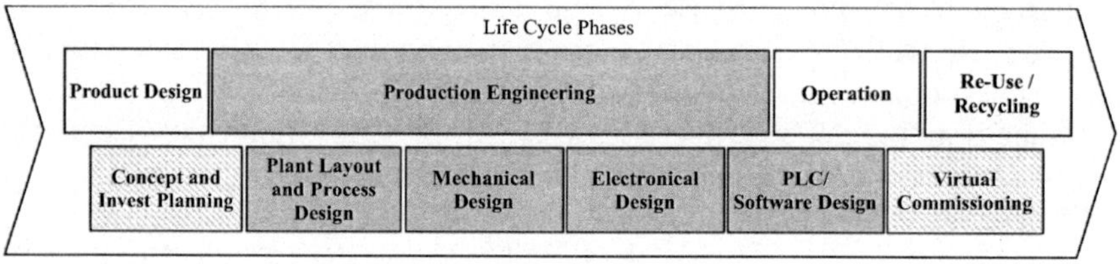

Fig. 1. Phases in production system engineering, in reference to [2]

With the finishing of the electrical/fluidic design, the PLC/software design phase begins. In this phase, the software (control program) of the production system is created. During this, a next important design and decision step is done, which is relevant for the evaluability of the data provided by the electrical components. The PLC is the hub that communicates with all electrical components and has access to all existing data in a frequency between 10 and 100 ms. In these sub-phases, important decisions are made for the system, which have significant effects on the subsequent design, functionalities, service life as well as the production quality of the product achieved during the operation of the system. A subsequent change is possible but usually associated with significant time and cost [2].

On the one hand, this results in enormous engineering data, which is usually not changed after the construction and commissioning of a production system and on the other hand, continuous system data is created during operation, such as error messages, style statuses, quality data, motion sequences, individual sensor data, etc. Depending on the recording rates, such recorded system data can reach large dimensions during system operation.

To merge and test all sub-phase, data without taking the risk of damaging the real plant as well as persons, the Virtual Commissioning (VC) is preferred as an ahead phase before the real commissioning [10, 11]. VC is therefore the first opportunity to check all data for logical errors and consistency as well as to perform the first checks on the data generated from the running system.

The simulation model required for VC is referred to as the mechatronic plant model in [12] and can be subdivided into the two sub-models, the extended 3D geometry model and the behavior model. The extended 3D geometry model includes the 3D geometry of the plants with mechanics, kinematics, part transport, sensors, and interfaces. This sub-model represents the plant mechanics for VC. The behavior model, in turn, is needed so that the mechatronic plant model can react to control outputs during VC in the same way as the real plant, pass information to other components and set appropriate control outputs [12].

Both sub-models are used as the basis for the digital twin of production plants. Among other things, the digital twin is used as the digital twin of the running production plant. In this case, the real PLC outputs are used. The output signals of the digital twin are not transmitted to the PLC. So, no PLC input signals of the real plant are manipulated.

In [13] an example of the file size of the behavior model and the generated run data is presented. The example shows that the required storage space for the behavior model, which consists of 106 individual behavior models, is only 8 MB, but for 8 h of running as a digital twin of the running real plant, it generates about 1 billion values with a data size of 11 GB at a sampling rate of 50 ms. This generated data directly depends on the number of input and output signals and the signal type, e.g., Boolean, Integer, Reals, etc. The file size depends on the used data format. In the example above, the behavior models were used as FMUs (10 electrical drives, 16 industrial robot, 80 pneumatic drives), according to the Functional Mock-Up Interface (FMI) [14, 15].

Based on expert discussions from the plant engineering industry, the biggest challenge is, which data to include and store during the development phase of a plant to achieve meaningful data recording during operation without collecting non-sincere, e.g., data over years of operation or, even worse, without recording data, which is of enormous importance. As a rule, experts decide in most cases to record and store all available data as a precaution. However, this creates a huge waste of data.

3 Digitalization and the Significance of Data

The importance of data in the digital age is expressed in headlines such as gold, raw material, the fuel of digital transformation or infinitely valuable.

Making data available became a "piece of cake", the challenge remains processing data for the right purpose, at the right time for the right people [16]. This aspect is even more important in a production environment, because "Manufacturing generates more data than any other sector of the economy. (…) More data than healthcare, more data than retail, finance. When we talk to manufacturers, they mostly throw away the data. Where they keep it, they don't know what to do" [17]. Even if generating data is a cost-effective process, the focus is on mastering the triad of data, information and knowledge and specific recommendations for action. Customer benefits, company benefits and above all resource efficiency are the guiding principles [5, 18].

In [19] it is brought to the point: "More recently, the digital era allows us to deal with lots of information, arising from sensors, CPS, IoT, and social networks. The challenge is to understand how to use these technologies to build on the fundamentals of lean thinking and create even more value and to improve industrial productivity".

4 Lean and a New Type of Waste

4.1 The Decisive Contribution of Lean

Efficiency is a masterpiece of lean, although the significance of lean in the digital age is a matter of debate. Representatives of the diversity of opinions are "Introducing Lean as a philosophy throughout the company with the traditional tools is dead, the Lean philosophy is alive" [5] or "Lean is the basis for digitalization and for "Industry 4.0". Waste must be eliminated before processes are automated and digitalized. Lean is therefore a prerequisite and basis for further progress" [6]. Lean arrived in the digital age.

If the system idea of lean is at the centre of attention, lean can still make a decisive contribution (Fig. 2). This aspect is summarized by Womack and Jones in lean thinking with the key messages: specify value from the end customer's perspective, identify value streams and stabilize them by eliminating waste, create a value flow, let customer pull value, and pursue perfection through continuous reduction of waste [7]. The essential role of people and leadership is also emphasized in lean enterprises [6, 20].

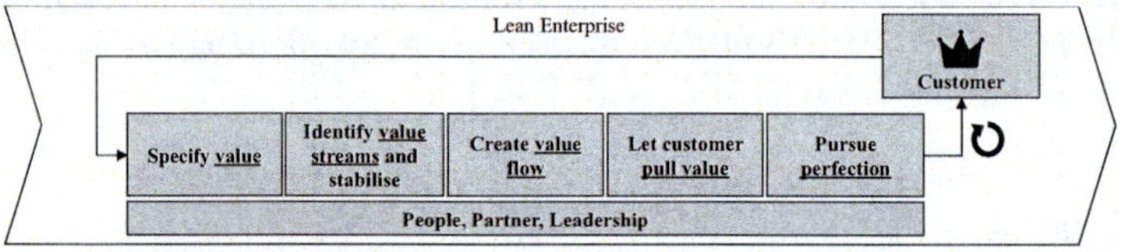

Fig. 2. System idea of lean, in reference to [6, 20, 21]

The reduction of waste in production or literally in all business processes is seen as the key principle of lean. Waste, or Muda in Japanese, means "to toil" or "pointless effort" and is any activity that does not add value to the customer's needs. It is found in overproduction, inventory, overprocessing, transport, waiting, movement/motion, defects, and unused intellect/skills in the production context [6, 20, 22]. In addition to waste, variability and inflexibility are also considered to be loss factors [6].

4.2 Waste in Engineering Processes

In the context of engineering, Muda arise in the planning processes itself and also in the processes to be planned, the later series production and affects engineering data as well as operation data (Fig. 3). Waste in the engineering process can be eliminated by taking the following levers into account, which finally lead to a reduction of the order throughput time [6].

- Reduced interfaces or at least the need for coordination and iteration loops.
- Structured data management and information exchange.
- Eliminated unnecessary work steps.

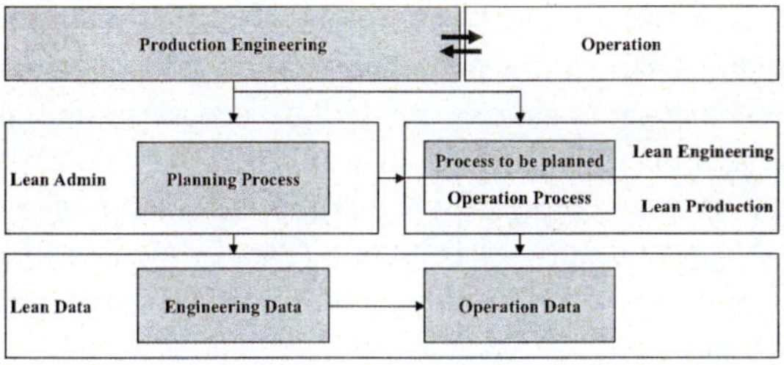

Fig. 3. Waste in engineering and data flow

The elimination of waste has its origin in product design and production planning with more potential influence than in ongoing production. Lean engineering provides planning principles, which can be applied to different types of workplaces and aim to create lean production processes [6].

- Ergonomic Workplace Design: Flow of horizontal material transport, work area with a best point area arranged in a radius of 80 cm, material supply always at the same place, position, and orientation (pick-optimized), and no set-up times of tools and machines.
- Machine Design: Machines' orientation in-depth and in an alignment reduces walking distances and required space.
- One-Piece Flow: A flow orientation of a layout or a cell is always the most important target, to reduce the lead time and stocks between stations.
- Human and Machine: The cooperation of machine and employee is to be designed in such a way that one employee can handle several machines and do not have to wait for the machine.
- Material Provision: Material is providing in the process with the help of chutes, rails, conveyors or hoppers, prepared in the smallest possible containers or units and the delivery schedule is organized via a Kanban system.
- Capacity: In the context of plant engineering, oversizing machine capacities can be defined as an additional type of waste [23].

If these principles are not already taken into consideration in the engineering, waste in automated production occurs as overproduction through lot sizes, overprocessing through production planning specifications or transport of goods through layouts without flow orientation [6].

4.3 Waste in the Information Flow

In addition to the traditional understanding of Muda, waste in relation to data has to be discussed to the increasing operationalization of industry 4.0 and usage of digital tools. It is generally agreed that the simple collection of data does not add value. The meaning of waste in the context of data and information is not finally defined.

In the sense of lean, the collection of data rather can be seen as an apparent performance, non-value-adding but necessary activity, which should be reduced to a minimum [6]. Waste in the information flow is less obvious and not immediately visible. There can be found parallels between lean thinking and waste in information management and defined as "the additional actions and any inactivity that arise as a consequence of not providing the information consumer immediate access to an adequate amount of appropriate, accurate and up-to-date information" [24].

The decision support is given priority and composed of the idea of a 5 C model in [16]: "Connection (sensor and networks), Cloud (data on-demand and anytime), Content (correlation and meaning), Community (sharing and social), and Customization (personalization and value)".

Alieva et al. propose to define a new form of waste, "Digital Muda", which can be found in uncollected, (partially) unprocessed or misinterpreted data in the production process also in the context of decision making [26].

A further literature review on the transfer of waste to information logistics can be found in [25]. Information logistics plans, manages, executes, controls, stores and prepares cross-process data flows with the purpose of decision support [27]. Meudt et al. set up a systematic approach to identify waste in information logistics according to the eight types of waste and the focus on brownfield processes (Fig. 4). The properties of information flow are considered, such as immateriality, compressibility, expandability and fast transportability. The model is based on the triad of data, information and knowledge and transfers these to data collection, data processing and data analysis [25].

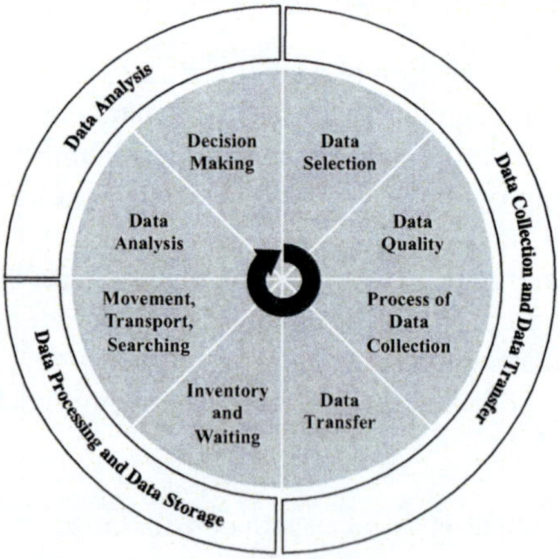

Fig. 4. A systematic approach to identify waste in information logistics, in reference to [25]

5 Fields of Action to a Systematic Reduction of Waste in Plant Engineering

5.1 Vision of Lean Aligned, Configurable Production Modules

In order to be able to transfer the lean data approach, the idea of template-based production modules is introduced. These modules aim to reduce complexity in plant engineering by creating pre-defined engineering artifacts in advance of a customer order.

The functionalities of individual components of the production cells, e.g., robot grippers, conveyor belts or sensors, are encapsulated, so that a distinction can be made between internal data, e.g., material or bearing of a conveyor belt, and user-specific engineering data, e.g., dimensions of a conveyer belt.

The idea follows a procedure model for project tasks as well as the project-independent activities of plant engineering, in particular the development of a continuous data model, reuse of artifacts, a standardized description language and the definition of reference models to describe knowledge formally [28, 29].

The approach uses a configurator [29], templates and a set of configuration rules to manage, for example, right of access, interdependencies of changes or configure customer orders. Project experiences are taken over in the knowledge management for further development of the standard based on previous configurations and best practices.

In the lean admin context, the implementation of configurable production modules increases efficiency in plant engineering in the first step by:

- Stakeholder [2] and interface management to reduce the need for clarification and iteration loops,
- Standardization as a starting point to document the knowledge of previous engineering activities, reduce redundancy of data and raise the quality of engineering artifacts and,
- Reduction of the order throughput time in particular through the predefined engineering artifacts.

In particular, the possibility of user-specific data exchange in the engineering chain through the encapsulation of functionalities establishes the conditions for a pull-oriented lean data implementation.

The production modules and as a result, the processes to be planned, follow the restrictions of lean engineering. Figure 5 demonstrates a draft of a configurable production module in the context of a production process to be automated for hydrogen electrolyzers. The shape of the casing is often comparable to a control cabinet, why the accessibility is focused on one side.

With the help of predefined templates and only a few data (variable and not variable), the parameters and the size of the plant can be customized and scaled, e.g., dimensions and positions of production cell, conveyors or robots. In addition, robot trajectories can already be pre-designed in a robot simulation program, depending on the variables of the product specific design. The template, especially if it is based on a neutral data exchange format, enables a lean information engineering and eliminates waste in information logistics starting with data selection and quality.

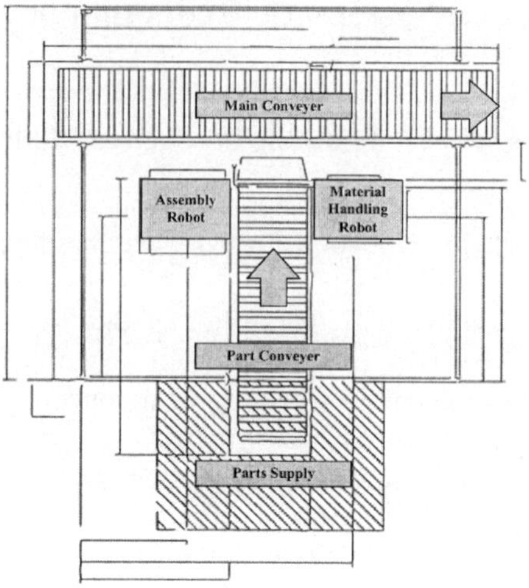

Fig. 5. Draft of a template-based configurable production module

The customer-specific scalability is to prevent waste due to under- or oversizing of the plant and eliminate waste in the movement of robots. The configurable production modules also ensure flexibility to arrange the plants according to diverse target layouts in a flow without stocks and in an optimized machine orientation.

A challenge is still to find an optimum of production modules considering variance and variability in product and process, not to risk a systematic under- or oversizing of plants, overproduction as the most critical form of waste, a lot of setups or waiting time in the process. Also, the number of different predefined production modules has to remain clear and manageable.

In the context of lean information management, the templates specify data requirements, processing and analysis as well as reduce data waste. The approach of configurable production modules is the baseline for a continuous data flow.

5.2 Levers to Create Lean Data Flow from the Beginning of a Lifecycle

Value in plant engineering results from the engineering artifacts as well as from the engineering data. Table 1 provides an overview of Data Muda Potentials in the context of plant engineering, derived from the current processes and considering the findings on lean information management and Digital Muda. The potentials can have different characteristics in relation to their maturity level and vary from Data Muda Potential not recognized to recognized and actions installed, running, or working.

A first indication is done to the effects on engineering processes or processes to be planned as well as to the superordinated category of waste according to [25]. All Data Muda Potentials impact the engineering processes, the impact on the processes to be planned is not always obvious at the first sight. Also, the range and intensity of Muda are indicated but have to be evaluated in detail.

Table 1. Data muda potential in plant engineering

	Data Muda Potentials	Affects		Waste [25]			Subject field [29]				
		Engineering Process	Engineering Artifacts	Data Collection	Data Processing	Data Analysis	workflow	Methods	Tools	OrganizationaStructur	Economics
Upstream Processes	Communication and coordination between engineering and product design or other stakeholders [2], such as maintenance.	x	x	x	x	x	x	x	x	x	x
	Input data/information does not correspond to the needs of engineering (e.g. digital product twin).	x		x			(x)	(x)		(x)	
	Documentation of requirements and questions. The same topics are discussed several times with same or even different results.	x		x				x			
Mechanical Design	Variety of native data up to engineering results are created and processed in numerous tools.	x		x	x	x			x		
	Data is only used for specific tasks or in specific tools and has no relevance for next process step.	x		x	x	x			x		
	Data is transferred with data/information loss due to transformation of file formats, e.g. CAD drawing to PDF.	x		x	x			x	x		
Electronical Design	Data availability and accuracy through selection of electrical components e.g. pneumatic drivers or sensors.	x	x	x					(x)		
	Frequency of data collection from the components by the communication technology, e.g. OPC UA.	x	x		x				(x)		
Software Design	Evaluability in PLC programming (input and output signals).	x	x	x	x				(x)		
	Frequency (between 10 and 100 ms) of evaluation.	x	x	x	x				(x)		
	Influence of the system onto the running process, e.g. production quality.	x	x	x	x				(x)		
Virtual Commissioning (VC)	Interaction of robot programme and control programme is not confirmed until VC.	x	x	x					(x)		
	Waste in simulation model as a basis for digital twin, transfers waste immediately to digital twin, e.g. library elements.	x	x	x	x	x			x		
	Conclusions about performance of the system (oversizing/undersizing) will not be transparent until VC.	x	x	x			(x)				
General topics in Engieering	Transparency of data and information flow and corresponding sources and sinks.	x		x					x		
	Push of data and information without focussing on selected needs of the user.	x			x		x		x	x	
	Changes from downstream process steps do not flow back in a structured or needs-oriented way.	x		x	x		x				
	Selected data formats influence file sizes.	x		x				x			
	Loss of information through the use of unsuitable media/tools for transfer e.g. email.	x			x			x			
	Information to all persons and to right persons.	x		x			x			x	
	Responsibility or acess rigth.	x		x	x	x	x			x	
	Handling of data during its life cycle (generation to deletion).	x	x	x	x		x			x	
	Optimum of overall process in context of the multidisziplinary engineering.	x	x	x	x	x				x	
	Data requirements from Simultaneous Engineering (e.g. work statuses, versioning, missing or changed information).	x		x			x	x	x	x	
Commissioning and Series Production	Recording and processing of plant data e.g. error messages.	x	x	x	x		x				
	Collection of relevant data during series production is not planded in plant enginering.	x	x	x			(x)			(x)	
Re-Use and Recycling	Data do not correspond to the needs of reuse and recycling (analogy: product design influences reuse).	x	x	x	x	x	(x)			(x)	

x: applicable, (x): partially applicable

In the context of plant engineering, the state-of-the-art target status for engineering organizations including technical, organizational and economic aspects is described in [29]. The assignment of the Data Muda Potentials to the defined topics gives a hint to the current situation and also an initial indication of starting points for eliminating waste. These findings lead to further key questions and the following research activities.

- How does pull of data from customer or decision-maker influence data waste?
- How does data waste, in particular, data processing and analyzing, depend on repetitive (e.g., KPIs) or irregular decision needs (variability)?
- How can data waste in specific engineering tasks be identified, measured (qualitative or quantitative) and rated according to the influences as well as the effort benefit ratio (reference model)?
- How should data waste be considered in the context of data economics, innovation with not conclusively defined customer value, or Big Data and AI applications (flexibility)?
- How can the lean paradigm Genchi Gembutsu, "go to the place of value creation" [6], be rethought in the context of data and information flow?

6 Summary and Outlook

In this paper, an overview of the data flow in plant engineering processes is given in the context of the increasing importance of data and information in the digital age. Lean is proven to give orientation to establish effective and efficient solutions and processes in information flow, plant engineering as well as engineering artifacts in a multidisciplinary environment. A first approach to reduce complexity and addresses lean data is presented with the templated-based configurable production module (research question I). The next step is to finalize and evaluate the modules concept especially focusing on data waste. In addition, the value stream-oriented arrangement in the layout and the control concept with a vision of one-piece flow have to elaborate.

The Data Muda Potentials in plant engineering are elaborated in the context of current processes. Specific levers and key questions to eliminate data waste and increase effectivity and efficiency are defined (research question II). Following research activities are dedicated to how data waste in specific engineering tasks can be identified, measured (qualitative or quantitative) and rated according to the influences as well as the effort benefit ratio and finally eliminated.

Acknowledgment. Parts of this work were supported by the Federal Ministry of Education and Research (BMBF) within the research project H2Giga – FertiRob.

References

1. Wiegand, B.: Der Weg aus der Digitalisierungsfalle: Mit Lean Management erfolgreich in die Industrie 4.0, 1st edn. Springer Fachmedien Wiesbaden, Wiesbaden (2018)
2. Biffl, S., Lüder, A., Gerhard, D. (eds.): Multi-Disciplinary Engineering for Cyber-Physical Production Systems: Data Models and Software Solutions for Handling Complex Engineering Projects. Springer, Cham (2017)
3. Burggräf, P., Schuh, G.: Fabrikplanung: Handbuch Produktion und Management, vol. 4. Springer Berlin Heidelberg, Berlin, Heidelberg (2021)
4. Webseite der Bundesregierung | Startseite, Rede von Bundeskanzlerin Merkel beim Publishers' Summit des Verbands Deutscher Zeitschriftenverleger (VDZ) am 2. November 2015. [Online]. Available: https://www.bundesregierung.de/breg-de/aktuelles/rede-von-bundeskanzlerin-merkel-beim-publishers-summit-des-verbands-deutscher-zeitschriftenverleger-vdz-am-2-november-2015-390088. Accessed 23 Apr 2022

5. Kieviet, A.: Digitalisierung der Wertschöpfung: Auswirkung auf das Lean Management. In: Künzel, H. (ed.) Erfolgsfaktor Serie, Erfolgsfaktor lean management 2.0: Wettbewerbsfhige verschlankung auf. [Place of publication not identified]: Gabler, pp. 41–59 (2016)
6. Bertagnolli, F.: Lean Management. Springer Fachmedien Wiesbaden, Wiesbaden (2022)
7. Womack, J.P., Jones, D.T.: Lean Thinking - Ballast abwerfen, Unternehmensgewinn steigern. Campus Verlag, Frankfurt/New York (2013)
8. Drescher, B., Stich, P., Kiefer, J., Strahilov, A., Reinhart, G.: Physikbasierte Simulation im Anlagenentstehungsprozess – Einsatzpotenziale bei der Entwicklung automatisierter Montageanlagen im Automobilbau pp. 271–281 (2013)
9. Groover, P.: Automation, Production Systems, and Computer-Integrated Manufacturing
10. Kiefer, J.: Mechatronikorientierte Planung automatisierter Fertigungszellen im Bereich Karosserierohbau. LFT, Univ, Saarbrücken. [Online]. Available: http://nbn-resolving.de/urn:nbn:de:bsz:291-scidok-14686 (2007)
11. Wünsch, G.: Methoden für die virtuelle Inbetriebnahme automatisierter Produktionssysteme. H. Utz, München (2008)
12. Kiefer, J., Ollinger, L., Bergert, M.: Virtuelle Inbetriebnahme - Standardisierte Verhaltensmodellierung mechatronischer Betriebsmittel im automobilen Karosserierohbau **07**, 40–46 (2009)
13. Strahilov, A., (ed.): Digitaler Schatten von Produktionsanlagen als Big Data Quelle – Herausforderungen & Potential. Big Data Technologien in der Produktion von der Datenerfassung bis zur Prozessoptimierung (2017)
14. Süß, S., Hauf, D., Strahilov, A., Diedrich, C.: Standardized classification and interfaces of complex behaviour models in virtual commissioning. Procedia CIRP **52**, 24–29 (2016). https://doi.org/10.1016/j.procir.2016.07.049
15. Functional Mock-up Interface. [Online]. Available: https://fmi-standard.org/. Accessed 3 May 2022
16. Lee, J., Lapira, E., Bagheri, B., Kao, H.: Recent advances and trends in predictive manufacturing systems in big data environment. Manuf. Lett. **1**(1), 38–41 (2013). https://doi.org/10.1016/j.mfglet.2013.09.005
17. "The digital-manufacturing revolution: How it could unfold," McKinsey & Company, 10 Jan 2015. https://www.mckinsey.com/business-functions/operations/our-insights/the-digital-manufacturing-revolution-how-it-could-unfold. Accessed 23 Apr 2022
18. Kieviet, A.: Lean Digital Transformation: Geschäftsmodelle transformieren, Kundenmehrwerte steigern und Effizienz erhöhen. Springer Berlin Heidelberg; Springer Gabler, Berlin, Heidelberg. [Online]. Available: https://ebookcentral.proquest.com/lib/kxp/detail.action?docID=5741612 (2019)
19. Cattaneo, L., Rossi, M., Negri, E., Powell, D., Terzi, S.: Lean thinking in the digital era. In: Ríos, J., Bernard, A., Bouras, A., Foufou, S., (eds.) IFIP Advances in Information and Communication Technology, vol. 517, Product Lifecycle Management and Industry of the Future: 14th IFIP WG 5.1 International Conference, PLM 2017, Seville, Spain, 10–12 July 2017: Revised Selected Papers. Springer, Cham, pp. 371–381 (2017)
20. Liker, J.K., Braun, A.: Der Toyota Weg: 14 Managementprinzipien des weltweit erfolgreichsten Automobilkonzerns, 8th edn. FBV, München. [Online]. Available: http://site.ebrary.com/lib/hsalbsig/docDetail.action?docID=10684046 (2013)
21. Womack, J.P., Jones, D.T.: Lean Thinking: Ballast abwerfen, Unternehmensgewinn steigern, 3rd edn. Campus Verlag, Frankfurt am Main. [Online]. Available: http://swb.eblib.com/patron/FullRecord.aspx?p=1219872 (2013)
22. Ohno, T.: Das Toyota Produktionssystem. Campus Verlag (1993)
23. Takeda, H.: The Synchronized production system – going beyond just-in-time through Kaizen. Kogan Page (2006)

24. Hicks, B.J.: Lean information management: understanding and eliminating waste. Int. J. Inf. Manage. **27**(4), 233–249 (2007). https://doi.org/10.1016/j.ijinfomgt.2006.12.001
25. Meudt, T., Leipoldt, C., Metternich, J.: Der neue Blick auf Verschwendungen im Kontext von Industrie 4.0. Zeitschrift für wirtschaftlichen Fabrikbetrieb **111**(11), 754–758 (2016). https://doi.org/10.3139/104.111617
26. Alieva, J., Haartman, R.: Digital Muda - the new form of waste by industry 4.0. OSCM: An Int. J. 269–278 (2020). https://doi.org/10.31387/oscm0420268
27. Winter, R., Schmaltz, M., Dinter, B., Bucher, T.: Das St. Galler Konzept der Informationslogistik. In: Töpfer, J., Winter, R. (eds.) Active enterprise intelligence: Unternehmensweite Informationslogistik als Basis einer wertorientierten Unternehmenssteuerung. Springer, Berlin, pp. 43–58 (2008)
28. Unverdorben, S.: Architecture framework concept for definition of system architectures based on reference architectures within the domain manufacturing (2021)
29. VDI/VDE 3695 - Engineering of industrial plants Evaluation and optimisation of the engineering Fundamentals and procedure (2020)

Sustainable Personnel Development Based on Production Plans

J. Möhle[1(✉)], L. Nörenberg[1], F. Shabanaj[1], M. Motz[2], P. Nyhuis[1], and R. Schmitt[2,3]

[1] Institute of Production Systems and Logistics IFA, An der Universität 2, 30823 Garbsen, Germany
moehle@ifa.uni-hannover.de

[2] Fraunhofer Institute for Production Technology IPT, Steinbachstr. 17, 52074 Aachen, Germany

[3] Laboratory for Machine Tools and Production Engineering WZL, Campus-Boulevard 30, 52074 Aachen, Germany

Abstract. The production environment is in a constant state of change. This results in a continuous change of production processes. A key factor in mastering change is to increase flexibility. To achieve this, the targeted training of employees is essential. Within the framework of the research project "reQenrol", research is being conducted to sustainably design personnel development based on the competence and tasks of the employees. Manufacturing companies face the challenge to efficiently training their personnel for an increasing and dynamic range of tasks. Training measures must be adapted to personal skill level of employees as well as to requirements of individual tasks in production. As a basis for a competence-based workforce deployment and the realization of targeted training measures, a survey was conducted on the current training situation and the relevance of competences in production. The results are placed into the context of the concept for an assistance system that enables manufacturing companies to perform a dynamic, competence-based workforce scheduling and realize targeted employee training.

Keywords: Workforce scheduling · Personnel training · Quality · Survey · Workforce flexibilization · Competence development

1 The Need for Sustainable Personnel Development

Nowadays, the shortage of skilled workers, sickness-related absences, and a high product variance are forcing manufacturing companies to deploy their production personnel in a highly flexible manner [1, 2]. As a result, employees must handle a growing and frequently changing range of tasks with consistently high quality. In his vision of sustainable productivity, Boos 2021 identifies this social perspective of production as one of the four key dimensions that manufacturing companies must address to achieve sustainable economic success [3]. Accordingly, manufacturing companies should increase the flexibility of their workforce scheduling and adapt it to the individual competence levels of employees, this applies in particular to small and medium-sized enterprises (SMEs). At the same time, employees must be trained through targeted, quality-oriented

M. Liewald et al. (Eds.): WGP 2022, LNPE, pp. 677–685, 2023.
https://doi.org/10.1007/978-3-031-18318-8_67

qualification measures in order to continuously adapt their competences to a growing and dynamic range of tasks [4].

However, the extent to which specific competences impact the quality of products appears to depend on a specific product and its quality requirements. This raises the question of which competences have a particularly large impact on quality and thus should be prioritized for training. The authors conducted a survey on the current training situation and the relevance of competences in production to investigate this question.

2 Fundamentals of Quality and Competences

According to the normative definitions, quality is defined as the "*[…] degree to which a set of inherent characteristics of an object meets requirements*" [5]. However, this does not allow conclusions about the factors directly influencing quality. Definitions, such as the entrepreneurial quality concept, refer to the employees' authority as one of the critical influences on fulfilling those quality requirements [6]. The relevance of competences is also emphasized in other common quality-related approaches. For example, DIN EN ISO 9001 demands that quality management systems ensure essential employee competences. Total quality management concepts list an employee focus as a key success factor for achieving quality [7, 8]. As a result, competent employees represent a source of competitive advantages that are difficult to imitate [9].

In the context of production, a direct link between quality and employee competence exists for product characteristics that are directly created within manual tasks by employees. Examples are the error-free assembly of components or setting a welding seam by requirements. Typically, the requirements for these tasks are documented in worker instructions. In [10], this understanding of quality is referred to as the so-called quality of execution and will be referred to in the following as quality for short.

Just like the requirements placed by customers on products, the general understanding of quality is also subject to a high degree of dynamic [6, 11]. The resulting change in quality requirements leads to changes in the demands placed on employees to provide the required quality. To meet the corresponding requirements, a broad spectrum of competences is necessary. The term competence is defined as a system of skills necessary for successfully completing a task [12]. This system includes qualifications, knowledge, skills, behaviors, and attitudes [8, 13, 14]. In the literature, there are various of possible classifications for competences [14–17]. A classification suitable for production is provided by Bughin et al. 2018, who distinguish competences between physical and manual, cognitive, social and emotional, as well as technological skills [16].

Within the research project reQenrol, an assistance system for competence-based workforce scheduling and adaptive provision of training materials based on the personal skill level of employees is developed [18]. The basis of the approach is the comparison of actual and target competences for individual production tasks. However, some production-relevant competences such as reliability and the ability to work in a team are difficult to measure and quantify. In practice, such competences are usually assessed utilizing a self-assessment or a peer assessment [14].

3 Survey on Competence Development in Production

In March 2022, an online survey was conducted with production managers from different sectors of the German manufacturing industry. 16 companies from the sectors mechanical engineering (40%), electronics (20%), and vehicle manufacturing (20%) participated in the survey.

The survey was divided into three sections. In the first section, the participants answered whether and which aids training is carried out in their companies. Furthermore, they were asked to rate how satisfied they were with the existing training materials. The latter was measured on a discrete scale ranging from satisfied (1), neutral (2), and not satisfied (3). The second section examined the current and future relevance of selected competences for production tasks. Finally, the third section covered the influence of selected competences on quality based on a five-point Likert scale [19], ranging from elementary importance (1) to no relevance (5), which is the standard for self-assessments and assessments of competences by others [20]. To pre-select production-relevant competences for the survey, 132 job postings from manufacturing companies for production workers with manual tasks have been analyzed regarding the most frequently required skills.

The results show that half of the companies surveyed actively implement competence development measures. Companies that already carry out competence development measures indicated various methods for this. These include classic measures such as initial training, instruction, internal training, further education, further development and standardization of processes together with employees, comprehension training during production, skills matrices, learning platforms such as LinkedIn Learning, leading as a coach, and 'training on the job'. Training videos, images, and text provided in print or digitally serve as the primary media used for competence development. In addition, face-to-face training is also a standard method for communicating training content.

Figure 1 shows the assessment of companies' satisfaction with their existing training materials. Satisfaction was assessed regarding the aspects of creation effort, acceptance by employees, dissemination, learning support, content, topicality, and format.

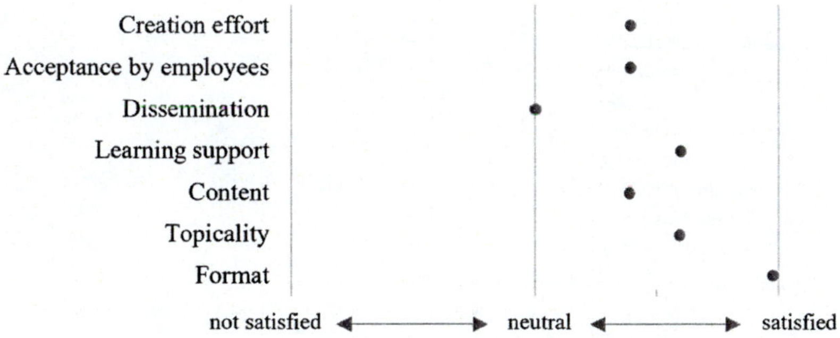

Fig. 1. Satisfaction with training materials

From the responses, it can be concluded that the most significant potential for improving training materials lies in the degree of dissemination. There is further potential for improvement in the level of acceptance, the effort required to create training materials, and their content design.

The diagram in Fig. 2 shows the assessment of the relevance of competences today. Here, the competences on the y-axis are sorted in descending order according to how frequently they are mentioned in job postings for manual production tasks. In Fig. 3, the assessment of the influence of the competences on quality is shown, while Fig. 4 depicts the expected change in the relevance of competences over the next ten years.

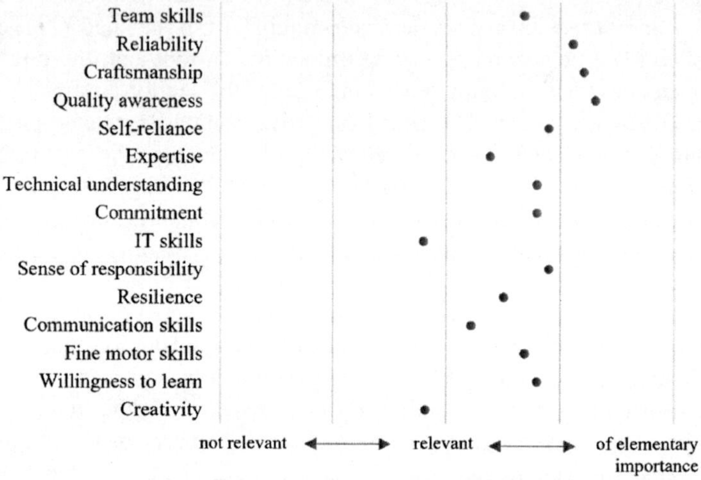

Fig. 2. Today's relevance of competences

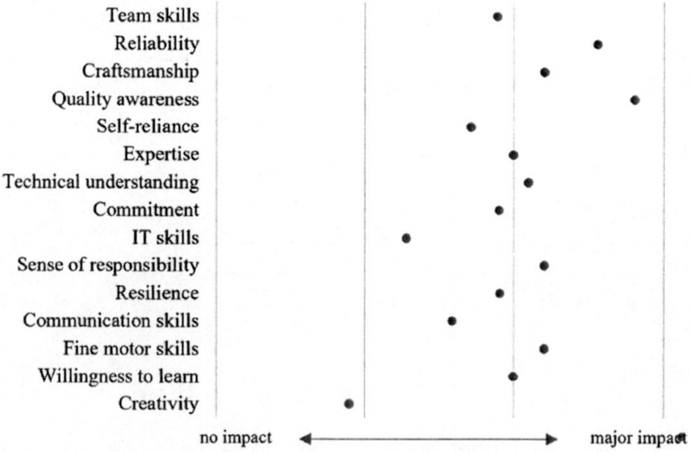

Fig. 3. Influence of competences on quality

The evaluation of the current relevance of competences (Fig. 2) shows a discrepancy between the competences requested in job postings and those relevant for production.

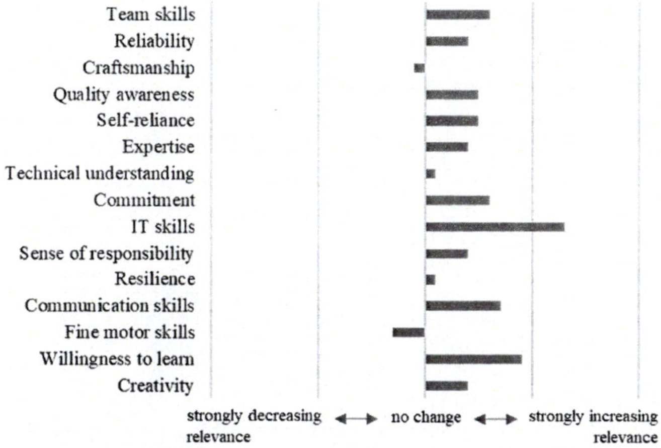

Fig. 4. Expected change in relevance of competences over the next ten years

Although quality awareness, craftsmanship, and reliability are rated as very important by the production managers, IT skills and creativity are currently assigned a relatively low relevance for production. Similarly, the competences of team skills, reliability and craftmanship are particularly often required in job postings, while fine motor skills, willingness to learn and creativity are among the less frequently required competences.

However, a comparison of diagrams in Figs. 2 and 3 underlines a high degree of agreement regarding the criteria most relevant to quality–quality awareness, reliability, and craftsmanship—and the criteria of IT skills and creativity, which are rated as having less influence. It can be concluded that competences with a great influence on quality are rated as particularly relevant by production managers. Figure 4 depicts the respondents expected extent to which a change in the relevance of production-relevant competences is expected over the next ten years. An exceptionally high increase in relevance is expected for IT skills and willingness to learn.

Overall, the survey results indicate that quality awareness, reliability, IT skills, and willingness to learn are key competences that manufacturing companies should prioritize in training to ensure sustainable personnel development aiming at maintaining a high level of quality in production. However, the number of participants is too small to derive statistically reliable statements about the whole manufacturing sector, so the survey would have to be expanded to achieve general validity. Nevertheless, the five-point Likert scale proved to be a suitable instrument for assessing competences in production that are difficult to quantify, as it enables an intuitive yet differentiated evaluation.

4 Competence Development Concept

Especially in SMEs, the lack of time, capacity, and budget often lead to a lack of adequate training measures, even though the need and benefits are well known [21, 22]. Therefore, forms of work-based learning are particularly attractive on the shop floor because of the reduced financial and time burden [19]. Furthermore, the selection of training measures is often subject to a certain degree of randomness, as the proportion of companies with a documented training strategy is low, especially among SMEs [21, 23]. This behaviour is the primary motivation for the research project reQenrol. The project aims to support manufacturing companies with an assistance system for targeted training of employees on the shop floor. In particular, the goal is to systematize the selection of training measures and align them with the employees' individual training needs. Figure 5 outlines the logic based on which competence development measures are prioritized and selected.

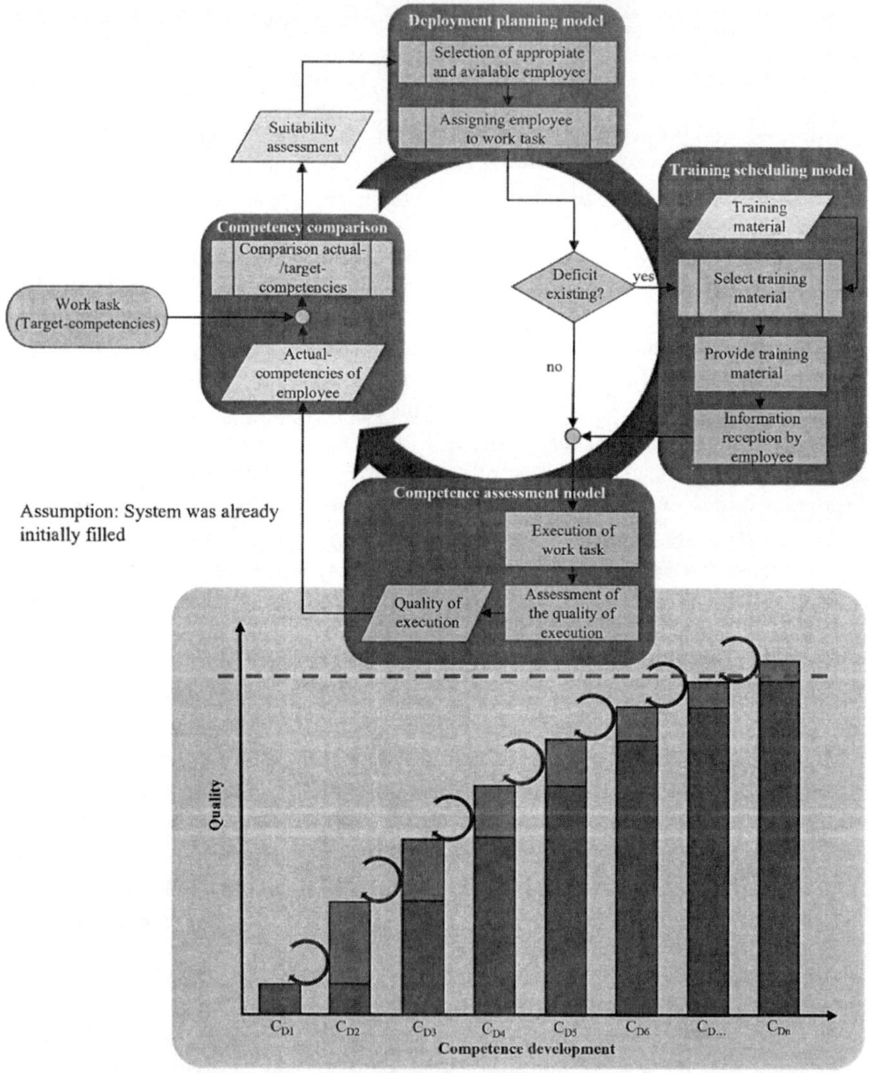

Fig. 5. Competence development concept

Figure 5 is divided into two parts. The diagram on the bottom illustrates that the training of specific competences has a varying influence on quality. For this reason, the sequence of competences to be developed is chosen so that the required quality (shown as a dashed line in Fig. 5) is achieved as quickly as possible by the sum of training measures $C_{D,n}$. The more influence a competence deficit has on quality, the earlier it should be trained. To achieve this, training materials are provided by the assistance system in a work-integrated manner. The training materials are tailored to the individual training needs, i.e. they address the competences that have the most significant impact on quality based on the personal skill profile. This way, employees should quickly be enabled to perform a task in production independently and in accordance with quality requirements.

The left side of Fig. 5 shows the competence development concept. This is divided into four sub-models. The starting point is the work task to be completed with the associated target competences necessary to complete the task successfully. This information, together with the actual skill profile of the employees, serves as the input for the first sub-model, the competence comparison model. When a work task is created, not only the competences required for it are assigned, but they are also weighted according to their influence on quality. In this way, comparing weighted target and actual competences results in a suitability assessment for individual employees. The calculated suitability level serves as an input for the deployment planning model. The second sub-model aims to select a specific employee for a production task from the set of available and suitable workers. Depending on the time urgency of a task, the model decides whether to select employees who can further develop their competences by absolving the task or select an employee with a higher skill level set and more experience. The former will typically require more time to complete the task. Furthermore, deploying an employee with less experience increases the risk of quality defects. Once the task has been assigned, a more detailed assessment of the skill deficit occurs (training planning model). Training materials appropriate to the individual skill level are selected and provided if there is a deficit. Depending on the type and criticality of training materials, the employees must either finish training before starting a task or can use the training material during the task execution. The execution of the task is followed by the quality assessment (competence assessment model). For this purpose, the assistance system records and analyzes quality defects reported by production employees or during scheduled quality inspections and analyzes them to update the skill profiles of employees. This is the basis for adapting the training materials to change employees' skills and systematically increase their deployment flexibility.

5 Conclusion

The results of the survey answered by production managers presented in this paper indicate that a significant change in the relevance of competences in production is expected in the next ten years and that the different competences of production employees have a varying influence on processing quality. To increase reliability, the terms used were defined at the outset. Since only 16 persons participated in the survey, who cannot be assigned to the sectors of manufacturing companies according to the job advertisements evaluated to determine the required competences, the survey result cannot be assumed to be representative. Consequently, there is no general validity of the survey result. Nevertheless, the result indicates that continuous development of employees on the shop floor is necessary to be able to meet future challenges.

This is where the presented competence development model comes in. This model aims at increasing the flexibility of the employees through targeted competence-oriented qualification measures and forms the basis for sustainable personnel development with consideration of the increasing flexibility requirements. The competence development concept presented is based on four sub-models: Competency comparison, deployment planning, training planning and competency assessment. Embedded in an overarching assistance system, the models enable manufacturing companies to carry out competence-based personnel deployment planning, derive targeted qualification measures, and continuously measure and update the competence level of their employees in production. The competence development concept presented is based on four sub-models: competence comparison, deployment planning, training planning, and competence assessment. Embedded in an overall assistance system, the models enable manufacturing companies to carry out a competence-based workforce scheduling, derive targeted training measures, and continuously measure and update the competence level of their employees in production.

The tools of self-assessment and peer-assessment based on a five-point Likert scale have proven effective for assessing competences that are difficult to measure. Therefore, in addition to data-driven measurements based on quality data, these will be used for competence assessment in the further course of the research project reQenrol. In this way, companies are enabled by a work-integrated learning environment to ensure high quality and sustainably anchor knowledge even with increasing demands on staff flexibility.

Acknowledgments. The CORNET project (No. 21842N) of the Federation of Quality Research and Science (FQS), August-Schanz-Straße 21A, 60433 Frankfurt am Main, Germany, was funded by the Federal Ministry for Economic Affairs and Climate Action through the German Federation of Industrial Research Associations (AiF) under the Industrial Collective Research (IGF) programme on the basis of a decision by the German Bundestag.

References

1. acatech (Hrsg.): Kompetenzen für Industrie 4.0. Qualifizierungsbedarfe und Lösungsansätze (acatech POSITION). Herbert Utz Verlag, München. https://www.acatech.de/wp-content/uploads/2018/03/161202_POS_Kompetenz_Industrie40_Web.pdf (2016). Last accessed 12 Mar 2022
2. Bundesministerium für Wirtschaft und Klimaschutz (BMWi): Fachkräfte für Deutschland, https://www.bmwk.de/Redaktion/DE/Dossier/fachkraeftesicherung.html. Last accessed 04 May 2022
3. Boos, W.: Production Turnaround—Turning Data into Sustainability. Through the Internet of Production towards sustainable production and operation (2021)
4. Ast, J., Möhle, J., Bleckmann, M., Nyhuis, P.: Preliminary study in a learning factory on functional flexibility on workforce. In: 12th Conference on Learning Factories, CLF2022 (2022)

5. DIN EN ISO 9000:2015a: Qualitätsmanagementsysteme—Grundlagen und Begriffe (ISO 9000:2015)
6. Schmitt, R., Pfeifer, T.: Qualitätsmanagement. Strategien—Methoden—Techniken. 5th edn. Hanser, München, Wien, p. 49, 108 (2015)
7. DIN EN ISO 9001:2015b. Qualitätsmanagementsysteme—Anforderungen (ISO9001:2015)
8. Miller, T.: Empowerment konkret! Handlungsentwürfe und Reflexionen aus der psychosozialen Praxis. De Gruyter, Stuttgart (2016)
9. Kolb, M.: Personalmanagement. Grundlagen und Praxis des Human Resource Management. Gabler Verlag, Wiesbaden (2010). ISBN 978-3-8349-1853-6
10. Masing, W.: Fertigungsplanung und Fabrikeinrichtungen. Automatisierung der betrieblichen Qualitätssicherung. Zeitschrift für wirtschaftlichen Fabrikbetrieb. vol. 71, 7, p. 275 ff. (1976)
11. Geiger, W.: Qualitäts Lehre. Einführung, Systematik, Technologien. Springer Vieweg. Wiesbaden, p. 43 ff. (1994)
12. Weinert, F.E.: Definition and Selection of Competencies. Concepts of Competence. Max Planck Institute for Psychological Research, Munich, Germany (1999)
13. Lambert, B., Plank, R., Reid, D., Fleming, D.: A competency model for entry level business-to-business services salespeople. Serv. Mark. Q. **35**(1), 84–103 (2014)
14. Erpenbeck, J., Rosernstiel, L., Grote, S., Sauter, W.: Handbuch Kompetenzmessung. Schäffler-Poeschel, Stuttgart (2017)
15. Thiele, P., Müller, W.: ESCO—Entwicklung einer europäischen Taxonomie für Berufe, Kompetenzen und Qualifikationen. BWP. vol. H4, p. 37 f. (2011)
16. Bughin, J., Hazan, E., Lund, S., Dahlström, P., Wiesinger, A., Subramaniam, A.: SKILL SHIFT. Automation and the future of the workforce. McKinsey&Company (2018)
17. The Future of Jobs Report. Centre of the New Economy and Society. World Economic Forum, Cologny, Genenva (2018)
18. Motz, M., et al.: Smarte Einsatzplanung und Schulung zur Qualitätssteigerung. Zeitschrift für wirtschaftlichen Fabrikbetrieb. **116**(12), 945–950 (2021)
19. Pfeiffer, S.: Montage und Erfahrung. Warum ganzheitliche Produktionssysteme menschliches Arbeitsvermögen brauchen, 1st edn. Rainer Hampp Verlag, München (2007)
20. Völkl, K., Korb, C.: Deskriptive Statistik. Springer, Halle, p. 20 (2018)
21. Weiterbildung für die digitale Arbeitswelt. Eine repräsentative Untersuchung von Bitkom Research im Auftrag des VdTÜV e.V. und des Bitkom e.V. https://www.bitkom.org/sites/default/files/2018-12/20181221_VdTU%CC%88V_Bitkom_Weiterbildung_Studienbericht.pdf. Last accessed 8 Apr 2022
22. Senderek, R.: Lernen in der digitalisierten Arbeitswelt, pp. 195–203. Bde. Tagungsband Bigital Engineering zum Planen, Testen und Betreiben technischer Systeme (2015)
23. Abel, J., Wagner, S.: Industrie 4.0. Mitabeiterqualifizierung in KMU. wt Werkstattstechnik online. 107, vol. H. 3, pp. 134–140 (2017)

Check for updates

Very Short-Term Electric Load Forecasting with Suitable Resolution Quality – A Study in the Industrial Sector

Lukas Baur[1,2](✉), Can Kaymakci[1,2], and Alexander Sauer[1,2]

[1] Institute for Energy Efficiency in Production EEP, University of Stuttgart, Nobelstraße 12, 70569 Stuttgart, Germany
lukas.baur@ipa.fraunhofer.de

[2] Fraunhofer Institute for Manufacturing Engineering and Automation IPA, Nobelstraße 12, 70569 Stuttgart, Germany

Abstract. To reduce energy costs in manufacturing, load forecasting plays a decisive role, for example, as a central instrument for load management or in the proactive marketing of demand-side flexibility. To allow for accurate forecasts and the ability to incorporate the latest available information, very short-term load forecasting is applied. While a 15 min time resolution standard is predominant in the energy sector, much higher input resolutions are available due to modern sensors, declining data storage prices, and increasing computing power. Nevertheless, an increase in the resolution does not have to improve the forecasting accuracy since higher-resolution data series are subject to stronger stochastic fluctuations. It is unclear up to what level of resolution refinement forecast improvements can be expected. This paper systematically examines the effects of varying the input data resolution with respect to the prediction performance of the resulting load forecasting models. We propose a method for identifying the optimal resolution for very short-term load forecasting applications and validate it on electrical load data of three companies from the industrial sector, each sampled at different resolutions.

Keywords: Data resolution · Load forecasting · Forecasting accuracy · Downsampling · Manufacturing sector

1 Introduction

Electricity load forecasting has shown great importance in various fields of application like energy purchasing, transmission and distribution planning, load peak shaving, and demand-side management since wrong decisions based on forecasting errors are directly connected to financial risks [1, 2]. Especially in the context of modern manufacturing, load forecasting plays a key central role since most energy efficiency optimization measures rely on accurate forecasts [3].

The literature distinguishes between four typical horizons used for load forecasting differing in time span and the resulting requirements on data resolution. While long-term and medium long-term forecasting span horizons longer than three years and up to one

M. Liewald et al. (Eds.): WGP 2022, LNPE, pp. 686–695, 2023.
https://doi.org/10.1007/978-3-031-18318-8_68

year, short-term load forecasting horizons are not longer than two weeks [3]. Especially very short-term load forecasting (VSTLF), focusing on load forecasting horizons up to one day, motivates the use of high-resolution input data to train robust forecasting models. Other than long-load forecasting, where coarse temporal resolutions, e.g., in the order of a month [4], are used, and curves are mainly influenced by seasonal and longterm trends, forecasting tasks of short length focus on features of a fine temporal scale. According to Setiawan et al. [2], VSTLF is influenced by the three main factors *time*, date-specific *irregularities*, and *random effects*. In contrast to long-term forecasting, single working shifts and irregular days such as public holidays or weekends must be identified and modelled correctly. VSTLF is subject to high volatile fluctuations and randomness since no dampening of single outlier events or irregularities occur, because it is applied during the downsampling process of long-term forecasting using averaging. For example, system outtakes change the expected electric load profile for VSTLF completely, even if they only last a few minutes, but only play a minor role when predicting the load of a whole month.

With the advent of powerful smart meters, well-developed IT infrastructures in today's companies, and increased bandwidths, high-resolution data can be recorded, transmitted, and processed [5]. From a data analysis perspective, it is advantageous to store sensor-measured data in the highest resolution available since coarser data instances can be calculated subsequently as needed.

In contrast, transferring data of higher quality than needed is a waste of data storage, energy consumption, and calculation time. For example, using recorded energy data on a nanosecond time scale will give no significantly better results in forecasting energy consumption on an hourly scale than using an instance of minute-wise aggregated data. Furthermore, refining the resolution by a factor of 0.5 doubles the number of input samples for preprocessing and training. Additionally, high-resolution sampling requires more complex models as the number of input features increases. Lastly, downsampled time series fluctuate less, have smaller variance and dampen outliers due to the averaging of neighbouring measurements.

Research is necessary to balance this tradeoff between high-resolution data and acceptable costs in terms of data storage, training, and processing times: A reasonable sampling resolution choice should represent the time series fine enough to accomplish the desired task within the required quality bounds but should be as coarse as possible to save resources.

In the literature, many different resolutions have been used, ranging from a second scale to an hourly resolution [6]. In the industrial sector of Germany, a temporal resolution of 15 min is obligatory for companies exceeding 100,000 kWh annual electricity consumption [7].

1.1 Related Work

In the context of downsampling, [8] artificially created time series to research on the forecast quality with respect to different pre-smoothing and downsampling levels. To include time series of different resolutions, load forecasting was performed in [9] using a multi-resolution approach combining low-level characteristics from hourly with high-resolution features from half-hourly data.

Wavelet transform or its discrete variant (DWT), as well as Fourier Analysis, has been applied in several load forecast preprocessing pipelines to extract time series features on various scales, as used in [10–13]. In [14], a hybrid load forecasting framework is presented that makes use of DWT to split the load input into several levels of resolution based on an entropy criterion. Reference [15] uses the Hurst exponent for measuring the predictability in a multi-resolution setup and showed that in a wind speed forecasting setup a data elimination of 29% of the input signal components does not reduce the prediction performance.

1.2 Research Gaps and Contributions

While the mentioned multi-scale methods extract features on different resolution scales implicitly using frequency decomposition during preprocessing of the high-resolution data set, to our best knowledge, none of the methods directly uses a subset of differently sampled instances, which would provide insight into the storing and processing resolution to be proposed. Nevertheless, from a data acquisition perspective, it is unclear which resolution should be chosen to still expect satisfactory forecasting results for a specific use case.

This work presents a framework for analyzing historical load series data to help determine a resolution choice used for further storing and modelling and supports identifying the optimal resolution for forecasting, if available. Because of its high relevance in electricity applications (compare [2]), this paper focuses on VSTLF for electric loads to evaluate the framework. For the evaluation, energy consumption data of three companies from the German industrial sector are used.

2 Methodology

This chapter first introduces the forecasting setup with its inputs, expected outputs, and meta parameter definitions. The setup defines the base on which the framework presented in the second section is built on.

2.1 Forecasting Setup

The overall forecasting setup is sketched in Fig. 1 and will be explained in more detail in the following two subsections. Wherever applicable, the notation was adopted from [7].

Reference System. On an abstract view of forecasting, each model receives a fixedlength historical input data slice (Fig. 1: highlighted in red) to predict a fixed number of forecasts in the future (Fig. 1: highlighted in green). According to the forecasting task, a temporal target resolution $\Delta t^{\max}$ is defined first, which has to be an integer multiple of the measured load resolution $\Delta t^{\min}$:

$$\Delta t^{\max} := \Delta \mathrm{t_k},\ \text{fixed}\, k \in \mathbb{N} \tag{1}$$

using the more general notation

$$\Delta t_i = i \cdot \Delta t^{\min} \tag{2}$$

The forecasting period z is defined by s_{out}-many target resolution time steps. Consequently, the length of the forecasting period equals $z = s_{\text{out}} \cdot \Delta t^{\max}$. The input period has a fixed length p, which is again defined with respect to the target resolution:

$$p = S_{\text{in}} \cdot \Delta t^{\max}. \tag{3}$$

Given the original load data series S^1 with time resolution $\Delta t^{\min}$, coarser load data series can be generated using downsampling. A downsampled version of S^1 is defined as S^i if it has an equidistant time resolution of Δt_i. There is an upper bound on i, because increasing i further than k would imply that the output resolution is of higher detail than the input resolution of the model, which cannot give better results. The output resolution $\Delta t^{\max}$ exactly matches the input resolutions in the boundary case of i being k, as it is shown with the blue curve in Fig. 1.

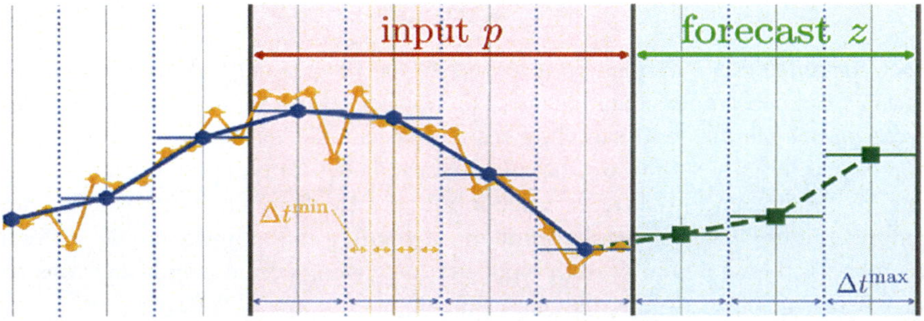

Fig. 1. Illustration showing the forecasting setup including the original load measurement series S^1 (orange) recorded with equidistant time resolution $\Delta t^{\min}$, its resampled version S^4 (blue) and the forecast (green) with target-resolution $\Delta t^{\max}$. Here, $s_{out} = 3$, $s_{in} = 4$, and $k = 4$.

Time Series Slicing. The framework presented in the upcoming chapter requires the splitting and combining of different time series into a list of windowed time slices. This section defines such a window.

When a forecast is made, data from the historical input of length p is used to forecast values representing a future time period of length z and load measurement data is used for model training. The number of output values s_{out} is constant for all iterations, whereas the number of input values differs depending on the choice of input resolution.

In case of an input resolution Δt_i, $\frac{p}{\Delta t_i} \overset{(2)}{=} \frac{p}{i \cdot \Delta t^{\min}} \overset{(3)}{=} \frac{S_{\text{in}} \Delta t^{\max}}{i \cdot \Delta t^{\min}} \overset{(1,2)}{=} \frac{S_{\text{in}}(k \cdot \Delta t^{\min})}{i \cdot \Delta t^{\min}} = \frac{S_{\text{in}} \cdot k}{i}$ many steps from the corresponding load data series S^i are used.

To sum up, for training the model with input resolution Δt_i, windows with $\frac{S_{\text{in}} \cdot k}{i}$ many input values from S^i, and s_{out} many output values from S^k are used.

2.2 Framework

A framework for the systematic analysis of the influence of varying the sampling resolution on the forecasting quality is presented with the pseudo-code in Program Code 1 using the notation introduced in the previous subsection.

Program Code 1. Abstract framework for analyzing the influence of varying the sampling resolution on the forecasting quality.

1: Define a target resolution Δt^{max} with respect to the forecasting task and obtain S^k
2: Select input resolution candidates $R \subseteq [\Delta t^{\text{min}}, \Delta t^{\text{max}}]$
3: Calculate downsampled load series S^i for resolutions $\Delta t_i \in R$
4: Select a model-family $\mathcal{M}$
5: **for each** $\Delta t_i \in R$ **do**
6: Train a model $m \sim \mathcal{M}$ using lagged input values from S^i and target values from S^k
7: Evaluate m with metric d: $\rho_{M,\Delta t_i,1}$
8: Repeat (6) and (7) w times to obtain $\rho_M(\Delta t_i) = [\rho_{M,\Delta t_i,1}, \dots, \rho_{M,\Delta t_i,w}]$
9: **end for**
10: Repeat (5) - (9) for different model family $\mathcal{M}'$
11: Analyze the set of ρ_M scores for minima, monotonicity, convexity, and common characteristics

---

Meta-parameters are defined with respect to the forecasting task in a setup phase (lines 1–2) and subsampled data instances are generated using averaging (line 3). For different model families and sampling resolutions, model instances are trained, and the forecasting quality scores are measured (lines 4–10). Training and evaluation are repeated w times each to reduce variance introduced by the nondeterministic learning procedures. Figure 2 visualizes the intuition of training model instances on different inputs for a fixed model family: Both input and forecasting period length stay constant, but increasing the data resolution increases the number of model features.

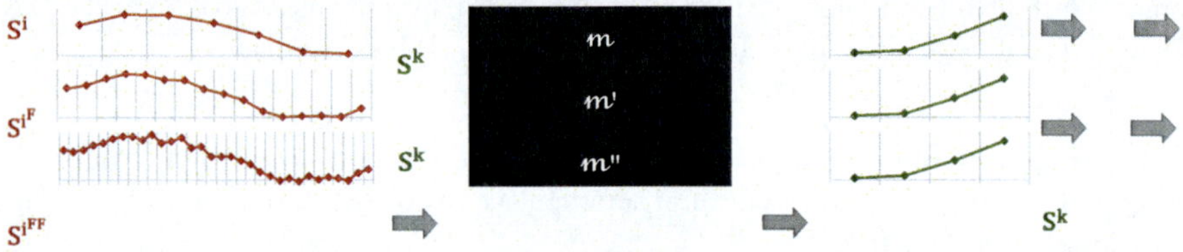

Fig. 2. Abstract sketch for training models of one common model family on different scales: the input data resolution Δt_i (red) is varied while the output resolution (green) remains constant. All models (black) are of the same model family.

3 Use Case Study

To show the framework's potential, it is applied and evaluated on load data of three companies in the industrial sector using Python.

3.1 Use Case Setup

Company Descriptions. The presented method is applied to the load data of three manufacturing companies from the industrial sector located in Germany denoted as

Company 1 to 3 in the following. For the evaluation, two years of electric load data has been gathered at an equidistance temporal resolution of one minute ($\Delta t^{\min} = 1$ min). Data gaps have been closed in a preprocessing step using linear interpolation. Table 1 gives an overview of the company data used in the study.

Table 1. Case study company load data characteristics

Company	Manufacturing sector	Recorded time	% Missing data	Peak/average load (MW)
1	Enclosures for electrics	2019–07–01 to 2021–06–30, 1,051,200 samples	0.4	0.332/0.159
2	Radial fan impellers		0.0	0.657/0.299
3	Glass		2.5	1.059/0.387

All use cases are from the manufacturing and processing industries, but they differ essentially in terms of power consumption. Figure 3 shows the companies' load profiles. Companies 1 and 2 process metal, while company 3 processes glass. Company 1 has a photovoltaic system that generates electricity for use in production.

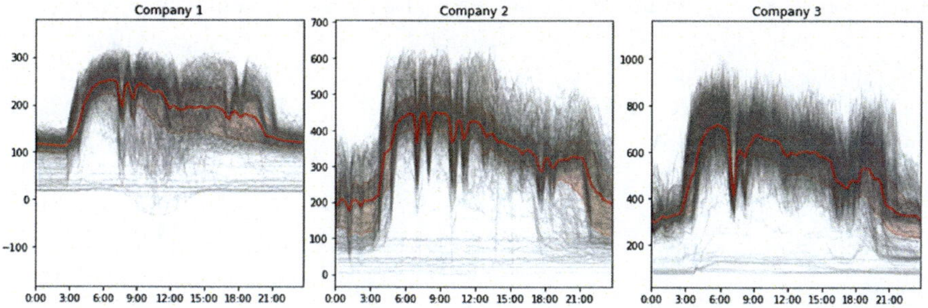

Fig. 3. Typical weekday electric load profiles for the three use case companies. The thick red line corresponds to the median load curve, the thin lines mark the 25 and 75 quantile transitions.

Forecast Requirements and Input Features. The task is to forecast the next hour ($z = 60$ min) in one step ($s_{out} = 1$) given the load data of the last 24 h ($p = 24 \cdot 60$ min). Hence, the maximum resolution equals 1 h ($\Delta t^{\max} = 60$ min). Since the history interval length stays constant for all input resolutions, the number of data points increases as the resolution decreases. Besides the endogenous lagged data inputs, eight exogenous timestamp features of the target time such as *day of the week*, *week of the year*, *hour of day*, etc., are provided to the models at forecasting time. Besides the minimum resolution of one minute from the input data and the maximum resolution of 60 min given by the target resolution, there are ten more resolutions dividing the target resolution evenly:

$$\mathrm{R} = [1, 2, 3, 4, 5, 6, 10, 12, 15, 20, 30, 60].$$

The total numbers of input features are listed in Table 2.

Table 2. Number of features for each resolution

Δt_i	1	2	3	4	5	6	10	12	15	20	30	60
#time	1440	720	480	360	288	240	144	120	96	72	48	24
#total	1448	728	488	368	296	248	152	128	104	80	56	32

Model Families. In the literature, short-term forecasting models are classified into *classical* and *machine learning models* [16]. To cover both classes, two representative classical model families and two machine learning families have been chosen. First, the classical set is covered by a multivariate linear repressor (MLR) and its regularized Lasso [17] variant, since they are intuitive and scale well with the number of features. To estimate reasonable regularization parameters, a parameter search with cross-validation is performed (LassoCV). The best model is returned and used for forecasting. Similarly, a Decision Tree Regressor model (DTR) is learned in a grid-search crossvalidation fashion varying the maximum leaf nodes (GSCV-DTR). It was chosen because of its straightforward interpretation. Finally, a multilayer perceptron (MLP) is trained as a second machine learning representative, as feedforward artificial neural networks are still popular on the application of forecasting [6].

Experiment Setup. The mean absolute error (MAE) is used to evaluate the model forecasts, because of its intuitive interpretation and its wide acceptance in electric load forecast research [6]. A relative metric was not applicable since target values $y_i \in y$ of the ground truth series can be zero, which would have resulted in undefined values. The MAE is defined between the two vectors y and its corresponding forecast $\hat{y}$, both having the same size [2]:

$$MAE(y, \hat{y}) = \frac{1}{|y|} \sum_{i=1}^{|y|} |y_i - \hat{y}_i|.$$

The repetition parameter was chosen to be $w = 50$ to cover model initialization fluctuations. To find a reasonable value, this parameter was increased until stable lower and upper prediction bounds were obtained. While repetition plays an important role for neuronal networks and partially for cross-validation, it does not change the results for deterministic methods like MLR. Data is split into train and test sets with $\alpha = 0.75$, i.e., the last quarter was used for evaluation. This choice of α is typical in this context [6].

3.2 Experimental Results

The results of applying the framework to the three companies are shown in Fig. 4.

For Company 1 and 2, error curves monotonically decrease for all linear models and MLP until resolution fall below 6 and 2 min respectively. For Company 1 the smallest error was measured with a 30 min resolution using a regression tree. Company 2 reaches its minimum using neuronal network at finest resolution. For Company 3, DTR and MLP curves are saturated for values between 30 and 3 min.

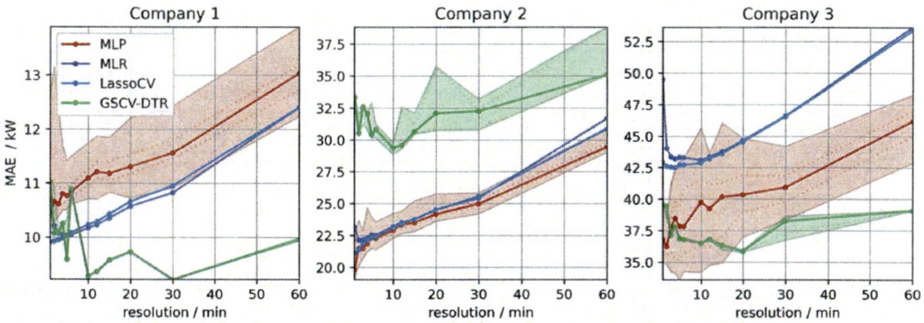

Fig. 4. Test set errors for each company: median (solid), upper and lower quartiles (dashed), minimum, and maximum evaluation errors for the three use-case companies.

The model families show similar behaviors across all companies: LassoCV and MLP strongly correlate with the input resolution. While the linear MLR is very similar to its regularized version, its error again increases for values smaller than 2–5 min, showing nearly convex shapes with single minima. The DTR instances perform worst for high and low resolution choices and have at least one optimal resolution value between 10 and 30 min.

The measured median errors differ for the three use cases in a magnitude and variance, ranging from 9–14 kW, 19–36 kW, and 35–54 kW for Company 1 to 3 respectively. Taking only the best model for each resolution into account, the errors vary between 10.0–11.1 kW, 20–29kW, and 36–39kW. The differences in prediction qualities indicate an increasing forecasting difficulty across companies 1–3. All models have in common that the error curves decrease for resolutions smaller than the reference resolution of 60 min.

Based on the median performances of the models used in this setup, choosing a high resolution is recommended for use case 2, whereas lower resolutions are sufficient for use cases 1 and 3: Company 1 reaches its lowest error on 30-min resolution input, Company 3 at 20 min, but single 5-min trained MLP instances perform better. This indicates more abstractly that the resolution must be chosen use-case specific.

4 Conclusion

In the context of load forecasting, there is a tradeoff between expensive storing, time- and energy-consuming model training, and preprocessing when using high-resolution load measurement data. On the other hand, information loss through downsampling high-resolution data leads to worse forecasting qualities. To address this problem, an abstract framework for analyzing the use-case-specific dependency of measurement resolution on the prediction quality using benchmark models is presented and evaluated, based on the data of three companies from the industrial sector.

Although the specific model choice typically influences the forecasting error more than the resolution choice, depending on the use-case-specific time series characteristics, it is highly recommended to analyze the resolution once a model class is fixed. The framework helps to identify a reasonable resolution choice. To our findings, for decision tree regressors, the input resolution plays a minor role, and for linear and neuronal network models, the resolution choice matters. The study shows that naively choosing the highest available resolution does not give the best results in general, but choosing a resolution smaller than 60 min in the given hour-ahead forecasting setup is recommended. Once a suitable resolution has been found using this method, more accurate forecasts can be expected, which in turn improves the energy efficiency measures used for manufacturing optimization.

Choosing higher resolutions increases the number of input features resulting in more complex models. In this work, optimizing the models with respect to their complexity has been applied only partially using Grid Search. Exploring this potential, hidden complexity influence in full detail is part of future research.

References

1. Hong, T.: Short term electric load forecasting. North Carolina State University (2010)
2. Setiawan, A., Koprinska, I., Agelidis, V.G.: Very short-term electricity load demand forecasting using support vector regression. In: 2009 International Joint Conference on Neural Networks. 2009 International Joint Conference on Neural Networks (IJCNN 2009—Atlanta), Atlanta, Ga, USA, 14.06.2009–19.06.2009, pp. 2888–2894. IEEE (2009). https://doi.org/10.1109/IJCNN.2009.5179063
3. Walther, J., Spanier, D., Panten, N., Abele, E.: Very short-term load forecasting on factory level—a machine learning approach. Procedia CIRP (2019). https://doi.org/10.1016/j.procir.2019.01.060
4. Mamun, M.A., Nagasaka, K.: Artificial neural networks applied to long-term electricity demand forecasting. In: Fourth International Conference on Hybrid Intelligent Systems (HIS'04), pp. 204–209. IEEE
5. Kabalci, Y.: A survey on smart metering and smart grid communication. Renew. Sustain. Energy Rev. (2016). https://doi.org/10.1016/j.rser.2015.12.114
6. vom Scheidt, F., Medinová, H., Ludwig, N., Richter, B., Staudt, P., Weinhardt, C.: Data analytics in the electricity sector—a quantitative and qualitative literature review. Energy AI (2020). https://doi.org/10.1016/j.egyai.2020.100009
7. Walser, T., Reisinger, M., Hartmann, N., Dierolf, C., Sauer, A.: Readiness of short-term load forecasting methods for their deployment on company level. In: Proceedings of GSM 2020, pp. 89–103
8. Romanuke, V.: Time series smoothing improving forecasting. Appl. Comput. Syst. (2021). https://doi.org/10.2478/acss-2021-0008
9. Amara-Ouali, Y., Fasiolo, M., Goude, Y., Yan, H.: Daily peak electrical load forecasting with a multi-resolution approach. https://arxiv.org/pdf/2112.04492 (2021)
10. Bashir, Z.A., El-Hawary, M.E.: Applying wavelets to short-term load forecasting using PSO-based neural networks. IEEE Trans. Power Syst. (2009). https://doi.org/10.1109/tpwrs.2008.2008606
11. Pandey, A.S., Singh, D., Sinha, S.K.: Intelligent hybrid wavelet models for short-term load forecasting. IEEE Trans. Power Syst. (2010). https://doi.org/10.1109/tpwrs.2010.2042471
12. RochaReis, A.J., AlvesdaSilva, A.P.: Feature extraction via multiresolution analysis for short-term load forecasting. IEEE Trans. Power Syst. (2005). https://doi.org/10.1109/tpwrs.2004.840380

13. Eljazzar, M.M., Hemayed, E.E.: Enhancing electric load forecasting of ARIMA and ANN using adaptive Fourier series. In: 2017 IEEE 7th Annual Computing and Communication Workshop and Conference (CCWC). IEEE (2017). https://doi.org/10.1109/ccwc.2017.7868457
14. Ghayekhloo, M., Menhaj, M.B., Ghofrani, M.: A hybrid short-term load forecasting with a new data preprocessing framework. Electric Power Syst. Res. (2015). https://doi.org/10.1016/j.epsr.2014.09.002
15. Doucoure, B., Agbossou, K., Cardenas, A.: Time series prediction using artificial wavelet neural network and multi-resolution analysis: application to wind speed data. Renew. Energy (2016). https://doi.org/10.1016/j.renene.2016.02.003
16. Walser, T., Sauer, A.: Typical load profile-supported convolutional neural network for short-term load forecasting in the industrial sector. Energy AI (2021). https://doi.org/10.1016/j.egyai.2021.100104
17. Tibshirani, R.: Regression shrinkage and selection via the Lasso. J. Roy. Stat. Soc.: Ser. B (Methodol.) (1996). https://doi.org/10.1111/j.2517-6161.1996.tb02080.x

Approach to Develop a Lightweight Potential Analysis at the Interface Between Product, Production and Material

S. Zeidler(✉), J. Scholz, M. Friedmann, and J. Fleischer

wbk Institute of Production Science, Karlsruhe Institute of Technology (KIT), 76131 Karlsruhe, Germany

simon.zeidler@kit.edu

Abstract. In this article, a methodology for estimating both the product and the production-side lightweight design potential is presented, which can be used at an early stage of the product development process due to the limited amount of data required. This can help companies to increase the performance of their production facilities through the proper use of their potential and, on the other hand, to identify the lightweight construction potential in their products. This allows for faster integration of lightweight construction in sectors not typically associated with lightweight construction due to the reveal of hidden possibilities in production and ultimately leads to resource savings in industry. For this purpose, possible influencing factors and existing potential analyses are examined first, the requirements for a methodology in the early phase of product development are analyzed and the use cases of calculation for a given component and calculation without a determined component are identified. From the information obtained, a linkage and relevance analysis is used to derive key factors influencing the lightweight design potential of the product and production. The methodology is developed on the basis of these key factors, with a division into potentials of geometric and material lightweight design. Parameters from both areas and their effects on product design and production were taken into account. The lightweight design potential of the production equipment and products is then given as a percentage of the optimal degree of fulfillment.

Keywords: Lightweight design · Methodology · Linkage and relevance analysis

1 Motivation

Environmental awareness is increasing in the society for years. Lightweight construction is a key technology for greater resource efficiency, but apart from the aviation and automotive industries, the advantages of lightweight design are not directly measurable. The relevance in other fields of industry like machine and plant engineering, medical technology or leisure industry has started to increase by a single digit percentage only in the recent past and is still faced with skepticism [1, 2]. The named industries have the missing readiness of the customer in common to pay more for lighter products. Therefore, lightweight solutions are often discarded due to their high material and manufacturing

M. Liewald et al. (Eds.): WGP 2022, LNPE, pp. 696–704, 2023.
https://doi.org/10.1007/978-3-031-18318-8_69

costs. To avoid late expensive changes in the product development, an adjusted V-Model was developed withing the research project "SyProLei". With this methodology, the development is an interaction between the domains product, material, production and joining technology from the beginning of the design process until the end of the process [3]. To support the interaction between the domains, a method is missing to identify the potential of manufacturing processes for producing lighter products in an early stage of development and without huge expert knowledge. In Sect. 2, existing analysis approaches in manufacturing are described, followed by the development approach for the presented method and the lightweight potential analysis itself (Sect. 3). At the end, a conclusion is given and an outlook regarding further development of the method.

2 State of the Art

For potential analyses exist a couple of well-established methods. A method known to a large number of applicants is, for example, the spider web diagram. To analyze potentials on a deeper and more specific level there exists a broad variety of specialized tools. In this paper, two potential analyses approaches are presented.

Schmidt [4] focuses on possible weight reduction of parts due to the geometric design freedom when produced by additive manufacturing processes. To achieve this, the two factors of minimal weight and utilization factor are introduced. The utilization factor is hereby derived via a FE-simulation. Additionally, Schmidt takes functional and monetary advantages into account.

In [5, 6], an automation potential analysis is presented which primarily targets the assembly. For this purpose, a systematic analysis of the used processes is used to identify processes technically as well as economically suitable for automation. This basis allows for a derivation of suitable fully or partly automated processes by the means of predefined characteristics. The automation potential analysis shows a possible solution on how an industry-ready solution can be realized. As seen, only specific methods to estimate the lightweight design potential exist as shown by Schmidt. Other potential analysis identified primarily target the automation of production plants. These already proven methods can be used as a basis for a transfer to method to estimate the general lightweight design potential.

To develop a methodology, research has shown that parameters for the assessment have to be identified [7, 8]. A thorough literature review was conducted for this purpose.

3 Approach

As the presented methods in the state of the art shows great potential for supporting the engineer in analyzing the specific cases a methodology is developed to identify the potential of manufacturing processes for the design of lighter products.

3.1 Requirements and Key Factor Identification

First, requirements of the early phase of product development like limited availability of information which restricts the possible parameters, have been identified and noted as boundaries to be considered. Then, 50 factors with an influence on the lightweight design potential have been identified. These can be clustered into production-, lightweight-strategic- and part-specific factors. The aspect of recyclability has been added due to its increasing relevance. In Table 1 some factors of the categories are shown as an example to get an idea which parameters are used. The recyclability is not mentioned in the table because it is used as one factor.

Table 1. Excerpt of the 50 evaluated factors

Production	Lightweight-strategic	Part specific
Flexibility	Conditional lightweight construction	Temperature
Geometry	Material lightweight design	Stress
Manufacturing process	Geometric lightweight design	Material
Volume	Concept lightweight design	Geometry

As research showed have strong interdependencies, the product and production system and therefore the lightweight design potential shows a highly dynamic reaction on changes of one parameter. As a direct result, key factors have to be identified to make the system complexity manageable. The aim is a reduction on the system relevant factors which determine the lightweight design potential. To a achieve a systematic reduction and analysis of the system parameters and to consider the system dynamic created by the interdependencies of the factors, an assessment with a linkage- and a relevance-analysis was chosen [9]. For this, all identified factors span a *m*x*m* matrix where *m* is the number of factors. An excerpt is shown in Fig. 1.

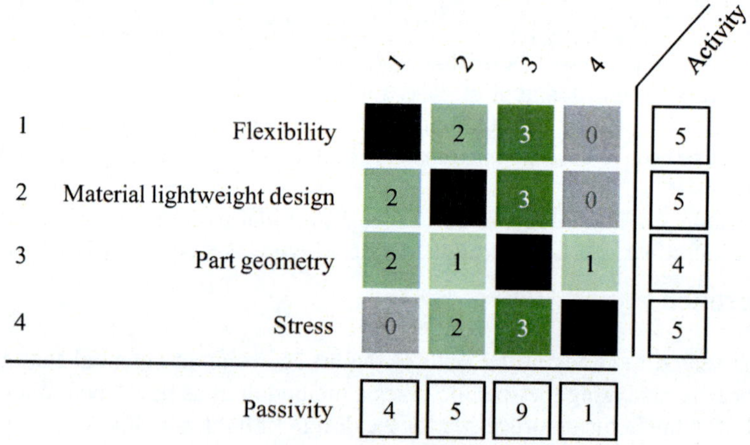

Fig. 1. Example of the linkage analysis

All factors are evaluated on their influence on the other factors following the convention row influences column. For this evaluation, the influence has been classified in four stages. A "0" is given, when no direct influence is occurring, while "1" represents a weak and delayed influence. Value "2" indicates an influence while "3" indicates a strong and direct influence. With this, an initial assessment of the parameters influence is possible. In the shown example the flexibility has influence on the parameter material lightweight design and strong influence on the part geometry. Additionally, the flexibility is influenced by the two factors. Yet, indirect effects have not been taken into account. These can be considered by an effect chain analysis, where "closed loop" influences are assessed. The sum of a row is called the activity and represents the influence of the investigated factor on the system. The sum of a column is called passivity and represents the influence of the system on the investigated factor. With these metrics, the parameters' role in the system can be identified. In the graphical representation a ranking is used rather than the absolute values of the activity and passivity to achieve a quadratic grid. Figure 2 shows the factors in the activity-passivity-grid.

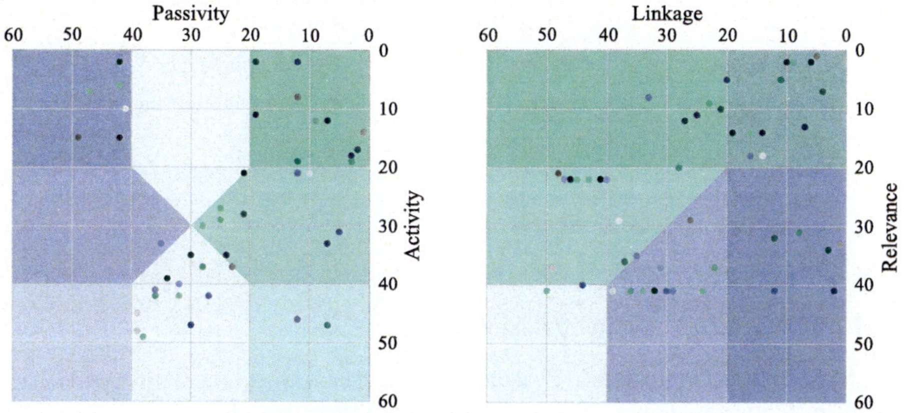

Fig. 2. Resulting activity-passivity-grid (left) and linkage-relevance-grid (right) of the impact factors of lightweight design potential

Depending on their position in the left grid, the factors can be classified. Elements with a high activity and a low passivity are called system levers and are represented by the dark blue area in the upper left corner. Elements with a high activity and a high passivity are called system knots and are represented by the dark green area in the upper right corner. Elements with a low activity and high passivity are called system indicators and are represented by the light green area in the bottom right corner. Elements with a low activity and low passivity are called independent factors and are represented by the light blue area in the bottom left corner. Examples for levers are boundary conditions like the load. For knots, there are the lightweight strategy and the type of production. An indicator is for example, recyclability. Key factors should fulfill three criteria:

1. Have a strong linkage (system knots) to depict a large part of the system dynamic.
2. Target the central topic and show a high relevance (Fig. 2 right) for the design field.
3. Include the central levers (system levers in Fig. 2 left).

Out of the activity-passivity-grid the linkage-relevance-grid is build up with the parameter of relevance, that represents the distance to the central topic. The relevance is identified via an interview with experts in the lightweight area and again consists of integers in the area [1 to m] with the addition of 0, which represents an elimination. The linkage is calculated by a multiplication of the activity and passivity. Again, a ranking is used to achieve a quadratic grid that is shown in Fig. 2 (right). The highest linked (green areas) and most relevant (blue areas) factors are found in the upper right corner and are called safe key factors for lightweight design. These safe key factors are supplemented by factors with a high leverage and linkage. The method will be built up with these in Table 2 depicted factors. The boundary conditions consist of load type, temperature, design space, tolerances and permitted stress or strain.

Table 2. The resulting safe key factors and factors possessing a strong leverage or linkage

Strong leverage	Safe key factors	Strong linkage
Boundary conditions	Process type	Material lightweight design
	Geometry	Geometric lightweight design
	Machine flexibility	
	Material	

3.2 Lightweight Design Potential Analysis

With these key factors identified the methodology will be build up out of three main parts. Part and machine parameters have been identified to strongly influence the lightweighting potential. Therefore, information about them will be gathered by a questionnaire. The needed information regarding machines and materials are already classified and researched for example in [10–13], this information can be stored in a database. Lastly, calculations of the lightweight design potential have to be developed.

Table 3. Excerpt of parameters asked in the questionnaire

General	Machines	Part
Production volume	Process	Material
Part considered	Maximum dimensions	Dimensions
Lightweight design strategy	Degrees of freedom	Volume of material
Recyclability considered		Machining directions
Costs considered		Load type
		Design space
		Operating temperature
		Tolerances

The information about the part to be produced and the machines to produce it vary between possible use cases. Therefore, a questionnaire has to be developed that allows for a collection of all relevant parameters, which are presented in Table 3.

Due to the identified complexity of the lightweight design potential, an approach was chosen to part the potential into subpotentials as seen in Fig. 3. These subpotentials are calculated independently. These subpotentials and the further subordinate calculations will be added together by a weighted mean. The weighting was determined by a survey among experienced engineers within the project consortium. Hereby, a consensus across different industries was observed.

As every potential need a reference, two usecases have been identified: Firstly, the case of a comparison between the potential of different machines. For example, this is the case when the decision between an invest into a machine for future products has to be made. Secondly, the potential of the already owned machines is not fully understood and a certain product should be optimized for this machine. Therefore, depending on the case, two references have been identified. For the first case, the reference is defined per property as the maximum of the machine properties amongst all possible machines. In the second case, the reference is simply the available machine. Equation 1 shows the calculation exemplary for the geometric flexibility. Equation 2 shows the calculation of a potential out of several subpotentials.

$$Potential_{geom} = \frac{\text{complexity}_{geom,part}}{capability_{geom,machine}} \tag{1}$$

$$Potential = \sum_{i=1}^{n} w_i * Subpotential_i \tag{2}$$

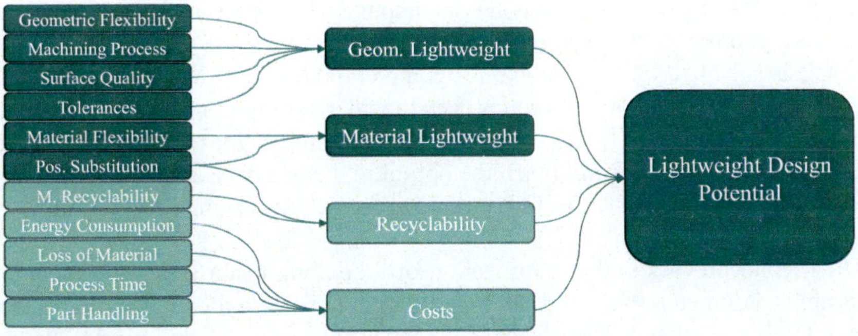

Fig. 3. Structure of the parameters for the lightweight design potential calculation

To verify this methodology, a prototype of an automated tool has been built in excel. Hereby the user adds the parameters already described in Table 3 into a mask. On another sheet, a database was built. This database contains information about selected machinery that has been investigated and evaluated for material and geometrical flexibility, tolerances, processable materials, undercuts, symmetry, surface quality, process and energy consumption. While the material database can be implemented by the means of existing material databases, a simplified one has been built up for testing, consisting of parameters Young's modulus, shear modulus, strength, stiffness, density, operating temperature as well as costs and recyclability. A material substitution can be calculated as in Eq. 3 for bending with regard to stiffness, depending on whether stiffness or strength is the relevant design parameter.

$$\Delta V_B = \frac{\sqrt{E_{original}}}{\sqrt{E_{subst}}} \tag{3}$$

For complex load, a direct estimation is not possible without a thorough analysis. Therefore, the assumption was implemented, that a complex load leads to the worst-case change in volume out of the considered parameters tension/compression, bending and denting.

The results are displayed by a percentage of fulfillment compared to the corresponding reference. There also the two worst fulfilled factors per subpotential are displayed and recommendations for action to tackle these flaws are given for these factors.

4 Case Study and Discussion

Two cases with which the method is to be tested will be discussed. First a CNC-milling process and a SLM process are to be compared. The comparison is viable, as both processes propose a high lightweight design potential. For both processes a maximum dimension of 1000 × 1000 × 1000 mm has been assumed. For the resolution, typical resolutions of IT6 for milling and 0.02 mm for the SLM process were chosen. While the milling process possesses 5 degrees of freedom, the SLM process possesses 6. While both processes showed a high suitability for lightweight design, the results showed the milling process (90.91%) to have a higher potential than the SLM process (82.98%), when only considering geometric and material lightweight design. This is mainly caused by the difference in the material lightweight design potential as the SLM process is limited to metals, while the milling process can process a wider variety of materials. As expected, the potential of geometric lightweight design is a bit higher for the SLM process, even though with 96.25% it does not reach the optimum there. This is due to the needed post processing of functional surfaces by a cutting process. The milling process follows with 94.69%.

The second process is the optimization of a bending beam as a well-known part. The beam is assumed to be 100 × 500 × 100 mm, have no undercuts and be made out of steel. Design parameter is stiffness. Tolerances and surfaces don't need to be highly accurate. Again, the milling process is compared to the SLM process. The potential of lightweight design is used to 40.78% on the milling process and to 33.78% with the SLM process. Recommended actions are an increased usage of the geometric and resolution capabilities of the processes for the geometric lightweight design. For the material lightweight design, a substitution and a check for the usage of fiber reinforced materials is recommended. Following the recommendations, a topology optimization (Fig. 4) increases the usage of the geometric potential drastically.

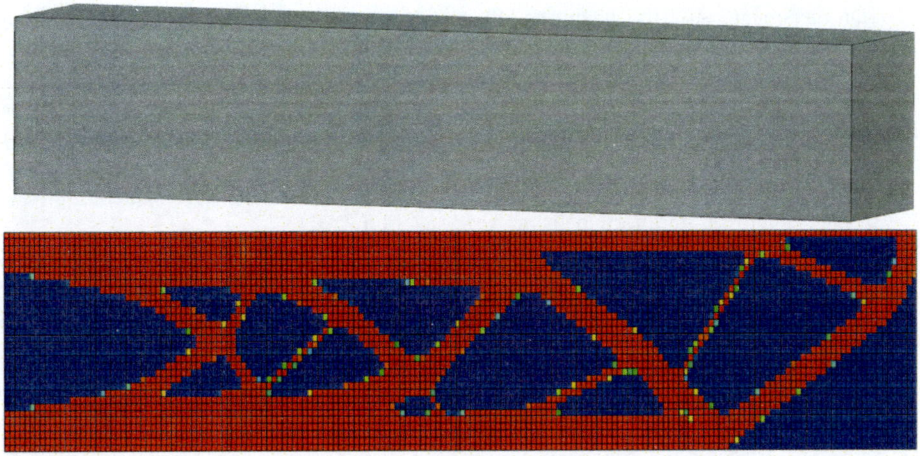

Fig. 4. CAD-model of the beam (top) and its 2D-topology optimization (bottom) with red representing part material and blue representing removed material

Even though the percentages suggest an exact potential, the results are in reality to be interpreted by the user and strongly case dependent. Nevertheless, the guidelines to increase lightweight design are functional and fulfill their purpose of leading the engineer to a more thought-out solution by pointing out unused potential.

5 Summary and Outlook

The paper presented an approach to develop a lightweight potential analysis at the interface between product, production and material. Based on a literature research and activity-passivity-analysis and a linkage-relevance-analysis key factors for the lightweight potential were identified. Based on the key factors a questionnaire for the interaction with the user were developed. Together with the information regarding product structure and manufacturing process the lightweight design potential is calculated based on a database regarding manufacturing processes and material. The developed questionnaire requires expert knowledge regarding the analyzed parts and some of the information is difficult to determine manually. Furthermore, the calculation of the part stress is inaccurate. For this purpose, a connection of the CAD-Model of the analyzed part as well as a FE-Simulation with the presented method would lead to an easier usage of the tool. Additionally, adding cost models to the manufacturing processes will also represent the economic effects of lightweight design in the model.

References

1. Fleischer, J., et al.: Leichtbau—Trends und Zukunftsmärkte und deren Bedeutung für Baden-Württemberg: Eine Studie im Auftrag der Leichtbau BW GmbH Koordination Fraunhofer-Institut für System- und Innovationsforschung ISI. Accessed 13 Sep 2021

2. Hansmersmann, A., Birenbaum, C., Burkhardt, J., Schneider, M., Stroka, M., Angabe, K.: Leichtbau im Maschinen-, Anlagen- und Gerätebau: Herausforderungen—Potenziale—Mehrwerte—Beispiele. Accessed 13 Sep 2021
3. Scholz, J., et al.: Konzept eines systemischen Entwicklungsprozesses zur Hebung von Leichtbaupotenzialen. Zeitschrift für wirtschaftlichen Fabrikbetrieb **116**(11), 797–800 (2021). https://doi.org/10.1515/zwf-2021-0182
4. Schmidt, T.: Potentialbewertung generativer Fertigungsverfahren für Leichtbauteile. Springer Berlin Heidelberg, Berlin, Heidelberg. Accessed 5 July 2021
5. Burger, N., Demartini, M., Tonelli, F., Bodendorf, F., Testa, C.: investigating flexibility as a performance dimension of a manufacturing value modeling methodology (MVMM): a framework for identifying flexibility types in manufacturing systems. Procedia CIRP **63**, 33–38 (2017). https://doi.org/10.1016/j.procir.2017.03.343
6. Neb, A., Schoenhof, R., Briki, I.: Automation potential analysis of assembly processes based on 3D product assembly models in CAD systems. Procedia CIRP **91**, 237–242 (2020). https://doi.org/10.1016/j.procir.2020.02.172
7. Prüß, H., Stechert, C., Thomas, V.: Methodik zur Auswahl von Fügetechnologien in Multimaterialsystemen (2010)
8. Kerbrat, O., Mognol, P., Hascoet, J.-Y.: Manufacturability analysis to combine additive and subtractive processes. Rapid Prototyp. J. **16**(1), 63–72 (2010). https://doi.org/10.1108/13552541011011721
9. Fink, A., Siebe, A.: Handbuch Zukunftsmanagement: Werkzeuge der strategischen Planung und Früherkennung, 2nd edn. Campus, Frankfurt am Main (2011)
10. Brecher, C., Weck, M.: Werkzeugmaschinen Fertigungssysteme, vol. 1. Springer Berlin Heidelberg, Berlin, Heidelberg (2019). Accessed 11 Oct 2021
11. Henning, F., Moeller, E., (eds.): Handbuch Leichtbau: Methoden, Werkstoffe, Fertigung. Carl Hanser Verlag GmbH & Co. KG, München (2011). Accessed 2 July 2021
12. Klocke, F., König, W.: Fertigungsverfahren 1: Drehen, Fräsen, Bohren, 8th edn. Springer-Verlag, Berlin, Heidelberg (2008)
13. Fritz, A.H., Schulze, G. (eds.): Fertigungstechnik, 11th edn. Springer Vieweg, Berlin, Heidelberg (2015)

Improving Production System Flexibility and Changeability Through Software-Defined Manufacturing

S. Behrendt[1,1(✉)], M. Ungen[2], J. Fisel[2], K.-C. Hung[2], M.-C. May[1], U. Leberle[2], and G. Lanza[1]

[1] Wbk Institute of Production Science, Karlsruhe Institute of Technology, Kaiserstraße 12, 76131 Karlsruhe, Germany
sebastian.behrendt@kit.edu

[2] Robert Bosch GmbH, Robert-Bosch-Campus 1, 71272 Renningen, Germany

Abstract. Caused by the trend of shorter product lifecycles, higher numbers of product variants and volatile markets, production systems face increasingly short periods with unchanged requirements. Therefore, the capability of manufacturing systems to reconfigure fast and cost-efficiently to changed requirements becomes a crucial factor for companies to maintain their competitiveness. Currently, reconfigurations of manufacturing systems are, on the one hand, limited due to technical constraints of the used hardware and software. On the other hand, reconfigurations require a lot of time due to manual engineering processes, planning procedures and inefficient deployment of changed production system configurations. Well-known response mechanisms for reducing reconfiguration efforts are the concepts of flexibility and changeability. This paper shows how the challenges of applying these concepts, such as managing complex modular systems or handling high reconfiguration frequencies, can be addressed with introduction of a new approach. With the paradigm shift towards software-defined manufacturing, the full potential of flexibility and changeability can be accessed. Software-defined manufacturing allows to largely decouple the production task from the operating production hardware and to manage the configuration of the production system via a continuous and highly digitized adaption process. By exploiting technologies like data mining and digital twins, the digital planning process determines new configurations of the production that fulfill changed requirements. Subsequently, the new configuration can be validated and procedures for the deployment to the production system can be determined.

Keywords: Flexibility · Changeability · Software-defined manufacturing · Digital twin · Production system · Modularity · Adaptability

M. Liewald et al. (Eds.): WGP 2022, LNPE, pp. 705–716, 2023.
https://doi.org/10.1007/978-3-031-18318-8_70

1 Introduction

Currently, manufacturing companies face a highly competitive market environment that is characterized by the VUCA world [1–3]. The VUCA world describes the challenging conditions for companies caused by volatility, uncertainty, complexity and ambiguity [4]. This demanding environment requires manufacturing companies to design their productions strategically, in order to be responsive in case of market shifts [3, 5–7]. Two enablers for a responsive production design are the concepts of flexibility and changeability [8]. Both concepts describe the capability of manufacturing systems to reconfigure in order to satisfy changed conditions concerning change dimensions such as a change in product portfolio or production quantities [3, 8, 9]. A reconfiguration refers in this context to a change in a manufacturing system's configuration regarding the utilized hardware, i.e. the production and transport resources, and the control software of the system and its resources [9]. The implementation of both flexibility and changeability is in practice, however, limited mainly by two constraints. Firstly, the foundations for changeable production hardware and software are mostly not fulfilled [5, 9–11]. These foundations are, according to Wiendahl et al. [8], the change enablers modularity, scalability, universality, compatibility and mobility. Although there exist several design guidelines for production hardware [5, 7, 9, 10, 12] that aim satisfy the change enablers, only little research focuses on the implementation of these concepts in the control software of production systems and resources. A second constraint is caused by the high degree of manual procedures in planning and deployment of production reconfigurations [5, 10, 13]. Domain experts are required to design new production configurations and to implement them on the shop floor. Both constraints lead to the fact that reconfigurations are time consuming and expensive.

A new paradigm that shows the potential to overcome these constraints by realizing production systems that reconfigure fast, cheap and autonomously is software-defined manufacturing (SDM) [14, 15]. Motivated by software-defined networking, SDM scopes to decouple production hardware and software with the goal of increasing the flexibility and changeability of both production hardware and software. Thereby, the development, planning and operation of a production can be integrated in a continuous adaption process that allows the production to reconfigure fast according to its requirements [16]. This paper introduces an approach for a continuous adaption process for production reconfigurations based on the SDM paradigm. Moreover, the required hardware and software infrastructure to realize this approach is specified.

2 Theoretical Foundation

Production reconfigurations can be realized, as explained earlier, by exploitation of the concepts of flexibility and changeability. The terms flexibility and changeability are widely referred to and defined in the literature on industrial production. A discussion of the terms can be found, for example, in [8, 9] or [17]. Figure 1 illustrates the distinction between the concepts flexibility and changeability according to different change dimensions such as volume, variants, costs, delivery time or process quality [18]. All changed requirements that are located within the flexibility corridor can be implemented by the inherent flexibility of the system in a short time, with little effort [19] and without a reconfiguration of the system [12]. The decision of the location and range of a flexibility corridor is already determined with the planning and design of the system [20]. If a changed requirement exceeds the defined flexibility corridor, the system has to be changed. The solution space envisaged for this purpose is referred to as a change corridor [12] and is shaped by proactive planning of potential change measures.

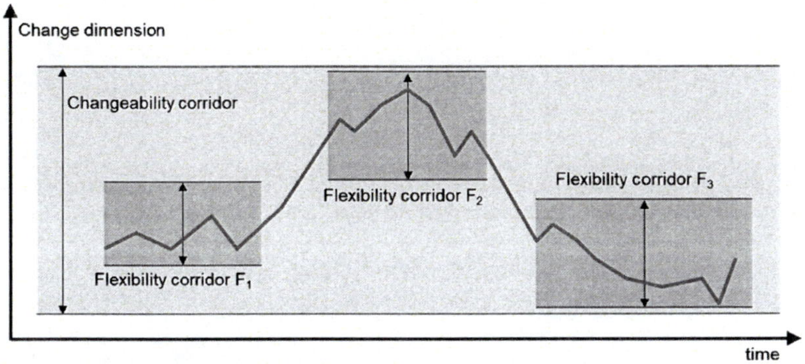

Fig. 1. Distinction of flexibility and changeability based on Wiendahl, Reichardt & Nyhuis[18], Nyhuis [12] and Zäh, Möller & Vogl [20]

Since modern production systems are due to the rise of Industry 4.0 highly digitized and the operation of production hardware is controlled by software, the realization of both flexibility and changeability is limited by rigid control software and manual planning procedures [10, 11, 13]. A potential solution for this problem is SDM.

SDM is a fairly new research area that builds on the vision of Industry 4.0 [15]. With SDM, a paradigm for industrial production is emerging that is based on two pillars. Firstly, the decoupling of software and hardware: Abstraction layers make it possible to control the functions of the hardware via software, or to adapt the properties of the hardware via software. The product manufacturing process thus becomes adaptable through software [14]. The second pillar of SDM is a new form of production system management [16]: The term continuity is omnipresent in the IT world. This is especially apparent in its manifestations of continuous integration, continuous delivery and continuous deployment (CI/CD) - i.e. the constant integration of new software fragments into the product stack, the continuous provision of a current compilation and the automated installation of the latest compilation on a defined target system. This results in fast and inexpensive software adaptions due to modular systems, lean processes and routine in their execution. These advantages are to be realized by a transfer of the continuity concept to the industrial production [16].

The reconfiguration of production systems is a planning problem associated with production planning and control (PPC). Different methodologies for PPC can be found in the literature, such as the Aachener model for PPC [21] or the hierarchical PPC model [22]. Both models, however, state that multiple distinct planning tasks have to be completed in PPC. In case of a reconfiguration, the most important tasks are configuration planning, production program planning and production control [9]. There exist multiple approaches in the literature that exploit digital technologies to fully automate these tasks [23]. This includes, for example, the use of optimization algorithms for configuration planning [10, 24] and production scheduling [25], production control by reinforcement learning agents [26, 27] and the extensive use of simulation in all planning phases [6, 28]. Although automation of individual tasks is available, a continuous process for production reconfigurations that allows the integration and combination of individual approaches is missing. This gap in the literature is targeted by the upcoming sections.

3 Interaction Between Flexibility or Changeability and SDM

Flexibility, changeability and SDM show intersections in their respective orientation: They aim at a simple and fast possibility to modify a production system according to changed requirements. Therefore, we will discuss in the following how they are related and how flexibility and changeability can be supported by SDM.

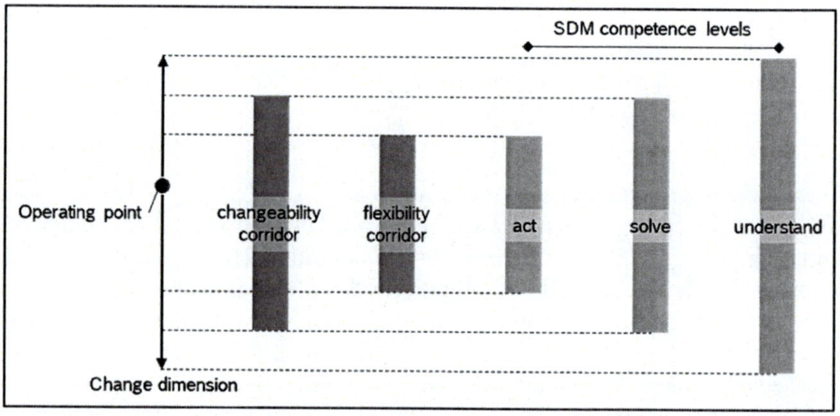

Fig. 2. Matching of flexibility and changeability corridors with SDM competence levels.

To this end, Fig. 1 is adopted and expanded upon, as shown in Fig. 2. The corridors of flexibility and changeability are shown on the left-hand side as ranges of a change dimension. The operating point indicates the value of the change dimension that is realized with the current configuration of the production system. On the right-hand side, the hierarchical competence levels "understand" (lowest), "solve" and "act" (highest) of SDM are introduced. Competence models for Industry 4.0, such as the I4.0 maturity index [29], demonstrate a degree of overlap, but pursue a different objective with the generic assessment of Industry 4.0 and are therefore not applied. The execution of the levels can comprise both manual and automated work. The north star, i.e. the long-term vision of an SDM system, is the fully-featured level "act", meaning a completely software-defined, automated reconfiguration of the production system with respect to changed requirements.

3.1 SDM Competence Level "Understand"

The level "understand" describes the ability to match requirements of the operating point with the capabilities of the production system. The effect of changed requirements on the operating point in the change dimensions is not part of the paper. For this purpose, models such as the receptor model [3] can be considered.

The basic requirement of this level is a description and interaction model of the elements involved, such as workpiece, production process and machine. Based on this, the specifically required model can be created from these sub models. An example of a "product portfolio" change dimension is the issue of whether an upcoming change of a product can be handled by the existing production system. A known approach is to describe production resources with provided production capabilities and products with required production capabilities and to compare them [30, 31] to each other. Different change dimensions require different information from the description of the elements. To increase the applicability in industrial practice, the description of required and provided capabilities can be automated: For example, the presence of a glue in the bill of material of a workpiece can be interpreted as a requirement of the capability "gluing".

The level "understand" represents the foundation of the levels solve and act, but already offers an added value on its own. The alignment of product and production system enables "design-to-line", i.e. target-oriented product development for a production system with existing production resources.

The ecosystem of SDM offers a variety of solution elements for mastering the level "understand". An essential element is the application of problem-specific digital twins. For the representation of descriptive and functional aspects of an asset, the concept of the asset administration shell including a series of submodels could be applied [31]. The relationships of the modelled entities such as components, machines, and workpieces can be modeled with ontologies [30].

3.2 SDM Competence Level "Solve"

The level "solve" extends the level "understand" by the ability to identify configurations of the production system that fulfill the requirements of an operating point. It refers to a preconceived solution space and therefore corresponds to the changeability displayed in Fig. 1.

More information is required to successfully implement the level "solve". The required information does not only include the status quo of the production system, but also three additional aspects. Firstly, its options for change, such as interchangeable process modules, shift planning, material flow and warehouse dimensioning. Secondly, its limits, such as restrictions by area or the assembly precedence graph. Thirdly, the target system of the production system reconfiguration such as the minimization of the reconfiguration costs, the reconfiguration duration or the resulting lead time of a product.

Substantial added value results from reduced planning activities as well as a minimization of possible planning errors. Challenges such as the preferred use of depreciated machines can be addressed by appropriate modeling of the objective functions of optimization problems.

The already widely used tools simulation and optimization are fundamental solution elements of SDM systems. They can be used effectively to control complex systems if models are largely generated or at least configured automatically instead of being created manually. This requires the connection and management of heterogeneous data sources, which can be implemented through cloud architectures and a middle layer for centralized data querying. In addition, virtualized components and systems increase and refine the solution space of simulation and optimization tasks. Hardware and software-in-the-loop approaches can be utilized, for example, to include and exchange components in system simulations without being physically available. Moreover, modular machine concepts can be used that allow the manufacturing process to be changed by exchanging set-up parts (e.g. exchanging the tool head on a robot). Thus, production equipment-as-a-service is realized, which can be used via a central management system.

3.3 SDM Competence Level "Act"

The highest level of an SDM system can be defined as "act". This extends the levels "understand" and "solve" by the ability to adjust the operating point of a production system. In regard of the corridors shown in Fig. 2, an SDM system of the "act" level corresponds to the flexibility corridor.

To implement the level "act" in a fully automated mode, challenges concerning production resources and the processes involved have to be addressed. For example, only manufacturing processes that can be fully controlled or set up by software are to be used. Manual work only takes place on the system, but no longer in the system. With regard to the processes involved, two different approaches are available. On the one hand, involved processes can be automated: These include, for example, the control of the material flow including supplied parts, the release of production processes and safety-related acceptance, spare parts management or shift planning. On the other hand, ranges of the operating point can be defined, in which a renewed run of a process is not necessary. For example, modular machines could be released once with different process modules, so that a replacement of the process module does not require a new release.

The key added value lies in the control of complexity and the associated reduction in planning and reconfiguration costs. With increasing automation of production system planning and management, the ability to react to changing requirements increases accordingly. Moreover, the level "act" offers the possibility of controlling the operating point of production systems.

To realize these potentials, a high degree of automation is required in the creation and management of software and digital twins. Container virtualization can be used to isolate applications in order to enable switching between physical and virtual assets as seamlessly as possible. A further solution component is the utilization of available data from a wide variety of data sources, such as sensors or MES systems, to allow status monitoring of production systems.

4 SDM-Driven Approach for Production Reconfigurations

The approach presented by this work, builds upon the SDM paradigm and aims at the targeted development, planning and operation of reconfigurable production systems. The following descriptions are based on the continuous SDM process introduced by Neubauer et al. [16]. Moreover, the different tasks associated with planning and deployment of production reconfigurations are classified according to the previously explained SDM competence levels.

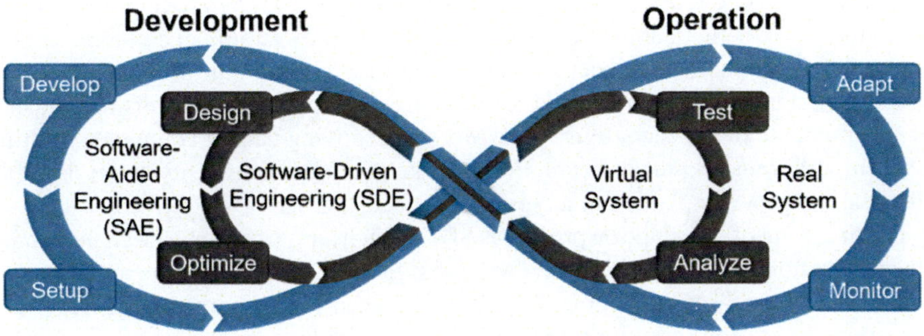

Fig. 3. Visualization of the continuous adaption process for reconfigurations of production systems, based on Neubauer et al. [16].

As motivated earlier, a new approach is needed to allow the hardware and software of production systems to be easily and rapidly reconfigured to satisfy new internal or external requirements. This new approach is based on a target-oriented, evolutionary adaption that changes the system not incrementally but continuously. This goal is achieved by closing the gap between development and operation activities with the help of the SDM pillars using CI/CD in the planning, development, deployment and operation of production systems (see Fig. 3).

Ideally, the tasks in these continuous processes can be fully automated. However, some procedures, e. g. the design of manufacturing modules, do not allow a full automation. Therefore, human interaction and support will still be required here. Besides, an essential aspect of a fully digital planning procedure is the use of a single data source for all planning steps and tools to ensure the continuity and consistency of the planning procedure.

A distinction in term of the operation of the virtual and the physical production system is also made in this approach. The virtual system allows testing, performance analysis and commissioning of new configurations without the need for physical changeover. The real system is only required during the physical adaption, operation and monitoring of the solution.

The development is also divided into two wings, namely software-aided engineering (SAE) and software-driven engineering (SDE). SDE summarizes design and optimization algorithms that generate new configurations by using available production assets. Contrarily, SAE aims to develop new production assets that extend the capabilities of the production system. The distinction in the development field is made to enable a human interaction with the system in SAE, if the automated procedures of SDE are not sufficient to create a solution that fulfills the needs of the production environment.

While the competence levels have a different impact on the left and right wings of the loop, a direct assignment is not conclusive. For example, the application of hardware-in-the-loop, described in Sect. 3.2, affects both the software-driven engineering wing and the "virtual system" wing. The competence levels rather describe the fundamental skills required for implementing the continuous adaption process.

A typical workflow in the approach is triggered by a change in requirements of the production systems that can be described and classified by an influence on the change dimensions. It should be noted here, that the tasks of the four wings of the approach are not completed in a defined order, but are completed in an order that suits the use case and requirements of the reconfiguration. Moreover, a wing can be completed multiple times with different degrees of detail. For the sake of comprehensibility, the tasks of the individual wings are explained in a typical sequence.

The first wing of the adaption process is SDE, which aims to control the changeability of the production system by generating new configurations that satisfy the requirements. To achieve this goal, optimization algorithms are used to automatically generate new production system configurations and production schedules. The design of new configurations is thereby constrained by a solution space that defines the degrees of freedom of a reconfiguration. Possible degrees of freedom are, for example, the utilized production resources, the layout of the production system or the mapping of processes and resources. Besides the requirement constraints, there is a multitude of technical and economical constraints that have to be satisfied by the new configuration. The design and optimization of the configuration can take place with different degrees of detail. For example, one could generate in a first iteration a set of rough configurations that get later optimized and enriched.

The virtual system is used to test and analyze the generated configurations. As SDE is performed with different degrees of detail, testing has to be performed with a suitable virtual system. It can be sufficient, for example, to use material flow simulations for testing and analysis in an early planning phase whereas highly detailed 3D simulations are used in a later planning phase for visualization or collision control. Testing aims to increase the credibility of new configurations by searching for potential errors before deployment. The analysis of the configurations is done with respect to the requirements of the production systems and defined KPIs that allow to assess how well a configuration performs. These KPIs can be economical, ecological or technological measures but should be in line with the goal of the new configuration.

It might happen, that it is not possible to generate a new configuration from the solution space that fulfills the requirements. In this case, the SAE wing is responsible to develop and setup production resources that resolve these requirements. Thereby, the solution space is expanded and allows to find a valid solution. SAE requires not only to generate a physical machine or component but also its virtual counterpart. This enables other wings to automatically adapt to the expanded solution space and to consider the new resource.

After deciding for a configuration based on a quantitative assessment or a human selection, the last wing gets passed which comprises of the adaption, operation and monitoring of the physical system. By using the digitally specified configurations, all information is derived to plan and implement the new configuration. This includes hardware modification sequences, order data for new hardware and the automatic deployment of new control software. During the operation of the new configuration, production data is gathered to evaluate the new configuration. Moreover, production data is used to synchronize the virtual and real system, serving as the basis for further SDM-driven reconfigurations of the production system.

5 Testbed and Case Study for Research on SDM

The previous section summarized the key ideas of SDM based on the presented concept using the SDM-Loop. To enable further research and development activities and to validate our concept, a basic testbed will be implemented. This testbed for SDM will be built upon existing technologies applied for a highly changeable production system that serves our case study. The integral technologies, characterizing the testbed, are described in the following paragraphs.

Highly convertible modular production machinery. A substantial component of the testbed is based on the concept CESA^{3}R [32]. Modular functional modules called "Mechatronic Objects" that are based on the Plug-and-Produce technology will be used. These Mechatronic Objects can be arranged arbitrarily within production cells and offer different production functionalities. Challenges that arise through the easy system reconfigurations provided by this concept will be identified and addressed using SDM.

Flexible intralogistics technology. Highly changeable material flows can be realized by e. g. automated guided vehicles (AGVs) or modular conveyor systems to establish the interlinkage of the individual production cells on the shop floor. This intralogistics technology will represent a significant part of the SDM testbed to realize interconnections between standalone production cells.

Intelligent floor infrastructure. The shop floor itself provides reconfigurability at the system level, as described in [33]. The infrastructure comprises tiles with wireless power transfer, which makes arbitrary positioning of production machinery possible. With this technology as the foundation for the infrastructure, validation of the presented SDM concept can be done more effectively.

Asset Administration Shell (AAS) Middleware. A common communication infrastructure for the connection and orchestration of the physical testbed components is offered by the approach in [34]. The application of this middleware in our testbed allows physical components to be integrated seamlessly and interoperable into the case study for validation.

By having the aforementioned technologies as the foundation for our testbed, several case studies will be performed in the future to validate our SDM approach. The inner part (i.e., SDE and virtual system operation) of the SDM loop and its transition to the real system operation will be considered at first. The key topics, which are planned for the coming future are the following:

1. Decision making regarding the suitability of implemented production system configurations to fulfill new production requirements (i.e., increased production output,

new product variants) will be studied. An important research aspect is the matching of production system capabilities with production requirements. This involves determination of performable production processes or process chains as well as achievable value stream performances for specific production system configurations.
2. Configuration planning based on production requirements and available resources represents another important research focus. The planning will involve the selection of resources, their arrangement as a production system and the definition of explicit workflows to fulfill the production task. In SDM, the focus will lie on the use of simulation-based optimization to solve this problem. Apart from finding an efficient optimization procedure, automated generation of simulation models will also be subject of the research.
3. Deployment preparation aims to support the changeover of the production system. This will involve the validation of planning results and the preparation of artefacts required for the implementation process. Research activities within this step will focus on automated reconfiguration of informational systems, generation of control code as well as the automated evaluation of safety and security aspects.

6 Conclusion

In regard of the VUCA world, increasing the ability of a production system to reconfigure fast and cost-efficient is critical to maintain a manufacturing companies' competitiveness. As Industry 4.0 drives the digitalization of production, reconfigurations are not solely hardware related but also impact the associated production software. In order to increase flexibility and changeability of production systems, a continuous adaption process is introduced that exploits the new production paradigm SDM. At first, this paradigm is explained and it is elaborated how it interacts with and differentiates itself from the concepts of flexibility and changeability. As a result, a model for the classification of SDM competences is presented that comprises of three levels, i.e. "understand", "solve" and "act". With reference to the competence model and the goal of achieving fully automated, fast and cost-efficient production system reconfigurations, a continuous adaption process for planning and deployment of production reconfigurations is introduced that builds upon SDM. The process integrates SDE, SAE and production planning processes with the operation of the physical and virtual production in an iterative process. Lastly, a testbed for this adaption process is presented which realizes the required software and hardware infrastructure for the concept's demonstration. It is explained that this testbed will be used in future research to study and demonstrate the continuous adaption process by the use of SDM elements concerning decision making, configuration planning and deployment preparation.

Acknowledgement. We extend our sincere thanks to the German Federal Ministry for Economic Affairs and Climate Action (BMWK) for supporting this research project 13IK001ZF "Software-Defined Manufacturing for the automotive and supplying industry https://www.sdm4fzi.de/".

References

1. Abele, E., Reinhart, G.: Zukunft der Produktion – Herausforderungen, Forschungsfelder, Chancen.1st end. Carl Hanser Verlag, München (2011)
2. Mehrabi, M.G., Ulsoy, A.G., Koren, Y., Heytler, P.: Trends and perspectives in flexible and reconfigurable manufacturing systems. J. Intell. Manuf. **13**, 135–146 (2002)
3. Cisek, R., Habicht, C., Neise, P.: Gestaltung wandlungsfähiger Produktionssysteme. Zeitschrift für wirtschaftlichen Fabrikbetrieb **97**(9), 441–445 (2002). https://doi.org/10.3139/104.100566
4. Bennett, N., Lemoine, G.J.: What a difference a word makes: Understanding threats to performance in a VUCA world. Bus. Horiz. **57**(3), 311–317 (2014). https://doi.org/10.1016/j.bushor.2014.01.001
5. Koren, Y., et al.: Reconfigurable Manufacturing Systems. CIRP Ann. **48**(2), 527–540 (1999). https://doi.org/10.1016/S0007-8506(07)63232-6
6. Benfer, M., Peukert, S., Lanza, G.: A Framework for Digital Twins for Production Network Management. Procedia CIRP **104**, 1269–1274 (2021). https://doi.org/10.1016/j.procir.2021.11.213
7. Heilala, J., Voho, P.: Modular reconfigurable flexible final assembly systems. Assem. Autom. **21**(1), 20–30 (2001). https://doi.org/10.1108/01445150110381646
8. Wiendahl, H.-P., et al.: Changeable Manufacturing - Classification. Design and Operation. CIRP Annals **56**(2), 783–809 (2007). https://doi.org/10.1016/j.cirp.2007.10.003
9. ElMaraghy, H.A.: Flexible and reconfigurable manufacturing systems paradigms. Int. J. Flex. Manuf. Syst. **17**(4), 261–276 (2006). https://doi.org/10.1007/s10696-006-9028-7
10. Stähr, T.J.: Methodik zur Planung und Konfigurationsauswahl skalierbarer Montagesysteme - Ein Beitrag zur skalierbaren Automatisierung. Dissertation, Karlsruher Institut für Technologie (KIT), Karlsruhe (2020)
11. ElMaraghy, H. A.: Reconfigurable Process Plans For Responsive Manufacturing Systems. In: P. F. Cunha & P. G. Maropoulos (Eds.), Digital enterprise technology: Perspectives and future challenges (pp.35–44). Springer. https://doi.org/10.1007/978-0-387-49864-5_4 (2007)
12. Nyhuis, P.: Wandlungsfähige Produktionssysteme, GITO Verlag, Berlin. ISBN: 9783942183154 (2010)
13. Leachman, R.C., Benson, R.F., Liu, C., Raar, D.J.: Impress: An Automated Production-Planning and Delivery-Quotation System at Harris Corporation - Semiconductor Sector. Interfaces **26**(1), 6–37 (1996). https://doi.org/10.1287/inte.26.1.6
14. Lechler, A., Kircher, C., Verl, A.: SDM – Software Defined Manufacturing. wt Werkstatttechnik online 2018(5), 307–312. https://doi.org/10.37544/1436-4980-2018-05-33 (2018)
15. Thames, L., & Schaefer, D.: Software-defined cloud manufacturing for industry 4.0. Procedia cirp, 52, 12–17 (2016)
16. Neubauer, M.; Ellwein, C.; Frick, F.; Fisel, J.; Kampert, D.; Leberle, U.; May, M.; Behrendt, S.; Esslinger, E.; Pfeifer, D.; Zahn, P.: Kontinuität als neues Paradigma. Computer&Automation (04) (2022)
17. Fisel, J.: Veränderungsfähigkeit getakteter Fließmontagesysteme: Planung der Fließbandabstimmung am Beispiel der Automobilmontage (2019)
18. Wiendahl, H.-P.; Reichardt, J. & Nyhuis, P.: Handbuch Fabrikplanung: Konzept, Gestaltung und Umsetzung wandlungsfähiger Produktionsstätten, Carl Hanser, München. ISBN: 9783446437029 (2014)
19. Abele, E.; Liebeck, T. & Wörn, A.: „Flexibilität im Investitionsentscheidungsprozess", wt Werkstattstechnik online, Vol. 97, 1/2, 85 f. (2007)

20. Zäh, M. F.; Möller, N. & Vogl, W.: Symbiosis of changeable and virtual production - the emperor's new clothes or key factor for future success. In: Proceedings (CD) of the international conference on changeable, agile, reconfigurable and virtual production, München (2005)
21. Schuh, G.: Produktionsplanung und -steuerung. Grundlagen, Gestaltung und Konzepte, Springer-Verlag Berlin Heidelberg, Berlin, Heidelberg. ISBN: 978–3–540–40306–7 (2006)
22. Hax A.C., H.C. Meal: Hierarchical integration of production planning and scheduling M.A. Geisler (Ed.), Studies in Management Sciences, Vol. 1: Logistics, Elsevier, Cambridge, MA (1975)
23. Usuga Cadavid, J.P., Lamouri, S., Grabot, B., Pellerin, R., Fortin, A.: Machine learning applied in production planning and control: a state-of-the-art in the era of industry 4.0. J. Intell. Manuf. **31**(6), 1531–1558 (2020). https://doi.org/10.1007/s10845-019-01531-7
24. Balzereit, K., & Niggemann, O.: Gradient-based Reconfiguration of Cyber-Physical Production Systems. In: 2021 4th IEEE International Conference on Industrial Cyber-Physical Systems (ICPS): Online, 10–13 May, 2021 (pp. 125–131). IEEE (2021)
25. Graves, S.C.: A Review of Production Scheduling. Oper. Res. **29**(4), 646–675 (1981). https://doi.org/10.1287/opre.29.4.646
26. Kuhnle, A.: Adaptive Order Dispatching based on Reinforcement Learning - Application in a Complex Job Shop in the Semiconductor Industry. Dissertation, Karlsruher Institut für Technologie (KIT), Karlsruhe (2020)
27. Waschneck, B., Reichstaller, A., Belzner, L., Altenmuller, T., Bauernhansl, T., Knapp, A., & Kyek, A.: Deep reinforcement learning for semiconductor production scheduling. In: 2018 29th Annual SEMI Advanced Semiconductor Manufacturing Conference (ASMC). IEEE. https://doi.org/10.1109/asmc.2018.8373191 (2018).
28. Kritzinger, W., Karner, M., Traar, G., Henjes, J., Sihn, W.: Digital Twin in manufacturing: A categorical literature review and classification. IFAC-PapersOnLine **51**(11), 1016–1022 (2018). https://doi.org/10.1016/j.ifacol.2018.08.474
29. Schuh, G., Anderl, R., Gausemeier, J., Ten Hompel, M., & Wahlster, W. (Eds.).: Industrie 4.0 Maturity Index: Die digitale Transformation von Unternehmen gestalten. Herbert Utz Verlag (2017)
30. Julius Pfrommer, Jan-Felix Klein, Marco Wurster, Simon Rapp, Patric Grauberger, Gisela Lanza, Albert Albers, Sven Matthiesen, and Jürgen Beyerer: An Ontology for Remanufacturing Systems. Automatisierungstechnik 06/2022. https://doi.org/10.1515/auto-2021-0156 (2022)
31. Tantik, E., & Anderl, R.: Integrated data model and structure for the asset administration shell in industrie 4.0. Procedia Cirp, 60, 86–91 (2017)
32. Vorderer, M., Junker, S., Lechler, A., & Verl, A.: CESA 3 R: highly versatile plug-and-produce assembly system. In: *2016 IEEE International Conference on Automation Science and Engineering (CASE)* (pp. 745–750). IEEE (2016)
33. Stillig, J., & Parspour, N.: Advanced manufacturing based on the intelligent floor: An infrastructure platform for the convertible production in the factory of the future. In: *2020 IEEE 20th Mediterranean Electrotechnical Conference (MELECON)* (pp. 248–253). IEEE (2020)
34. Ewert, D., Jung, T., Tasci, T., Stiedl, T.: Assets2036 – Lightweight Implementation of the Asset Administration Shell Concept for Practical Use and Easy Adaptation. In: Weißgraeber, P., Heieck, F., Ackermann, C. (eds.) Advances in Automotive Production Technology – Theory and Application. A, pp. 153–161. Springer, Heidelberg (2021). https://doi.org/10.1007/978-3-662-62962-8_18

Improvement of Personnel Resources Efficiency by Aid of Competency-Oriented Activity Processing Time Assessment

A. Keuper(✉), M. Kuhn, M. Riesener, and G. Schuh

Laboratory for Machine Tools and Production Engineering WZL, RWTH Aachen University, Campus-Boulevard 30, 52074 Aachen, Germany
a.keuper@wzl.rwth-aachen.de

Abstract. Manufacturing companies in high-wage countries face a variety of challenges. International competition exposes them to constantly increasing pressure to offer new products that meet customer requirements in increasingly shorter intervals and at competitive prices. Particularly in high-wage countries, where personnel resources are the second highest cost factor it is important to utilize these resources efficiently in order to be competitive in global markets. An important lever to increase the efficiency of personnel resources is the specific allocation of employees according to their competencies and not according to their function in the company. Employees with a set of competencies that matches well with the characteristics of an activity or a process are able to achieve the same results in less time compared to employees with a less matching set of competencies. Therefore, the goal of this paper is to develop a methodology for the improvement of personnel resources efficiency by assessing the time needed for specific activities based on the competencies of personnel resources. This enables a better resource management since activities and processes can be assigned to the employees who are able to finish them fastest.

Keywords: Competencies · Resource management · Work efficiency · Processing time assessment

1 Introduction

1.1 Motivation

Manufacturing companies face many challenges in today's VUCA world (volatility, uncertainty, complexity and ambiguity), especially in high-wage countries [1]. To be successful, they must offer new products that meet requirements in higher frequency and at marketable prices [2]. At the same time, customer demands have increased [3]. This combination leads to a strong need for innovation [1]. Projects are increasingly parallelized in order to save development time and costs [4], but this increases the coordination and planning effort [5]. Despite project management, many projects fail to achieve their goals, 45% of projects miss their time target and 38% miss their budget

M. Liewald et al. (Eds.): WGP 2022, LNPE, pp. 717–726, 2023.
https://doi.org/10.1007/978-3-031-18318-8_71

target [6]. In light of these challenges, the goal must be to achieve the best possible use of the resources employed and to minimize waste due to poor project performance. One approach to increase the personnel resource efficiency is to improve the project and task allocation by finding best-suited employees or teams [7]. Thus, the resource management should be competency-oriented in order to identify suitable personnel resources for the activities, to reduce bottlenecks and to boost efficiency. It can be assumed that employees with a set of competencies that matches well with the characteristics of an activity are able to achieve the same results in less time compared to employees with a less matching set of competencies. Moreover, competencies are a decisive factor in ensuring the innovative capability, effectiveness and target achievement of companies [8]. Especially in product development, with a high share of human labour, the efficient use of personnel resources is crucial for success. Therefore, the goal of this paper is to develop a methodology that enables a resource-specific and competency-based assessment of an activities' processing time. In order to do so the available competencies in a company provided by their employees are described and the influence between these competencies and the processing time of activities is evaluated.

The paper is structured as follows. After the introduction, Sect. 1 continues with a description of the relevant terminology. Section 2 analyzes related research and identifies a research deficit. Section 3 provides an overview of the research approach. Afterwards the results of the research are presented in Sect. 4 and applied on an exemplary use case in Sect. 5. The last section draws a conclusion and gives an impulse for further research.

1.2 Definitions

In this section, a short definition of resource management in product development and competencies in the context of resource management is given to establish a common understanding.

Resource Management in product development

Resource management is a central element in project management and the most important lever for achieving objectives in multi-project environments [9, 10]. Usually, resource management in the context of project management considers human resources [11]. Resource management is responsible for resource planning, including the identification and allocation of appropriate resources [12]. The aim of resource management is to bring the available resources and the resource demand into balance [11]. It should be emphasized that the correct use of resources and resource management is cited in many project management studies as a key success factor for successful project execution. Human resources in particular are attributed an important role in this regard [12].

Competencies in the context of resource management

A competency can be understood as the ability to act appropriately in a given situation. It is the ratio between demands made on a person or group and their ability to meet these demands. In this context, it goes beyond pure knowledge, because it describes the totality of knowledge, skills and abilities that people use to solve problems [13]. The four basic types of competency are personal, activity and action-oriented, technical-methodical and social-communicative competencies [14].

After the definition of the most relevant terms, the next section of the paper will analyze related work in the field of competency-based resource management.

2 Related Work

In the context of this paper, the analysis of related work will focus on the following approaches: Approaches in the field of competency-based resource management, approaches to determine the fit between competency and activity and approaches to determine processing times for (project) activities.

Wysocki is dealing with the competency-based resource management. Since employees are the most difficult resources to plan, several approaches are summarized in this work. On the one hand, competency matrices according to Wysocki contain the competency requirements of activities and, on the other hand, the competencies availability through personnel resources and are connected through a company-specific definition of competencies. The identification of qualified personnel can be done with this approach, but the influence of the competency fit between resource and activity is not further discussed [15].

An approach to determine the fit between competency and activity is a qualification matrix, as shown by Kosar and Biedermann. By aid of an employee qualification matrix, various skills of employees can be analyzed, reviewed and expanded at regular intervals. Thus, the qualification status of the employees can be read off the matrix at any time. Qualifications are recorded annually and in five different qualification levels. From these, further training needs can be derived. Again, the influence of the competency fit between resource and activity is not further discussed [16].

The next approach originates in the field of personnel psychology. A competency model according to Krumm et al. consists of competencies of different areas and the corresponding characteristics, which are divided into several levels. In companies, this can be used to create competency requirement profiles for various positions in the company and thus select people with a suitable competency profile. The match between position and person is feasible with this model, but the deployment of human resources in projects based on the competency model is not addressed [17].

Haroune et al. deal with human resource deployment in multi-project environments from a common resource pool with different competencies. Each combination of activity and human resource is assigned to different efficiency values, which influence the processing time of the activities. This is used within an optimization model to generate a project schedule in which the projects can be finished as fast as possible. However, it is not described how the efficiency values can be assessed and how the processing time of an activity changes based on the competency profiles of the resource [7].

The four exemplary approaches presented above as well as further analyzed approaches show, that a competency-oriented use of human resources in the context of multi-project management has not yet been intensively investigated and, above all, the consideration of how a resource with an individual competency profile influences the processing time of an activity has not yet been carried out. Therefore, this paper will develop a methodology to consider systematically the influence of competencies on activities' processing times.

3 Research Approach

The goal of this paper is to develop a methodology to improve the estimation of project activity processing time and thereby the quality of project planning and resource efficiency. In order to be able to improve project activity processing time estimations it is necessary to consider the competencies of allocated human resources. Raising the following research question: "*How can the influence of an allocated human resource on the processing time of a project activity be determined?*".

To give an answer to this research question a four-step methodology in the product development context is developed. However, it is feasible that a competency-based allocation approach with individual process times can also be used in other areas of manufacturing companies, for example in shop-floor management. Step one of the methodology develops a description model for the competencies of human resources in the product development context. In step two, a description model for the activities within a development project is derived. Step three focuses on the determination of relations between competencies and the processing time of an activity. Step four shows how individual activity processing times can be calculated. Finally, the four steps will be applied to a simplified use case to demonstrate how the methodology can be used in practice.

4 Development of the Methodology

4.1 Step 1—Description of Competencies of Human Resources

The goal of the first step is to describe the competencies of human resources in a standardized manner in order to be able to evaluate their proficiency level of these competencies objectively. For this purpose, it is necessary to first identify the competencies, which are relevant for the company applying this methodology. This can be done based on numerous approaches already existing in scientific literature. For example the approach of Krumm et al. describes how a company specific competency model can be developed [17]. Other approaches already include comprehensive lists of competencies and can be used to select competencies relevant to the use case [18, 19]. For the product development context, an analysis of scientific literature and 150 job descriptions of different manufacturing companies was conducted in order to create a longlist of 163 generally relevant competencies. Afterwards the selected competencies need to be operationalized by defining different levels of proficiency and describing the necessary skills and knowledge, which are required for each proficiency level. This procedure is common when developing competency models [17]. It is advised to choose skill- and knowledge-elements that can be assessed objectively. An example for competencies and the operationalization of a competency with skill- and knowledge-elements is shown in Fig. 1. After the company-specific competencies are selected and operationalized it is necessary to assess the proficiency level for these competencies of all employees. Numerous methods exist to do so, some examples like interviews, surveys or behavioral observations are described by Krumm et al. [17].

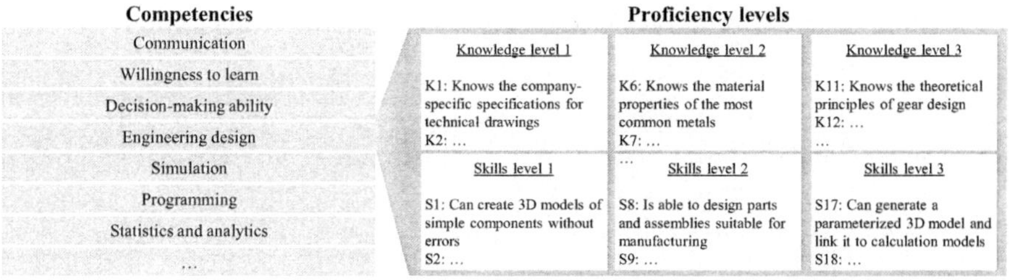

Fig. 1. Exemplary competencies and proficiency levels in product development

4.2 Step 2—Description of Activities

In the second step activities are described by features and characteristics in order to be able to identify activities with similar competency requirements. The requirement for the description is that it must be possible to distinguish activities that differ in terms of their competency requirements. At the same time, activities that require the same or very similar competencies should be described by the same or very similar characteristics. By conducting a systematic literature review it was possible to identify 111 features of development project activities. These features were subjected to a plausibility check and narrowed down to the most relevant by checking whether an adjustment of the characteristic of a feature causes a change in the required competencies to perform the activity. The description model was developed for the context of product development of manufacturing companies and is shown in Fig. 2.

Features	Characteristics					
Novelty	repetitive	similar to standard	modified	unknown		
Variability	planned, anticipatory	unplanned, reactive				
Interdiciplinarity	subject-specific	interface topic	interdisciplinary			
Cooperation	individual work	local team	global team	company-network		
Use of methods	synthesis	management	analysis	evaluation		
Responsibility	organizational	supporting	executing	representative		
IT affinity	low	medium	high			
Focus area of expertise	mechanics	electrics	software	sales & services	business	production
Process interdependencies	few	some	many			

Fig. 2. Description model for project activities in product development

When describing an activity with this description model, a vector with nine dimensions, one for each feature, can represent the description. The value in every dimension is determined by the respective characteristic that is chosen to describe the activity. By representing the activities and their description through vectors, it is possible to cluster them into groups of similar activities using common clustering algorithms [20]. As mentioned above it is assumed that the competency requirements within these groups are homogeneous.

4.3 Step 3—Determination of Relations Between Competencies and Activities

After competencies of the resources and activities have been described, the next step is to determine the relation between them. The goal is to take the influence of competencies of the resource on the processing time of an activity into account. For this purpose, it is first determined in which way the proficiency-level of a competency can influence the processing time of an activity. Based on expert interviews five types of relations were identified (see Fig. 3).

1. Basic competency: this competency must be at maximum level to meet the planned activity processing time. The greater the deviation from the maximum level of proficiency, the greater the slowdown in activity completion.
2. Bidirectional competency: this competency can speed up activity execution when a high level of proficiency is present and might slow it down when a low level of proficiency is present.
3. Optional competency: this competency does not lead to a slowdown if it is not present, but can accelerate the activity if it is present.
4. Insignificant competency: the presence or absence of this competency does not affect the process time.
5. Counterdirectional competency: competency acts in reversed direction. A higher proficiency level of this competency leads to a slowdown of the activity.

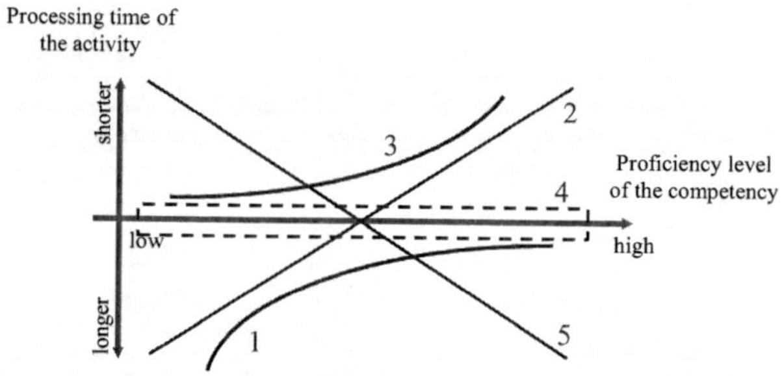

Fig. 3. Types of relations between competencies and processing time of activities

A closer examination of these basic types reveals that they behave very similarly to the attributes and customer satisfaction in the Kano-model [21]. For this reason, a method from the application of the Kano-model can be converted for the evaluation of the relations between competencies and activities. The method is the functional and dysfunctional questioning, which becomes the qualifying and disqualifying questioning in the context of this paper. In this method, two questions are asked, "What is the effect on the processing time of the activities of cluster X if proficiency level of competency Y is high?" and "What is the effect on the processing time of the activities of cluster X if proficiency level of competency Y is low?" By combining the answers, a basic type of relation can be assigned to a competency/activity pair as shown in Fig. 4.

Qualifying questioning	Disqualifying questioning: A	B	C	D	E
A: This significantly reduces the processing time	Contradiction	Contradiction	Optional competency	Bidirectional competency	Bidirectional competency
B: This reduces the processing time	Counterdirectional	Contradiction	Insignificant competency	Basic competency	Basic competency
C: No impact on the processing time	Counterdirectional	Insignificant competency	Insignificant competency	Insignificant competency	Basic competency
D: This increases the processing time.	Counterdirectional	Counterdirectional	Insignificant competency	Contradiction	Basic competency
E: This significantly increases the processing time.	Counterdirectional	Counterdirectional	Counterdirectional	Contradiction	Contradiction

Fig. 4. Evaluation matrix for the qualifying/disqualifying questioning in accordance with [22]

Alternatively, it is possible to evaluate the competency influence on the duration of a project activity by analyzing historical project data. If the planned duration, actual time stamps and allocated employees of activities can be extracted from a company's database it can be combined with a retrospective assessment of these employees' competencies. The resulting data set can be analyzed in order to identify the impact of the competencies on the processing time of the different clusters of activities. A similar approach is presented by Riesener et al. [23].

4.4 Step 4—Calculation of Resource-Specific Activity Duration

In the final step, the relations from the previous step are used to calculate resource-specific activity processing times. This enables an improved project planning and scheduling since processing times can be estimated more precisely. As an additional input, an estimation of the processing time according to the Program Evaluation Review Technique (PERT) is required [24]. With the PERT optimistic, realistic and pessimistic estimates for the processing times are gathered. Depending on the competency fit between resource and activity, it can then be determined whether the processing time for an activity allocated to a specific resource is closer to the optimistic or pessimistic estimate (see Fig. 5). First, the activity-cluster for an activity is determined based on the description model from (see Sect. 4.1). In Fig. 5, the activity "create technical drawing" is part of the activity-cluster "mechanical design". Then for each relevant competency for this activity-cluster, the proficiency level of the resource indicates whether that competency influences the processing time towards the optimistic or pessimistic PERT-estimate. In the example, a high proficiency-level of the competency "analytical thinking" means according to the relation type "bidirectional competency" (see Sect. 4.3) that this competency influences the processing time towards the optimistic estimate. By forming a weighted average over all relevant competencies for that activity, a resource-specific processing time can be calculated.

By applying the presented methodology, it is possible to improve project planning by more precise resource-specific activity processing times as well as resource efficiency by allocating the resources with the best competency fit.

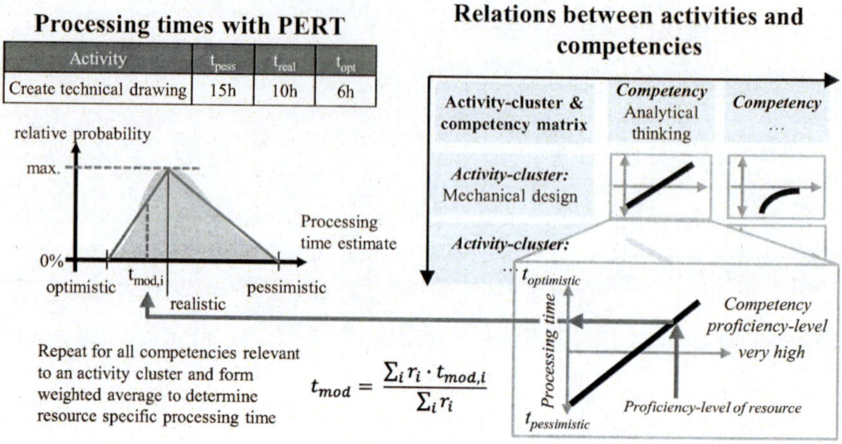

Fig. 5. Calculation of resource-specific competency-based processing times

5 Exemplary Application of the Methodology

This section will show the application of the developed methodology based on an exemplary simplified use case. In the use case, five employees (ID 1–5, upper matrix in Fig. 6) of a development department are considered. The relevant competencies are reduced to four competencies for simplification (C1–C4). A project (design of a brake disc) with five activities is considered. The activities can be assigned to three different activity clusters (middle left matrix in Fig. 6). Accordingly, the relationships between four competencies and three clusters of activities must be determined (middle right matrix in Fig. 6). In addition, the planned activity process times are recorded for all five activities using

Resource ID	Competencies (Proficiency-levels, five levels)			
	C1 Mechanical design	C2 Simulation	C3 Creativity	C4 Analytical thinking
1	high	low	very low	high
2	very high	very low	medium	medium
3	very low	very high	low	very high
4	medium	medium	low	medium
5	medium	low	very high	very low

Activity	t_{pess}	t_{real}	t_{opt}
Concept design	6h	4h	3h
Design brake disc	18h	12h	8h
Create technical drawing	15h	10h	6h
Perform thermal simulation	30h	20h	12h
Perform mechanical simulation	24h	16h	10h

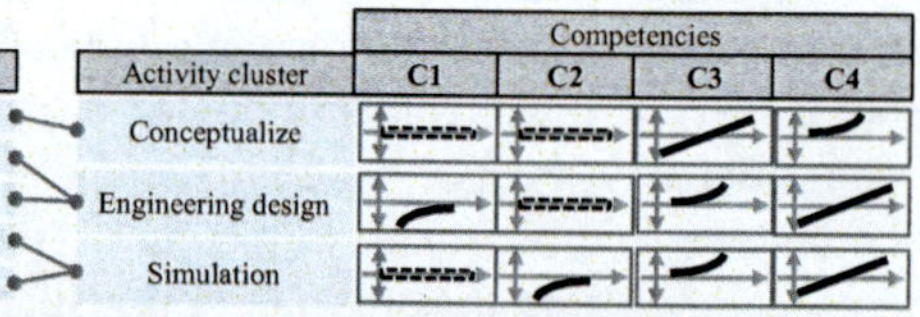

Resource ID	Individual processing times for the project activities				
	Concept design	Design brake disc	Create technical drawing	Perform thermal simulation	Perform mech. simulation
1	4.1h	11.8h	9.8h	21.2h	17h
2	3.8h	11.3h	9.3h	22h	17.7h
3	4h	12.3h	10h	16.7h	13.5h
4	4.3h	12.7h	10.5h	21h	16.8h
5	3.5h	13.7h	11.2h	23.2h	18.7h

Fig. 6. Exemplary application of the methodology

PERT. By applying the calculation method presented in Sect. 4.4, a matrix is generated in which the resource-individual processing times of the five activities are shown (lower matrix in Fig. 6).

The results of this use case show, that it is possible calculate individual processing times based on the competencies of the allocated resource. Furthermore, the results for the individual processing times depending on the competencies can be estimated as realistic according to the expectations based on empirical experience. It must be noted that these are artificially generated data sets, which is why transferability to practice must be examined separately. However, the necessary data input as well as the calculation procedure were successfully developed in this methodology.

6 Conclusion, Limitations and Further Research

The objective of the presented paper was to develop a methodology for determining resource-specific processing times to improve project planning. Furthermore, resource efficiency can be increased by selecting the resources with the best fit and thus the shortest processing time for an activity. A four-step approach was presented in which competencies and activities are described, followed by the identification of correlations and subsequent conversion to resource-specifically adjusted processing time. So far, the methodology was successfully tested with artificially generated data sets, a validation with actual company-specific data sets needs to be carried out in further research. Additionally, empirical surveys can help to quantify the influence of competencies on processing times even more precisely. Another limitation of the methodology is that so far only the processing time of activates, not the quality of the result is considered. The underlying hypothesis for this is that even with lower competencies a high-quality result can be achieved, but the processing time for this is significantly higher. Thus, if the quality of the result is fixed, there is only a time dependency. Nevertheless, the methodology could be expanded in the future to include the aspect of result quality. Besides this, future research activities can use the presented approach and integrate it into automated project scheduling with optimization methods.

Acknowledgement. Funded by the Deutsche Forschungsgemeinschaft (DFG, German Research Foundation) under Germany's Excellence Strategy — EXC-2023 Internet of Production — 390621612.

References

1. Schuh, G., Dölle, C.: Sustainable Innovation, 2nd edn. Springer, Berlin, Heidelberg (2021)
2. Schuh, G., Rudolf, S., Riesener, M., Kantelberg, J.: Application of highly-iterative product development in automotive and manufacturing industry. In: ISPIM Innovation Symposium, S.1 (2016)
3. Zafirov, R.: Produktionsmodelle und Simulation (DiFa). In: Eigner, M., Roubanov, D., Zafirov, R. (eds.) Modellbasierte Virtuelle Produktentwicklung, pp. 197–226. Springer, Heidelberg (2014). https://doi.org/10.1007/978-3-662-43816-9_10

4. Dammer, H.: Multiprojektmanagement. Zugl.: Berlin, Technology University, Dissertation, 2007, 1st ed., Gabler, Wiesbaden (2008)
5. Ogura, M., Harada, J., Kishida, M., Yassine, A.: Resource optimization of product development projects with time-varying dependency structure. Res. Eng. Des. **30**(3), 435–452 (2019). https://doi.org/10.1007/s00163-019-00316-6
6. Project Management Institute: Pulse of the Profession (R) 2021. Beyond Agility. (2021)
7. Haroune, M., Dhib, C., Neron, E., Soukhal, A., Babou, H. M., Nanne, M.: Multi-project scheduling problems with shared multi-skill resource constraints. In: PMS-2020 17th International Workshop on Project Management and Scheduling, Toulouse, France (2021)
8. Snauwaert, J., Vanhoucke, M.: A new solution procedure for multi-skilled resources in resource-constrained project scheduling. In: 17th International Workshop on Project Management and Scheduling 2020/21 (2020)
9. Steinle, C., Eßeling, V., Eichenberg, T.: Handbuch Multiprojektmanagement und-controlling: Projekte erfolgreich strukturieren und steuern, Erich Schmidt Verlag (2010)
10. Fiedler, R.: Controlling von Projekten, Springer Fachmedien Wiesbaden, Wiesbaden (2020)
11. Scheuring, H.: Ressourcenmanagement endlich in den Griff bekommen. In: projektManagement aktuell (2016)
12. Redaktion Can Do: Whitepaper Ressourcenmanagement (2015)
13. North, K., Reinhardt, K., Sieber-Suter, B.: Was ist Kompetenz ? In: Kompetenzmanagement in der Praxis (North, K., Reinhardt, K. & Sieber-Suter, B., eds), pp. 35–110, Springer Fachmedien Wiesbaden, Wiesbaden (2018)
14. Kuhlmann, A., Sauter, W.: Innovative Lernsysteme. Kompetenzentwicklung mit Blended Learning und Social Software, Springer Berlin Heidelberg, Berlin, Heidelberg (2008)
15. Wysocki, R.K.: Effective project management. Traditional, agile, extreme, 7th ed., Wiley, Indianapolis, Indiana (2014)
16. Kosar, G., Biedermann, H.: Wissen intern vernetzen. In: Wissen schafft Neues (Petra Wimmer, ed), pp. 71–78 (2017)
17. Krumm, S., Mertin, I., Dries, C.: Kompetenzmodelle. Hogrefe, Göttingen (2012)
18. Heyse, V., Erpenbeck, J.: Kompetenz training. Informations- und Trainings programme, 2nd ed., Schäffer-Poeschel, Stuttgart (2010)
19. Leslie, C.: Engineering competency model. In: 2016 ASEE Annual Conference (2016)
20. Backhaus, K., Erichson, B., Plinke, W., Weiber, R.: Multivariate Analysemethoden. Eine anwendungsorientierte Einführung, 15th ed., Springer Gabler, Berlin (2018)
21. Kano, N., Seraku, N., Takahashi, F., ichi Tsuji, S.: Attractive quality and must-be quality. J. Japan. Soc. Qual. Control **14**, 147–156 (1984)
22. Schlößer, E. S.: Auslegung prototypischer Produktinkremente im Kontext agiler Entwicklungsprojekte. Dissertation, 1st ed
23. Riesener, M., Kuhn, M., Keuper, A., Schuh, G.: Framework for FAMD-based identification of RCPSP-constraints for improved project scheduling. In: Design Conference (2022)
24. Project Management Institute: A guide to the project management body of knowledge. (PMBOK® guide); an American National Standard ANSI-PMI 99-001-2013, 5th ed., PMI, Newtown Square, Pa. (2013)

An Efficient Method for Automated Machining Sequence Planning Using an Approximation Algorithm

S. Langula(✉), M. Erler, and A. Brosius

Chair of Forming and Machining Processes, Technische Universität Dresden, 01062 Dresden, Germany

sebastian.langula1@tu-dresden.com

Abstract. Machining sequence planning for milling, also called operation sequence planning, can be considered one of the most important tasks of manufacturing process planning. Computer-Aided Process Planning (CAPP) is one of the application areas of machining sequence planning and is also an important interface between computer-aided design and computer-aided manufacturing. The planning tasks are multidimensional, but they are often handled in a linear way, which is one of the problems of conventional CAPP systems. This problem leads to limited solution space. The solution can be far away from the optimum or even not represented in reality due to resections caused by technical reasons. Multiple planning tasks cannot be combined in every way. They are restricted by the technological properties of the machining process, which makes the solution even more complicated. In contrast to the conventional approach, this paper generates valid sequences of operations based on a graph (Hamiltonian path) using a Simulated Annealing algorithm. Simulated Annealing is meta-heuristic, which finds global extrema within a graph by approximation. The algorithm's goal is to minimize the number of setups and tool changes. To evaluate the validity of the method, the Simulated Annealing algorithm was tested on parts with known experimental machining sequence optimum.

Keywords: Machining sequence planning · Simulated Annealing · Computer-aided process planning (CAPP)

1 Introduction

With the advent of Computer-Aided Design (CAD) and Computer-Aided Process Planning (CAPP), the product development process, which extends from the concept phase to the design and the manufacture of a part, has become more complex and multi-layered [1]. Information relevant to the manufacture of a part is already available during the product development process and can be used to improve the product at an early stage. An essential part of this development process is the planning of the production phases as it has a great influence on the quality and costs of a later product [2]. To create a manufacturing plan, two things are needed. A system for parameterization of production

M. Liewald et al. (Eds.): WGP 2022, LNPE, pp. 727–736, 2023.
https://doi.org/10.1007/978-3-031-18318-8_72

parts and logic for process planning [3]. The system for parameterizing production parts provides information such as machining features, tools, tool paths, and their tool access direction (TAD.) Two technologies play an overriding role here: feature recognition and feature-based design [4].

Subsequently, it is the task of the machining sequence planning (logic for process planning) to define in which order the different steps will be realized. Technical restrictions of the manufacturing technologies must be taken into account. A component can reach its final geometric shape in countless different processing sequences. In addition, the part to be manufactured constantly changes its geometric shape during the manufacturing process. Consequently, the task of planning the sequence of manufacturing features is very complex. The challenge is to find an optimal solution from the complex solution space given.

This paper focuses on the problem of machining sequencing and aims at minimizing the cost of machine tools- and setup changes.

2 Literature Review

Many state-of-the-art approaches use graph-based methods to address the problem of machining sequence planning. In the search for a valid and optimal solution (sequence), various graph-based algorithms have been developed [3].

Genetic algorithms generate a mutation of the solution sequence according to a previously defined fitting function [5]. A low runtime complexity of the algorithm is opposed by the disadvantage that the algorithm does not always find the global extrema [6].

Rule-[7] and knowledge-based [8] approaches are capable of finding high-performance solutions for a specially defined use case. The algorithm is only as good as the implemented knowledge. If the set of the base rule is incomplete or not applicable to the use case, poor or no solution sequences are found.

Artificial neural networks (ANN) are also used in machining sequencing [9]. The advantage of ANN is that they provide reasonable results for large amounts of data and many data dimensions. Nevertheless, the problem is that enormous amounts of data are required to train the algorithm. In addition, ANN behaves like a black box, thus making the verification of the result extremely difficult.

Other approaches like the ant algorithm [10] or particle swarm algorithms [11] offer the possibility of very high success rates at high efforts for the parameterization of the efforts.

Approximative algorithms like Simulated Annealing randomly swap elements within the sequence and check the resulting solution concerning predefined optimization parameters [12]. There is no guarantee of having found the optimal sequence even with this algorithm [13]. However, the runtime complexity of the algorithm can be adjusted by parameterization [14].

The probability of very good optimization results can be significantly increased by combining the various algorithms presented at the expense of runtime complexity [15].

3 Machining Sequence Problem Modeling

Known algorithms for automated machining sequence planning use TAD as a data basis. A machining sequence planning based on TAD does not fully consider the change of the manufacturing part during machining. The approaches imply that the manufacturing part can be spanned so that the feature is accessible without investigating this assertion. To provide a more realistic CAPP, this paper presents an alternative model in conjunction with a Simulated Annealing algorithm that has been extensively researched in the field.

The model for automated machining sequence planning which this work is based on is visualized in Fig. 1. The input is provided by a system used for the parameterization of production parts. Specific data such as manufacturing features to be machined with geometric specifications and technological information on type, shape, dimensions, and position are determined for the finished part.

Suitable tools and their access direction as well as possible clamping situations based on realistic clamping surfaces are explicitly assigned to each manufacturing feature to generate the individual manufacturing characteristics.

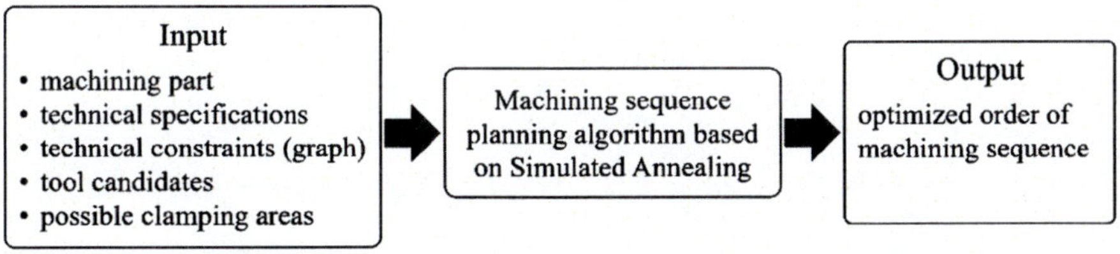

Fig. 1. The basic framework of the method for automated machining sequence planning.

Subsequently, the information obtained for each manufacturing feature is mapped in a directed graph. The individual nodes of the graph represent manufacturing features, and the edges of the graph contain manufacturing information for machining (tools and their access direction, as well as setups). For the sake of completeness and clarity, an output node representing the raw part is added to the graph. The graph represents the complete solution space of the machining sequencing problem. However, choosing an arbitrary machining sequence is not possible due to manufacturing restrictions. To represent these restrictions, they are introduced by precedence relations between the nodes of the graph. A machining sequence in the graph is only valid if it satisfies the priority relationships between the machining operations of the individual manufacturing features.

Constraints between features arise from the practice of machining. Accordingly, a constraint exists if a production feature prevents the processing of another production feature. This means that one production feature conceals another one. Thus, the constraint is that the upper production feature can only be machined first, and afterward the hidden production feature, e.g. roughing before finishing or core drilling before tapping.

Based on this, the sequence of processing the manufacturing features is represented as a Hamiltonian path [16] of least cost. The generation of valid plans has the structure of the well-known traveling salesman problem from graph theory. It is a combinatorial optimization problem and has its roots in theoretical computer science. The traveling salesman problem is considered to be an NP-hard problem. Those are decision problems that can be solved with non-deterministic algorithms in polynomial runtime concerning the input size.

Exact solution methods harbor the problem that the computation time to find the solution increases proportionally with the complexity of the solution. An additional challenge arises from the ever-changing geometry of the part during the process. Approximate algorithms are used to deal with complex issues such as machining sequencing.

In this approach, Simulated Annealing is applied to the problem. The result of the algorithm is a valid sequence of features that have been optimized according to the number of mold and fixture changes.

4 Searching for a Better Machining Plan

4.1 Simulated Annealing

After generating the solution space by modeling the problem in a directed graph and defining the optimization criteria, the next step is to determine a sequence from a large number of variants that is optimal according to both optimization criteria. The approach of this paper is the approximate optimization method Simulated Annealing. Figure 2 shows the flowchart of the algorithm tailored to the machining sequencing problem, where S depicts the currently best cost and C the cost of the actual sequence.

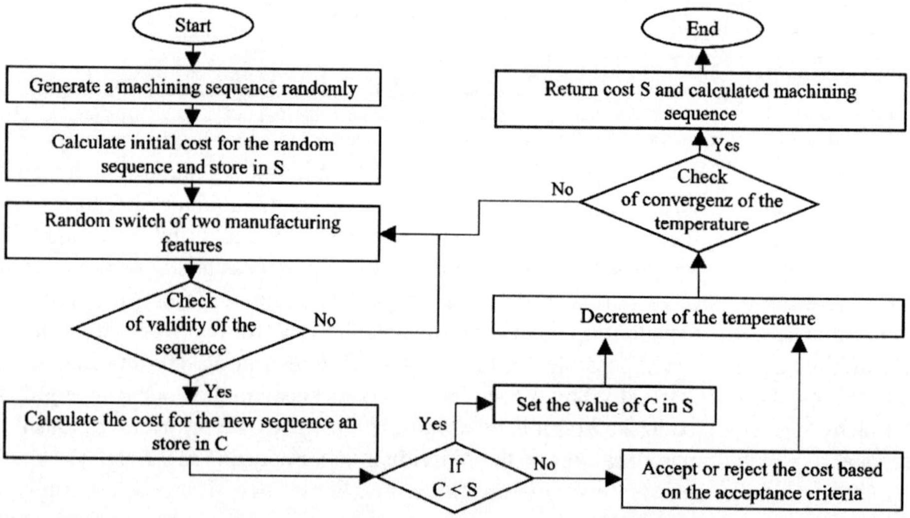

Fig. 2. Flow chart for the algorithm for automated machining sequencing planning based on Simulated Annealing.

Specifically, it is a meta-heuristic algorithm for approximating global optimization in a large search space for an optimization problem. The algorithm is applied to problems where it is more important to find an approximate global optimum than a precise local optimum in a specified period.

4.2 Determination of Costs and Optimization Criteria

The criteria for the evaluation and optimization of machining sequence plans used in this work are the number of setups and tool changes. These should be reduced to a minimum. The cost C for the tool changes C_T and reclamping operations C_S are calculated according to Eq. 1, where j is the number of manufacturing features.

$$C = \sum_{i=1}^{j} (C_{S_i} + C_{T_i}) \tag{1}$$

A pool of tools and setups has been developed to determine the effort. Each manufacturing feature has its tool and setup pool. This pool contains tools and setups suitable for the explicit feature to be processed. The functionality of the pools is identical. The functionality of the pools is exemplarily shown for the tools in Fig. 3. The toolpool represents a list of the tools suitable for processing a feature.

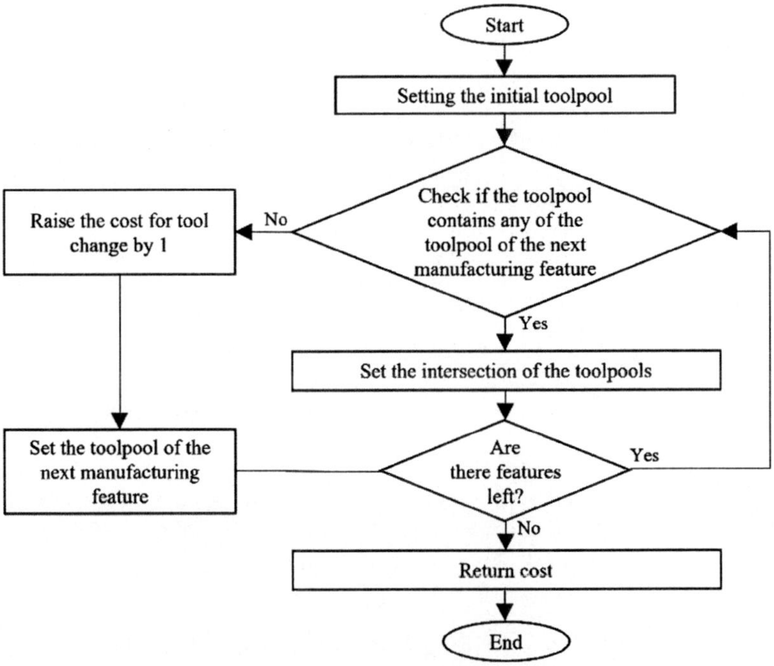

Fig. 3. Flowchart to determine the cost of tool changes.

The setup pool represents explicit clamping situations that are mapped by realistic geometries of clamping surfaces. Opposite clamping surfaces in combination result in possible setups. This approach takes the constantly changing shape of the production part during machining into account, thus enabling realistic machining sequence planning.

4.3 Tolerance Criteria

Simulated Annealing works by accepting higher costs for some time, i.e., steps upwards are allowed based on tolerance criteria avoiding local extrema (at the same time). The temperature determines the maximum tolerance between iterations of the sequence accepted so that sometimes there are higher costs allowed than the currently known optimum. It authorizes the algorithm to deviate from local optimal solutions and to search for global optimal solutions. The basic requirement is that the temperature is not reduced too quickly, otherwise a global search may not be possible. By decreasing the temperature increasing iterations, the tolerance steadily becomes smaller, converging the probability of accepting higher costs to zero. The convergence of the temperature T is shown in Eq. 2, where c depicts the change in sequence, t the current temperature, and r a random number between 0 and 1 [17].

$$T = exp\left[-\left(\frac{c}{t}\right)\right]r \tag{2}$$

5 Case Studies

A prismatic fabrication part as shown in Fig. 4 is used to test the algorithm. The constraints of the manufacturing features are visualized on the right side of Fig. 4.

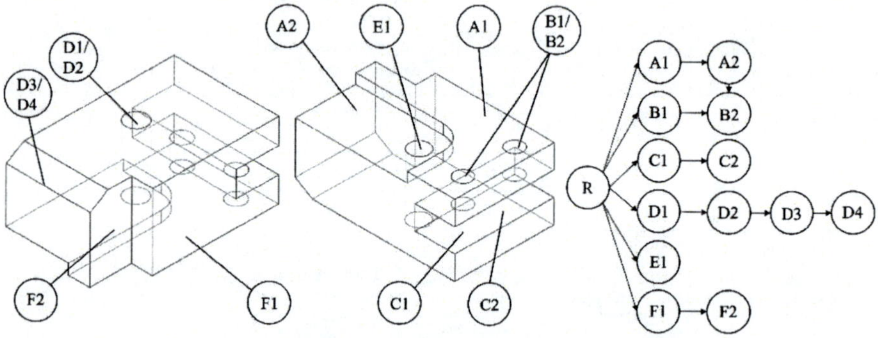

Fig. 4. Sample part and its constraints

The part consists of 13 manufacturing features, e.g. steps, a hole, threaded holes, and a chamfer. The features of the manufacturing part are explained in Table 1. In addition, the clamping possibilities compatible with the respective feature and the tools that can be used for machining are assigned.

Figure 5 depicts an example of the clamping surfaces given by the cubature for the production part. The results are shown in the matching tables according to Chap. 4.2.

The case study was performed with the parameters shown in Table 2.

Table 1. Manufacturing features of the sample part with their parameters

Feature	Feature descriptions	Setup candidates	Tool candidates
B	Blank	1 up to 24	T1 up to T9
A1	Rough face milling	1, 2, 3, 4, 18, 19, 20, 21, 22, 23, 24	T1, T6, T8
A2	Finish step milling	2, 3, 4, 9, 10, 11, 12, 13, 14, 15, 16, 18, 19, 20, 21, 22, 23, 24	T2, T7, T9
B1	Drilling	1, 2, 3, 4, 5, 6, 7, 8, 13, 14, 15, 16, 17, 18, 19, 20, 21, 22, 23, 24	T3
B2	Thread cutting	1, 2, 3, 4, 5, 6, 7, 8, 13, 14, 15, 16, 17, 18, 19, 22, 23, 24	T4
C1	Rough milling	5, 6, 7, 8, 12, 13, 14, 15, 16, 17, 18, 19, 20, 21, 23, 24	T1, T6, T8
C2	Slot milling	5, 6, 7, 8,12, 13, 14, 15, 16, 17, 18, 19, 20,21, 23, 24	T2, T7, T9
D1	Rough face milling	1, 3, 4, 18, 19, 20, 21, 22, 23, 24	T1, T6, T8
D2	Finish face milling	1, 3, 4, 18, 19, 20, 21, 22, 23, 24	T2, T7, T9
D3	Chamfer rough milling	3, 4, 5, 6, 7, 8, 9, 10, 11, 12, 13, 14, 15, 16, 17, 18, 19, 20, 21, 22, 23, 24	T1, T6, T8
D4	Chamfer finish milling	3, 4, 5, 6, 7, 8, 9, 10, 11, 12, 13, 14, 15, 1, 17, 18, 19, 22, 23, 24	T2, T7, T9
E1	Drilling	1, 2, 3, 4, 5, 6, 7, 8, 9, 10, 11, 12, 13, 14, 15, 16, 18, 19, 20, 21, 22, 23, 24	T5
F1	Rough face milling	1, 2, 5, 6, 7, 8, 9, 10, 11, 12, 18, 19	T1, T6, T8
F2	Finish step milling	1, 2, 5, 6, 7, 8, 9, 10, 11, 12, 13, 14, 15, 16, 20, 21, 22, 23, 24	T2, T7, T9

6 Results and Discussion

The presented method for automated machining sequence planning based on the approximation algorithm Simulated Annealing was applied to an example part. The results are compared with the known optimum in Table 3. Optimization criteria were the number of tools and chipping changes.

The method presented found the optimum in all test runs. This was proven by the calculation of all possible sequences. The method requires 3911 iterations with 6.2 billion possible combinations. The runtime behavior of the method is shown in Fig. 6. The local minima at a small number of iterations are left again by the large tolerance at the initial high temperature. Only with decreasing temperature, the global minimum is approached step by step.

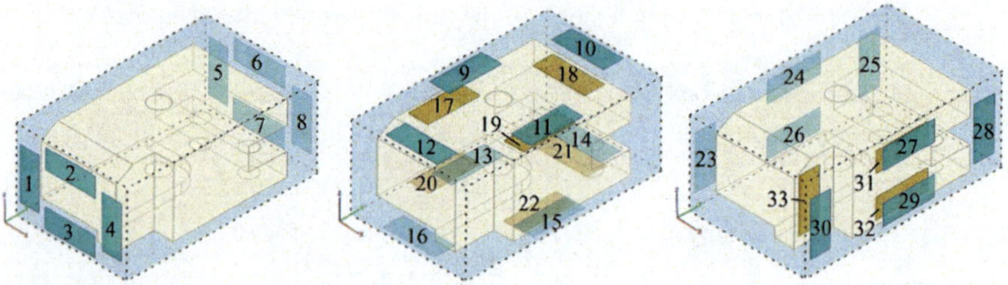

Possible Setups

Clamping Area	Setup	Clamping Area	Setup	Clamping Area	Setup	Clamping Area	Setup	Clamping Area	Setup
1 and 5	A	9 and 13	E	18 and 14	L	23 and 30	R	25 and 28	V
2 and 6	B	9 and 20	F	18 and 21	M	23 and 33	S	26 and 29	W
3 and 7	C	17 and 13	G	11 and 15	N	24 and 27	T	26 and 32	X
4 and 8	D	17 and 20	H	11 and 22	O	24 and 31	U		
		10 and 14	I	19 and 15	P				
		10 and 21	J	19 and 22	Q				
		12 and 16	K						

Fig. 5. Clamping areas and the resulting possible setups of the sample part.

Table 2. Parameter setting

Feature	Value
Cost changing setup C_S	100
Cost changing tool C_T	10
Initial temperature T	1
Decrement of temperature	0.99
Static iterations	100

Table 3. Comparison of the algorithm and the known optimum

Feature	Simulated Annealing	Calculation of all sequences
Minimum cost	159	159
Iterations	3911	$13! = 6.227.020.800$
Valid sequences found	3280	8108100

7 Conclusion and Outlook

In this paper, a method for optimization of machining sequence planning for milling based on the approximation algorithm Simulated Annealing is proposed. The method has been applied to a sample part. The approach of the paper expands the state of the art with realistic setups. The results are compared with the known optimum.

The proposed method can determine nourishing solutions to NP-hard problems such as machining sequencing problems. From the results obtained, it is clear to say that

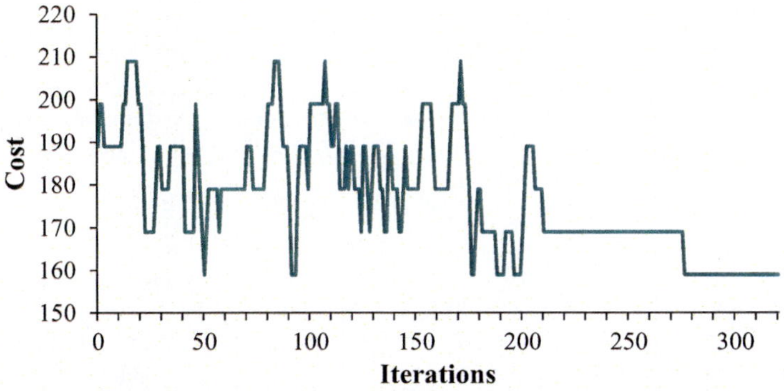

Fig. 6. The runtime behavior of the algorithm

the algorithm produces improved optimal sequences combined with less computational time. In the topic of machining sequence planning and computer automated process planning, efficient heuristic search is required to explore the large solution space of potential machining sequences. In the case studies, the algorithm has always found a valid machining sequence. Hence, the algorithm found the real optimum for the problem addressed. It seems that the algorithm can be considered quite effective as in most cases it finds a solution representing a good approximation to the optimum in a reasonable number of iterations and runtime.

Gefördert durch:

Bundesministerium für Wirtschaft und Energie

aufgrund eines Beschlusses des Deutschen Bundestages

Acknowledgements. The IGF project 21808 BR/2 of the Bundesvereinigung Logistik (BVL) e.V. is funded via the AiF as part of the program to promote joint industrial research (IGF) by the Federal Ministry of Economics and Energy based on a resolution of the German Bundesta.

References

1. Ehrlenspiel, K., Kiewert, A., Lindemann, U.: Cost-Efficient Design, 1st edn. Springer-Verlag, Heidelberg (2007)
2. Wang, H.: A fault feature characterization-based method for remanufacturing process planning optimization. J. Clean. Prod. **161**, 708–719 (2017)
3. Yusof, Y., Latif, K.: Survey on computer-aided process planning. Int. J. Adv. Manuf. Technol. **75**(1–4), 77–89 (2014). https://doi.org/10.1007/s00170-014-6073-3
4. Wang, J., Wu, X., Fan, X.: A two-stage ant colony optimization approach based on a directed graph for process planning. Int. J. Adv. Manuf. Technol. **80**(5–8), 839–850 (2015). https://doi.org/10.1007/s00170-015-7065-7
5. Zhang, F.: Using genetic algorithms in process planning for job shop machining. IEEE Trans. Evol. Comput. **1**(4), 278–289 (1997)

6. Su, Y., Chu, X., Chen, D., Sun, X.: A genetic algorithm for operation sequencing in CAPP using edge selection based encoding strategy. J. Intell. Manuf. **29**(2), 313–332 (2015). https://doi.org/10.1007/s10845-015-1109-6
7. Wang, W., Li, Y., Huang, L.: Rule and branch-and-bound algorithm based sequencing of machining features for process planning of complex parts. J. Intell. Manuf. **29**(6), 1329–1336 (2016). https://doi.org/10.1007/s10845-015-1181-y
8. Xuwen, J.: Intelligent generation method of 3D machining process based on process knowledge. Int. J. Comput. Integr. Manuf. **33**(1), 38–61 (2020)
9. Natarajan, K.: Application of artificial neural network techniques in computer aided process planning — a review. Int. J. Process Manage. Benchmarking **11**(1), 80–100 (2021)
10. Ha, C.: Evolving ant colony system for large-sized integrated process planning and scheduling problem considering sequence-dependent setup times. Flex. Serv. Manuf. J. **32**(3), 523–560 (2019). https://doi.org/10.1007/s10696-019-09360-9
11. Dou, J.: A discrete particle swarm optimisation for operation sequencing in CAPP. Int. J. Prod. Res. **56**(11), 3795–3814 (2018)
12. Li, W.D.: A simulated annealing-based optimization approach for integrated process planning and scheduling. Int. J. Comput. Integr. Manuf. **20**(1), 80–95 (2007)
13. Ingber, L.: Simulated annealing: practice versus theory. Math. Comput. Model. **18**(11), 29–57 (1993)
14. Haddadzade, M., Razfar, M.R., Zarandi, M.H.F.: Integration of process planning and job shop scheduling with stochastic processing time. Int. J. Adv. Manuf. Technol. **71**(1–4), 241–252 (2013). https://doi.org/10.1007/s00170-013-5469-9
15. Salehi, M.: Optimization process planning using hybrid genetic algorithm and intelligent search for job shop machining. J. Intell. Manuf. **22**, 643–653 (2011)
16. Garrod, C.: Hamiltonian path-integral method. Rev. Mod. Phys. **38**(3), 483–494 (1966)
17. Kirkpatrick, S.: Optimization by simulated annealing. Science **220**(4598), 671–680 (1983)

Early Detection of Rejects in Presses

J. Koß(✉), A. Höber, R. Krimm, and B.-A. Behrens

Institut für Umformtechnik und Umformmaschinen, Leibniz Universität Hannover, An der Universität 2, 30823 Garbsen, Germany
koss@ifum.uni-hannover.de

Abstract. Various production parameters such as inhomogeneous material properties or varying lubrication lead to deviations in manufacturing. Quality management must ensure that the required geometric dimensions and tolerances are maintained. In many cases, the inspection is carried out randomly, manually and at the end of the production chain, which prevents early detection of rejects and intervention. The problem complexifies due to the increasing demand for 100% testing of workpieces, e.g. for safety-relevant components in the automotive industry. This leads to additional effort regarding time, personnel and logistics. One solution to these problems is the inline measurement of the workpiece geometry. Due to rough environmental conditions in forming machines, the implementation presents particular challenges. In this publication, the disturbance variables occurring in presses are described and requirements are derived which result for the applied sensor technology. Based on this, a methodology for measuring small rotationally symmetrical workpieces is presented.

Keywords: Inline measurement · Reject detection · Quality management

1 Introduction

Due to a high output rate and efficient use of material, multi-stage presses (Fig. 1) are established in many industries for the production of small to medium-sized components with large batch sizes [4]. One aspect that offers potential for further development is the achievable manufacturing accuracy using multi-stage presses. The uncontrolled variation of workpiece dimensions has already been investigated and documented in various scientific papers [1, 5–9]. In general, the causes for the undesired variation of workpiece dimensions are the forming tool, the positioning of the workpieces in the tool, the lubrication, the raw material as well as the forming machine itself. In order to reduce weight and costs while increasing functionality and safety, the requirements on workpieces are getting higher and higher. A defective workpiece may have serious consequences and result in considerable economic damage to a company. Therefore, a demand for 100% good parts is no longer a rarity. Continuously dimensionally accurate, defect-free workpieces at the end of the production chain are the desired optimum required for profitable and competitive production [1, 10].

Even today, manual inspection of the workpieces at the end of the production chain is a common method to monitor the component quality. This requires a huge effort

M. Liewald et al. (Eds.): WGP 2022, LNPE, pp. 737–746, 2023.
https://doi.org/10.1007/978-3-031-18318-8_73

of logistics, personnel and time. The inspection costs are very high due to the high investment and operating costs of measuring equipment, the additional facilities required, and the need for specially trained personnel. In addition, component quality is often only checked after the entire batch has been produced. It is then no longer possible to intervene or change production parameters [3]. Thus, ensuring consistently high component quality at low production costs remains a major challenge for all manufacturing companies. Automated in-process monitoring of component dimensions is therefore intended to supplement or, in some cases, replace the manual inspection.

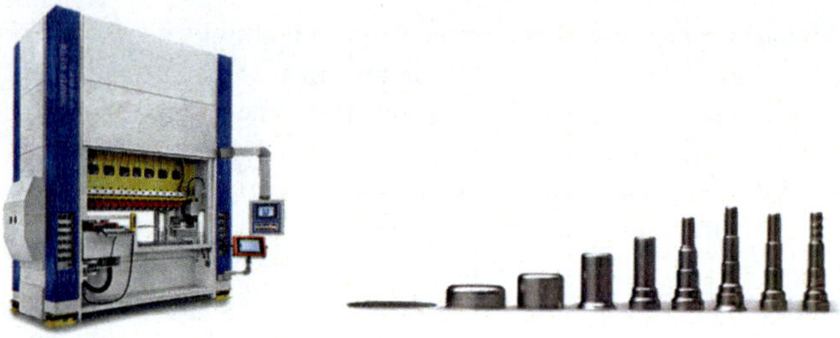

Fig. 1. Multi-stage press and exemplary stage sequence [11]

2 State of Research

Quality management

Although quality management has no direct impact on the value generation of a company, a wrong strategy can cause high costs and, hence, cut the profit. The time required for manual measurement of the workpieces after the production processs and, thus, the personnel costs are high. Once rejects have been detected, the machine must be stopped in order to identify and fix the cause. This procedure is also referred to as reactive maintenance. Another approach is the concept of preventive maintenance, in which active elements are replaced and readjusted after a certain time regardless of their condition. However, the potential to produce more quality parts with the existing components and the previous settings are lost [12]. In [13] a simulation method was presented in which the wear of all quality-relevant components in multi-stage presses is monitored in order to predict the quality loss of the final products and the system failure rate. From this, a decision can be made as to whether preventive maintenance is appropriate. A third variant is condition-based maintenance, in which intervention immediately takes place as soon as the production of rejects occurs. This approach requires that the workpieces and/or the active elements of the tools are monitored during the process.

In-process control of component dimensions

The inline, machine-integrated measurement of forming parameters has been a topic of academic discussion for many years. Ever more accurate and robust sensor technologies are constantly offering new application possibilities. However, such systems are still rarely found in forming plants [2].

The reasons for this include the following:

- High demands on the measuring technology to be applied.
- Lack of guidelines and planning instructions for users.
- Financial investment without prior assessability of the actual benefit.

According to the current state of the art, only indirect systems are available for the automated, in-process control of sheet metal components manufactured using multi-stage presses. In these systems, for example, the forming force or noise emission are measured and irregularities or process errors are inferred [14–17]. However, the accurate and direct measurement of the component geometry of small components during ongoing production is not possible with these systems. Systems providing 100% in-process monitoring only exist for large automotive body components, which are produced with a low number of cycles [18, 19].

To provide direct measurement of small rotationally symmetrical components in multi-stage presses and multi-stage dies, the topic is subject of the already finished AiF project 19904N [20] and the still ongoing project 21554N at IFUM. The measurement is supposed to take place during the process in the last, free stage of the multi-stage die (Fig. 2). The disturbance variables in presses that are relevant for sensors, will be investigated. Based on this, the use of different sensor systems will be discussed.

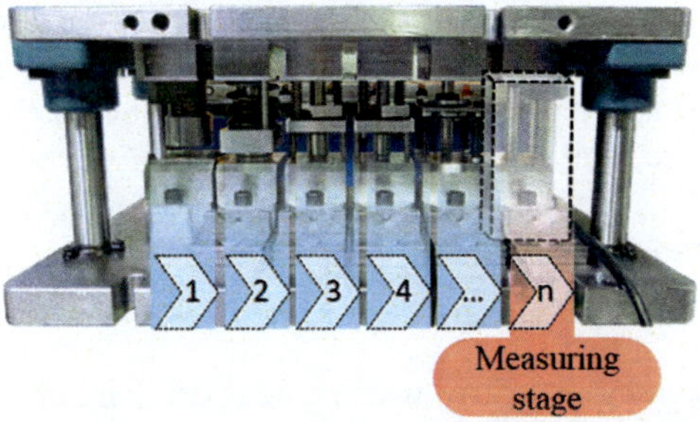

Fig. 2. Measuring system in the last stage of a multi-stage press or a multi-stage die

3 Measurement Conditions in Production Environments

The lack of available measuring systems for automated, in-process control of sheet metal components produced by multi-stage presses is mainly a result of the difficult environment conditions in the installation space of a press [20].

- thermal interactions and high temperatures,
- changing lubrication condition of the workpieces,

- shock and vibrations,
- short measuring time window due to high stroke rates,
- impurities,
- limited installation spaces for measuring systems.

As part of the AiF project 19904N, the effects of disturbance influences on the measurement of the workpiece geometry during production were investigated and requirements for sensor technology were derived. For this purpose, an industrially used multi-stage press was equipped with various sensors and the possibility of integrating a measuring system in the last, free-standing stage was investigated.

Temperature

Measurements with a thermal imaging camera showed that the maximum temperatures in the multi-stage press (~70 °C) were locally limited to the component and the temperature in the press room (~35 °C) and the die stages were lower (~45–50 °C). In addition, the temperatures of both component (~45 °C) and installation space (~35–45 °C) were lower as well in the last stage, where the measuring stage is to be integrated. This is because no forming takes place at this point and accordingly no additional thermal energy is released. Depending on the processed material, the stroke rate, the number of stages, the forming operations, the dimensions of the components and the installation situation, may vary.

However, the measurements show that even in cold forming processes the temperature conditions certainly play a role. Sensor systems that have to be used in close proximity to the workpiece as a result of a small measuring range can reach their limits or not be used at all, since adhesive layers between sensor components can soften, for example. The use of sensors with a greater distance between sensor and measured object (e.g. laser profile sensors) is less critical in this respect. Nevertheless, in all cases it must be kept in mind that the workpiece to be measured in a warm state still undergoes a reduction in volume as a result of the cooling process, which can certainly play a role depending on the accuracy requirements [20].

Lubrication

Lubrication can be challenging in two ways: One is dripping oil or aerosol can get into the interior of sensor systems and cause damage or cause dirtying of relevant parts such as lenses. For another, oiling of the workpiece can lead to measurement errors, which especially affects the use of optical sensor systems. In the latter, two superimposed effects occur. First, the laser beam is refracted when it enters the oil layer so that it hits a different point on the surface of the workpiece than intended (Fig. 3a). Second, the incoming light is reflected by the oil layer, which leads to blurring or scattering of the measured values (Fig. 3b). The resulting error increases with the entrance angle of the laser relative to the surface normal and the oil layer thickness. Common oiling quantities are in the range of 1–5 g/m^2, which, depending on the component surface and assuming homogeneous distribution, results in layer thicknesses in the range of one to several tens of micrometers. Depending on the entrance angle, this leads to measurement errors of the same order. This must be taken into account in particular if the desired manufacturing and measurement accuracy is in this range. An inhomogeneous distribution of the oil layer thickness makes this aspect even more challenging.

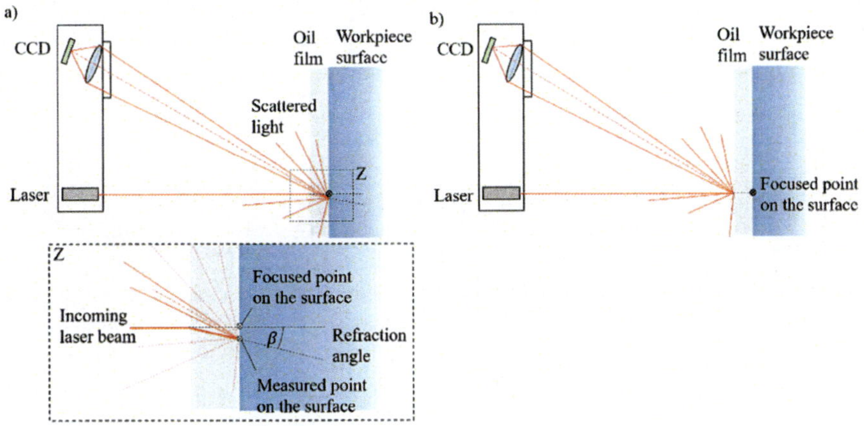

Fig. 3. Measurement error due to oiled surfaces

These effects can be prevented to a certain extent by the following approaches:

- Orientation of the sensor so that the entrance angle with respect to the surface normal is as small as possible,
- Off-center exposure of the measured object to reduce the light reflected by the oil layer,
- Removal of the oil layer in the measuring area by applying compressed air.

Vibration

The shocks and vibrations that occur in presses can result in a particular challenge for the applied measurement technology. They can lead to damages in the sensor and loosen adhesive surfaces or screw connections. As a result, measurement values become faulty and a readjustment of the setup is necessary. Further, measurement errors can occur because of the vibrations. The occurrence of errors even increases when the target and sensor are subject to different vibrational movements.

Hence, when using measurement technology in vibrating systems, particular attention must be paid to

- shielding the measuring system from vibrations by means of damping elements,
- designing the measuring system in such a way that the measured object and the sensor vibrate in the same manner, and
- designing the measuring system so that the natural frequency is higher than that of the press.

In order to quantify the requirements on the measurement technology in the case of the investigated multi-stage press, measurements were carried out with acceleration sensors at various positions (ram, transfer bars, machine block, bottom die, press frame, transport level). The measurements showed that vibrations of more than 100g occurred at the slide and the lower dies. On the machine block and press frame, however, significantly lower values occurred. The permanently occurring vibrations were in the range up to a maximum of 10 g [20]. Again, it should be noted that the results are highly dependent on the machine, the process, the die, and other aspects.

Measuring Time

The processes on multi-stage presses and multi-stage dies are designed for high performance and are carried out with the highest possible number of strokes. The time window available for measurement is therefore accordingly short. Its definition depends on various factors, such as object visibility, object position, and disturbance variables. To determine the available time, the transfer movement, the ram movement, the gripper activity and the occurrence of vibrations and shocks in particular must be considered in detail. On the one hand, it must be ensured that the measured object is not in the grip of the transfer system and that the grippers do not cover the measured object after setting down or before picking up. On the other hand, the measurement time window must be chosen so that the vibrations are as low as possible in order to reduce or avoid the measurement errors caused by this. The aspects mentioned above must be considered for every production case, as well as shock and vibration.

Installation space

One of the great benefits of multi-stage presses and dies is the efficient use of space. The stages are therefore kept as compact as possible, which makes the integration of measuring systems more difficult. Particularly in the case of small components, the available installation space is another limiting factor in the integration of sensors in press systems. The characteristics of the transfer movement are also an important factor. Depending on the measuring range of the sensors, they must be positioned more or less close to the workpiece. To avoid collisions, it is therefore necessary to match the positioning and mounting of the sensor to the conditions of the transfer system.

4 Suitable Sensor Systems and Their Limitations

Depending on the measuring task, application point, temperature of the workpieces and materials used, there are different types of sensors best suited for the task. Sensors that have a high measuring rate and high accuracy while being robust against vibrations and shock are laser-optical systems and eddy current sensors. The latter are the least susceptible to disturbances. However, compared to laser-optical systems, the sensors are often more expensive and only allow point measurement. In addition, the sensor heads are large compared to the dimensions of the available measuring spot. Thus, they can only be inserted into narrow openings such as inner diameters of rotationally symmetrical components to a limited extent. Another challenge is that these sensors must be mounted very close to the workpieces due to the small measuring range. Especially in complex systems such as multi-stage presses or multi-stage dies, difficult installation conditions arise because of the transfer of parts between the stages. On the other hand, laser-optical systems are more susceptible to disturbances, especially when reflective surfaces are to be measured in oily environments. The use of laser profile sensors, however, offers the possibility of measuring workpieces in their entirety with just a few sensor units. For workpieces with particularly narrow inner diameters, the measuring range of a laser profile sensor may not be sufficient, or the angle of incidence may be too wide. In this case, a compact sensor could be moved into the interior of the workpiece. However, this requires a measurement offset by 90° relative to the longitudinal axis. Such models can be found among the confocal chromatic sensors.

Table 1. Suitable sensor systems

	 [21]Laser profile sensor [21]	 [21]Confocal displacement sensor	 [21]Eddy current sensor [21]
Measurement			
Type	Profile	Point	Point
Range	++	–	–
Rate	+	0	++
Accuracy	+	++	++
Special feature	/	90° version	/
Disturbance resistance			
Vibration/shock	+	+	++
Temperature	+	+	++
Material surface/lubrication	–	0	++
Mounting space/integration			
Size	–	++	+
Distance sensor-target	++	–	–
Costs	+	–	–

5 Measuring Device in Multistage Forming Tools

In AiF project 21554N, a measuring stage is currently under development that enables an integration into the final stage of a multi-stage die. Here, the use of two laser profile sensors will be tested, which are used to measure the entire workpiece geometry of a flange housing (Fig. 4). The measurement methodology involves rotating the workpiece once. During rotation, angle-dependent contour profiles are measured. In order to relate the two sensors to each other locally, a reference object is placed under the workpiece. This object is manufactured with high precision and the position of key features and contour elements in relation to each other is known very precisely. Since the sensors measure certain parts of the reference object, the measured contour profiles of the workpiece can also be assigned to a specific location.

In order to perform the measurement reliably, the component has to be centered, fixed and rotated in the measuring stage. A pneumatically actuated centric gripper was selected for centering and fixing on the inner side of the workpiece (Fig. 5). This allows a fast reaction time and a high force transmission while minimizing the covering of the workpiece. In addition, the three gripper fingers are mechanically coupled to each other via the wedge-hook principle, which allows very precise centering of the component. However, this gripper requires a compressed air supply for operation. The rotary unit therefore requires the possibility of a compressed air feed-through. A servo-electrically operated rotary unit with a hollow shaft is thus used in the measuring stage.

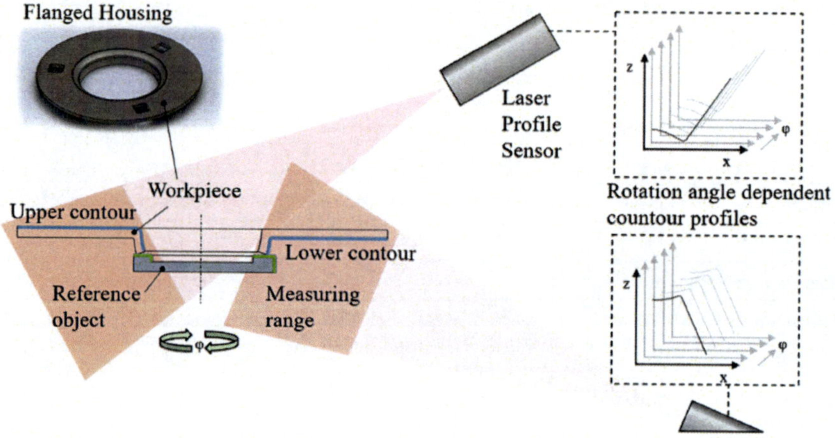

Fig. 4. Measurement of workpiece geometry with two laser profile sensors

In order to precisely measure the angle of rotation, the rotary unit has an integrated encoder. Despite the extremely compact design, the dynamics of the motor are sufficient to completely rotate both the gripper and the workpiece in 200 ms. The measuring stage is currently being implemented within a test setup and will subsequently be subjected to a measurement capability analysis. For this purpose, the measuring stage is also tested under different lubrication and temperature conditions of the component in a vibration test rig available at IFUM.

6 Summary

Today, there exist no systems for direct, in-process measurement of component geometry in press systems, which is mainly due to the challenging environmental conditions in press systems. However, advances in sensor technology are constantly opening up new

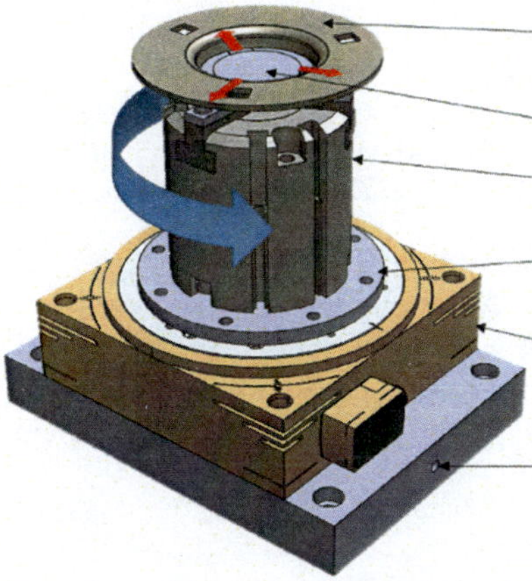

Fig. 5. Model of the measuring stage

possibilities for implementing appropriate systems. In various completed and ongoing projects, the IFUM is working on the direct measurement of components in multi-stage presses and multi-stage dies. In this context, the disturbance variables occurring in presses have been measured and analyzed, and requirements for sensors that could potentially be used have been derived. On the basis of the knowledge gained, a measuring system is currently under construction which will be used in the final stage of a test die to measure the entire geometry of small rotationally symmetrical workpieces during the process.

Acknowledgement. The IGF project "In-process reject detection by means of comprehensive measurement of the workpiece geometry in multi-stage presses" (funding number 21554 N) of the research association EFB e.V. was funded by the German Federal Ministry of Economics and Climate Protection via the German Federation of Industrial Research Associations (AiF) within the framework of the program for the promotion of joint industrial research (IGF) on the basis of a resolution of the German Bundestag.

References

1. Endelt, B., Nielsen, K.B.; Danckert, J.: Adaptive shimming control for the ultimate deep drawing process. In: Proceedings of the IDDRG 2008 International Conference, pp. 569–580, Olofström, Sweden (2008)
2. Maier, S.J.: Inline-Qualitätsprüfung im Presswerk durch intelligente Nachfolgewerkzeuge. Dissertation Technische Universität München. Schriftenreihe Umformtechnik und Gießereiwesen, TUM.University Press, Munich, Germany (2018)
3. Gevatter, H.-J., Grünhaupt, U.: Handbuch der Mess- und Automatisierungstechnik in der Produktion, Springer-Verlag, Heidelberg, Germany (2006)
4. Herlan, T.: Optimaler Energieeinsatz bei der Fertigung durch Massivumformung. Dissertation Universität Stuttgart, Springer-Verlag, Berlin Heidelberg New York London Paris Tokyo, 1989
5. Behrens, B.-A., Javadi, M.: Exakte und kostengünstige Qualitätskontrolle an Pressen in der Blechverarbeitungsindustrie. UTFScience **2**, 1–5 (2009)

6. Doege, E., Strasser, D.: Instabilitäten bei Pressen für den Karosseriebau. Idee — Vision — Innovation. Verlag Meisenbach, Bamberg, Germany (2001)
7. Doege, E., Derenthal, M.J., Großmann, K., Jungnickel, G.: Analyse der Werkzeug- und Maschinenerwärmung während der Anlaufphase von Anlagen der Blechverarbeitung. In: Research Report on EFB Project 208 (AIF 12721BG), Europäische Forschungsgesellschaft für Blechverarbeitung, Hannover, Germany (2003)
8. Malikov, V., Ossenbrink, R., Viehweger, B., Michailov, V.: Experimental study of the change of stiffness properties during deep drawing of structured sheet metal. J. Mater. Process. Technol. **213**(11), 1811–1817 (2013)
9. Weck, M., Brecher, C.: Werkzeugmaschinen 5, Messtechnische Untersuchung und Beurteilung, dynamische Stabilität. 7 edn. Springer Verlag, Berlin, Heidelberg, New York (2006)
10. N.N.: Sensing the quality. Sheet Metal Industries, July 2001, pp. 4–5 (2001)
11. Hubert Stüken GmbH & Co. KG
12. Reichel, J., Müller, G., Mandelartz, J.: Betriebliche Instandhaltung. Springer, Berlin Heidelberg, Germany (2009)
13. Lu, B., Zhou, X.: Quality and reliability oriented maintenance for multistage manufacturing systems subject to condition monitoring. J. Manuf. Syst. **52**(A), 76–85 (2019)
14. Doege, E., Behrens, B.-A.: Handbuch Umformtechnik: Grundlagen, Technologien, Maschinen, 2nd edn. Springer-Verlag, Berlin (2010)
15. MARPOSS Monitoring Solutions GmbH Homepage. https://brankamp.com. Last accessed 27 Apr 2022
16. TRsystems GmbH Homepage. https://www.unidor.info. Last accessed 27 Apr 2022
17. Yun, J.W.: Stoffflussregelung beim Tiefziehen mittels eines optischen Sensors und eines Fuzzy — Reglers. Dissertation, University of Hannover, Hannover, Germany (2005)
18. GOM — Gesellschaft für Optische Messtechnik mbH Homepage. https://www.gom.com/de. Last accessed 27 Apr 2022
19. Carl Zeiss Optotechnik GmbH Homepage. https://optotechnik.zeiss.com. Last accessed 27 Apr 2022
20. Behrens, B.-A., Krimm, R., Höber, A.: Prozessbegleitende Bauteilvermessung in Stufenpressen. In: Research report on EFB project 541, Europäische Forschungsgesellschaft für Blechverarbeitung e.V., Hannover, Germany (2020)
21. Micro Epsilon Messtechnik GmbH & Co. KG Homepage. https://www.micro-epsilon.de/. Last accessed 27 Apr 2022

Aspects of Resilience of Production Processes

Optimal Selection of Decarbonization Measures in Manufacturing Using Mixed-Integer Programming

C. Schneider(✉), S. Büttner, and A. Sauer

Fraunhofer IPA, Institute for Energy Efficiency in Production University of Stuttgart, Nobelstraße 12, 70569 Stuttgart, Germany
christian.schneider@ipa.fraunhofer.de

Abstract. Scholars have highlighted the importance of decarbonizing manufacturing industries for several years already. Industry accounts for about 20% of the EU's greenhouse gas emissions. In order to meet the targets set in the Paris Agreement, industry must reduce emissions to almost zero by 2050. A wide range of measures can be taken to achieve climate neutrality consisting of three categories: reducing greenhouse gases by adapting business models, substituting products or offsetting the emitted greenhouse gases. Companies have to determine the optimal set of measures taking into account their individual situation as well as available resources. From this, a complex optimization problem arises and the proposed decision model offers significant sup-port for the selection of decarbonization measures. By using the decision model, companies can achieve the greatest possible emissions reduction with a minimal set of resources according to their target system, thus taking into account net present value, benefits, and risks. This paper introduces a novel modeling of measures that incorporates relevant evaluation criteria. The arising decision model is solved by using Mixed-Integer Programming. The presented approach was validated in a case study with an industrial corporation.

Keywords: Operations research · Decarbonization · Optimization

1 Motivation

Rising energy and CO_2 prices increase the financial pressure on companies to reduce their emissions [1]. One of the most relevant groups in the energy transition is the industrial sector. Not only does it account for a large proportion of most countries' energy consumption, but also for associated energy- and process-related emissions [2]. Bauer et al. [26] present pathways for decarbonising different emission-intensive sectors such as production and end-use optimization. Available measures to reduce CO_2-emissions in industrial companies include [3]. Reduction of energy consumption through energy efficiency measures, reduction of process-related or process-induced emissions, for instance by substituting (metallurgical) coke with green hydrogen in steel production as well as

M. Liewald et al. (Eds.): WGP 2022, LNPE, pp. 749–760, 2023.
https://doi.org/10.1007/978-3-031-18318-8_74

the self-generation of renewable energies and their storage. Making decisions to decarbonize goes along with the need to identify an optimal mix of measures for a company. For stakeholders in general, but also for a company in particular, it makes sense to pursue pathways to achieve what is needed in an optimal way.

The ideal mix cannot be taken off the shelf, as each company's situation varies, even if the difference appears marginal [4].

Despite the importance of decarbonization, companies still face a lack of decision support methods to help identify the optimal mix of measures. By developing a decision support system, barriers can be lowered and the decarbonization of the industry can be advanced.

Therefore, the research question arises, how companies can identify their optimal selection of decarbonization measures. The focus of this paper is to present an approach using mixed-integer optimization to determine a company's ideal set of decarbonization measures on the basis of strategic priorities, predefined measures and boundaries. The paper is structured as follows: First, the related work is presented. This is followed by a definition of the problem, the measures and the mathematical formulation. Then, a case study is presented followed by a conclusion.

2 Related Work

Buettner et al. [2] focusses on questions that need to be answered to determine one's ideal decarbonization strategy and present a literature review that is condensed hereafter [2]. A number of studies explore pathways for decarbonization. Many of them focused much on the technological pathways and less on organizational frameworks [5]. Bataille et al. [6] present an "integrated [policy] strategy for a managed transition" in energy intensive industries, also including technology options. Rissman et al. [7] review policy options, sociological, technological, and practical solutions in detail. These studies address decarbonization of industry from either a policy, a supply-side, or technology perspective but are short of giving corporate concrete advice on how to get started from an individual company's perspective. Similarly, studies such as the one by Johnson et al. [8] analyze and compare national roadmaps for decarbonizing the heavy industry on a global scale, alongside factors such as ambition, financial effort, and mitigation measures. Nevertheless, this approach again leaves a gap when it comes to company-tailored advice. One effective way to develop decarbonization roadmaps involves applying approaches from the backcasting framework literature. This concept, established by Robinson [9], refers to a strategy where stakeholders/policymakers set up a target (energy consumption/emissions) and work backwards from this target to reach it in the future. This framework is widely applied in designing emission-reduction pathways.

Despite the importance of decarbonizing manufacturing industry, the optimal selection of decarbonization measures is an area of research that has received little attention to date. However, approaches from the field of energy efficiency provide valuable information and can be generalized to other decarbonization measures. Bayata et al. [10] develop a model for energy efficiency optimization in the design process of buildings using multi objective optimization. There are three objective functions to be minimized: building energy consumption, investment level, and CO_2 emissions. Bre et al. [11] present an optimization model for the optimal determination of building design parameters. For example, window type, roof type, and wall type are considered and variants with three to eight different states are modeled for individual parameters. The solution is based on the evolutionary algorithm for multi-objective optimization NSGA-2. Diakaki et al. [12] utilize multi-criteria optimization for increasing the energy efficiency of buildings including but not only focused on the design phase compared to Bayata et al. [10]. The model consists of different decision variables related to the building envelope and technologies. Energy demand, investment costs and CO_2 emissions are the objective functions. Eskander et al. [13] focus on the optimization of energy efficiency measures in Portuguese households. The optimization model identifies optimal retrofit solutions in different regions of Portugal. Six different energy efficiency measures are considered. The optimal energy efficiency measures are selected using a genetic algorithm with the inclusion of a constrained investment budget. The investment budget can only take the forms low, medium and high. Kontogirgos et al. [14] present a model for mixed-integer evaluation of residential energy conservation measures under uncertainty.

In addition to approaches in energy efficiency there are other optimization approaches focusing on decarbonization in industry. Maigret et al. [15] present a multi objective optimization using an evolutionary algorithm to minimize annual costs and CO_2-emissions in a refinery. Hu et al. [16] develop a multi-objective decision-making method to evaluate correlated decarbonization measures using a pareto optimal and marginal cost-effectiveness criterion.

In summary, it can be seen that approaches for the optimized selection of decarbonization measures in industry are still an understudied field of research. The formulation of the selection of decarbonization measures in the form of a decision model has not yet been carried out. However, preliminary work in the field of energy efficiency provides valuable information for the development of a solution procedure.

3 Methodology

3.1 Problem Definition

The following chapter describes the optimization problem that companies face when selecting decarbonization measures.

Choosing what measures must be prioritized requires determining the target criteria. Targets are the foundation to allow assessing which states or results are desirable and how their quality is to be measured [17]. For the development of a decision model, it is therefore crucial to determine the relevant dimensions of these preferences and to derive how these are to be measured. Cooreman [18] examines strategic dimensions of energy efficiency measures and concludes that the three strategic competitive advantages are cost, risk, and value of the measure. Adapting this approach, the goal criteria used for the optimization problem are cost, risks, and avoided CO_2 emissions.

With regard to the available measures Buettner et al. [3] describe measure categories that can be pursued by companies on site (internal measures).

- reduction of energy consumption (and of the connected load) through energy efficiency measures, including utilizing waste energy and passive resources such as passive ventilation;
- reduction of process-related or process-induced emissions, for instance, by substituting coke by green hydrogen in steel production or emissions released by the process itself;
- self-generation of renewable energies and their storage, for instance, solar-, wind-, hydro- or geothermal energy, including means for flexibilising the energy demand.

In addition to the optimized selection of decarbonization measures, constraints may exist that need to be incorporated into the optimization problem in the form of equations and inequalities. Constraints mentioned in the literature for the selection of measures include the investment budget [19–21] and emission targets [19, 22]. Based on these literature, the investment budget and CO_2 emission targets are used in the optimization problem in the form of constraints.

3.2 Assessment of Decarbonization Measures

In order to derive relevant evaluation criteria and thereby create an evaluation system, it is first necessary to define what characterizes the relevance of the evaluation system. Based on the research of Cooreman [18] referenced in the previous chapter, three strategic dimensions, cost, risk, and avoided CO_2 emissions, are to be used to assess a decarbonization measure.

With regard to the costs, three methods are most frequently used to assess the profitability of an investment [23]: Payback period, net present value (NPV), and internal rate of return (IRR). For the cost evaluation of decarbonization measures, the NPV is selected in the context of this work because it is described as an important lever for energy efficiency, which again can be generalized to decarbonization measures [24].

The second dimension is based on the avoided CO_2 emissions represented by $\text{tonesCO}_{2\text{eq. avoided}}$. Due to the importance of other greenhouse gases, the assessment is based on CO_2 equivalents.

Using the previously defined target system, a method for assessing risks in implementing decarbonization measures utilizing an approach presented by Schneider et al. [25] is used. Using the method, each measure is assigned a risk value representing the risk assigned connected to the measure's implementation. Thus, there is a ranking for the risk of all considered decarbonization measures.

3.3 Mathematical Formulation

The following section describes the previously defined problem in a mathematical form. The decision variables of the optimization problem in this case represent the decision for a set of certain decarbonization measures, which can be represented in a form where 1 stands for a selection of the measure and 0 for a non-selection.

$$A_i = \{a_{i,0}, a_{i,1}\}$$

where:

$$a_{i,0} = \text{Do not implement decarbonization measure } i$$

$$a_{i,1} = \text{Implement decarbonization measure } i$$

The basic decision to select a set of decarbonization measures is reduced to a series of individual decisions on whether or not to implement an available measure. Since a measure can be selected or not selected in binary form, it is a mixed integer problem. Integer constraints are, for example, compliance with the available investment budget.

The set of available action plans (1) of the decision maker is corresponding to the measures previously given as input variables and is formed by the cross product of the individual measures. This means that each combination of measures represents a possible investment plan.

$$A = \prod_{i=1}^{n} A_i = A_1 \times \ldots \times A_n \quad (1)$$
$$n = \text{Number of decarbonization measures}$$

A preference function is used to order given alternatives based on their relative utility for the user. Within the optimization problem the function consists of the weighted sum of the individual factors presented in the chapter above. These consist of the tons of CO_2 emissions avoided, the net present value associated with the measure and represented by the NPV and the risks of implementation. The individual preference functions are first transposed to the interval [0, 1]. U represents the value function, while Φ is used to describe the sum of the individual value functions.

$$U_{\text{CO}_2\text{avoided}}(A) := \text{CO}_2\text{emissions avoided}\left[t\text{CO}_2eq\right]$$

$$U_{NPV}(A) := \text{Net present value [\$]}$$

$$U_{Risk}(A) := \text{ Risk value [dimensionless]}$$

The value Φ(A) present in (2) represents the valuation of an alternative according to the preference of the decision maker. The individual preference dimensions are summarized to the overall weighted preference function.

$$\begin{aligned}\phi(A) = U_{CO_2\text{avoided}}(A) * q_{CO_2\text{avoided}} \\ + U_{NPV}(A) * q_{NPV} - U_{\text{Risk}}(A) * q_{\text{Risk}}\end{aligned} \quad (2)$$

$$\text{with} q_{\text{CO}_2\text{savings}} + q_{\text{Cost}} + q_{\text{Risk}} = 1$$

The weighting can be set directly by the decision-maker in a simple form. However, this method, which is often used in practice, is criticized from a scientific point of view and alternative methods are recommended [26]. Alternatively, the determination of the weights can follow the pairwise comparison of the Analytic Hierarchy Process by Saaty [27].

Constraints of the problem arise from the limitations of the investment budget and possible requirements on the CO_2 savings achieved.

$$g_{\text{Budget}}(A) \leq 0$$

$$g_{CO_2\ \text{savings}}(A) \leq 0$$

Equations can occur when exact savings (possibly with deviations) are to be fulfilled. Next, the individual components are combined to form an optimization problem represented in a maximization form. The following optimization problem (3) arises, where $h_i(A)$ and $g_i(A)$ and represent constraints in the form of equations and inequalities.

$$\max \phi(A) \text{subject to} \begin{cases} h_i(A) = 0, & i \in \{1, \dots m\} \\ g_j(A) \leq 0, & i \in \{1, \dots m\} \end{cases} \tag{3}$$

For the solution of the defined problem under the selected conditions, different optimization algorithms can be used fulfilling the requirements of the problem class of the optimization problem.

Figure 1 summarizes the process of determining the optimal investment plans. The starting point is a set of decarbonization measures that are already available in evaluated form, i.e. the criteria necessary for decision-making have been evaluated. For the risk assessment for example a separate risk assessment method was presented by Schneider [25]. For the optimization, it is furthermore necessary to determine the relevant constraints and the weighting of the target dimensions. Once this is done, the mixed-integer problem can be solved and the optimal set of measures can be determined.

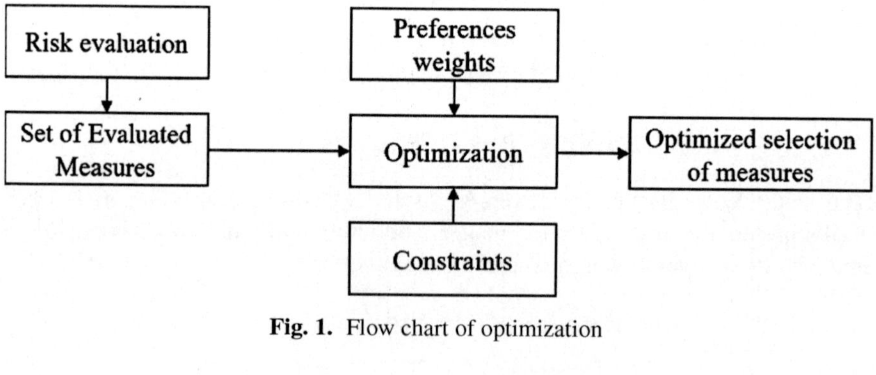

Fig. 1. Flow chart of optimization

4 Case Study

In the following, the presented optimization is applied in a case study. The aim is to identify the optimal action plans for selected investment budgets, i.e. to decide which of the available measures should be implemented. By determining relevant economic evaluation criteria for the action plans, this represents a key decision-making tool for management. The implementation of the case study requires a data set of measures. The case presented below is founded on a database of proposed decarbonization measures from a large industrial company in Germany. The dataset includes 32 decarbonization measures, most of which relate to energy efficiency measures. Individual measures also include the self-generation of renewable energy.

The measures included have already been evaluated by the company's experts in terms of cost, annual energy saved, the measures' service life, and other parameters. Due to the confidentiality of the data, only excerpts of selected measures are presented as examples in Table 1.

Table 1. Exemplary measures of the data set

Measure category	Investment (€)	Service life (years)	CO_2 avoidance (tCO_2e/a)
Energy efficiency	20000	10	39.8
Self generation of renewable energy	294492	20	196.0
Energy efficiency	125000	15	13

A number of algorithms are available for solving mixed-integer optimization problem with most of those based on the branch-and-bound method [28]. Within the selected programming environment *Matlab* the branch-and-bound algorithm *intlinprog* is applied due to the already available implementation.

To calculate avoided CO_2 emissions from energy efficiency measures, the specific greenhouse gas emissions were estimated at 200 g/kWh in CO_2equivalents per kilowatt-hour of electricity. The number used is based on assumptions taking into account numbers published by the German Umweltbundesamt [29] specifying greenhouse gas emissions in CO_2equivalents (CO_2eq) per kilowatt hour of electricity with 428 g/kWh as well as estimated reductions by around 50% compared to today's level to meet the ambition level of the Climate Protection Plan by 2030 [30]. For energy efficiency measures, the amount of avoided CO_2equivalents is calculated by multiplying the saved energy per year with the lifetime of the measure and the CO_2equivalent.

Avoiding CO_2 emissions from self-generation of energy with the aid of PV systems is based on the assumptions made above and on figures from Fraunhofer ISE [31] and the Umweltbundesamt, who assume a greenhouse gas potential for PV electricity of 56 g CO_2 eq./kWh for system operation in Germany. The limitations of the estimation have to be pointed out.

An exact calculation on a yearly level and the company-specific emission data was not carried out because the data was not made available by the company the case study is based on. Therefore the presented case study is based on data derived from the sources stated above. This represents a weakness of the results. Moreover, no reliable range can be given because the specific CO_2 savings vary so much from company to company. For example, a company with a 100% self-supply of renewable energy would represent one extreme value, while a company that is supplied with electricity from coal power would represent another extreme value.

The calculation of the net present value is based on a discount rate of 6%. The discount rate used for the calculation of the NPV is based on numbers given by the company the case study is based on and represents the value and represents the rate at which future cash flows are discounted. The discount rate, represents the return that could be earned per unit of time on an investment with similar risk.

The following illustration in Fig. 1 represents a central result of the optimization algorithm. The best possible selection of decarbonization measures was calculated for available investment budgets in steps of fifty thousand euros each. Accordingly, each data point represents an iteration of the optimization algorithm. In a figurative sense, the performed mixed-integer optimization corresponds to the so-called knapsack problem.

A single data point represents an optimal investment programme for decarbonization measures. I.e. for all available measures it is decided whether to implement the measures or not. The curve flattens out at the end as the investment budget is fully utilized. Accordingly, no further measures can be integrated as there are no more available.

The objective function is based on a weighting in which the avoided CO_2 emissions are weighted with 0.4, the present value with 0.4 and the risk with 0.2. The figure supports a decision-maker on which savings are possible depending on the investment budget.

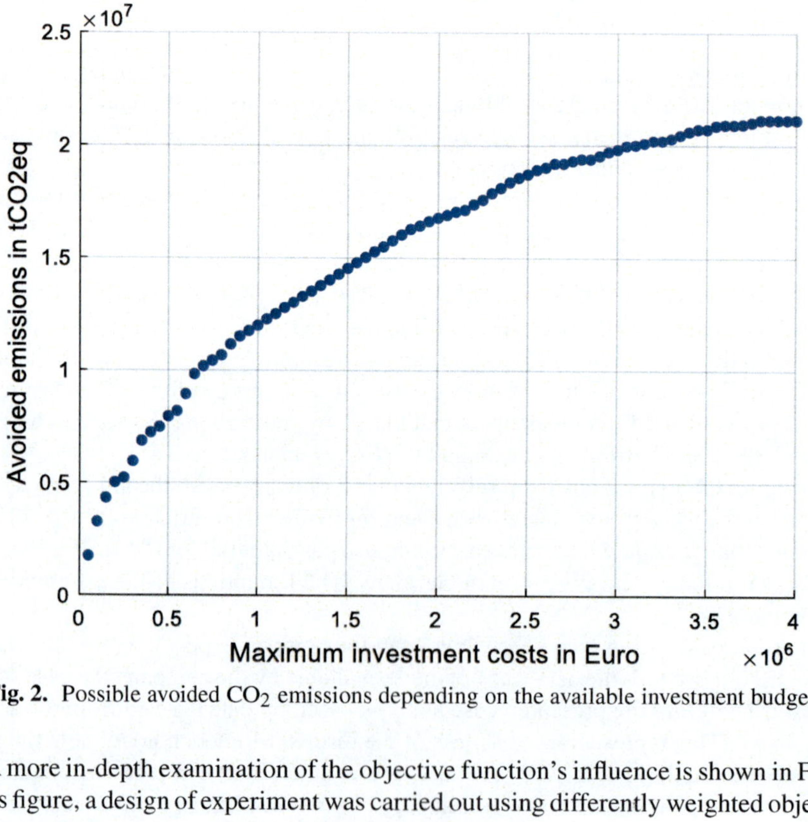

Fig. 2. Possible avoided CO_2 emissions depending on the available investment budget.

A more in-depth examination of the objective function's influence is shown in Fig. 2. In this figure, a design of experiment was carried out using differently weighted objective functions and the maximum CO_2 emissions that can be avoided. The used objective function was evaluated in each case. The aim of this presentation is to gain a more precise understanding of how the weighting of the target dimensions by the decision-maker influences the achievable CO_2 savings.

The achievable CO_2 reduction clearly depends on how the risk is evaluated. The more a decision-maker weights the importance of the risk of decarbonization measures, the lower the achievable savings. In the extreme case of weighting it 1, i.e., a decision based exclusively on the risk associated with the implementation of measures, the achievable savings converge towards 0. On the other hand, there is a high correlation between NPV and avoided emissions. This is particularly due to the cost savings associated with energy efficiency measures (Fig. 3).

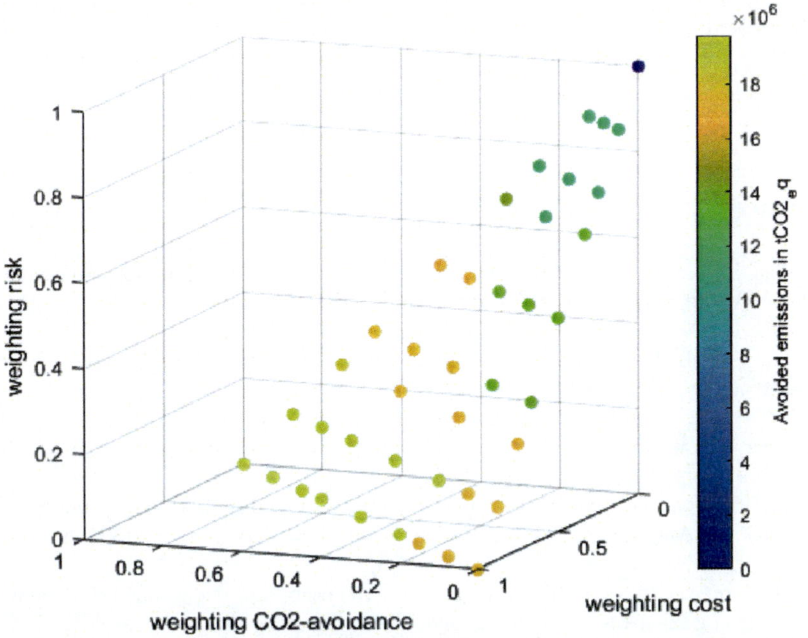

Fig. 3. Design of Experiment using different weights for the goal function

5 Conclusion

Due to the increasing importance of reducing CO_2 emissions in the industrial sector, companies must decide on measures and determine the optimal quantity of measures. Especially for large companies, there is a great variety of possible measures and the decision is correspondingly complex. Using the presented mixed-integer optimization approach, companies can determine the set of optimal decarbonization measures in their individual situation. Decision-makers are faced with the complex challenge of determining the optimal mix of measures for their company. Yet, due to the individual situation of each company, optimal action plans cannot be determined generically and each case must be considered individually.

For the user, the proposed decision model provides the benefit of scientifically sound decision support that determines the best possible actions for the selected preference.

The approach is based on the assumption of individually characterized utility functions, which are represented by an objective function individually weighted by the decision maker. The presented investigation of the objective function with the help of an experiment design illustrates the high dependence of the achievable CO_2 savings on the target weighting, in particular the weighting of the risk.

Further research on the approach focuses on a more precise determination of avoided CO_2 emissions, since at the company level, there is a complex interaction between, for example, energy efficiency measures and the company's own generation of renewable energies. This leads to a complex determination of the actually achieved CO_2 avoidance with an individual measure because the implementation of for example renewable energies leads to lower avoidance of CO_2 emissions in energy efficiency measures. Possible further research can be in this consideration of interactions between decarbonization measures.

References

1. Büttner, S.M., Wang, D., Schneider, C.: Der Weg zur Klimaneutralität. Bausteine einer neuen Methodik zur Bestimmung eines wirtschaftlichen Maßnahmenmix. In: Digitalisierung im Kontext von Nachhaltigkeit und Klimawandel, pp. 89–106 (2021)
2. Buettner, S.M.: Roadmap to neutrality—what foundational questions need answering to determine one's ideal decarbonisation strategy. Energies **15**, 3126 (2022)
3. Buettner, S.M., Schneider, C., König, W., Mac Nulty, H., Piccolroaz, C., Sauer, A.: How do German manufacturers react to the increasing societal pressure for decarbonisation? Appl. Sci. **12**(2022), 543 (2022)
4. Eco, U., Schick, W.: Wie man eine wissenschaftliche Abschlußarbeit schreibt: Doktor-, Diplom- und Magisterarbeit in den Geistes- und Sozialwissenschaften, 7, unveränd. Aufl. der dt. Ausg, Facultas Univ.-Verl., Wien (1998)
5. Nurdiawati, A., Urban, F.: Towards deep decarbonisation of energy-intensive industries: a review of current status, technologies and policies. Energies **14**, 2408 (2021)
6. Bataille, C., Åhman, M., Neuhoff, K., Nilsson, L.J., Fischedick, M., Lechtenböhmer, S., et al.: A review of technology and policy deep decarbonization pathway options for making energy-intensive industry production consistent with the Paris Agreement. J. Clean. Prod. **187**, 960–973 (2018)
7. Rissman, J., Bataille, C., Masanet, E., Aden, N., Morrow, W.R., Zhou, N., et al.: Technologies and policies to decarbonize global industry: review and assessment of mitigation drivers through 2070. Appl. Energy **266**, 114848 (2020)
8. Johnson, O.W., Mete, G., Sanchez, F., Shawoo, Z., Talebian, S.: Toward climate-neutral heavy industry: an analysis of industry transition roadmaps. Appl. Sci. **11**, 5375 (2021)
9. Robinson, J.B.: Energy backcasting a proposed method of policy analysis. Energy Policy **10**, 337–344 (1982)
10. Bayata, Ö., Temiz, İ: Developing a model and software for energy efficiency optimization in the building design process: a case study in Turkey. Turk. J. Electr. Eng. Comput. Sci. **25**, 4172–4186 (2017)

11. Bre, F., Fachinotti, V.D.: A computational multi-objective optimization method to improve energy efficiency and thermal comfort in dwellings. Energy Build. **154**, 283–294 (2017)
12. Diakaki, C., Grigoroudis, E., Kabelis, N., Kolokotsa, D., Kalaitzakis, K., Stavrakakis, G.: A multi-objective decision model for the improvement of energy efficiency in buildings. Energy **35**, 5483–5496 (2010)
13. Eskander, M.M., Sandoval-Reyes, M., Silva, C.A., Vieira, S.M., Sousa, J.M.: Assessment of energy efficiency measures using multi-objective optimization in Portuguese households. Sustain. Cities Soc. **35**, 764–773 (2017)
14. Kontogiorgos, P., Chrysanthopoulos, N., Papavassilopoulos, G.: A mixed-integer programming model for assessing energy-saving investments in domestic buildings under uncertainty. Energies **11**, 989 (2018)
15. de Maigret, J., Viesi, D., Mahbub, M.S., Testi, M., Cuonzo, M., Thellufsen, J.Z., et al.: A multi-objective optimization approach in defining the decarbonization strategy of a refinery. Smart Energy **6**, 100076 (2022)
16. Hu, H., Yuan, J., Nian, V.: Development of a multi-objective decision-making method to evaluate correlated decarbonization measures under uncertainty — the example of international shipping. Transp. Policy **82**, 148–157 (2019)
17. Göbel, E.: Entscheidungstheorie, 2., durchgesehene Auflage; Studienausgabe, UVK Verlagsgesellschaft mbH; UVK Lucius; UTB GmbH, Konstanz, München, Stuttgart (2018)
18. Cooremans, C., Energy-efficiency investments and energy management: an interpretative perspective. In: EECB'12 Proceedings (2012)
19. He, Y., Liao, N., Bi, J., Guo, L.: Investment decision-making optimization of energy efficiency retrofit measures in multiple buildings under financing budgetary restraint. J. Clean. Prod. **215**, 1078–1094 (2019)
20. Tan, B., Yavuz, Y., Otay, E.N., Çamlıbel, E.: Optimal selection of energy efficiency measures for energy sustainability of existing buildings. Comput. Oper. Res. **66**, 258–271 (2016)
21. Malatji, E.M., Zhang, J., Xia, X.: A multiple objective optimisation model for building energy efficiency investment decision. Energy Build. **61**, 81–87 (2013)
22. Cano, E.L., Moguerza, J.M., Ermolieva, T., Ermoliev, Y.: Energy efficiency and risk management in public buildings: strategic model for robust planning. CMS **11**(1–2), 25–44 (2013). https://doi.org/10.1007/s10287-013-0177-3
23. Cooremans, C.: Investment in energy efficiency: do the characteristics of investments matter? Energ. Effi. **5**, 497–518 (2012)
24. Müller, E., Engelmann, J., Löffler, T., Strauch, J.: Energieeffiziente Fabriken planen und betreiben. Springer-Verlag, Berlin Heidelberg, Berlin, Heidelberg (2009)
25. Schneider, C., Burkert, M., Weise, P., Sauer, A.: Risikobewertung von Energieeffizienzmaßnahmen/Risk assessment of energy efficiency measures, wt Werkstattstechnik online **111** (2021), 44–48
26. Rommelfanger, H.J., Eickemeier, S.H.: Entscheidungstheorie: Klassische Konzepte und Fuzzy-Erweiterungen. Springer, Berlin Heidelberg, Berlin, Heidelberg, s.l. (2002)
27. Saaty, R.W.: The analytic hierarchy process—what it is and how it is used. Math. Model. **9**, 161–176 (1987)
28. Achterberg, T., Wunderling, R.: Mixed integer programming: analyzing 12 years of progress. In: Jünger, M., Reinelt, G. (eds.) Facets of Combinatorial Optimization: Festschrift for Martin Grötschel, pp. 449–481. Springer, Berlin Heidelberg, Berlin, Heidelberg, s.l. (2013)
29. Umweltbundesamt, Strom- und Wärmeversorgung in Zahlen. 05 Apr 2022. https://www.umweltbundesamt.de/themen/klima-energie/energieversorgung/strom-waermeversorgung-in-zahlen#Kraftwerke
30. Umweltbundesamt, Klimaschutz im Stromsektor 2030 — Vergleich von Instrumenten zur Emissionsminderung. 05 Apr 2022. https://www.umweltbundesamt.de/sites/default/files/medien/1/publikationen/2017-01-11_cc_02-2017_strommarkt_endbericht.pdf

31. Fraunhofer-Institut für Solare Energiesysteme ISE. Aktuelle Fakten zur Photovoltaik in Deutschland. https://www.ise.fraunhofer.de/de/veroeffentlichungen/studien/aktuelle-fakten-zur-photovoltaik-in-deutschland.html

Concept for Increasing the Resilience of Manufacturing Companies

J. Tittel(✉), M. Kuhn, M. Riesener, and G. Schuh

Laboratory for Machine Tools and Production Engineering (WZL), RWTH Aachen University, Campus-Boulevard 30, 52074 Aachen, Germany
j.tittel@wzl.rwth-aachen.de

Abstract. In the context of sustainable management, organizational resilience is gaining importance. Manufacturing companies are increasingly exposed to external disturbances. Crisis-resistant product development is of particular importance, as innovative products offer a promising opportunity to create competitive advantages and thus secure the company's existence or even enable a company to increase its market share in the event of a crisis. At the same time, corporate functions today are usually geared towards efficient execution. In this context, the paper presents a concept for the alignment of product development in the conflict between efficient goal achievement and the prevention of the impacts of disturbances. For this purpose, the design elements, goals and relevant disturbances of product development are taken into account. Based on the interdependencies of these elements, a methodological approach for a company-specific determination of the target characteristics of the design elements is presented, in order to enable an alignment in the conflict between efficient goal achievement and resilience. The concept is designed to the alignment of product development, but can be transferred to other corporate functions and corporate divisions.

Keywords: Resilience · Product development · Efficiency

1 Introduction

Modern value chains are usually organized towards efficiency and productivity. However, the effects of the Corona pandemic exemplarily illustrate the vulnerability of these value chains to disturbances and crises. There is a conflict between the efficient achievement of corporate goals and the prevention of the impacts of disturbances. Efficiency-oriented companies perform better than resilience-oriented companies do in times of no crisis. In the event of a crisis, however, the performance of efficiency-oriented companies decreases rapidly and their existence may be threatened. Science, politics and industry have recognized the need for action. [1, 2] The goal of a resilient design of value chains has emerged [3]. In particular the corporate function of product development plays a decisive role here. Innovative products that meet the market requirements better than competitor products enable competitive advantages [4] and thus ensure the probability of survival or even enable a company to increase its market share in the event of a crisis.

M. Liewald et al. (Eds.): WGP 2022, LNPE, pp. 761–770, 2023.
https://doi.org/10.1007/978-3-031-18318-8_75

This paper defines organizational resilience as a company's ability to anticipate, deal with and adapt in the aftermath of disturbances [5, 6]. Crises are typically the result of a combination of multiple disturbances [7]. Disturbances are more frequent and more difficult to predict and deal with in volatile, uncertain, complex and ambiguous times [8]. Product development, as with the entire company, is under the influence of external and internal disturbances. Examples of disturbances can be the extreme situation of the Corona crisis, but weaker disturbances such as new regulations or emerging technologies of competitor companies can also have a disruptive influence. [6]

Both a sufficient level of efficiency and resilience must be ensured. Within product development, there is usually a reactive, short-term handling of the impacts of disturbances [9], which must be replaced by a preventive approach. For this reason, answering the following question promises relevant information: How can product development be aligned within the conflict of efficient goal achievement and prevention of the impacts of disturbances? The concept presented in this paper is intended to present answers to this question and enable companies in the manufacturing industry to position themselves at the desired level of resilience through the alignment of product development. The concept is based on [6] but includes relevant changes and concretizations in all submodels.

2 Fundamentals

The purpose of this section is to provide relevant basics and definitions of resilience, crisis management and product development. According to [5], organizational resilience describes the company's ability to anticipate, deal with and adapt in the aftermath of disturbances. Organizational resilience can therefore also be interpreted as the ability of a system to withstand crises. Resilience makes a system more capable of acting in crises, but it is not meant to overcome all negative influences of a crisis. Accordingly, the added value of a resilient organization is particularly advantageous in times of crisis, but is often neglected in times of abundance [2, 6]. Corporate crises are unplanned, undesirable and temporary processes with ambivalent results, which can threaten the existence of a company in the long term [10]. Three categories can be distingusted. Strategic crises, success crises and liquidity crises are usually passed through in sequence and increase in their negative effects [11]. The extent of the threat and the need for action increases continuously and the scope for action for the company decreases gradually with each category [12].

The corporate function of product development focuses on the design of products and variants for specific customers or the anonymous market based on requirements, applications and specifications using available technology [13]. Product development is an interdisciplinary corporate function with multiple interfaces. A distinction is made between two viewpoints on product development. On the one hand, product development controls the processes of development activities and the actions of the associated employees and teams. [14] A distinction can be made between classical, plan-oriented, sequential, iterative and hybrid processes. On the other hand, product development is an organizational unit, which describes the structure of the necessary workspace, the allocation of the unit into subsystems and the assignment of subtasks to the respective subsystems [15]. [6, 16]. The concept follows the presented basics and definitions.

3 Related Work

The following section discusses existing approaches to organizational resilience in the context of product development.

Fundamental principles of organizational resilience and related attributes is provided by ISO 22316 [17]. The norm provides generally applicable principles for organizations regardless of size, industry or sector. Due to the general approach, the norm does not provide concrete approaches for an implementation and does not go into details of different business functions.

DIN ISO 31000 [18] provides guidelines for successful risk management in organizations regardless of the nature of the risk. The norm provides principles for organizations regardless of size, industry or sector. Risk management is presented as a multi-phase process consisting of communication and consultation, context setting, risk identification, analysis, assessment and treatment, as well as monitoring and review. In [19], literature regarding risk management in product development is reviewed using the framework of DIN ISO 31000 [18]. This shows that the applicability of the norm is given, but there are shortcomings in the implementation.

A practical approach is taken in [20]. In the paper, an exploratory study of crises in product development is conducted with designers from industrial practice. It included 15 examples of crises in product development to develop a total of 9 contextual factors to characterize product development crises and 56 success factors for managing crises.

Reference [21] provides an overview of resilient technology strategies in volatile environments by deriving requirements for long-term strategic positioning in VUCA times. It is stated that a successful technology strategy in VUCA environments requires context-adaptive technology planning through clear and consistent strategic goals.

In summary, the approaches presented emphasize the importance of organizational resilience. However, the trade-off between efficient goal achievement and resilience is not adequately addressed. Additionally, it has to be noted that none of the approaches simultaneously considers product development and its contribution to the resilience of the overall company. This paper aims to address the identified weaknesses in the aforementioned literature.

4 Methodology

This paper presents a concept for increasing the resilience of manufacturing companies through the alignment of product development in the conflict between efficient goal achievement and prevention of the impacts of disturbances. The concept is divided into five submodels (see Fig. 1). In submodel one, the procedure for deriving goals of product development from corporate goals is presented. Submodel two describes the procedure for the determination of relevant disturbances for product development. The approach for developing a morphology of the design elements of product development results from submodel three. In submodel four, a procedure for explaining interdependencies between goals, disturbances and design elements are determined. Submodel five presents a methodical approach for a company-specific determination of the target characteristics of the design elements of product development, in order to enable an alignment in the conflict between efficient goal achievement and resilience. The concept is geared to the sensible alignment of product development, but can in principle be transferred to other corporate functions and corporate divisions.

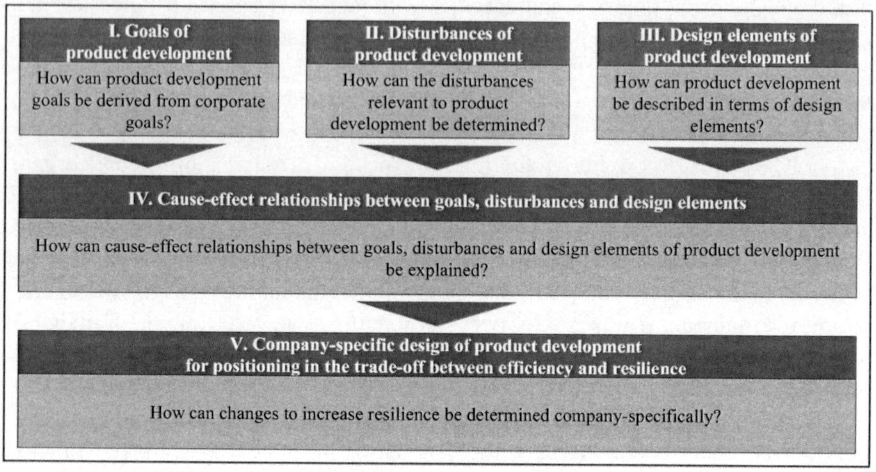

Fig. 1. Concept for the alignment of product development in the conflict between efficient goal achievement and the prevention of the impacts of disturbances

4.1 I. Derivation of Product Development Goals from Corporate Goals

A resilient company strives to achieve its goals in the best possible way even in times of crises. The subject of the first submodel is therefore to derive product development goals from overall corporate goals for the manufacturing industry. This is a prerequisite for the explanation of the cause-effect relationships between goals, disturbances and design elements of product development in the fourth submodel.

The first step is to describe and hierarchize corporate goals. With the help of an analytical deductive analysis of existing approaches in the scientific literature and an empirical inductive analysis of case studies of manufacturing companies, the relevant goals of companies in the manufacturing industry are elaborated. Subsequently, the identified corporate goals are arranged in a hierarchical structure. Financial, environmental, social and regulatory corporate goals such as productivity, environmental performance and workplace safety must be taken into account.

The second step is to describe and hierarchize product development goals. Analogous to the derivation of corporate goals, relevant product development goals are elaborated with the help of an analytical deductive analysis of existing approaches in the scientific literature and an empirical inductive analysis of case studies. This is followed by their hierarchization, taking into account both the result-side and the cost-side goals of product development.

Finally, an assignment of corporate goals and product development goals is made based on evaluation criteria, which enable an examination of possible correlations between the respective lowest levels of corporate and product development goals. The result of submodel one is a model for describing product development goals based on corporate goals.

4.2 II. Derivation of Disturbances Relevant to Product Development from Corporate Disturbances

The subject of submodel two is the derivation of disturbances that are relevant to product development. This is a prerequisite for the explanation of the cause-effect relationships between goals, disturbances and design elements of product development in the fourth submodel. The description here is limited to exogenous disturbances, which by definition cannot be influenced by a company. Endogenous disturbances can be influenced by the company and are therefore interpreted as potential design elements in the third submodel.

The first steps are a literature review and expert interviews on exogenous disturbances. For this purpose, first an analytical deductive analysis of existing approaches in the scientific literature and an empirical inductive analysis of case studies of manufacturing companies are conducted. Redundancies are then eliminated. This is followed by further consolidation using a design structure matrix. Substitution technology, entry of new competitors, change in legislation and esource scarcity are examples for disturbances.

In the second step, the disturbances relevant to product development are selected. The concept is focused on the corporate function of product development. Therefore, the total amount of exogenous disturbances has to be reduced to the subset of disturbances relevant for product development. For this purpose, suitable criteria for assessing the product development relevance of exogenous company disturbances are defined and applied to the total set of disturbances.

In the third step, the effect-relevant disturbances are selected. The totality of exogenous disturbances can be divided into cause-related disturbances and effect-related disturbances. Relevant, however, are the effect-related disturbances, which can become apparent due to the impact at the company and must be addressed. For this purpose, an effect network is constructed, which sorts the elaborated disturbances into the cause level or the effect level. From the effect network, conclusions can be drawn about the concatenation of the disturbances. The result of submodel two is a model for describing the disturbances relevant to product development.

4.3 III. Description of Product Development Based on Design Elements and Characteristics

Submodel three describes the corporate function of product development based on design elements as well as their possible characteristics. It represents a further prerequisite for the explanation of the cause-effect relationships between goals, disturbances and design elements of product development in the fourth submodel. The design elements are the parameters for aligning product development with the conflicting goals of efficient goal achievement and prevention of the impacts of disturbances.

The first step is a collection of design elements of the corporate function product development. The design elements are derived analytically deductively from existing approaches and models in the scientific literature. For example, publications such as VDI Norm 2221 [14] are used to describe the product development process. Other relevant publications describe the object area of product development, the product development organization and culture. The collection is supplemented by empirically inductively derived design elements from industry projects as well as structured expert interviews with participants of industry working groups of the authors.

In the next step, the design elements are consolidated by eliminating redundancies and by means of a design structure matrix. To structure the design elements in a meaningful way, superordinate dimensions are formed in which the design elements can be classified. In this way, a systematic presentation can be ensured.

Finally, the design elements of product development are operationalized by defining potential characterisitcs for every design element. For this purpose, the results of the scientific literature and the industry exchange are used. The result of submodel three is a model for the morphological description of product development (see Fig. 2).

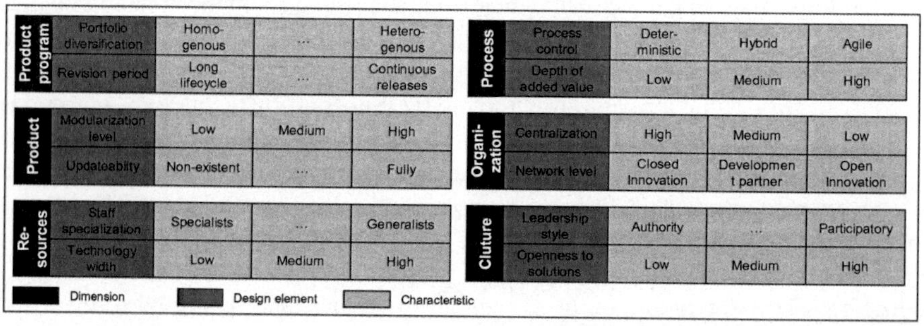

Fig. 2. Description of product development based on exemplary design elements and characteristics

4.4 IV. Explanation of the Cause-Effect Relationships Between Goals, Disturbances and Design Elements of Product Development

Submodel four explains the cause-effect relationships between design elements and goals, design elements and disturbance, and within design elements of product development. It has to be noted that no direct cause-effect relationships between goals and disturbances of product development are considered, since goals cannot influence exogenous disturbances. Although disturbances have an influence on the subsequent achievement of goals, they do not effect the initial definition of the goals.

A conformity matrix is created from the characteristics of the design elements and the goals of product development. For each cell of this matrix, the following question has to be answered: "How well is the characteristic of the design element separately suited for achieving the goal?" An additional conformity matrix is formed from the characteristics of the design elements and the disturbances of product development. For each cell of this matrix, the following question has to be answered: "How well is the characteristic of the design element separately suited for the prevention of the impact of the disturbance?" In addition, an influence matrix is created in which the characteristics of the design elements of product development are compared. For each cell of the matrix, the following question is to be answered: "How does a horizontal characteristic of the design elements influence a vertical characteristic of the design elements? Decisive prerequisites for reliable answers to the formulated questions are the precise and comprehensible formulation of the design elements including their characteristics, goals, as well as disturbances of product development. In addition, a Likert scale that is understood uniformly by all interview partners is constructed.

In the next step, the cause-effect relationships are logically derived through expert interviews and literature research. In the course of preparing the interviews, the interview guidelines are developed and tested in advance. In addition, managers and experts in product development are selected as suitable interview partners. For this purpose, the network of scientists and participants of the authors' industry working groups are used. This is followed by a structured and recorded interview process and systematic interview evaluation.

Based on the results, both goal-related and disturbance-related conformity sums can be calculated for the characteristics of the design elements. Thus, it becomes clear which characteristics positively influence the achievement of goals and which characteristics positively influence the prevention of the impacts of disturbances. In the influence matrix of the design elements, active and passive sums can be determined for each characteristic of the design elements. This shows which characteristics of a design element positively influence or are influenced by other characteristics of other design elements. The result of submodel four is a model for explaining the interdependencies between goals, disturbances and design elements of product development. Figure 3 shows the matrices with exemplary design elements, goals and disturbances.

4.5 V. Company-Specific Design of Product Development for Positioning in the Trade-off Between Efficiency and Resilience

Submodel five defines the company-specific changes for positioning in the trade-off between efficiency and resilience. From the totality of disturbances relevant to product development, one company specifically chooses the disturbances relevant for the company. For handling these relevant disturbances, an adjustment of the characteristics of the design elements is to be examined. Potential negative effects of an adjustment on the goal achievement in times without disturbances are to be accepted. This results in an increase of the company's resilience.

First, the company-specific product development goals and disturbances are derived, prioritized and weighted. To determine the relevant goals and disturbances, the company applies the results from submodels one and two and weights these elements.

Subsequently, the conformity matrices are reduced accordingly by deleting the rows with goals and disturbances that are not relevant for the company. After checking the numerical values from submodel four for the specific use case of the company, the specific goal-related and disturbance-related conformity sums can be determined for each design element characteristic. Both sums are connected with each other via the risk affinity factor. The risk affinity factor is the ratio of the specific weighting of goals to the prevention of impacts of disturbances. This results in the company-specific, combined conformity sum for each characteristic of the design elements. These values are used to weight the numerical values of the influence matrix of the design elements from

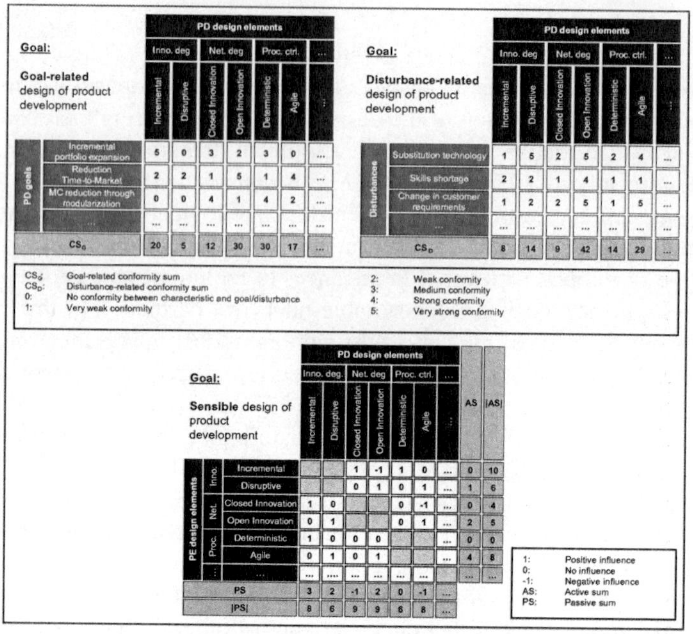

Fig. 3. Cause-effect relationships between goals, disturbances and design elements of product development

submodel four. Therefore, the results are the conformity sum-specific active and passive sums per design element characteristic.

In the last step, the adaptation measures are determined, taking into account the conformity of the overall system. For this purpose, a portfolio is created in which the characteristics of the design elements are classified according to their respective conformity sum-specific active and passive sums. Based on the calculated sums, critical, active, inert and reactive characteristics can be distinguished. The structure of the portfolio is based on the method of networked thinking [22]. The relationships shown in the portfolio can be used to derive optimization approaches for the company-specific positioning in the trade-off between efficient goal achievement and prevention of impacts of disturbances. Taking into account the current characteristics of each design element in the company and estimating the required time and financial transformation effort, target characteristics can be defined for each design element. The result of submodel four is a model for the company-specific definition of the characteristics of the design elements.

5 Conclusion

The manufacturing industry faces major challenges. Exogenous, combined disturbances are gaining in importance. The presented concept enables the positioning of corporate functions and corporate divisions in the area of conflict between the efficient achievement of goals and the prevention of the impacts of disturbances. On the one hand, the paper is targeted at researchers in the fields of product development and organizational resilience.

However, the methodology is transferable to other corporate functions and divisions. On the other hand, the paper is aimed at managers of manufacturing companies who want to make their company resilient to disturbances.

Based on the derivation of product development goals from corporate goals, the derivation of product development relevant disturbances from corporate disturbances, and the description of product development based on design elements and characteristics, the cause-effect relationships between goals, disturbances and design elements of product development are explained. This allows the design elements of product development to be configured company-specifically, thus enabling the positioning in the trade-off between efficiency and resilience.

The presented topic is currently being researched as part of a dissertation at the Chair of Production Engineering of the Laboratory for Machine Tools and Production Engineering WZL at RWTH Aachen University. It will be applied in consulting projects and industry working groups to ensure its feasibility in industrial applications.

Acknowledgment. Funded by the Deutsche Forschungsgemeinschaft (DFG, German Research Foundation) under Germany's Excellence Strategy — EXC-2023 Internet of Production — 390621612.

References

1. Kagermann, H., Süssenguth, F., Körner J., Liepold A., Behrens J.H.: Resilienz als wirtschafts- und innovationspolitische Gestaltungsziel. In: Acatech IMPULS (2021)
2. Reeves, M., Nanda, S., Whitaker, K., Wesselink, E.: Becoming an all-weather company. BCG Henderson Institute (2020)
3. Boos, W., Trauth, D., Arntz, K., Prümmer, M., Niemietz, M., Wilms, M., Lürken, C., Mayer, J.: Wettbewerbsfaktor Resilienz. Handlungsfelder für den krisensicheren Werkzeugbau 2021. 1st edn. WBA Aachener Werkzeugbau Akademie GmbH (2021)
4. Porter, M.E.: What is strategy? Harvard Bus. Rev., Nov–Dec (1996)
5. Duchek, S.: Organizational resilience: a capability-based conceptualization. Bus. Res. **13**(1), 215–246 (2019). https://doi.org/10.1007/s40685-019-0085-7
6. Riesener, M., Kuhn, M., Tittel, J., Schuh, G.: Concept for enhancing the contribution of product development to organizational resilience of manufacturing companies. In: 2021 IEEE International Conference on Industrial Engineering and Engineering Management (IEEM). IEEE, Singapore (2021)
7. Krystek, U., Moldenhauer, R.: Handbuch Krisen- und Restrukturierungsmanagement. Kohlhammer Verlag, Stuttgart (2007)
8. Meyer, G., Knüppel, K., Möhwald, H., Nyhuis P.: Kompetenzorientiertes Störgrößenmanagement (2014)
9. Repenning, N.P.: Understanding fire-fighting in new product development. J. Prod. Innov. Manag. **18**, 285–300 (2001)
10. Krystek, U.: Unternehmungskrisen. Gabler, Wiesbaden (1987)
11. Müller, R.: Krisenmanagement in der Unternehmung (1986)
12. Bickhoff, N., et al.: Die Unternehmenskrise als Chance. Springer-Verlag, Berlin Heidelberg (2004)
13. Schloske, A.: Innovation und Produktentwicklung. In: Fabrikbetriebslehre 1, ch. 3, pp. 67–102. Springer Vieweg, Berlin (2020)

14. VDI 2221 Part 1: Entwicklung technischer Produkte und Systeme — Modell der Produktentwicklung (2019)
15. Kosiol, E.: Organisation der Unternehmung. Gabler-Verlag (1976)
16. Ponn, J., Lindemann, U.: Konzeptentwicklung und Gestaltung technischer Produkte. Springer-Verlag, Berlin Heidelberg (2011)
17. ISO 22316: Security and resilience- organizational resilience principles and attributes (2017)
18. DIN ISO 31000: Risikomanagement-Leitlinien (2018)
19. Oehmen, J., Ben-Daya, M., Seering, W., Al-Salamah, M.: Risk management in product design: current state, conceptual model and future research. In: ASME International (2010)
20. Münzberg, C., Gericke, K., Oehmen, J., Lindemann, U.: An exploratory study of crises in product development. Presented at the International Design Conference, Dubrovnik, Croatia (2016)
21. Schuh G., Patzwald M., Imhäuser Cardoso M.C.: Resilient technology strategy in volatile environments (2019)
22. Probst, G.J.B., Gomes, P.: Vernetztes Denken, 2nd edn. Gabler, Wiesbaden (1991)

Industrialization of Remanufacturing in the Highly Iterative Product and Production Process Development (HIP^3D)

A. Hermann(✉), S. Schmitz, A. Gützlaff, and G. Schuh

Laboratory for Machine Tools and Production Engineering (WZL), RWTH Aachen University, Campus-Boulevard 30, 52074 Aachen, Germany
a.hermann@wzl.rwth-aachen.de

Abstract. In view of the increasing scarcity of resources and global efforts to reduce CO_2 emissions, production management approaches are focusing on enabling a circular economy. The remanufacturing of used products to the quality standards of a new product is one key enabler. Remanufacturing offers economic and ecologic advantages by reducing the amount of resources used in production. Thus, associated manufacturing costs are reduced and the dependence on imports of critical raw materials decreases. To do so, remanufacturing requirements must be considered in the early development phases of products and production processes. In practice, companies focus on the economic perspective in the development phase, methodically supported by highly iterative product and production process development (HIP^3D) approaches. However manufacturing companies neglect the inclusion of the ecological perspective in the development phases, partly due to missing methodical support. This paper presents a framework for the industrialization of remanufacturing in the HIP^3D. Since the feasibility of remanufacturing is defined at the early stages of product and production process development, this paper aims at integrating remanufacturing requirements in the development phase. First, the requirements arising from remanufacturing are identified through a systematic literature review. Subsequently, it is examined to what extent HIP^3D already covers these requirements. For non-fulfilled remanufacturing requirements, adaptions and extensions to the HIP^3D approach are derived and described in design guidelines. This results in a framework for the industrialization of remanufacturing in the HIP^3D, enabling manufacturing companies to exploit their economic and ecological potential.

Keywords: Remanufacturing · Highly iterative product and process development · Industrialization

1 Introduction

As resources become scarcer and political environmental regulations are becoming stricter, companies are encouraged to comply with these new demands. This leads to a rethinking among consumers and investors [1]. Companies need to change from a

M. Liewald et al. (Eds.): WGP 2022, LNPE, pp. 771–780, 2023.
https://doi.org/10.1007/978-3-031-18318-8_76

linear economy, following the "take-make-waste"-principle, to a circular economy. The concept of circular economy aims to separate economic growth from the use of finite resources and thus reduce the accompanying negative impact on the environment [2].

To reach a circular economy various strategies are presented. For manufacturing companies, a key strategy to meet these challenges is remanufacturing, the return of used products or components, and their reprocessing into as-new or better condition [3, 4]. The extension of the lifetime of products or their components through reuse in the same or similar application scenario leads to a reduction or avoidance of waste. The extended product and component cycles are characteristic of the circular economy and replace traditional end-of-life recycling strategies. A product design that allows non-destructive disassembly of the products or components enables the reuse and return into the product cycle [5]. Through remanufacturing, more than half of the resources in comparison to a new product can be saved, whereby resources are not only considered to be the raw materials of the product but also energy and water consumption [6]. Remanufacturing has already been practiced in some industries for several years, but it is still characterized by cost-intensive and manual processes. Furthermore, there are many challenges that prevent the comprehensive implementation of remanufacturing in the industry today. These challenges range from lack of technology or product knowledge to legal restrictions [7]. Uncertain product conditions as well as inconsistent quality and fluctuating availability of end-of-life products also prevent remanufacturing [1]. Most products and manufacturing processes are not designed for a circular economy use of the products. To counteract that considerations and measures that significantly facilitate remanufacturing are already required during product and process development [8]. Also, to utilize the economic benefits of remanufacturing, it is necessary to automatize and industrialize the disassembly as well as the remanufacturing process in general. Due to the high interdependencies between the product, the production process and remanufacturing, it is important to consider these dimensions holistically. Consequently, due to the interactions that occur between the three dimensions of product, production process and remanufacturing process, it is necessary to adapt an integrated approach to product and production process development regarding remanufacturing requirements. Wlecke et al. developed a concept for a highly iterative product and production process development (HIP^3D), which allows considering the interactions between product and production process development interactions [9]. This integrated development approach ensures to address the dimensions of product and production process simultaneously and their interactions with each other are subsequently addressed. However, this approach does not take into account the integration of remanufacturing. Therefore, an extension of the concept is needed to industrialize remanufacturing.

This paper aims to present a framework for the industrialization of remanufacturing in the HIP^3D, enabling manufacturing companies to exploit their economic and ecological potential. Chapter two analyses existing approaches that provide insight into remanufacturing challenges. Chapter three uses the results of systematic literature research to derive the requirements for remanufacturing that influence product and production process development. In chapter four, the extension of the concept of HIP^3D is described. The extension serves to cover the previously identified requirements of remanufacturing.

Chapter five concludes the findings and results as well as an outlook on relevant future research.

2 Requirements and Prior Research in Product and Process Development for Remanufacturing

2.1 Requirements

Remanufacturing has already been practiced for several years, but the remanufacturing process is characterized as time-consuming and cost-intensive. Furthermore, the return of products and components cannot be planned due to the lack of information about the lifetime and condition of the products. To industrialize remanufacturing, it is consequently essential *to incorporate its requirements and constraints into the development of products and production processes* and, simultaneously, to *design an associated remanufacturing process* [10]. In addition, disassembly is a key aspect of the design of the remanufacturing process [1]. Currently, this process is hardly automated and cannot be planned due to uncertainties, which complicates *the industrialization of both the disassembly and the remanufacturing process.* Since remanufacturing has a significant influence on the product and the production process, it is important to *consider these interactions throughout the entire development cycle and to generate insights for development.*

2.2 Prior Research in Product and Process Development for Remanufacturing

In the following, existing approaches are presented, which focus either on the highly iterative and integrated development of products [9], describe the inclusion of remanufacturing in development [10–13], or approaches to industrialize disassembly [1, 14].

The approach presented in [9] considers a holistic industrialization concept for physical products using a highly iterative product development (HIPD). Based on the analysis of the conflict between HIPD and plan-driven production process development, product- and production-specific design guidelines are developed and six steps for the industrialization of highly iteratively developed products are defined. This approach uses hypothesis-based development as well as the validation of hypothetical increments. In addition, investment decisions are considered in the early phases. Overall, this process is the most advanced in terms of highly iterative product and process development (HIP^{3}D). The method presented by Wlecke [9] serves as a tool for realizing the industrialization of remanufacturing in the HIP^{3}D. The consideration of remanufacturing does not take place in this approach.

Furthermore, it is essential to analyze the influences that remanufacturing has on development and the requirements that arise as a result. The requirements that must be considered for the inclusion of remanufacturing in the development phase are described by Prendeville and Bocken [10]. The described key factors for design for remanufacturing are the consideration of technology integration, the inclusion of reverse logistics, the description of a detailed design, the selection of materials to be used and the selection of a standardized design to reduce complexity and simplify the production process.

The approach presented by Boorsma [11] focuses on the requirements that arise in the development of a remanufacturable product. A framework is developed explaining different roles in the design for remanufacturing based on Balance Scorecard Model. Here, different perspectives of design management are extracted and subsequently related to remanufacturing. In this approach, the perspectives design as differentiator, the customer perspective design as integrator, the process perspective design as transformer, the learning perspective design as good business, and the financial perspective are considered. The influence of a design for remanufacturing and the different perspectives that have to be considered are shown.

In [12] a method is provided for the strategic and practical implementation of feedback from remanufacturing to design for OEMs. This first involves an assessment of the current situation. The results of the assessment are incorporated into the future vision. After this measures are defined for the step-by-step implementation of feedback from remanufacturing to design. These actions are prioritized and a timeline is established. The final step is an evaluation of the impact of the implemented measures and their impact on the design for remanufacturing.

An approach that addresses remanufacturing in the early stages of product development is described by Yang [13]. In this approach a tool is presented supporting the decision-making process in the early phases of development with respect to the product design under influence of the requirements of remanufacturing.

The industrialization of disassembly is one main challenge in the industrialization of remanufacturing. An approach to industrialize the disassembly is described by Sprenger [1]. The focus of this approach is the automation of the disassembly and modularity of products. This is to contribute to the effective realization of Remanufacturing. For the realization of an automated disassembly constant transparency is necessary regarding the condition of the product, so the return and the disassembly can be planned accordingly. Complete information transparency facilitates decisions throughout the entire product life cycle. Furthermore, human-robot collaboration in disassembly is an essential factor. The disassembly process should be designed of flexible cells, for reaction to different product conditions.

In [14], a concept for an update factory is presented. According to the principle of remanufacturing, the update factory renew or upgrade products after their use and return as-new product to the customer. Framework conditions for product development are derived to address the core elements of the circular economy and remanufacturing. The approach exposes that the conditions of a circular economy use of products must be created in the development. In addition, the approach stresses that a suitable modular product architecture should be considered to ensure that products can be updated. The approach aims to extend the use of the products. This requires an intensive interaction between development and production. A strong focus should be placed on future customer requirements as early as the product development stage. This involves forward-looking optimization for future behavior and the preferences of users.

Summarizing, there is no approach in the literature that provides remanufacturing for HIP^{3}D. Thus, the research already conducted is concerned with a consideration of influences through remanufacturing on product and process development.

3 Design Guidelines for Remanufacturing

To identify further requirements of remanufacturing in terms of product and production process development, a systematic literature review (SLR) was conducted. For this purpose, the method by Borrego [15] was applied. The goal is to reduce the subjectivity and fallibility, while allowing efficient handling of the amount of available literature in different databases.

3.1 Methodology of the Systematic Literature Review.

At first, search terms were defined to describe the topic as precisely as possible. For the literature search, the following terms were used in combination: *Remanufacturing product, Remanufacturing goods, Remanufacturing commodity, Remanufacturing process, Remanufacturing system, Remanufacturing operation, Remanufacturing requirements, Remanufacturing demands, Remanufacturing requests, Design for Remanufacturing, Remanufacturing in product development, Remanufacturing in process development.* These search terms are based on four components, each describing a different aspect of the field of interest and connected with the boolean operator "AND". For each component a selection of synonyms of the respective aspect were integrated with the boolean operator "OR" in the search string. This approach was executed both in English and in German. The databases used for the SLR were ScienceDirect, Springer Link, IEEE Xplore, Scopus and Google Scholar. An overview of the databases, components and synonyms used for the literature search is depicted in Fig. 1.

Databases			
• Science Direct (www.sciencedirect.com) • Springer Link (www.link.springer.com)	• IEEE Xplore (www.ieeexplore.com) • Scopus (www.scopus.com)	• Google Scholar (www.scholar.google.com)	
Search string components			
English	Compon. 1	(„Remanufacturing product" OR „Remanufacturing goods" OR „Remanufacturing commodity")	AND
	Compon. 2	(„Remanufacturing process" OR „Remanufacturing system" OR „Remanufacturing operation")	AND
	Compon. 3	(„Remanufacturing requirements" OR „Remanufacturing demands" OR „Remanufactruing requests")	AND
	Compon. 4	(„Design for Remanufacturing" OR „Remanufactruing in product development" OR „Remanufactruing in process development")	
German	Compon. 1	(„Produkt Refabrikation" OR „Refabrikation von Ware")	AND
	Compon. 2	(Refabrikationsprozess OR Refabrikationssystem OR Refabrikationsvorgang)	AND
	Compon. 3	(„Anforderungen an die Refabrikation OR „Rahmenbedingungen für die Refabrikation")	AND
	Compon. 4	(„Entwicklung für die Refabrikation" OR „Refabrikation in der Produktentwicklung" OR „Refabrikation in der Prozessentwicklung")	
	Compon. 5	(„Produkt Remanufacturing" OR „Remanufacturing von Ware")	AND
	Compon. 6	(Remanufacturingprozess OR Remanufacturingsystem OR Remanufacturingvorgang)	AND
	Compon. 7	(„Anforderungen an Remanufacturing" OR „Rahmenbedingungen für Remanufacturing")	AND
	Compon. 8	(„Entwicklung für Remanufacturing" OR „Remanufacturing in der Produktentwicklung" OR „Remanufacturing in der Prozessentwicklung")	

Fig. 1. Used databases and search strings.

In the following, the requirements for remanufacturing are derived based on the identified papers that influence product and production process development.

3.2 Results and Identified Requirements of Remanufacturing

From the SLR, several requirements are identified when considering remanufacturing in early phases. The identified requirements either influence the development of the product, the development of the production process, but also overarching requirements that arise from the integration of product and production process development under consideration of remanufacturing.

Considering the influences concerning the product development, it is important that *the components of the product intended for remanufacturing are suitable for all phases of the remanufacturing process (1)* and not only for some phases [16]. Furthermore, the *modularity of the products (2)* is decisive for the realization of remanufacturing [1, 17]. The necessary modularity must be provided in the product development to enable upgradeability of the products and creating a clear delimitation between remanufacturing and reuse.

Additionally, remanufacturing influences the development of production processes, as the production process is supposed to ensure a complete disassembly for successful remanufacturing. Therefore, it is important to consider and enable disassembly in addition to the actual manufacturing process. Currently, disassembly is considered a separate process and is not already considered during development. For the efficient and effective application of remanufacturing, it is essential to *design disassembly in early stages of the product and production process development (3)* [8, 14]. To provide this, a *flexible design of the production process in the form of changeable production concepts (4)* is important to react to different conditions of the returning products [1, 5]. Consequently, the focus lays on the *industrialization of disassembly (5)* [5].

Furthermore, when designing the disassembly, automation should be preferred and integrated into the production process and considered as part of it. In addition to the requirements assigned to the product and production process, further requirements occur generally in the overall development process. Since remanufacturing must take into account the disassembly process and not only the actual production process, *the influence of fluctuating quality uncertainties and the degree of damage to the components* (6) must be considered [1].

Consideration of the constraints imposed by the approach to a remanufacturing process and the impact on product and production process development (7) [14]. Nowadays, the condition of returned components or products is uncertain and cannot be planned. For the successful implementation of remanufacturing in the development process for product and production process, it is important to use *information from previous product phases to incorporate these into the development of future products and processes (8)* [14].

The identified requirements, which are incorporated in the consideration of remanufacturing in the development phase of product and production, must be addressed in the development process. In Sect. 4, the identified requirements are compared with the design characteristics of HIP^{3}D (see Fig. 2) and the concept is extended to consider product, production process and remanufacturing simultaneously.

4 Concept for Industrialization of Remanufacturing in the Highly Iterative Product and Production Process Development

The model of HIP^{3}D described by Wlecke [9] provides a development approach that includes a parallel and highly iterative development of product and production process and thus offers the possibility to consider the influence of remanufacturing on product and production process. However, this approach does not meet all the requirements identified in Chapter 3.2 (see Fig. 2).

Requirements Remanufacturing	(1)	(2)	(3)	(4)	(5)	(6)	(7)	(8)
Fullfilled by the concept of HIP³D	X	O	X	O	X	X	✔	✔

X not fullfilled
O partly fullfilled
✔ fullfilled

Fig. 2. Requirements of Remanufacturing fulfilled by the concept of HIP^{3}D.

In this chapter, the concept of HIP^{3}D is extended to meet the requirements identified in Sect. 3.2 for integrating remanufacturing into development. Therefore, an eight-step concept (see Fig. 3) is presented in the following (I–VIII).

Due to the challenges facing the development as well as the additional requirements occur from remanufacturing, in a first step (I) of the development process, customer's requirements and potential constraints of remanufacturing on product and production process must be identified and their influence on each other has to be evaluated. The constraints are described, for example, by product geometry, the product materials and the manufacturing processes.

The identified customer requirements and restrictions due to remanufacturing are the basis for the initial formulation of the product hypothesis (II), production process hypothesis (III) and a hypothesis describing the remanufacturing process (IV). The hypotheses serve as a basis for the initial development of product prototypes and manufacturing concepts for the production process as well as manufacturing concepts and information requirements for remanufacturing. The formulated hypotheses are further concretized and evaluated in the course of the development in iterations, to ensure the reaction to changes in customer requirements as well as the consideration of reprocessing is given during the development. The interaction of the three dimensions implies a constant simultaneous analysis of those to realize this consultation, a continuous exchange of information between the three dimensions is necessary.

The development phase is followed by the transition to the validation phase. In the validation phase, the prototypes of the product cycle (V) and the manufacturing concepts from the production process cycle (VI) as well as the manufacturing concepts and information requirements from the reprocessing cycle (VII) are validated. The prototypes of the product phase are first manufactured at component level so the most important functionalities can be tested initially. The goal of the last iteration is the completion of a holistic prototype. In the validation of the prototypes (V), it is essential to determine whether compliance with the constraints of remanufacturing in terms of material or component geometry has been observed. In the validation of (VI), it is necessary to provide the flexibility of the process. In the validation of (VII), new input is generated for the next product hypothesis, which describes the necessary information that must be integrated into the product e.g. by sensor technology. The information collected through the integration of sensors is used for planning the return of the products and components and provides conclusions about the assembly and the input required for disassembly. The results of the validation are evaluated. Based on the evaluation, it is decided whether the development moves to the next phase or if the current phase has to be iterated. The decisive factor for the evaluation is the maturity of the prototype on the product side and the degree of freedom of production or reprocessing on the process side. Development and validation phase are carried out within the cycles of production (PC), production process (PPC) and remanufacturing (RemC). The RemC is concerned with the development of a concept for the industrialization of disassembly, as this is the core process of remanufacturing. This enables the remanufacturing process to be economically advantageous.

At the same time, by considering the requirements of remanufacturing from the start of development, it is ensured that both the developed product and the production concept allow remanufacturing. The progress of the development process is continuously monitored by means of the synchronization points provided at the phase transitions. Once the product maturity required by the customer has been reached and the degree of freedom of the production has been sufficiently reduced, the iterative development ends and a remanufacturable product, a production concept and disassembly concept is derived and production begins (VIII).

Figure 3 summarizes the eight steps of the hypothesis-based approach for the industrialization of remanufacturing in the highly iterative product and production process development.

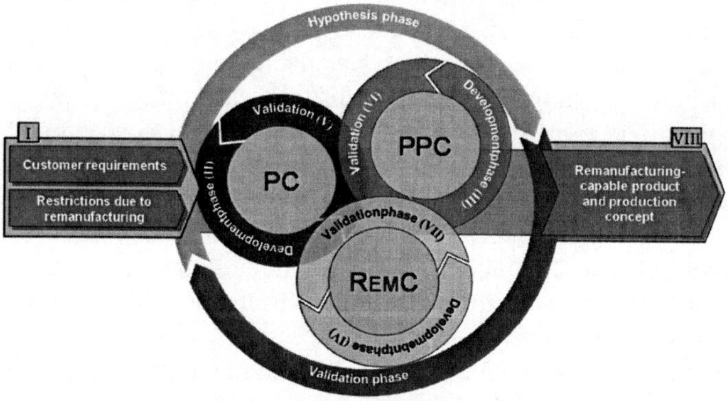

Fig. 3. Concept for industrialization of remanufacturing in the (HIP^{3}D)

Due to the parallel development of product, production process and remanufacturing, it is ensured that both the customer requirements and the remanufacturing requirements are integrated in the development. As a consequence, a concept for the industrialization of the remanufacturing, in particular the disassembly, can be realized. Furthermore, by identifying the required information for data-based support, it is possible to control the subsequent return of the components for remanufacturing. The integration of customer requirements and the constraints resulting from remanufacturing in the first step (I) of the development process ensures that both product modularity (1) and initial influences on the product and the associated production process as well as a subsequent remanufacturing process (7) are taken into account.

The addition of a separate cycle which considers the development of the remanufacturing process (IV and VII), enables the simultaneous development of the remanufacturing process in parallel with the product and production process development. The industrialization of disassembly can be provided on this basis (3 and 5). Furthermore, by developing the remanufacturing process in early phases of development, information needs can be incorporated into product and production process development, which are necessary at the end of the product life cycle to enable remanufacturing to be planned (6 and 8). The integrated development and consideration of remanufacturing (II–VII) ensures that all components of the product and all process steps enable remanufacturing (1) while also realizing a sufficiently flexible production concept that can respond to quality and quantity uncertainties (4).

5 Conclusion and Further Research

As resources become increasingly scarce, companies are forced to rethink and move from a linear economy to a circular economy. A decisive strategy for manufacturing companies is remanufacturing. For the industrialization of remanufacturing, it is essential to address this in early stages of development in order to design the remanufacturing process. For these reasons, a concept for the industrialization of remanufacturing is presented in this paper. An integrated approach described by HIP^3D is suitable as a basis because of the parallel development and consideration of the interactions between product and production. Through adaptations, which are necessary due to the requirements of remanufacturing, the approach is suitable for the development of remanufacturable products as well as the associated production and remanufacturing process. Further research is required in formulating the initial requirements of remanufacturing and in describing the remanufacturing process in detail. In particular, the consideration of the derived information requirements requires detailed consideration.

Acknowledgement. Funded by the Deutsche Forschungsgemeinschaft (DFG, German Research Founda-tion) under Germany's Excellence Strategy—EXC-2023 Internet of Production—390621612.

References

1. Sprenger, K., Klein, J.-F., Wurster, M., et al.: Industrie 4.0 im Remanufacturing. I40M 2021(4), 37–40 (2021). https://doi.org/10.30844/I40M_21-4_S37-40
2. Ministerium für Umwelt, Forsten und Verbraucherschutz (ed) (2008) Kreislaufwirtschaftsland Rheinland-Pfalz (2008)
3. Steinhilper, R.: Remanufacturing-the ultimate form of recycling. Fraunhofer IRB Verlag (1998)
4. Ijomah, W.L., McMahon, C.A., Hammond, G.P., et al.: Development of robust design-for-remanufacturing guidelines to further the aims of sustainable development. Int. J. Prod. Res. **45**(18–19), 4513–4536 (2007)

5. Hollah, A.M., Kreisköther, K.D., Kampker, A., et al.: Electromobile remanufacturing—Nutzenpotenziale für batterieelektrische Fahrzeuge. In: 5th Conference on Future Automotive Technology Focus Electromobility, COFAT, 2016-10-12–2016-10-13, Fürstenfeld, Germany, 8 Seiten (2016)
6. Lange, U: Ressourceneffizienz durch Remanufacturing - Industrielle Aufarbeitung von Altteilen. VDI ZRE Publikationen(18) (2017)
7. Kurilova-Palisaitiene, J., Sundin, E., Poksinska, B.: Remanufacturing challenges and possible lean improvements. J. Clean. Prod. **172**, 3225–3236 (2018). https://doi.org/10.1016/j.jclepro.2017.11.023
8. Matsumoto, M., Masui, K., Fukushige, S., et al. (eds.): Sustainability Through Innovation in Product Life Cycle Design. Springer eBook Collection Earth and Environmental Science. Springer, Singapore (2017)
9. Wlecke, S., Prote, J.-P., Molitor, M., et al.: Concept for the industrialization of physical products in the highly iterative product development. In: Production at the Leading Edge of Technology, pp. 583–592. Springer, (2019)
10. Prendeville, S., Bocken, N.: Design for remanufacturing and circular business models. In: Sustainability Through Innovation in Product Life Cycle Design, pp. 269–283. Springer, (2017)
11. Boorsma, N., Balkenende, R., Bakker, C., Tsui, T., Peck, D.: Incorporating design for remanufacturing in the early design stage: a design management perspective. J. Remanuf. **11**(1), 25–48 (2020). https://doi.org/10.1007/s13243-020-00090-y
12. Lindkvist Haziri, L., Sundin, E.: Supporting design for remanufacturing—a framework for implementing information feedback from remanufacturing to product design. J. Remanuf. **10**(1), 57–76 (2020). https://doi.org/10.1007/s13243-019-00074-7
13. Yang, S.S., Ong, S.K., Nee, A.Y.C.: A decision support tool for product design for remanufacturing. Proc. CIRP **40**, 144–149 (2016). https://doi.org/10.1016/j.procir.2016.01.085
14. Schulze, V., Aurich, J.C.: Update-Factory für ein industrielles Produkt-Update (2021)
15. Borrego, M., Foster, M.J., Froyd, J.E.: Systematic literature reviews in engineering education and other developing interdisciplinary fields. J. Eng. Educ. **103**(1), 45–76 (2014)
16. Salah, B., Ziout, A., Alkahtani, M., et al.: A Qualitative and quantitative analysis of remanufacturing research. Processes **9**(10), 1766 (2021). https://doi.org/10.3390/pr9101766
17. Wurster, M., Häfner, B., Gauder, D., et al.: Fluid Automation—a definition and an application in remanufacturing production systems. Proc. CIRP **97**, 508–513 (2021). https://doi.org/10.1016/j.procir.2020.05.267

Determining the Product-Specific Energy Footprint in Manufacturing

P. Pelger[1,2](✉), C. Kaymakci[1,2], S. Wenninger[3,4], L. Fabri[3,4], and A. Sauer[1,2]

[1] Fraunhofer Institute for Manufacturing Engineering and Automation IPA, 70569 Stuttgart, Germany
philipp.pelger@ipa.fraunhofer.de
[2] Institute for Energy Efficiency in Production, University of Stuttgart, 70569 Stuttgart, Germany
[3] FIM Research Center, University of Applied Sciences Augsburg, 86159 Augsburg, Germany
[4] Project Group Business and Information Systems Engineering of the Fraunhofer FIT, 86159 Augsburg, Germany

Abstract. In the energy transition context, the manufacturing industry moves into the spotlight, as it is responsible for significant proportions of global greenhouse gas emissions. The consequent pressure to decarbonize leads to suppliers needing to report and continuously reduce the energy consumption incurred in manufacturing supplied goods. To track the energy footprint of their products, manufacturing companies need to integrate energy data with process and planning data, enabling the tracing of the product-specific energy consumption on the shop floor level. Since manufacturing processes are prone to disturbances such as maintenance, the energy footprint of each product differs. Meanwhile, the demand for energy-efficiently produced products is increasing, supporting the development of a sustainability-focused procurement by OEMs. This paper addresses this development and outlines the technical requirements as well as how companies can identify product-specific energy consumption. Furthermore, a case study is conducted detailing how to determine the product-specific energy footprint.

Keywords: Energy footprint · Energy transparent products · Data analytics

1 Introduction

Climate change drastically alters the world with diverse impacts on nature, society, and the economy [1, 2]. To counteract this, the European Union has set the goal of becoming the first climate-neutral continent by 2050 [3]. Furthermore, the German government has tightened its climate protection targets by amending the Climate Protection Act, thereby setting the goal of greenhouse gas neutrality by 2045 [4]. As the manufacturing industry is responsible for a significant amount of greenhouse gas emissions (GHG), decarbonization pressure increases, necessitating manufacturing value chains to develop towards more sustainable, energy-efficient, and digitalized structures [5, 6]. In particular, this development focuses on the entire supply chain since the industrial sector

M. Liewald et al. (Eds.): WGP 2022, LNPE, pp. 781–790, 2023.
https://doi.org/10.1007/978-3-031-18318-8_77

plays a significant role in achieving long-term emission reduction [7]. Moreover, new decision-making criteria of OEMs, evaluating their entire supply chain based on energy and GHG-saving potential, particularly exert pressure on suppliers in the manufacturing sector [8]. While the demand for low-carbon, energy-efficient products is rising, identifying the product-specific energy footprint in manufacturing is an important enabler for industrial processes to contribute to the decarbonization of the economy [9]. To meet the requirements of the OEMs and remain competitive, suppliers in the manufacturing sector need to create transparency regarding the energy consumed during product manufacture. Thus, determining the product-specific energy footprint of manufacturing processes is the first step toward transparency. However, existing approaches to identify the energy footprint in manufacturing comprise static analysis like one-time simulations that do not consider the dynamic environment of manufacturing processes and the continuous improvement processes [10, 11]. To address this gap in research and practice, we describe the technical requirements and the procedure of how companies can switch from static analyses to a dynamic, data-based determination of the energy footprint of their products utilizing digital solutions such as data analytics, laying the foundation for a smart, interconnected, and sustainable manufacturing ecosystem [12–14].

The transparency gained enables companies to respond to OEMs' increasingly strong pull effect concerning sustainable supply chains and political or customer requirements, thus reflecting efficiency improvements at the product level and remaining competitive. In other words, our work aims to address the dynamic manufacturing environment and provides a data-based and real-time solution for identifying the product-specific energy footprint.

This paper is structured as follows: Sect. 2 presents the related work for manufacturing products' energy footprint, laying the theoretical foundations. Section 3 presents the technical requirements, the procedure, and the scope for data-driven determination of the product-specific energy footprint in manufacturing. For better comprehensibility, we provide an exemplary determination of the product-specific energy footprint in Sect. 4. Last, Sect. 5 concludes our work and presents opportunities for further research.

2 Related Work

This section describes the theoretical foundations of determining the product-specific energy footprint. Furthermore, the related work assessing the energy footprint in manufacturing systems and processes will be evaluated to identify the gaps in current research. Jeon et al. [11] introduce a method for determining and analyzing the energy footprint of manufacturing systems on four different aggregation levels—product, machine, plant, and industry. Therefore, information from the lowest levels, such as CAD models, is used for extracting key energy parameters from product design. The machine-level focuses on the different states determining energy consumption levels. The plant and industrial levels are simulated environments where experiments are designed and optimized. Although Jeon et al.'s work comprises a holistic approach to analyzing the energy footprint in manufacturing, the work focuses on model-based simulation, not considering real-world manufacturing data. Xie et al. [10] present an integrated model for the specific energy consumption of machine tools used in manufacturing processes. The authors model the specific energy consumption from spindle systems by abating established and theory-based power equations from motor driving and transmission systems. Limitations of the study conducted by Xie et al. include a static approach, requiring adjustment of the model coefficients in order to predict the specific energy consumption of different machines, complicating the determination of the product-specific energy footprint. Peng et al. [15] use a feature-based approach for evaluating the energy consumption in manufacturing. The solution considers three different aspects of energy consumption in the manufacturing plant—product, machining, and peripheral. Product-associated energy consumption is related to the raw workpiece or geometric dimensioning. Machining energy consumption considers all machine-related data, such as the machining parameters or cutting fluids. Peripheral energy consumption comprises all data generated from the manufacturing infrastructure. Furthermore, the authors define six reliability and accuracy confidence levels to classify the available data into confidence levels. However, the applicability of the feature-based approach is based on simplified calculations for modeling the energy consumption. Actual manufacturing data, e.g., from manufacturing execution systems, are not considered.

In contrast to related work, this paper presents a dynamic, data-based approach for identifying the product-specific energy footprint in manufacturing. Whereby most of the methods in the literature primarily identify the footprint for a particular product type but not for the individual product instance manufactured at a particular time, our approach for dynamic determination of the product-specific energy footprint in manufacturing, presented in the next chapter, is intended to close this research gap.

3 Determining the Product-Specific Energy Footprint in Manufacturing

3.1 Technical Requirements and Procedure

This chapter presents the technical requirements and how companies can identify the product-specific energy consumption in manufacturing, describing the functional components and relations concerning data and information flow. In the context of this work, the technical requirements and the procedure introduced in this chapter are based on interviews with three IT and manufacturing system experts with expertise in data processing in a manufacturing environment. Capturing the real-world energy footprint of manufactured products compels companies to meet specific requirements regarding digital information systems as well as physical sources such as sensors or machine data, thereby overcoming the challenge of integrating different data sources [16–18].

First, companies need to collect **energy data on a machinery level** to monitor the energy consumption of manufacturing systems, enabling energy data evaluation via time-series analysis [19]. Therefore, integrating sensors such as smart meters is necessary, enabling ingestion of energy time series data to the company's energy management system (EMS) and integration with other data from manufacturing or information systems that can be correlated to the energy consumption [16, 19]. Second, collecting and ingesting **machine-specific process data** such as the processing time is needed to link the machine state to the energy data. Third, **organizational planning data** such as the

order data is needed to allocate the energy footprint to a specific product and order. To do so, **traceability of products** on the shop floor level is necessary. Depending on the underlying product, process, and IT, traceability requires either a machine-induced assignment of a manufacturing step to a specific product and order by automatically labeling the process data or a product-induced assignment of an order to a manufacturing process by using RFID tags or other technologies enabling product-specific tracking on shop floor level. Therefore, linking planning data from the enterprise resource planning (ERP) system with process data managed in the manufacturing execution system (MES) makes monitoring and updating the order status based on the actual manufacturing progress possible. The resulting traceability of products allows allocating the energy consumption incurred during machine-related processing times to one specific order. As shown in Fig. 1, the acquired data are extracted from their respective database and integrated, thereby merging the data based on their time stamps, thus preparing it for further analysis. Therefore, linking process and planning data with energy data enables constructing an **energy footprint data array**, integrating all necessary information for determining the product-specific energy footprint.

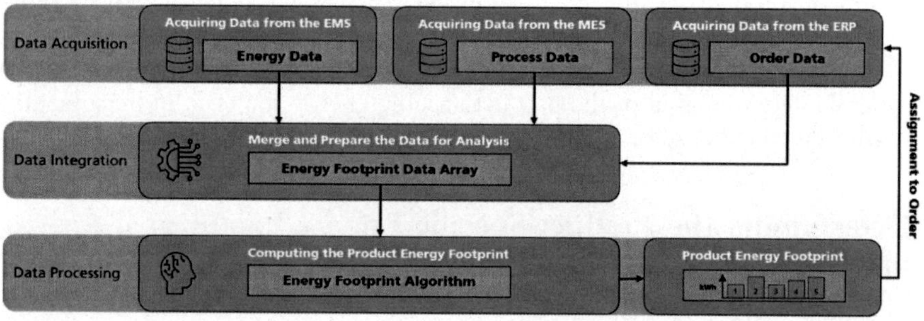

Fig. 1. Procedure for the determination of the product-specific energy footprint

For manufacturing processes, where only one product is processed per station, the product-specific energy footprint is determined by assigning the energy consumption incurred during a product's processing period to its specific instance. Since the order execution duration depends on various machine states, such as conveying, tool change, or disruptions, the power utilized during these states must be considered and assigned to the respective product when determining the energy footprint. Consequently, the **product energy footprint** on the machinery level can be calculated following Eq. 1:

$$\textit{Product Energy Footprint} = \sum_{i=1}^{M}\sum_{j=1}^{N}(\textit{Power } P_i * \textit{Time } t_i)_j \tag{1}$$

$$\textit{whereby } i = \textit{product related machine state and } j = \textit{machine}$$

As the equation indicates, the product-specific energy footprint on the machine level results from the sum of the power induced by the product-related machine states multiplied by the time of the corresponding machine state. Accumulating the resulting energy consumption induced by a manufacturing line's various machines and machine states allows for identifying the product-specific energy footprint on a plant level. Once all the necessary data have been acquired and processed into the proposed data array, the product-specific energy footprint can be determined. Therefore, an **energy footprint algorithm** can classify the energy data regarding the machine and the processed product, accumulating the machine-induced energy consumption and assigning it to the respective order, thereby determining the **product energy footprint**.

3.2 Defining the Scope of the Product-Specific Energy Footprint

While the effort of determining the product-specific energy footprint depends on the complexity of the product and the associated manufacturing system, the procedure is customizable regarding the process boundaries. As indicated in Fig. 2, the scope of the product-specific energy footprint in manufacturing comprises three different scaling options:

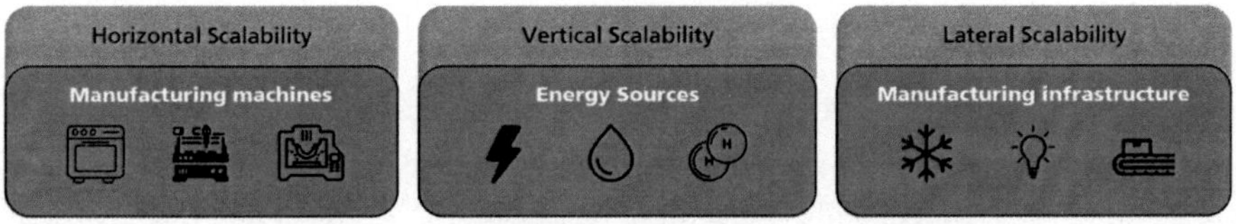

Fig. 2. Scope of the product-specific energy footprint

The **horizontal scalability** represents the scope of the considered machines when determining the energy footprint. Although a holistic review of a product's manufacturing process is reasonable, narrowing the scope to monitor one sub-process, thereby adjusting Eq. 1 accordingly, enables product-specific energy efficiency benchmarking for one machine. Consequently, efficiency monitoring facilitates total productive maintenance, enabling continuous optimization of manufacturing processes.

While this paper focuses on electricity-based energy consumption, **vertical scalability** enables extending the scope of the product-specific energy footprint by integrating other energy sources such as the water or hydrogen consumed during the manufacturing of one specific product. Vertical adaptation of the energy footprint also requires an adaptation of the physical infrastructure, as flow meters have to be installed, and machine-specific hydrogen or water consumption data must be integrated into the MES. When determining the product-specific energy footprint regarding different energy sources, Eq. 1 must be adjusted, focusing on the product-related water or hydrogen consumption on a machine level rather than the power utilized. A combined energy footprint from different energy sources is also conceivable. For example, electricity and water consumption can be reported for a product.

Lateral scalability represents the further possibility of extending the process boundaries of the product-specific energy footprint by integrating the energy consumption induced by peripheral components. Peripheral energy consumption consists of all energy data generated from the manufacturing infrastructure, such as intralogistics or cooling

[15]. While allocating the peripheral energy consumption remains rather challenging, one approach for auxiliary equipment is incorporating it as a fixed energy share regarding manufacturing time and output during this period. The more comprehensive the energy footprint scope, the more complex the digital and physical infrastructure requirements and the more implementation effort are needed. In particular, vertical scaling and the associated consideration of different energy sources and lateral scaling and the accompanying integration and product-specific assignment of peripheral energy data require significant domain knowledge and substantial implementation effort.

4 Exemplary Determination of the Product Energy Footprint

Determining the energy footprint is demonstrated by applying the procedure outlined in chapter three to an exemplary product. The manufacturing process comprises three different machines. A plastic component is manufactured using an injection molding machine, picked up by a robot, and conveyed to a laser, which applies a logo to the component. The scope of the product-specific energy footprint considered in this use case covers the electricity-based energy consumption of the three machines, measured by near-machine smart meters on the shop floor level. In addition, the process data accrued during manufacturing are captured via a machine interface and logged in a time-series database. Since a wide range of heterogenic process data arise, domain knowledge is necessary when determining which data are necessary to identify the product-related energy consumption at the machinery level. Regarding the injection molding process, the cycle time is decisive, as it provides information on how long it took for a component to be manufactured. By linking the cycle time with the start timestamp of the cycle and the cycle id, it is possible to determine the cycle end timestamp by pre-processing the historical data. Consequently, mapping this period to the energy time series data enables allocating the energy consumption induced by the injection molding process to one product and its unique order identification number. An exemplary segment of the pre-processed energy footprint data array for the injection molding machine is demonstrated in Fig. 3. Information regarding the planning data is acquired by the administrative IT system, providing information about which order was manufactured at what time. The data array illustrates the result of integrating energy and process data with planning data, thereby assigning the power values of the injection molding machine to a specific cycle and order ID, enabling determining the product-specific energy footprint on a machinery level.

index	_orderid	_machineid	_cycleid	_cyclestart	_cyclestop	_time	_value	_id
0	99000154…	Injection_Molding	452387785…	07:30:07.361	07:30:44.1…	07:30:07.…	0.238999…	Power
1	99000154…	Injection_Molding	452387785…	07:30:07.361	07:30:44.1…	07:30:08.…	0.459999…	Power
2	99000154…	Injection_Molding	452387785…	07:30:07.361	07:30:44.1…	07:30:10.…	0.829999…	Power
3	99000154…	Injection_Molding	452387785…	07:30:07.361	07:30:44.1…	07:30:12.…	0.329999…	Power
4	99000154…	Injection_Molding	452387785…	07:30:07.361	07:30:44.1…	07:30:14.…	0.501000…	Power
5	99000154…	Injection_Molding	452387785…	07:30:07.361	07:30:44.1…	07:30:15.…	0.216979…	Power
6	99000154…	Injection_Molding	452387785…	07:30:07.361	07:30:44.1…	07:30:17.…	0.524640…	Power

Fig. 3. Energy footprint data array excerpt of the injection molding machine

Since the machines and the underlying processes are interrelated, the stop timestamp of an injection molding cycle is the start timestamp of the subsequent handling process of the robot. As the available process data of the robot do not provide the same information as the injection molding machine, domain knowledge is required to identify the relevant datasets that allow the assignment of the robot-specific energy consumption to one specific product. Linking the start timestamp of the handling process with the travel speed of the robot and its arm enables identifying the stop timestamp of the handling process as the travel speed value drops to zero. As a result, mapping the robot processing time to the energy time series data enables identifying the product-specific energy consumption, thus connecting it to the order ID previously processed by the injection molding machine. The end timestamp of the robot handling again signals the start timestamp of the subsequent laser process. The product-specific processing time and the subsequent energy consumption result from the start timestamp and the value of the traverse speed of the laser, which drops to zero after the process is over. The resulting product-specific energy consumption can be connected to the order ID previously processed by the injection molding machine and the robot.

After determining the energy footprint data array for all three machines, an algorithm can process the resulting arrays according to formula one, thus determining the product-specific energy footprint on a machinery level. As shown in Fig. 4, summing up the individual energy footprint of the machines enables determining the product-specific energy footprint in the considered scope of the respective manufacturing line.

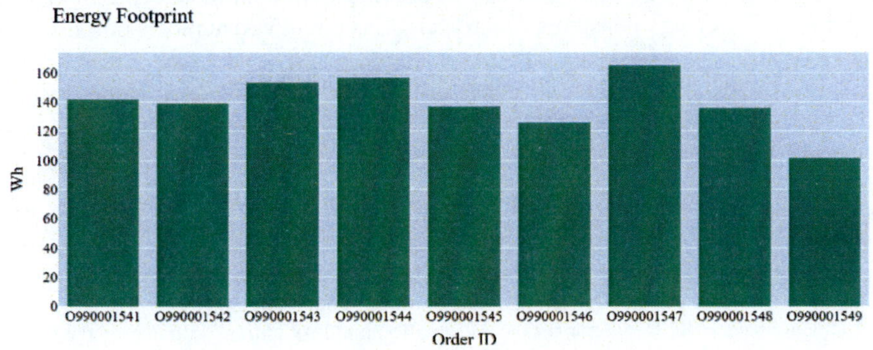

Fig. 4. Product-specific energy footprint

As the results indicate, determining the product-specific energy footprint creates transparency regarding energy use in manufacturing, facilitating product-specific energy-efficiency benchmarking and optimization, enabling efficiency-oriented supply chain monitoring of OEMs.

5 Conclusion and Future Work

Our study outlined technical requirements and presented a procedure enabling companies to identify the product-specific energy footprint in manufacturing. In contrast to existing work in research and practice, we offer a novel approach that enables data-based determination of product-specific energy consumption by rigorously recording energy and process-related data and combining them, thereby considering the dynamic environment of manufacturing processes. With the approach presented in this paper, continuous tracking of the product-specific energy footprint and energy efficiency monitoring of manufacturing processes are enabled, differentiating this procedure from existing static approaches. Furthermore, it is essential to consolidate manufacturing and energy management domains to achieve transparency in energy consumption, which have often been considered separately. This involves extracting the data from the respective information systems and combining it into a common solution.

Naturally, our study is subject to limitations and prospects for further research. First, we focused on the measurable key figure of energy consumption. Aiming to quantify and reduce product-specific GHG emissions might interest further studies [20]. Here, the information systems perspective is increasingly crucial concerning electricity, as electricity can be procured via various contracts, each with a different electricity mix, and must be matched to the respective consumers. Second, while we focus on extracting product-specific energy consumption, we neglect to ensure tamper-proof data management and storage approaches. Research could build on our study to develop solutions for tamper-proof traceability and verification of product-specific energy consumption, enabling seamless documentation along supply chains. Third, our research is at a relatively early stage of development, so validation and testing in the field are central to a viable solution. However, considering the high costs for sensors and measurement devices might limit the scalability and reproducibility of our study. This could incentivize less costly approaches, such as non-intrusive load monitoring to quantify machine-specific energy consumption instead of installing vast numbers of sensors and meters.

Despite these limitations, we are convinced that this study provides an important first step towards traceable and transparent energy consumption in manufacturing to achieve the industry's climate targets.

Acknowledgements. The industrial data examined in this publication were provided from CUNA production by Fraunhofer IOSB-INA. CUNA Production consists of an injection molding production facility set up in the SmartFactoryOWL in Lemgo in a cooperative of 10 industrial partners and has been operating since 2021. The framework for this cooperative was set by the research and development project "KI-Reallabor für die Automation und Produktion" initiated by the German initiative "Plattform Industrie 4.0", funded by the German Federal Ministry for Economic Affairs and Climate Action (BMWK) and managed by the VDI Technologiezentrum (VDI TZ). Under the leadership of Fraunhofer IOSB-INA, the project pursues the central objective of making industrial datasets available to a broad community of AI developers via an open data platform. Fraunhofer IOSB-INA generates and processes the data arising from CUNA production to enable the training of models. One focus of the "KI Reallabor" is the development of an energetic footprint by Fraunhofer IPA based on the data presented in this publication.

References

1. Kara, M., Ghadge, A., Bititci, U.: Modelling the impact of climate change risk on supply chain performance. Int. J. Prod. Res. **59**(24), 7317–7335 (2021)
2. Mitchell, G.: Climate change and manufacturing. Proc. Manuf. **12**, 298–306 (2017)
3. European Commission: A European green deal. https://ec.europa.eu/info/strategy/priorities-2019-2024/european-green-deal_en. Last Accessed 20 Apr 2022
4. Umweltbundesamt: Treibhausgasminderungsziele Deutschlands, https://www.umweltbundesamt.de/daten/klima/treibhausgasminderungsziele-deutschlands#internationale-vereinbarungen-weisen-den-weg. Last Accessed 08 Apr 2022
5. Stock, T., Seliger, G.: Opportunities of Sustainable Manufacturing in Industry 4.0. Proc. CIRP **40**, 536–541 (2016)
6. Buettner, S., Schneider, C., König, W., Mac Nulty, H., Piccolroaz, C., Sauer, A.: How do German manufacturers react to the increasing societal pressure for decarbonisation? Appl. Sci. **12**(2), 543 (2022)
7. Fais, B., Sabio, N., Strachan, N.: The critical role of the industrial sector in reaching long-term emission reduction, energy efficiency and renewable targets. Appl. Energy **162**, 699–712 (2016)
8. Centobelli, P., Cerchione, R., Esposito, E.: Environmental sustainability and energy-efficient supply chain management: a review of research trends and proposed guidelines. Energies **11**(2), 275 (2018)
9. Di Foggia, G.: Energy-efficient products and competitiveness in the manufacturing sector. J. Open Innov. Technol. Market Complex. **7**(1), 33 (2021)
10. Xie, J., Liu, F., Qiu, H.: An integrated model for predicting the specific energy consumption of manufacturing processes. Int. J. Adv. Manuf. Technol. **85**(5–8), 1339–1346 (2015). https://doi.org/10.1007/s00170-015-8033-y
11. Jeon, H., Taisch, M., Prabhu, V.: Modelling and analysis of energy footprint of manufacturing systems. Int. J. Prod. Res. **53**(23), 7049–7059 (2015)
12. Bauer, D., Maurer, T., Henkel, C., Bildstein, A.: Big-data-analytik: Datenbasierte Optimierung Produzierender Unternehmen. Zenodo (2017)
13. Donnelly, J., John, A., Mirlach, J., Osberghaus, K., Rother, S., Schmidt, C., Voucko-Glockner, H., Wenninger, S.: Enabling the smart factory—a digital platform concept for standardized data integration. CPSL 2021 (2021)
14. Pauli, T., Marx, E., Matzner, M.: Leveraging industrial IoT platform ecosystems: insights from the complementors' perspective. ECIS (2020)
15. Peng, T., Xu, X.: Energy consumption evaluation for sustainable manufacturing: a feature-based approach, pp. 2310–2315. WCICA (2014)
16. Kaymakci, C., Sauer, A.: Automated profiling of energy data in manufacturing. In: Behrens, B.-A., Brosius, A., Hintze, W., Ihlenfeldt, S., Wulfsberg, J.J. (eds.) WGP 2020. LNPE, pp. 559–567. Springer, Heidelberg (2021). https://doi.org/10.1007/978-3-662-62138-7_56
17. Harding, J., Shahbaz, M., Srinivas, Kusiak, A.: Data mining in manufacturing: a review. J. Manuf. Sci. Eng. **128**(4), 969–976 (2006)

18. Westkämper, E., Löffler, C.: Visionen und strategische Konzepte für das System Produktion. In: Westkämper E, Löffler C (Hrsg) Strategien der Produktion, 71–237 (2016)
19. Bränzel, J.: Energiemanagement. Praxisbuch Für Fachkräfte, Berater und Manager (2019)
20. Laurent, A., Olsen, S., Hauschild, M.: Carbon footprint as environmental performance indicator for the manufacturing industry. CIRP **59**(1), 37–40 (2010)

A Service-Oriented Sustainability Platform—Basic Considerations to Facilitate a Data-Based Sustainability Management System in Manufacturing Companies

D. Koch[1(✉)], L. Waltersmann[1], and A. Sauer[1,2]

[1] Fraunhofer Institute for Manufacturing Engineering and Automation IPA, 70569 Stuttgart, Germany
david.koch@ipa.fraunhofer.de

[2] University of Stuttgart, Institute for Energy Efficiency in Production EEP, 70569 Stuttgart, Germany

Abstract. Sustainability is important aspect of management. Public awareness for climate change and other aspects of sustainability (resource efficiency, energy efficiency, and social responsibility) has forced companies to integrate sustainability considerations into their day-to-day management activities and overall enterprise strategy. Historically, economic aspects have been the focus of all management activities and thus controlling mechanisms and management systems have evolved to provide all kinds of data to facilitate management activities and decision-making. Sustainability data, however, is currently not available for management with comparable ease. Therefore, this paper describes a service-oriented hub (EcoHub) to enable a sustainability management system and to facilitate management decision-making based on sustainability data in manufacturing companies. The focus is on data and service requirements of real use cases and the respective requirements for a data system based on the Asset Administration Shell (AAS).

Keywords: Sustainable production · Technology management · Sustainability · Sustainability management · Data management · Asset Administration Shell · Digitalization

1 Introduction

Due to recent developments in society and science, public awareness for sustainability of companies has significantly increased. This is due to both an increased intrinsic interest of stakeholders in the topic and to more stringent legal requirements regarding companies' sustainability and sustainability management [1]. Sustainable business practices have therefore become a competitive advantage for companies [2]. These changes in markets and public awareness have forced companies to integrate sustainability considerations into their day-to-day management activities and overall enterprise strategy [3].

M. Liewald et al. (Eds.): WGP 2022, LNPE, pp. 791–798, 2023.
https://doi.org/10.1007/978-3-031-18318-8_78

Historically, the sole focus of management has been on the economic situation of a business with sustainability entering the picture only as a side consideration when sustainability and economics happen to be aligned. Accordingly, controlling mechanisms and management systems have evolved to provide all kinds of data to facilitate business management activities and decision-making, while sustainability data is currently not available for management with comparable ease [4].

While some of the sustainability data is available within the company, some is not and needs to be specifically collected. When there is data, it is often not available electronically or in formats that are easily interchangeable across platforms. Also, the data is often not automatically gathered and stored at all. [5].

One of the keys to credible communication and improvement is the collection of comprehensible sustainability data. Due to the complexity and the variety of data sources (e.g., ERP system, machine data collection, external sustainability databases), digital solutions are essential for supporting sustainability management. A service-oriented sustainability platform "EcoHub" thus can help to create the necessary transparency and to manage the complexity [6, 7].

Therefore, this paper seeks to identify the data requirements and types of data that manufacturing companies need for a successful sustainability management and propose a possible way of data acquisition and processing as an outlook.

2 Current Status of the Use of Digitalization for Sustainability Management

It is accepted and recognized, that data collection and processing are key aspects of sustainability management to satisfy information requirements for both internal and external stakeholders [8]. In order to manage the increased complexity, digital approaches can be helpful [8]. Holistic Industrie 4.0 approaches are expected to provide transparency regarding resource consumption and efficiency [9]. A common issue is the low data quality and availability of data in heterogeneous systems. This is a key obstacle in optimizing sustainability aspects such as resource efficiency [10]. First attempts have been made to integrate sustainability data along the supply chain and to exchange data between suppliers and manufacturers [11].

However, availability of tools to organize the acquisition of sustainability data in a structured manner is very limited [12]. There are approaches to manage and automate the data collection (e.g. use of Digital Twin [13], Asset Administration Shell (AAS) [13, 14]) in manufacturing companies. Some are already implemented for classical process optimizations, e.g. for machine parameters [15]. While first steps have been taken to set standards regarding the environmental performance [16], the research and use of these approaches for the improvement of sustainability is still in its beginning [12, 15]. Especially, there is a lack of holistic approaches to collect and exchange environmental impact and sustainability data along the supply chain [17, 18]. The comprehensive view, however, is imperative in order to prevent having information systems that focus on single sustainability aspects [19].

Currently, there is no comprehensive sustainability platform for the use in manufacturing companies. There is a mismatch in the desire and necessity to communicate sustainability information and use it for management decisions and the available sustainability accounting information systems [20]. Hence, the goal of the EcoHub project is to create such a sustainability platform that allows for the integrated storage and analysis of sustainability data, based on the requirements of real-life use cases. The following section will describe how we derived the requirements from the project partners. Based on these requirements, the data structure, data analysis capabilities, and user interface definitions will be derived.

3 Methodology for Deriving the Requirements

The requirements for an integrated data-based sustainability management system were established with experts from manufacturing companies that represent a wide range of possible use cases and industries. This ensures capturing a divers set of requirements that will allow for general applicability of the sustainability platform. The companies and use cases are described in detail below. All use cases are relevant for the requirement specification of the sustainability platform.

3.1 Use Cases

Optimization of Energy Consumption. The company is a wire manufacturer and the production site features a solar power plant as well as a block heat and power station. The management would like to use the sustainability platform to integrate production planning data and energy availability data from the two company-owned power sources to optimize energy efficiency in production.

Process Optimization in Machining. A manufacturer of metal cutting machinery would like to integrate production process data across multiple production steps and production parts in order to obtain an overview of Good Parts vs. Bad Parts per production step and part number in production. In case an anomaly occurs during production, process and machinery data is to be stored on the sustainability platform for post-mortem analysis. The analysis and subsequent optimization of the machining process is expected to yield better resource and energy efficiency.

Optimized Plastics Recycling and Data Exchange. A producer of injection molded parts and a manufacturer of 3D-printers are working together to determine the specification of residual raw material of plastics production and production waste for further use as raw material in 3D-printing. Collected material properties data based on material history (i.e. prior use) is exchanged with the printer manufacturer for printer setup and optimization of the printing process. In this use case, both data collection and analysis and data exchange between two companies play an integral role.

Transparency Regarding Scrap Quantities and Resource and Energy Use and Equipment Control. A manufacturer of corrugated board manufacturing equipment is implementing optimized control strategies for its machinery. This should result in efficiency improvements for their customer. Again, in this use case the purpose of the sustainability platform is twofold: create transparency for the customer regarding machinery's process efficiency through data collection regarding scrap quantities and use of energy and resources. For the manufacturer of the machinery, environmental data and optical data captured on equipment-internal storage shall be transferred to the sustainability platform, once the internal storage is exhausted. This data should be accessible to the equipment manufacturer.

Optimization of Useful Life and Recycling of Cutting Tools. A manufacturer and distributor of machine cutting tools would like to optimize remanufacturing operations and increase the number of recycled tools. To this end, the entire part history needs to be documented on the sustainability platform, including the documentation of the costumer's usage.

Transparent Sustainability Reporting. A producer of paint and varnish would like to streamline the management of environmental data both within the company and along its supply chain. The data is used to generate a thorough overview of the company's environmental impact as well as for sustainability reporting and certification purposes.

3.2 Establishing the Requirements for the Sustainability Platform

The requirements for the sustainability platform were established with the use cases' representatives. First, questionnaires were distributed to the project partners and completed by them. The questionnaires were structured into five sections in order to provide a detailed understanding: (1.) Understanding of the use-case challenges (2.) understanding of the expected analysis services (3.) input data required and available to be used for the services (4.) interface requirements (5.) requirements regarding data security and integrity. Subsequently, follow-up workshops have been conducted with all representatives to clarify any ambiguities. The process that was followed is depicted in Fig. 1.

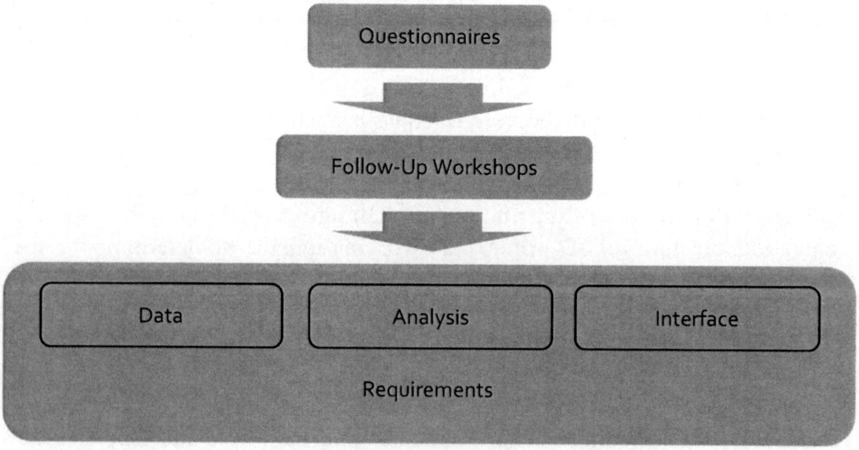

Fig. 1. Process for establishing the requirements for the sustainability platform

The analysis of the questionnaires and workshops gives an overview of the anticipated benefit of an integrated data-based sustainability platform. It highlights data required and differences in expected data analysis and interfaces to meet the expectation.

4 Requirements for a Data-Based Sustainability Management System

As described in the previous section, expert interviews were conducted with representatives from the associated companies. The result is a detailed overview regarding the different specifics of the use cases and the resulting requirements regarding the sustainability platform. The use cases differ widely and represent different aspects of sustainability targets that companies aim to achieve. They range from primarily reporting aspects to optimizing production processes in order to increase energy efficiency or to increase the use of recyclate or optimizing the remanufacturing/maintenance activities for tools.

Naturally, these differences in the nature of the use cases results in a variety of requirements regarding the data interface and sustainability services. Sustainability management requirements vary significantly depending on the use case. The requirements include: (a) collection and storage of sustainability data from various areas within the company for reporting purposes (b) using the accumulated data for process analysis (c) sharing of sustainability data across companies (d) using sustainability data for process optimization.

The identified requirements are now explained in further detail. The requirements are clustered by area of use and not by specific use case. The reason is that some of the use cases have requirements for multiple areas of use and this paper focuses primarily on the design and requirements for the sustainability platform, regardless of how they pertain to one or more particular use cases.

4.1 Data Acquisition Requirements

Data Requirements for Reporting and Data Transparency. As discussed in the introduction, a common challenge for sustainability management is the structured availability of the required data. Hence, one requirement for the sustainability platform is to be a structured enterprise data repository for sustainability data. Energy consumption data, logistics data, carbon emissions data, material flow data, waste data, and production quality data needs to be readily accessible for sustainability management.

Data Requirements for Production Process Control. The extent to which production process control is anticipated using the sustainability platform varies across use cases. However, common requirements exist regarding types of data. Data to be considered include quality data (number of good parts vs. bad parts), process data pertaining to machine control setpoints, energy use data, energy supply data, and tooling data.

Data Requirements for Optimization of Recycling and Maintenance. Recycling and maintenance require primarily information regarding material flow. Both in the supply chain and within the company. Besides material flow data (quantities, materials, etc.), types and frequency of use of tools and materials involved need to be tracked and made available.

Data Requirements for Data Exchange between Companies. Generally, the data stored on the sustainability platform must only be accessible to the company that owns the data. It is first and foremost a proprietary data storage solution. Data safety and security as a common concern for all use-case representatives involved. However, as supply chain data is part of the information that needs to be provided, naturally the standardized platform lends itself to being used as a means of data exchange between supplier and customer.

This is recognized and requested by some of the use cases that are part of the EcoHub working group. Therefore, a cross-company interface for selected data—to be defined by the respective owners—shall be implemented in order to efficiently facilitate such exchange of information.

4.2 Data Processing Requirements

In order to implement the requirements above that were established as described in Chapter 3.2 the sustainability platform will provide a standardized interface that can be addressed by neighboring IT-systems to provide the sustainability data. It is imperative that the interface and data structure is complied with since the absence of a common standard for both data format and data quality is currently a major obstacle in analyzing and using sustainability data that is in principle available in many companies, but not readily accessible (see Chapter 2).

As with data acquisition, the requirements regarding the processing of the acquired data varies widely among the use cases. On one end of the spectrum, the storage of the acquired data is sufficient for some use cases as long as data access is possible. Some use cases require simple analysis in order to compute KPIs.

At the other end of the requirements spectrum, there is the desire to analyze the process data to optimize production planning and increase production efficiency and energy and resource consumption.

Another requirement closely related to data processing is the interface design. Again, the requirements greatly vary, from simple download requirements to displaying process KPIs on operator devices to detailed production planning.

5 Conclusion and Outlook

This paper discusses the challenges of data-based sustainability management. It was shown that there is currently no holistic solution for integrated sustainability data acquisition and management. The EcoHub project is set up to develop a sustainability platform that fulfills the need for modern sustainability management. Using approaches from literature and expert interviews with representatives of various manufacturing companies, the requirements for numerous use cases regarding sustainability management are identified.

After having identified data transfer and storage requirements further research should be devoted to developing analysis and visualization tools for the accumulated data. The challenge is the wide range of use cases. Most use cases require some sort of key performance indicator (KPI) calculation so that this might be one application that should be integrated into a data hub software solution.

To that end, the next steps within the EcoHub project will involve continuous and frequent feedback loops in order to be able to integrate optimizations and requirement clarifications and extensions, as they arise while working with prototyped versions of the EcoHub solution within the various use-cases.

Further, the AAS [13, 14] seems to be a promising specification of a digital twin. Therefore, further research will also consider the possibility of using AAS as a basis for data collection and exchange.

The users currently focus on company-internal acquisition and use of data. However, as discussed above, there are use cases where data exchange between users across the platform maybe very beneficial. Exchange can enhance the potential of the platform if it satisfies user expectations regarding data safety and security. This is another topic to be further investigated.

Acknowledgements. We gratefully acknowledge the support of the Federal Ministry of Education and Research within the Project "Serviceorientierter Hub zur Verwertung von Nachhaltigkeitsinformationen für produzierende Unternehmen" (FKZ 02J20E528).

References

1. Seidel, S.; Recker, J.; vom Brocke, J.: Sensemaking and Sustainable Practicing: Functional Affordances of Information Systems in Green Transformations. In: MISQ 37 (4), p. 1275–1299. DOI: https://doi.org/10.25300/MISQ/2013/37.4.13. (2013)
2. Löser, F.: Strategic information systems management for environmental sustainability. Enhancing firm competitiveness with Green IS. Zugl.: Berlin, Techn. Univ., Diss., Berlin: Universitätsverlag der TU Berlin (Schriftenreihe Informations- und Kommunikationsmanagement der Technischen Universität Berlin, (2015). Online available at http://nbn-resolving.de/urn:nbn:de:kobv:83-opus4-65987. Last Accessed 03 May 2022
3. Schaltegger, S., Hörisch, J., Windolph, S. E., Harms, D.: Corporate Sustainability Barometer 2012: Praxisstand und Fortschritt des Nachhaltigkeitsmanagements in den größten Unternehmen Deutschlands. p. 19f Centre for Sustainability Management (CSM). Universität Lüneburg. Lüneburg. (2012)
4. Burritt, R., Christ, K.: Industry 4.0 and environmental accounting: a new revolution? Asian J. Sustain. Social Responsib. **1**(1), 23–38 (2016). https://doi.org/10.1186/s41180-016-0007-y
5. Friedrich, R., Ploner, F., Schäfer, C., Disselhoff, T., Petkau, A., Hennemann, C., Moecke, J., Wätzig, T., Zimmert, O., Waltersmann, L., Kiemel, S., Miehe, R., Sauer, A.: Potenziale der schwachen künstlichen Intelligenz für die betriebliche Ressourceneffizienz. Berlin: VDI Zentrum Ressourceneffizienz GmbH (VDI ZRE) (VDI ZRE Publikationen) (2021). Available online at: https://www.ressource-deutschland.de/fileadmin/user_upload/downloads/studien/VDI-ZRE_Studie_KI-betriebliche-Ressourceneffizienz_Web_bf.pdf. Last Accessed 03 May 2022
6. Teuteberg, F., Marx Gómez, J.C. (Hg.): Corporate environmental management information systems. In: Advancements and Trends. Business Science Reference (Premier reference source), Hershey, New York (2010)
7. Perl-Vorbach, E.: Communicating environmental information on a company and inter-organizational level. In: Organizational Communication and Sustainable Development. Hershey, Pa. [u.a.]: Information Science Reference (2010)
8. Geibler, J., Brandt, J., Waltersmann, L., Miehe, R., Tesch, R.: Digitales Nachhaltigkeitsmanagement in Unternehmen. In: Industrie 4.0 Management 38 (Nr. 1), p. 45–47 (2022)

9. Schebek, L., Kannengießer, J., Campitelli, A.: Ressourceneffizienz durch Industrie 4.0. Potenziale für KMU des verarbeitenden Gewerbes. Berlin: VDI Zentrum Ressourceneffizienz (VDI ZRE) (VDI ZRE Publikationen) (2017). Available online at: https://www.ressource-deutschland.de/fileadmin/Redaktion/Bilder/Newsroom/Studie_Ressourceneffizienz_durch_Industrie_4.0.pdf. Last Accessed 03 May 2022
10. Böhner, J., Scholz, M., Franke, J., Sauer, A.: Integrating digitization technologies into resource efficiency driven industrial learning environments. Proc. Manuf. **23**, 39–44 (2018). https://doi.org/10.1016/j.promfg.2018.03.158
11. Papetti, A., Marconi, M., Rossi, M., Germani, M.: Web-based platform for eco-sustainable supply chain management. Sustain. Prod. Consumption **17**, 215–228 (2019). https://doi.org/10.1016/j.spc.2018.11.006
12. Al Assadi, A., Waltersmann, L., Miehe, R., Fechter, M., Sauer, A.: Automated environmental impact assessment (EIA) via asset administration shell. In: Philipp Weißgraeber, Frieder Heieck und Clemens Ackermann (Hg.): Advances in Automotive Production Technology—Theory and Application. Stuttgart Conference on Automotive Production (SCAP2020), pp. 45–52. Springer Vieweg (ARENA2036), Berlin, Heidelberg (2021)
13. Jacoby, M.; Usländer, T.: Digital twin and internet of things—current standards landscape. Appl. Sci. **10**(18), 6519 (2020). https://doi.org/10.3390/app10186519
14. Plattform Industrie 4.0: Details of the asset administration shell—Part 1—The exchange of information between partners in the value chain of Industrie 4.0 (Version 3.0RC01). https://www.plattform-i40.de/SiteGlobals/IP/Forms/Listen/Downloads/DE/Downloads_Formular.html?gtp=1022694_list%253D2&cl2Categories_TechnologieAnwendungsbereich_name=Verwaltungsschale. Last Accessed 10 Mar 2022
15. Miehe, R., Waltersmann, L., Sauer, A., Bauernhansl, T.: Sustainable production and the role of digital twins–basic reflections and perspectives. J. Adv. Manuf. Process **3**(2) (2021). https://doi.org/10.1002/amp2.10078
16. ISO 20140-5:2017 Automation systems and integration—evaluating energy efficiency and other factors of manufacturing systems that influence the environment—Part 5: Environmental performance evaluation data
17. Cudre-Mauroux, P., Ceravolo, P., Gašević, D. (eds.): SIMPDA 2012. LNBIP, vol. 162. Springer, Heidelberg (2013). https://doi.org/10.1007/978-3-642-40919-6
18. Schiffleitner, A., Bley, T., Schneider, R., Wimpff, D.: Stakeholder perspectives on business model requirements for a sustainability data exchange platform across supply chains. In: Electronics Goes Green 2012: Taking Green to the Next Level. Joint International Conference and Exhibition, p. 5, Sept 9–12, 2012, Berlin, Germany. Proceedings.Fraunhofer Verlag, Stuttgart (2012)
19. Junker, H., Farzad, T.: Towards sustainability information systems. Proc. Comput. Sci. **64**, 1130–1139 (2015). https://doi.org/10.1016/j.procs.2015.08.587
20. Dagiliene, L., Šutiene, K.: Corporate sustainability accounting information systems: a contingency-based approach. SAMPJ **10**(2), 260–289 (2019). https://doi.org/10.1108/SAMPJ-07-2018-0200

Leveraging Peripheral Systems Data in the Design of Data-Driven Services to Increase Resource Efficiency

T. Kaufmann[1(✉)], P. Niemietz[1], and T. Bergs[1,2]

[1] Laboratory for Machine Tools and Production Engineering (WZL) of RWTH Aachen University, Campus-Boulevard 30, 52074 Aachen, Germany
t.kaufmann@wzl.rwth-aachen.de

[2] Fraunhofer Institute for Production Technology (IPT), Steinbachstraße 17, 52074 Aachen, Germany

Abstract. Production and sustainability represent a challenge that still exists today. The demand for more efficient use of resources and operating materials is clear, and possible through the pragmatic integration of digital technologies and the approach of the circular economy along the entire process chain. However, when leaving individual processes, the complexity of data increases since causal effects between the process steps and their impact on the resulting KPIs must be considered simultaneously. This is where data-driven analysis unfolds its full potential. For this purpose, in addition to the manufacturing process itself, it is imperative to consider the often-neglected peripheral systems, including the provision of raw materials, consumables and supplies. In addition to the necessary consistent and cross-process-step data, manufacturing companies and especially small and medium-sized companies lack usable digital services for the demand-oriented control of process and periphery and for event-based instead of time-controlled recommendations for action for staff, maintenance, and management to achieve an increase in resource efficiency. This work provides an approach that addresses prediction, classification, and anomaly detection using modular machine learning models trained on heterogeneous data in the electroplating industry and gives a conceptual outlook for transforming these models into robust edge services for control systems in manufacturing.

Keywords: Demand-oriented operating resource supply · Data-driven modeling · Data-driven services · Electro chemical plating · Circular economy

1 Introduction and Motivation

Increasingly rare raw materials, drastic price increases, political constraints and capital market aspects pose new challenges for manufacturing companies [1]. The steel and metal processing industry has many energy-intensive production processes and is one of the largest consumers of resources in Germany. Subsequently summarised under raw materials and supplies (RHB), depending on the technology, they represent a medium

M. Liewald et al. (Eds.): WGP 2022, LNPE, pp. 799–809, 2023.
https://doi.org/10.1007/978-3-031-18318-8_79

(cutting) to considerable (electroplating and surface coating technology) share of the resource consumption of manufacturing plants. With an average material cost share of 45% of the gross production value in the manufacturing industry, the optimised, efficient, and sustainable use of resources is of socio-political and ecological, but also strategic relevance. Approaches for the optimisation and adaptive design of manufacturing processes and process chains make an important contribution to this. In addition, peripheral systems offer significant efficiency potential. Process-related peripheral systems in the example of electroplating are, including energy efficiency potential, rectifiers (32%), bath temperature control (23%), drying devices (2%), drives (12%) and compressed air supply (1%) [2]. Unknown are the additional efficiency potentials possible through material savings, order sequence and coating scenarios (order sequence, rinsing cascade control, demand-oriented chemical bath control, process control with prediction of quality parameters) and the resulting change in the CO_2 footprint. Nevertheless, the identification and utilisation of the potential of these systems has received little attention so far. This is also due to the complexity of the resulting overall system. However, data-driven methods are suitable for recording and using these. So far, however, there is still a lack of the necessary data-driven models and digital services that implement the potential of data science in companies in a timely manner.

This paper therefore pursues the goal of a methodology suitable for SMEs for the prediction, classification, and detection of anomalies with the help of modular machine learning models. The results of automated data pre-processing, clustering and event-based prediction for production and maintenance are to be brought back to manufacturing in the form of services.

2 State of the Art

The coating produced by electroplating has functional properties that are defined by the essential process steps of pre-treatment, coating, and post-treatment [3]. With a 35–40% share of sales, the automotive industry is the largest customer for electroplated parts [4]. Due to the large number of resources required for tempering, cleaning and electrolytes, rising raw material and energy prices as well as a shortage of resources have a direct impact on electroplating [5]. With a value of 40%, zinc and zinc alloys are among the most required raw materials for coating [5].

In electroplating, the environmental impact and thus the conservation of resources is an important aspect. The disclosure of potentials for the improvement of the ecological evaluation and the operating balance required for environmental reports requires in all cases a condition monitoring system of the production [5].

Thanks to scientific-technical developments, valid data from almost every process is available in theory, down to the highest level of detail, and can be used to improve efficiency. A study conducted by the VDI Zentrum Ressourceneffizienz GmbH [6] revealed a crucial deficit in the use of data: In none of the companies surveyed was data specifically processed and used to optimise the resource consumption of individual processes and process chains using the possibilities of digitalisation. Today, process actual data are already processed in a few digitalized companies in the industry, but mainly by visualization and simple threshold value consideration. Furthermore, although several concepts are available which analyse the process of electroplating regarding electrolyte management and layer thickness optimization using simulation [7] and data-driven methods—an overview is given by Soergel et al. [8]—the practical and operational use with reference to the entire process chain and system periphery has not yet been developed into digital services which can be used by SMEs. A current challenge also lies in the processing of the high-dimensional data streams and volumes. Although the potential of using data-driven analyses has already become clear, even in small companies, there is a lack of practicable and low-threshold solutions from university research to leverage the potential and justify the investment, as well as qualification to enable plant and personnel.

In addition, it was found that the SMEs were aware of the benefits of individual accounting tools, such as the CO_2 footprint, but that their application was too complex for their own use. This implies a great need for qualification regarding the connection between resource efficiency, sustainability management in companies and the digital transformation through available and standardised solutions, and at the same time calls for available digital services to utilise the potential of data science tools. Overall, it became clear that the data basis for quantifying the effects of the digital transformation on corporate resource efficiency is currently still insufficient [6].

Schmitt identified the following three potentials with data collection that is integrated into production as much as possible [1]: (1) Increasing the degree of utilisation (e.g., based on the reduction of malfunctions, predictive maintenance). For example, fluid systems in machine tools cause 13% of the entire system to fail. (2) Increasing the degree of performance. (3) Identification and management usage of KPI to evaluate resource efficiency. The Fraunhofer Institute for Systems and Innovation Research ISI identified a potential for optimising the power consumption of compressed air supply equipment (lines, couplings, valves) by 30% [9]. Furthermore, compressed air generation itself can be used to increase overall resource efficiency by converting up to 95% of the energy used to generate the expensive medium into usable heat output (heat recovery) [10].

Sievers, König and Sessler summarise an overview of tools and software tools, some of which are freely available, for analysing, modelling, and advising medium-sized manufacturing companies with the aim of improving eco-efficiency [11].

Due to often very heterogeneous data sources and a lack of networking or homogenisation, relevant data on product, energy and material consumption are not continuously recorded. The development of such dynamic value and material flow models for the identification of resource efficiency potentials in manufacturing is still in its infancy [12] and is therefore not available to companies.

Due to the interactions between material, process and subsequent process, dependencies arise in the process or quality characteristics between successive manufacturing processes. This dependence of the effect of a variable on the characteristics of another variable is also referred to in the literature as the interaction effect [13]. Especially for the targeted provision of required resources, a great potential for optimisation and efficiency increase is hidden in the cross-process-step correlation analysis. However, even with ML models with high precision, it is not guaranteed that the recognised interaction effects are consistent with the interactions of the real mechanisms of action [13]. The interaction effects must therefore be plausibly with the help of domain knowledge or understood as a starting point for further experimental investigation. Learning correct cause-effect relationships from data is implemented with methods of causal inference [13].

So far, there is only a very limited number of published papers dealing with the application of machine learning methods to energy-related goals in production engineering. The number of scientific publications on the application of machine learning methods has been increasing since 2018 [14], but there is still a need for the pragmatic implementation or integration of these methods in manufacturing companies.

3 Methodology

The approach is based on a structured methodology for integrating data-driven methods into digitally transformed production, such as electrolyte flow and anode flow, including supply peripherals such as rinsing cascades, compressed air supply and chemical supply. The central parameter is the CO_2 footprint, related to the energy and material input. By extending the balance envelope to the entire system, the complexity for increasing resource efficiency increases considerably. Such an internal balance sheet for energy and material is currently not available but is urgently needed by manufacturing companies.

The methodology for achieving the goal is divided into (1) a preliminary data benefit analysis, (2) an extended plant networking with sensors for the peripheral systems and energy management, (3) an exploratory data analysis and (4) a technology selection for digital services or machine learning methods, so that models for energy and RHB use can be trained in (5) and fed back into production and control in the form of services in (6) (Fig. 1).

The digital services to be developed should, on the one hand, provide improved parameters for the process-controlled use of chemicals, energy consumption and fresh water consumption in rinsing cascades by investigating mechanisms of action for the influence of these parameters on the quality characteristics of layer thickness and layer thickness distribution, and, on the other hand, provide event-based decision-supporting action measures for the personnel for control and predictive maintenance.

The basis for the digital services is (a) a complete, networked and system-differentiated condition monitoring system (sensor infrastructure) including an IT infrastructure for processing, storing, providing, and visualizing data and results, (b) a system-differentiated resource flow balancing system for deriving ecological and economic key indicators and (c) trained machine learning models for improving the resource efficiency of the overall system.

Steps 3–5 focus on the application of data science methods to production engineering field data. The achievement of fundamentals (a) and (b) are explained in Chap. 4. For the development of prediction algorithms (c), primarily supervised learning methods are used. Depending on the labels of the field data of the data set to be acquired, classification or regression models are trained. Due to their high flexibility, especially with time series features, different artificial neural networks (ANN) are evaluated and compared, as well as random forests and support vector machines. The advantage of random forests is that they are non-parametric and therefore very flexible. It is also analysed whether prediction models can be trained that can be reliably used for the anticipatory avoidance of negative

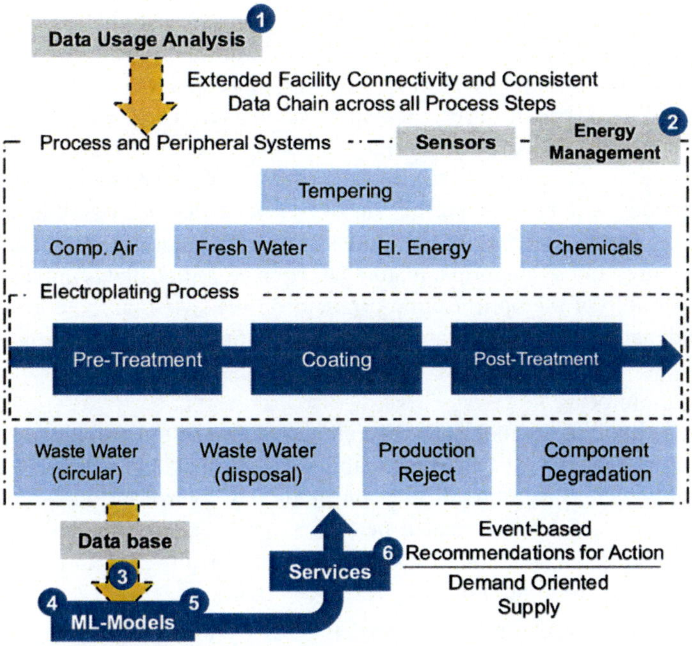

Fig. 1. Methodological approach for leveraging peripheral systems data in data driven services for efficient use of resources in production

events. This is conceptually possible in time-dependent contexts but requires traces of events before they occur in the data and the presence of such situations in training data. Another focus is the development of anomaly detection methods. This can be based on dimensionality reduction by autoencoders using an analysis of the time series of reconstruction errors. In addition to the pure detection of occasional anomalies, it is being investigated (i) whether unfavourable events such as failures can be announced in the anomaly detection time series and proactively detected accordingly, as well as (ii) the possibility of detecting the change to stable other process states.

Dialogue-based approaches and measures for operating and maintenance personnel are also considered when deriving measures for action. The derivation of control signals to increase resource efficiency represents an optimisation problem consisting of high output, at least constant quality, and low resource input.

In the context of decision support, a decision model is developed that models the influence of the characteristics on the named target variables. This model is used as the objective function of the optimisation problem to derive new manipulated/controlled variables from the optimum point and to get as close as possible to the improved state. Similarly, the simultaneous optimisation of all objective variables is considered unrealistic. Therefore, regarding modular services, an attempt is made to determine a Pareto front and to present this to the operating personnel in a comprehensible way.

The relationship of the characteristics to the target variable can be mapped in the case of categorical variables by means of classification procedures, and in the case of metric variables by means of regression procedures. It can be assumed that a manageable number of characteristics of a categorical target variable will be suitable. To calculate the Pareto front, the possible values can therefore be considered fixed. Here, the classification procedure is used to restrict the space of characteristics. The continuous target variables are studied under these restrictions. With the help of classification, all categorical variables are successively fixed at all possible values and optimised with respect to the continuous variables. From this, an overview of the best possible values for all variables can be found. Furthermore, it must be assumed that a correlation between the identified features and a categorical feature has been identified that is not directly related to a target variable of the optimisation problem. In the case that the correlation or the feature is independently assessed as relevant (for example, in the context of exploratory cluster analysis), this feature can be predicted and displayed by means of a classification in a process monitoring.

With the help of these models, the overarching optimisation problem of productivity in terms of time, quality, costs, and sustainability is to be tackled within the framework of the methodology. This means identifying and predicting control and adjustment variables for the necessary and process-related periphery using existing model approaches for the electroplating process, so that maximum output is achieved with the best possible, required quality and minimum primary resource input.

4 Preliminary Results

The electroplating process under consideration is a fully automated drum coating line for approx. 80–120 kg/batch of bulk material for binding elements, which are batched into a different number of drums after weighing for each order. For the in-depth data benefit analysis by the domain experts and data scientists, a diagram of the electroplating line including all peripheral systems was first prepared. This showed the status quo of the data situation. Already recorded and available data streams of the rectifiers, target dosage of chemicals and all bath temperatures were connected to a powerful TimeScale DB locally at the application partner. Important quality parameters are the layer thickness, the layer thickness distribution, the depth scattering and the ductility. These are recorded after coating and integrated into the database so that labelled data is available for analysis and improvement. All other job-related data was provided from the ERP system via.csv-based interfaces.

The data benefit analysis resulted in the following objectives: (1) demand-based fresh water control of the rinsing bath cascades in the multi-step pre-treatment on the basis of the status variables pH value, filling level and predicted carry-over of the subsequent batch; (2) event-controlled injection of compressed air into the coating baths; (3) order-controlled bath temperature control; (4) demand-based control of the extraction system; (5) prediction of anode wear; (6) event-controlled maintenance measures through anomaly detection in the overall system; (7) resource-efficient adjustment of the (projected) coating thickness and coating thickness distribution. The results showed that in order to achieve the objectives, an energy management system with differentiated power recording for the end users, flow sensors for all fluids for the actual recording of fresh water consumption, waste water flows and compressed air, as well as recording of selected chemicals were required. The process-specific data for coating, such as weights, amperage, voltage, temperature, and time, were digitally recorded from the plant control software using an MQTT interface. The concentrations of the process solutions were digitally entered into the system after manual sampling and laboratory evaluation. With the help of the extended and process-step-spanning data chain, a transparent production was achieved, which forms the basis for the identification of efficiency-increasing potentials and further analysis.

In the course of further sensor technology to achieve this transparency, a comprehensive energy management system from JANITZA was put into operation. In addition to the recording of individual consumers (broken down into individual supplies) in the form of time series, this also enabled the system-internal creation of virtual sensors that function as an interconnection of several lines or devices. In this way, various process stations and individual peripheral devices could each be combined as a virtual element, so that the modelling for each batch was available in terms of electrical energy in the form of time series data.

Another challenge was to record the volume flows of the RHB (fluids). The solution was the systematic integration of non-invasive, ultrasound-based volume flow sensors on all supply and discharge lines. After the measurement uncertainty had been successfully checked, actual and total values were available for the fresh water supply, as well as for the waste water lines for treatment and discharge into the sewer system, the compressed air supply and the supply and return flows of heating and cooling lines for temperature control. The actual values for the addition of chemicals were also recorded.

Based on the expanded data, the electroplating line was modelled in software using processes and resource flows with the aim of creating a cradle-to-gate scenario by mapping the complete process chain of an electroplating metal deposition. The modelling was carried out using the UMBERTO LCA + software. This offers a graphical modelling environment with a network structure and the integration of the ECOINVENT 3.8 database, with the help of which CO_2 equivalents of the substances were available. Umberto also offers the possibility to import and export (live) data using the Python and Excel API. The modelling, which was also used to calculate a CO_2 footprint, served as a starting point for an evaluation of different process and supply scenarios regarding the CO_2 footprint. The entire production process of the line consisted of 97 individual process steps and is described in the following Fig. 2. Part of the representation is the display of resource flows by modelling them as Sankey diagrams. In the present modelling, the functional unit was a batch of fasteners that was zinc electroplated. The automatic system for lowering and lifting the drums into the baths and the procedure between the individual process steps was not taken into account, nor was the loading, transfer and holding of the goods in the storage facility.

From the system modelling, shares of the CO_2 footprint of the entire system could be quantified. In addition to the relative shares, a weighting of particularly critical RHB could also be established by means of an impact assessment on the environment (Fig. 3). The share of electrical energy is almost four times as high as that of zinc, oil, dirt, tinder, and sewage sludge. Hydrochloric acid and nitric acid, natural gas, waste water, water and caustic soda have only a small share in the CO_2 footprint, due to their functional

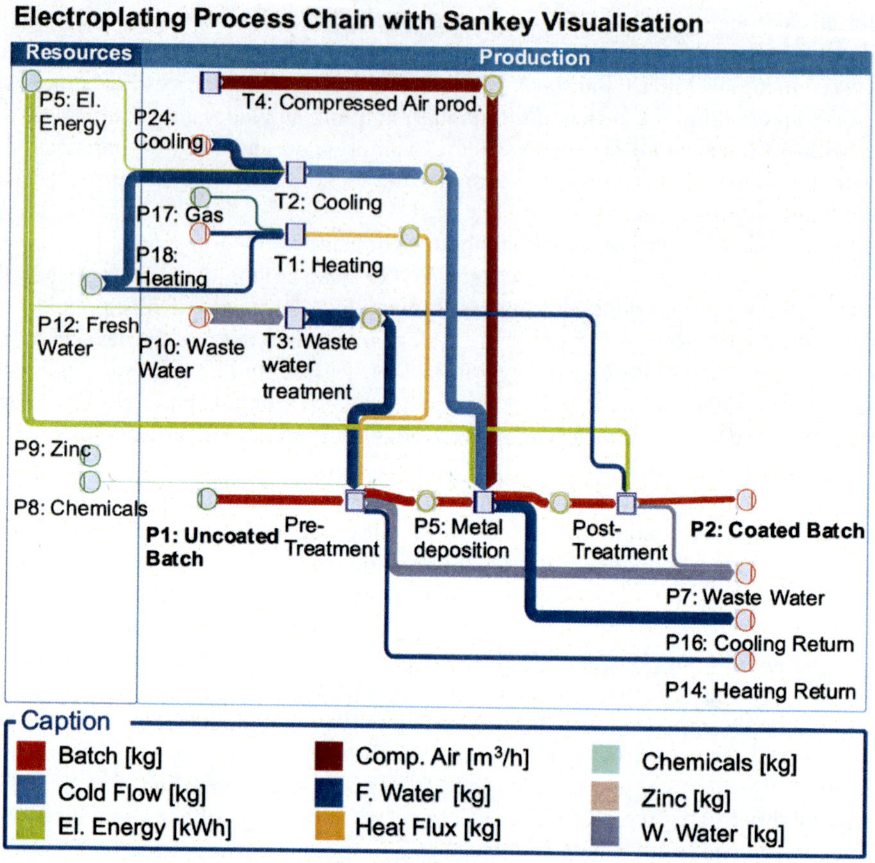

Fig. 2. Sankey visualization of the electroplating process chain

unity. Regarding their environmental impact, the chemicals for pre-treatment as well as production of the electrolyte are to be weighted highly.

Based on an initial consideration of the resource flows, the modelling primarily served to secure the data basis. For the intended training of machine learning models in accordance with the objective mentioned in Chap. 3, it is essential that the target variables mentioned can be determined in the data. With this step, the complete recording of process data can be started. This is done considering all available data streams in the highest resolution in a time interval of $t = 1.5$ s. The aim of the data collection is to contain enough, on the one hand, repetitions of the same articles and, on the other hand, different orders or articles for the demand-oriented control of orders. Upcoming analyses will initially be performed on approx. 60–100 batches in discussion with experts.

Initial results from the existing test data set provide eight scenarios with three parameters in a first investigation. The time-dependent variables here are the water consumption of the baths; the temperature of the baths for metal separation; the compressed air requirement for mixing the electrolyte; and the mass of clarifying sludge due to carry-over. The cooling capacity to be considered for metal separation had to be omitted due to

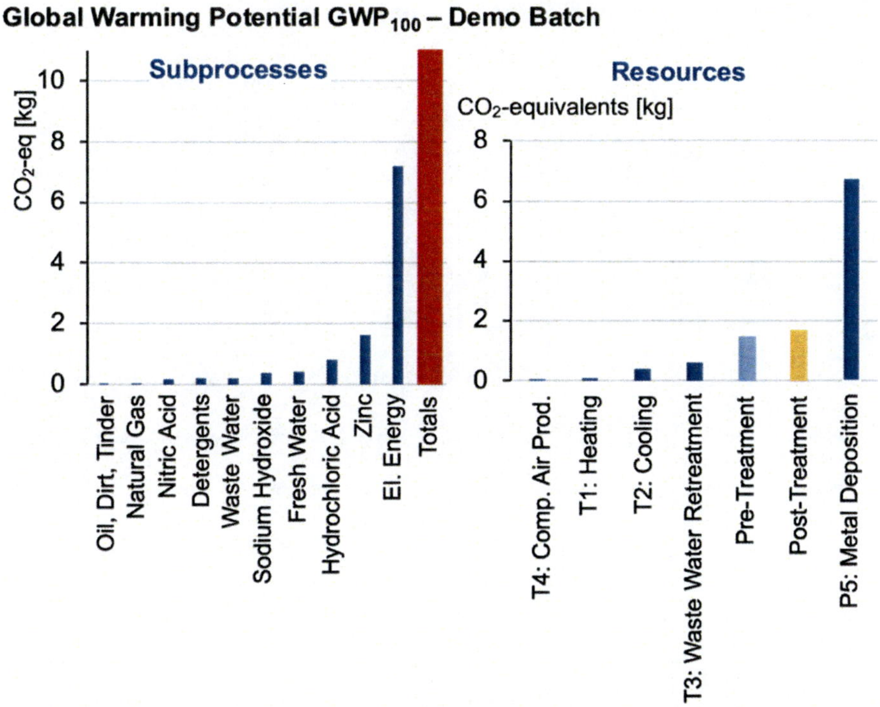

Fig. 3. Differentiated consideration of the factors influencing the GWP

the still insufficient data available. To achieve a possible increase in resource efficiency, the demo batch was varied via the states of the holding time in the bath (target-actual), the temperature control of the remaining baths (target-actual-lower bound-upper bound) and a demand-controlled compressed air injection. Figure 4 shows the results. Overall, the result of scenario VIII with all parameters in the measured actual state (CO_2-eq = 15.2 kg) is nearly equal to the calculated target state (scenario VII) with a CO_2-eq = 15.5 kg). A notable difference is the demand-oriented (predicted) supply of compressed air (to homogenize the electrolyte bath), which results in an average difference of 28%. Further deviations, which are already evident with the prevailing data basis of low and undifferentiated level, exist in the bath temperature control. Here, there is considerable potential for job- and thus demand-oriented temperature control of the process bath and pre-treatment. However, it must be considered that a constant component quality regarding layer thickness and layer thickness distribution could not yet be validated but is provided in the more in-depth analysis.

5 Future Work

The future work is based on the validated data situation and the application of the methodology for the feedback of digital services into production, which makes trained machine learning models (steps 3–5) usable for SMEs. The methodology described in this

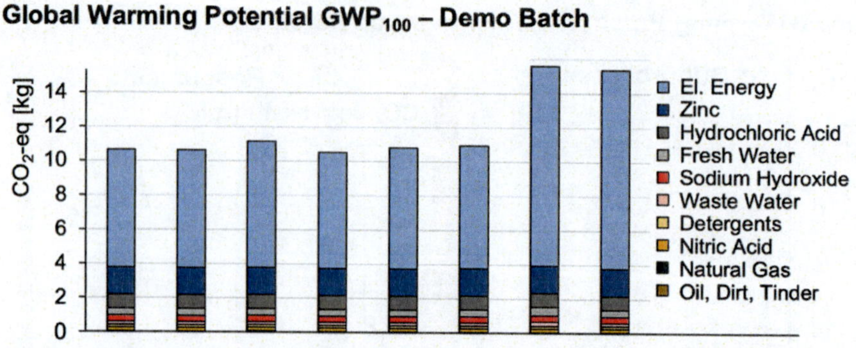

I	II	III	IV	V	VI	VII	VIII	Scenario
actual	target	upper bound	lower bound	target	actual	target	actual	Temperature
actual	actual	actual	actual	target	target	target	actual	Time
demand-oriented supply						continuous supply		Comp. Air

Fig. 4. Scenario analysis for first resource efficiency potentials

paper for demand-driven control of peripheral systems in the context of industrial series production, using electroplating as an example, was developed together with technology experts and data scientists and is being tested with an extensive field data set. In particular, the preliminary results show the application of the data science methods to the overall process to increase resource efficiency to be promising. A dynamic resource flow model is being integrated to look at the potential identified so far in more detail.

Acknowledgement. The presented methodology and the preliminary results, as well as the scientific-technical outlook of this article, were developed within the framework of the project BeStPeri (funding code: 03EI5008) funded by the Federal Ministry of Economic Affairs and Climate Action (BMWK) in close cooperation with the companies B+T Oberflächentechnik GmbH and DiTEC GmbH.

References

1. Schmitt, R., Brecher, C., Nau-Hermes, M., Berners, T.: Material- und Energieeffizienzpotenziale durch den Einsatz von Fertigungsdatenerfassung und -verarbeitung. In: VDI Zentrum Ressourceneffizienz GmbH Publikationen. Berlin (2015)
2. Zimmer, M.-M.: Prozess- und anlagentechnisch optimierte Auslegung, Konstruktion, Planung und Installationsvorbereitung einer galvanischen Hartchromanlage. Abschlussbericht über ein Entwicklungsprojekt gefördert unter dem Az: 25418-21/2 (2009). https://www.dbu.de/ab/DBU-Abschlussbericht-AZ-25418.pdf. Last Accessed 06 May 2022
3. Kanani, N.: Galvanotechnik. Grundlagen, Verfahren, Praxis. München, Wien: Hanser (2000)
4. N.N.: Oberflächentechnik: Impulsgeber des technischen Fortschritts (2005)

5. Lampke, T., Steger, H., Zacher, M., Steinhäuser, S., Wielage, B.: Status quo und Trends der Galvanotechnik. In: Materialwissenschaft und Werkstofftechnik. 39. Jg., 1, S. 52–57 (2008)
6. VDI Zentrum Ressourceneffizienz GmbH: Ressourceneffizienz durch Industrie 4.0. Potenziale für KMU des verarbeitenden Gewerbes (2017). https://www.ressource-deutschland.de/themen/industrie-40/studie-industrie-40/. Last Accessed 06 May 2022
7. Leiden, A., Kölle, S., Thiede, S., Schmid, K., Metzner, M., Herrmann, C.: Model-based analysis, control and dosing of electroplating electrolytes. Int. J. Adv. Manuf. Technol. **111**(5–6), 1751–1766 (2020). https://doi.org/10.1007/s00170-020-06190-0
8. Soergel, T., Buettner, R., Baumgartl, H., Seifert, T., Metzner, M., Feige, K., Ispas, A., et al.: The need for digitalisation in electroplating—how digital approaches can help to optimize the electrodeposition of chromium from trivalent electrolytes. J. Electrochem. Plating Technol. (2021)
9. Heyde, S.: Energieeffiziente Druckluftsysteme. Ein Merkblatt der IHK Projekte Hannover GmbH (2012). https://www.hannover.ihk.de/fileadmin/data/Dokumente/Themen/Energie/Energie-Merkblaetter/120823_Merkblatt_DruckluftHey_.pdf. Last Accessed 07 Mar 2021
10. Sperling, B.: Effizienzpotenziale in der Drucklufttechnik. Veranstaltungsreihe - Kostensenkung durch Ressourceneffizienz. IHK Nordrheinwestfalen, Effizienz-Agentur NRW, Energie-Agentur NRW, HWK Münster, VDI Münsterländer Bezirksverein e.V. (2022). https://www.ihk.de/nordwestfalen/system/vst/3498908?id=368479&terminId=630592. Last Accessed 18 July 2022
11. Sievers, U.; König, U.; Zimmer, M.-M.: Ressourceneffiziente Fertigung - Erfahrungen und Handlungsempfehlungen zur Verbesserung von Rohstoff- und Energie-Effizienz in der verarbeitenden Industrie. 3. Aufl.: WOTech GbR (2013)
12. Institut für Werkzeugmaschinen und Betriebswissenschaften (iwb) TU München (Editor.). Industrie 4.0 als Chance für Energie-und Ressourceneffizienz für die Galvanik München (2018)
13. Bergs, T., Stauder, L., Beckers, A., Grünebaum, T., Barth, S.: Adaptive design of manufacturing process sequences in case of short-term disruptions in the production process. Manuf. Lett. **27**. Jg., S. 92–95 (2021)
14. Narciso, D.A.C., Martins, F.G.: Application of machine learning tools for energy efficiency in industry: a review. Energy Reports. **6**. Jg., S. 1181–1199 (2020)

Potential for Stamping Scrap Reduction in Progressive Processes

S. Rosenthal[1](✉), T. -S. Hainmann[1], M. Heuse[2], H. Sulaiman[2], and A. -E. Tekkaya[1]

[1] Institute of Forming Technology and Lightweight Components (IUL), TU Dortmund University, 44227 Dortmund, Germany
Stephan.Rosenthal@iul.tu-dortmund.de
[2] Faurecia Automotive Seating and Interior, Hannover, Germany

Abstract. The reduction of CO_2-emissions is an essential need in the automotive industry. Progressive die stamping offers a large potential to reduce CO_2 emissions due to potential savings in material scrap after the punching operations. This work presents a methodology to calculate the material loss in progressive die stamping and illustrates strategies to rearrange the stamped parts to determine the potential of saving material. The share of scrap in the conventional process ranges from 16 to 60%. The average savings potential with a redesigned component layout in sheet metal band lies between 28 and 41%, revealing a high potential to reduce material loss and CO_2-emissions.

Keywords: Progressive die stamping · CO_2-reduction · Material efficiency

1 Introduction

The global climate change is largely driven by the greenhouse gas effect, which is why avoiding greenhouse gas emissions as a climate protection measure is also becoming increasingly relevant in industrial processes. Not only the growing interest of stakeholders in sustainable action and the increasing social pressure but also political demands—for example in the context of the European Union's "Green Deal"—require sustainable action from industrial companies. Due to the high CO_2 emissions during energy-intensive steel production, efficiency improvement measures are particularly suitable for steel processing operations.

As Milford et al. [1] show, up to 16% of CO_2 emissions could be saved by reducing the generation of scrap during production processes and thus the demand for raw materials across the steel sector. Cooper et al. [2] analyze the environmental impact of stamping processes for sheet metal components and highlight the importance of reducing stamping scrap, especially for high batch sizes. Horton et al. [3] estimate that stamping scrap accounts for 44% of material requirements in automotive production. In a follow-up paper, realistic opportunities for material savings are identified, and a concrete improvement process is presented, which should be considered in both product and process development [4].

M. Liewald et al. (Eds.): WGP 2022, LNPE, pp. 810–820, 2023.
https://doi.org/10.1007/978-3-031-18318-8_80

For sheet metal working processes, progressive die manufacturing has proven its worth especially due to the high output rate and the high degree of automation for large quantities. The progressive die process is schematically shown in Fig. 1 with its relevant features such as pilot holes. It can be seen that the part is formed while passing through several stations.

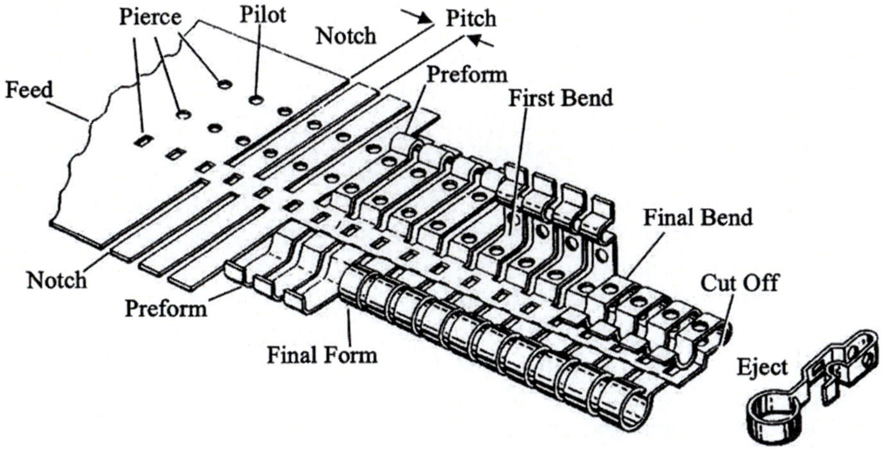

Fig. 1. Schematic illustration of a conventional progressive stamping sequence [5]

However, due to the connection of the workpieces to the sheet metal strip, often only low material utilization rates are achieved. Material efficiency can be increased by carefully arranging the workpieces on the sheet metal strip.

An algorithm for maximizing material utilization in stamping processes is presented by Nye [6]. Here, the optimal placement of the components on the sheet metal strip is determined, and the planar anisotropy of the sheet metal is also taken into account as an additional technological constraint. Licari and Lo Valvo [7] provide an extended solution that, in addition to considering the available bandwidth as an additional parameter, also considers the placement of different geometries on a sheet metal strip. The nesting system of Peng and Zhao [8] is designed for practical applicability and offers the possibility of nesting optimization for different single-row to multi-row blank layouts.

Other papers explicitly address ways to increase the material efficiency of progressive die processes. In their work, Ghatrehnaby and Arezoo [9] develop software for minimizing punching scrap in progressive die processes which focus on geometric optimization. The nesting of components and the placement of guide holes are identified as particularly critical for optimal material utilization. Therefore, two software modules are presented which suggest the optimal placement based on lines and arcs of the component geometry.

Although some of the above-mentioned work provides a solid data basis for the material utilization of stamping processes, it does not explicitly consider progressive die processes. This paper examines how large the proportion of stamping scrap in common parts from progressive die manufacturing currently is, in order to determine the potential for reducing said scrap. Furthermore, the extent to which optimized nesting could increase material efficiency is examined. While previous works have only considered larger parts for transfer processes, a representative sample of parts from progressive processes will be assessed in this work. It will also be evaluated whether stamping layouts without a permanent connection of the components to the sheet metal strip would lead to an improvement.

2 Analysis of Material Savings Potential

2.1 Characterization of Sample Part Stamping Scrap

As the first step of this work, components manufactured via the conventional progressive process are characterized. A total of 20 parts are identified in the course of a literature review (Fig. 2) and it is assumed that the investigated components are a representative sample. Only the contours are used as a basis for the further analysis but no data regarding the scrap proportions is available at this stage. For the analysis of the stamping scrap, it is of importance that the station sequence of the parts is known. After determining the stamping scrap, the scrap to which the production of the 20 parts could be reduced is determined if the initial geometries for the forming process were to be stamped out independently, without consideration of the carrying web. Finally, it will be discussed whether it is feasible to simplify industrially manufactured components using primitive geometries and to evaluate the potential material savings for all components from progressive dies. The parts are characterized based on their geometric properties using the following criteria:

- Symmetry of the stamping geometry,
- Number and position of parallel edges of the stamping geometry,
- Number of perforations in the finished part,
- Number of web connections to the sheet metal strip of the progressive die, and
- Layout of the workpieces in the conventional progressive process as a symmetrical pair.

The above-mentioned criteria are selected to allow a generally valid comparison of the geometries. The stamping geometries are described 1st. in the case of axial symmetry via the number of symmetry axes (Fig. 2, #9); and 2nd. in the case of cyclic symmetry via the number of cyclic duplicates n, in which the appearance matches its initial state (Fig. 2, #1). Out of the 20 stamped geometries shown in Fig. 2, twelve parts are asymmetric, five are axisymmetric with one axis of symmetry, one part is axisymmetric with two axes of symmetry, and two parts are rotationally symmetric with $n = 3$ and $n = 9$.

In addition, 60% of the parts have parallel body edges. In ten of these twelve geometries, the parallel body edges are located approximately in the center of the workpiece or in close proximity to the center of gravity, which could be advantageous in a new transfer system for gripping the workpieces. In the case of the remaining 40% of the stamped parts, there are no contours in the form of parallel body edges for gripping.

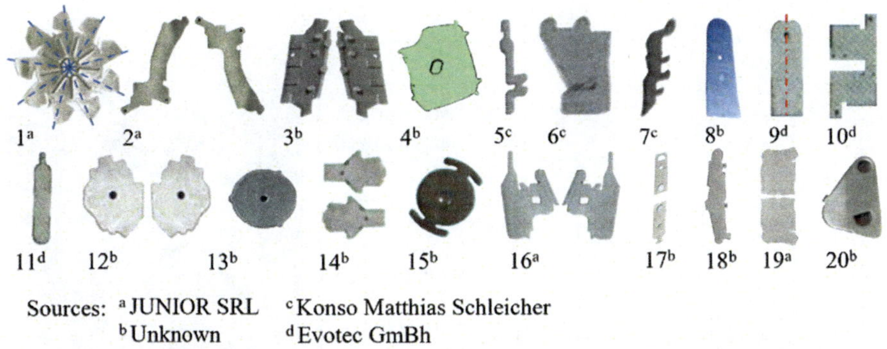

Fig. 2. Listing of the 20 selected parts from conventional progressive processes

Three-quarters of the components are characterized by two or more perforations of different shapes in their final stage. The remainder has at least one circular hole. For these parts, 44% of all perforations are located near the edge of the stamped part. In addition, each of the 20 parts is connected to the sheet metal strip via one to four connections. On average, a component has two connections to the strip. Finally, it should be noted that six of the 20 formed parts are arranged together as a mirrored pair in the stamping sequence on the sheet metal strip. The two parts #17 share a progressive process and are presumably arranged and produced in parallel on the same strip due to their high similarity and common functionality.

The analysis of the parts is carried out on the basis of the available photos. All the parts are shown in their sequence of stages in the progressive process, without a matching perspective of the photos. To correct the distortion, the photographs are first subjected to a perspective correction in an image processing program so that they can be viewed perpendicular to the sheet metal plane. Next, the stage in the progressive process is examined, in which the component is first completely stamped out. The connection to the sheet metal strip is also considered. For this process stage, the part surface area and the surface of connection areas are calculated. Stamped-out geometries within the component contour, such as holes and other perforations, are not evaluated as scrap, since these would also occur in any other process in which the workpieces are stamped from the sheet metal. Subsequently, the proportional stamping scrap for a workpiece in relation to the total sheet area per part is determined.

The results of the stamping scrap analysis arithmetically averaged for all evaluated parts, are shown as boxplots in Fig. 3a. The diagram shows three different categories of proportional stamping scrap for the 20 parts from Fig. 2. These include the stamping scrap resulting solely from the carrying web of the respective progressive die process, the stamping scrap without the carrying web, and the total stamping scrap, which represents cumulative stamping scrap. Figure 3b exemplarily shows the underlying methodology.

It is noticeable that the median of the stamping scrap proportions without consideration of the carrying webs of the progressive process is 28%, and therefore three times as high as for the stamping scrap proportions resulting from the carrying webs. This means that the total proportion of stamping scrap results particularly from unfavorable positioning of the formed parts in the sheet and to a lesser extent from the connection to

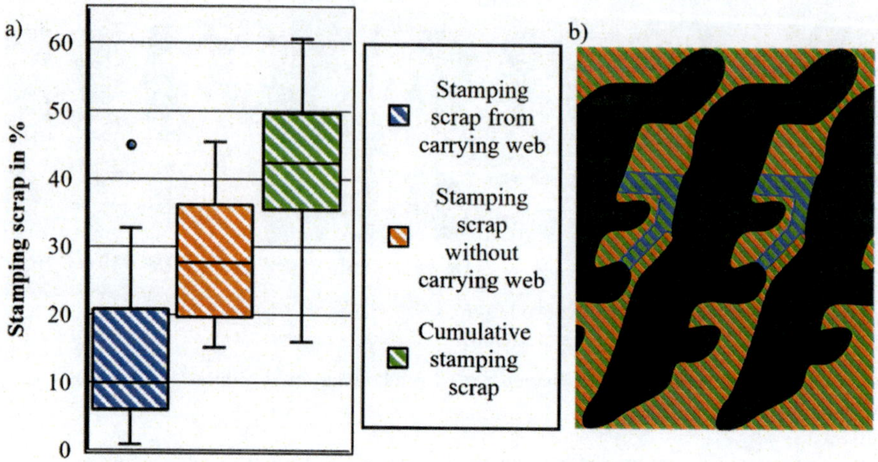

Fig. 3. a) Stamping scrap results averaged for all investigated geometries b) Visualization of the methodology (exemplarily shown for part #7, based on Fig. 2).

the strip for the progressive process. However, it should be noted that this unfavorable positioning may be due to the necessary connection to the sheet metal strip and is thus directly related to the kinematic boundary conditions of the transfer system.

In Fig. 3a, an outlier can be noticed, which is part #13 with three-fold rotational symmetry. The circular part #1 with nine-fold rotational symmetry ($n = 9$) also belongs to the upper quartile of the sample considered here, with stamping scrap amounting to 23% due to the carrying web alone. The two rotationally symmetrical components are found in the upper quartile in the analysis of the total amount of stamping scrap. Thus, the proportion of stamping scrap in the conventional progressive die process appears to be generally high for circular parts.

Since the lowest value of the stamping scrap fraction without considering the carrying webs and the lowest value for the total stamping scrap are approximately the same, it can be concluded that the lowest value in Fig. 3a of the stamping scrap fraction due to the carrying webs must also belong to the same part. This tendency that low stamping scrap due to carrying webs also means low additional stamping scrap can be observed for yet another part. Stamping scrap fractions of this part also lie in the lower quartile of the boxplots in all three categories. These are the parts #9 and #19, which have a largely rectangular geometry. Similarly, for the upper quartile of the boxplots, it is possible to check whether parts with a high proportion of stamping scraps from carrying webs also have a high proportion of additional stamping scraps without considering the carrying webs. However, this is not the case for any of the parts in the upper quartile of either category. All parts that appear in the upper quartile of the boxplot for the category of the total stamping scrap are either also represented in the upper quartile of the stamping scrap from sheet metal carrying webs or in the upper quartile of the stamping scrap without consideration of carrying webs.

Parts #1, #2, #13, #16 and #20 account for the largest total stamping scrap, at 49% to 61%, while the smallest total stamping scrap, at 18% to 33%, is produced by parts #7, #8,

#11, #18 and #19. The parts with the largest proportion of scrap are characterized by the fact that they include the two rotationally symmetrical parts, #1 and #13, as well as the two most complex geometries (due to their irregular, non-symmetrical outer contour), #2 and #16. The parts with a small total amount of stamping scrap can be characterized by the fact that they are narrow (parts #7, #8, #11, and #18) and/or resemble a rectangle in their outer contour (parts #8, #11, #18, #19).

2.2 Analysis of Material Savings Potential for Optimal Stamping Layouts

The stamping scrap fractions for the conventional progressive process of the individual parts shown in Fig. 2 have been calculated in the previous chapter (Fig. 3). Next, the potential for increasing material efficiency is determined. As a preparation for this, the geometries must first be arranged on the sheet metal strip in the most material-efficient way possible. The following assumptions are made for this procedure.

The conventional progressive process starts with the unwinding of the sheet coil from the coiler. This initial state should also be retained for a new stamping process since the coil is well suited for series production of sheet metal parts due to its easily automated handling. The sheet metal coil can be simplified as a rectangle with width W and infinite length $L \to \infty$. Thus, the stamping scrap that originates at the beginning and at the end of the nesting sequence can be neglected. For the design of the stamping sequences on the sheet metal strip, a metal coil optimized in width will be assumed so that exactly the minimum edge width a can be set. The minimum bridge width e and edge width a (Fig. 4) of the stamping sequence must be considered to ensure the proper function of the transfer tool, since the compressive force of the punching tool can lead to deformation of the sheet surface. If the webs and edges are too narrow, they may twist and hinder the further progressive process.

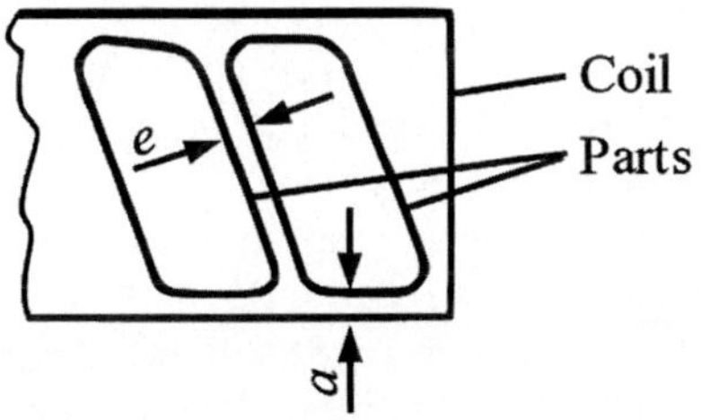

Fig. 4. Visualization of edge width a and bridge width e

Since the sheet thickness t is unknown for most parts, it is assumed that minimum distances that can be stamped within a part contour can also be realized between two parts and are taken as a reference for the bridge and edge widths. For simplicity, $e = a = t$ is assumed for the parts at hand. The potential for reducing the stamping scrap is primarily to be determined independently from the transfer system and forming process, the parts can be arranged and nested as desired. However, only identical geometries or geometries that were also previously processed together in the conventional progressive forming process should be nested together. If a wide variety of geometries were positioned on a strip, it might be possible to achieve lower stamping scrap proportions, but the parts could then not be produced in independent quantities. If two mirrored or other similar parts are produced on one strip in the original progressive process, they should also be produced in equal proportions in the new process. Taking these boundary conditions and simplifications into account, it is now possible to determine the stamping scrap proportion of each part with the most material-efficient nesting possible. The procedure of this analysis is shown schematically in Fig. 5 as an example for a two-row nesting of part #7. The part analysis takes place in the punching sequence when only the strip remains as a transfer feature, utilizing image processing software. The complete part surface is colored black. Then the components are arranged as space-efficient as possible in a single row, two rows, three rows, and four rows on a white background, taking into account the estimated minimum bridge width e in their stamping sequence.

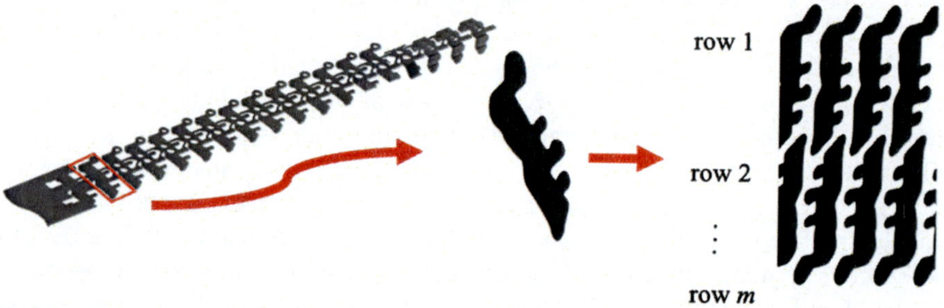

Fig. 5. Procedure of the analysis for determining the stamping scrap in idealized layouts where m is the number of rows

Individual rows may overlap. After positioning the parts in a stamping sequence, a periodically repeating area of the punching sequence is generated, considering the estimated minimum edge width a. The black and white proportions will be determined for this periodically repeating area. The proportion of the white background then corresponds to the scrap proportion.

Finally, the results of the stamping scrap analysis of all the considered parts and arrangements are illustrated in Fig. 6.

The bar chart in Fig. 7 shows the average stamping scrap of all 20 parts in relation to the total material consumption per part for different layouts.

The green columns in Fig. 7 are stamping scrap percentages that can only be achieved under ideal conditions. The potential for reducing the stamping scrap P_{St} is defined in Eq. 1.

$$P_{St} = 1 - \frac{\textit{Stamping waste in the new stamping process}}{\textit{Total stamping waste in the original progressive process}} \quad (1)$$

The boxplots in Fig. 8 show the potential for reducing the stamping scrap of all 20 components for the various ideal stamping layouts. It is noticeable that all potential savings span large ranges from 68 to 75%. In addition, the arithmetic mean and the median of the savings potential of all parts are minimally larger for the two-row stamping

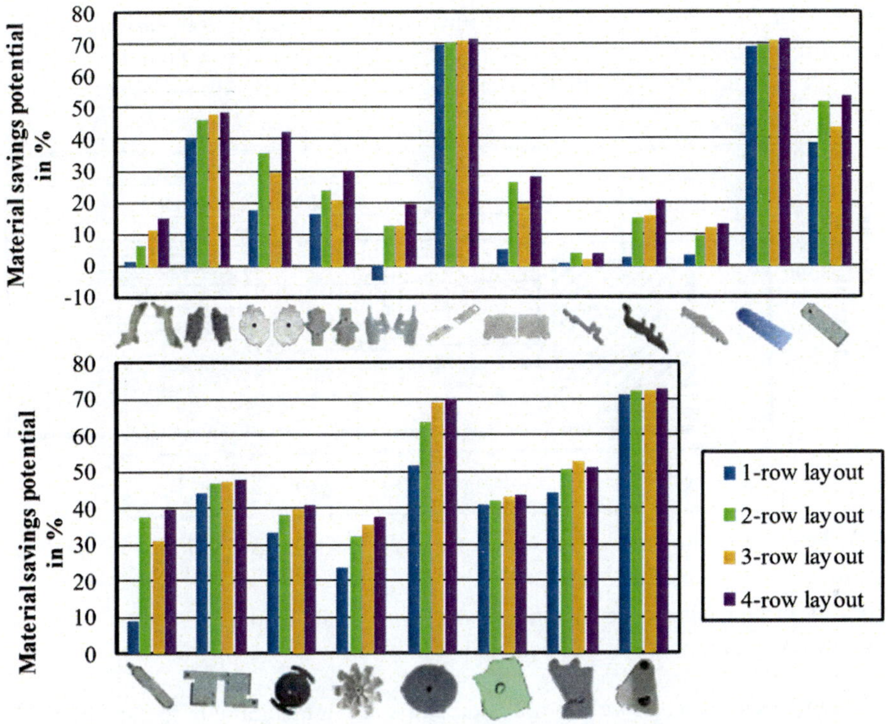

Fig. 6. Comparison of the punching scrap reduction potentials for all considered parts

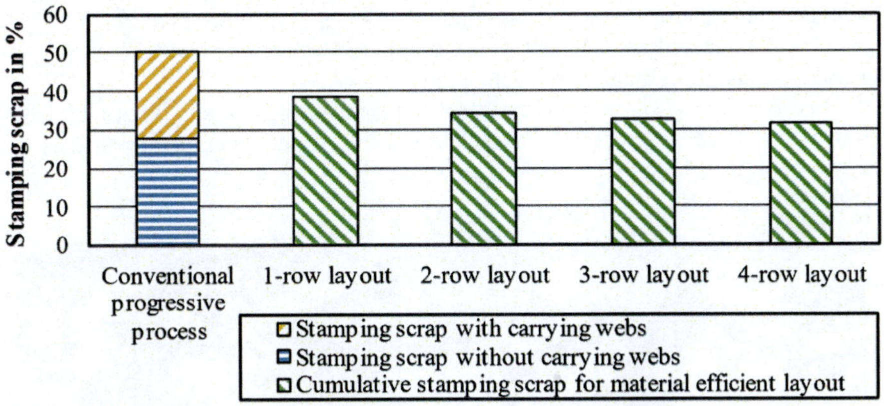

Fig. 7. Representation of the average scrap proportions for all investigated geometries in various stamping layouts

process than for the three-row stamping process. However, both values are around 37%, so that the potential for reducing stamping scrap is roughly the same. Otherwise, it is evident that a higher number of rows in the stamping process has a positive effect on the reduction of material scrap on average. Finally, when considering the arithmetic mean, the two-row and three-row layouts are about 9%, and the four-row layout is about 12% above the potential for reducing the stamping scrap of the single-row layout.

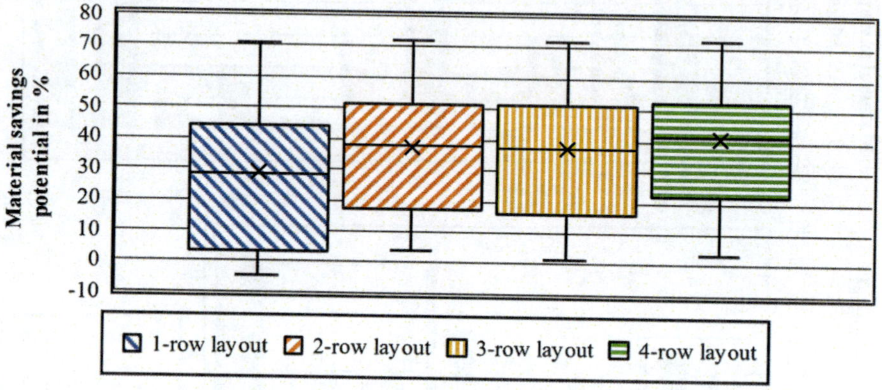

Fig. 8. Representation of the potential for scrap savings by switching to idealized punching layouts

From an analysis of the material savings potentials in the one-, two-, three- and four-row layout, the following conclusions can be drawn:

- Asymmetrical parts, as well as mirrored pairs, are unfavorably processed in a single row configuration due to the inefficient usage of the sheet metal strip (Fig. 9a).
- A switch from the conventional progressive layout to the two-row layout can result in material savings for all investigated parts. Such a change is particularly beneficial for parts where the overlap effect as shown in Fig. 9b can be utilized.
- The three-row layout is disadvantageous compared to the four-row layout due to the global asymmetric arrangement of the parts.
- The four-row layout slightly increases the material savings potential but may lead to an overall increase in processing complexity due to wider sheet metal strips.

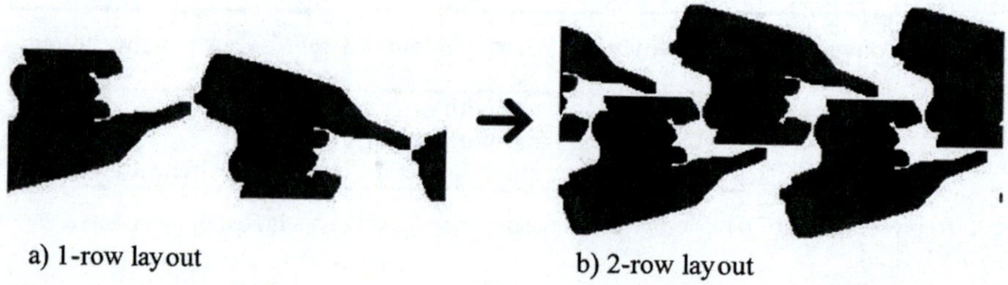

Fig. 9. Switch from a single-row to a double-row arrangement of the parts with overlapping effect in the double-row case

Generally, it can be stated that the material savings potential from an *n*- to an $n + 1$-row punching layout increases significantly if the overlapping effect shown in Fig. 9

occurs between all existing arrangement rows. Otherwise, only the influence of material waste between the parts and the sheet edge decreases, and only very small improvements in material efficiency can be expected.

3 Discussion and Conclusions

To assess the proportion of stamping scrap from progressive stamping processes, a representative sample has been generated using 20 real parts. The scrap proportions, ranging from 16 to 60% have been determined by analyzing the individual part and nesting geometries. In a second step, the material savings potential has been individually determined for each part geometry for different idealized stamping layouts. The results show an average material savings potential ranging from 28 to 41%, depending on the investigated layout. The results from the stamping scrap analysis reinforce the importance of material usage reduction and can be used as a quantified base for specific improvements in the progressive process. The results can also be utilized as a guidance to review scrap savings potential in existing progressive processes. Further studies should focus on the technical application of improved stamping layouts. The investigated ideal layouts in this work are not immediately applicable in conventional progressive processes but require different technical approaches. Since the highest material utilization rates can only be achieved without using carrying webs as a means of transfer, especially new transfer systems are needed. The challenge here is to retain all advantages such as the high productivity and level of automation that the conventional progressive process offers.

Acknowledgments. The authors would like to thank Dr.-Ing. Till Clausmeyer for his support in proofreading this manuscript.

References

1. Milford, R.L., Allwood, J.M., Cullen, J.M.: Assessing the potential of yield improvements, through process scrap reduction, for energy and CO_2 abatement in the steel and aluminum sectors. Resour. Conserv. Recycl. **55**(12), 1185–1195 (2011). https://doi.org/10.1016/j.resconrec.2011.05.021
2. Cooper, D.R, Rossie, K.E, Gutowski, T.G.: An environmental and cost analysis of stamping sheet metal parts, vol. 3: Joint MSEC-NAMRC Symposia, Blacksburg, Virginia, USA. https://doi.org/10.1115/MSEC2016-8880
3. Horton, P.M., Allwood, J.M.: Yield improvement opportunities for manufacturing automotive sheet metal components. J. Mater. Process. Technol. **249**, 78–88 (2017). https://doi.org/10.1016/j.jmatprotec.2017.05.037
4. Horton, P.M., Allwood, J.M., Cleaver, C.: Implementing material efficiency in practice: a case study to improve the material utilization of automotive sheet metal components. Resour. Conserv. Recycl. **145**, 49–66 (2019). https://doi.org/10.1016/j.resconrec.2019.02.012
5. Tripar Inc. Homepage. https://www.triparinc.com/wp-content/uploads/2018/10/tripar-tech-progressive-dies-nov-2018_eng.pdf. Last Accessed 04 May 2022
6. Nye, T.J.: Stamping strip layout for optimal raw material utilization. J. Manuf. Syst. **19**(4), 239–248 (2000). https://doi.org/10.1016/S0278-6125(01)80003-0

7. Licari, R., Lo Valvo, E.: Optimal positioning of irregular shapes in stamping die strip. Int. J. Adv. Manuf. Technol. **52**(5–8), 497–505. https://doi.org/10.1007/s00170-010-2772-6
8. Peng, Y., Zhao, Z.: Development of a practical blank layout optimisation system for stamping die design. Int. J. Adv. Manuf. Technol. **20**(5), 357–362. https://doi.org/10.1007/s001700200163
9. Ghatrehnaby, M., Arezoo, B.: A fully automated nesting and piloting system for progressive dies. J. Mater. Process. Technol. **209**(1), 525–535 (2009). https://doi.org/10.1016/j.jmatprotec.2008.02.049

Creating Digital Twins for Production

Digital Twins in Battery Cell Production

J. Krauß[1(✉)], A. Kreppein[2], K. Pouls[1], T. Ackermann[1], A. Fitzner[1], A. D. Kies[2], J. -P. Abramowski[2], T. Hülsmann[1], D. Roth[1], A. Schmetz[1], and C. Baum[1,2]

[1] Fraunhofer Research Institution for Battery Cell Production FFB, Bergiusstraße 8, 48165 Münster, Germany
jonathan.krauss@ffb.fraunhofer.de

[2] Fraunhofer Institute for Production Technology IPT, Steinbachstr. 17, 52074 Aachen, Germany

Abstract. A digital twin enables the accessibility of data, information, models, and simulations for a physical object. Therefore, digital twins become increasingly relevant for different areas of production. In particular, the production of battery cells with its high complexity could benefit from digital representations such as digital twins. Still, there is no coherent definition for digital twins in the battery cell production yet. In this paper we introduce the first concept of digital twins in battery cell production. For this we combine existing ideas for the digital twin with the characteristics of the battery cell production. The concept consists of digital twins for buildings, products, and machines or assets, which we validate based on different use cases. By this, we demonstrate the benefits that arise from the implementation of digital twins in battery cell production—from increased productivity, faster ramp-ups, to increased sustainability.

Keywords: Digital twin · Battery production · Battery cell production · Digital building twin · Digital product twin · Digital machine twin

1 Introduction

Current estimates forecast a growth in demand for lithium-ion batteries from currently 200 GWh to 1.5–3 TWh per year in 2030 [1]. One of the main drivers for this increase is the move towards electric mobility, which will account for up to 80% of the battery demand [2]. To meet this growing market, manufactures have announced many new battery cell production plants. Until 2030 there are approx. 40 new plants planned in Europe [3], which can produce about a third of the forecasted battery demand [4, 5].

However, battery cell manufacturing is highly complex and poses several challenges still to be solved such as ensuring high cell quality while achieving high process stability and efficiency. In addition to this, the ecological footprint of cells needs to be reduced to ensure sustainability according to all Environmental, Social, and Governance (ESG) criteria. This can be achieved through reduction of energy needs in production, avoiding toxic and environmental or socially unsustainable materials [6–8] as well as designing the cells with recycling and second life in mind [9].

To ensure sustainability and a transparent supply chain, the EU has introduced the "battery passport". It will become mandatory for all batteries with a capacity above 2

M. Liewald et al. (Eds.): WGP 2022, LNPE, pp. 823–832, 2023.
https://doi.org/10.1007/978-3-031-18318-8_81

kWh from 2026 onwards [10]. Among other things, it will include the carbon footprint of the battery. In addition to this, national regulation like the "supply chain act" in Germany [11] set rules for supply chain transparency.

In the context of Industrie 4.0, new concepts have been emerging to approach the aforementioned challenges. One approach lies in Digital Twins. We provide an overview of the state of the art of Digital Twins (DT) in production and introduce the first concept for DT in battery cell production. Since high quality battery cells require complex alignment between the (sub-) products, the machines and the building, we introduce three forms of DT, which consider the individual requirements of the respective physical asset. To coordinate the three forms of DT we introduce the Digital Twin of Battery Cell Production, following a modular approach. The three forms of the DT, the DT of Battery Cell Production, and their components, structures, and relations are presented. In addition, we present applications and discuss the challenges and benefits.

2 State of the Art

In the 1970s NASA used physical copies of spacecrafts and satellites to replicate issues and their solutions on earth [12]. Thereafter, physical twins were improved or replaced with cost efficient digital copies to gain insights into the behavior and state of the original spacecraft. The term "Digital Twin" was coined by NASA in 2010 and describes these digital copies [13]. Thereafter, DTs gained traction in academia and various industries and additional connecting terminologies, such as the digital shadow [14] or the digital model [15] were introduced. However, there is no consensus on the definition what a DT is. While industrial companies often proclaim new or existing hard- and software components as DT, academical research proposes various definitions. While some say that a DT is a digital replica of a physical asset others speak of DTs of processes and other immaterial things. They also vary in other points, e.g. some focus on the inclusion of simulations and models [13] while others are purely data driven [16], as well as in terms of direction and automation of information flow. To avoid this confusion there are attempts to use different nomenclatures like Digital Shadow and Digital Twin for the integration level of automated information flow [17], however this is not consistently used in academia or the industry. An overview of the various definitions, misconceptions and differences can be found in Sjarow et al. [14] and Fuller et al. [15].

While there is no DT for a battery cell production implemented and published there are works towards it [18], and DTs are used in other industries and research. Examples include the design and implementation of a DT in a networked Micro Smart Factory [19], the use of DTs for anomaly detection [20], for adaptive process planning and optimization [21], or the implementation of a DT of the battery for usage optimization and evaluation of degradation [22]. Reviews of other publications on the use of DT are provided by Kritzinger et al. [23], Fuller et al. [15], and Liu et al. [24].

3 The Digital Twin in the Battery Cell Production

Based on the previously presented definitions, in particular building on the widely used definition of Stark et al. [25], we define the DT as a digital representation of a physical object. The DT includes the properties, states, and the behavior of the object via data, models, and information. This enables optimization using physical simulations as well as data-based models with a structured and traceable data basis including all relevant information of the physical asset throughout the whole life cycle. To achieve the DT in battery cell production, we identified three main forms of the DT each corresponding to certain physical assets. This enables a use-case oriented design, while keeping complexity, with three forms, limited. The identification took place in cooperation with all relevant research partners from the BMBF-funded project FoFeBat:

- *Digital Product Twin (DPT)* incorporates information on raw materials and all intermediate or end products. This includes parameters of various processing operations.
- *Digital Machine Twin (DMT)* includes information of all production-relevant machines in the factory.
- *Digital Building Twin (DBT)* contains all components and information necessary for the construction and operation of the factory building.

Through direct communication and interconnections, these individual DTs form the DT of the complete battery cell production. This approach decreases complexity, thus allowing a holistic optimization as well as optimizing the individual twins for their use cases. It is important to note, that the aim of the DT is not to replace existing IT systems but to interact with and aggregate information from them.

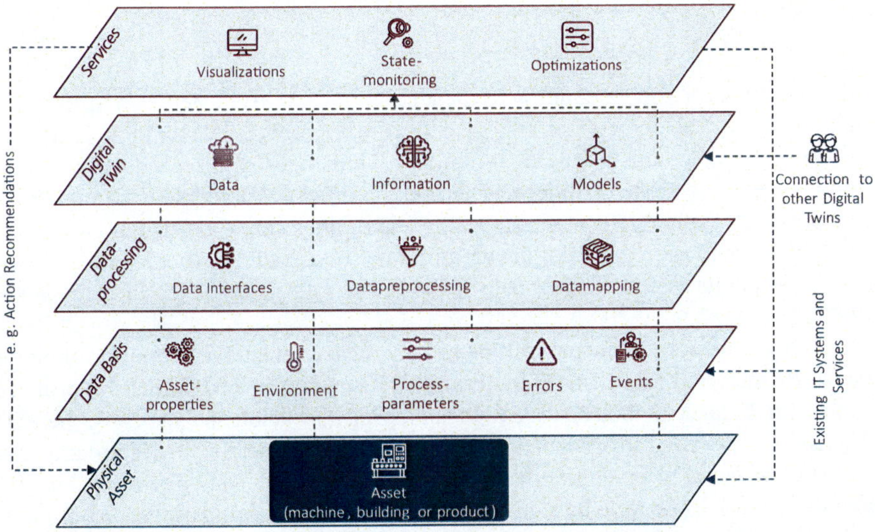

Fig. 1. Structure of a Digital Twin and its embedding into existing systems

As shown in Fig. 1, each DT represents a single physical asset like a machine, building, or battery cell. Among other things asset properties, environmental conditions, and information about events create the data basis for DTs. This data is then processed before it gets transferred and used in the DT, e.g. for models and simulations. Each DT can connect to other DTs and IT-Systems, so that services can be built on top of it. These services can be optimizations or visualizations but also create a feedback loop to the asset with, for example, action recommendations.

3.1 The Digital Product Twin

DPT in battery cell manufacturing enables the structured consolidation and management of data, information, and models associated with a specific instance of a physical intermediate or end product, e. g. an electrode coil or a battery cell. It evolves dynamically across the entire process chain.

A particular feature of the DPT is the large number and variety of intermediate products that must be referenced among each other. As depicted in Fig. 2, these include the electrode pastes and electrode rolls for the anode and cathode, which each can have their own DT and are referenced in the digital twin of the battery.

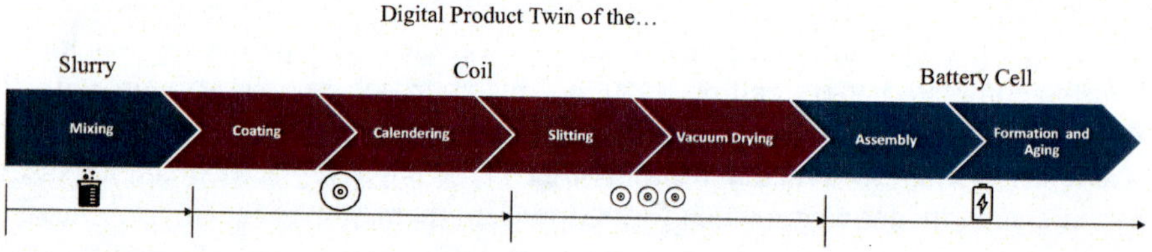

Fig. 2. Evolution of the intermediate products and their respective digital product twins along the process chain

For a successful implementation of the DPT, a traceability system is required which assigns quality measurements and process data to a product, like a single battery cell or an electrode coil. This is one of the biggest challenges as the production includes continuous, e.g. electrode coating, and discrete, e.g. aging, processes and one product unit is not easily definable in some steps, e.g. continuous slurry extrusion. However the correct linking and aggregation of the various data sources at different levels is essential for the DPT and its uses cases, like data analysis, accurate machine learning models and predictions.

The DPT will store inline and offline quality measurements, information about precursors, raw materials as well as references to where and when the product was produced. This structured data can then be used to visualize each product, analyze cause and effect relationships, optimize the product design and process parameters, feed simulations, and train machine learning models. In addition to this, the DPT can generate value in the rest of the products life cycle by storing usage data and optimizing usage, recycling, and second life.

3.2 The Digital Machine Twin

The overall aim of the DMT is to decrease downtime, improve user friendliness, and increase the efficiency of the machine and the respective process to achieve a more economical and sustainable production. In addition to this, the linking of all DMT along the highly complex process chain offers potential for data-driven, cross-machine optimizations in all steps from electrode production to the formation of the finished cell.

In order to add value to the DMT, a comprehensive data basis must be created, which contains, among other things, process data, information on machine states as well as data of events. The quality of the event data, such as information and times of maintenance and special machine states, often depends on input from maintenance staff and machine operators. These inputs—in addition to the process and condition data of the machine as well as the information of the machine manufacturer resulting from remote maintenance—are the basis of subsequent predictive models of the machine [26]. To represent the condition of the machine as accurately as possible, it is necessary to sensitize process participants to record event data. If the data is generated manually, the challenge is the lack of standardization of the inputs. Therefore, various standardized input methods must be included in the design to enable a fast and user-friendly utilization for the DMT.

After a comprehensive data basis is established, the states, properties, and behavior of the machine are linked together in the DMT. The data basis can now detect relevant information in raw data or aggregate this data to relevant key performance indicators. Also, relevant machine information can be enriched with information from other DTs, which allow data driven recommendations for the production line (see Chapter 4).

3.3 The Digital Building Twin

In addition to the shop-floor production, which is mapped by the Digital Product and Machine Twin, the factory building also plays a central role in battery cell manufacturing. The construction and building sector accounted for close to 40% of global CO_2 emissions in 2020 [27]. Construction and operation of production facilities for battery cell manufacturing are associated with high emissions. For example, there is a great need for high-performance clean and dry rooms, the operation of which is associated with high energy consumption [28].

The Building Information Modeling (BIM) is often used as a basis for optimizing construction planning and operation. The resulting 3D models can partly be equated with DTs, but do not meet the requirements for DTs in battery cell production.

To enable the potential of the DBT, challenges have to be overcome. In addition to the cross-domain challenges that arise during the design and implementation of any DT, there are several other characteristics related to the DBT that need to be considered. One of these characteristics stems from the multitude of partners involved in the construction and operation of complex buildings such as factories. From initial ground surveys to architectural design to handover and commissioning of the building, a single major project can easily involve dozens of different companies. Not all relevant information is always available to all partners in the process. While this is where the great potential of the DBT as a one-stop shop for all building-related data becomes apparent, at the same time it is also one of the greatest challenges.

During operation, the degree of networking of the various building components and systems is of high importance. Not every system is "smart" and provides its data via a freely accessible interface. This also applies to actuators such as shading systems or ventilation systems, which cannot always be controlled centrally and mapped by the DBT. Furthermore, not all types of data can be captured automatically. Compared to the DPT, a large amount of information regarding defects, conversions, maintenance, or repairs has to be captured and updated manually. This is prone to errors and jeopardizes the DBT's claim to always be an exact replica of its real counterpart. Clear process models are therefore needed for such non-digitized processes.

3.4 Interaction of the Digital Twins

The DT of the machine, product, or building offer different services and benefits. A digital twin exists for each physical object—e.g. every coating machine has its own digital twin. In addition, there are use-cases in which the various forms of the DT must be considered in combination. This is brought together in the DT of battery cell production, which uses standardized interfaces to access the subordinate DTs and create a comprehensive exchange of data and information.

The DT of battery cell production is a group of interconnected modules of which each is a digital machine, product or building twin. The modular structure reduces complexity and facilitates manageability at the technical level. New functionalities, for example based on new sensor technology, can be implemented easily so that all resulting data and information are available to other applications. Furthermore, changes to the building, product or machine can also be taken into account in the digital twin of battery cell production by simple adaptation in the corresponding module. This flexibility in design allows the whole DT or just individual modules to be transferred to other battery manufacturing sites and implemented there.

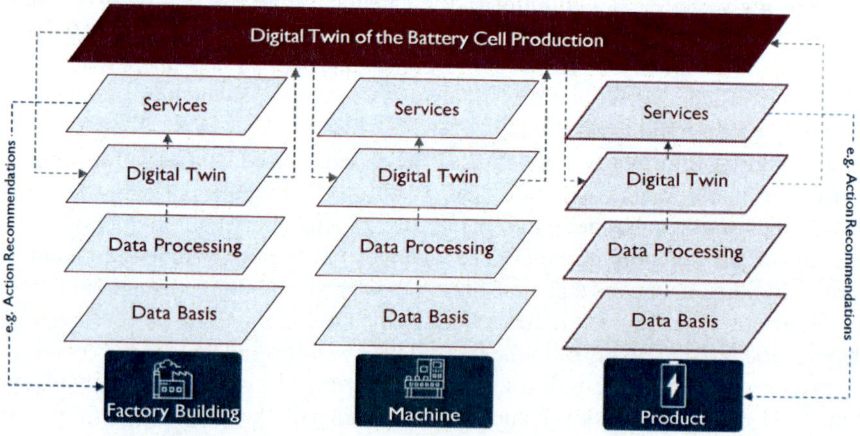

Fig. 3. Interaction of the different forms of digital twins

Networking the individual forms to build a DT of battery cell production opens up further opportunities to increase both sustainability and efficiency. Questions with regard to energy requirements, management and flexibility can be answered by information from the machines and the building. Particularly when considering the operation of the dry rooms associated with the DBT, there exists a strong dependency on production. This represents a clear and necessary interface between the DMT and the DBT. Furthermore, the product quality is strongly influenced by the machine settings and parameters, such as the influence of the slot die settings of the coater on the layer thickness of the electrode foil. For an analysis of the cause-effect relationships, machine data from the DMT must therefore be linked with quality data from the DPT. The linking of all forms of expression is also necessary for the creation of the so-called "Battery Passport". Figure 3 schematically shows the interaction of the forms. In the superordinate digital twin of battery cell production, the digital twins of machine, product and building are accessed, and information, data and models can be viewed together.

4 Application of the Digital Twin

The **Digital Product Twin** is a structured data basis of all quality and process data for each individual product. This structured data can be used in correlation analysis, simulations, and machine learning models. The first goal of this is to understand the cause-and-effect relationships along the full production chain for example using correlation analysis or linear regression methods. This knowledge can then be used to improve the quality of cells as well as the design of the machines.

The second goal is to optimize the product design, like material composition, and process parameters, coming from the respective DMT, such as drying parameters, using data-based approaches. This can for example be achieved by using clustering algorithms to identify good parameter combinations or by estimating optimal parameters using neural networks.

The availability of structured and connected (historical) data in the **Digital Machine Twin** enables the creation of models that recognize the circumstances that lead to unplanned failures, e.g. due to tool wear, and can predict such events [29]. Based on the current data, the condition as well as the remaining operating time of machine wear can be estimated through the same model. This information can be used to create more intelligent intervals for the replacement of machine components, which can prevent expensive, unplanned downtimes in the future, and thus increase production efficiency. At the same time—in contrast to the periodic replacement of machine components—the utilization period of the individual parts is optimized, thus saving costs, and making more sustainable use of machine components.

The **Digital Building Twin** can be used to monitor and manage energy needs ensuring stable operations while reducing peak loads and increasing energy flexibility. This is achieved using demand forecasts for all technical building equipment, energy production forecasts from photovoltaic installations, and the state of energy storage facilities form the DBT as well as energy forecasts from the DMT. All this data can then be used to optimize production plans and energy usage of technical building equipment with data based models. Optimizations goals could be scheduling peak loads during times of high energy production and free capacity in the grid as well as optimal usage of renewables, energy recuperation, and storage capacity.

The **Digital Twin of the Battery Cell Production** enables use-cases which involve more than one DT, such as the adaptive control of a processes. On the one hand, adaptive process control allows optimization of process variables in the machine, based on the quality of the output—e.g. the residual solvent content after drying—to improve following products, see Fig. 4. On the other hand, the process parameters, like the calendar gap, can also be adjusted based on the input—e.g. the coating thickness and damages after drying—to reduce reject and bring product qualities back into optimal ranges as well as reducing wear on the machine. This is achieved by linking data from the DPT with the current states of the DMTs and using predictive models to suggest optimized parameter settings. Adaptive process control enables more efficient processes with increasing quality and a reduction of waste. This in turn reduces production costs.

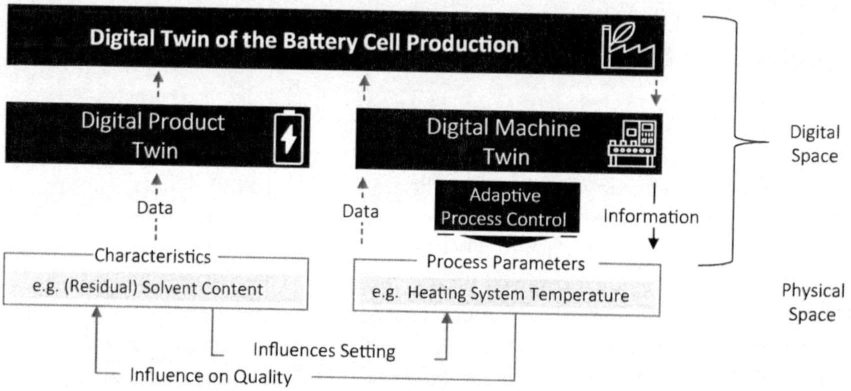

Fig. 4. Schematic illustration of adaptive process control through the DPT and DMT

5 Conclusion and Outlook

The DT in battery cell production sets the fundamental course for an economically and ecologically optimized factory. At the same time, the DT can be the foundation for further applications, research and development projects along the life cycle of the battery cell [27].

In this paper, a definition of the digital twin in battery cell manufacturing is presented based on the scientific discourse. As a main result of this paper, we distinguish between three forms of Digital Twins (DT): The Digital Product Twin (DPT), the Digital Machine Twin (DMT), and the Digital Building Twin (DBT). The foundation of each DT are data-based models with a structured and traceable data basis including all relevant information of the physical asset throughout the whole life cycle. Distinguishing between three forms of DT creates a modularity, which enables a use-case oriented approach, while keeping complexity limited. This allows for the specific consideration of various use cases in the battery cell production, while interactions between the forms also allow generating a holistic DT that is able to represent the whole battery cell production. Nevertheless, resources are required to implement a DT. The cost-benefit of DT use-cases were not considered in this paper and should therefore be discussed in future research.

DTs can help to enable a more sustainable production by reducing waste, increasing product quality, and optimizing energy consumption. The DT can also make a decisive contribution in the future to fulfilling the Supply Chain Sourcing Obligations Act [11], which has already been passed by the German government, and to the creation of a "Battery Passport" as part of the Green Deal [30] envisaged by the European Union. Depending on the use-case under consideration, the benefits will be visible in every life-cycle phase. Nevertheless, a rebound effect must be prevented. A holistic consideration of the environmental benefits should therefore be carried out in future research.

Acknowledgements. "FoFeBat—Forschungsfertigung Batteriezelle Deutschland" is funded by the Federal Ministry of Education and Research. Reference number: 03XP0256, 03XP0416. We want to thank our colleagues in the project that supported in creating the results of this paper. We look forward to publishing an extended version in form of a white paper.

References

1. Jinasena, A., Burheim, O.S., Strømman, A.H.: A Flexible Model for Benchmarking the Energy Usage of Automotive Lithium-Ion Battery Cell Manufacturing. Batteries **7** (2021)
2. World Economic Forum: A Vision for a Sustainable Battery Value Chain in 2030 Unlocking the Full Potential to Power Sustainable Development and Climate Change Mitigation. World Economic Forum (2019). http://www3.wefo-rum.org/docs/WEF_A_Vision_for_a_Sustainable_Battery_Value_Chain_in_2030_Report.pdf. Accessed 22 Nov 2021
3. Bockey, G.: Batterie-Projekte in Europa (Stand: Januar 2022). https://battery-news.de/in-dex.php/2022/01/14/batterie-projekte-in-europa-stand-januar-2022/
4. MWi: atterien "ma e in Germany" – ein Beitrag zu nachhaltigem Wachstum und kli- mafreundlicher Mobilität (2021). https://www.bmwi.de/Redak-tion/DE/Dossier/batteriezellfertigung.html
5. Rasch, M.: Tesla, Volkswagen, Porsche – Deutschland wird zum Zentrum der europäi- schen Batteriezellen-Produktion (2021). https://www.nzz.ch/wirtschaft/deutschland-wird-zum-zentrum-der-batterieproduktion-ld.1631548. Accessed 22 Nov 2021
6. Romare, M., Dahllöf, L.: The life cycle energy consumption and greenhouse gas emissions from lithium-ion batteries. A study with focus on current technology and Batteries for light-duty vehicles (2017). http://www.energimyndigheten.se/globalassets/for-skning--innovation/transporter/c243-the-life-cycle-energy-consumption-and-co2-emis-sions-from-lithium-ion-batteries-.pdf. Accessed 22 Nov 2021
7. Kwade, A.: Advances in battery cell production. Energy Technol. **8**, 1900751 (2020)
8. Michaelis, S., Rahimzei, E., Kampker, A., Heimes, H., Offermanns, C., Locke, M., Löbberding, H., Sarah, W., Thielmann, A., Hettesheimer, T., Neef, C.: Roadmap Batterie-Produktionsmittel 2030. Update 2020, Frankfurt am Main (2020)
9. Emilsson, E., Dahllöf, L.: Lithium-ion vehicle battery production. Status 2019 on energy use, CO_2 emissions, use of metals, products environmental footprint, and recycling (2022). https://www.diva-portal.org/smash/get/diva2:1549551/FULLTEXT01.pdf
10. European Commission: Proposal for a regulation of the European Parliament and of the council concerning batteries and waste batteries, repealing Directive (2020)
11. BUNDESMINISTERIUM FÜR WIRTSCHAFT UND KLIMASCHUTZ: Lieferketten-sorgfaltspflichtengesetz. LkSG (2021)
12. Rosen, R., von Wichert, G., Lo, G., Bettenhausen, K.D.: About the importance of autonomy and digital twins for the future of manufacturing. IFAC-PapersOnLine **48**, 567–572 (2015)

13. Shafto, M., et al.: Modeling, simulation, information technology & processing roadmap. Nat. Aeronaut. Space Adm. **32**, 1–38 (2010)
14. Kunath, M., Winkler, H.: Integrating the Digital Twin of the manufacturing system into a decision support system for improving the order management process. Proc. CIRP **72**, 225–231 (2018)
15. Brauner, P., Dalibor, M., Jarke, M., Kunze, I., Koren, I., Lakemeyer, G., Liebenberg, M., Michael, J., Pennekamp, J., Quix, C., Rumpe, B., van der Aalst, W., Wehrle, K., Wortmann, A., Ziefle, M.: A computer science perspective on digital transformation in production. ACM Trans. Internet Things **3**, 1–32 (2022)
16. Sjarov, M., Lechler, T., Fuchs, J., Brossog, M., Selmaier, A., Faltus, F., Donhauser, T., Franke, J.: The Digital Twin concept in industry—a review and systematization. In: 2020 25th IEEE International Conference on Emerging Technologies and Factory Automation (ETFA), pp. 1789–1796 (2020)
17. Fuller, A., Fan, Z., Day, C., Barlow, C.: Digital Twin: enabling technologies, challenges and open research. IEEE Access **8**, 108952–108971 (2020)
18. Ngandjong, A.C., et al.: Investigating electrode calendering and its impact on electrochemical performance by means of a new discrete element method model: towards a digital twin of Li-Ion battery manufacturing. J. Power Sources (2021). https://doi.org/10.1016/j.jpowsour.2020.229320
19. Park, K.T., et al.: Design and implementation of a digital twin application for a connected micro smart factory. Int. J. Comput. Integr. Manuf. **32**, 596–614 (2019)
20. Chhetri, S.R., Faezi, S., Canedo, A., Faruque, M.A.A.: QUILT: Quality inference from living Digital Twins in IoT-enabled manufacturing systems. In: Landsiedel, O., Nahrstedt, K. (eds.) Proceedings of the International Conference on Internet of Things Design and Implementation. IoTDI '19: International Conference on Internet-of-Things Design and Implementation, Montreal Quebec Canada, 15 04 2019 18 04 2019, pp. 237–248. ACM, New York, NY, USA (2019)
21. Liu, J., et al.: Dynamic evaluation method of machining process planning based on Digital Twin. IEEE Access **7**, 19312–19323 (2019)
22. Singh, S., Weeber, M., Birke, K.P.: Implementation of battery Digital Twin: approach, functionalities and benefits. Batteries **7**, 78 (2021)
23. Kritzinger, W., Karner, M., Traar, G., Henjes, J., Sihn, W.: Digital Twin in manufacturing: a categorical literature review and classification. IFAC-PapersOnLine **51**, 1016–1022 (2018)
24. Liu, M., Fang, S., Dong, H., Xu, C.: Review of digital twin about concepts, technologies, and industrial applications. J. Manuf. Syst. **58**, 346–361 (2021)
25. Stark, R., Kind, S., Neumeyer, S.: Innovations in digital modelling for next generation manufacturing system design. CIRP Ann. **66**, 169–172 (2017)
26. Singh, S., Weeber, M., Birke, K.-P.: Advancing digital twin implementation: a toolbox for modelling and simulation. Procedia CIRP **99**, 567–572 (2021)
27. Kies, D.A., Krauß, J., Schmetz, A., Baum, C., Schmitt, H.R., Brecher, C.: Der digitale Zwilling in der Batteriezellfertigung/Digital Twin in battery cell production—from data management and traceability system to target-oriented application. wt **111**, 286–290 (2021)
28. Lehne, J., Preston, F.: Making concrete change innovation in low-carbon cement and concrete. Chatham House Report (2018)
29. Jardine, A.K., Lin, D., Banjevic, D.: A review on machinery diagnostics and prognostics implementing condition-based maintenance. Mech. Syst. Signal Process. **20**, 1483–1510 (2006)
30. European Commission: Green deal: sustainable batteries for a circular and climate neutral economy (2020)

Use Cases for Digital Twins in Battery Cell Manufacturing

S. Henschel(✉), S. Otte, D. Mayer, and J. Fleischer

wbk–Institute of Production Science at Karlsruhe Institute of Technology (KIT), Kaiserstraße 12, 76131 Karlsruhe, Germany
sebastian.henschel@kit.edu

Abstract. Increasing concerns for a more sustainable future have led to a fast-growing demand for high quality lithium-ion batteries. In order to expand available manufacturing capacities to the desired magnitudes within a reasonable timeframe, the concept of Digital Twins is seen as a possible solution. With the purpose of better understanding the abilities of this concept and showcasing how it can be used to accelerate the ramp-up process of manufacturing technology, this paper presents an analysis of existing approaches to Digital Twins within battery cell manufacturing. Available case-studies and scientific publications along with novel concepts will be used to identify and cluster the potentials and benefits of the technology. The resulting framework provides an overview for possible Digital Twin implementations as well as an opportunity to identify future areas of research. Based on the performed analysis, two conceptual use cases will be presented. It is shown, that Digital Twins can help transform both the mixing of electrode slurry as well as the production sequence of separating and stacking battery electrodes from traditional discrete processes to continuous production flows.

Keywords: Digital Twin · Battery cell manufacturing · Process modelling

1 Introduction

Driven by the transformation towards a more sustainable future and the expansion of electric mobility, the demand for lithium-ion batteries is growing rapidly [1]. To cope with the demand, new manufacturing capabilities are sought to be established especially in Europe and North America. Whilst these regions have an extensive background in traditional manufacturing operations, such as the automotive industry, experience with battery cell manufacturing is often limited. It is for this reason, that manufacturing capabilities have to be set up with a limited knowledge and in parallel to a necessary fast ongoing learning curve regarding battery cell manufacturing [2].

A possible solution, for dealing with the aforementioned challenges can be found in the concept of Digital Twins (DTs). It was first used by Michael Grieves in the manufacturing context in 2003 and allows for the optimization of products and processes by the aid of a virtual counterpart [3].This digital representation can be used to further study cause and effect relationships, thus decreasing development lead times and increasing process accuracy for ongoing productions for example.

M. Liewald et al. (Eds.): WGP 2022, LNPE, pp. 833–842, 2023.
https://doi.org/10.1007/978-3-031-18318-8_82

In order to better understand the opportunities associated with the DT technology in regards to battery manufacturing, this paper will present a structured study of existing use cases. The state of the art on current DT developments will be discussed, in order to establish a common ground for the following description of possible implementations. The existing use cases will be transferred in a cluster according to their focus level and the level of integration that was chosen. In addition to research literature, two new use cases will be presented as well. They will be used to showcase how the idea of a DT can be used, not only to analyze and fine tune existing processes, but also change existing processes at a more fundamental level. With the help of the DTs, the traditional discrete processes of mixing and stacking can be transformed to continuous production flows.

2 State of the Art

2.1 Definition of Digital Twins

As with any new occurring technology, it is important to create a common baseline of understanding for the given technology. In the case of DTs this is especially important, due to the variety of concepts existing behind this term. In addition, the abundance of use cases makes this step even more necessary. Extensive research on the definition of the terminology itself has been conducted by Jones et al., showing that the common ground between different approaches can be found in the idea, that a virtual entity is created of a physical object [4]. However, especially the connection or communication between the two counterparts both from the virtual entity to the physical entity and vice versa is not uniformly implemented or regarded as necessary in many DT applications, as was shown by Kuehner et al. for example [5]. In this paper, the terms introduced by Kritzinger et al. will therefore be used to determine the level of integration, distinguishing between a Digital Model, a Digital Shadow and a DT [6]. The digital model has no continuous connection at all to its physical version. While the digital shadow receives data from its physical counterpart automatically, only the actual DT has an automated connection in both directions, allowing the virtual model to influence the physical object.

Depending on the level of integration chosen for each twin, a number of different functionalities can be incorporated. According to Pires et al. these can include process optimization, process monitoring and process control [7]. While the optimization could be performed solemnly based upon a Digital Model, the process monitoring requires at least a Digital Shadow in order to display real time information from the physical instance. A DT, representing the highest level of integration, then allows for the full process control, since it is also able to send automated information and commands to the physical machine.

2.2 Classification of Digital Twins

In order to further distinguish DT applications in the field of battery production from one another, the different physical objects, such as individual products or factories, being subject to the creation of a virtual counterpart can also be used. Separations here can be made according to the focus level with which a given production operation is regarded. A common use case for DTs is to create a virtual representation of an entire factory, or even a global production network. This corresponds to a broad focus level and allows for material flows and potential bottlenecks of the system to be analyzed. Moving further along the focus levels, one can also set up an individual twin for an individual machine within a production system. At this medium focus level, the intention can be to analyze the performance of the machine or predict maintenance activities. The approach focusing on the narrow end of the focus spectrum is to analyze the individual materials being processed during the production. At this level, a DT is set up for a finished end-product. This is often done alongside the manufacturing process in order to create a solid understanding of the finished product, as well as its individual instances. Hereby it can be made possible to implement advanced quality control mechanisms. Similar to the final end product, DTs with a very narrow focus level can be set up to simulate and represent the processed materials up to a microscopic level. This way, material performance can be better understood and quality parameters of the intermediate product can be predicted for example. The different focus levels for implemented DTs, as well as examples for the added value they provide can be found in Table 1.

Table 1. Focus levels for creating DTs in the context of production engineering.

Focus level	Physical object	Added value (exemplary)
Very broad	Global production system	Determining bottle necks
Broad	Individual factory	Optimizing material flow
Medium	Individual or connected machine(s)	Predicting process quality
Narrow	Finished product	Analyzing product quality
Very narrow	Individual material	Simulating material behavior

2.3 Classification of Existing Use Cases

Using both the focus level, and the level of integration as a possible means for distinguishing between use cases for DTs in the battery cell production, a literature research has been conducted. Research papers published within the last three years from both Scopus and Science Direct were utilized and categorized, as can be seen in Table 2.

As can be seen from the research, numerous case-studies and use cases have been conducted for both ends of the spectrum regarding the focus level. The broad to very broad focus level corresponds to research being done at the factory level. Here multiple process steps are mirrored into one digital entity allowing a conclusion for quality assessments across the entire production process for example. The narrow to very narrow focus level can be found mostly in projects, that either focus on the battery as a final end product, or the individual material steps during individual production phases. Most use cases implement the DT technology to simulate the battery over the course of its lifecycle and thus generate additional information about the state of charge for example. Depending on the level of integration implemented, this information can either be used to plan

Table 2. Existing literature on the implementation of DTs in the context of battery cell manufacturing.

Focus level	Digital Model	Digital Shadow	Digital Twin
(Very) broad	[8, 9]	[10–12]	[13]
Medium	[14, 15]	[16]	
(Very) narrow	[17–21]	[22–26]	[27–29]

scheduled maintenance activities for the system, giving it a very limited amount of feedback, or in case of a full DT provide live feedback to the system, allowing for immediate correctional measures.

During the research, it became apparent, that very few implementations of a DT at a medium focus level have been taken place for the battery cell manufacturing industry. This is especially true for higher levels of integration. In order to help bridge this gap, two novel concepts for applying DTs to an individual process step are presented in this paper.

3 Own Approach

3.1 Continuous Slurry Production

Currently, the first steps in battery cell production, the mixing and dispersing of the electrode slurry, is mainly being carried out in a batch process [30–32]. This is to be replaced by a continuous process using a twin-screw extruder. The aim is to achieve an optimization of the mixing process and its result. In addition, an increase in throughput and a reduction in mixing time is thought after. In this innovative process, it has not yet been possible to identify all process dependencies quantitatively and qualitatively. In order to achieve the optimum process conditions and parameters quickly, a DT is to be used. With this it should be possible to determine the optimum parameters in advance in the event of a recipe change, thus minimizing a trail-and-error phase. Therefore, the use of a DT seems particularly suitable and its concept is presented in the following.

The structure of the DT is based upon the real layout of the twin-screw extruder, see Fig. 1. The real CAD data are used to set up the DT in virtual, which means that a geometric comparability exists. The Software Siemens NX is used for this process step. The mechatronic limitations are also considered. This means that all necessary mechatronic functions, required by the real machine, are specified in the DT. Here, the software Siemens Mechatronic Concepted Designer was used. By following this approach, it should be possible to add or remove single components of the machine without having to rebuild the DT.

Furthermore, the DT should have the necessary interfaces so that data can be transmitted from the machine control to the twin and vice versa, as can be seen in Fig. 1. In addition, an interface with the real extruder is to be implemented so that a data exchange is possible. The relevant parameters such as screw speed, torque, imputed solids content and solvent content, temperature during the extrusion section and pressure at the

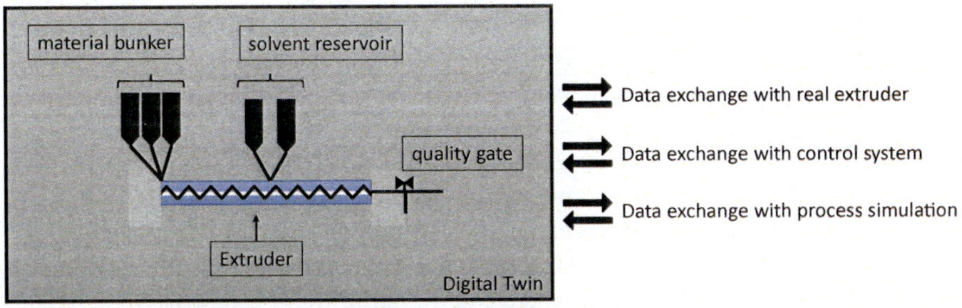

Fig. 1. The DT of the extruder as a schematic visualization

screw tip can be registered by the system control (see Fig. 2). In addition, in-line measurement of viscosity, density, temperature and pressure is possible with downstream measurement technology. An in-line measurement of particle size distribution is currently under development. All these data can be transmitted to the digital twin via a communication interface. The DT can manipulate the variables solids content, solvent content, temperature and screw speed and send them back to the extruder and the control system.

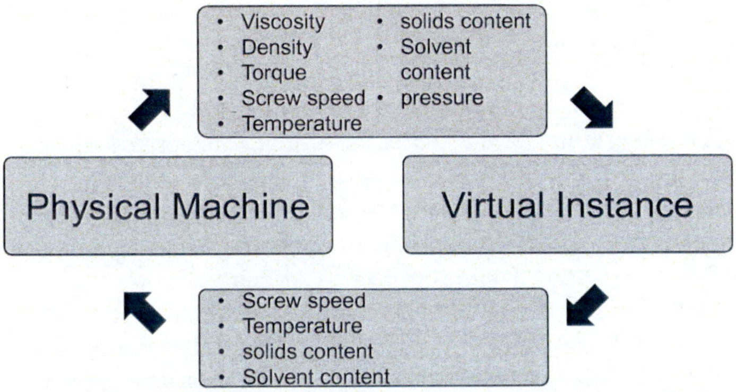

Fig. 2. Concept for the DT for an extruder with necessary information exchanged between the two entities

Furthermore, an interface with process simulation is conceivable. This allows a comprehensive simulation of the process and the machine-process interaction. By doing so, the process data will be made available to a database and have an interface for the use of Artificial Intelligence algorithms. On the one hand, various AI methods are to be used to be able to simulate the mixing process as realistically as possible. On the other hand, the AI approaches should serve to optimize machine and process control.

The structure and functionality of the DT supports its intended use and the operation of the extruder. One goal is to use the DT to optimize and speed up start up of the extruder. Due to its structure, the extruder control can be digitally simulated and validated in advance. The main focus is on the controllability of the process and the relevant parameters of viscosity, screw refractive index and pressure. The DT thus supports the virtual start up of the real machine. The DT can also be used to validate customized controls during later operation. This reduces the probability of an unexpected failure of the real machine.

At the same time the DT allows effective validation of the simulation models. Via the interface to an external process simulation model, a simulation of the machine process interactions can be generated and carried out from the results of the optimization. It is also possible to mirror the real machine status and the current process parameters. This creates a live image of the process and easier monitoring. This has the great advantage that, in the context of battery cell production, the relationships with regard to viscosity can be better understood and material waste is reduced by running simulative optimized experiments.

3.2 Virtual Sensing for a Continous Separation and Stacking Process

In order to further demonstrate the opportunities of a full DT on the medium focus level, a second concept is presented here. During the separation and stacking of electrodes, it is of high importance, to assure the cutting quality and the correct size of the produced electrodes. This is because deviations from the norm can have a significant impact on the final cell performance [32]. A DT can therefore be used to combine live measured data with digitally computed values to automatically determine the dimensional accuracy of a cut electrode. By providing this feedback to the machine, it is possible to exclude the faulty electrode from further processing, without having to perform all following production steps on the faulty part and only recognizing the mistake at a later stage.

The DT is to be set up for the successor of a new integrated cutting and stacking machine described in [33, 34]. The new machine will be capable of separating two continuous coils of electrode material, one for the anode and one for the cathode and afterwards gluing them to a continuous separator, creating a finished electrode composite. Due to the continuous nature of this process, it is especially important to ensure the correct size of the electrodes, since a quality check of the individual sheets is not possible at a later time. Among others, the DT is therefore to allow for the control of the dimensional accuracy, without having to incorporate any additional measuring or control equipment within the process. The required information to be exchanged between the two entities can be seen in Fig. 3.

In order to perform the necessary computations within the virtual instance of the machine, the data from the physical twin will be combined with available models of the material, as well as the electromechanical behavior of the machine. With the help of advanced simulation tools, the electronic behavior of electric drives for example can also be incorporated into the modelling, thus further increasing the accuracy of the digitally computed values. As an additional benefit, it will become possible to validate the modeled data in real time with physical measurements, since both the physical and the virtual entities are operating simultaneously. This setup therefore represents a true DT, as described at the start of this paper.

Since measuring the exact alignment of the continuous electrode web at every point within the process and especially when performing the cut is both not technically possible, nor economically feasible, a conventional control approach will always be limited

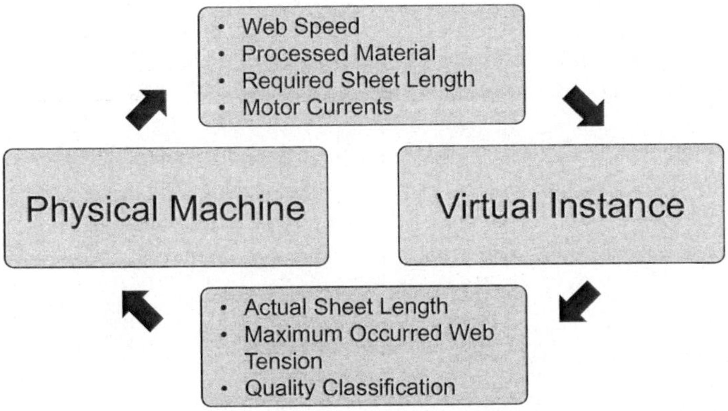

Fig. 3. Concept for the DT with necessary information exchanged between the two entities.

in its accuracy. In a virtual model of the machine however, performing all the required measurements becomes possible by recreating the mechanical interdependencies and allowing for a virtual sensing process during the ongoing operation of the machine. In case of the web alignment, the accuracy of the computed values is thought to be further improved, by combining the limited physical measurements with the virtual model. By doing so, deviations of the calculated alignment from the measured values are to be recognized and compensated in real time, thus further increasing the accuracy of the model, even at points within the process, where physical measurements are not taking place.

Due to this extra degree of accuracy, the modelling results of the virtual twin are to be used to draw real time quality conclusions about the produced part, in this case the cut electrode.

4 Conclusion and Outlook

It was shown in this paper, that there are numerous implementations of DTs in the area of battery production. A closer analysis however showed that true DTs at the medium focus level barely exist. Most implementations either focus on the production process as a whole, modelling material flows and optimizing intralogistics, or they have a narrow focus level and create virtual representations of the batteries being build, or even just individual components, such as electrodes. Only few reports of DTs being implemented for the scope of individual machines exist. The potential, that resides within this research gap has been shown by the two novel use cases presented in this paper.

Especially when a high level of integration is reached, the optimization and even transformation of individual process steps becomes possible with the help of DTs. Both the mixing of electrode slurry and the separation of electrodes can be taken from a discrete to a continuous process. This becomes possible through the added information that can be calculated within the virtual instance of the machine and therefore allowing for greater accuracy and precision, both of which are necessary for the continuous operation of the processes.

In the future, the introduced concepts will be implemented, allowing for the collection of data from the ongoing operation. This will help to further quantify the advantages of a highly integrated DT on the medium focus level for battery production.

Acknowledgements. The presented research is funded by the *German Federal Ministry of Education and Research* under the funding code 03XP0343A (Project: *IntelliPast*) and 03XP0252B (Project: *E-Qual*), the *Center for Electrochemical Energy Storage Ulm | Karlsruhe (CELEST)* and the *Battery Technology Center of KIT*.

References

1. EV-Volumes—The Electric Vehicle World Sales Database (2022)
2. Michaelis, S., Rahimzei, E., Kampker, A., Heimes, H., Huang, Z.: Roadmap Batterie-Produktionsmittel 2030—Update 2020, Frankfurt am Main (2021)
3. Grieves, M.: Digital twin: manufacturing excellence through virtual factory replication. White Paper **1**, 1–7 (2014)
4. Jones, D., Snider, C., Nassehi, A., Yon, J., Hicks, B.: Characterising the Digital Twin: a systematic literature review. CIRP J. Manuf. Sci. Technol. **29**, 36–52 (2020). https://doi.org/10.1016/j.cirpj.2020.02.002
5. Kuehner, K.J., Scheer, R., Strassburger, S.: Digital Twin: finding common ground—a meta-review. Proc. CIRP **104**, 1227–1232 (2021). https://doi.org/10.1016/j.procir.2021.11.206
6. Kritzinger, W., Karner, M., Traar, G., Henjes, J., Sihn, W.: Digital Twin in manufacturing: a categorical literature review and classification. IFAC-PapersOnLine **51**(11), 1016–1022 (2018). https://doi.org/10.1016/j.ifacol.2018.08.474
7. Pires, F., Cachada, A., Barbosa, J., Moreira, A.P., Leitao, P.: Digital Twin in industry 4.0: technologies, applications and challenges. In: 2019 IEEE 17th International Conference on Industrial Informatics (INDIN). 2019 IEEE 17th International Conference on Industrial Informatics (INDIN), Helsinki, Finland, 22.07.2019–25.07.2019, pp. 721–726. IEEE (2019). https://doi.org/10.1109/INDIN41052.2019.8972134
8. Rohkohl, E., Schönemann, M., Bodrov, Y., Herrmann, C.: A data mining approach for continuous battery cell manufacturing processes from development towards production. Adv. Ind. Manuf. Eng. **4**, 100078 (2022). https://doi.org/10.1016/j.aime.2022.100078
9. Xin, X.: Research on digital manufacturing of lithium battery pilot production line based on virtual reality. J. Phy.: Conf. Series **1996**(1) (2021). https://doi.org/10.1088/1742-6596/1996/1/012007
10. Reynolds, C.D., Slater, P.R., Hare, S.D., Simmons, M.J., Kendrick, E.: A review of metrology in lithium-ion electrode coating processes. Mater. Des. **209**, 109971 (2021). https://doi.org/10.1016/j.matdes.2021.109971
11. Sommer, A., Leeb, M., Haghi, S., Günter, F.J., Reinhart, G.: Marking of electrode sheets in the production of lithium-ion cells as an enabler for tracking and tracing. Procedia CIRP **104** (2021). https://doi.org/10.1016/j.procir.2021.11.170
12. Assad, F., Konstantinov, S., Ahmad, M.H., Rushforth, E.J., Harrison, R.: Utilising web-based digital twin to promote assembly line sustainability. In: Proceedings—2021 4th IEEE International Conference on Industrial Cyber-Physical Systems, ICPS 2021 (2021). https://doi.org/10.1109/ICPS49255.2021.9468209
13. Kies, A.D., Krauß, J., Schmetz, A., Baum, C., Schmitt, R.H., Brecher, C.: Digital twin in battery cell production—from data management and traceability system to target-oriented application [Vom datenmanagement über das traceability-system zur zielgerichteten nutzung:

Der digitale zwilling in der batteriezellfertigung]. WT Werkstattstechnik **111**(5), 286–290 (2021). https://doi.org/10.37544/1436-4980-2021-05-20
14. Arcelus, O., Franco, A.A.: Perspectives on manufacturing simulations of Li-S battery cathodes. J. Phys. Energy **4**(1) (2022). https://doi.org/10.1088/2515-7655/ac4ac3
15. Ngandjong, A.C., et al.: Investigating electrode calendering and its impact on electrochemical performance by means of a new discrete element method model: towards a digital twin of Li-Ion battery manufacturing. J. Power Sources **485**, 229320 (2021). https://doi.org/10.1016/j.jpowsour.2020.229320
16. Husseini, K., Schmidgruber, N., Weinmann, H.W., Maibaum, K., Ruhland, J., Fleischer, J.: Development of a Digital Twin for improved ramp-up processes in the context of li-ion-battery-cell-stack-formation. Proc. CIRP **106**, 27–32 (2022). https://doi.org/10.1016/j.procir.2022.02.150
17. Prifling, B., Neumann, M., Hlushkou, D., Kübel, C., Tallarek, U., Schmidt, V.: Generating digital twins of mesoporous silica by graph-based stochastic microstructure modeling. Comput. Mater. Sci. **187**, 109934 (2021). https://doi.org/10.1016/j.commatsci.2020.109934
18. Drakopoulos, S.X., et al.: Formulation and manufacturing optimization of lithium-ion graphite-based electrodes via machine learning. Cell Reports Phys. Sci. **2**(12), 100683 (2021). https://doi.org/10.1016/j.xcrp.2021.100683
19. Heinrich, F., Noering, F.-D., Pruckner, M., Jonas, K.: Unsupervised data-preprocessing for long short-term memory based battery model under electric vehicle operation. J. Energy Storage **38** (2021). https://doi.org/10.1016/j.est.2021.102598
20. Lombardo, T., Ngandjong, A.C., Belhcen, A., Franco, A.A.: Carbon-binder migration: a three-dimensional drying model for lithium-ion battery electrodes. Energy Storage Mater. **43**, 337–347 (2021). https://doi.org/10.1016/j.ensm.2021.09.015
21. Park, J., Bae, K.T., Kim, D., Jeong, W., Nam, J., Lee, M.J., Shin, D.O., Lee, Y.-G., Lee, H., Lee, K.T., Lee, Y.M.: Unraveling the limitations of solid oxide electrolytes for all-solid-state electrodes through 3D digital twin structural analysis. Nano Energy **79** (2021). https://doi.org/10.1016/j.nanoen.2020.105456
22. Li, J., Zhou, Q., Williams, H., Xu, H., Du, C.: Cyber-physical data fusion in surrogate-assisted strength Pareto evolutionary algorithm for Phev energy management optimization. IEEE Trans. Industr. Inf. **18**(6), 4107–4117 (2022). https://doi.org/10.1109/TII.2021.3121287
23. Lizaso-Eguileta, O., Martinez-Laserna, E., Rivas, M., Miguel, E., Iraola, U., Cantero, I.: Module-level modelling approach for a cloudbased Digital Twin platform for li-ion batteries. In: 2021 IEEE Vehicle Power and Propulsion Conference, VPPC 2021—Proceedings (2021). https://doi.org/10.1109/VPPC53923.2021.9699271
24. Merkle, L., Pöthig, M., Schmid, F.: Estimate e-golf battery state using diagnostic data and a digital twin. Batteries **7**(1), 1–22 (2021). https://doi.org/10.3390/batteries7010015
25. Sancarlos, A., Cameron, M., Abel, A., Cueto, E., Duval, J.-L., Chinesta, F.: From ROM of electrochemistry to AI-Based Battery Digital and Hybrid Twin. Archives Comput. Methods Eng. **28**(3), 979–1015 (2020). https://doi.org/10.1007/s11831-020-09404-6
26. Tang, H., Wu, Y., Cai, Y., Wang, F., Lin, Z., Pei, Y.: Design of power lithium battery management system based on digital twin. J. Energy Storage **47** (2022). https://doi.org/10.1016/j.est.2021.103679
27. Guardabascio, V., Pesce, M., Magistrali, S., Marcigliano, F., Fonti, G., Valesano, F., Dimitrakopoulos, P., Framke, N.-H., Papadimitriou, I.: Model-based control development using real-time 1D thermal management in co-simulation for high performance BEV Digital Twin. SAE Technical Papers (2022). https://doi.org/10.4271/2022-01-0200
28. Wang, W., Wang, J., Tian, J., Lu, J., Xiong, R.: Application of Digital Twin in smart battery management systems. Chinese J. Mech. Eng. **34**(1), 1–19 (2021). https://doi.org/10.1186/s10033-021-00577-0

29. Xu, Z., Xu, J., Guo, Z., Wang, H., Sun, Z., Mei, X.: Design and optimization of a novel microchannel battery thermal management system based on Digital Twin. Energies **15**(4) (2022). https://doi.org/10.3390/en15041421
30. Kwade, A., Haselrieder, W., Leithoff, R., Modlinger, A., Dietrich, F., Droeder, K.: Current status and challenges for automotive battery production technologies. Nat. Energy **3**(4), 290–300 (2018). https://doi.org/10.1038/s41560-018-0130-3
31. Borzutzki, K., Börner, M., Eckstein, M., Wessel, S., Winter, M., Tübke, J.: Kontinuierliche und Batch-basierte Prozessierung von Batterieelektroden für Lithium-Ionen-Batterien (2022)
32. Zhang, G., Wei, X., Tang, X., Zhu, J., Chen, S., Dai, H.: Internal short circuit mechanisms, experimental approaches and detection methods of lithium-ion batteries for electric vehicles: a review. Renew. Sustain. Energy Rev. **141**, 110790 (2021). https://doi.org/10.1016/j.rser.2021.110790
33. Weinmann, H.W., Eichelkraut, M., Da Woke Silva, L., Fleischer, J.: Batteriezellenfertigung vom Coil zum Stack: Integriert, automatisiert und dadurch hoch flexibel. C2 Coating & Converting(4), 21–24 (2020)
34. Weinmann, H.W., Töpper, H.-C., Fleischer, J.: Coil2Stack: Ein innovatives Verfahren zur formatflexiblen Batteriezellherstellung. Zeitschrift für wirtschaftlichen Fabrikbetrieb **115**(4), 241–243 (2020). https://doi.org/10.3139/104.112192

Author Index

M. Liewald et al. (Eds.): WGP 2022, LNPE, pp. 843–846, 2023.
https://doi.org/10.1007/978-3-031-18318-8